Name	Symbol	Atomic Number	Atomic Mass[a]	Name	Symbol	Atomic Number	Atomic Mass[a]
Actinium	Ac	89	(227)[b]	Molybdenum	Mo	42	95.94
Aluminum	Al	13	26.98	Neodymium	Nd	60	144.2
Americium	Am	95	(243)	Neon	Ne	10	20.18
Antimony	Sb	51	121.8	Neptunium	Np	93	(237)
Argon	Ar	18	39.95	Nickel	Ni	28	58.69
Arsenic	As	33	74.92	Niobium	Nb	41	92.91
Astatine	At	85	(210)	Nitrogen	N	7	14.01
Barium	Ba	56	137.3	Nobelium	No	102	(259)
Berkelium	Bk	97	(247)	Osmium	Os	76	190.2
Beryllium	Be	4	9.012	Oxygen	O	8	16.00
Bismuth	Bi	83	209.0	Palladium	Pd	46	106.4
Bohrium	Bh	107	(264)	Phosphorus	P	15	30.97
Boron	B	5	10.81	Platinum	Pt	78	195.1
Bromine	Br	35	79.90	Plutonium	Pu	94	(244)
Cadmium	Cd	48	112.4	Polonium	Po	84	(209)
Calcium	Ca	20	40.08	Potassium	K	19	39.10
Californium	Cf	98	(251)	Praseodymium	Pr	59	140.9
Carbon	C	6	12.01	Promethium	Pm	61	(145)
Cerium	Ce	58	140.1	Protactinium	Pa	91	231.0
Cesium	Cs	55	132.9	Radium	Ra	88	(226)
Chlorine	Cl	17	35.45	Radon	Rn	86	(222)
Chromium	Cr	24	52.00	Rhenium	Re	75	186.2
Cobalt	Co	27	58.93	Rhodium	Rh	45	102.9
Copernicium	Cn	112	(285)	Roentgenium	Rg	111	(272)
Copper	Cu	29	63.55	Rubidium	Rb	37	85.47
Curium	Cm	96	(247)	Ruthenium	Ru	44	101.1
Darmstadtium	Ds	110	(271)	Rutherfordium	Rf	104	(261)
Dubnium	Db	105	(262)	Samarium	Sm	62	150.4
Dysprosium	Dy	66	162.5	Scandium	Sc	21	44.96
Einsteinium	Es	99	(252)	Seaborgium	Sg	106	(266)
Erbium	Er	68	167.3	Selenium	Se	34	78.96
Europium	Eu	63	152.0	Silicon	Si	14	28.09
Fermium	Fm	100	(257)	Silver	Ag	47	107.9
Fluorine	F	9	19.00	Sodium	Na	11	22.99
Francium	Fr	87	(223)	Strontium	Sr	38	87.62
Gadolinium	Gd	64	157.3	Sulfur	S	16	32.07
Gallium	Ga	31	69.72	Tantalum	Ta	73	180.9
Germanium	Ge	32	72.64	Technetium	Tc	43	(99)
Gold	Au	79	197.0	Tellurium	Te	52	127.6
Hafnium	Hf	72	178.5	Terbium	Tb	65	158.9
Hassium	Hs	108	(265)	Thallium	Tl	81	204.4
Helium	He	2	4.003	Thorium	Th	90	232.0
Holmium	Ho	67	164.9	Thulium	Tm	69	168.9
Hydrogen	H	1	1.008	Tin	Sn	50	118.7
Indium	In	49	114.8	Titanium	Ti	22	47.87
Iodine	I	53	126.9	Tungsten	W	74	183.8
Iridium	Ir	77	192.2	Uranium	U	92	238.0
Iron	Fe	26	55.85	Vanadium	V	23	50.94
Krypton	Kr	36	83.80	Xenon	Xe	54	131.3
Lanthanum	La	57	138.9	Ytterbium	Yb	70	173.0
Lawrencium	Lr	103	(262)	Yttrium	Y	39	88.91
Lead	Pb	82	207.2	Zinc	Zn	30	65.41
Lithium	Li	3	6.941	Zirconium	Zr	40	91.22
Lutetium	Lu	71	175.0	—	—	113	(284)
Magnesium	Mg	12	24.31	—	—	114	(289)
Manganese	Mn	25	54.94	—	—	115	(288)
Meitnerium	Mt	109	(268)	—	—	116	(292)
Mendelevium	Md	101	(258)	—	—	118	(294)
Mercury	Hg	80	200.6				

[a]Values for atomic masses are given to four significant figures.
[b]Values in parentheses are the mass number of an important radioactive isotope.

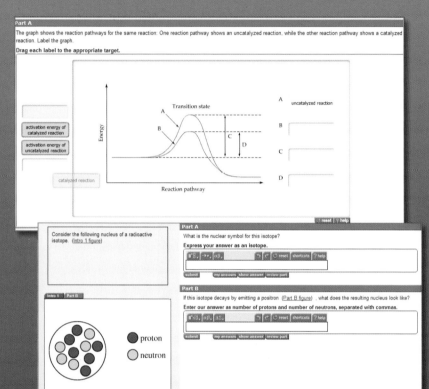

Basic Chemistry

THIRD EDITION

Basic Chemistry

Karen Timberlake

Los Angeles Valley College

William Timberlake

Los Angeles Harbor College

Prentice Hall

Boston Columbus Indianapolis New York San Francisco Upper Saddle River
Amsterdam Cape Town Dubai London Madrid Milan Munich Paris Montréal Toronto
Delhi Mexico City São Paulo Sydney Hong Kong Seoul Singapore Taipei Tokyo

Library of Congress Cataloging-in-Publication Data
Timberlake, Karen.
    Basic chemistry / Karen C. Timberlake, William Timberlake.—3rd ed.
    p. cm.
    ISBN-13: 978-0-13-703841-1
    ISBN-10: 0-13-703841-0
    1. Chemistry—Textbooks   I. Timberlake, William E.   II. Title.

QD31.3.T54 2011
540—dc22

                                    2009044403

Acquisitions Editor: *Terry Haugen*
Editor in Chief, Chemistry and Geosciences: *Nicole Folchetti*
Marketing Manager: *Erin Gardner*
Assistant Editor: *Jessica Neumann*
Editorial Assistant: *Lisa Tarabokjia*
Marketing Assistant: *Nicola Houston*
VP/Executive Director, Development: *Carol Trueheart*
Development Editor: *Ray Mullaney*
Managing Editor, Chemistry and Geosciences: *Gina M. Cheselka*
Project Manager, Production: *Beth Sweeten*
Senior Media Producer: *Angela Bernhardt*
Associate Media Producer: *Victoria Prather*
Senior Media Production Supervisor: *Liz Winer*
Media Editor: *Shannon Kong*
Art Editor: *Connie Long*
Art Studio: *Precision Graphics*
Art Director: *Mark Ong*
Interior Design: *Jill Little*
Cover Design: *Derek Bacchus*
Senior Manufacturing and Operations Manager: *Nick Sklitsis*
Operations Specialist: *Maura Zaldivar*
Image Permissions Coordinator: *Elaine Soares*
Photo Researcher: *Eric Shrader*
Production Supervision/Composition: *Prepare, Inc.*
Cover Illustrator: *Blakely Kim*
Front cover photograph: *magnetcreative/iStockphoto.com*
Back cover photograph: *chrislitwin/iStockphoto*

Credits and acknowledgments borrowed from other sources and reproduced, with permission, in this textbook appear on p. C-1.

Printed in the United States
10 9 8 7 6 5 4 3

ISBN 10: 0-321-66310-1
ISBN 13: 978-0-321-66310-8

Prentice Hall
is an imprint of

www.pearsonhighered.com

BRIEF CONTENTS

CONTENTS

5

Electronic Structure and Periodic Trends   121

6

Inorganic and Organic Compounds: Names and Formulas   154

7

Chemical Quantities   193

8

Chemical Reactions of Inorganic and Organic Compounds   221

9

Chemical Quantities in Reactions   254

10

Structures of Solids and Liquids   285

11

Gases   326

12
Solutions   364

13
Chemical Equilibrium   406

14
Acids and Bases   441

18

Biochemistry
e-chapter   603

APPLICATIONS AND ACTIVITIES

CHEMISTRY AND INDUSTRY

Guide to Problem Solving

ABOUT THE AUTHORS

Karen Timberlake is Professor Emerita of Chemistry at Los Angeles Valley College, where she taught chemistry for allied health and preparatory chemistry for 36 years. She received her bachelor's degree in chemistry from the University of Washington and her master's degree in biochemistry from the University of California at Los Angeles.

Professors Timberlake and Timberlake have been writing chemistry textbooks for 33 years. During that time, their names have become associated with the strategic use of pedagogical tools that promote student success in chemistry and the application of chemistry to real-life situations. More than 1 million students have learned chemistry using texts, laboratory manuals, and study guides written by Karen and William Timberlake. In addition to *Basic Chemistry*, Karen Timberlake is also the author of *General, Organic, and Biological Chemistry, Structures of Life*, third edition, with the accompanying "Study Guide" and "Selected Solutions Manual" and *Chemistry: An Introduction to General, Organic, and Biological Chemistry*, tenth edition, with the accompanying *Study Guide and Selected Solutions, Laboratory Manual*, and *Essential Laboratory Manual*.

Professors Timberlake and Timberlake belong to numerous science and educational organizations including the American Chemical Society (ACS) and the National Science Teachers Association (NSTA). In 1987, Karen Timberlake was the Western Regional Winner of Excellence in College Chemistry Teaching Award given by the Chemical Manufacturers Association. In 2004, she received the McGuffey Award in Physical Sciences from the Textbook Author Association for her textbook *Chemistry: An Introduction to General, Organic, and Biological Chemistry*, eighth edition, which has demonstrated excellence over time. In 2006, she also received the "Texty" Textbook Excellence Award from the Textbook Authors Association for the first edition of *Basic Chemistry*. She has participated in education grants for science teaching including the Los Angeles Collaborative for Teaching Excellence (LACTE) and a Title III grant at her college. She speaks at conferences and educational meetings on the use of student-centered teaching methods in chemistry to promote the learning success of students.

William Timberlake is Professor Emeritus of Chemistry at Los Angeles Harbor College where he taught preparatory and organic chemistry for 36 years. He received his bachelor's degree in chemistry from Carnegie Mellon University and his master's degree in organic chemistry from the University of California at Los Angeles. When the Professors Timberlake are not writing textbooks, they relax by hiking, traveling to Mexico, Europe, and Asia, trying new restaurants, cooking, and playing tennis.

PREFACE

Welcome to the third edition of *Basic Chemistry*. This text was written to prepare students with little or no chemistry experience for future chemistry courses or careers in a health-related profession, such as nursing, dietetics, respiratory therapy, environmental or agricultural science, or as laboratory technicians. Our main objective in writing this text is to make the study of chemistry an engaging and positive experience for you by relating the structure and behavior of matter to real-life applications. This new edition introduces rich problem-solving strategies, including new concept checks, more problem-solving guides, conceptual, challenge, and combined problem sets.

It is our goal to help you become a critical thinker by understanding scientific concepts that will form a basis for making important decisions about issues concerning health and the environment. Thus, we have utilized materials that

- help you to learn and enjoy chemistry.
- develop problem-solving skills that lead to success in your chemistry course.
- promote learning and success in your chosen career.

New to This Edition

New features have been added throughout this third edition, including the following:

- The most advanced chemistry homework and tutorial system, MasteringChemistry®, providing an online homework and tutoring system, can be packaged with the third edition of *Basic Chemistry*.
- New *Concept Checks* with Answers help students focus on their understanding of newly introduced chemical terms and ideas.
- New photos and updated diagrams improve clarity and provide visual understanding of chemical reactions.
- New captions on photos identify photo contents.
- More Guides to Problem Solving (GPS) illustrate step-by-step problem-solving strategies.
- Chapter 5 is now **Electronic Structure and Periodic Trends**.
- Chapter 6, now **Inorganic and Organic Compounds: Names and Formulas**, includes two new sections, "Organic Compounds" and "Names and Formulas of Hydrocarbons: Alkanes."
- Chapter 8, now **Chemical Reactions of Inorganic and Organic Compounds**, includes two new sections, "Functional Groups in Organic Compounds" and "Reactions of Organic Compounds."
- New Chemistry and Health boxes, "Brachytherapy", "Polycyclic Aromatic Hydrocarbons (PAHs)", and "Breathing Mixtures for Scuba", have been added.
- New Chemistry and the Environment boxes—"Energy-Saving Fluorescent Bulbs", "Vanilla", and "Biodiesel as an Alternative Fuel"—have been added.
- New material about the discovery of the electron by J. J. Thomson using the cathode ray tube has been added.
- "Oxidation of Alcohols" is now part of Chapter 15, **Oxidation and Reduction**.
- New Understanding the Concepts problems add more visual examples to conceptual learning.
- Learning Goals are now included in the Chapter Review.
- Revised interchapter Combining Ideas problem sets provide problems with greater depth using concepts from several chapters.

Chapter Organization

In each textbook we write, we consider it essential to relate every chemical concept to real-life issues of health and environment. Because a course of chemistry may be taught in different time frames, it may be difficult to cover all the chapters in this text. However, each chapter is a complete package, which allows some chapters to be skipped or the order of presentation to be changed. In this edition, we have incorporated many topics of organic chemistry from Chapter 17 and some of biochemistry from Chapter 18 into the early chapters of the text to integrate general chemistry with organic chemistry.

Chapter 1, Chemistry in Our Lives, introduces the concepts of chemicals and chemistry, describes the goals and achievements of scientists, discusses the scientific method in everyday terms, and guides students in developing a study plan for learning chemistry.

- A new section discusses the "Branches of Chemistry."
- More discoveries were added to Table 1.2, "Some Important Scientific Discoveries, Laws, Theories, and Technological Innovations".

Chapter 2, Measurements, looks at measurement and emphasizes the need to understand numerical structures of the metric system in the sciences. Table 2.6 has been reworked to include more prefixes and equalities. An explanation of scientific notation and working with a calculator is included in the chapter.

- New photos include the standard kilogram for the United States, hard-disk drive, an ophthalmologist, and fertilizer application.
- New Table 2.7 gives "Daily Values for Selected Nutrients".
- A new Chemistry and the Environment box, "Density of Crude Oil"—including oil spill cleanup—is discussed in Section 2.8.
- New Concept Checks on metric prefixes, problem solving, and density expand the everyday application of density.
- The Chemistry and Health feature "Bone Density" is now included in Chapter 2.
- A new Sample Problem and Solution, "Calculating Density", describes the calculation of the density of high-density lipoprotein (HDL).

Chapter 3, Matter and Energy, classifies matter and states of matter, describes temperature measurement, and discusses energy and its measurement. Physical and chemical changes and physical and chemical properties are now discussed in more depth. The section on forms of energy has been deleted. Combining Ideas from Chapters 1, 2, and 3 follows as an interchapter problem set.

- A new Chemistry and Health feature, "Breathing Mixtures for Scuba", has been added.
- New illustrations utilize macro-to-micro diagrams to visualize pure substances, elements, and compounds.
- A discussion and photos of physical methods of separating mixtures, filtration, and chromatography have been added.
- Section 3.2 expands the content of "States and Properties of Matter" and includes more macro-to-micro diagrams of the atomic arrangement in matter.
- New Concept Checks on potential and kinetic energy, temperature scales, specific heat, and physical and chemical properties clarify conceptual content.
- The Chemistry and the Environment box, "Carbon Dioxide and Global Warming", has been rewritten to include a graph of the changes in carbon dioxide levels for the years from 1000 C.E. to 2000 C.E.
- "Energy and Nutrition" has now been written as separate Section 3.6.

Chapter 4, Atoms and Elements, looks at elements, atoms, subatomic particles, atomic numbers, and mass numbers. Using the naturally occurring isotopes and abundances, atomic mass is calculated. The types of compounds formed by the element carbon is discussed. The periodic table emphasizes the numbering of groups from 1–18. Elements with atomic numbers 116 and 118 have been added to the periodic table and atomic masses of elements. All isotopic symbols now include the atomic number. Section 4.6, "Electron Energy Levels", has been deleted.

- A new Chemistry and Industry box, "Many Forms of Carbon", has been added.
- A new Chemistry and Health box, "Elements Essential to Health", has been added.
- A new discussion of the discovery of electrons by J. J. Thomson using the cathode ray tube has been added.
- New macro-to-micro illustrations update the concepts of isotopes, percent abundances, and atomic masses.
- New Concept Checks include the periodic table, subatomic particles, and calculating atomic mass.

Chapter 5, Electronic Structure and Periodic Trends, uses the electromagnetic spectrum to explain atomic spectra and develop the concept of energy levels and sublevels. Electrons in sublevels and orbitals are represented using orbital diagrams and electron configurations. Periodic properties of elements including atomic radius and ionization energy are related to their valence electrons. Section 5.3 is now entitled "Sublevels and Orbitals".

- Chapter 5, previously Chapter 9, now appears earlier in the text.
- The calculations in Section 5.1 that convert frequency to wavelength have been deleted.
- A new Chemistry and Environment box, "Energy-Saving Fluorescent Bulbs", has been added.
- The sublevel diagram has been updated to clarify the order of filling sublevels.
- New Concept Checks include the electromagnetic spectrum, orbital diagrams, and atomic radius.

Chapter 6, Inorganic and Organic Compounds: Names and Formulas, describes how atoms form ionic and covalent bonds. Chemical formulas are written, and ionic compounds—including those with polyatomic ions—and covalent compounds are named. An introduction to the three-dimensional shape of carbon molecules provides a basis for the shape of organic and biochemical compounds. Organic chemistry is introduced with the properties of inorganic and organic compounds and condensed structural formulas of alkanes.

- New photos provide visual images of ionic and covalent compounds.
- A new Section 6.6, "Organic Compounds", discusses the properties of inorganic and organic compounds, three-dimensional shapes of carbon compounds, and the condensed structural formulas of alkanes.
- A new Section 6.7, "Names and Formulas of Hydrocarbons: Alkanes", introduces the condensed structural formulas and names of alkanes with 1 to 10 carbon atoms.
- New Concept Checks include the names and formulas of ionic and covalent compounds, as well as properties of organic compounds.

Chapter 7, Chemical Quantities, discusses Avogadro's number, the mole, and molar masses of compounds, which are used in calculations to determine the mass or number of particles in a quantity of a substance. The percent composition of a compound is calculated and used to determine its empirical and molecular formula. Combining Ideas from Chapters 4, 5, 6, and 7 follows as an interchapter problem set.

- The section "Atomic Mass and Formula Mass" has been deleted.
- A new diagram illustrates the concept of percent composition.
- A new Guide to Problem Solving (GPS), "Calculating the Atoms or Molecules of a Substance", has been added.
- A new Guide to Problem Solving (GPS), "Calculating the Moles (or Grams) of a Substance from Grams (or Moles)", has been added.
- New Concept Checks including the number of particles, molar mass, moles, and empirical formula have been added.

Chapter 8, Chemical Reactions of Inorganic and Organic Compounds, looks at the interaction of atoms and molecules in chemical reactions. Chemical equations are balanced and organized into combination, decomposition, single replacement, and double replacement reactions. A new Section 8.4, "Functional Groups in Organic Compounds", classifies compounds according to their structures, which allows predictions of their reactions and properties.

- The material on physical change has been deleted.
- A new Section 8.4, "Functional Groups in Organic Compounds", has been added.
- Molecular models of organic compounds have been included in the section on functional groups.
- A new Section 8.5, "Reactions of Organic Compounds", has been added.
- The material on combustion—including the combustion of alcohols—has been moved to Section 8.5, "Reactions of Organic Compounds".
- Material on functional groups in biomolecules has been added.
- The Chemistry and Health features, "Amines in Health and Medicine" and "Fatty Acids", and the Chemistry and Industry feature, "Hydrogenation of Unsaturated Fats", are now included in Chapter 8.
- The section "Energy in Chemical Reactions" is now in Chapter 9.
- New Concept Checks include reaction types, balancing equations, and the identification of functional groups in organic and biochemical molecules.

Chapter 9, Chemical Quantities in Reactions, describes the mole and mass relationships among the reactants and products and provides calculations of percent yields and limiting reactants. A section on endothermic and exothermic reactions and a new section on "Energy in the Body" complete the chapter.

- Mole and mass relationships among the reactants and products are examined along with calculations of percent yield and limiting reactants.
- The Chemistry and Health features, "ATP Energy and Ca^{2+} Needed to Contract Muscles" and "Stored Fat and Obesity", are now included in Chapter 9.
- Section 9.5, "Energy in Chemical Reactions", is now included in this chapter.
- A new Section 9.6, "Energy in the Body", has been added.
- New Concept Checks include calculations of mass of reactants and products, percent yield, and limiting reactant.

Chapter 10, Structures of Solids and Liquids, introduces electron-dot formulas for molecules and ions with single and multiple bonds as well as resonance structures. Electronegativity leads to a discussion of the polarity of bonds and molecules. Electron-dot formulas and VSEPR theory illustrate covalent bonding and the three-dimensional shapes of molecules and ions. The attractive forces between particles and their impact on states of matter and changes of state are described. Combining Ideas from Chapters 8, 9, and 10 follows as an interchapter problem set.

- New illustrations of molecular models provide visual representations of three-dimensional structures.
- The section "Electron Configuration of Ionic Compounds" has been deleted.
- A new Chemistry and Health box, "Attractive Forces in Biological Compounds", has been added.
- New Concept Checks include electron-dot formulas of compounds and polyatomic ions with single and double bonds as well as resonance structures, VSEPR and shapes of molecules and polyatomic ions, and polarity of bonds and molecules.

Chapter 11, Gases, describes the properties of a gas and calculates changes in gases using the gas laws and the ideal gas law. The amounts of gases required or produced in chemical reactions are calculated.

- Descriptions of properties of gases and gas laws have been rewritten to clarify concepts.
- A new Chemistry and the Environment box, "Greenhouse Gases", has been added.
- New Concept Checks include properties of gases, gas laws, and applications of gas laws—including quantities of gases in reactions.

Chapter 12, Solutions, describes solutions, saturation and solubility, concentrations, and colligative properties. The volumes and molarities of solutions are used in calculations of reactants and products in chemical reactions, as well as dilutions and titrations.

- Sections on "Formation of Solutions" and "Electrolytes" have been rewritten for improved clarity.

- A new Section 12.7, "Properties of Solutions", now discusses properties of solutions and includes types of solutions and the impact of particle concentration on boiling point, freezing point, and osmotic pressure.
- A new Chemistry and Health box, "Dialysis by the Kidneys and the Artificial Kidney", has been added.
- New Concept Checks include electrolytes and nonelectrolytes, solubility, dilution, and colligative properties.

Chapter 13, Chemical Equilibrium, looks at the rates of reactions and the equilibrium condition when forward and reverse rates for a reaction become equal. Equilibrium expressions for reactions are written and equilibrium constants are calculated. Using equilibrium constants, reactions are evaluated to determine if the equilibrium favors the reactants or the products and the concentrations of components are calculated. Le Châtelier's principle is used to evaluate the impact on concentrations when a stress is placed on the equilibrium system. The equilibrium of dissolving and crystallizing in saturated solutions is evaluated using solubility product constants.

- Diagrams and illustrations have been updated.
- New Concept Checks include rate of reactions, equilibrium constant, Le Châtelier's principle, and solubility product constant.

Chapter 14, Acids and Bases, discusses acids and bases and their strengths, conjugate acid–base pairs, the dissociation of weak acids and bases and water, pH and pOH, and buffers. Acid–base titration uses the neutralization reactions between acids and bases to calculate quantities of acid in a sample. Combining Ideas from Chapters 11, 12, 13, and 14 follows as an interchapter problem set.

- Section 14.1 now includes condensed structural formulas and names of organic acids.
- New problems with weak organic bases have been added.
- Section 14.6, "The pH Scale", has been updated to clarify calculations of pH and hydronium ion concentration.
- The Chemistry and the Environment box, "Acid Rain", has been updated.
- New problems related to acid rain have been added.
- New Concept Checks include the names of acids and bases, conjugate acid–base pairs, pH of solutions, acid dissociation constants, and buffers.

Chapter 15, Oxidation and Reduction, looks at the characteristics of oxidation and reduction reactions. Oxidation numbers are assigned to the atoms in elements, molecules, and ions to determine the components that lose electrons during oxidation and gain electrons during reduction. Changes in oxidation numbers and the half-reaction method are both utilized to balance oxidation–reduction reactions. The production of electrical energy in voltaic cells and the requirement of electrical energy in electrolytic cells are diagrammed using half-cells. The activity series is used to determine the spontaneous direction of an oxidation–reduction reaction.

- A new Section 15.4, "Oxidation of Alcohols", has been added.
- Table 15.1, "Rules for Assigning Oxidation Numbers", has been rewritten for clarity.
- The material on "Oxidizing and Reducing Agents" has been rewritten for clarity.
- New photos provide more visual images of compounds that undergo oxidation and reduction.
- The Chemistry and Health box, "Oxidation of Alcohol in the Body", is now included in Chapter 15.
- New material on "Balancing Oxidation–Reduction Equations in Basic Solutions" has been added.
- New Concept Checks include loss and gain of electrons, oxidation numbers, classification of alcohols, batteries, and spontaneous reactions.

Chapter 16, Nuclear Radiation, looks at the type of radioactive particles that are emitted from the nuclei of radioactive atoms. Equations are written and balanced for both naturally occurring radioactivity and artificially produced radioactivity. The half-lives of radioisotopes are discussed and the amount of time for a sample to decay is calculated. Radioisotopes important in the field of nuclear medicine are described. Combining Ideas from Chapters 15 and 16 follows as an interchapter problem set.

- Section 16.1, "Natural Radioactivity", and Section 16.2, "Nuclear Equations", have been rewritten for clarity.
- The imaging methods computed tomography and magnetic resonance imaging have been added.
- A new Chemistry and Health box, "Brachytherapy", has been added.
- New Concept Checks include radiation particles, alpha decay, bombardment, nuclear equations, measurement of activity, and half-lives.

Chapter 17, Organic Chemistry, discusses each family of organic compounds, thus forming a basis for understanding the biomolecules of living systems. In this third edition, topics such as functional groups, reactions of organic compounds, and naming organic acids are now included in earlier chapters on naming covalent compounds and chemical reactions. This organic chapter is now streamlined with an emphasis on naming and drawing formulas of organic compounds from alkanes with substituents to amines and amides. Chapter 17 is available as an online chapter.

- A new list, "Possible Number of Isomers for Alkanes with 1–10 Carbon Atoms", has been added.
- The material "The Tetrahedral Structure of Carbon" has been moved to an earlier chapter.
- New material, "Haloalkanes", has been added.
- The section "Functional Groups" has been moved to an earlier chapter.
- The material "Solubility and Density of Alkanes", "Combustion of Alkanes", and "Hydrogenation" has been moved to earlier chapters in the text.
- The sections "Alkenes and Alkynes" and "Polymers" have been combined as Section 17.3.
- The material "Classification of Alcohols" and the "Oxidation of Primary and Secondary Alcohols" has been moved to an earlier chapter.
- New Chemistry and the Environment boxed features, "Polycyclic Aromatic Hydrocarbons (PAHs)" and "Vanilla", have been added.
- New Guides, "Naming Aldehydes", "Naming Ketones", and "Naming Esters", have been added.
- Material on "Naming Carboxylic Acids" has been moved to an earlier chapter.
- New Chemistry and Health boxed features, "Common Uses of Haloalkanes" and "Carboxylic Acids in Metabolism", have been added.
- New Concept Checks—including naming alkanes with substituents, naming alkenes and alkynes, naming aromatic compounds, naming ethers, naming isomers, naming amines, and amidation have been added.
- New Challenge Questions with visual art have been added.

Chapter 18, Biochemistry, looks at the chemical structures and reactions of chemicals that occur in living systems. We focus on four types of biomolecules—carbohydrates, lipids, proteins, and nucleic acids—as well as their biochemical reactions. The shape of proteins is related to the activity and regulation of enzyme activity. Combining Ideas from Chapters 17 and 18 follows as an interchapter problem set.

- A new Chemistry and the Environment box, "Biodiesel as an Alternative Fuel", has been added.
- The Chemistry and Health box "How Sweet Is My Sweetener?" now includes information about Neotame.
- The Chemistry and Health box "Trans Fatty Acids and Hydrogenation" has been updated.
- A discussion of the genetic code and Table 18.10, "mRNA Codons: The Genetic Code for Amino Acids", have been added to Section 18.8, "Protein Synthesis".
- New Concept Checks include fatty acids, peptides, structures of proteins, and codons in protein synthesis.

Features of This Text

You may wonder why your career path includes a class in chemistry. A common view is that chemistry is just a lot of facts to be memorized. To change this perception, we have included many features to help you learn about chemistry in your life and career choice and to give you the skills to learn chemistry successfully. These features include

- Connections to health and the environment.
- Visual guides to problem solving.
- In-chapter problem sets to work immediately that reinforce the learning of new concepts.

New to this edition, organic chemistry and biochemistry are now introduced in the early chapters. A successful learning program in this text provides you with many learning tools, which are now discussed.

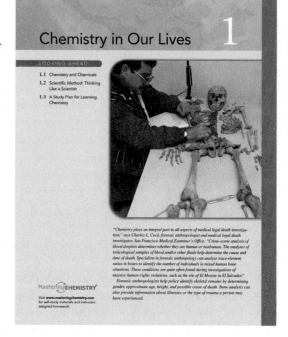

Chemistry in Our Lives   1

LOOKING AHEAD

1.1 Chemistry and Chemicals
1.2 Scientific Method: Thinking Like a Scientist
1.3 A Study Plan for Learning Chemistry

MasteringCHEMISTRY

Visit www.masteringchemistry.com for self-study materials and instructor-assigned homework.

"Chemistry plays an integral part in all aspects of medical legal death investigation," says Charles L. Cecil, forensic anthropologist and medical legal death investigator, San Francisco Medical Examiner's Office. "Crime-scene analysis of blood droplets determines whether they are human or nonhuman. The analyses of toxicological samples of blood and/or other fluids help determine the cause and time of death. Specialists in forensic anthropology can analyze trace-element ratios in bones to identify the number of individuals in mixed human bone situations. These conditions are quite often found during investigations of massive human-rights violations, such as the site of El Mozote in El Salvador."

Forensic anthropologists help police identify skeletal remains by determining gender, approximate age, height, and possible cause of death. Bone analysis can also provide information about illnesses or the type of trauma a person may have experienced.

Career Focus and Real-World Applications

This Text Was Designed to Help Students Attain Their Career Goals

Chapter Opening Interviews with Scientists

Each chapter begins with an interview with a professional in a career such as medicine, forensic anthropology, nuclear medicine, dentistry, and oceanography. These interviews discuss the importance of chemistry in their careers.

Career Focus

Within the chapters are additional interviews with allied health professions using chemistry.

CAREER FOCUS

Geologist

"Chemistry underpins geology," says Vic Abadie, consulting geologist. "I am a self-employed geologist consulting in exploration for petroleum and natural gas. Predicting the occurrence of an oil reservoir depends in part on understanding chemical reactions of minerals. This is because over geologic time, such reactions create or destroy pore spaces that host crude oil or gas in a reservoir rock formation. Chemical analysis can match oil in known reservoir formations with distant formations that generated oil and from which oil migrated into the reservoirs in the geologic past. This can help identify target areas to explore for new reservoirs.

"I evaluate proposals to drill for undiscovered oil and gas. I recommend that my clients invest in proposed wells that my analysis suggests have strong geologic and economic merit. This is a commercial application of the scientific method: The proposal to drill is the hypothesis, and the drill bit tests it. A successful well validates the hypothesis and generates oil or gas production and revenue for my clients and me. I do this and other consulting for private and corporate clients. The risk is high, the work is exciting, and my time is flexible."

On the Web

The **MasteringChemistry® Study Area** features in-depth resources for each of the health professions featured in the book and takes students through interactive case studies.

Students Will Learn Chemistry Using Real-World Examples

Chemistry and Health

The many **Chemistry and Health boxes** in each chapter apply chemical concepts to relevant topics of health and medicine. These topics include trans fats, sweeteners, scuba breathing mixtures, effects of alcohol on the body, and pheromones.

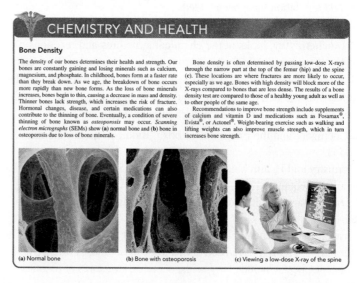

CHEMISTRY AND HEALTH

Bone Density

The density of our bones determines their health and strength. Our bones are constantly gaining and losing minerals such as calcium, magnesium, and phosphate. In childhood, bones form at a faster rate than they break down. As we age, the breakdown of bone occurs more rapidly than new bone forms. As the loss of bone minerals increases, bones begin to thin, causing a decrease in mass and density. Thinner bones lack strength, which increases the risk of fracture. Hormonal changes, disease, and certain medications can also contribute to the thinning of bone. Eventually, a condition of severe thinning of bone known as *osteoporosis* may occur. *Scanning electron micrographs* (SEMs) show (a) normal bone and (b) bone in osteoporosis due to loss of bone minerals.

Bone density is often determined by passing low-dose X-rays through the narrow part at the top of the femur (hip) and the spine (c). These locations are where fractures are more likely to occur, especially as we age. Bones with high density will block more of the X-rays compared to bones that are less dense. The results of a bone density test are compared to those of a healthy young adult as well as to other people of the same age.

Recommendations to improve bone strength include supplements of calcium and vitamin D and medications such as Fosamax®, Evista®, or Actonel®. Weight-bearing exercise such as walking and lifting weights can also improve muscle strength, which in turn increases bone strength.

(a) Normal bone   (b) Bone with osteoporosis   (c) Viewing a low-dose X-ray of the spine

Chemistry and the Environment

Chemistry and the Environment boxes highlight the practical applications of chemistry that impact the environment and provide sustainability. They delve into issues such as global warming, ozone depletion, radon in the home, acid rain, and biodiesel fuels.

CHEMISTRY AND THE ENVIRONMENT

Fertilizers

Every year in the spring, homeowners and farmers add fertilizers to the soil to produce greener lawns and larger crops. Plants require several nutrients, but the major ones are nitrogen, phosphorus, and potassium. Nitrogen promotes green growth, phosphorus promotes strong root development for strong plants and abundant flowers, and potassium helps plants defend against diseases and weather extremes. The numbers on a package of fertilizer give the percentages each of N, P, and K by mass. For example, the set of numbers 30–3–4 describes a fertilizer that contains 30% N, 3% P, and 4% K.

Nitrogen (N) 30%
Phosphorus (P) 3%
Potassium (K) 4%

Lawn Fertilizer 30-3-4

The label on a bag of fertilizer states the percentages of N, P, and K.

The major nutrient, nitrogen, is present in huge quantities as N_2 in the atmosphere, but plants cannot utilize nitrogen in this form. Bacteria in the soil convert atmospheric N_2 to usable forms by nitrogen fixation. To provide additional nitrogen to plants, several types of nitrogen-containing chemicals, including ammonia, nitrates, and ammonium compounds, are added to soil. The nitrates are absorbed directly, but ammonia and ammonium salts are first converted to nitrates by the soil bacteria.

The percent nitrogen depends on the type of nitrogen compound used in the fertilizer. The percent nitrogen by mass in each type is calculated using percent composition.

Type of Fertilizer		Percent Nitrogen by Mass
NH_3	$\dfrac{14.01 \text{ g N}}{17.03 \text{ g } NH_3}$	$\times\ 100\% = 82.27\%$ N
NH_4NO_3	$\dfrac{28.02 \text{ g N}}{80.05 \text{ g } NH_4NO_3}$	$\times\ 100\% = 35.00\%$ N
$(NH_4)_2SO_4$	$\dfrac{28.02 \text{ g N}}{132.15 \text{ g } (NH_4)_2SO_4}$	$\times\ 100\% = 21.20\%$ N
$(NH_4)_2HPO_4$	$\dfrac{28.02 \text{ g N}}{132.06 \text{ g } (NH_4)_2HPO_4}$	$\times\ 100\% = 21.22\%$ N

The choice of a fertilizer depends on its use and convenience. A fertilizer can be prepared as crystals or a powder, in a liquid solution, or as a gas such as ammonia. The ammonia and ammonium fertilizers are water soluble and quick-acting. Other forms may be made to slow-release by enclosing water-soluble ammonium salts in a thin plastic coating. The most commonly used fertilizer is NH_4NO_3 because it is easy to apply and has a high percent of N by mass.

Chemistry and History

The **Chemistry and History box** describes the historical development of chemical ideas.

CHEMISTRY AND HISTORY

Early Chemists: The Alchemists

For many centuries, chemists have studied changes in matter. From the time of the ancient Greeks to about the sixteenth century, early scientists, called alchemists, described matter in terms of four components of nature: earth, air, fire, and water. These components had the qualities of hot, cold, damp, or dry. By the eighth century, alchemists believed that they could rearrange these qualities to change metals such as copper and lead into gold and silver. They searched for an unknown substance called a *philosopher's stone*, which they thought would turn metals into gold as well as prolong youth and postpone death. Although these efforts failed, the alchemists did provide information on the processes and chemical reactions involved in the extraction of metals from ores. During the many centuries that alchemy flourished, alchemists made observations of matter and identified the properties of many substances. They also designed some of the first laboratory equipment and developed early laboratory procedures.

Chemistry and Industry

The boxed feature **Chemistry and Industry** describes industrial and commercial applications such as the oil industry and the commercial production of margarine and solid shortening.

CHEMISTRY AND INDUSTRY

Many Forms of Carbon

Carbon, which has the symbol C and atomic number 6, is a central element located in Group 4A (14) in Period 2 on the periodic table. However, its atoms can be arranged in different ways to give several different substances. Two forms of carbon—diamond and graphite—have been known since prehistoric times. In diamond, each carbon atom is attached to four other carbon atoms in a rigid structure. A diamond is transparent and harder than any other substance, whereas graphite is black and soft. In graphite, carbon atoms are arranged in hexagonal rings that form planes that slide over each other. Graphite is used as pencil lead, as lubricants, and as carbon fibers for the manufacture of lightweight golf clubs and tennis rackets.

Two other forms of carbon have been discovered more recently. In the form called Buckminsterfullerene, or buckyball, 60 carbon

Student-Friendly Approach

Keeping Students Engaged Is the Ultimate Goal

Student-Friendly Writing Style

To enhance student understanding, we try to use an accessible writing style, based on a carefully paced and simple development of chemical ideas, suited to your background. All terms are precisely defined, and clear goals are set for each section of the text. Clear analogies help you visualize and understand key chemical concepts.

Learning Goals

At the beginning of each section, a **Learning Goal** clearly identifies the key concept of the section, providing a road map for studying. All information contained in that section relates back to the Learning Goal. The Learning Goals for each section are also repeated in the Chapter Review so you can make sure you have mastered the key concepts.

New Concept Checks

The many **Concept Checks** throughout each chapter allow students to check their understanding of new chemical terms and ideas.

9.2 Mass Calculations for Reactions

LEARNING GOAL
Given the mass in grams of a substance in a reaction, calculate the mass in grams of another substance in the reaction.

When you perform a chemistry experiment in the laboratory, you use a laboratory balance to measure out a certain mass of the reactant. From the mass in grams, you can determine the number of moles of reactant. By using mole–mole factors, you can predict the moles of product that can be produced. Then the molar mass of the product is used to convert the moles back into mass in grams. The process in Concept Check 9.3 can be used to set up and solve problems that involve calculations of quantities for substances in chemical reactions.

CONCEPT CHECK 7.5

■ **Empirical and Molecular Formulas**

A compound has an empirical formula of CH_2O. Indicate if each of the following would be a possible molecular formula. Explain your answer.

a. CH_2O
b. C_2H_4O
c. $C_2H_4O_2$
d. $C_5H_{10}O_5$
e. $C_6H_{12}O$

ANSWER

a. The empirical formula and molecular formula have a ratio of 1:1, which gives an empirical formula that is the same as the molecular formula.

b. Because only the C and H are twice their moles in the empirical formula, this cannot be a corresponding molecular formula.

c. This can be a possible molecular formula because the C, H, and O are twice the moles of the empirical formula.

d. This can be a possible molecular formula because the C, H, and O are five times the moles of the empirical formula.

e. Because only the C and H are six times their moles in the empirical formula, this cannot be a corresponding molecular formula.

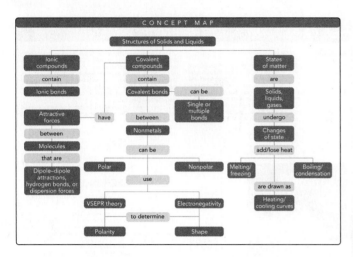

Clear Illustrations

The **art program** is not only beautifully rendered but also pedagogically effective to help students visualize chemistry.

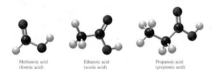

Methanoic acid
(formic acid)

Ethanoic acid
(acetic acid)

Propanoic acid
(propionic acid)

FIGURE 14.1 The IUPAC names of carboxylic acids use the alkane names but replace e with oic acid.
Q What is the IUPAC and common name of a carboxylic acid with a chain of four carbons?

Concept Maps

Each chapter ends with a **Concept Map** that reviews the key concepts of each chapter and how they fit together.

Macro-to-Micro Art

The photographs and drawings portray the atomic structure of recognizable objects, putting chemistry in context and connecting the atomic world to the macroscopic world. A question with each figure challenges you to think critically about photos and illustrations. Many new photos expand visual connections.

Propane, C_3H_8, is an organic compound, whereas sodium chloride, NaCl, is an inorganic compound.

Problem Solving

Many Tools Show Students How to Solve Problems

A Visual Guide to Problem Solving

As part of a comprehensive learning program, the **Guides to Problem Solving (GPS)** illustrate the steps you need to solve problems. We clearly understand the learning challenges facing students in this course, so we walk students through the problem-solving process step-by-step. For each type of problem, we use a unique, color-coded flow chart that is coordinated with parallel worked examples as a visual guide for each problem-solving strategy.

Sample Problems with Study Checks

Numerous **Sample Problems** appear throughout the text to demonstrate the application of each new concept. The worked-out solutions give step-by-step explanations, provide a problem-solving model, and illustrate required calculations. Each Sample Problem is followed by a **Study Check Question** that allows students to test their understanding of the problem-solving strategy.

Integrated Questions and Problems in Every Section

Questions and Problems at the end of each section encourage students to apply concepts and begin problem solving after each section. **Paired Problems** of each even-numbered problem with a matching odd-numbered problem guide students as they solve the problems. **Answers** to odd-numbered problems are given at the end of each chapter.

Guide to Problem Solving Using Conversion Factors

STEP 1
State the given and needed quantities.

STEP 2
Write a plan to convert the given unit to the needed unit.

STEP 3
State the equalities and conversion factors needed to cancel units.

STEP 4
Set up problem to cancel units and calculate answer.

SAMPLE PROBLEM 7.7

■ **Converting Grams to Particles**

A 10.00-lb bag of table sugar contains 4536 g of sucrose, $C_{12}H_{22}O_{11}$. How many molecules of sugar are present?

SOLUTION

STEP 1 **Given** 4536 g of sucrose   **Need** moles of sucrose

STEP 2 **Plan** Convert the grams of sugar to moles of sugar using the molar mass of $C_{12}H_{22}O_{11}$. Then use Avogadro's number to calculate the number of molecules of sugar.

g of sugar Molar mass factor moles Avogadro's number of molecules

STEP 3 **Equalities/Conversion Factors** The molar mass of $C_{12}H_{22}O_{11}$ is the sum of the masses of 12 mol of C + 22 mol of H + 11 mol of O:

(144.1 g/mol) + (22.18 g/mol) + (176.0 g/mol) = 342.3 g/mol

$$1 \text{ mol of } C_{12}H_{22}O_{11} = 342.3 \text{ g of } C_{12}H_{22}O_{11}$$

$$\frac{342.3 \text{ g } C_{12}H_{22}O_{11}}{1 \text{ mol } C_{12}H_{22}O_{11}} \quad \text{and} \quad \frac{1 \text{ mol } C_{12}H_{22}O_{11}}{342.3 \text{ g } C_{12}H_{22}O_{11}}$$

$$1 \text{ mol of } C_{12}H_{22}O_{11} = 6.022 \times 10^{23} \text{ molecules of } C_{12}H_{22}O_{11}$$

$$\frac{6.022 \times 10^{23} \text{ molecules}}{1 \text{ mol } C_{12}H_{22}O_{11}} \quad \text{and} \quad \frac{1 \text{ mol } C_{12}H_{22}O_{11}}{6.022 \times 10^{23} \text{ molecules}}$$

STEP 4 **Set Up Problem** Using the molar mass as a conversion factor, convert the grams of sucrose to moles of sucrose and then use Avogadro's number to convert moles of sucrose to molecules.

$$4536 \text{ g } C_{12}H_{22}O_{11} \times \frac{1 \text{ mol } C_{12}H_{22}O_{11}}{342.3 \text{ g } C_{12}H_{22}O_{11}} \times \frac{6.022 \times 10^{23} \text{ molecules}}{1 \text{ mol } C_{12}H_{22}O_{11}}$$

$$= 7.980 \times 10^{24} \text{ molecules of sucrose}$$

STUDY CHECK

In the body, caffeine, which has a formula of $C_8H_{10}N_4O_2$, acts as a stimulant and diuretic. How many moles of nitrogen are in 2.50 g of caffeine?

End-of-Chapter Questions

Understanding the Concepts questions encourage students to think about the concepts they have learned. **Additional Questions and Problems** integrate the topics from the entire chapter to promote understanding and critical thinking. **Challenge Questions** are designed for group work in inquiry-based learning environments. **Combining Ideas Problem Sets** appear after every two to four chapters as a set of integrated problems designed to test students' cumulative understanding of the previous chapters.

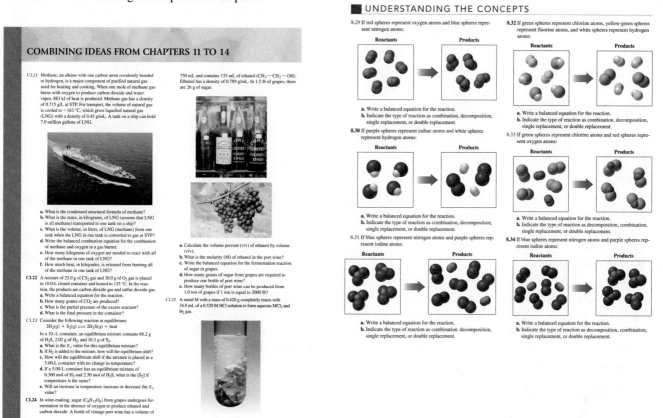

Make Learning Part of the Grade® (www.masteringchemistry.com)

MasteringChemistry® emulates the instructor's office-hour environment, coaching students on problem-solving techniques by asking questions that reveal gaps in understanding and giving students the power to answer questions on their own. It tutors students *individually*— with feedback specific to their errors, offering optional simpler steps.

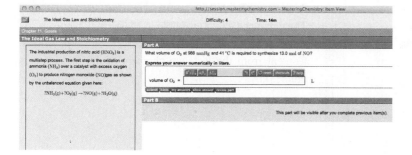

Student Tutorial

MasteringChemistry® is the only system to provide instantaneous feedback specific to the most-common wrong answers. Students submit an answer and receive immediate, error-specific feedback. Simpler sub-problems—"hints"—are provided upon request.

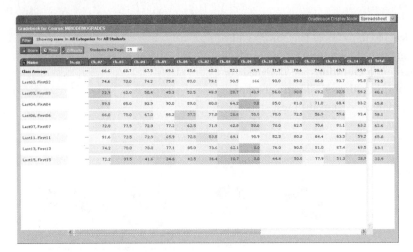

Unmatched Gradebook Capability

MasteringChemistry® is the only system to capture the step-by-step work of each student, including wrong answers submitted, hints requested, and time taken on every step. This data powers an unprecedented gradebook.

At-a-glance Diagnostics

Spot students in trouble at a glance with the color-coded gradebook, effortlessly identify the most difficult problem (and step within that problem) in your last assignment, or critique the detailed work of any one student who needs more help. Compare results on any problem and any step with a previous class, or the national average.

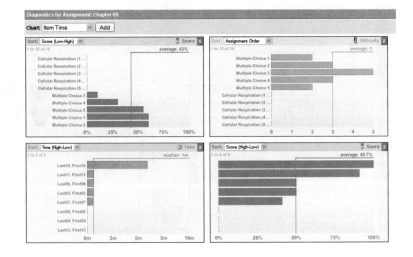

Unparalleled Study Tools

Robust Study Area

This area within MasteringChemistry® provides a unique approach to engaging students using Learning Goals, Self-study Activities, Career Focus, and Case Studies.

Pearson eText

The enhanced eText available in MasteringChemistry® is fully searchable with note delivering capabilities. Content from the last two chapters of the text appears online only.

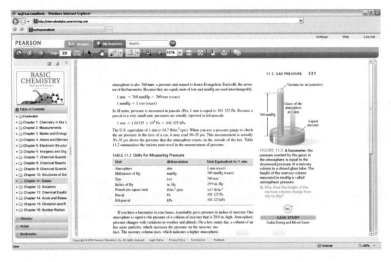

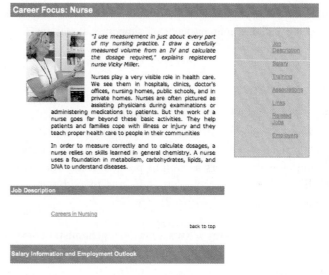

Career Focus

The text's career focus is reinforced with MasteringChemistry® activities.

Case Studies

Case studies in the book correspond to activities within MasteringChemistry®.

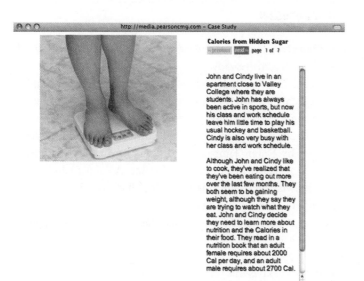

Self Study Activities

These activities appear within the book and in MasteringChemistry®.

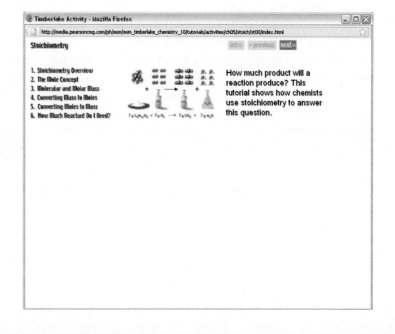

Instructional Package

Basic Chemistry, third edition, is the nucleus of an integrated teaching and learning package of support material for both students and professors.

For Students

The Study Guide for *Basic Chemistry*, third edition, by Karen Timberlake is keyed to the learning goals in the text and designed to promote active learning through a variety of exercises with answers as well as practice tests. The **Study Guide** also contains complete solutions to odd-numbered problems. (ISBN 0-321-67626-2)

MasteringChemistry® (www.masteringchemistry.com)
The most advanced, most widely used online chemistry tutorial and homework program is available for the third edition of *Basic Chemistry*. MasteringChemistry® is a homework and tutoring system that utilizes the Socratic method to coach students through problem-solving techniques, offering hints and simpler questions on request to help students *learn,* not just practice. A powerful gradebook with diagnostics that gives instructors unprecedented insight into their students' learning is also available.

Pearson eText
myeBook is integrated within MasteringChemistry® giving students, with new books, easy access to the electronic text. myeBook offers students the power to create notes, highlight text in different colors, create bookmarks, zoom, and view single or multiple pages.

Media Icons
Media Icons in the margins throughout the text direct students to tutorials and case studies in the Study Area located within MasteringChemistry® for *Basic Chemistry,* third edition.

Laboratory Manual by Karen Timberlake
This best-selling lab manual coordinates 42 experiments with the topics in *Basic Chemistry,* third edition, and uses new terms during the lab and explores chemical concepts. Laboratory investigations develop skills of manipulating equipment, reporting data, solving problems, making calculations, and drawing conclusions. (ISBN 0-321-66310-1)

Essential Laboratory Manual by Karen Timberlake
This lab manual contains 25 experiments for the chemical topics in *Basic Chemistry*, third edition. (ISBN 0-321-66310-1)

For Instructors

MasteringChemistry® (www.masteringchemistry.com)
MasteringChemistry® is the first adaptive-learning online homework and tutorial system. Instructors can create online assignments for their students by choosing from a wide range of items, including end-of-chapter problems and research-enhanced tutorials. Assignments are automatically graded with up-to-date diagnostic information, helping instructors pinpoint where students struggle either individually or as a class as a whole.

Instructor Solutions Manual
The ***Instructor Solutions Manual,*** by Karen and William Timberlake with contributions by Mark Quirie, includes answers and solutions for all questions and problems in the text. (ISBN 0-321-67662-9)

Instructor Resource Center on CD/DVD for *Basic Chemistry*, third edition
This CD/DVD includes all the art, photos, and tables from the book in high-resolution format for use in classroom projection or when creating study materials and tests. In addition, the instructors can access over 2000 PowerPoint lecture outlines, written by Karen Timberlake, or create their own easily with the art provided in PowerPoint format. Also available on the Media Manager are downloadable files of the *Instructor Solutions Manual* and *Test Bank*, as well as a set of "clicker questions" designed for use with classroom-response systems. (ISBN 0-321-67625-4)

Printed Test Bank
The **Test Bank**, by William Timberlake, includes more than 1600 multiple-choice, matching, and true/false questions. (ISBN 0-321-67624-6)

Online Instructor Manual for Laboratory Manual
This manual contains answers to report pages for the *Laboratory Manual* and the *Essential Laboratory Manual*. (ISBN 0-321-66310-1)

Also visit the Prentice Hall catalog page for Timberlake's *Basic Chemistry*, third edition, at **www.pearsonhighered.com** to download available instructor supplements.

ACKNOWLEDGMENTS

The preparation of a new text is a continuous effort of many people. As in our work on other textbooks, we are thankful for the support, encouragement, and dedication of many people who put in hours of tireless effort to produce a high-quality book that provides an outstanding learning package. The editorial team at Pearson Publishing has done an exceptional job. We want to thank Nicole Folchetti, editor in chief, and acquisitions editor, Terry Haugen, who supported our vision of this third edition and the development of a new topic sequence including organic chemistry in the early chapters as well as the addition of new Concept Checks and more Guides to Problem Solving, new Chemistry and Health, Chemistry and History, Chemistry and Industry, and Chemistry and the Environment boxes, and new problems in Understanding the Concepts and Combining Ideas. We much appreciate all the wonderful work of Jessica Neumann, assistant editor, who was like an angel encouraging us at each step, while skillfully coordinating reviews, art, Web site materials, and all the things it takes to make a book come together. We are grateful to Ray Mullaney, editor in chief of science book development, for his watchful eyes during the writing and production of this new edition. I appreciate the work of Beth Sweeten, project manager and Simone Lukashov of Preparé, who brilliantly coordinated all phases of the manuscript to the final pages of a beautiful book. Thanks to Mark Quirie, reviewer, who precisely analyzed and edited the initial and final manuscripts and pages to make sure the words and problems were correct to help students learn chemistry.

We are especially proud of the art program in this text, which lends beauty and understanding to chemistry. We would like to thank Maureen Eide, art director, Mark Ong, book designer, and Travis Amos, photo editor, whose creative ideas provided the outstanding design for the cover and pages of the book. Eric Schrader, photo researcher, was invaluable in researching and selecting vivid photos for the text so that students can see the beauty of chemistry. Thanks also to Bio-Rad Laboratories for their courtesy and use of KnowItAll ChemWindows, drawing software that helped us produce chemical structures for the manuscript. The macro-to-micro illustrations designed by Production Solutions and Precision Graphics give students visual impressions of the atomic and molecular organization of everyday things and are a fantastic learning tool. We want to thank Donna Mulder for the hours spent proofreading all the pages. We also appreciate all the hard work put in by the marketing team in the field and by Erin Gardner, marketing manager.

We are extremely grateful to an incredible group of peers for their careful assessment of all the new ideas for the text; for their suggested additions, corrections, changes, and deletions; and for providing an incredible amount of feedback about improvements for the book. In addition, we appreciate the time scientists took to let us take photos and discuss their work with them. We admire and appreciate every one of you.

If you would like to share your experience with chemistry, or have questions and comments about this text, we would appreciate hearing from you.

Karen and Bill Timberlake
Email: khemist@aol.com

Reviewers

Maher Atteya
Georgia Perimeter College

Pamela Goodman
Moraine Valley Community College

David Nachman
Mesa Community College

MaryKay Orgill
University of Nevada, Las Vegas

Mark Quirie
Algonquin College

Ben Rutherford
Washington State Community College

Previous Edition Reviewers

Michelle Driessen
University of Minnesota

Wesley Fritz
College of DuPage

Amy Waldman Grant
El Camino College

Richard Lavallee
Santa Monica College

MaryKay Orgill
University of Nevada, Las Vegas

Cyriacus Chris Uzomba
Austin Community College

Chemistry in Our Lives

1

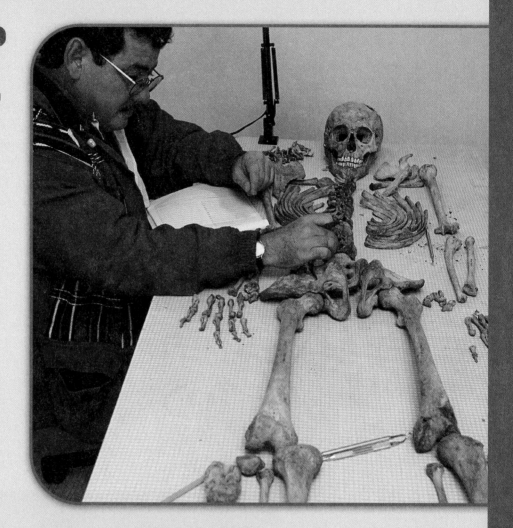

"*Chemistry plays an integral part in all aspects of medical legal death investigation,*" says Charles L. Cecil, forensic anthropologist and medical legal death investigator, San Francisco Medical Examiner's Office. "*Crime-scene analysis of blood droplets determines whether they are human or nonhuman. The analyses of toxicological samples of blood and/or other fluids help determine the cause and time of death. Specialists in forensic anthropology can analyze trace-element ratios in bones to identify the number of individuals in mixed human bone situations. These conditions are quite often found during investigations of massive human-rights violations, such as the site of El Mozote in El Salvador.*"

Forensic anthropologists help police identify skeletal remains by determining gender, approximate age, height, and possible cause of death. Bone analysis can also provide information about illnesses or the type of trauma a person may have experienced.

Visit **www.masteringchemistry.com** for self-study materials and instructor-assigned homework.

Now that you are in a chemistry class, you may be wondering what you will be learning. What are some questions in science you have been curious about? Perhaps you are interested in how smog is formed, what causes ozone depletion, how nails form rust, or how aspirin relieves a headache. Just like you, chemists are curious about the world we live in.

- How does car exhaust produce the smog that hangs over our cities? One component of car exhaust is nitrogen oxide (NO), which forms in car engines where high temperatures convert nitrogen gas (N_2) and oxygen gas (O_2) to NO. In chemistry, reactions like this are written in the form of an equation: $N_2(g) + O_2(g) \longrightarrow 2NO(g)$. Then, the product NO reacts with the oxygen in the air, producing nitrogen dioxide, or NO_2: $2NO(g) + O_2(g) \longrightarrow 2NO_2(g)$. It is NO_2 that causes smog to have a characteristic reddish brown color.

- Why has the ozone layer been depleted in certain parts of the atmosphere? During the 1970s, scientists discovered that substances called *chlorofluorocarbons* (CFCs) were associated with the depletion of ozone. As CFCs are broken down by ultraviolet (UV) light, chlorine (Cl) is released and reacts rapidly with ozone (O_3) in the atmosphere to form chlorine oxide gas (ClO) and oxygen: $Cl(g) + O_3(g) \longrightarrow ClO(g) + O_2(g)$. This reaction causes the breakdown of ozone molecules and the destruction of the protective ozone layer, which has been linked to an increased risk of skin cancer.

Molecules of NO_2

The chemical reaction of NO with oxygen in the air forms NO_2, which produces the reddish brown color of smog.

- Why does an iron nail rust when exposed to air and rain? When solid iron (Fe) in a nail reacts with oxygen gas in the air, the oxidation of iron forms rust (Fe_2O_3): $4Fe(s) + 3O_2(g) \longrightarrow 2Fe_2O_3(s)$.

- Why does aspirin relieve a headache? When a part of the body is injured, substances called *prostaglandins* are produced, which cause inflammation and pain. Aspirin acts to block the production of prostaglandins, thereby reducing inflammation, pain, and fever.

Chemists perform many different kinds of research. Some design new fuels and more efficient ways to use them. Researchers in the medical field look for evidence that will aid in developing new treatments for diabetes, genetic defects, cancer, AIDS, and other diseases. Researchers in the environmental field study the ways in which human development impacts the environment and develop processes that help reduce environmental degradation. For the researcher in the laboratory, the physician in the dialysis unit, the environmental chemist, or the agricultural scientist, chemistry plays a central role in understanding problems, assessing possible solutions, and making important decisions.

The formation of rust on an iron nail is an example of a chemical reaction.

1.1 Chemistry and Chemicals

Chemistry is the study of the composition, structure, properties, and reactions of matter. *Matter* is another word for all the substances that make up our world. Perhaps you imagine that chemistry takes place only in a laboratory where a chemist is working in a white coat and goggles. Actually, chemistry happens all around you every day and has a big impact on everything you use and do. You are doing chemistry when you cook food, add bleach to your laundry, or start your car. A chemical reaction has taken place when silver tarnishes or an antacid tablet fizzes when dropped into water. Plants grow because chemical reactions convert carbon dioxide, water, and energy to carbohydrates. Chemical reactions take place when you digest food and break it down into substances that you need for energy and health.

LEARNING GOAL

Define the term *chemistry* and identify substances as chemicals.

Antacid tablets undergo a chemical reaction when dropped into water.

Branches of Chemistry

The field of chemistry is divided into branches such as organic, inorganic, and general chemistry. Organic chemistry is the study of substances that contain the element carbon. Inorganic chemistry is the study of all other substances except those that contain carbon. General chemistry is the study of the composition, properties, and reactions of matter.

Today chemistry is often combined with other sciences, such as geology, biology, and physics, to form cross-disciplines such as geochemistry, biochemistry, and physical chemistry. Geochemistry is the study of the chemical composition of ores, soils, and minerals of the surface of the Earth and other planets. Biochemistry is the study of the chemical reactions that take place in biological systems. Physical chemistry is the study of the physical nature of chemical systems including energy changes.

Chemicals

All the things you see around you are composed of one or more chemicals. A **chemical** is a substance that always has the same composition and properties wherever it is found. When a chemical undergoes a chemical change, a new substance with a new composition and properties is formed. Chemical processes take place in chemistry laboratories, manufacturing plants, and pharmaceutical labs as well as every day in nature and in our bodies. Often the terms *chemical* and *substance* are used interchangeably to describe a specific type of matter.

A biochemist studies the chemistry of biological substances.

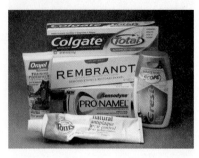

Toothpaste is a combination of many chemicals.

TABLE 1.1 Chemicals Commonly Used in Toothpaste

Chemical	Function
Calcium carbonate	An abrasive used to remove plaque
Sorbitol	Prevents loss of water and hardening of toothpaste
Carrageenan (seaweed extract)	Keeps toothpaste from hardening or separating
Glycerin	Makes toothpaste foam in the mouth
Sodium lauryl sulfate	A detergent used to loosen plaque
Titanium dioxide	Makes toothpaste white and opaque
Triclosan	Inhibits bacteria that cause plaque and gum disease
Sodium fluorophosphate	Prevents formation of cavities by strengthening tooth enamel with fluoride
Methyl salicylate	Gives toothpaste a pleasant wintergreen flavor

Every day, you use products containing substances that were developed and prepared by chemists. Soaps and shampoos contain chemicals that remove oils on your skin and scalp. When you brush your teeth, the chemicals in toothpaste clean your teeth, prevent plaque formation, and stop tooth decay. Some chemicals commonly contained in toothpaste are listed in Table 1.1.

In cosmetics and lotions, chemicals are used to moisturize, prevent deterioration of the product, fight bacteria, and thicken the product. Your clothes may be made of natural materials such as cotton or synthetic substances such as nylon or polyester. Perhaps you wear a ring or watch made of gold, silver, or platinum. Your breakfast cereal is probably fortified with iron, calcium, and phosphorus, while the milk you drink is enriched with vitamins A and D. Antioxidants are chemicals added to food to prevent it from spoiling. Some of the chemicals you may encounter when you cook in the kitchen are shown in Figure 1.1.

Silicon dioxide (glass)
Chemically treated water
Metal alloy
Natural polymers
Natural gas
Fruits grown with fertilizers and pesticides

FIGURE 1.1 Many of the items found in a kitchen are chemicals or products of chemical reactions.

Q What are some other chemicals found in a kitchen?

CONCEPT CHECK 1.1

■ **Chemicals**

Why is copper wire an example of a chemical, while sunlight is not?

ANSWER Copper wire is a substance that has the same composition and properties wherever it is found. Sunlight is energy given off by the Sun. Thus, sunlight is not a substance, and it does not contain matter.

SAMPLE PROBLEM | 1.1

■ **Everyday Chemicals**

Identify the chemical within each of the following statements:

a. Soda cans are made from aluminum.
b. Salt (sodium chloride) is used to preserve meat and fish.
c. Sugar (sucrose) is used as a sweetener.

SOLUTION

a. aluminum   **b.** salt (sodium chloride)
c. sugar (sucrose)

STUDY CHECK

Which of the following are chemicals?

a. iron   **b.** tin
c. a low temperature   **d.** water

The answers to all of the *Study Checks* can be found at the end of each chapter.

QUESTIONS AND PROBLEMS

Chemistry and Chemicals

The answers to all the magenta, odd-numbered *Questions and Problems* can be found at the end of each chapter. Check your answers to determine if you understand the material.

1.1 Write a one-sentence definition for each of the following:
     a. chemistry   **b.** chemical

1.2 Ask two of your friends (not in this class) to define the terms in Problem 1.1. Do their answers agree with the definitions you provided?

1.3 Obtain a vitamin bottle and observe its ingredients. List four. Which ones are chemicals?

1.4 Obtain a box of breakfast cereal and observe its ingredients. List four. Which ones are chemicals?

1.5 Read the labels on some items found in your medicine cabinet. What are the names of some chemicals contained in those items?

1.6 Read the labels on products used to wash and clean your car. What are the names of some chemicals contained in those products?

1.7 A "chemical-free" shampoo includes the following ingredients: water, cocomide, glycerin, and citric acid. Is the shampoo truly "chemical-free"?

1.8 A "chemical-free" sunscreen includes the following ingredients: titanium dioxide, vitamin E, and vitamin C. Is the sunscreen truly "chemical-free"?

1.9 Pesticides are chemicals. Give one advantage and one disadvantage of using pesticides.

1.10 Sugar is a chemical. Give one advantage and one disadvantage of eating sugar.

CHEMISTRY AND HISTORY

Early Chemists: The Alchemists

For many centuries, chemists have studied changes in matter. From the time of the ancient Greeks to about the sixteenth century, early scientists, called alchemists, described matter in terms of four components of nature: earth, air, fire, and water. These components had the qualities of hot, cold, damp, or dry. By the eighth century, alchemists believed that they could rearrange these qualities to change metals such as copper and lead into gold and silver. They searched for an unknown substance called a *philosopher's stone*, which they thought would turn metals into gold as well as prolong youth and postpone death. Although these efforts failed, the alchemists did provide information on the processes and chemical reactions involved in the extraction of metals from ores. During the many centuries that alchemy flourished, alchemists made observations of matter and identified the properties of many substances. They also designed some of the first laboratory equipment and developed early laboratory procedures.

The alchemist Paracelsus (1493–1541) thought that alchemy should be about preparing new medicines, not about producing gold. Using observation and experimentation, he proposed that a healthy body was regulated by a series of chemical processes that could be unbalanced by certain chemical compounds and rebalanced by using minerals and medicines. For example, he determined that inhaled dust, not underground spirits, caused lung disease in miners. He also thought that goiter was a problem caused by contaminated water, and he treated syphilis with compounds of mercury. His opinion of medicines was that the right dose makes the difference between a poison and a cure. Today this idea is part of the risk assessment of medicines. Paracelsus changed alchemy in ways that helped to establish modern medicine and chemistry.

Alchemists in the Middle Ages developed laboratory procedures.

Swiss alchemist Paracelsus (1493–1541) identified the properties of many substances.

LEARNING GOAL

Describe the activities that are part of the scientific method.

Linus Pauling, Nobel Prize winner in chemistry, 1954

1.2 Scientific Method: Thinking Like a Scientist

When you were very young, you explored the things around you by touching and tasting. As you grew, you asked questions about the world in which you live. What is lightning? Where does a rainbow come from? Why is water blue? As an adult, you may have wondered how antibiotics work or why vitamins are important to your health. Every day, you ask questions and seek answers to organize and make sense of the world around you.

When the late Nobel Laureate Linus Pauling described his student life in Oregon, he recalled that he read many books on chemistry, mineralogy, and physics. "I mulled over the properties of materials: why are some substances colored and others not, why are some minerals or inorganic compounds hard and others soft?" He said, "I was building up this tremendous background of empirical knowledge and at the same time asking a great number of questions." Linus Pauling won two Nobel Prizes: the first, in 1954, was in chemistry for his work on the nature of chemical bonds and the determination of the structures of complex substances; the second, in 1962, was the Peace Prize.

Scientific Method

Although the process of trying to understand nature is unique to each scientist, a set of general principles, called the **scientific method**, helps to describe how a scientist thinks.

1. **Observations.** The first step in the scientific method is to observe, describe, and measure an event in nature. Observations based on measurements are called *data*.

2. **Hypothesis.** After sufficient data are collected, a *hypothesis* is proposed, which states a possible interpretation of the observations. The hypothesis must be stated in a way that it can be tested by experiments.

3. **Experiments.** Experiments are tests that determine the validity of the hypothesis. Often, many experiments are performed to test the hypothesis, and a large amount of information is collected. Many experiments are needed to support the original hypothesis. However, if just one experiment produces a different result than predicted by the hypothesis, a modified hypothesis must be proposed. Then new experiments are conducted to test the new hypothesis.

4. **Theory.** When experiments are repeated by many scientists with consistent results, the hypothesis may be confirmed. Consequently, that hypothesis may become a *theory*. Even then, a theory continues to be tested and, based on new experimental results, may need to be modified or replaced. Then the cycle of the scientific method begins again with the proposal of a new hypothesis.

TUTORIAL
The Scientific Method

Students make observations in the chemistry laboratory.

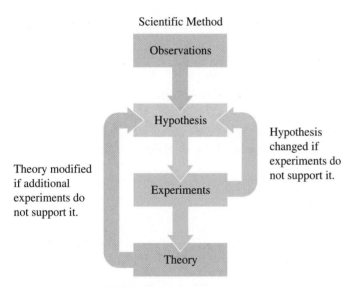

The scientific method develops a theory using observations, hypotheses, and experiments.

Using the Scientific Method in Everyday Life

You may be surprised to realize that you use the scientific method in your everyday life. Suppose you visit a friend in her home. Soon after you arrive, your eyes start to itch and you begin to sneeze. Then you observe that your friend has a new cat. Perhaps you ask yourself why you are sneezing and form the hypothesis that you are allergic to cats. To test your hypothesis, you leave your friend's home. If the sneezing stops, perhaps your hypothesis is correct. You test your hypothesis further by visiting another friend who also has a cat. If you start to sneeze again, your experimental results support your hypothesis that you are allergic to cats. However, if you continue sneezing after you leave your friend's home, your hypothesis is not supported. Now you need to form a new hypothesis, which could be that you have a cold.

Through observation you may determine that you are allergic to cat hair and dander.

A silver tray tarnishes when exposed to the air.

Tomato plants grow faster when placed in the sun.

CONCEPT CHECK 1.2

■ **Scientific Method**

Identify each of the following as an observation (O), a hypothesis (H), or an experiment (E):

a. Drinking coffee at night keeps me awake.
b. If I stop drinking coffee in the afternoon, I will be able to sleep at night.
c. I will try drinking coffee only in the morning.

ANSWER

a. Describing what happens when I drink coffee is an observation (O).
b. Describing what may happen if I stop drinking coffee in the afternoon is a hypothesis (H).
c. Changing the time for drinking coffee is an experiment (E).

SAMPLE PROBLEM | **1.2**

■ **Scientific Method**

Identify each of the following statements as an observation (O) or a hypothesis (H):

a. A silver tray turns a dull gray color when left uncovered.
b. North of the equator, it is warmer in summer than in winter.
c. Ice cubes float in water because they are less dense.

SOLUTION

a. observation (O)   **b.** observation (O)   **c.** hypothesis (H)

STUDY CHECK

The following statements are found in a student's notebook. Identify each of the following as an observation (O), a hypothesis (H), or an experiment (E):

a. "Today I placed two tomato seedlings in the garden and two more in a closet. I will give all the plants the same amount of water and fertilizer."
b. "After 50 days, the tomato plants in the garden are 3 feet high with green leaves. The plants in the closet are 8 inches tall and yellow."
c. "Tomato plants need sunlight to grow."

Science and Technology

When scientific information is applied to industrial and commercial uses, it is called technology. Such uses have made the chemical industry one of the largest industries in the United States. Every year, technology provides new materials or procedures that produce more energy, cure diseases, improve crops, and produce new kinds of synthetic materials. Table 1.2 lists some of the important scientific discoveries, laws, theories, and technological innovations that have been made over the past 300 years.

Not all scientific discoveries have been positive, however. The production of some substances has contributed to the development of hazardous conditions in our environment. We have become concerned about the energy requirements of new products and how some

TABLE 1.2 Some Important Scientific Discoveries, Laws, Theories, and Technological Innovations

Discovery, Law, Theory, or Innovation	Date	Discoverer or Inventor	Country
Law of gravity	1687	Isaac Newton	England
Oxygen	1774	Joseph Priestley	England
Electric battery	1800	Alessandro Volta	Italy
Atomic theory	1803	John Dalton	England
Anesthesia, ether	1842	Crawford Long	United States
Nitroglycerin	1847	Ascanio Sobrero	Italy
Germ theory	1865	Louis Pasteur	France
Antiseptic surgery	1865	Joseph Lister	England
Discovery of nucleic acids	1869	Friedrich Miescher	Switzerland
Radioactivity	1896	Henri Becquerel	France
Discovery of radium	1898	Marie and Pierre Curie	Poland, France
Quantum theory	1900	Max Planck	Germany
Theory of relativity	1905	Albert Einstein	Germany
Identification of components of RNA and DNA	1909	Phoebus Theodore Levene	United States
Insulin	1922	Frederick Banting, Charles Best, John Macleod	Canada
Penicillin	1928	Alexander Fleming	England
Nylon	1937	Wallace Carothers	United States
Discovery of DNA as genetic material	1944	Oswald Avery	United States
Synthetic production of transuranium elements	1944	Glenn Seaborg, Arthur Wahl, Joseph Kennedy, Albert Ghiorso	United States
Determination of DNA structure	1953	Francis Crick, Rosalind Franklin, James Watson	England, United States
Polio vaccine	1954	Jonas Salk	United States
	1957	Albert Sabin	
Laser	1958	Charles Townes	United States
	1960	Theodore Maiman	
Cellular phones	1973	Martin Cooper	United States
MRI	1980	Paul Lauterbur	United States
Prozac	1988	Ray Fuller	United States
World Wide Web available to the public	1993	Tim Berners-Lee	Switzerland
HIV protease inhibitor	1995	Joseph Martin, Sally Redshaw	United States
DVD	1996	Many contributors	Japan
Human genome mapped	2007	Craig Venter	United States

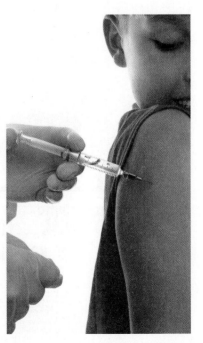

The polio vaccine protects a child from this harmful disease.

materials may cause changes in our oceans and atmosphere. We want to know if the new materials can be recycled, how they are broken down, and whether there are processes that are safer. The ways in which we continue to utilize scientific research will strongly impact our planet and its community in the future. These decisions can best be made if every citizen has an understanding of science.

CHEMISTRY AND THE ENVIRONMENT

DDT—Good Pesticide, Bad Pesticide

DDT (**D**ichloro**d**iphenyl**t**richloroethane) was once one of the most commonly used pesticides. DDT is an example of organic compounds known as hydrocarbons, which typically are composed of the elements of carbon (C) and hydrogen (H). In a molecule of DDT, there are 14 carbon atoms and 9 hydrogen atoms, as well as 5 chlorine atoms. The hydrocarbon portion makes DDT insoluble in water, and the Cl atoms make DDT difficult to break down.

Although DDT was first synthesized in 1874, it was not used as an insecticide until 1939. Before DDT was widely used, insect-borne diseases such as malaria and typhus were rampant in many parts of the world. Paul Müller, who discovered that DDT was an effective pesticide, was recognized for saving many lives and received the Nobel Prize in Physiology or Medicine in 1948. DDT was considered the ideal pesticide because it was toxic to many insects, had a low toxicity to humans and animals, and was inexpensive to prepare.

In the United States, DDT was extensively used in homes as well as on crops, such as cotton and soybeans. Because of its stable chemical structure, DDT did not break down quickly in the environment, which meant that it did not have to be applied as often. At first, everyone was pleased with DDT as crop yields increased and diseases such as malaria and typhus were controlled.

However, by the early 1950s, problems attributed to DDT began to surface. Insects were becoming more resistant to the pesticide. At the same time, the public was increasingly concerned about the long-term impact of a substance that could remain in the environment for

The Brown Pelican was once an endangered species due to the use of DDT.

many years. The metabolic systems of humans and animals cannot break down DDT, which is soluble in fats but not in water and is stored in the fatty tissues of the body. Although the concentration of DDT applied to crops was very low, runoff containing DDT reached the oceans, where the DDT was absorbed by fish.

When birds such as the Brown Pelicans in Florida and California consumed fish contaminated with DDT, the amount of calcium in their eggshells was significantly reduced. As a result, incubating eggs cracked open early, causing offspring to die. Due to this difficulty with reproduction, the populations of birds such as the Brown Pelican dropped significantly and they became endangered.

By 1972, DDT was banned in the United States. Since then, the population of Brown Pelicans has increased and they are no longer considered endangered. Today new types of pesticides, which are more water soluble and break down faster in the environment, have replaced the long-lasting pesticides such as DDT. However, these new pesticides are much more toxic to humans.

A 1947 advertisement recommends the household use of DDT.

A field is sprayed with pesticide.

QUESTIONS AND PROBLEMS

Scientific Method: Thinking Like a Scientist

1.11 Define each of the following terms of the scientific method:
 a. hypothesis   **b.** experiment
 c. theory   **d.** observation

1.12 Identify each of the following activities in the scientific method as an observation (O), a hypothesis (H), an experiment (E), or a theory (T):
 a. Formulate a possible explanation for your experimental results.
 b. Collect data.
 c. Design an experimental plan that will give new information about a problem.
 d. State a generalized summary of your experimental results.

1.13 At a popular restaurant, where Chang is the head chef, the following events occurred:
 a. Chang determined that sales of the house salad had dropped.
 b. Chang decided that the house salad needed a new dressing.
 c. In a taste test, Chang prepared four bowls of lettuce, each with a new dressing: sesame seed, olive oil and balsamic vinegar, creamy Italian, and blue cheese.
 d. The tasters rated the sesame seed salad dressing as the favorite.
 e. After two weeks, Chang noted that the orders for the house salad with the new sesame seed dressing had doubled.
 f. Chang decided that the sesame seed dressing improved the sales of the house salad because the sesame seed dressing enhanced the taste.

The favorite salad has sesame seed dressing.

Identify each activity, **a–f**, as an observation (O), a hypothesis (H), an experiment (E), or a theory (T).

1.14 Lucia wants to develop a process for dyeing shirts so that the color will not fade when the shirt is washed. She proceeds with the following activities:
 a. Lucia notices that the dye in a design fades when the shirt is washed.
 b. Lucia decides that the dye needs something to help it combine with the fabric.
 c. She places a spot of dye on each of four shirts and then places each one separately in water, salt water, vinegar, and baking soda and water.
 d. After one hour, all the shirts are removed and washed with a detergent.
 e. Lucia notices that the dye has faded on the shirts in water, salt water, and baking soda, while the dye did not fade on the shirt soaked in vinegar.
 f. Lucia thinks that the vinegar binds with the dye so it does not fade when the shirt is washed.

Identify each activity, **a–f**, as an observation (O), a hypothesis (H), an experiment (E), or a theory (T).

1.3 A Study Plan for Learning Chemistry

Here you are taking chemistry, perhaps for the first time. Whatever your reasons for choosing to study chemistry, you can look forward to learning many new and exciting ideas.

LEARNING GOAL

Develop a study plan for learning chemistry.

Features in This Text Help You Study Chemistry

This text has been designed with a variety of study aids to complement different learning styles. On the inside of the front cover is a periodic table of the elements that make up matter. On the inside of the back cover are tables that summarize useful information needed throughout the study of chemistry. Each chapter begins with *Looking Ahead*, which outlines the topics in the chapter. A *Learning Goal* at the beginning of each section previews the concepts you are to learn. A comprehensive *Glossary and Index* is included at the end of the text.

Before you begin reading, obtain an overview of the chapter by reviewing the topics in *Looking Ahead*. As you prepare to read a section of the chapter, look at the section title and turn it into a question. For example, for Section 1.1, "Chemistry and Chemicals," you could write a question that asks "What is chemistry?" or "What are chemicals?" When you are ready to read through that section, review the *Learning Goal*, which tells you what to expect in that section and what you need to accomplish. As you read, try to answer the questions you considered. Throughout the chapter, you will find *Concept Checks* that will help you understand key ideas. When you come to a *Sample Problem*, take the time to work it through, and try the associated *Study Check*. Many *Solutions* are now accompanied

Studying in a group can be beneficial to learning.

by a visual *Guide to Problem Solving (GPS)*. At the end of each section, you will find a set of *Questions and Problems* that allows you to apply problem solving immediately to the new concepts.

When you finish each section, immediately work through the *Questions and Problems* for practice. Throughout the chapters, boxes titled "Chemistry and Health," "Chemistry and History," "Chemistry and Industry," and "Chemistry and the Environment" help you connect the chemical concepts you are learning to real-life situations. Many of the figures and diagrams throughout the text use macro-to-micro illustrations to depict the atomic level of organization of ordinary objects. These visual models illustrate the concepts described in the text and allow you to "see" the world in a microscopic way.

At the end of each chapter, you will find several study aids that complete the chapter. The *Key Terms* are boldfaced in the text and listed again with their definitions at the end of the chapter. *Concept Maps* and *Chapter Reviews* show the connections between important concepts and provide a summary. *Understanding the Concepts*, a set of questions that use art and structures, help you visualize concepts. *Additional Questions and Problems* and *Challenge Problems* provide additional opportunities to test your understanding of the topics in the chapter. All the problems are paired, and the answers to the *Study Checks* and to the odd-numbered *Selected Questions and Problems* are provided at the end of the chapter. If your answers match, you most likely understand the topic; if not, you need to study the section again.

After some chapters, problem sets called *Combining Ideas* test your ability to solve problems containing material from more than one chapter.

Using Active Learning to Learn Chemistry

A student who is an active learner thinks about the new chemical concepts while reading the text and attending lectures. Let's see how this is done.

As you read and practice problem solving, you remain actively involved in studying, which enhances the learning process. In this way, you learn small bits of information at a time and establish the necessary foundation for understanding the next section. You may also note questions you have about the reading to discuss with your professor and laboratory instructor. Table 1.3 summarizes these steps for active learning. The time you spend in a lecture can also be useful as a learning time. By keeping track of the class schedule and reading the assigned material before a lecture, you become aware of the new terms and concepts you need to learn. Some questions that occur during your reading may be answered during the lecture. If not, you can ask for further clarification from your professor.

Many students find that studying with a group can be beneficial to learning. In a group, students motivate each other to study, fill in gaps, and correct misunderstandings by teaching and learning together. Studying alone does not allow the process of peer correction. In a group, you can cover the ideas more thoroughly as you discuss the reading and problem solve with other students. It is easier to retain new material and new ideas if you study in small sessions throughout the week rather than all at once. Waiting to study until the night before an exam does not give you time to understand concepts and practice problem solving.

TABLE 1.3 Steps in Active Learning

1. Read each *Learning Goal* for an overview of the material.
2. Form a question from the title of each section before reading.
3. Read the section looking for answers to your question.
4. Self-test by working *Concept Checks*, *Sample Problems*, and *Study Checks*.
5. Complete the *Questions and Problems* that follow that section and check the answers for the magenta odd-numbered problems.
6. Work the exercises in the *Study Guide* and go to *http://www.masteringchemistry.com* for self-study materials and instructor-assigned homework (optional).
7. Proceed to the next section and repeat the steps.

Thinking Scientifically about Your Study Plan

As you embark on your journey into the world of chemistry, think about your approach to studying and learning chemistry. You might consider some of the ideas in the following list. Check those ideas that will help you successfully learn chemistry. Commit to them now. *Your* success depends on *you*.

My study of chemistry will include the following:

_____ reviewing the *Learning Goals*

_____ keeping a problem notebook

_____ reading the text as an active learner

_____ self-testing by working the *Chapter Problems* and checking *Solutions* in the text

_____ reading the chapter before lecture

_____ going to lecture

_____ being an active learner in lecture

_____ organizing a study group

_____ seeing the professor during office hours

_____ completing exercises in the *Study Guide*

_____ working through the tutorials at *http://www.masteringchemistry.com*

_____ attending review sessions

_____ organizing my own review sessions

_____ studying in small doses as often as I can

Students discuss a chemistry problem with their professor during office hours.

CONCEPT CHECK 1.3

■ A Study Plan for Chemistry

What are some advantages to studying in a group?

ANSWER In a group, students motivate and support each other, fill in gaps, and correct misunderstandings. Ideas are discussed while reading and problem solving together.

SAMPLE PROBLEM | 1.3

■ A Study Plan for Learning Chemistry

Which of the following activities would you include in a study plan for learning chemistry successfully?

a. skipping a lecture
b. forming a study group
c. keeping a problem notebook
d. waiting to study until the night before the exam
e. becoming an active learner

SOLUTION

Your success in chemistry can be improved by

b. forming a study group
c. keeping a problem notebook
e. becoming an active learner

STUDY CHECK

Which of the following will help you learn chemistry?

a. skipping review sessions
b. working assigned problems
c. attending the professor's office hours
d. staying up all night before an exam
e. reading the assignment before a lecture

QUESTIONS AND PROBLEMS

A Study Plan for Learning Chemistry

1.15 What are four things you can do to help you to succeed in chemistry?

1.16 What are four things that would make it difficult for you to learn chemistry?

1.17 A student in your class asks you for advice on learning chemistry. Which of the following might you suggest?
  a. Form a study group.
  b. Skip a lecture.
  c. Visit the professor during office hours.

  d. Wait until the night before an exam to study.
  e. Become an active learner.
  f. Work the exercises in the *Study Guide*.

1.18 A student in your class asks you for advice on learning chemistry. Which of the following might you suggest?
  a. Do the assigned problems.
  b. Don't read the text; it's never on the test.
  c. Attend review sessions.
  d. Read the assignment before a lecture.
  e. Keep a problem notebook.
  f. Do the tutorials at *http://www.masteringchemistry.com*.

CONCEPT MAP

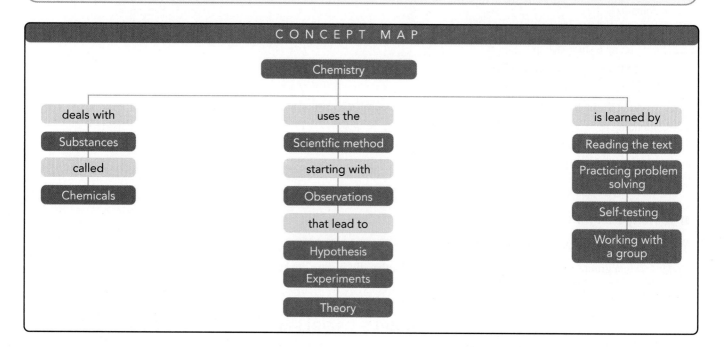

CHAPTER REVIEW

1.1 Chemistry and Chemicals
LEARNING GOAL: *Define the term* chemistry *and identify substances as chemicals.*
Chemistry is the study of the composition, structure, properties, and reactions of matter. A chemical is any substance that always has the same composition and properties wherever it is found.

1.2 Scientific Method: Thinking Like a Scientist
LEARNING GOAL: *Describe the activities that are part of the scientific method.*
The scientific method is a process of explaining natural phenomena beginning with making observations, forming a hypothesis, and per-

forming experiments. A theory may be proposed when repeated experimental results by many scientists support the hypothesis. Technology involves the application of scientific information to industrial and commercial uses.

1.3 A Study Plan for Learning Chemistry
LEARNING GOAL: *Develop a study plan for learning chemistry.*
A study plan for learning chemistry utilizes the features in the text and develops an active learning approach to study. By using the *Learning Goals* in the chapter and working the *Concept Checks*, *Sample Problems*, and *Study Checks*, and the *Questions and Problems* at the end of each section, the student can successfully learn the concepts of chemistry.

KEY TERMS

chemical A substance that has the same composition and properties wherever it is found.

chemistry A science that studies the composition of substances and the way they interact with other substances.

experiment A procedure that tests the validity of a hypothesis.

hypothesis An unverified explanation of a natural phenomenon.

observation Information determined by noting and recording a natural phenomenon.

scientific method The process of making observations, proposing a hypothesis, testing the hypothesis, and developing a theory that explains a natural event.

theory An explanation of an observation that has been validated by experiments that support a hypothesis.

UNDERSTANDING THE CONCEPTS

1.19 According to Sherlock Holmes, "One must follow the rules of scientific inquiry, gathering, observing, and testing data, then formulating, modifying, and rejecting hypotheses, until only one remains." Did Sherlock use the scientific method? Why or why not?

Sherlock Holmes is a fictional detective in novels written by Arthur Conan Doyle.

Aluminum melts at 660 °C.

1.20 In "A Scandal in Bohemia," Sherlock Holmes receives a mysterious note. He states, "I have no data yet. It is a capital mistake to theorize before one has data. Insensibly one begins to twist facts to suit theories, instead of theories to suit facts." What do you think Sherlock meant?

1.21 Classify each of the following statements as an observation (O) or a hypothesis (H):
  a. Aluminum melts at 660 °C.
  b. Dinosaurs became extinct when a large meteorite struck Earth and caused a huge dust cloud that severely decreased the amount of light reaching Earth.
  c. The 100-yard dash was run in 9.8 seconds.

1.22 Classify each of the following statements as observation (O) or a hypothesis (H):
  a. Analysis of ten ceramic dishes showed that four dishes contained lead levels that exceeded federal safety standards.
  b. Marble statues undergo corrosion in acid rain.
  c. Statues corrode in acid rain because the acidity is sufficient to dissolve calcium carbonate, the major substance of marble.

ADDITIONAL QUESTIONS AND PROBLEMS

For instructor-assigned homework, go to
www.masteringchemistry.com.

1.23 Why does the scientific method include a hypothesis?

1.24 Why is experimentation an important part of the scientific method?

1.25 Select the correct phrase(s) to complete the following statement: If experimental results do not support your hypothesis, you should
  a. pretend that the experimental results support your hypothesis.
  b. write another hypothesis.
  c. do more experiments.

1.26 Select the correct phrase(s) to complete the following statement: A hypothesis becomes a theory when
  a. one experiment proves the hypothesis.

  b. many experiments by many scientists validate the hypothesis.
  c. you decide to call it a theory.

1.27 Which of the following will help you develop a successful study plan?
  a. Skip lecture and just read the text.
  b. Work the *Sample Problems* as you go through a chapter.
  c. Go to your professor's office hours.
  d. Read through the chapter, but work the problems later.

1.28 Which of the following will help you develop a successful study plan?
  a. Study all night before the exam.
  b. Form a study group and discuss the problems together.
  c. Work problems in a notebook for easy reference.
  d. Copy the answers to homework from a friend.

CHALLENGE QUESTIONS

1.29 Classify each of the following as an observation (O), a hypothesis (H), or an experiment (E):
  a. The bicycle tire is flat.
  b. If I add air to the bicycle tire, it will expand to the proper size.
  c. When I added air to the bicycle tire, it was still flat.
  d. The bicycle tire must have a leak in it.

1.30 Classify each of the following as an observation (O), a hypothesis (H), or an experiment (E):
  a. A big log in the fire does not burn well.
  b. If I chop the log into smaller wood pieces, it will burn better.
  c. The small wood pieces burn brighter and make a hotter fire.
  d. The small wood pieces are used up faster than burning the big log.

ANSWERS

ANSWERS TO STUDY CHECKS

1.1 **a**, **b**, and **d**

1.2 **a.** experiment (E) **b.** observation (O)
c. hypothesis (H)

1.3 **b**, **c**, and **e**

ANSWERS TO SELECTED QUESTIONS
AND PROBLEMS

1.1 **a.** Chemistry is the science of the composition and properties of matter.
b. A chemical is a substance that has the same composition and properties wherever it is found.

1.3 Many chemicals are listed on a vitamin bottle such as vitamin A, vitamin B_3, vitamin B_{12}, vitamin C, folic acid, etc.

1.5 Typical items found in a medicine cabinet and some of the chemicals they contain:

Antacid tablets: calcium carbonate, cellulose, starch, stearic acid, silicon dioxide

Mouthwash: water, alcohol, thymol, glycerol, sodium benzoate, benzoic acid

Cough suppressant: menthol, beta-carotene, sucrose, glucose

1.7 No. All of the ingredients are chemicals.

1.9 An advantage of a pesticide is that it gets rid of insects that bite or damage crops. A disadvantage is that a pesticide can destroy beneficial insects or be retained in a crop that is eventually eaten by animals or humans.

1.11 **a.** A hypothesis proposes a possible explanation for a natural phenomenon.
b. An experiment is a procedure that tests the validity of a hypothesis.

c. A theory is a hypothesis that has been validated many times by many scientists.
d. An observation is a description or measurement of a natural phenomenon.

1.13 **a**, **d**, and **e** are observations (O).
b is a hypothesis (H).
c is an experiment (E).
f is a theory (T).

1.15 There are several things you can do to help you successfully learn chemistry, including forming a study group, going to lecture, working *Sample Problems* and *Study Checks*, working *Questions and Problems*, and checking answers, reading the assignment ahead of class, keeping a problem notebook, etc.

1.17 **a**, **c**, **e**, and **f**

1.19 Yes. Sherlock's investigation includes making observations (gathering data), formulating a hypothesis, testing the hypothesis, and modifying it until one of the hypotheses is validated.

1.21 **a.** observation (O)
b. hypothesis (H)
c. observation (O)

1.23 A hypothesis, which is a possible explanation for an observation, can be tested with experiments.

1.25 **b** and **c**

1.27 **b** and **c**

1.29 **a.** observation (O)
b. hypothesis (H)
c. experiment (E)
d. hypothesis (H)

Measurements

2

"*I use measurement in just about every part of my nursing practice,*" *says registered nurse Vicki Miller.* "*When I receive a doctor's order for a medication, I have to verify that order. Then I draw a carefully measured volume from an IV or a vial to create that particular dose. Some dosage orders are specific to the size of the patient. I measure the patient's weight and calculate the dosage required for the weight of that patient.*"

Nurses use measurement each time they take a patient's temperature, height, weight, or blood pressure. Measurement is used to obtain the correct amounts for injections and medications and to determine the volumes of fluid intake and output. For each measurement, the amounts and units are recorded in the patient's records.

Visit **www.masteringchemistry.com** for self-study materials and instructor-assigned homework.

Your weight on a bathroom scale is a measurement.

Chemists working in research laboratories test new products and develop new pharmaceuticals.

Chemistry and measurements are an important part of our everyday lives. Levels of toxic materials in the air, soil, and water are discussed in our newspapers. We read about radon gas in our homes, holes in the ozone layer, trans fatty acids, global warming, and DNA analysis. Understanding chemistry and measurement helps us make proper choices about our world.

Think about your day. You probably took some measurements. Perhaps you checked your weight by stepping on a bathroom scale. If you made some rice for dinner, you added 2 cups of water to 1 cup of rice. If you did not feel well, you may have taken your temperature. Whenever you take a measurement, you use a measuring device such as a balance, a measuring cup, or a thermometer. Over the years, you have learned to read the markings on each device to take a correct measurement.

Scientists measure the amounts of the materials that make up everything in our universe. An engineer determines the amount of metal in an alloy or the volume of seawater flowing through a desalination plant. A physician orders laboratory tests to measure substances in the blood such as glucose or cholesterol. An environmental chemist measures the levels of pollutants such as lead and carbon monoxide in our soil and atmosphere.

By learning about measurement, you develop skills for solving problems and how to work with numbers in chemistry. An understanding of measurement is essential to evaluate our health and surroundings.

2.1 | Units of Measurement

Suppose today you walk 2.1 km to campus carrying a backpack that has a mass of 12 kg. Perhaps it is 8:30 A.M. and the temperature is 22 °C. You have a mass of 58.2 kg and a height of 165 cm. If these measurements seem unusual to you, it is because the *metric system* was used to measure each one. Perhaps you are more familiar with these measurements stated in the U.S. system of measurement; then you would walk 1.3 mi carrying a backpack that weighs 26 lb. The temperature at 8:30 A.M. would be 72 °F. You have a weight of 128 lb and a height of 65 in.

8:30 A.M.
165 cm (65 in.)
22 °C (72 °F)
58.2 kg (128 lb)
12 kg (26 lb)
2.1 km (1.3 mi)

There are many measurements in everyday life.

LEARNING GOAL

Write the names and abbreviations for the metric or SI units used in measurements of length, volume, mass, temperature, and time.

Scientists and health professionals throughout the world use the **metric system**. It is also the common measuring system in all but a few countries of the world. In 1960, scientists adopted a modification of the metric system called the International System of Units (**SI**), *Le Système*

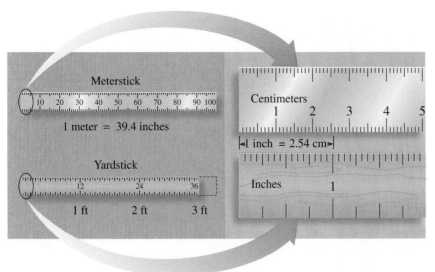

FIGURE 2.1 Length in the metric (SI) system is based on the meter, which is slightly longer than a yard.

Q How many centimeters are in a length of 1 in.?

International d'Unités, which provides additional uniformity of units throughout the world. In chemistry, we use metric and SI units for length, volume, mass, temperature, and time.

Length

The metric and SI unit of length is the **meter (m)**. The standard for length is the distance that light travels in 1/299 792 458 s. A meter is 39.37 in., which makes it slightly longer than a yard (1.094 yd). The **centimeter (cm)**, a smaller unit of length, is commonly used in chemistry and is about equal to the width of your little finger. For comparison, there are 2.54 cm in 1 in. (see Figure 2.1).

1 m	=	1.094 yd
1 m	=	39.37 in.
1 m	=	100 cm
2.54 cm	=	1 in.

Volume

Volume (*V*) is the amount of space a substance occupies. The SI unit of volume, the **cubic meter (m³)**, is the volume of a cube that has sides that measure 1 m in length. In a laboratory or a hospital, the cubic meter is too large for practical use. Instead, chemists work with metric units of volume that are smaller and more convenient, such as the **liter (L)** and **milliliter (mL)**. A liter is slightly larger than a quart (1 L = 1.057 qt) and contains 1000 mL, as shown in Figure 2.2. A cubic meter is the same volume as 1000 L.

1 m³	=	1000 L
1 L	=	1000 mL
1 L	=	1.057 qt
1 qt	=	946.3 mL

Mass

The **mass** of an object is the quantity of material it contains. The standard for mass, the international prototype kilogram (IPK), is a cylinder that is made of a platinum-iridium alloy. The SI unit of mass, the **kilogram (kg)**, is used for larger masses, such as body mass. In the metric system, the unit for mass is the **gram (g)**, which is used for smaller masses. There are 1000 g in one kilogram.

The standard kilogram for the United States is stored at the National Institute of Standards and Technology (NIST).

FIGURE 2.2 Volume is the space occupied by a substance. In the metric system, volume is based on the liter, which is slightly larger than a quart.

Q How many quarts are in 1 L?

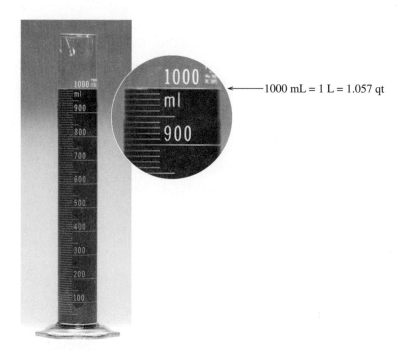

1000 mL = 1 L = 1.057 qt

FIGURE 2.3 On an electronic balance, a nickel has a mass of 5.01 g in the digital readout.

Q What is the mass of 10 nickels?

FIGURE 2.4 A thermometer is used to determine temperature.

Q What kinds of temperature readings have you made today?

You may be more familiar with the term *weight*. Weight is a measure of the gravitational pull on an object measured at a specific location. On Earth, an astronaut with a mass of 75.0 kg has a weight of 165 lb. On the moon, where the gravitational pull is one-sixth that of Earth, the astronaut has a weight of 27.5 lb. However, the mass of the astronaut is the same as on Earth, 75.0 kg. Scientists measure mass rather than weight because mass does not depend on gravity.

A balance in a chemistry laboratory measures the mass, in grams, of a substance, not its weight (see Figure 2.3). In comparison to the U.S. system, the mass of 1 kg is equivalent to 2.205 lb, and 1 lb is equivalent to 453.6 g.

$$1 \text{ kg} = 1000 \text{ g}$$
$$1 \text{ kg} = 2.205 \text{ lb}$$
$$453.6 \text{ g} = 1 \text{ lb}$$

Temperature

Temperature tells us how hot something is and how cold it is outside (see Figure 2.4). It is also used to determine if we have a fever. A typical laboratory thermometer consists of a glass bulb with a liquid in it that expands as the temperature increases. The metric system uses the *Celsius temperature scale* for measuring temperature. On the **Celsius (°C) scale**, water freezes at 0 °C and boils at 100 °C, while on the Fahrenheit (°F) scale, water freezes at 32 °F and boils at 212 °F. In the SI system, temperature is measured using the **Kelvin (K) scale**, on which the lowest temperature possible is assigned a value of 0 K. Note that the units on the Kelvin scale are called kelvins (K) and are not given degree signs.

Time

We typically measure time in units such as years, days, hours, minutes, or seconds. Of these, the SI and metric unit of time is the **second (s)**. The standard second for the United States is determined using an atomic clock called NIST-F1, located in Boulder, Colorado. A comparison of metric and SI units for measurement is shown in Table 2.1.

TABLE 2.1 Units of Measurement

Measurement	Metric	SI
Length	meter (m)	meter (m)
Volume	liter (L)	cubic meter (m³)
Mass	gram (g)	kilogram (kg)
Temperature	degree Celsius (°C)	kelvin (K)
Time	second (s)	second (s)

CONCEPT CHECK 2.1

■ **Units of Measurement**

State the type of measurement indicated by each of the following units:

a. gram   **b.** liter
c. centimeter   **d.** degree Celsius

ANSWER

a. A gram is a unit of mass.   **b.** A liter is a unit of volume.
c. A centimeter is a unit of length.   **d.** A degree Celsius is a unit of temperature.

SAMPLE PROBLEM   2.1

■ **Units of Measurement**

State the type of measurement indicated in each of the following:

a. 45.6 kg   **b.** 1.895 m^3
c. 14 s   **d.** 45 m
e. 315 K

SOLUTION

a. mass   **b.** volume
c. time   **d.** length
e. temperature

STUDY CHECK

Give the SI unit and abbreviation that would be used to express the following measurements:

a. length of a football field   **b.** daytime temperature
c. mass of salt in a shaker

A stopwatch is used to measure the time of a race.

QUESTIONS AND PROBLEMS

Units of Measurement

2.1 State the name of the unit and the type of measurement indicated for each of the following quantities:
    a. 4.8 m   **b.** 325 g   **c.** 1.5 L
    d. 480 s   **e.** 28 °C

2.2 State the name of the unit and the type of measurement indicated for each of the following quantities:
    a. 0.8 L   **b.** 3.6 m   **c.** 14 kg
    d. 35 g   **e.** 373 K

2.3 State the name of the unit in each of the following, and identify that unit as an SI unit, a metric unit, both, or neither:
    a. 5.5 m   **b.** 45 kg   **c.** 16 in.
    d. 25 s   **e.** 22 °C

2.4 State the name of the unit in each of the following, and identify that unit as an SI unit, a metric unit, both, or neither:
    a. 8 m^3 **b.** 245 K   **c.** 45 °F
    d. 125 L   **e.** 125 g

2.5 State the name of the unit in each of the following, and identify that unit as an SI unit, a metric unit, both, or neither:
    a. 25.2 g   **b.** 1.5 L   **c.** 15 °F
    d. 8.2 lb   **e.** 15 s

2.6 State the name of the unit in each of the following, and identify that unit as an SI unit, a metric unit, both, or neither:
    a. 24 °C   **b.** 268 K   **c.** 0.48 qt
    d. 28.6 m   **e.** 4.2 m^3

2.2 Scientific Notation

In chemistry, we use numbers that are extremely small and extremely large. We might measure something as tiny as the width of a human hair, which is 0.000 008 m. Or perhaps we want to count the number of hairs on the average human scalp, which is about 100 000 (see Figure 2.5). In this text, we add a space between sets of three digits when it helps make the places easier to count. However, it is more convenient to write large and small numbers in *scientific notation*.

LEARNING GOAL

Write a number in scientific notation.

FIGURE 2.5 Humans have an average of 1×10^5 hairs on their scalps. Each hair is about 8×10^{-6} m wide.

Q Why are large and small numbers written in scientific notation?

8×10^{-6} m

TUTORIAL

Scientific Notation

Item	Standard Number	Scientific Notation
Width of a human hair	0.000 008 m	8×10^{-6} m
Hairs on a human scalp	100 000 hairs	1×10^5 hairs

Writing a Number in Scientific Notation

A number written in **scientific notation** consists of three parts: a coefficient, a power of 10, and a measurement unit. For example, the quantity 2400 m is written in scientific notation as 2.4×10^3 m. The coefficient is 2.4, and the value 10^3 shows the power of 10 is 3. The coefficient was determined by moving the decimal point to give a number from 1 to 9. Because we moved the decimal point three places to the left, the power of 10 is a positive 3, which is written as 10^3. For any quantity greater than 1, the power of 10 will be positive.

$$2400. \, \text{m} \underset{\longleftarrow 3 \text{ places}}{} = 2.4 \times 1000 = \underset{\text{Coefficient}}{2.4} \times \underset{\substack{\text{Power} \\ \text{of 10}}}{10^3} \, \text{m}$$

When a number less than 1 is written in scientific notation, the power of 10 will be negative. For example, the number 0.000 86 g is written in scientific notation by moving the decimal point to give a coefficient of 8.6. Because the decimal point was moved four places to the right, the power of 10 becomes a negative 4, written as 10^{-4}.

$$0.00086 \, \text{g} \underset{4 \text{ places} \longrightarrow}{} = \frac{8.6}{10\,000} = \frac{8.6}{10 \times 10 \times 10 \times 10} = \underset{\text{Coefficient}}{8.6} \times \underset{\substack{\text{Power} \\ \text{of 10}}}{10^{-4}} \, \text{g}$$

Table 2.2 gives some examples of numbers written as positive and negative powers of 10. The powers of 10 are a way of keeping track of the decimal point in the decimal number. Table 2.3 gives several examples of writing standard numbers in scientific notation.

TABLE 2.2 Some Powers of 10

Number	Multiples of 10		Scientific Notation	
10 000	$10 \times 10 \times 10 \times 10$		1×10^4	
1000	$10 \times 10 \times 10$		1×10^3	
100	10×10		1×10^2	Some positive powers of 10
10	10		1×10^1	
1	0		1×10^0	
0.1	$\dfrac{1}{10}$		1×10^{-1}	
0.01	$\dfrac{1}{10} \times \dfrac{1}{10}$	$= \dfrac{1}{100}$	1×10^{-2}	
0.001	$\dfrac{1}{10} \times \dfrac{1}{10} \times \dfrac{1}{10}$	$= \dfrac{1}{1000}$	1×10^{-3}	Some negative powers of 10
0.0001	$\dfrac{1}{10} \times \dfrac{1}{10} \times \dfrac{1}{10} \times \dfrac{1}{10}$	$= \dfrac{1}{10\,000}$	1×10^{-4}	

TABLE 2.3 Some Measurements Written in Scientific Notation

Measured Quantity	Standard Number	Scientific Notation
Volume of gasoline used in United States each year	550 000 000 000 L	5.5×10^{11} L
Diameter of Earth	12 800 000 m	1.28×10^7 m
Time for light to travel from the Sun to Earth	500 s	5×10^2 s
Mass of a typical human	68 kg	6.8×10^1 kg
Mass of a hummingbird	0.002 kg	2×10^{-3} kg
Diameter of a chickenpox (varicella zoster) virus	0.000 000 3 m	3×10^{-7} m
Mass of bacterium (mycoplasma)	0.000 000 000 000 000 000 1 kg	1×10^{-19} kg

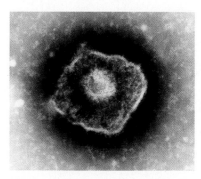

A chickenpox virus has a diameter of 3×10^{-7} m.

Scientific Notation and Calculators

You can enter a number written in scientific notation on many calculators using the EE or EXP key. After you enter the coefficient, press the EXP (or EE) key and enter only the power of 10, because the EXP function key already includes the $\times$ 10 value. To enter a negative power of 10, press the plus/minus ($+/-$) key or the minus ($-$) key (depending on your calculator), but *not* the key for subtraction. Some calculators require entering the sign before the power.

Number to Enter	Method	Display Reads
4×10^6	4 EXP (EE) 6	4 06 or 4 ⁰⁶ or 4 E06
2.5×10^{-4}	2.5 EXP (EE) +/-4	2.5−04 or 2.5⁻⁰⁴ or 2.5 E−04

When a calculator answer is given in scientific notation, it is usually shown in the display as a coefficient from 1 to 9 followed by a space and the power of 10. To express this display in scientific notation, write the coefficient, insert "$\times$ 10", and use the power of 10 as an exponent.

Calculator Display	Expressed in Scientific Notation
7.52 04 or 7.52⁰⁴ or 7.52 E04	7.52×10^4
5.8−02 or 5.8⁻⁰² or 5.8 E−02	5.8×10^{-2}

On many scientific calculators, a number can be converted into scientific notation using the appropriate keys. For example, the number 0.000 52 can be entered followed by hitting the second or third function key and the SCI key. The scientific notation appears in the calculator display as a coefficient and the power of 10.

0.000 52 [2nd or 3rd function key] [SCI] = 5.2−04 or 5.2 ⁻⁰⁴ or 5.2 E−04 = 5.2×10^{-4}

Key   Key   Calculator display

Converting Scientific Notation to a Standard Number

When a number written in scientific notation has a positive power of 10, the standard number is obtained by moving the decimal point to the right for the same number of places as the power of 10. Placeholder zeros are used, as needed, to give additional decimal places.

$$4.3 \times 10^2 = 4.3 \times 10 \times 10 = 4.3 \times 100 = 430$$

For a number with a negative power of 10, the standard number is obtained by moving the decimal point to the left for the same number of places as the power of 10. Placeholder zeros are added in front of the coefficient as needed.

$$2.5 \times 10^{-5} = 2.5 \times \frac{1}{10^5} = 2.5 \times 0.000\,01 = 0.000\,025$$

CONCEPT CHECK 2.2

■ **Scientific Notation**

Indicate whether the power of 10 is positive or negative when each of the following is written in scientific notation:

a. 75 000 m   **b.** 0.0092 g
c. 143 mL

ANSWER

a. To make a coefficient from 1 to 9, the decimal point is moved four places to the left, which gives a positive power of 10 (7.5×10^4 m).
b. To make a coefficient from 1 to 9, the decimal point is moved three places to the right, which gives a negative power of 10 (9.2×10^{-3} g).
c. To make a coefficient from 1 to 9, the decimal point is moved two places to the left, which gives a positive power of 10 (1.43×10^2 mL).

SAMPLE PROBLEM | 2.2

■ **Scientific Notation**

1. Write each of the following in scientific notation:

    a. 350 g   **b.** 0.000 16 L   **c.** 5 220 000 m

2. Write each of the following as a standard number:

    a. 2.85×10^2 L   **b.** 7.2×10^{-3} m   **c.** 2.4×10^5 g

SOLUTION

1. a. 3.5×10^2 g   **b.** 1.6×10^{-4} L   **c.** 5.22×10^6 m
2. a. 285 L   **b.** 0.0072 m   **c.** 240 000 g

STUDY CHECK

Write each of the following in scientific notation:

a. 425 000 m   **b.** 0.000 000 8 g

QUESTIONS AND PROBLEMS

Scientific Notation

2.7 Write each of the following in scientific notation:
  a. 55 000 m   **b.** 480 g   **c.** 0.000 005 cm
  d. 0.000 14 s   **e.** 0.007 85 L   **f.** 670 000 kg

2.8 Write each of the following in scientific notation:
  a. 180 000 000 g   **b.** 0.000 06 m   **c.** 750 000 g
  d. 0.15 mL   **e.** 0.024 s   **f.** 1500 m^3

2.9 In each of the following pairs, which number is larger?
  a. 7.2×10^3 cm or 8.2×10^2 cm
  b. 4.5×10^{-4} kg or 3.2×10^{-2} kg
  c. 1×10^4 L or 1×10^{-4} L
  d. 0.000 52 m or 6.8×10^{-2} m

2.10 In each of the following pairs, which number is smaller?
  a. 4.9×10^{-3} s or 5.5×10^{-9} s
  b. 1250 kg or 3.4×10^2 kg
  c. 0.000 000 4 m or 5.0×10^2 m
  d. 2.50×10^2 g or 4×10^5 g

2.11 Write each of the following as a standard number:
  a. 1.2×10^4 s   **b.** 8.25×10^{-2} kg
  c. 4×10^6 g   **d.** 5.8×10^{-3} m^3

2.12 Write each of the following as a standard number:
  a. 3.6×10^{-5} L   **b.** 8.75×10^4 cm
  c. 3×10^{-2} mL   **d.** 2.12×10^5 kg

2.3 Measured Numbers and Significant Figures

When you make a measurement, you use some type of measuring device. For example, you may use a meterstick to measure your height, a scale to check your weight, or a thermometer to take your temperature. **Measured numbers** are the numbers you obtain when you measure a quantity such as your height, weight, or temperature.

Measured Numbers

Suppose you are going to measure the lengths of the objects in Figure 2.6. You would select a ruler with a scale marked on it. By observing the lines on the scale, you determine the measurement for each object. Perhaps the divisions on the scale are marked as 1 cm. Another ruler might be marked in divisions of 0.1 cm. To report the length, you would first read the numerical value of the marked line. Finally, you *estimate* by visually dividing the space between the smallest marked lines. This estimated number is the final digit in a measured number.

For example, in Figure 2.6a, the end of the object falls between the lines marked 4 cm and 5 cm. That means that the length is 4 cm plus an estimated digit. If you estimate that the end is halfway between 4 cm and 5 cm, you would report its length as 4.5 cm. However, someone else might estimate the length as 4.6 cm. The last digit in a measured quantity can differ because people do not estimate in the same way. The ruler shown in Figure 2.6b is marked with lines at 0.1 cm. With this ruler, you can estimate the value of the hundredths place (0.01 cm). Perhaps you would report the length of the object as 4.55 cm, while someone else may report its length as 4.56 cm. Both results are acceptable.

There is always uncertainty in every measurement. When a measurement ends on a marked line, a zero is written as the estimated digit. For example, in Figure 2.6c, the measurement for length is written as 3.0 cm, not 3 cm. This means that the uncertainty of the measurement (the last digit) is in the tenths place.

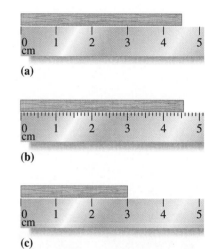

(a)

(b)

(c)

FIGURE 2.6 The lengths of the rectangular objects are measured as (a) 4.5 cm and (b) 4.55 cm.

Q What is the length of the object in (c)?

Significant Figures

In a measured number, the **significant figures (SFs)** are all the digits including the estimated digit. All *nonzero* numbers are always counted as significant figures. Zeros may or may not be significant, depending on their position in a number. Table 2.4 gives the rules and examples for counting significant figures.

When one or more zeros in a large number are significant, they are shown by writing the number using scientific notation. For example, if the first zero in the measurement 500 m is significant, it is written as 5.0×10^2 m. In this text, we will place a decimal point at the

TABLE 2.4 Significant Figures in Measured Numbers

Rule	Measured Number	Number of Significant Figures
1. A number is a *significant figure* if it is		
a. not a zero	4.5 g	2
	122.35 m	5
b. a zero between nonzero digits	205 m	3
	5.082 kg	4
c. a zero at the end of a decimal number	50. L	2
	25.0 °C	3
	16.00 g	4
d. in the coefficient of a number written in scientific notation	5.5×10^{-9} kg	2
	3.00×10^2 m^3	3
2. A zero is *not a significant figure* if it is		
a. at the beginning of a decimal number	0.0004 s	1
	0.075 cm	2
b. used as a placeholder in a large number without a decimal point	850 000 m	2
	1 250 000 g	3

end of a number if all the zeros are significant. For example, a measurement written as 500. g indicates that the zeros are all significant. It could also be written as 5.00×10^2 g. We will assume that zeros at the end of large standard numbers without a decimal point are not significant. Therefore, we would read 400 000 g as having one significant figure, which would be written in scientific notation as 4×10^5 g.

■ **Significant Zeros**

Identify the zeros as significant or not significant in each of the following measured numbers:

a. 0.000 250 m   **b.** 70.040 g   **c.** 1 020 000 L

ANSWER

a. The zeros preceding the first nonzero digit of 2 are not significant. The zero in the last decimal place following the 5 is significant.
b. Zeros between nonzero digits or at the end of decimal numbers are significant. All zeros in 70.040 g are significant.
c. Zeros between nonzero digits are significant. Zeros at the end of a large number with no decimal point are placeholders but not significant. The zero between 1 and 2 is significant, but the four zeros following the 2 are not significant.

Exact Numbers

Exact numbers are obtained by counting items or using a definition that compares two units in the same measuring system. Suppose a friend asks you how many classes you are taking this term. You would answer by counting the number of classes. It is not necessary for you to use any type of measuring tool. Suppose you are asked to state the number of seconds in 1 minute. Without using any measuring device, you would give the definition: 60 seconds is 1 minute. Exact numbers are not measured, do not have a limited number of significant figures, and do not affect the number of significant figures in a calculated answer. For more examples of exact numbers, see Table 2.5.

TABLE 2.5 Examples of Some Exact Numbers

Counted Numbers	Defined Equalities	
	U.S. System	Metric System
8 doughnuts	1 ft = 12 in.	1 L = 1000 mL
2 baseballs	1 qt = 4 cups	1 m = 100 cm
5 caps	1 lb = 16 ounces (oz)	1 kg = 1000 g

The number of baseballs is counted, which means 2 is an exact number.

■ **Significant Figures**

Identify each of the following numbers as measured or exact and give the number of significant figures (SFs) in each of the measured numbers:

a. 42.2 g   **b.** three eggs   **c.** 5.0×10^{-4} m
d. 450 000 kg   **e.** 3.500×10^5 s

SOLUTION

a. measured, three SFs   **b.** exact   **c.** measured, two SFs
d. measured, two SFs   **e.** measured, four SFs

STUDY CHECK

State the number of significant figures in each of the following measured numbers:
a. 0.000 35 g   **b.** 2000 m   **c.** 2.0045 L

QUESTIONS AND PROBLEMS

Measured Numbers and Significant Figures

2.13 What is the estimated digit in each of the following measured numbers?
a. 8.6 m   **b.** 45.25 g   **c.** 25.0 °C

2.14 What is the estimated digit in each of the following measured numbers?
a. 125.04 g   **b.** 5.057 m   **c.** 525.8 °C

2.15 Identify the numbers in each of the following statements as measured or exact:
a. A person weighs 155 lb.
b. The basket holds 8 apples.
c. In the metric system, 1 kg is equal to 1000 g.
d. The distance from Denver, Colorado, to Houston, Texas, is 1720 km.

2.16 Identify the numbers in each of the following statements as measured or exact:
a. There are 31 students in the laboratory.
b. The oldest-known flower lived 1.2×10^8 years ago.
c. The largest gem ever found, an aquamarine, has a mass of 104 kg.
d. A laboratory test shows a blood cholesterol level of 184 mg/dL.

2.17 In each of the following pairs of numbers, identify the measured number(s), if any:
a. 3 hamburgers and 6 oz of hamburger
b. 1 table and 4 chairs
c. 0.75 lb of grapes and 350 g of butter
d. 60 s equals 1 min

2.18 In each of the following pairs of numbers, identify the exact number(s), if any:
a. 5 pizzas and 50.0 g of cheese
b. 6 nickels and 16 g of nickel
c. 3 onions and 3 lb of onions
d. 5 miles and 5 cars

2.19 For each of the following measurements, indicate if the zeros are significant or not:
a. 0.0038 m   **b.** 5.04 cm   **c.** 800. L
d. 3.0×10^{-3} kg   **e.** 85 000 g

2.20 For each of the following measurements, indicate if the zeros are significant or not:
a. 20.05 °C   **b.** 5.00 m   **c.** 0.000 02 L
d. 120 000 years   **e.** 8.05×10^2 g

2.21 How many significant figures are in each of the following measured quantities?
a. 11.005 kg   **b.** 0.000 32 m³   **c.** 36 000 000 m
d. 1.80×10^4 g   **e.** 0.8250 L   **f.** 30.0 °C

2.22 How many significant figures are in each of the following measured quantities?
a. 20.60 mL   **b.** 1036.48 g   **c.** 4.00 m
d. 20.88 °C   **e.** 60 800 000 kg   **f.** 5.0×10^{-3} L

2.23 In which of the following pairs do both numbers contain the same number of significant figures?
a. 11.0 m and 11.00 m   **b.** 405 K and 504.0 K
c. 0.000 12 s and 12 000 s   **d.** 250.0 L and 2.500×10^{-2} L

2.24 In which of the following pairs do both numbers contain the same number of significant figures?
a. 0.005 75 g and 5.75×10^{-3} g
b. 0.0250 m and 0.205 m
c. 150 000 s and 1.50×10^4 s
d. 3.8×10^{-2} L and 7.5×10^5 L

2.25 Write each of the following in scientific notation with two significant figures:
a. 5000 L   **b.** 30 000 g
c. 100 000 m   **d.** 0.000 25 cm

2.26 Write each of the following in scientific notation with two significant figures:
a. 5 100 000 g   **b.** 26 000 s
c. 40 000 m   **d.** 0.000 820 kg

2.4 Significant Figures in Calculations

In the sciences, we measure many things: the length of a bacterium, the volume of a gas sample, the temperature of a reaction mixture, or the mass of iron in a sample. The numbers obtained from these types of measurements are often used in calculations. The number of significant figures in the measured numbers determines the number of significant figures in the calculated answer.

Using a calculator will help you perform calculations faster. However, calculators cannot think for you. It is up to you to enter the numbers correctly, press the correct function keys, and adjust the calculator display to give an answer with the correct number of significant figures.

Rounding Off

To calculate the area of a carpet that measures 5.5 m by 3.5 m, multiply 5.5 times 3.5 to obtain the number 19.25, the area in square meters. However, four digits cannot be used in the answer, because they are not all significant figures. Each measurement of length and width has only two significant figures. This means that the calculated result must be rounded off to give an answer that also has two significant figures, 19 m². When you obtain a calculator result, determine the number of significant figures for the answer and round off using the following rules:

LEARNING GOAL

Adjust calculated answers to give the correct number of significant figures.

TUTORIAL
Significant Figures in Calculations

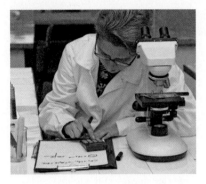

A technician uses a calculator in the laboratory.

Rules for Rounding Off

1. If the first digit to be dropped is *4 or less*, then it and all following digits are simply dropped from the number.

2. If the first digit to be dropped is *5 or greater*, then the last retained digit of the number is increased by 1.

	Three Significant Figures	Two Significant Figures
Example 1: 8.4234 rounds off to	8.42	8.4
Example 2: 14.780 rounds off to	14.8	15
Example 3: 3256 rounds off to	3260 (3.26×10^3)	3300 (3.3×10^3)

Note: The value of a large number is retained by using placeholder zeros to replace dropped digits.

CONCEPT CHECK 2.4

■ Rounding Off

Select the correct value when 2.8456 m is rounded off to each of the following:

a. three significant figures: 2.84 m   2.85 m   2.8 m   2.90 m
b. two significant figures: 2.80 m   2.85 m   2.8 m   2.90 m

ANSWER

a. To round off 2.8456 m to three significant figures, drop the final digits 56 and increase the last retained digit by 1 to give 2.85 m.
b. To round off 2.8456 m to two significant figures, drop the final digits 456 to give 2.8 m.

SAMPLE PROBLEM | 2.4

■ Rounding Off

Round off each of the following numbers to three significant figures:

a. 35.7823 m
b. 0.002 627 L
c. 3.8268×10^3 g
d. 1.2836 kg

SOLUTION

a. 35.8 m
b. 0.002 63 L
c. 3.83×10^3 g
d. 1.28 kg

STUDY CHECK

Round off each of the numbers in Sample Problem 2.4 to two significant figures.

Multiplication and Division

In multiplication and division, the final answer is written so that it has the same number of significant figures (SFs) as the measurement with the *fewest* SFs.

Example 1

Multiply the following measured numbers: 24.65 × 0.67

   To perform this problem on a calculator, enter the number and then press the operation key. In this case, we might press the keys in the following order:

24.65	⊠	0.67	⊟	*16.5155*	→	17
4 SFs		2 SFs		Calculator display		Final answer, rounded off to 2 SFs

A calculator is helpful in working problems and doing calculations faster.

The answer in the calculator display has more digits than the data allow. The measurement 0.67 has the fewer number of significant figures, which is two. Therefore, the calculator answer is rounded off to two significant figures.

Example 2

Solve the following:

$$\frac{2.85 \times 67.4}{4.39}$$

On a calculator, we might press the keys in the following order:

2.85   ⊠   67.4   ÷   4.39   =   *43.756264*   ⟶   43.8

3 SFs   3 SFs   3 SFs   Calculator   Final answer, rounded
                                    display   off to 3 SFs

All of the measurements in this problem have three significant figures. Therefore, the calculator result is rounded off to give an answer, 43.8, that has three significant figures.

Adding Significant Zeros

Sometimes a calculator display gives a small whole number. To report an answer with the correct number of significant figures, you need to write significant zeros after the calculator value. For example, suppose the calculator display is 4 but the measurements each have three significant numbers. Then the number 4 is written as 4.00 by adding two significant zeros.

$$\frac{8.00}{2.00} = \quad 4 \quad \longrightarrow \quad 4.00$$

3 SFs   Calculator display   Final answer, two zeros
                                              added to give 3 SFs

---

SAMPLE PROBLEM | **2.5**

■ Significant Figures in Multiplication and Division

Perform the following calculations of measured numbers. Give the answers with the correct number of significant figures.

a. 56.8×0.37 **b.** $\dfrac{71.4}{11.0}$

c. $\dfrac{(2.075)(0.585)}{(8.42)(0.0045)}$ **d.** $\dfrac{25.0}{5.00}$

SOLUTION

a. 21   **b.** 6.49
c. 32   **d.** 5.00 (add significant zeros)

STUDY CHECK

Perform the following calculations of measured numbers and give the answers with the correct number of significant figures:

a. $45.26 \times 0.010\ 88$ **b.** $2.60 \div 324$

c. $\dfrac{4.0 \times 8.00}{16}$

Addition and Subtraction

In addition or subtraction, the answer is written so it has the same number of decimal places as the measurement having the *fewest* decimal places.

Example 3

Add:

	2.045	Three decimal places
$\boxed{+}$	34.1	One decimal place
	36.145	Calculator display
	36.1	Answer rounded off one decimal place (tenths place)

Example 4

Subtract:

	255	Ones place
$\boxed{-}$	175.65	Two decimal places
	79.35	Calculator display
	79	Answer rounded off to the ones place

When numbers are added or subtracted to give an answer ending in zero, the zero does not appear after the decimal point in the calculator display. For example, 14.5 g − 2.5 g = 12.0 g. However, if you do the subtraction on your calculator, the display shows 12. To write the correct answer, a significant zero is written in the tenths place.

Example 5

Subtract:

	14.56 g	Two decimal places
$\boxed{-}$	4.16 g	Two decimal places
	10.4	Calculator display
	10.40 g	Answer; a significant zero is written to give two decimal places (hundredths place)

SAMPLE PROBLEM | 2.6

■ Significant Figures in Addition and Subtraction

Perform the following calculations and give the answers with the correct number of decimal places:

a. 27.8 cm + 0.235 cm   **b.** 104.45 mL + 0.838 mL + 46 mL

c. 153.247 g − 14.82 g

SOLUTION

a. 28.0 cm   **b.** 151 mL   **c.** 138.43 g

STUDY CHECK

Perform the following calculations and give the answers with the correct number of decimal places:

a. 82.45 g + 1.245 g + 0.000 56 g   **b.** 4.259 L − 3.8 L

QUESTIONS AND PROBLEMS

Significant Figures in Calculations

2.27 Why do we usually need to round off calculations that use measured numbers?

2.28 Why do we sometimes add a zero to a number in a calculator display?

2.29 Round off each of the following to three significant figures:
a. 1.854 kg   **b.** 88.0238 L   **c.** 0.004 738 265 cm
d. 8807 m   **e.** 1.8329×10^3 s

2.30 Round off each of the numbers in Problem 2.29 to two significant figures.

2.31 Round off or add zeros to each of the following to give an answer with three significant figures:
a. 56.855 m   **b.** 0.002 282 5 g
c. 11 527 s   **d.** 8.1 L

2.32 Round off or add zeros to each of the following to give an answer with two significant figures:
a. 3.2805 m   **b.** 1.855×10^2 g
c. 0.002 341 mL   **d.** 2 L

2.33 For each of the following, give an answer with the correct number of significant figures:
a. 45.7×0.034 **b.** $0.002\,78 \times 5$

c. $\dfrac{34.56}{1.25}$ **d.** $\dfrac{(0.2465)(25)}{1.78}$

e. $(2.8 \times 10^4)(5.05 \times 10^{-6})$

f. $\dfrac{(3.45 \times 10^{-2})(1.8 \times 10^5)}{(8 \times 10^3)}$

2.34 For each of the following, give an answer with the correct number of significant figures:

a. 400×185 **b.** $\dfrac{2.40}{(4)(125)}$

c. $0.825 \times 3.6 \times 5.1$ **d.** $\dfrac{(3.5)(0.261)}{(8.24)(20.0)}$

e. $\dfrac{(5 \times 10^{-5})(1.05 \times 10^4)}{(8.24 \times 10^{-8})}$ **f.** $\dfrac{(4.25 \times 10^2)(2.56 \times 10^{-3})}{(2.245 \times 10^{-3})(56.5)}$

2.35 For each of the following, give an answer with the correct number of decimal places:

a. 45.48 cm $+ 8.057$ cm

b. 23.45 g $+ 104.1$ g $+ 0.025$ g

c. 145.675 mL $- 24.2$ mL

d. 1.08 L $- 0.585$ L

2.36 For each of the following, give an answer with the correct number of decimal places:

a. 5.08 g $+ 25.1$ g

b. 85.66 cm $+ 104.10$ cm $+ 0.025$ cm

c. 24.568 mL $- 14.25$ mL

d. 0.2654 L $- 0.2585$ L

2.5 Prefixes and Equalities

The special feature of the SI as well as the metric system is that a **prefix** can be placed in front of any unit to increase or decrease its size by some factor of 10. For example, the prefixes *milli* and *micro* are used to make the smaller units milligram (mg) and microgram (μg). Table 2.6 lists some of the SI and metric prefixes, their symbols, and their numerical values.

The U.S. Food and Drug Administration (FDA) has determined the daily values (DVs) of nutrients for adults and children age 4 or older. Examples of these recommended daily values, some of which use prefixes, are listed in Table 2.7.

LEARNING GOAL

Use the numerical values of prefixes to write a metric equality.

SELF STUDY ACTIVITY
The Metric System

TUTORIAL
SI Prefixes and Units

TABLE 2.6 Metric and SI Prefixes

Prefix	Symbol	Numerical Value	Scientific Notation	Equality
Prefixes That Increase the Size of the Unit				
peta	P	1 000 000 000 000 000	10^{15}	$1\ \text{Pg} = 1 \times 10^{15}\ \text{g}$ $1\ \text{g} = 1 \times 10^{-15}\ \text{Pg}$
tera	T	1 000 000 000 000	10^{12}	$1\ \text{Ts} = 1 \times 10^{12}\ \text{s}$ $1\ \text{s} = 1 \times 10^{-12}\ \text{Ts}$
giga	G	1 000 000 000	10^{9}	$1\ \text{Gm} = 1 \times 10^{9}\ \text{m}$ $1\ \text{m} = 1 \times 10^{-9}\ \text{Gm}$
mega	M	1 000 000	10^{6}	$1\ \text{Mg} = 1 \times 10^{6}\ \text{g}$ $1\ \text{g} = 1 \times 10^{-6}\ \text{Mg}$
kilo	k	1 000	10^{3}	$1\ \text{km} = 1 \times 10^{3}\ \text{m}$ $1\ \text{m} = 1 \times 10^{-3}\ \text{km}$
Prefixes That Decrease the Size of the Unit				
deci	d	0.1	10^{-1}	$1\ \text{dL} = 1 \times 10^{-1}\ \text{L}$ $1\ \text{L} = 10\ \text{dL}$
centi	c	0.01	10^{-2}	$1\ \text{cm} = 1 \times 10^{-2}\ \text{m}$ $1\ \text{m} = 100\ \text{cm}$
milli	m	0.001	10^{-3}	$1\ \text{ms} = 1 \times 10^{-3}\ \text{s}$ $1\ \text{s} = 1 \times 10^{3}\ \text{ms}$
micro	μ	0.000 001	10^{-6}	$1\ \mu\text{g} = 1 \times 10^{-6}\ \text{g}$ $1\ \text{g} = 1 \times 10^{6}\ \mu\text{g}$
nano	n	0.000 000 001	10^{-9}	$1\ \text{nm} = 1 \times 10^{-9}\ \text{m}$ $1\ \text{m} = 1 \times 10^{9}\ \text{nm}$
pico	p	0.000 000 000 001	10^{-12}	$1\ \text{ps} = 1 \times 10^{-12}\ \text{s}$ $1\ \text{s} = 1 \times 10^{12}\ \text{ps}$
femto	f	0.000 000 000 000 001	10^{-15}	$1\ \text{fs} = 1 \times 10^{-15}\ \text{s}$ $1\ \text{s} = 1 \times 10^{15}\ \text{fs}$

TABLE 2.7 Daily Values for Selected Nutrients

Nutrient	Amount Recommended
Protein	44 g
Vitamin C	60 mg
Vitamin B_{12}	6 μg
Vitamin B_6	2 mg
Calcium	1000 mg
Iron	18 mg
Iodine	150 μg
Magnesium	400 mg
Niacin	20 mg
Potassium	3500 mg
Sodium	2400 mg
Zinc	15 mg

The prefix *centi* is like cents in a dollar. One cent would be a centidollar, or 0.01 of a dollar. That also means that one dollar is the same as 100 cents. The prefix *deci* is like dimes in a dollar. One dime would be a decidollar, or 0.1 of a dollar. That also means that one dollar is the same as 10 dimes.

The relationship of a prefix to a unit can be expressed by replacing the prefix with its numerical value. For example, when the prefix *kilo* in kilometer is replaced with its value of 1000, we find that a kilometer is equal to 1000 m. Other examples follow:

1 **kilo**meter (1 km) = **1000** meters   (1000 m = 10^3 m)

1 **kilo**liter (1 kL)   = **1000** liters   (1000 L = 10^3 L)

1 **kilo**gram (1 kg) = **1000** grams   (1000 g = 10^3 g)

A terabyte hard-disk drive stores 10^{12} bytes of information.

CONCEPT CHECK 2.5

■ **Prefixes**

The storage capacity for a hard-disk drive (HDD) is specified using prefixes such as megabyte (MB), gigabyte (GB), or terabyte (TB). Indicate the storage capacity in bytes for each of the following hard-disk drives. Suggest a reason for describing an HDD storage capacity in gigabytes or terabytes.

a. 5 MB   **b.** 2 GB   **c.** 1 TB

ANSWER

a. 5 MB = 5 000 000 (5×10^6) bytes
b. 2 GB = 2 000 000 000 (2×10^9) bytes
c. 1 TB = 1 000 000 000 000 (1×10^{12}) bytes

Expressing HDD capacity in gigabytes or terabytes gives a more reasonable number to work with than a number with many zeros or a large power of 10.

SAMPLE PROBLEM | 2.7

■ **Prefixes**

Fill in the blanks with the correct prefix:

a. 1000 g = 1 _____ g   **b.** 1×10^{-9} m = 1 _____ m
c. 1×10^6 L = 1 _____ L

SOLUTION

a. The prefix for 1000 is *kilo*; 1000 g = 1 kg.
b. The prefix for 1×10^{-9} is *nano*; 1×10^{-9} m = 1 nm.
c. The prefix for 1×10^6 is *mega*; 1×10^6 L = 1 ML.

STUDY CHECK

Write the correct prefix in the blanks:

a. 1 000 000 000 s = 1 _____ s   **b.** 0.01 m = 1 _____ m

Using a retinal camera, an ophthalmologist photographs the retina of an eye.

Measuring Length

An ophthalmologist may measure the retina of the eye in centimeters (cm), whereas a surgeon may need to know the length of a nerve in millimeters (mm). When the prefix *centi* is used with the unit meter, it becomes centimeter, a length that is one-hundredth of a meter (0.01 m). When the prefix *milli* is used with the unit meter, it becomes millimeter, a length that is one-thousandth of a meter (0.001 m). There are 100 cm and 1000 mm in a meter.

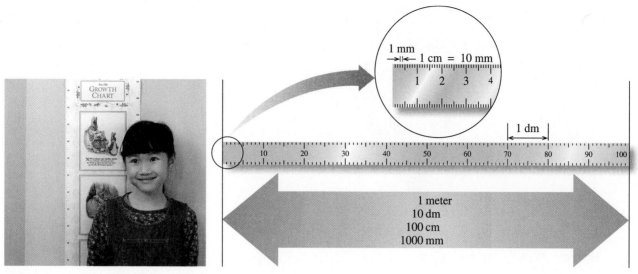

FIGURE 2.7 The metric length of 1 m is the same length as 10 dm, 100 cm, and 1000 mm.
Q How many millimeters (mm) are in 1 cm?

If we compare the lengths of a millimeter and a centimeter, we find that 1 mm is 0.1 cm; there are 10 mm in 1 cm. These comparisons are examples of **equalities**, which show the relationship between two units that measure the same quantity. For example, in the equality 1 m = 100 cm, each quantity describes the same length but in a different unit. In an equality, each quantity has both a number and a unit.

First Quantity   Second Quantity

$$1 \quad m \quad = \quad 100 \quad cm$$

Number + unit   Number + unit

Some Length Equalities

$$1 \text{ m} = 100 \text{ cm} = 1 \times 10^2 \text{ cm}$$
$$1 \text{ m} = 1000 \text{ mm} = 1 \times 10^3 \text{ mm}$$
$$1 \text{ cm} = 10 \text{ mm} = 1 \times 10^1 \text{ mm}$$

Some metric units for length are compared in Figure 2.7.

Measuring Volume

In the sciences, volumes smaller than 1 L are commonly used. When a liter is divided into 10 equal portions, each portion is a deciliter (dL). There are 10 dL in 1 L. Laboratory results for blood work are often reported in mass per deciliter. Table 2.8 lists typical laboratory test values for some substances in the blood.

When a liter is divided into a thousand parts, each of the smaller volumes is a milliliter (mL). In a 1-L container of physiological saline, there are 1000 mL of saline solution.

Some Volume Equalities

$$1 \text{ L} = 10 \text{ dL} = 1 \times 10^1 \text{ dL}$$
$$1 \text{ L} = 1000 \text{ mL} = 1 \times 10^3 \text{ mL}$$
$$1 \text{ dL} = 100 \text{ mL} = 1 \times 10^2 \text{ mL}$$

The **cubic centimeter (cm^3 or cc)** is the volume of a cube whose dimensions are 1 cm on each side. A cubic centimeter has the same volume as a milliliter, and the units are often used interchangeably.

$$1 \text{ cm}^3 = 1 \text{ cc} = 1 \text{ mL}$$

When you see *1 cm*, you are reading about length; when you see *1 cc* or *1 cm³* or *1 mL*, you are reading about volume. A comparison of units of volume is illustrated in Figure 2.8.

A laboratory technician transfers small volumes using a micropipette.

TABLE 2.8 Some Typical Laboratory Test Values

Substance in Blood	Typical Range
Albumin	3.5–5.0 g/dL
Ammonia	20–150 μg/dL
Calcium	8.5–10.5 mg/dL
Cholesterol	105–250 mg/dL
Iron	80–160 μg/dL (male)
Protein (total)	6.0–8.0 g/dL

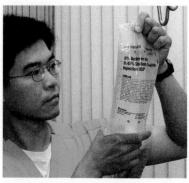

A plastic container contains 1000 mL of intravenous fluid.

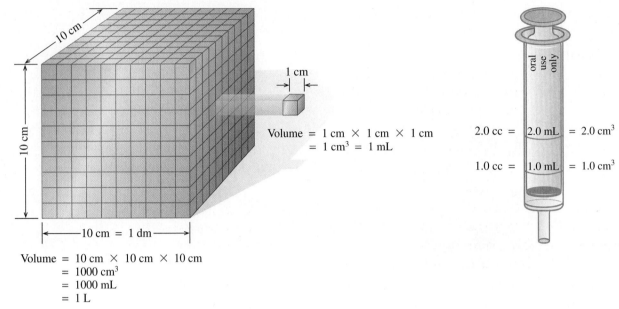

Volume = 1 cm × 1 cm × 1 cm
= 1 cm³ = 1 mL

2.0 cc = 2.0 mL = 2.0 cm³

1.0 cc = 1.0 mL = 1.0 cm³

Volume = 10 cm × 10 cm × 10 cm
= 1000 cm³
= 1000 mL
= 1 L

FIGURE 2.8 A cube measuring 10 cm on each side has a volume of 1000 cm³, or 1 L; a cube measuring 1 cm on each side has a volume of 1 cm³ (cc) or 1 mL.

Q What is the relationship between a milliliter (mL) and a cubic centimeter (cm³)?

Measuring Mass

When you go to the doctor for a physical examination, your mass is recorded in kilograms, whereas the results of your laboratory tests are reported in grams, milligrams (mg), or micrograms (μg). A kilogram equals 1000 g. One gram represents the same mass as 1000 mg, and 1 mg equals 1000 μg.

Some Mass Equalities

$$1 \text{ kg} = 1000 \text{ g} = 1 \times 10^3 \text{ g}$$
$$1 \text{ g} = 1000 \text{ mg} = 1 \times 10^3 \text{ mg}$$
$$1 \text{ mg} = 1000 \text{ } \mu\text{g} = 1 \times 10^3 \text{ } \mu\text{g}$$

SAMPLE PROBLEM 2.8

■ **Writing Metric Relationships**

1. Identify the larger unit in each of the following pairs:
   a. centimeter or kilometer **b.** L or dL **c.** mg or μg

2. Complete the following list of metric equalities:
   a. 1 L = _____ dL **b.** 1 km = _____ m
   c. 1 cm = _____ m **d.** 1 cm³ = _____ mL

SOLUTION

1. **a.** kilometer **b.** L **c.** mg

2. **a.** 10 dL **b.** 1000 m
   c. 0.01 m **d.** 1 mL

STUDY CHECK

Complete the following equalities:

a. 1 kg = _____ g **b.** 1 mL = _____ L

QUESTIONS AND PROBLEMS

Prefixes and Equalities

2.37 The speedometer is marked in both km/h and mi/h or mph. What is the meaning of each abbreviation?

2.38 In a French car, the odometer reads 2250. What units would this be? What units would it be if this were an odometer in a car made for the United States?

2.39 How does the prefix *kilo* affect the gram unit in *kilogram*?

2.40 How does the prefix *centi* affect the meter unit in *centimeter*?

2.41 Write the abbreviation for each of the following units:
a. milligram   **b.** deciliter
c. kilometer   **d.** femtogram
e. microliter   **f.** nanosecond

2.42 Write the complete name for each of the following units:
a. cm   **b.** kg   **c.** ms
d. Gm   **e.** μg   **f.** pg

2.43 Write the numerical value for each of the following prefixes:
a. centi   **b.** tera   **c.** milli
d. deci   **e.** mega   **f.** nano

2.44 Write the complete name (prefix + unit) for each of the following numerical values:
a. 0.10 g   **b.** 10^{-6} g   **c.** 1000 g
d. 0.01 g   **e.** 0.001 g   **f.** 10^{-12} g

2.45 Complete each of the following equalities:
a. 1 m = _____ cm   **b.** 1 nm = _____ m
c. 1 mm = _____ m   **d.** 1 L = _____ mL

2.46 Complete each of the following equalities:
a. 1 Mg = _____ g   **b.** 1 μL = _____ L
c. 1 g = _____ kg   **d.** 1 g = _____ mg

2.47 For each of the following pairs, which is the larger unit?
a. milligram or kilogram   **b.** milliliter or microliter
c. cm or pm   **d.** kL or dL
e. nanometer or picometer

2.48 For each of the following pairs, which is the smaller unit?
a. mg or g   **b.** centimeter or millimeter
c. mm or μm   **d.** mL or dL
e. mg or Mg

2.6 Writing Conversion Factors

Many problems in chemistry require a change of units. You make changes in units every day. For example, suppose you worked 2.0 hours (h) on your homework, and someone asked you how many minutes that was. You must have multiplied 2.0 h × 60 min/h because you know the equality (1 h = 60 min) that relates the two units. When you expressed 2.0 h as 120 min, you did not change the amount of time. You changed only the unit of measurement for time. You can write any equality in the form of a fraction called a **conversion factor**, in which one of the quantities is the numerator and the other is the denominator. Be sure to include the units when you write the conversion factors. Two conversion factors are possible from any equality.

Two Conversion Factors for the Equality 1 h = 60 min

$$\frac{\text{Numerator} \longrightarrow}{\text{Denominator} \longrightarrow} \quad \frac{60\ \text{min}}{1\ \text{h}} \quad \text{and} \quad \frac{1\ \text{h}}{60\ \text{min}}$$

These factors are read as "60 minutes per 1 hour," and "1 hour per 60 minutes." The term *per* means "divide." Some common relationships are given in Table 2.9. It is important that the equality you select to form a conversion factor is a true relationship.

When an equality shows the relationship for two units from the same system, it is considered a definition and exact. Thus, the numbers in that definition are not used to determine significant figures. When an equality shows the relationship of units from two different systems, the number is measured and counts toward the significant figures in a calculation. For example, in the equality 1 lb = 453.6 g, the measured number 453.6 has four significant figures. The number 1 in 1 lb is considered exact. An exception is the relationship of 1 in. = 2.54 cm; the value 2.54 has been defined as exact.

LEARNING GOAL

Write a conversion factor for two units that describe the same quantity.

TABLE 2.9 Some Common Equalities

Quantity	U.S.	Metric (SI)	Metric–U.S.
Length	1 ft = 12 in. 1 yd = 3 ft 1 mi = 5280 ft	1 km = 1000 m 1 m = 1000 mm 1 cm = 10 mm	2.54 cm = 1 in. (exact) 1 m = 39.37 in. 1 km = 0.6214 mi
Volume	1 qt = 4 cups 1 qt = 2 pints 1 gal = 4 qt	1 L = 1000 mL 1 dL = 100 mL 1 mL = 1 cm^3	1 L = 1.057 qt 946.3 mL = 1 qt
Mass	1 lb = 16 oz	1 kg = 1000 g 1 g = 1000 mg	1 kg = 2.205 lb 453.6 g = 1 lb
Time		1 h = 60 min 1 min = 60 s	

Metric Conversion Factors

We can write metric conversion factors for any of the metric relationships. For example, from the equality for meters and centimeters, we can write the following factors:

Metric Equality	**Conversion Factors**
1 m = 100 cm	$\dfrac{100\ cm}{1\ m}$ and $\dfrac{1\ m}{100\ cm}$

Both are proper conversion factors for the relationship; one is just the inverse of the other. The usefulness of conversion factors is enhanced by the fact that we can turn a conversion factor over and use its inverse.

CONCEPT CHECK 2.6

■ Conversion Factors

Identify one or more correct conversion factors for the equality of gigagrams and grams.

a. $\dfrac{1\ Gg}{1 \times 10^9\ g}$

b. $\dfrac{1 \times 10^{-9}\ g}{1\ Gg}$

c. $\dfrac{1 \times 10^9\ Gg}{1\ g}$

d. $\dfrac{1 \times 10^9\ g}{1\ Gg}$

ANSWER In Table 2.6, the equality for gigagrams and grams is 1 Gg = 1×10^9 g. Answers **a** and **d** are correctly written conversion factors.

Metric–U.S. System Conversion Factors

Suppose you need to convert from pounds, a unit in the U.S. system, to kilograms in the metric (or SI) system. A relationship you could use is

1 kg = 2.205 lb

The corresponding conversion factors would be

$$\frac{2.205\ lb}{1\ kg} \quad \text{and} \quad \frac{1\ kg}{2.205\ lb}$$

Figure 2.9 illustrates the contents of some packaged foods in both U.S. and metric units.

Conversion Factors with Powers

Sometimes we need to use a factor that is squared or cubed. This is the case when we need to calculate an area or a volume.

Distance = length

Area = length × length = $length^2$

Volume = length × length × length = $length^3$

FIGURE 2.9 The mass or volume of a packaged food is listed in U.S. and metric units.

Q What are some advantages of using the metric system?

To obtain the necessary conversion factor, we can use a known equality and raise both sides of the equality to the same power.

Measurement	Equality	Conversion Factors	
Length	1 in. = 2.54 cm	$\dfrac{1 \text{ in.}}{2.54 \text{ cm}}$ and	$\dfrac{2.54 \text{ cm}}{1 \text{ in.}}$
Area	$(1 \text{ in.})^2 = (2.54 \text{ cm})^2$	$\dfrac{(1 \text{ in.})^2}{(2.54 \text{ cm})^2}$ and	$\dfrac{(2.54 \text{ cm})^2}{(1 \text{ in.})^2}$
Volume	$(1 \text{ in.})^3 = (2.54 \text{ cm})^3$	$\dfrac{(1 \text{ in.})^3}{(2.54 \text{ cm})^3}$ and	$\dfrac{(2.54 \text{ cm})^3}{(1 \text{ in.})^3}$

Suppose you want to write the equality and the conversion factors for the relationship between an area in square centimeters and in square meters.

Equality: 1 m = 100 cm

Area = length × length = $(1 \text{ m})^2 = (100 \text{ cm})^2$

Conversion factors: $\dfrac{(1 \text{ m})^2}{(100 \text{ cm})^2}$ and $\dfrac{(100 \text{ cm})^2}{(1 \text{ m})^2}$

SAMPLE PROBLEM | 2.9

■ **Writing Conversion Factors**

Write conversion factors for the relationship between the following pairs of units:

a. milligrams and grams   **b.** quarts and liters

c. square inches and square feet

SOLUTION

Equality	Conversion Factors	
a. 1 g = 1000 mg	$\dfrac{1 \text{ g}}{1000 \text{ mg}}$ and	$\dfrac{1000 \text{ mg}}{1 \text{ g}}$
b. 1 L = 1.057 qt	$\dfrac{1 \text{ L}}{1.057 \text{ qt}}$ and	$\dfrac{1.057 \text{ qt}}{1 \text{ L}}$
c. $(1 \text{ ft})^2 = (12 \text{ in.})^2$	$\dfrac{(1 \text{ ft})^2}{(12 \text{ in.})^2}$ and	$\dfrac{(12 \text{ in.})^2}{(1 \text{ ft})^2}$

STUDY CHECK

A zeptosecond (zs) is a very small quantity of time. As an equality, it is written

$1 \text{ zs} = 1 \times 10^{-21} \text{ s}$

Write the conversion factors for this equality.

Conversion Factors Stated Within a Problem

Many times, an equality is specified within a problem that applies only to that problem. It might be the price of 1 kg of oranges or the speed of a car in kilometers per hour. Such equalities are easy to miss when you first read a problem. However, we can write conversion factors for relationships stated within a problem.

1. The motorcycle was traveling at a speed of 85 km/h.

Equality: 1 h = 85 km

Conversion Factors: $\dfrac{85 \text{ km}}{1 \text{ h}}$ and $\dfrac{1 \text{ h}}{85 \text{ km}}$

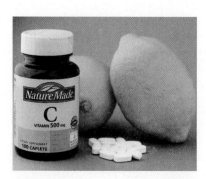

Vitamin C, an antioxidant needed by the body, is found in fruits such as lemons.

2. One tablet contains 500 mg of vitamin C.

Equality: 1 tablet = 500 mg of vitamin C

Conversion Factors: $\dfrac{500 \text{ mg vitamin C}}{1 \text{ tablet}}$ and $\dfrac{1 \text{ tablet}}{500 \text{ mg vitamin C}}$

Conversion Factors from a Percentage, ppm, and ppb

Sometimes a percentage is given in a problem. The term *percent* (%) means parts per 100 parts. To write a percentage as a conversion factor, we choose a unit and express the numerical relationship of the parts to 100 parts of the whole. For example, an athlete might have 18% (18 percent) body fat by mass (see Figure 2.10). The percent quantity can be written as 18 mass units of body fat in every 100 mass units of body mass. Different mass units, such as grams (g), kilograms (kg), or pounds (lb) can be used, but both units in the factor must be the same.

Percent quantity:	18% body fat by mass
Equality:	18 kg of body fat = 100 kg of body mass
Conversion factors:	$\dfrac{100 \text{ kg body mass}}{18 \text{ kg body fat}}$ and $\dfrac{18 \text{ kg body fat}}{100 \text{ kg body mass}}$

or

Equality:	18 lb of body fat = 100 lb of body mass
Conversion factors:	$\dfrac{100 \text{ lb body mass}}{18 \text{ lb body fat}}$ and $\dfrac{18 \text{ lb body fat}}{100 \text{ lb body mass}}$

When scientists want to indicate ratios with particularly small percentage values, they use numerical relationships called parts per million (ppm) or parts per billion (ppb). The ratio of parts per million is the same as the milligrams of a substance per kilogram (mg/kg). The ratio of parts per billion equals the micrograms per kilogram (μg/kg). For example, the maximum amount of lead allowed by FDA in glazed pottery bowls is 5 ppm, which is 5 mg/kg.

ppm quantity:	5 ppm of lead in glaze
Equality:	5 mg of lead = 1 kg of glaze
Conversion factors:	$\dfrac{5 \text{ mg lead}}{1 \text{ kg glaze}}$ and $\dfrac{1 \text{ kg glaze}}{5 \text{ mg lead}}$

FIGURE 2.10 The thickness of the skin fold at the waist measured in millimeters (mm) is used to determine the percent of body fat.

Q What is the percent body fat of an athlete with a body mass of 120 kg and 18 kg of body fat?

(MC)

TUTORIAL

Using Percentage
as a Conversion Factor

SAMPLE PROBLEM | 2.10

■ Conversion Factors Stated in a Problem

Write possible conversion factors for each of the following statements:

a. There are 325 mg of aspirin in 1 tablet.
b. At the grocery store, 1 kg of bananas costs $1.25.
c. The EPA has set the maximum level for mercury in tuna at 0.1 ppm.

SOLUTION

a. $\dfrac{325 \text{ mg of aspirin}}{1 \text{ tablet}}$ and $\dfrac{1 \text{ tablet}}{325 \text{ mg of aspirin}}$

b. $\dfrac{\$1.25}{1 \text{ kg of bananas}}$ and $\dfrac{1 \text{ kg of bananas}}{\$1.25}$

c. $\dfrac{0.1 \text{ mg of mercury}}{1 \text{ kg of tuna}}$ and $\dfrac{1 \text{ kg of tuna}}{0.1 \text{ mg of mercury}}$

STUDY CHECK

What conversion factors can be written for the following statements?

a. A cyclist in the Tour de France bicycle race rides at an average speed of 62.2 km/h.
b. The permissible level of arsenic in water is 10 ppb.

The maximum amount of mercury allowed in tuna is 0.1 ppm.

CHEMISTRY AND THE ENVIRONMENT

Toxicology and Risk–Benefit Assessment

Every day, we make choices about what we do or what we eat, often without thinking about the risks associated with these choices. We are aware of the risks of cancer from smoking, and we know there is a greater risk of having an accident if we cross a street where there is no light or crosswalk.

A basic concept of toxicology is the statement of Paracelsus that the right dose is the difference between a poison and a cure. To evaluate the level of danger from various substances, natural or synthetic, a risk assessment is made by exposing laboratory animals to the substances and monitoring the effects. Often, doses very much greater than humans might ordinarily encounter are given to the test animals.

Many hazardous chemicals or substances have been identified by these tests. One measure of toxicity is the LD_{50} or "lethal dose," which is the concentration of the substance that causes death in 50% of the test animals. A dosage is typically measured in milligrams per kilogram of body mass (mg/kg) or micrograms per kilogram (μg/kg).

Dosage	Units
parts per million (ppm)	milligrams per kilogram (mg/kg)
parts per billion (ppb)	micrograms per kilogram (μg/kg)

Other evaluations need to be made, but it is easy to compare LD_{50} values. Parathion, a pesticide, with an LD_{50} of 3 mg/kg, would be highly toxic. That means that 3 mg of parathion per kg of body mass would be fatal to half of the test animals. Table salt (sodium chloride) with an LD_{50} of 3000 mg/kg has a much lower toxicity. You would need to ingest a huge amount of salt before any toxic effect would be observed. However, increased salt intake may increase renal and blood pressure disorders. Although the risk to animals based on dose can be evaluated in the laboratory, it is more difficult to determine the impact in the environment since there is a difference between continuous exposure and a single, large dose of the substance.

Table 2.10 lists some LD_{50} values and compares substances in order of increasing toxicity.

TABLE 2.10 Some LD_{50} Values for Substances Tested in Rats

Substance	LD_{50} (mg/kg)
Table sugar	29 700
Baking soda	4220
Table salt	3000
Ethanol	2080
Aspirin	1100
Caffeine	192
DDT	113
Sodium cyanide	6
Parathion	3

The LD_{50} of caffeine is 192 ppm.

QUESTIONS AND PROBLEMS

Writing Conversion Factors

2.49 Why can two conversion factors be written for an equality like 1 m = 100 cm?

2.50 How can you check that you have written the correct conversion factors for an equality?

2.51 What equality is expressed by the conversion factor

$$\frac{1000 \text{ g}}{1 \text{ kg}}?$$

2.52 What equality is expressed by the conversion factor

$$\frac{1 \text{ L}}{1 \times 10^6 \, \mu\text{L}}?$$

2.53 Write the equality and conversion factors for each of the following:
a. One yard is 3 ft.
b. One liter is 1000 mL.
c. One minute is 60 s.
d. A car travels 27 miles on 1 gal of gasoline.

2.54 Write the equality and conversion factors for each of the following:
a. One gallon is 4 quarts.
b. At the store, lemons are $1.29 per lb.
c. There are 7 days in 1 week.
d. One dollar has four quarters.

2.55 Write the equality and conversion factors for the following pairs of units:
a. centimeters and meters
b. grams and nanograms
c. kiloliters and liters
d. kilograms and milligrams
e. cubic meters and cubic centimeters

2.56 Write the equality and conversion factors for the following pairs of units:
a. centimeters and inches
b. pounds and kilograms
c. pounds and grams
d. quarts and milliliters
e. square centimeters and square inches

2.57 Write the conversion factors for each of the following statements:
    a. A bee flies at an average speed of 3.5 m/s.
    b. One milliliter of gasoline has a mass of 0.74 g.
    c. An automobile traveled 46.0 km on 1.0 gal of gasoline.
    d. Sterling silver is 93% by mass silver.
    e. The pesticide level in plums was 29 ppb.

2.58 Write the conversion factors for each of the following statements:
    a. The highway gas mileage was 28 mi/gal.
    b. There are 20 drops in 1 mL of water.
    c. The nitrate level in well water was 32 ppm.
    d. A DVD contains 17 GB of information.
    e. The price of a gallon of gas is $2.29.

2.7 Problem Solving

The process of problem solving in chemistry often requires one or more conversion factors to change a given unit to the needed unit.

Given unit × one or more conversion factors = needed unit

Suppose a problem requires the conversion of 165 lb to kilograms. One part of this statement (165 lb) is the given unit, while another part (kilograms) is the unit needed for the answer. Once you identify these units, you can determine which equalities you need in order to convert the given unit to the needed unit.

STEP 1 **Given** 165 lb   **Need** kilograms

STEP 2 **Plan** Write out a sequence of units that starts with the given unit and progresses to the needed unit for the answer. We can see that the given unit is in the U.S. system of measurement and the needed unit is a metric unit. Therefore, the connecting conversion factor is an equality that relates the U.S. unit of lb to the metric unit of kg.

$$\text{lb} \quad \boxed{\begin{array}{c}\text{U.S.–Metric} \\ \text{factor}\end{array}} \quad \text{kg}$$

STEP 3 **Equalities/Conversion Factors** For each change of unit in your plan, state the equality and corresponding conversion factors.

$$1 \text{ kg} = 2.205 \text{ lb}$$
$$\frac{2.205 \text{ lb}}{1 \text{ kg}} \quad \text{and} \quad \frac{1 \text{ kg}}{2.205 \text{ lb}}$$

STEP 4 **Set Up Problem** Now we can write the setup to solve the problem using the plan and one or more conversion factors. First, write down the given information, 165 lb. Then we multiply by the conversion factor that has the unit lb in the denominator (bottom number) to cancel out the given unit lb in the numerator.

Unit for answer goes here

$$165 \cancel{\text{ lb}} \quad \times \quad \frac{1 \text{ kg}}{2.205 \cancel{\text{ lb}}} \quad = \quad 74.8 \text{ kg}$$

Given unit   Conversion factor
(cancels given unit)   Answer
(needed unit)

Look at how the units cancel. The given unit lb cancels out and the needed unit kg is in the numerator (top number). The unit that you want in the answer is the one that remains after all the other units have canceled out. This is a helpful way to check that a problem is set up properly.

$$\cancel{\text{lb}} \times \frac{\text{kg}}{\cancel{\text{lb}}} = \text{kg} \quad \text{Unit needed for answer}$$

The calculation done on a calculator gives the numerical part of the answer, which is then adjusted to give a final answer with the proper number of significant figures (SFs).

$$165 \times \frac{1}{2.205} = 165 \; \boxed{\div} \; 2.205 = \mathit{74.829932} = 74.8$$

3 SFs   4 SFs   Calculator   3 SFs
                                               display   (rounded off)

The value of 74.8 combined with the needed unit, kg, gives the answer of 74.8 kg. With few exceptions, answers to numerical problems contain a number and a unit.

CONCEPT CHECK 2.7

■ Canceling Units

Cancel the units in the following setup and give the unit of the final answer:

$$3.5 \, \text{L} \times \frac{1 \times 10^3 \, \text{mL}}{1 \, \text{L}} \times \frac{0.48 \, \text{g}}{1 \, \text{mL}} \times \frac{1 \times 10^3 \, \text{mg}}{1 \, \text{g}} =$$

ANSWER

All matching units cancel to give mg in the numerator as the needed unit for the answer.

$$3.5 \, \cancel{\text{L}} \times \frac{1 \times 10^3 \, \cancel{\text{mL}}}{1 \, \cancel{\text{L}}} \times \frac{0.48 \, \cancel{\text{g}}}{1 \, \cancel{\text{mL}}} \times \frac{1 \times 10^3 \, \text{mg}}{1 \, \cancel{\text{g}}} = \text{needed unit is mg}$$

SAMPLE PROBLEM | 2.11

■ Problem Solving Using Metric Factors

The daily recommended amount of potassium in the diet is 3500 mg. How many grams of potassium is that?

SOLUTION

STEP 1   **Given** 3500 mg   **Need** grams

STEP 2   **Plan** When we look at the given unit (mg) and the needed unit (g), we see that both are metric units. Therefore, the connecting conversion factor must relate the metric units of g and mg.

     mg   Metric factor   g

STEP 3   **Equalities/Conversion Factors** For the equality for grams and milligrams, we can write two conversion factors:

$$1 \, \text{g} = 1000 \, \text{mg}$$

$$\frac{1 \, \text{g}}{1000 \, \text{mg}} \quad \text{and} \quad \frac{1000 \, \text{mg}}{1 \, \text{g}}$$

STEP 4   **Set Up Problem** The setup is written starting with the given 3500 mg and the conversion factor that converts mg to g. The needed unit (g) is obtained by using the conversion factor that cancels the unit mg. Then the calculation is done and the answer adjusted to give the proper number of significant figures.

Unit for answer goes here

$$3500 \, \cancel{\text{mg}} \times \frac{1 \, \text{g}}{1000 \, \cancel{\text{mg}}} = 3.5 \, \text{g}$$

2 SFs   Exact   2 SFs

STUDY CHECK

If 1890 mL of orange juice is prepared from orange juice concentrate, how many liters of orange juice is that?

Guide to Problem Solving Using Conversion Factors

> **STEP 1**
> State the given and needed quantities.

> **STEP 2**
> Write a plan to convert the given unit to the needed unit.

> **STEP 3**
> State the equalities and conversion factors needed to cancel units.

> **STEP 4**
> Set up problem to cancel units and calculate answer.

Forensic Toxicologist

"I never have a boring day at work, and I never stop learning new things, which may help me deliver better science to the criminal justice system," says Dr. Nikolas P. Lemos, Chief Forensic Toxicologist and Forensic Laboratory Director of the San Francisco Office of the Chief Medical Examiner. "Primarily, I work on post-mortem cases to determine the absence or presence of drugs and their metabolites, chemicals such as ethanol and other volatile substances, carbon monoxide and other gases, metals, and other toxic chemicals in human fluids and tissues, and to evaluate their role as a determinant or contributory factor in the cause and manner of death. Additionally, I am involved in human-performance cases to determine the absence or presence of drugs and chemicals in blood, breath, or other appropriate specimen(s), and to evaluate their role in modifying human performance or behavior. My work also encompasses urine drug testing used in workplace drug-screening procedures to determine the absence or presence of drugs and their metabolites in urine to demonstrate prior use or abuse."

Lava flows from an eruption of Mauna Loa volcano, Hawaii.

Using Two or More Conversion Factors

In many problems, two or more conversion factors are needed to complete the change of units. In setting up these problems, one factor follows the other. Each factor is arranged to cancel the preceding unit until the needed unit is obtained. Once the problem is set up, the calculations can be done without writing intermediate values. Then the answer is written with the correct number of significant figures. This process is worth practicing until you understand unit cancellation and the mathematical calculations.

If you work a problem in single steps, be sure to keep one or two extra digits in the intermediate values. Then round off the calculator display to give the correct number of significant figures for the final answer.

SAMPLE PROBLEM | 2.12

■ Problem Solving Using Two Factors

During a volcanic eruption on Mauna Loa, Hawaii, the lava flowed at a rate of 33 m/min. At this rate, how far, in kilometers, can the lava travel in 45 min?

SOLUTION

STEP 1 **Given** 45 minutes; 33 m/min **Need** kilometers

STEP 2 **Plan**

min [Rate factor] m [Metric factor] km

STEP 3 **Equalities/Conversion Factors** In the problem, the information for the rate of lava flow is given as 33 m/min. We will use this rate as one of the equalities as well as the metric equality for meters and kilometers and write conversion factors for each.

$$1\ min = 33\ m \qquad\qquad 1\ km = 1000\ m$$

$$\frac{1\ min}{33\ m} \quad and \quad \frac{33\ m}{1\ min} \qquad \frac{1\ km}{1000\ m} \quad and \quad \frac{1000\ m}{1\ km}$$

STEP 4 **Set Up Problem** The problem can be set up using the rate as a conversion factor to cancel minutes, and then the metric factor that gives kilometers as the needed unit.

$$45\ \cancel{min}\ \times\ \frac{33\ \cancel{m}}{1\ \cancel{min}}\ \times\ \frac{1\ km}{1000\ \cancel{m}}$$

The calculation is done as follows:

45 [×] 33 [÷] 1000 = [1.485]

Calculator display

By counting the significant figures in the measured quantities, we write the needed answer with two significant figures.

$$45\ \underset{\text{2 SFs}}{\cancel{min}}\ \times\ \frac{33\ \cancel{m}}{1\ \underset{\text{2 SFs}}{\cancel{min}}}\ \times\ \frac{1\ km}{1000\ \underset{\text{Exact}}{\cancel{m}}}\ =\ \underset{\text{2 SFs}}{1.5\ km}$$

STUDY CHECK

How many hours are required for the lava to flow a distance of 5.0 km?

SAMPLE PROBLEM | 2.13

■ Using a Percent as a Conversion Factor

Bronze is 80.0% by mass copper and 20.0% by mass tin. A sculptor is preparing to cast a figure that requires 1.75 lb of bronze. How many kilograms of copper are needed for the bronze figure?

Molten bronze is poured into molds.

SOLUTION

STEP 1   Given 1.75 lb of bronze; 80.0% by mass copper **Need** kilograms of copper

STEP 2   Plan

lb of bronze → U.S.–Metric factor → kg of bronze → Percent factor → kg of copper

STEP 3   Equalities/Conversion Factors   Now we can write the equalities and conversion factors. One is the U.S.–metric factor for lb and kg. The second is the percent factor derived from the information given in the problem.

$$1 \text{ kg of bronze} = 2.205 \text{ lb of bronze}$$
$$\frac{1 \text{ kg bronze}}{2.205 \text{ lb bronze}} \text{ and } \frac{2.205 \text{ lb bronze}}{1 \text{ kg bronze}}$$

$$100 \text{ kg of bronze} = 80.0 \text{ kg of copper}$$
$$\frac{80.0 \text{ kg copper}}{100 \text{ kg bronze}} \text{ and } \frac{100 \text{ kg bronze}}{80.0 \text{ kg copper}}$$

STEP 4   Set Up Problem   We can set up the problem using conversion factors to cancel each unit, starting with lb of bronze, until we obtain the final unit, kg of copper, in the numerator. After we count the significant figures in the measured quantities, we write the needed answer with the proper number of significant figures.

$$\underset{3 \text{ SFs}}{1.75 \text{ lb bronze}} \times \underset{4 \text{ SFs}}{\frac{1 \text{ kg bronze}}{2.205 \text{ lb bronze}}} \times \underset{3 \text{ SFs}}{\frac{80.0 \text{ kg copper}}{100 \text{ kg bronze}}} = \underset{3 \text{ SFs}}{0.635 \text{ kg of copper}}$$

You could set up this problem in a different way by using the following relationships and conversion factors:

$$1 \text{ lb} = 453.6 \text{ g}$$
$$\frac{1 \text{ lb}}{453.6 \text{ g}} \text{ and } \frac{453.6 \text{ g}}{1 \text{ lb}}$$

$$100 \text{ g of bronze} = 80.0 \text{ g of copper}$$
$$\frac{80.0 \text{ g copper}}{100 \text{ g bronze}} \text{ and } \frac{100 \text{ g bronze}}{80.0 \text{ g copper}}$$

$$1 \text{ kg} = 1000 \text{ g}$$
$$\frac{1 \text{ kg}}{1000 \text{ g}} \text{ and } \frac{1000 \text{ g}}{1 \text{ kg}}$$

Then the setup would appear as follows:

$$\underset{3 \text{ SFs}}{1.75 \text{ lb bronze}} \times \underset{4 \text{ SFs}}{\frac{453.6 \text{ g bronze}}{1 \text{ lb bronze}}} \times \underset{3 \text{ SFs}}{\frac{80.0 \text{ g copper}}{100.0 \text{ g bronze}}} \times \underset{\text{Exact}}{\frac{1 \text{ kg copper}}{1000 \text{ g copper}}} = \underset{3 \text{ SFs}}{0.635 \text{ kg of copper}}$$

STUDY CHECK

Uncooked lean ground beef can contain up to 22% fat by mass. How many grams of fat would be contained in 0.25 lb of the ground beef?

QUESTIONS AND PROBLEMS

Problem Solving

2.59 When you convert one unit to another, how do you know which unit of the conversion factor to place in the denominator?

2.60 When you convert one unit to another, how do you know which unit of the conversion factor to place in the numerator?

2.61 Use metric conversion factors to solve each of the following problems:
a. The height of a student is 175 cm. How tall is the student in meters?
b. A cooler has a volume of 5500 mL. What is the capacity of the cooler in liters?
c. A hummingbird has a mass of 0.0055 kg. What is the mass of the hummingbird in grams?
d. A balloon has a volume of 350 cm³. What is the volume in m³?

2.62 Use metric conversion factors to solve each of the following problems:
a. The daily requirement of phosphorus is 800 mg. How many grams of phosphorus are recommended?
b. A glass of orange juice contains 0.85 dL of juice. How many milliliters of orange juice is that?
c. A package of chocolate instant pudding contains 2840 mg of sodium. How many grams of sodium is that?
d. A park has an area of 150 000 m². What is the area in km²?

2.63 Solve each of the following problems using one or more conversion factors (see Table 2.9):
a. A container holds 0.750 qt of liquid. How many milliliters of lemonade will it hold?
b. In England, a person is weighed in stones. If one stone is 14.0 lb, what is the mass, in kilograms, of a person who weighs 11.8 stones?
c. The femur, or thighbone, is the longest bone in the body. In a 6-ft-tall person, the femur is 19.5 in. long. What is the length of that femur in millimeters?
d. How many inches thick is an arterial wall that measures 0.50 μm?

2.64 Solve each of the following problems using one or more conversion factors (see Table 2.9):
a. You need 4.0 oz of a steroid ointment. If there are 16 oz in 1 lb, how many grams of ointment does the pharmacist need to prepare?
b. During surgery, a person receives 5.0 pints of plasma. How many milliliters of plasma were given?
c. Solar flares containing hot gases can rise to 120 000 miles above the surface of the Sun. What is that distance in kilometers?
d. A filled gas tank contains 18.5 gal of fuel. If a car uses 46 L, how many gallons of fuel remain in the tank?

2.65 The singles portion of a tennis court is 27.0 ft wide and 78.0 ft long.
a. What is the length of the court in meters?
b. What is the area of the court in square meters (m²)?
c. If a serve is measured at 185 km/h, how many seconds does it take for the tennis ball to travel the length of the court?
d. How many liters of paint are needed to paint the court if 1 gal of paint covers 150 ft²?

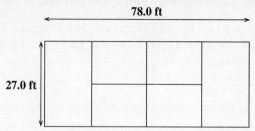

2.66 A football field is 160. ft wide and 300. ft long between goal lines.

a. How many meters does a player run if he catches the ball on his own goal line and scores a touchdown?
b. If a player catches the football and runs 45 yd, how many meters will he gain?
c. How many square meters of Astroturf are required to completely cover the playing field?
d. If a player runs at a speed of 36 km/h, how many seconds does it take to run from the 50-yd line to the 20-yd line?

2.67 **a.** Oxygen (O) makes up 46.7% by mass of Earth's crust. How many grams of oxygen are present if a sample of the crust has a mass of 325 g?
b. Magnesium (Mg) makes up 2.1% by mass of Earth's crust. How many grams of magnesium are present if a sample of the crust has a mass of 1.25 g?
c. A plant fertilizer contains 15% by mass nitrogen (N). In a container of soluble plant food, there are 10.0 oz of fertilizer. How many grams of nitrogen are in the container?

Agricultural fertilizers applied to a field provide nitrogen for plant growth.

d. In a candy factory, the nutty chocolate bars contain 22.0% by mass pecans. If 5.0 kg of pecans were used for candy last Tuesday, how many pounds of nutty chocolate bars were made?

2.68 **a.** Water is 11.2% by mass hydrogen (H). How many kilograms of water would contain 5.0 g of hydrogen?
b. Water is 88.8% by mass oxygen. How many grams of water would contain 2.25 kg of oxygen?
c. Blueberry fiber cakes contain 51% dietary fiber. If a package with a net weight of 12 oz contains six cakes, how many grams of fiber are in each cake?
d. A jar of crunchy peanut butter contains 1.43 kg of peanut butter. If you use 8.0% of the peanut butter for a sandwich, how many ounces of peanut butter did you take out of the container?

2.8 Density

Density is a physical property of matter that compares the mass of a substance to its volume. Every substance has a unique density, which distinguishes it from other substances. Density can be used to identify a specific substance. For example, silver has a density of 10.5 g/cm^3, whereas aluminum has a density of 2.70 g/cm^3. If we had 1.00 cm^3 of both metals, the 1.00 cm^3 of silver would have a mass of 10.5 g, and 1.00 cm^3 of aluminum would have a lower mass of 2.70 g. Density can be used to determine the purity of a substance. If we have 1.00 cm^3 of silver and find that it has a mass of 10.3 g, we would think that the silver is not 100% pure silver but has a contaminant. Density is also used to predict if an object will sink or float in a liquid or in the air. If an object is less dense than a liquid, the object floats when placed in the liquid. In Figure 2.11, the lead object sinks in water because the density of lead is greater than the density of water. However, the cork floats in water because the density of cork is less than that of water.

LEARNING GOAL

Calculate the density of a substance; use the density to calculate the mass or volume of a substance.

Calculating Density

Once the mass and volume of a substance are measured, we can place them in a relationship called the *density expression*:

$$\text{Density} = \frac{\text{mass of substance}}{\text{volume of substance}}$$

In substances with high densities, the particles tend to be close together, whereas in substances with low densities, the particles are farther apart. Thus, metals such as gold and lead tend to have higher densities because their atoms are close and tightly packed in small volumes, whereas gases have very low densities because their atoms are far apart in large volumes. In the metric system, the densities of solids and liquids are usually expressed as grams per cubic centimeter (g/cm^3) or grams per milliliter (g/mL). The density of a gas is usually stated as grams per liter (g/L). Table 2.11 gives the densities of some common substances.

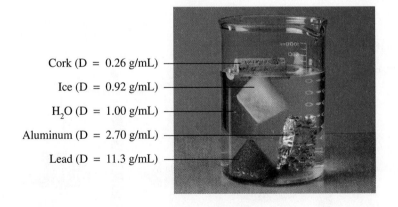

Cork (D = 0.26 g/mL)
Ice (D = 0.92 g/mL)
H$_2$O (D = 1.00 g/mL)
Aluminum (D = 2.70 g/mL)
Lead (D = 11.3 g/mL)

FIGURE 2.11 Objects that sink in water are more dense than water; objects float if they are less dense.

Q Why will an ice cube float and a piece of aluminum sink in water?

TABLE 2.11 Densities of Some Common Substances

Solids (at 25 °C)	Density (g/cm³ or g/mL)	Liquids (at 25 °C)	Density (g/mL)	Gases (at 0 °C)	Density (g/L)
Cork	0.26	Gasoline	0.74	Hydrogen	0.090
Ice (at 0 °C)	0.92	Ethyl alcohol	0.785	Helium	0.179
Sugar	1.59	Olive oil	0.92	Methane	0.714
Salt (NaCl)	2.16	Water (at 4 °C)	1.00	Neon	0.90
Aluminum	2.70	Milk (skim)	1.04	Nitrogen	1.25
Diamond	3.52	Mercury	13.6	Air (dry)	1.29
Copper	8.92			Oxygen	1.43
Silver	10.5			Carbon dioxide	1.96
Lead	11.3				
Gold	19.3				

CHEMISTRY AND THE ENVIRONMENT

Density of Crude Oil

Crude oil (also called petroleum or fossil fuel) is oil that has not been processed, which comes up from underneath the ground. The most typical compounds found in crude oil are hydrocarbons, which are compounds composed of carbon and hydrogen. The differences are in the number of carbon atoms that make up the carbon chains.

Crude oil may be transported by trucks, pipelines, or ships to oil refineries, where it is separated into compounds mostly used for energy. The compounds with 1 to 4 carbon atoms are the gases methane, ethane, propane, and butane, which are commonly used for heating and cooking. The hydrocarbons with 5 or more carbon atoms, such as octane, with 8 carbon atoms, are liquids. They are found in gasoline and other liquid fuels, such as diesel and jet fuel. Those with long carbon chains of about 25 carbon atoms or longer are processed as waxes and tars, such as asphalt.

When tanker ships transporting crude oil have accidents at sea, the spills of oil greatly impact the environment. Because the hydrocarbons typically have densities from 0.74 g/mL for gasoline to about 0.85 g/mL for diesel fuel, they are less dense than water (1.00 g/mL). When there is an oil spill, the crude oil components do not mix with water and thus form a thin layer about one millimeter thick on the surface, which spreads over a large area of the ocean.

In the *Exxon Valdez* oil spill in 1989, 40 million L of oil covered more than 25 000 km² of Prince William Sound in Alaska. Other oil spills occurred in Queensland, Australia (2009), the coast of Wales (1996), and in the Shetland Islands (1993). If the crude oil reaches the beaches and inlets, there can be considerable damage to shellfish, fish, birds, and wildlife habitats. When animals such as birds

In an oil spill, crude oil spreads out to form a thin layer on top of the ocean surface.

are covered with oil, they must be cleaned quickly because ingestion of the hydrocarbons when they try to clean themselves is fatal.

Cleanup of oil spills includes mechanical, chemical, and micro-biological methods. A boom may be placed around the ship to contain the leaking oil until it can be removed. Boats called skimmers then scoop up the oil and place it in tanks. In a chemical method, a substance that attracts oil is used to pick it up and then the oil is scraped off into recovery tanks. Certain bacteria that can ingest oil are also used to break it down into less harmful products.

CONCEPT CHECK 2.8

■ Density

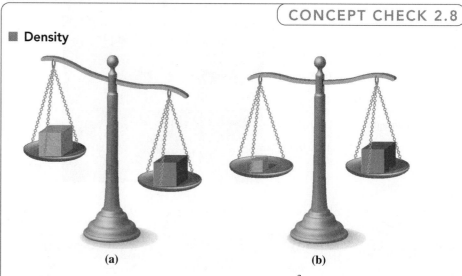

(a)   (b)

a. In drawing (a), the gray cube has a density of 4.5 g/cm³. Is the density of the green cube the same, less than, or greater than that of the gray cube?

b. In drawing (b), the gray cube has a density of 4.5 g/cm³. Is the density of the green cube the same, less than, or greater than that of the gray cube?

ANSWER

a. The green cube has the same volume as the gray cube but has a greater mass. Thus, the green cube has a density that is greater than that of the gray cube.

b. The green cube has the same mass as the gray cube, but the green cube has a greater volume. Thus, the green cube has a density that is less than that of the gray cube.

SAMPLE PROBLEM   2.14

■ Calculating Density

High-density lipoprotein (HDL) contains large amounts of proteins and small amounts of cholesterol. If a 0.258-g sample of HDL has a volume of 0.215 cm^3, what is the density of the HDL sample?

SOLUTION

STEP 1 **State the given and needed quantities.**
Given mass of HDL sample = 0.258 g; volume = 0.215 cm^3
Need density (g/cm^3)

STEP 2 **Write the density expression.**

$$\text{Density} = \frac{\text{mass of substance}}{\text{volume of substance}}$$

STEP 3 **Express mass in grams and volume in milliliters (mL) or cm^3.**

Mass of HDL sample = 0.258 g
Volume of HDL sample = 0.215 cm^3

STEP 4 **Substitute mass and volume into the density expression and calculate the density.**

$$\text{Density} = \frac{\overset{3\ \text{SFs}}{0.258\ \text{g}}}{\underset{3\ \text{SFs}}{0.215\ cm^3}} = \frac{1.20\ \text{g}}{1\ cm^3} = \underset{3\ \text{SFs}}{1.20\ \text{g/}cm^3}$$

STUDY CHECK

Low-density lipoprotein (LDL) contains small amounts of proteins and large amounts of cholesterol. If a 0.380-g sample of LDL has a volume of 0.362 cm^3, what is the density of the LDL sample?

Density of Solids

The density of a solid is calculated using its mass and volume. When a solid is completely submerged, it displaces a volume of water that is equal to the volume of the solid. In Figure 2.12, the water level rises from 35.5 mL to 45.0 mL after the zinc object is added. This means that 9.5 mL of water is displaced and that the volume of the object is 9.5 mL. The density of the zinc is calculated as follows:

$$\text{Density} = \frac{68.60\ \text{g Zn}}{9.5\ \text{mL}} = 7.2\ \text{g/mL}$$

Mass of zinc object

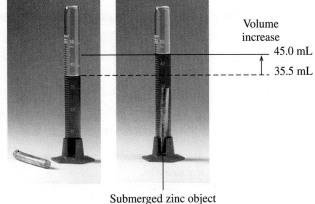

Volume increase
45.0 mL
35.5 mL

Submerged zinc object

FIGURE 2.12 The density of a solid can be determined by volume displacement because a submerged object displaces a volume of water equal to its own volume.

Q How is the volume of the zinc object determined?

Diamond is a crystalline form of carbon.

■ Calculating Density

Suppose you purchased an antique ring. The owner has told you that the gem in your ring is a diamond, a crystalline form of carbon. You ask the owner to remove the gem so you can test its density. You determine that the gem has a mass of 1.65 g. When you place it in a small cylinder containing 3.50 mL of water, the water level rises to 3.97 mL. When you look up the density values, you find the following: diamond 3.5 g/mL, cubic zirconia 4.6 g/mL, glass 2.5 g/mL

a. What is the density of the gem?
b. From your answer in **a**, do you think the gem is diamond, cubic zirconia, or glass?
c. If 5.00 carats have a mass of 1.00 g, how many carats is the gem?

ANSWER

a. The volume of the gem is calculated as the difference in the water levels: 3.97 mL − 3.50 mL = 0.47 mL. The density is calculated by the expression D = mass/volume.

$$\text{Density} = \frac{1.65 \text{ g}}{0.47 \text{ mL}} = 3.5 \text{ g/mL}$$

b. A density of 3.5 g/mL is the same density as diamond but not cubic zirconia or glass.
c. The number of carats in the ring is calculated using the conversion factor 5.00 carats/1.00 g.

$$1.65 \text{ g} \times \frac{5.00 \text{ carats}}{1.00 \text{ g}} = 8.25 \text{ carats}$$

That is a nice-sized diamond in your ring.

■ Using Volume Displacement to Calculate Density

A lead weight used in the belt of a scuba diver has a mass of 226 g. When the weight is carefully placed in a graduated cylinder containing 200.0 cm^3 of water, the water level rises to 220.0 cm^3. What is the density of the lead weight (g/cm^3)?

SOLUTION

STEP 1 State the given and needed quantities.
   Given mass of lead = 226 g; water level before object submerged = 200.0 cm^3; water level after object submerged = 220.0 cm^3
   Need density (g/cm^3)

STEP 2 Write the density expression.

$$\text{Density} = \frac{\text{mass of substance}}{\text{volume of substance}}$$

STEP 3 Express mass in grams and volume in milliliters (mL) or cm^3.

Mass of lead weight = 226 g
The volume of the lead weight is equal to the volume of water displaced, which is calculated as follows:

Water level after object submerged	=	220.0 cm^3
Water level before object submerged	=	−200.0 cm^3
Water displaced (volume of lead)	=	20.0 cm^3

STEP 4 Substitute mass and volume into the density expression and calculate the density.

Lead weights in a belt counteract the buoyancy of a scuba diver.

$$\text{Density} = \frac{\overset{\text{3 SFs}}{226 \text{ g}}}{\underset{\text{3 SFs}}{20.0 \text{ cm}^3}} = \underset{\text{3 SFs}}{11.3 \text{ g/cm}^3}$$

STUDY CHECK

A total of 0.50 lb of glass marbles is added to 425 mL of water. The water level rises to a volume of 528 mL. What is the density (g/cm³) of the glass marbles?

Problem Solving Using Density

Density can be used as a conversion factor. For example, if the volume and the density of a sample are known, the mass in grams of the sample can be calculated.

SAMPLE PROBLEM | 2.16

■ Problem Solving Using Density

If the density of a sample of skim milk is 1.04 g/mL, how many grams are in 0.50 qt of skim milk?

A sample of skim milk has a density of 1.04 g/mL.

SOLUTION

STEP 1 **State the given and needed quantities.**
  Given 0.50 qt; density = 1.04 g/mL   **Need** grams of milk

STEP 2 **Write a plan to calculate the needed quantity.**

qt ⟩ U.S.–metric factor ⟩ L ⟩ Metric factor ⟩ mL ⟩ Density factor ⟩ g

STEP 3 **Write equalities and their conversion factors including density.**

1 L = 1.057 qt	1 L = 1000 mL	1 mL = 1.04 g
$\dfrac{1 \text{ L}}{1.057 \text{ qt}}$ and $\dfrac{1.057 \text{ qt}}{1 \text{ L}}$	$\dfrac{1 \text{ L}}{1000 \text{ mL}}$ and $\dfrac{1000 \text{ mL}}{1 \text{ L}}$	$\dfrac{1 \text{ mL}}{1.04 \text{ g}}$ and $\dfrac{1.04 \text{ g}}{1 \text{ mL}}$

STEP 4 **Set up problem to calculate the needed quantity.**

$$\underset{\text{2 SFs}}{0.50 \text{ qt}} \times \underset{\text{4 SFs}}{\frac{1 \text{ L}}{1.057 \text{ qt}}} \times \underset{\text{Exact}}{\frac{1000 \text{ mL}}{1 \text{ L}}} \times \underset{\text{3 SFs}}{\frac{1.04 \text{ g}}{1 \text{ mL}}} = \underset{\text{2 SFs}}{490 \text{ g } (4.9 \times 10^2 \text{ g})}$$

Guide to Using Density

STEP 1
State the given and needed quantities.

STEP 2
Write a plan to calculate the needed quantity.

STEP 3
Write equalities and their conversion factors including density.

STEP 4
Set up problem to calculate the needed quantity.

STUDY CHECK

How many milliliters of mercury are in a thermometer that contains 20.4 g of mercury (see Table 2.11 for the density of mercury)?

CHEMISTRY AND HEALTH

Bone Density

The density of our bones determines their health and strength. Our bones are constantly gaining and losing minerals such as calcium, magnesium, and phosphate. In childhood, bones form at a faster rate than they break down. As we age, the breakdown of bone occurs more rapidly than new bone forms. As the loss of bone minerals increases, bones begin to thin, causing a decrease in mass and density. Thinner bones lack strength, which increases the risk of fracture. Hormonal changes, disease, and certain medications can also contribute to the thinning of bone. Eventually, a condition of severe thinning of bone known as *osteoporosis* may occur. *Scanning electron micrographs* (SEMs) show (**a**) normal bone and (**b**) bone in osteoporosis due to loss of bone minerals.

Bone density is often determined by passing low-dose X-rays through the narrow part at the top of the femur (hip) and the spine (**c**). These locations are where fractures are more likely to occur, especially as we age. Bones with high density will block more of the X-rays compared to bones that are less dense. The results of a bone density test are compared to those of a healthy young adult as well as to other people of the same age.

Recommendations to improve bone strength include supplements of calcium and vitamin D and medications such as Fosamax®, Evista®, or Actonel®. Weight-bearing exercise such as walking and lifting weights can also improve muscle strength, which in turn increases bone strength.

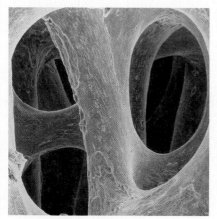

(**a**) Normal bone

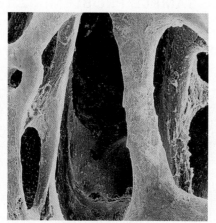

(**b**) Bone with osteoporosis

(**c**) Viewing a low-dose X-ray of the spine

QUESTIONS AND PROBLEMS

Density

2.69 In an old trunk, you find a piece of metal that you think may be aluminum, silver, or lead. In a lab, you find it has a mass of 217 g and a volume of 19.2 cm³. Using Table 2.11, what is the metal you found?

2.70 Suppose you have two 100-mL graduated cylinders. In each cylinder, there is 40.0 mL of water. You also have two cubes: one is lead, and the other is aluminum. Each cube measures 2.0 cm on each side. After you carefully lower each cube into the water of its own cylinder, what will the new water level be in each of the cylinders?

2.71 Determine the density (g/mL) for each of the following:
    a. A 20.0-mL sample of a salt solution has a mass of 24.0 g.
    b. A cube of butter weighs 0.250 lb and has a volume of 130.3 mL.
    c. A gem has a mass of 4.50 g. When the gem is placed in a graduated cylinder containing 2.00 mL of water, the water level rises to 3.45 mL.

Lightweight heads on the drivers of golf clubs are made of titanium.

    d. A lightweight head on the driver of a golf club is made of titanium. If the volume of a sample of titanium is 114 cm³ and the mass is 485.6 g, what is the density of titanium?

e. A syrup is added to an empty container with a mass of 115.25 g. When 0.100 pint of syrup is added, the total mass of the container and syrup is 182.48 g. (1 qt = 2 pt)

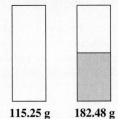

115.25 g   182.48 g

2.72 Determine the density (g/mL) for each of the following:
  a. A plastic material weighs 2.68 lb and has a volume of 3.5 L.
  b. The fluid in a car battery has a volume of 125 mL and a mass of 155 g.
  c. A 5.00-mL urine sample from a person suffering from diabetes mellitus has a mass of 5.025 g.
  d. An ebony carving has a mass of 275 g and a volume of 207 cm³.
  e. A 10.00-L sample of oxygen gas has a mass of 0.014 kg.

2.73 Use the density values in Table 2.11 to solve the following problems:
  a. How many liters of ethyl alcohol contain 1.50 kg of alcohol?
  b. How many grams of mercury are present in a barometer that holds 6.5 mL of mercury?

c. A sculptor has prepared a mold for casting a bronze figure. The figure has a volume of 225 mL. If bronze has a density of 7.8 g/mL, how many ounces of bronze are needed in the preparation of the bronze figure?
  d. What is the mass, in grams, of a cube of copper that has a volume of 74.1 cm³?
  e. How many kilograms of gasoline fill a 12.0-gal gas tank? (1 gal = 4 qt)

2.74 Use the density values in Table 2.11 to solve the following problems:
  a. A graduated cylinder contains 28.0 mL of water. What is the new water level after 35.6 g of silver metal is submerged in the water?
  b. A thermometer containing 8.3 g of mercury has broken. What volume of mercury spilled?
  c. A fish tank holds 35 gal of water. How many pounds (lb) of water are in the fish tank?
  d. The mass of an empty container is 88.25 g. The mass of the container and a liquid with a density of 0.758 g/mL is 150.50 g. What is the volume (mL) of the liquid in the container?
  e. An iron object has a volume of 115 cm³. If iron has a density of 7.86 g/cm³, what is the mass, in kilograms, of the object?

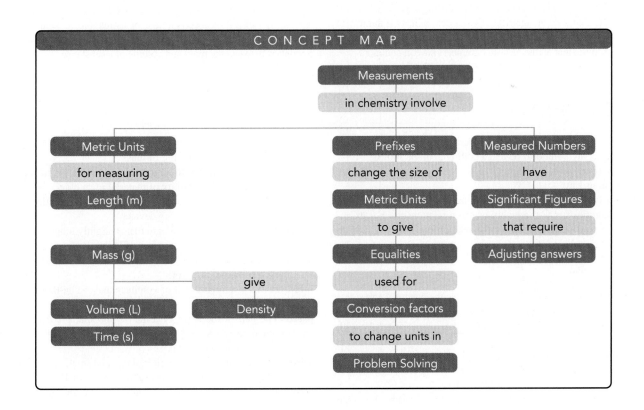

CONCEPT MAP

Measurements

in chemistry involve

Metric Units

for measuring

Length (m)

Mass (g)

give

Volume (L)   Density

Time (s)

Prefixes

change the size of

Metric Units

to give

Equalities

used for

Conversion factors

to change units in

Problem Solving

Measured Numbers

have

Significant Figures

that require

Adjusting answers

CHAPTER REVIEW

2.1 Units of Measurement

LEARNING GOAL: *Write the names and abbreviations for the metric or SI units used in measurements of length, volume, mass, temperature, and time.*
In science, physical quantities are described in units of the metric or International System of Units (SI). Some important units are meter (m) for length, liter (L) for volume, gram (g) and kilogram (kg) for mass, degree Celsius (°C) for temperature, and second (s) for time.

2.2 Scientific Notation

LEARNING GOAL: *Write a number in scientific notation.*
Large and small numbers can be written using scientific notation, in which the decimal point is moved to give a coefficient from 1 to 9 and the number of decimal places moved shown as a power of 10. A number greater than 1 will have a positive power of 10, while a number less than 1 will have a negative power of 10.

2.3 Measured Numbers and Significant Figures

LEARNING GOAL: *Identify a number as measured or exact; determine the number of significant figures in a measured number.*
A measured number is any number derived from using a measuring device to determine a quantity. An exact number is obtained by counting items or from a definition; no measuring device is used. Significant figures are the numbers reported in a measurement including the estimated digit.

2.4 Significant Figures in Calculations

LEARNING GOAL: *Adjust calculated answers to give the correct number of significant figures.*
In multiplication or division, the measured number in the calculation that has the smallest number of significant figures determines the number of significant figures in an answer. In addition or subtraction, the answer reflects the last place that all the measured numbers have in common.

2.5 Prefixes and Equalities

LEARNING GOAL: *Use the numerical values of prefixes to write a metric equality.*
Prefixes placed in front of a unit change the size of the unit by factors of 10. Prefixes such as *centi*, *milli*, and *micro* provide smaller units; prefixes such as *kilo* provide larger units. An equality relates two units that measure the same quantity of length, volume, mass, or time. Examples of metric equalities are: 1 m = 100 cm; 1 L = 1000 mL; 1 kg = 1000 g; 1 s = 1000 ms.

2.6 Writing Conversion Factors

LEARNING GOAL: *Write a conversion factor for two units that describe the same quantity.*
Conversion factors are used to express a relationship in the form of a fraction. Two factors can be written for any relationship in the metric or U.S. system.

2.7 Problem Solving

LEARNING GOAL: *Use conversion factors to change from one unit to another.*
Conversion factors are useful when changing a quantity expressed in one unit to a quantity expressed in another unit. In the process, a given unit is multiplied by one or more conversion factors that cancel units until the needed answer is obtained.

2.8 Density

LEARNING GOAL: *Calculate the density of a substance; use the density to calculate the mass or volume of a substance.*
The density of a substance is a ratio of its mass to its volume, usually expressed as g/mL or g/cm^3. The units of density can be used as a factor to convert between the mass and volume of a substance.

KEY TERMS

Celsius (°C) temperature scale A temperature scale on which water has a freezing point of 0 °C and a boiling point of 100 °C.

centimeter (cm) A unit of length in the metric system; there are 2.54 cm in 1 in.

conversion factor A ratio in which the numerator and denominator are quantities from an equality or given relationship. For example, the conversion factors for the relationship 1 kg = 2.205 lb are written as:

$$\frac{2.205 \text{ lb}}{1 \text{ kg}} \quad \text{and} \quad \frac{1 \text{ kg}}{2.205 \text{ lb}}$$

cubic centimeter (cm^3, cc) The volume of a cube that has 1-cm sides; 1 cubic centimeter is equal to 1 mL.

cubic meter (m^3) The SI unit of volume; the volume of a cube with sides that measure 1 m.

density The relationship of the mass of an object to its volume, expressed as grams per cubic centimeter (g/cm^3), grams per milliliter (g/mL), or grams per liter (g/L).

equality A relationship between two units that measure the same quantity.

exact number A number obtained by counting or definition.

gram (g) The metric unit used in measurements of mass.

Kelvin (K) temperature scale A temperature scale on which the lowest possible temperature is 0 K, which makes the freezing point of water 273 K and the boiling point of water 373 K.

kilogram (kg) A metric mass of 1000 g, equal to 2.205 lb. The kilogram is the SI standard unit of mass.

liter (L) A metric unit for volume that is slightly larger than a quart.

mass A measure of the quantity of material in an object.

measured number A number obtained when a quantity is determined by using a measuring device.

meter (m) A metric unit for length that is slightly longer than a yard. The meter is the SI standard unit of length.

metric system A system of measurement used by scientists and in most countries of the world.

milliliter (mL) A metric unit of volume equal to one-thousandth of a liter (0.001 L).

prefix The part of the name of a metric unit that precedes the base unit and specifies the size of the measurement. All prefixes are related on a decimal scale.

scientific notation A form of writing large and small numbers using a coefficient from 1 to 9, followed by a power of 10.

second (s) The standard unit of time in both the SI and metric systems.

SI An international system of units that modifies the metric system.

significant figures (SFs) The numbers recorded in a measurement.

temperature An indicator of the hotness or coldness of an object.

volume (V) The amount of space occupied by a substance.

UNDERSTANDING THE CONCEPTS

2.75 Indicate if each of the following is answered with an exact number or a measured number:

a. number of legs
b. height of table
c. number of chairs at the table
d. area of tabletop

2.76 State the temperature, including the estimated digit, on each of the Celsius thermometers.

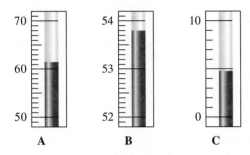

2.77 Measure the length and width, including the estimated digit, of the rectangle using a metric ruler.

a. What is the length and width of this rectangle measured in centimeters?
b. What is the length and width of this rectangle measured in millimeters?
c. How many significant figures are in the length measurement?
d. How many significant figures are in the width measurement?
e. What is the area of the rectangle in cm^2?
f. How many significant figures are in the calculated answer for area?

2.78 Measure the length of each of the objects in diagrams (a), (b), and (c) using the metric ruler shown. Indicate the number of significant figures for each and the estimated digit for each.

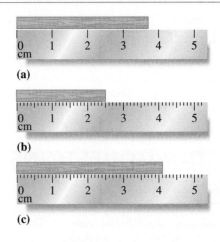

(a)

(b)

(c)

2.79 What is the density of the solid object that is weighed and submerged in water?

2.80 Each of the following diagrams represents a container of water and a cube. Some cubes float while others sink. Match diagrams 1, 2, 3, or 4 with one of the following descriptions and explain your choices:
a. The cube has a greater density than water.
b. The cube has a density that is 0.60–0.80 g/mL.
c. The cube has a density that is one-half the density of water.
d. The cube has the same density as water.

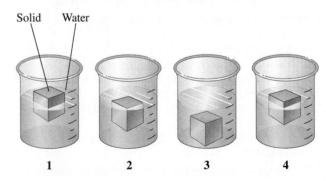

2.81 A graduated cylinder contains three liquids A, B, and C, which have different densities and do not mix: mercury (D = 13.6 g/mL), vegetable oil (D = 0.92 g/mL), and water (D = 1.00 g/mL). Identify the liquids A, B, and C in the cylinder.

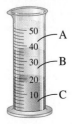

2.82 The solids A, B, and C represent aluminum, gold, and silver. If each has a mass of 10.0 g, what is the identity of each?

Density of aluminum = 2.70 g/mL

Density of gold = 19.3 g/mL

Density of silver = 10.5 g/mL

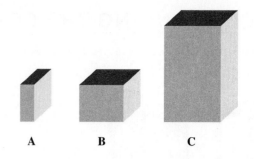

A   B   C

ADDITIONAL QUESTIONS AND PROBLEMS

For instructor-assigned homework, go to
www.masteringchemistry.com.

2.83 Round off or add zeros to the following calculated answers to give a final answer with three significant figures:
a. 0.000 012 58 L   **b.** 3.528 × 10² kg
c. 125 111 m³   **d.** 58.703 m
e. 3 × 10⁻³ s   **f.** 0.010 826 g

2.84 What is the total mass, in grams, of a dessert containing 137.25 g of vanilla ice cream, 84 g of fudge sauce, and 43.7 g of nuts?

2.85 During a workout at the gym, you set the treadmill at a pace of 55.0 m/min. How many minutes will you walk if you cover a distance of 7500 ft?

2.86 A fish company delivers 22 kg of salmon, 5.5 kg of crab, and 3.48 kg of oysters to your seafood restaurant.
a. What is the total mass, in kilograms, of the seafood?
b. What is the total number of pounds?

2.87 Bill's recipe for onion soup calls for 4.0 lb of thinly sliced onions. If an onion has an average mass of 115 g, how many onions does Bill need?

2.88 The price of 1 pound (lb) of potatoes is $1.75. If all the potatoes sold today at the store bring in $1420, how many kilograms (kg) of potatoes did grocery shoppers buy?

2.89 The following nutrition information is listed on a box of crackers:
Serving size 0.50 oz (6 crackers)
Fat 4 g per serving **Sodium** 140 mg per serving
a. If the box has a net weight (contents only) of 8.0 oz, about how many crackers are in the box?
b. If you ate 10 crackers, how many ounces of fat are you consuming?
c. How many grams of sodium are used to prepare 50 boxes of crackers?

2.90 An aquarium store unit requires 75 000 mL of water. How many gallons of water are needed? (1 gal = 4 qt)

2.91 In Mexico, avocados are 48 pesos per kilogram. What is the cost, in cents, of an avocado that weighs 0.45 lb if the exchange rate is 14.4 pesos to the dollar?

2.92 Celeste's diet restricts her intake of protein to 24 g per day. If she eats an 8.0-oz burger that is 15.0% protein, has she exceeded her protein limit for the day? How many ounces of a burger would be allowed by her diet?

2.93 A sunscreen preparation contains 2.50% by mass benzyl salicylate. If a tube contains 4.0 oz of sunscreen, how many kilograms of benzyl salicylate are needed to manufacture 325 tubes of sunscreen?

2.94 An object has a mass of 3.15 oz. When it is submerged in a graduated cylinder initially containing 325.2 mL of water, the water level rises to 442.5 mL. What is the density (g/mL) of the object?

2.95 What is a cholesterol level of 1.85 g/L in units of mg/dL?

2.96 If a recycling center collects 1254 aluminum cans and there are 22 aluminum cans in 1 lb, what volume, in liters, of aluminum was collected (see Table 2.11)?

2.97 The water level in a graduated cylinder initially at 215 mL rises to 285 mL after a piece of lead is submerged. What is the mass, in grams, of the lead (see Table 2.11)?

2.98 A graduated cylinder contains 155 mL of water. A 15.0-g piece of iron (density = 7.86 g/cm³) and a 20.0-g piece of lead are added. What is the new water level in the cylinder (see Table 2.11)?

2.99 How many cubic centimeters (cm³) of olive oil have the same mass as 1.00 L of gasoline (see Table 2.11)?

2.100 What is the volume, in quarts, of 1.50 kg of ethyl alcohol (C₂H₆O) (see Table 2.11)?

2.101 a. Some athletes have as little as 3.0% body fat. If such a person has a body mass of 65 kg, how many pounds of body fat does that person have?
b. In liposuction, a doctor removes fat deposits from a person's body. If body fat has a density of 0.94 g/mL and 3.0 liters of fat is removed, how many pounds of fat were removed from the patient?

2.102 A mouthwash is 21.6% ethyl alcohol by mass. If each bottle contains 0.358 pint of mouthwash with a density of 0.876 g/mL, how many kilograms of ethyl alcohol are in 180 bottles of the mouthwash?

A mouthwash may contain ethyl alcohol.

2.103 Sterling silver is 92.5% silver by mass with a density of 10.3 g/cm^3. If a cube of sterling silver has a volume of 27.0 cm^3, how many ounces of pure silver are present?

2.104 A typical adult body contains 55% water. If a person has a mass of 65 kg, how many pounds of water does she have in her body?

CHALLENGE QUESTIONS

2.105 A balance measures mass to 0.001 g. If you determine the mass of an object that weighs about 30 g, would you record the mass as 30 g, 32 g, 32.1 g, or 32.075 g? Explain your choice by writing two or three complete sentences that describe your thinking.

2.106 When three students use the same meterstick to measure the length of a paperclip, they obtain results of 5.8 cm, 5.75 cm, and 5.76 cm. If the meterstick has millimeter markings, what are some reasons for the different values?

2.107 A car travels at 55 miles per hour and gets 11 kilometers per liter of gasoline. How many gallons of gasoline are needed for a 3.0-h trip?

2.108 For a 180-lb person, calculate the quantity of each of the following that must be ingested to provide the LD$_{50}$ for caffeine given in Table 2.10:
   a. cups of coffee if one cup is 12 fluid ounces and there is 100 mg of caffeine per 6-fl oz of drip-brewed coffee
   b. cans of cola if one can contains 50 mg of caffeine
   c. tablets of No-Doz if one tablet contains 100 mg of caffeine

2.109 A package of aluminum foil is 66$\frac{2}{3}$ yd long, 12 in. wide, and 0.000 30 in. thick. If aluminum has a density of 2.70 g/cm^3, what is the mass, in grams, of the foil?

2.110 A circular pool with a diameter of 27 ft is filled to a depth of 50. in. Assume the pool is a cylinder ($V_{cylinder} = \pi r^2 h$).
   a. What is the volume of water in the pool in cubic meters?

   b. The density of water is 1.00 g/cm^3. What is the mass, in kilograms, of the water in the pool?

2.111 An 18-karat gold necklace is 75% gold by mass, 16% silver, and 9.0% copper.
   a. What is the mass, in grams, of the necklace if it contains 0.24 oz of silver?
   b. How many grams of copper are in the necklace?
   c. If 18-karat gold has a density of 15.5 g/cm^3, what is the volume in cubic centimeters?

2.112 In the manufacturing of computer chips, cylinders of silicon are cut into thin wafers that are 3.00 in. in diameter and have a mass of 1.50 g of silicon. How thick (mm) is each wafer if silicon has a density of 2.33 g/cm^3? (The volume of a cylinder is $V = \pi r^2 h$.)

2.113 A 50.0-g silver object and a 50.0-g gold object are both added to 75.5 mL of water contained in a graduated cylinder. What is the new water level in the cylinder?

2.114 The label on a 1-pint bottle of water lists the following components. If the density is the same as pure water and you drink three bottles of water in one day, how many milligrams of each component will you obtain?
   a. calcium, 28 ppm
   b. fluoride, 0.08 ppm
   c. magnesium, 12 ppm
   d. potassium, 3.2 ppm
   e. sodium, 15 ppm

ANSWERS

ANSWERS TO STUDY CHECKS

2.1 **a.** meter (m)   **b.** kelvin (K)   **c.** kilogram (kg)

2.2 **a.** 4.25×10^5 m   **b.** 8×10^{-7} g

2.3 **a.** two SFs   **b.** one SF   **c.** five SFs

2.4 **a.** 36 m   **b.** 0.0026 L
      c. 3.8×10^3 g   **d.** 1.3 kg

2.5 **a.** 0.4924   **b.** 0.008 02 or 8.02×10^{-3}
      c. 2.0

2.6 **a.** 83.70 g   **b.** 0.5 L

2.7 **a.** giga   **b.** centi

2.8 **a.** 1000 g   **b.** 0.001 L

2.9 Equality: 1 zs $= 1 \times 10^{-21}$ s

   Conversion Factors: $\dfrac{1 \text{ zs}}{1 \times 10^{-21} \text{ s}}$ and $\dfrac{1 \times 10^{-21} \text{ s}}{1 \text{ zs}}$

2.10 **a.** $\dfrac{1 \text{ h}}{62.2 \text{ km}}$ and $\dfrac{62.2 \text{ km}}{1 \text{ h}}$

   b. $\dfrac{10 \text{ } \mu\text{g arsenic}}{1 \text{ kg water}}$ and $\dfrac{1 \text{ kg water}}{10 \text{ } \mu\text{g arsenic}}$

2.11 1.89 L

2.12 2.5 h

2.13 25 g of fat

2.14 1.05 g/cm^3

2.15 2.2 g/cm^3

2.16 1.50 mL of mercury

ANSWERS TO SELECTED QUESTIONS AND PROBLEMS

2.1 **a.** meter, length   **b.** gram, mass
      c. liter, volume   **d.** second, time
      e. degree Celsius, temperature

2.3 **a.** meter, both   **b.** kilogram, both
      c. inch, neither   **d.** second, both
      e. degree Celsius, metric

2.5 **a.** gram, metric   **b.** liter, metric
      c. degree Fahrenheit, neither **d.** pound, neither
      e. second, both

2.7 **a.** 5.5×10^4 m   **b.** 4.8×10^2 g
      c. 5×10^{-6} cm   **d.** 1.4×10^{-4} s
      e. 7.85×10^{-3} L   **f.** 6.7×10^5 kg

2.9 **a.** 7.2×10^3 cm   **b.** 3.2×10^{-2} kg
      c. 1×10^4 L   **d.** 6.8×10^{-2} m

2.11 **a.** 12 000 s   **b.** 0.0825 kg
      c. 4 000 000 g   **d.** 0.0058 m^3

2.13 **a.** first decimal place or tenths place (0.6)
      b. second decimal place or hundredths place (0.05)
      c. first decimal place or tenths place (0.0)

2.15 **a.** measured   **b.** exact
c. exact   **d.** measured

2.17 **a.** 6 oz of hamburger   **b.** none
c. 0.75 lb, 350 g   **d.** none (definitions are exact)

2.19 **a.** not significant   **b.** significant
c. significant   **d.** significant
e. not significant

2.21 **a.** 5 SFs   **b.** 2 SFs
c. 2 SFs   **d.** 3 SFs
e. 4 SFs   **f.** 3 SFs

2.23 Both measurements in part **c** have two significant figures and both measurements in part **d** have four significant figures.

2.25 **a.** 5.0×10^3 L   **b.** 3.0×10^4 g
c. 1.0×10^5 m   **d.** 2.5×10^{-4} cm

2.27 The number of figures in the answer is limited by the measurements used in the calculation.

2.29 **a.** 1.85 kg   **b.** 88.0 L
c. 0.004 74 cm   **d.** 8810 m
e. 1.83×10^3 s

2.31 **a.** 56.9 m   **b.** 0.002 28 g
c. 11 500 s (1.15×10^4 s)   **d.** 8.10 L

2.33 **a.** 1.6   **b.** 0.01
c. 27.6   **d.** 3.5
e. 1.4×10^{-1} (0.14)   **f.** 8×10^{-1} (0.8)

2.35 **a.** 53.54 cm   **b.** 127.6 g
c. 121.5 mL   **d.** 0.50 L

2.37 km/h is kilometers per hour; mi/h is miles per hour.

2.39 The prefix *kilo* means to multiply by 1000; 1 kg is the same mass as 1000 g.

2.41 **a.** mg   **b.** dL   **c.** km
d. fg   **e.** μL   **f.** ns

2.43 **a.** 0.01   **b.** 10^{12} **c.** 0.001
d. 0.1   **e.** 10^6 **f.** 10^{-9}

2.45 **a.** 100 cm   **b.** 10^{-9} m   **c.** 0.001 m
d. 1000 mL

2.47 **a.** kilogram   **b.** milliliter   **c.** cm
d. kL   **e.** nanometer

2.49 A conversion factor can be inverted to give a second conversion factor.

2.51 1 kg = 1000 g

2.53 **a.** 1 yd = 3 ft; $\dfrac{3 \text{ ft}}{1 \text{ yd}}$ and $\dfrac{1 \text{ yd}}{3 \text{ ft}}$

b. 1 L = 1000 mL; $\dfrac{1000 \text{ mL}}{1 \text{ L}}$ and $\dfrac{1 \text{ L}}{1000 \text{ mL}}$

c. 1 min = 60 s; $\dfrac{60 \text{ s}}{1 \text{ min}}$ and $\dfrac{1 \text{ min}}{60 \text{ s}}$

d. 1 gal = 27 mi; $\dfrac{1 \text{ gal}}{27 \text{ mi}}$ and $\dfrac{27 \text{ mi}}{1 \text{ gal}}$

2.55 **a.** 1 m = 100 cm; $\dfrac{100 \text{ cm}}{1 \text{ m}}$ and $\dfrac{1 \text{ m}}{100 \text{ cm}}$

b. 1 g = 1×10^9 ng; $\dfrac{1 \text{ g}}{1 \times 10^9 \text{ ng}}$ and $\dfrac{1 \times 10^9 \text{ ng}}{1 \text{ g}}$

c. 1 kL = 1000 L; $\dfrac{1000 \text{ L}}{1 \text{ kL}}$ and $\dfrac{1 \text{ kL}}{1000 \text{ L}}$

d. 1 kg = 10^6 mg; $\dfrac{10^6 \text{ mg}}{1 \text{ kg}}$ and $\dfrac{1 \text{ kg}}{10^6 \text{ mg}}$

e. $(1 \text{ m})^3 = (100 \text{ cm})^3$; $\dfrac{(100 \text{ cm})^3}{(1 \text{ m})^3}$ and $\dfrac{(1 \text{ m})^3}{(100 \text{ cm})^3}$

2.57 **a.** $\dfrac{3.5 \text{ m}}{1 \text{ s}}$ and $\dfrac{1 \text{ s}}{3.5 \text{ m}}$

b. $\dfrac{0.74 \text{ g}}{1 \text{ mL}}$ and $\dfrac{1 \text{ mL}}{0.74 \text{ g}}$

c. $\dfrac{46.0 \text{ km}}{1.0 \text{ gal}}$ and $\dfrac{1.0 \text{ gal}}{46.0 \text{ km}}$

d. $\dfrac{93 \text{ g silver}}{100 \text{ g sterling}}$ and $\dfrac{100 \text{ g sterling}}{93 \text{ g silver}}$

e. $\dfrac{29 \text{ μg}}{1 \text{ kg}}$ and $\dfrac{1 \text{ kg}}{29 \text{ μg}}$

2.59 The unit in the denominator must cancel with the preceding unit in the numerator.

2.61 **a.** 1.75 m   **b.** 5.5 L
c. 5.5 g   **d.** 3.5×10^{-4} m^3

2.63 **a.** 710. mL   **b.** 74.9 kg
c. 495 mm   **d.** 2.0×10^{-5} in.

2.65 **a.** 23.8 m   **b.** 196 m^2
c. 0.463 s   **d.** 53 L of paint

2.67 **a.** 152 g of oxygen   **b.** 0.026 g of magnesium
c. 43 g of nitrogen   **d.** 50. lb of chocolate bars

2.69 lead, 11.3 g/cm^3

2.71 **a.** 1.20 g/mL   **b.** 0.870 g/mL
c. 3.10 g/mL   **d.** 4.26 g/mL
e. 1.42 g/mL

2.73 **a.** 1.91 L of ethyl alcohol   **b.** 88 g of mercury
c. 62 oz of bronze   **d.** 661 g of copper
e. 34 kg of gasoline

2.75 **a.** exact   **b.** measured
c. exact   **d.** measured

2.77 **a.** length = 6.96 cm, width = 4.75 cm. Each answer may vary in the estimated digit.
b. length = 69.6 mm, width = 47.5 mm
c. 3 significant figures
d. 3 significant figures
e. 33.1 cm^2
f. 3 significant figures

2.79 1.8 g/mL

2.81 A is vegetable oil, B is water, and C is mercury.

2.83 **a.** 1.26×10^{-5} L   **b.** 3.53×10^2 kg
c. 1.25×10^5 m^3 **d.** 58.7 m
e. 3.00×10^{-3} s   **f.** 0.0108 g

2.85 42 min

2.87 16 onions

2.89 **a.** 96 crackers   **b.** 0.2 oz of fat   **c.** 110 g of sodium

2.91 68 cents

2.93 0.92 kg

2.95 185 mg/dL

2.97 790 g

2.99 8.0×10^2 cm^3

2.101 **a.** 4.3 lb of body fat   **b.** 6.2 lb

2.103 9.07 oz of pure silver

2.105 Because the balance can measure mass to 0.001 g, the mass should be given to 0.001 g. You should record the mass of the object as 32.075 g.

2.107 6.4 gal

2.109 3.8×10^2 g of aluminum

2.111 **a.** 43 g of 18-karat gold   **b.** 3.9 g of copper   **c.** 2.8 cm^3

2.113 82.9 mL

Matter and Energy

3

LOOKING AHEAD

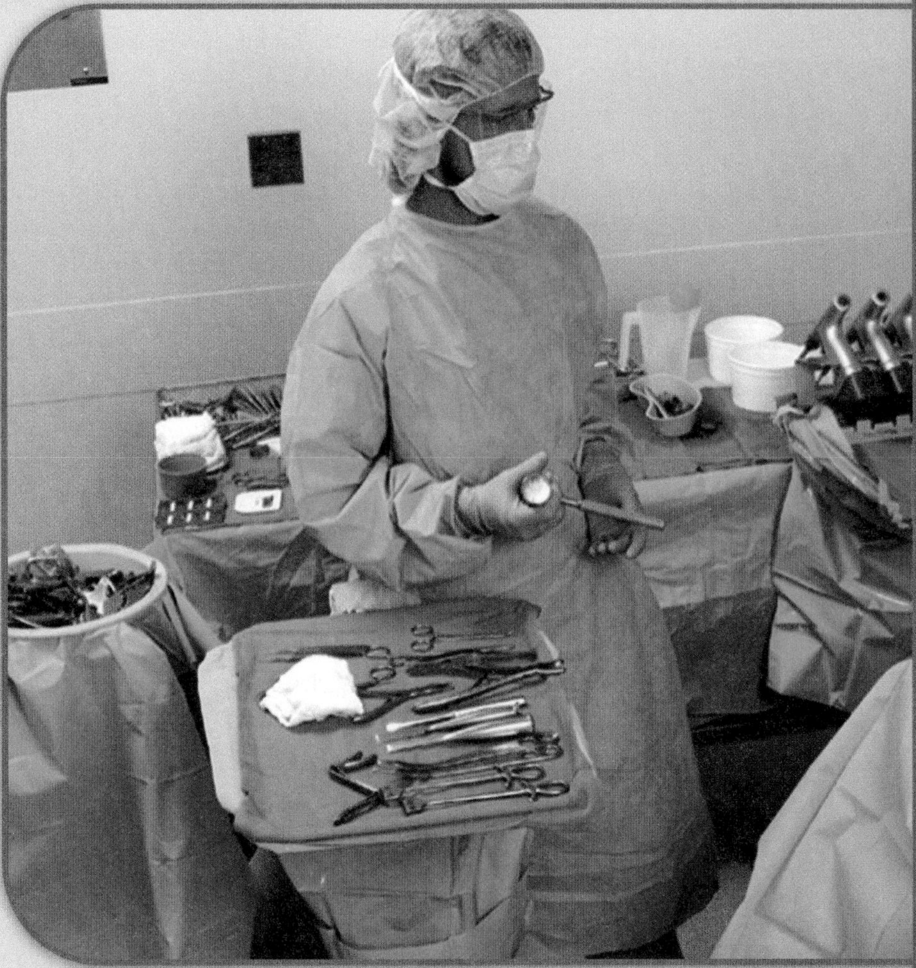

"As a surgical technologist, I assist the doctors during surgeries," says Christopher Ayars, a surgical technologist at Kaiser Hospital. "I am there to help during general or orthopedic surgery by passing instruments, holding retractors, and maintaining the sterile field. Our equipment for surgery is sterilized by steam that is heated to 270 °F, which is the same as 130 °C."

Surgical technologists assist with surgical procedures by preparing and maintaining surgical equipment, instruments, and supplies; providing patient care in an operating room setting; preparing and maintaining a sterile field; and ensuring that there are no breaks in aseptic technique. Instruments, which have been sterilized, are wrapped and sent to surgery, where they are checked again before they are opened.

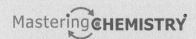

MasteringCHEMISTRY

Visit **www.masteringchemistry.com** for self-study materials and instructor-assigned homework.

Every day, we see a variety of materials with many different shapes and forms. To a scientist, all of this material is *matter*. Matter is everywhere around us: the orange juice we had for breakfast, the water we put in the coffeemaker, the plastic bag that holds our sandwich, our toothbrush and toothpaste, the oxygen we inhale, and the carbon dioxide we exhale.

When we look around, we see that matter takes the physical form of a solid, a liquid, or a gas. Water is a familiar example that we routinely observe in all three states. In the solid state, water can be an ice cube or a snowflake. It is a liquid when it comes out of a faucet or fills a pool. Water forms a gas, or vapor, when it evaporates from wet clothes or boils in a pan. In these examples, water changes state by losing or gaining energy. For example, energy is added to melt ice cubes and to boil water in a teakettle. Conversely, energy is removed to freeze liquid water in an ice-cube tray and to condense water vapor to liquid droplets.

Almost everything we do involves energy. We use energy when we walk, play tennis, study, and breathe. We use energy when we heat water, cook food, turn on lights, use computers, use a washing machine, or drive our cars. Of course, that energy has to come from something. In our bodies, the food we eat provides us with energy. Energy from burning fossil fuels or the Sun is used to heat a home or water for a pool.

3.1 Classification of Matter

LEARNING GOAL

Classify examples of matter as pure substances or mixtures.

TUTORIAL

Classification of Matter

An aluminum can consists of many atoms of aluminum.

Matter is anything that has mass and occupies space. Matter makes up all things we use, such as water, wood, plates, plastic bags, clothes, and shoes. The different types of matter are classified by their composition.

Pure Substances

A **pure substance** is matter that has a fixed or definite composition. There are two kinds of pure substances: elements and compounds. An **element**, the simplest type of a pure substance, is composed of only one type of material, such as silver, iron, and aluminum. Every element is composed of *atoms*, which are extremely tiny particles that make up each type of matter. For example, silver is composed of silver atoms, iron of iron atoms, and aluminum of aluminum atoms. A full list of the elements is found on the inside front cover of this text.

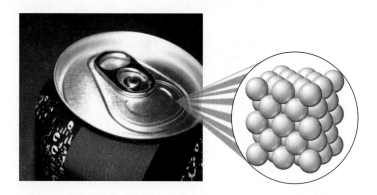

A **compound** is also a pure substance, but it consists of atoms of two or more elements chemically combined in the same proportion. In many compounds, the atoms are held together by attractions called *bonds*, which form small groups of atoms called molecules. For example, a molecule of the compound water has two hydrogen atoms for every one oxygen atom and is represented by the formula H_2O. The compound hydrogen peroxide is also a combination of hydrogen and oxygen, but it has two hydrogen atoms for every two oxygen atoms and is represented by the formula H_2O_2. Water (H_2O) and hydrogen peroxide (H_2O_2) are different compounds, which means they have different properties.

A molecule of water consists of two atoms of hydrogen (white) for one atom of oxygen (red) and has a formula of H_2O.

An important difference between compounds and elements is that compounds can be broken down by chemical processes into simpler substances, whereas elements cannot be broken down further. For example, ordinary table salt consists of the compound NaCl,

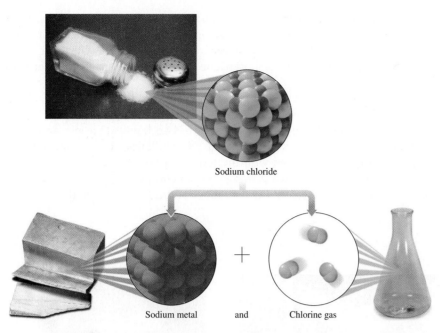

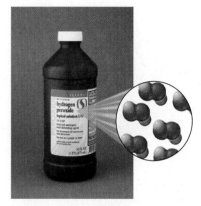

FIGURE 3.1 The decomposition of salt, NaCl, produces the elements sodium and chlorine.

Q How do elements and compounds differ?

Sodium chloride

Sodium metal   and   Chlorine gas

which can be separated by chemical processes into sodium metal and chlorine gas, as seen in Figure 3.1. However, compounds such as NaCl cannot be separated into simpler substances by using physical methods such as boiling or sifting.

Mixtures

Much of the matter in our everyday lives consists of mixtures. In a **mixture**, two or more substances are physically mixed but not chemically combined. The air we breathe is a mixture of mostly oxygen and nitrogen gases. The steel in buildings and railroad tracks is a mixture of iron, nickel, carbon, and chromium. The brass in doorknobs and fixtures is a mixture of zinc and copper. Tea, coffee, and ocean water are mixtures, too. In any mixture, the proportions of the components can vary. For example, two sugar–water mixtures may look the same, but the one with the higher ratio of sugar to water would taste sweeter. Different types of brass have different properties, such as color or strength, depending on the ratio of copper to zinc (see Figure 3.2).

A molecule of hydrogen peroxide consists of two atoms of hydrogen (white) for two atoms of oxygen (red) and has a formula of H_2O_2.

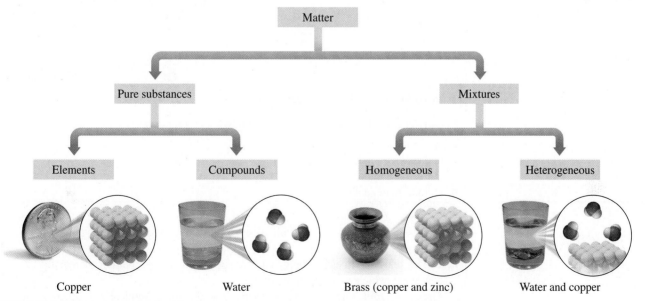

Matter

Pure substances

Mixtures

Elements

Compounds

Homogeneous

Heterogeneous

Copper

Water

Brass (copper and zinc)

Water and copper

FIGURE 3.2 Matter is organized by its components: elements, compounds, and mixtures. (a) The element copper consists of copper atoms. (b) The compound water consists of H_2O molecules. (c) Brass is a homogeneous mixture of copper and zinc atoms. (d) Copper metal in water is a heterogeneous mixture of Cu atoms and H_2O molecules.

Q Why are copper and water pure substances, but brass is a mixture?

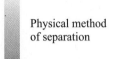
Physical method of separation

FIGURE 3.3 A mixture of spaghetti and water is separated using a strainer, a physical method of separation.

Q Why can physical methods be used to separate mixtures but not compounds?

Physical processes can be used to separate mixtures because there are no chemical interactions between the components. For example, different coins, such as nickels, dimes, and quarters, can be separated by size; iron particles mixed with sand can be picked up with a magnet; and water is separated from cooked spaghetti by using a strainer (see Figure 3.3).

In the chemistry laboratory, mixtures are separated by various methods. Solids are separated from liquids by filtration, which involves pouring a mixture through a filter paper set in a funnel. If a mixture consists of two liquids that mix or a liquid with a dissolved solid, they can be separated by distillation. Distillation involves boiling the mixture; the substance with the lower boiling temperature forms a gas that condenses back to a liquid in a cooling tube. In chromatography, different components of a liquid mixture separate as they move at different rates up the surface of a piece of paper.

A mixture of a liquid and a solid is separated by filtration.

Different substances are separated as they travel at different rates up the surface of chromatography paper.

CONCEPT CHECK 3.1

■ **Pure Substances and Mixtures**

Classify each of the following as a pure substance or a mixture:

a. sugar in a sugar bowl
b. nickels and dimes in a piggy bank
c. coffee with milk and sugar
d. aluminum in an aluminum soda can

ANSWER

a. Sugar would be a compound with one type of matter, which makes it a pure substance.
b. The nickels and dimes in a piggy bank are physically mixed, but not chemically combined, which makes them a mixture.
c. The coffee, milk, and sugar are physically mixed but not chemically combined, which makes it a mixture.
d. The aluminum in a soda can is an element, one type of matter, which makes it a pure substance.

CHEMISTRY AND HEALTH

Breathing Mixtures for Scuba

The air we breathe is composed mostly of the gases oxygen (21%) and nitrogen (79%). The homogeneous breathing mixtures used by scuba divers differ from the air we breathe, depending on the depth of the dive. Nitrox is a mixture of oxygen and nitrogen, but with more oxygen gas (up to 32%) and less nitrogen gas (68%) than air. A breathing mixture with less nitrogen gas decreases the risk of nitrogen narcosis, which is a decrease in mental and physical functions due to breathing nitrogen at high pressure. Heliox is a breathing mixture of oxygen and helium gases typically used for diving more than 200 feet. With deep dives, there is more chance of nitrogen narcosis, but by replacing nitrogen with helium, it does not occur. However, at dive depths over 300 ft, helium is associated with severe shaking and body temperature drop.

A breathing mixture used for dives over 400 ft is Trimix, which contains oxygen, helium, and some nitrogen. The addition of some nitrogen lessens the problem of shaking that comes with breathing high levels of helium. Both Heliox and Trimix are only used by professional, military, or highly trained divers.

A Nitrox mixture is used to fill scuba tanks.

Types of Mixtures

Mixtures are classified further as homogeneous or heterogeneous. In a *homogeneous mixture*, also called a *solution*, the composition is uniform throughout the sample. Examples of familiar homogeneous mixtures are air, which contains oxygen and nitrogen gases; and seawater, a solution of salt and water.

In a *heterogeneous mixture*, the components do not have a uniform composition throughout the sample. For example, a mixture of oil and water is heterogeneous because the oil floats on the surface of the water. Other examples of heterogeneous mixtures include the raisins in a cookie and the bubbles in a soda. Table 3.1 summarizes the classification of matter.

TABLE 3.1 Classification of Matter

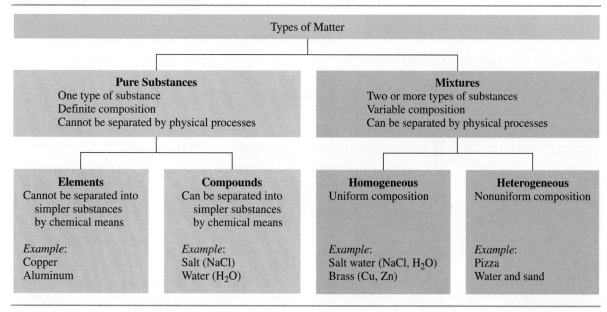

Types of Matter			
Pure Substances One type of substance Definite composition Cannot be separated by physical processes		**Mixtures** Two or more types of substances Variable composition Can be separated by physical processes	
Elements Cannot be separated into simpler substances by chemical means *Example:* Copper Aluminum	**Compounds** Can be separated into simpler substances by chemical means *Example:* Salt (NaCl) Water (H_2O)	**Homogeneous** Uniform composition *Example:* Salt water (NaCl, H_2O) Brass (Cu, Zn)	**Heterogeneous** Nonuniform composition *Example:* Pizza Water and sand

Oil and water form a heterogeneous mixture.

■ **Classifying Mixtures**

Classify each of the following as a pure substance (element or compound) or a mixture (homogeneous or heterogeneous):

a. copper in copper wire
b. a chocolate-chip cookie
c. Nitrox, a breathing mixture of oxygen and nitrogen used by scuba divers
d. carbon dioxide (CO_2)

SOLUTION

a. Copper is an element, which is a pure substance.
b. A chocolate-chip cookie does not have a uniform composition, which makes it a heterogeneous mixture.
c. The gases oxygen and nitrogen have a uniform composition in Nitrox, which makes it a homogeneous mixture.
d. Carbon dioxide, which contains one carbon atom chemically combined with two oxygen atoms, is a pure substance.

STUDY CHECK

A salad dressing is prepared with oil, vinegar, and chunks of blue cheese. Is this a homogeneous or heterogeneous mixture?

QUESTIONS AND PROBLEMS

Classification of Matter

3.1 Classify each of the following as a pure substance or a mixture:
a. baking soda ($NaHCO_3$)
b. a blueberry muffin
c. ice (H_2O)
d. zinc (Zn)
e. Trimix, oxygen, nitrogen, and helium, in a scuba tank

3.2 Classify each of the following as a pure substance or a mixture:
a. a soft drink   **b.** propane (C_3H_8)
c. a cheese sandwich   **d.** an iron (Fe) nail
e. salt substitute (KCl)

3.3 Classify each of the following pure substances as an element or a compound:
a. a silicon (Si) chip   **b.** hydrogen peroxide (H_2O_2)
c. oxygen (O_2)   **d.** rust (Fe_2O_3)
e. methane (CH_4) in natural gas

3.4 Classify each of the following pure substances as an element or a compound:
a. helium gas (He)
b. mercury (Hg) in a thermometer
c. sugar ($C_{12}H_{22}O_{11}$)
d. sulfur (S)
e. lye ($NaOH$)

3.5 Classify each of the following mixtures as homogeneous or heterogeneous:
a. vegetable soup   **b.** seawater
c. tea   **d.** tea with ice and lemon slices
e. fruit salad

3.6 Classify each of the following mixtures as homogeneous or heterogeneous:
a. nonfat milk
b. chocolate-chip ice cream
c. gasoline
d. peanut-butter-and-jelly sandwich
e. cranberry juice

3.2 States and Properties of Matter

LEARNING GOAL

Identify the states and the physical and chemical properties of matter.

TUTORIAL
Properties and Changes of Matter

One way to describe matter is to observe its properties. For example, if you were asked to describe yourself, you might list your characteristics such as the color of your eyes and skin or the length, color, and texture of your hair.

Physical Properties

Physical properties are those characteristics that can be observed or measured without affecting the identity of a substance. In chemistry, typical physical properties include the shape, color, melting point, boiling point, and physical state of a substance. For example, a

TABLE 3.2 Some Physical Properties of Copper

Color	Reddish orange
Odor	Odorless
Melting point	1083 °C
Boiling point	2567 °C
State at 25 °C	Solid
Luster	Shiny
Conduction of electricity	Excellent
Conduction of heat	Excellent

Copper, used in cookware, is a good conductor of heat.

penny has the physical properties of a round shape, an orange-red color, a solid state, and a shiny luster. Table 3.2 gives more examples of physical properties of copper found in pennies, electrical wiring, and copper pans.

States of Matter

On Earth, matter exists in one of three *physical forms* called the **states of matter**: *solids*, *liquids*, and *gases*. A **solid**, such as a pebble or a baseball, has a definite shape and volume. You can probably recognize several solids within your reach right now, such as books, pencils, or a computer mouse. In a *solid*, strong attractive forces hold particles such as atoms or molecules close together. The particles are arranged in such a rigid pattern they can only vibrate slowly in fixed positions. For many solids, this rigid structure produces a crystal such as that seen in amethyst.

Amethyst, a solid, is a purple form of quartz (SiO_2).

A **liquid** has a definite volume but not a definite shape. In a *liquid*, the particles move in random directions but are sufficiently attracted to each other to maintain a definite volume, although not a rigid structure. Thus, when water, oil, or vinegar is poured from one container to another, the liquid maintains its own volume but takes the shape of the new container.

Water as a liquid takes the shape of its container.

TABLE 3.3 A Comparison of Solids, Liquids, and Gases

Characteristic	Solid	Liquid	Gas
Shape	Has a definite shape	Takes the shape of the container	Takes the shape of the container
Volume	Has a definite volume	Has a definite volume	Fills the volume of the container
Arrangement of particles	Fixed, very close	Random, close	Random, far apart
Interaction between particles	Very strong	Strong	Essentially none
Movement of particles	Very slow	Moderate	Very fast
Examples	Ice, salt, iron	Water, oil, vinegar	Water vapor, helium, air

A **gas** does not have a definite shape or volume. In a *gas*, the particles are far apart, have little attraction to each other, and move at high speeds, taking the shape and volume of their container. When you inflate a bicycle tire, the air, which is a gas, fills the entire volume of the tire. The propane gas in a tank fills the entire volume of the tank. Table 3.3 compares the three states of matter.

A gas takes the shape and volume of its container.

Physical Changes

Water is a substance that is commonly found in all three states: solid, liquid, and gas. When matter undergoes a **physical change**, its state or its appearance will change, but its composition remains the same. The solid state of water—snow or ice—has a different appearance than its liquid or gaseous state, but all three states are water (see Figure 3.4).

The physical appearance of a substance can change in other ways, too. Suppose that you dissolved some salt in water. The appearance of the salt changes, but you could re-form the salt crystals by heating the mixture and evaporating the water. Thus, in a physical change, there are no new substances produced. Table 3.4 gives more examples of physical changes.

FIGURE 3.4 Water exists in an ice cube as a solid, water as a liquid, water vapor as a gas.

Q In what state of matter does water have a definite volume but not a definite shape?

Some Physical Changes of Water

$$\text{Solid water (ice)} \underset{\text{freezing}}{\overset{\text{melting}}{\rightleftarrows}} \text{liquid water} \underset{\text{condensing}}{\overset{\text{boiling}}{\rightleftarrows}} \text{water vapor (gas)}$$

TABLE 3.4 Examples of Some Physical Changes

Type of Physical Change	Example
Change of state	Water boiling
	Freezing of liquid water to solid water (ice)
Change of appearance	Dissolving sugar in water
Change of shape	Hammering a gold ingot into shiny gold leaf
	Drawing copper into thin copper wire
Change of size	Cutting paper into tiny pieces for confetti
	Grinding pepper into smaller particles

In a physical change, a gold ingot is hammered to form gold leaf.

CONCEPT CHECK 3.2

▧ **States of Matter**

Identify the state(s) of matter described in each of the following:

a. Its volume does not change in a different container.
b. It has a very low density.
c. Its shape depends on the container.
d. It has a definite shape and volume.

ANSWER

a. Both a solid and a liquid have their own volume that does not depend on the volume of their container.
b. In a gas, the particles are far apart, which gives a small mass per volume, or a low density.
c. Both a liquid and a gas take the shape of their containers.
d. A solid has a rigid arrangement of particles that gives it a definite shape and volume.

Chemical Properties and Chemical Changes

Chemical properties are those that describe the ability of a substance to change into a new substance. When a **chemical change** takes place, the original substance is converted into one or more new substances, which have different chemical and physical properties. For example, methane (CH_4) in natural gas can burn because it has the chemical property of being flammable. When methane burns in oxygen (O_2), it is converted to water (H_2O) and carbon dioxide (CO_2), which have different chemical and physical properties. Rusting or corrosion is a chemical property of iron. In the rain, an iron nail undergoes a chemical change when it reacts with oxygen in the air to form rust (Fe_2O_3), a new substance. Table 3.5 gives some examples of chemical changes, and Table 3.6 summarizes physical and chemical properties and changes.

Flan has a topping of caramelized sugar.

TABLE 3.5 Examples of Some Chemical Changes

Type of Chemical Change	Change in Chemical Properties
Silver tarnishes	Shiny, silver metal reacts in air to produce a black, grainy coating.
Methane burns	Methane burns with a bright flame, producing water vapor and carbon dioxide.
Sugar caramelizes	At high temperatures, white granular sugar changes to a smooth caramel-colored substance.
Iron rusts	Iron, which is gray and shiny, combines with oxygen to form orange-red rust.

TABLE 3.6 Summary of Physical and Chemical Properties and Changes

	Physical	Chemical
Property	A characteristic of a substance such as color, shape, odor, luster, size, melting point, and density.	A characteristic that indicates the ability of a substance to form another substance: paper can burn, iron can rust, and silver can tarnish.
Change	A change in a physical property that retains the identity of the substance: a change of state, a change in size, or a change in shape.	A change in which the original substance is converted to one or more new substances: paper burns, iron rusts, silver tarnishes.

CONCEPT CHECK 3.3

■ Physical and Chemical Properties

Classify each of the following as a physical or chemical property:

a. Water is a liquid at room temperature.
b. Gasoline burns in air.
c. Aluminum foil has a shiny appearance.

ANSWER

a. A liquid is a state of matter, which makes it a physical property.
b. When gasoline burns, it changes to different substances with new properties, which is a chemical property.
c. The shininess of a substance does not change the type of substance; it is a physical property.

SAMPLE PROBLEM   3.2

■ **Physical and Chemical Changes**

Classify each of the following as a physical or chemical change:

a. An ice cube melts to form liquid water.
b. Bleach removes a stain.
c. An enzyme breaks down the lactose in milk.
d. Garlic is chopped into small pieces.

SOLUTION

a. A physical change occurs when the ice cube changes state from solid to liquid.
b. A chemical change occurs when the color from the stain is removed.
c. A chemical change occurs when an enzyme breaks down lactose into simpler substances.
d. A physical change occurs when the size of an object changes.

STUDY CHECK

Which of the following are chemical changes?

a. Water freezes on a pond.
b. Gas bubbles form when baking powder is placed in vinegar.
c. A log is chopped for firewood.
d. A log burns in a fireplace.

QUESTIONS AND PROBLEMS

States and Properties of Matter

3.7 Indicate whether each of the following describes a gas, a liquid, or a solid:
a. This substance has no definite volume or shape.
b. The particles in a substance do not interact with each other.
c. The particles in a substance are held in a rigid structure.

3.8 Indicate whether each of the following describes a gas, a liquid, or a solid:
a. This substance has a definite volume but takes the shape of its container.
b. The particles in a substance are very far apart.
c. This substance occupies the entire volume of the container.

3.9 Describe each of the following properties as physical or chemical:
a. Chromium is a steel-gray solid.
b. Hydrogen reacts readily with oxygen.
c. Nitrogen freezes at −210 °C.
d. Milk will sour when left in a warm room.
e. Butane gas in an igniter burns in oxygen.

3.10 Describe each of the following properties as physical or chemical:
a. Neon is a colorless gas at room temperature.
b. Apple slices turn brown when they are exposed to air.
c. Phosphorus will ignite when exposed to air.
d. At room temperature, mercury is a liquid.
e. Propane gas is compressed to a liquid for placement in a small cylinder.

3.11 What type of change, physical or chemical, takes place in each of the following?
a. Water vapor condenses to form rain.
b. Cesium metal reacts explosively with water.
c. Gold melts at 1064 °C.
d. A puzzle is cut into 1000 pieces.
e. Sugar dissolves in water.

3.12 What type of change, physical or chemical, takes place in each of the following?
a. Gold is hammered into thin sheets.
b. A silver pin tarnishes in the air.
c. A tree is cut into boards at a saw mill.
d. Food is digested.
e. A chocolate bar melts.

3.13 Describe each property of the element fluorine as physical or chemical.
a. is highly reactive
b. is a gas at room temperature
c. has a pale, yellow color
d. will explode in the presence of hydrogen
e. has a melting point of −220 °C

3.14 Describe each property of the element zirconium as physical or chemical.
a. melts at 1852 °C
b. is resistant to corrosion
c. has a grayish white color
d. ignites spontaneously in air when finely divided
e. is a shiny metal

3.3 Temperature

Temperatures in science, and in most of the world, are measured and reported in *Celsius* (*°C*) units. In the United States, everyday temperatures are commonly reported in *Fahrenheit (°F)* units. A typical room temperature of 21 °C would be the same as 70 °F. A normal body temperature of 37.0 °C is 98.6 °F.

Celsius and Fahrenheit Temperatures

On the Celsius and Fahrenheit scales, the temperatures of melting ice and boiling water are used as reference points. On the Celsius scale, the freezing point of pure water is defined as exactly 0 °C and the boiling point as exactly 100 °C. On the Fahrenheit scale, pure water freezes at exactly 32 °F and boils at exactly 212 °F. On each scale, the temperature difference between freezing and boiling is divided into smaller units called *degrees*. The Celsius scale has 100 degrees between the freezing and boiling temperatures of water, compared with 180 degrees on the Fahrenheit scale. That makes a Celsius degree almost twice the size of a Fahrenheit degree: 1 °C = 1.8 °F (see Figure 3.5).

$$180 \text{ Fahrenheit degrees} = 100 \text{ Celsius degrees}$$

$$\frac{180 \text{ Fahrenheit degrees}}{100 \text{ Celsius degrees}} = \frac{1.8 \text{ °F}}{1 \text{ °C}}$$

In a chemistry laboratory, temperatures are measured in Celsius degrees. To convert to a Fahrenheit temperature, the Celsius temperature is multiplied by 1.8 and then 32 degrees is added, which adjusts the freezing point of 0 °C on the Celsius scale to 32 °F on the Fahrenheit scale. Both values, 1.8 and 32, are exact numbers. The equation for this conversion follows:

$$T_{\text{F}} = \underbrace{\frac{1.8 \text{ °F}(T_{\text{C}})}{1 \text{ °C}}}_{\substack{\text{Changes} \\ \text{°C to °F}}} + \underbrace{32}_{\substack{\text{Adjusts} \\ \text{freezing point}}} \quad \text{or} \quad T_{\text{F}} = 1.8(T_{\text{C}}) + 32$$

FIGURE 3.5 A comparison of the Fahrenheit, Celsius, and Kelvin temperature scales between the freezing and boiling points of water.

Q What is the difference in the freezing points of water on the Celsius and Fahrenheit temperature scales?

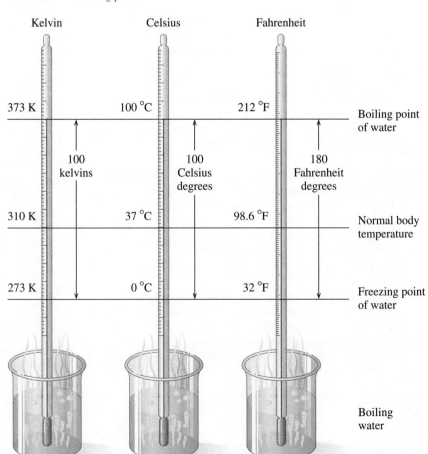

CONCEPT CHECK 3.4

■ **Temperature Scales**

A student in your chemistry class has designed a new temperature scale, which he has named degrees Zupa. The freezing point of water on this Zupa scale is 10 °Z, and the boiling point occurs at 130 °Z.

a. What is the relationship between degrees Zupa and degrees Celsius?
b. How would you adjust the freezing point?
c. Write an equation that relates degrees Zupa to degrees Celsius.
d. Convert a temperature of 35 °C to degrees Zupa.

ANSWER

a. On the Celsius scale, there are 100 °C between the freezing and the boiling points of water. On the Zupa scale, there are 120 °Z. Thus, 100 °C = 120 °Z and the conversion factor is 120 °Z/100 °C.
b. The freezing point is adjusted by adding 10 degrees.
c. The equation to convert °C to °Z would be written as

$$T_Z = \frac{120 \text{ °Z}(T_C)}{100 \text{ °C}} + 10$$
$$T_Z = 1.2(T_C) + 10$$

d. $T_Z = 1.2(35) + 10 = 42 + 10 = 52 \text{ °Z}$

SAMPLE PROBLEM | 3.3

■ **Converting Celsius to Fahrenheit**

The temperature of a room is set at 22 °C. If that temperature is lowered by 1 °C, it can save as much as 5% in energy costs. What temperature, in Fahrenheit degrees, should be set to lower the Celsius temperature by 1 °C?

SOLUTION

STEP 1   **Given** 22 °C − 1 °C = 21 °C   **Need** T_F

STEP 2   **Plan**

$$T_C \quad \boxed{\text{Temperature equation}} \quad T_F$$

STEP 3   **Equality/Conversion Factor**

$$T_F = 1.8(T_C) + 32$$

STEP 4   **Set Up Problem** Substitute the Celsius temperature into the equation and solve.

$$T_F = 1.8(21) + 32 \qquad \text{1.8 is exact; 32 is exact}$$

$$T_F = 38 + 32$$
$$= 70. \text{ °F} \qquad \text{Answer to the ones place}$$

In the temperature equation, *the values of 1.8 and 32 are exact numbers.* The answer is reported using significant figure rules.

STUDY CHECK

In the process of making ice cream, rock salt is added to the crushed ice. If the temperature drops to −11 °C, what is it in °F?

In a chemistry laboratory, temperatures are measured in Celsius degrees. To convert from Fahrenheit to Celsius, the temperature equation is rearranged for T_C. Start with

$$T_F = 1.8(T_C) + 32$$

Then subtract 32 from both sides.

$$T_F - 32 = 1.8(T_C) + 32 - 32$$
$$T_F - 32 = 1.8(T_C)$$

Solve the equation for T_C by dividing both sides by 1.8.

$$\frac{T_F - 32}{1.8} = \frac{\cancel{1.8}(T_C)}{\cancel{1.8}}$$

$$\frac{T_F - 32}{1.8} = T_C$$

SAMPLE PROBLEM | 3.4

■ **Converting Fahrenheit to Celsius**

In a type of cancer treatment called *thermotherapy,* temperatures as high as 113 °F are used to destroy cancer cells. What is that temperature in degrees Celsius?

SOLUTION

STEP 1   **Given** 113 °F   **Need** T_C

STEP 2   **Plan**

T_F [Temperature equation] T_C

STEP 3   **Equality/Conversion Factor**

$$\frac{T_F - 32}{1.8} = T_C$$

STEP 4   **Set Up Problem** To solve for T_C, substitute the Fahrenheit temperature into the equation and solve.

$$T_C = \frac{T_F - 32}{1.8}$$

$$T_C = \frac{(113 - 32)}{1.8}$$ 32 is exact; 1.8 is exact

$$= \frac{81}{1.8} = 45 \ °C$$ Answer to the ones place

STUDY CHECK

A child has a temperature of 103.6 °F. What is this temperature on a Celsius thermometer?

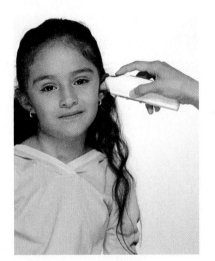

A digital ear thermometer is used to measure body temperature.

Kelvin Temperature Scale

Scientists have learned that the coldest temperature possible is −273 °C (more precisely, −273.15 °C). On the Kelvin scale, this temperature, called *absolute zero,* has the value of 0 K. Units on the Kelvin scale are called kelvins (K); no degree symbol is used. Because there are no lower temperatures, the Kelvin scale has no negative numbers. Between the freezing and boiling points of water, there are 100 kelvins, which makes a kelvin equal to a Celsius degree.

$$1 \ K = 1 \ °C$$

TABLE 3.7 Comparison of Temperatures

Example	Fahrenheit (°F)	Celsius (°C)	Kelvin (K)
Sun	9937	5503	5776
A hot oven	450	232	505
A desert	120	49	322
A high fever	104	40	313
Room temperature	70	21	294
Water freezes	32	0	273
An Alaska winter	−66	−54	219
Helium boils	−452	−269	4
Absolute zero	−459	−273	0

To calculate a Kelvin temperature, add 273 to the Celsius temperature:

$$T_K = T_C + 273$$

To calculate a Celsius temperature, 273 is subtracted from the Kelvin temperature:

$$T_K - 273 = T_C + 273 - 273$$
$$T_K - 273 = T_C$$
$$T_C = T_K - 273$$

Table 3.7 gives a comparison of some temperatures on the three temperature scales.

SAMPLE PROBLEM | 3.5

■ **Converting from Celsius to Kelvin Temperature**

A dermatologist may use cryogenic liquid nitrogen at −196 °C to remove skin lesions and some skin cancers. What is the temperature of the liquid nitrogen in K?

SOLUTION

STEP 1   **Given** −196 °C   **Need** T_K

STEP 2   **Plan**

$$T_C \quad \boxed{\text{Temperature equation}} \quad T_K$$

STEP 3   **Equality/Conversion Factor**

$$T_K = T_C + 273$$

STEP 4   **Set Up Problem** Substitute the Celsius temperature into the equation and solve.

$$T_K = T_C + 273$$
$$T_K = -196 + 273$$
$$= 77 \text{ K} \qquad \text{Answer to the ones place}$$

STUDY CHECK

On the planet Mercury, the average night temperature is 13 K and the average day temperature is 683 K. What are these temperatures in degrees Celsius?

CHEMISTRY AND HEALTH

Variation in Body Temperature

Normal body temperature is considered to be 37.0 °C, although it varies throughout the day and from person to person. Oral temperatures of 36.1 °C are common in the morning and climb to a high of 37.2 °C between 6 P.M. and 10 P.M. Temperatures above 37.2 °C for a person at rest are usually an indication of illness. Individuals who are involved in prolonged exercise may also experience elevated temperatures. Body temperatures of marathon runners can range from 39 °C to 41 °C as heat production during exercise exceeds the body's ability to lose heat.

Changes of more than 3.5 °C from the normal body temperature begin to interfere with bodily functions. At body temperatures above 41 °C, a condition called hyperthermia or heat stroke may occur in which sweat production stops, the pulse rate is elevated, and respiration becomes weak and rapid. A person with hyperthermia may become lethargic and lapse into a coma. In children, convulsions can occur, which may lead to permanent brain damage. Damage to internal organs is a major concern, and treatment, which must be immediate, may involve immersing the person in an ice-water bath.

At the low temperature extreme of hypothermia, body temperature can drop as low as 28.5 °C. The person may appear cold and pale and have an irregular heartbeat. Unconsciousness can occur if the body temperature drops below 26.7 °C. Respiration becomes slow and shallow, and oxygenation of the tissues decreases. Treatment involves providing oxygen and increasing blood volume with glucose and saline fluids. Injecting warm fluids (37.0 °C) into the peritoneal cavity may restore the internal temperature.

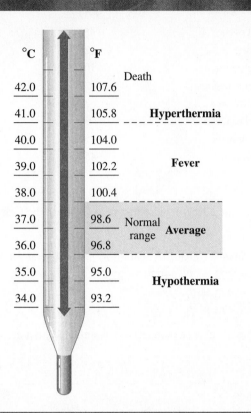

QUESTIONS AND PROBLEMS

Temperature

3.15 Your friend who is visiting from France just took her temperature. When she reads 99.8, she becomes concerned that she is quite ill. How would you explain the temperature to your friend?

3.16 You have a friend who is using a recipe for flan from a Mexican cookbook. You notice that he set your oven temperature at 175 °F. What would you advise him to do?

3.17 Solve the following temperature conversions:
   a. 37.0 °C = _____ °F   **b.** 65.3 °F = _____ °C
   c. −27 °C = _____ K   **d.** 62 °C = _____ K
   e. 114 °F = _____ °C   **f.** 72 °F = _____ K

3.18 Solve the following temperature conversions:
   a. 25 °C = _____ °F   **b.** 155 °C = _____ °F
   c. −25 °F = _____ °C   **d.** 224 K = _____ °C
   e. 145 °C = _____ K   **f.** 875 K = _____ °F

3.19 a. A person with hyperthermia has a temperature of 106 °F. What does this read on a Celsius thermometer?
   b. Because high fevers can cause convulsions in children, the doctor wants to be called if a child's temperature goes over 40.0 °C. Should the doctor be called if a child has a temperature of 103 °F?

3.20 a. Water is heated to 145 °F. What is the temperature of the hot water in °C?
   b. During extreme hypothermia, a young woman's temperature dropped to 20.6 °C. What was her temperature on the Fahrenheit scale?

3.4 Energy

LEARNING GOAL

Identify energy as potential or kinetic; convert between units of energy.

When you are running, walking, dancing, or thinking, you are using energy to do **work**, any activity that requires energy. In fact, **energy** is defined as the ability to do work. Suppose you are climbing a steep hill and you become too tired to go on. At that moment, you do not have sufficient energy to do any more work. Now suppose you sit down and have lunch. In a while, you will have obtained energy from the food, and you will be able to do more work and complete the climb (see Figure 3.6).

Kinetic and Potential Energy

Energy can be classified as kinetic or potential energy. **Kinetic energy** is the energy of motion. Any object that is moving has kinetic energy. **Potential energy** is determined by the position of an object or by the chemical composition of a substance. A boulder resting on top of a mountain has potential energy because of its location. If the boulder rolls down the mountain, the potential energy becomes kinetic energy. Water in a reservoir behind a dam has potential energy. When the water goes over the dam and falls to the stream below, its potential energy is converted to kinetic energy. Foods and fossil fuels have potential energy in their molecules. When you digest food or burn gasoline in your car, potential energy is converted to kinetic energy to do work.

FIGURE 3.6 Work is done as the rock climber moves up the cliff. At the top, the climber has more potential energy than when she started the climb.

Q What happens to the potential energy of the climber when she descends?

CONCEPT CHECK 3.5

■ Potential and Kinetic Energy

Identify each of the following as an example of potential or kinetic energy:

a. gasoline   **b.** skating
c. candy bar

SOLUTION

a. Gasoline is burned to provide energy and heat; it contains potential energy in its molecules.
b. A skater uses energy to move; skating is kinetic energy (energy of motion).
c. A candy bar has potential energy. When digested, its components provide energy for the body to do work.

Heat and Units of Energy

Heat is associated with the motion of particles. A frozen pizza feels cold because heat flows from your hand into the pizza. The faster the particles move, the greater the heat or thermal energy of the substance. In the frozen pizza, the particles are moving very slowly. As heat is added and the pizza becomes warmer, the motions of the particles in the pizza increase. Eventually, the particles have enough energy to make the pizza hot and ready to eat.

Units of Energy

The SI unit of energy and work is the **joule (J)** (pronounced "jewel"). The joule is a small amount of energy, so scientists often use the kilojoule (kJ), 1000 joules. When you heat water for one cup of tea, you use about 75 000 J or 75 kJ of heat. Table 3.8 shows a comparison of energy in joules for several energy sources.

You may be more familiar with the unit **calorie (cal)**, from the Latin *caloric,* meaning "heat." The calorie was originally defined as the amount of energy (heat) needed to raise the temperature of 1 g of water by 1 °C. Now one calorie is defined as exactly 4.184 J. This equality can also be written as a conversion factor:

1 cal = 4.184 J (exact)

$$\frac{4.184 \text{ J}}{1 \text{ cal}} \quad \text{and} \quad \frac{1 \text{ cal}}{4.184 \text{ J}}$$

One *kilocalorie* (kcal) is equal to 1000 calories, and one *kilojoule* (kJ) is 1000 joules.

1 kcal = 1000 cal
1 kJ = 1000 J

When water flows from the top of a dam, potential energy is converted to kinetic energy.

TUTORIAL
Energy Conversions

TABLE 3.8 A Comparison of Energy for Various Resources

Energy in joules

Energy value	Resource
10^{27}	
10^{24}	Energy radiated by Sun per second (10^{26})
10^{21}	World reserves of fossil fuel (10^{23})
10^{18}	Energy consumption for one year in US (10^{20})
10^{15}	Solar energy reaching the Earth per second (10^{17})
10^{12}	
10^9	Energy use per person in one year in US (10^{11})
10^6	Energy from one gallon of gasoline (10^8)
10^3	Energy from one serving of pasta, a doughnut, or needed to bicycle one hour (10^6)
10^0	Energy used to sleep one hour (10^5)

Octane, a component of gasoline, reacts in a car engine to produce energy.

SAMPLE PROBLEM | 3.6

■ **Energy Units**

When 1.0 g of octane fuel burns in an automobile engine, 48 000 J are released. Convert this quantity of energy to both kilojoules and calories.

SOLUTION

a. kilojoules

STEP 1 **Given** 48 000 J   **Need** kilojoules

STEP 2 **Plan** J → Energy factor → kJ

STEP 3 **Equalities/Conversion Factors**

$$1 \text{ kJ} = 1000 \text{ J}$$

$$\frac{1000 \text{ J}}{1 \text{ kJ}} \quad \text{and} \quad \frac{1 \text{ kJ}}{1000 \text{ J}}$$

STEP 4 **Set Up Problem** $48\ 000\ \cancel{J} \times \dfrac{1 \text{ kJ}}{1000\ \cancel{J}} = 48 \text{ kJ}$

b. calories

STEP 1 **Given** 48 000 J   **Need** calories (cal)

STEP 2 **Plan** J → Energy factor → cal

STEP 3 **Equalities/Conversion Factors**

$$1 \text{ cal} = 4.184 \text{ J}$$

$$\frac{1 \text{ cal}}{4.184 \text{ J}} \quad \text{and} \quad \frac{4.184 \text{ J}}{1 \text{ cal}}$$

STEP 4 **Set Up Problem**

$$48\,000\,\cancel{J} \times \frac{1\ cal}{4.184\,\cancel{J}} = 11\,000\ (1.1 \times 10^4)\ cal$$

STUDY CHECK

The burning of 1.0 g of coal produces 35 000 J of energy. How many kcal are produced?

QUESTIONS AND PROBLEMS

Energy

3.21 Discuss the changes in the potential and kinetic energy of a roller-coaster ride as the roller coaster climbs to the top and goes down the other side.

3.22 Discuss the changes in the potential and kinetic energy of a ski jumper taking the elevator to the top of the jump and going down the ramp.

3.23 Indicate whether each item describes potential or kinetic energy.
a. water at the top of a waterfall
b. kicking a ball
c. the energy in a lump of coal
d. a skier at the top of a hill

3.24 Indicate whether each item describes potential or kinetic energy.
a. the energy in your food
b. a tightly wound spring

c. an earthquake
d. a car speeding down the freeway

3.25 A burning match releases 1.1×10^3 J. Convert the energy released by 20 matches to the following energy units:
a. kilojoules   **b.** calories
c. kilocalories

3.26 A person uses 750 kcal to run a race. Convert the energy used for the race to the following energy units:
a. calories   **b.** joules
c. kilojoules

3.27 Convert each of the following energy units:
a. 3500 cal to kcal   **b.** 415 J to cal
c. 28 cal to J   **d.** 4.5 kJ to cal

3.28 Convert each of the following energy units:
a. 8.1 kcal to cal   **b.** 325 J to kJ
c. 2550 cal to kJ   **d.** 2.50 kcal to J

3.5 Specific Heat

Every substance has the ability to absorb or lose heat with temperature change. When you bake a potato, you place it in a hot oven. If you are cooking pasta, you add the pasta to boiling water. Some substances absorb more heat than others to reach a certain temperature. These energy requirements for different substances are described in terms of a physical property called specific heat. **Specific heat** (*SH*) is the amount of heat (*q*) added to raise the temperature of exactly 1 g of a substance by exactly 1 °C. This temperature change is written as ΔT (*delta T*), where the delta symbol means "a change in."

$$\text{Specific heat } (SH) = \frac{\text{heat } (q)}{\text{mass} \times \Delta T} = \frac{\text{J (or cal)}}{1\ g \times {}^{\circ}C}$$

Now we can write the specific heat for water using our definition of the calorie and joule.

$$\text{Specific heat } (SH) \text{ of } H_2O(l) = 4.184\ \frac{J}{g \times {}^{\circ}C} = 1.00\ \frac{cal}{g \times {}^{\circ}C}$$

If we look at Table 3.9, we see that 1 g of water requires 4.184 J to increase its temperature by 1 °C. Water has a large specific heat that is about five times the specific heat of aluminum. Aluminum has a specific heat that is about twice that of copper.

Therefore, 4.184 J (1 cal) will increase the temperature of 1 g of water by 1 °C. However, the same amount of heat (4.184 J or 1 cal) will increase the temperature of 1 g of aluminum by 5 °C and 1 g of copper by 10 °C. The high specific heat of water gives it the capacity to absorb or release large amounts of heat in the body, which maintains an almost constant body temperature. The low specific heats of aluminum and copper mean they transfer heat efficiently, which makes them useful in cookware.

LEARNING GOAL

Use specific heat to calculate heat loss or gain, temperature change, or mass of a sample.

TUTORIAL

Heat Capacity

TABLE 3.9 Specific Heats of Some Substances

	Substance	Specific Heat (J/g °C)
Elements	Aluminum, Al(s)	0.897
	Copper, Cu(s)	0.385
	Gold, Au(s)	0.129
	Iron, Fe(s)	0.452
	Silver, Ag(s)	0.235
	Titanium, Ti(s)	0.523
Compounds	Ammonia, $NH_3(g)$	2.04
	Ethanol, $C_2H_5OH(l)$	2.46
	Sodium chloride, NaCl(s)	0.864
	Water, $H_2O(l)$	4.184
	Water, $H_2O(s)$	2.03

CONCEPT CHECK 3.6

■ **Specific Heat**

A 1.0-g sample of iron has a temperature of 21 °C. Using Table 3.9, predict the final temperature of the iron sample when 4.184 J of heat is added.

a. 21 °C   **b.** 25 °C   **c.** 30. °C   **d.** 120 °C   **e.** 200 °C

Explain.

ANSWER

The specific heat of iron indicates that it absorbs less heat per gram than does water for a 1 °C increase. Thus, there will be a greater increase in the temperature of iron, which can be calculated using the ratio of the specific heat of water and iron:

$$\frac{\text{Specific heat of water 4.184 J/g °C}}{\text{Specific heat of iron 0.452 J/g °C}} = 9.26$$

Thus, the temperature of iron increase about 9 times the 1 °C of change in water, which increases the temperature of iron from 21 °C to 30. °C (answer **c**).

SAMPLE PROBLEM | **3.7**

■ **Calculating Specific Heat**

What is the specific heat of lead if 57.0 J are needed to raise the temperature of 35.6 g of lead by 12.5 °C?

SOLUTION

STEP 1   **Given** heat 57.0 J mass 35.6 g of Pb temperature change 12.5 °C
Need specific heat (J/g °C)

STEP 2   **Plan** The specific heat (*SH*) is calculated by dividing the heat by the mass (g) and by the temperature change (ΔT).

$$SH = \frac{\text{heat}}{\text{mass} \ \Delta T}$$

STEP 3   **Substitute the given values into the equation.**

$$\text{Specific heat } (SH) = \frac{57.0 \text{ J}}{35.6 \text{ g} \ 12.5 \text{ °C}} = 0.128 \frac{\text{J}}{\text{g °C}}$$

STUDY CHECK

What is the specific heat of sodium if 123 J are needed to raise the temperature of 4.00 g of sodium by 25.0 °C?

TUTORIAL
Heat

TUTORIAL
Specific Heat Calculations

Calculations Using the Heat Equation

When we know the specific heat of a substance, we can rearrange it to obtain the *heat equation*.

$$\text{Specific heat } (SH) = \frac{\text{heat}}{\text{mass} \times \Delta T}$$

$$\text{Specific heat } (SH) \times \text{mass} \times \Delta T = \frac{\text{heat}}{\cancel{\text{mass}} \times \cancel{\Delta T}} \times \cancel{\text{mass}} \times \cancel{\Delta T}$$

Heat Equation

Heat	=	mass	×	temperature change	×	specific heat
Heat (q)	=	mass	×	ΔT	×	SH
J (cal)	=	mass	×	°C	×	$\dfrac{\text{J (cal)}}{\text{g °C}}$

In heat calculations, the temperature change is always the difference between the final temperature and the initial temperature.

$$\Delta T = T_{\text{final}} - T_{\text{initial}}$$

The heat lost or gained is calculated by substituting the mass (in grams), the change in temperature, and specific heat into the heat equation. Canceling units gives heat in units of calories or joules. When a substance absorbs energy, the temperature rises. Then the sign of ΔT is positive ($+$), which gives a positive sign ($+$) for the heat (q) that is calculated. However, if a substance loses energy, the temperature drops. Then the sign of ΔT is negative ($-$), which gives a negative sign ($-$) for the heat (q). Thus, a positive ($+$) sign for heat (q) means that heat flows into the substance, whereas a negative ($-$) sign for heat (q) means that heat flows out of the substance. In the following sample problems, we will see how the heat equation is used to calculate heat and how it can be rearranged to solve for mass.

SAMPLE PROBLEM   3.8

■ Calculating Heat with Temperature Increase

How many joules are absorbed by 45.2 g of aluminum if its temperature rises from 12.5 °C to 76.8 °C (see Table 3.9)?

SOLUTION

STEP 1 **List given and needed data.**

Given mass = 45.2 g

  SH for aluminum = 0.897 J/g °C
  Initial temperature = 12.5 °C
  Final temperature = 76.8 °C

Need heat (q) in joules (J)

STEP 2 **Calculate the temperature change.** The temperature change ΔT is the difference between the two temperatures.

$$\Delta T = T_{\text{final}} - T_{\text{initial}} = 76.8\ °C - 12.5\ °C = 64.3\ °C$$

STEP 3 **Write the heat equation.**

$$\text{Heat } (q) = m \times \Delta T \times SH$$

STEP 4 **Substitute the given values and solve, making sure units cancel.**

$$\text{Heat } (q) = 45.2\ \cancel{g} \times 64.3\ \cancel{°C} \times \frac{0.897\ \text{J}}{\cancel{g}\ \cancel{°C}} = 2.61 \times 10^3\ \text{J}$$

STUDY CHECK

Some cooking pans have a layer of copper on the bottom. How many kilojoules are needed to raise the temperature of 125 g of copper from 22 °C to 325 °C if the specific heat of copper is 0.385 J/g °C?

Guide to Calculations Using Specific Heat

STEP 1
List given and needed data.

STEP 2
Calculate the temperature change (ΔT).

STEP 3
Write the heat equation
$q = m \times \Delta T \times SH$
and rearrange for unknown.

STEP 4
Substitute the given values and solve, making sure units cancel.

The copper on a cooking pan conducts heat rapidly to the food in the pan.

Hot tea cools by losing heat energy.

SAMPLE PROBLEM | 3.9

■ Calculating Heat Loss

A 225-g sample of hot tea cools from 74.6 °C to 22.4 °C. How much heat, in kilojoules, is lost, assuming that tea has the same specific heat as water?

SOLUTION

STEP 1 **List given and needed data.**

> **Given** mass = 225 g   *SH* for tea = 4.184 J/g °C
>
> Initial temperature = 74.6 °C final temperature = 22.4 °C
>
> **Equality/Conversion Factors**
>
> 1 kJ = 1000 J
>
> $\dfrac{1000\ J}{1\ kJ}$ and $\dfrac{1\ kJ}{1000\ J}$
>
> **Need** heat (q) in kilojoules (kJ)

STEP 2 **Calculate the temperature change.** The temperature change ΔT is the difference between the two temperatures.

$$\Delta T = T_f - T_i = 22.4\ ^\circ C - 74.6\ ^\circ C = -52.2\ ^\circ C$$

STEP 3 **Write the heat equation.**

$$\text{Heat } (q) = m \times \Delta T \times SH$$

STEP 4 **Substitute the given values and solve, making sure units cancel.**

$$\text{Heat } (q) = 225\ \cancel{g} \times (-52.2\ \cancel{^\circ C}) \times \frac{4.184\ \cancel{J}}{\cancel{g}\ \cancel{^\circ C}} \times \frac{1\ kJ}{1000\ \cancel{J}} = -49.1\ kJ$$

STUDY CHECK

How much heat, in joules, is lost when 15.5 g of gold cools from 215 °C to 35 °C? The specific heat of gold is 0.129 J/g °C.

SAMPLE PROBLEM | 3.10

■ Calculating Mass Using Specific Heat

Ethanol has a specific heat of 2.46 J/g °C. When 655 J are added to a sample of ethanol, its temperature rises from 18.2 °C to 32.8 °C. What is the mass of the ethanol sample?

SOLUTION

STEP 1 **List given and needed data.**

> **Given** Heat (q) = 655 J   *SH* for ethanol = 2.46 J/g °C
>
> Initial temperature = 18.2 °C final temperature = 32.8 °C
>
> **Need** mass of ethanol sample

STEP 2 **Calculate the temperature change.** The temperature change ΔT is the difference between the two temperatures.

$$\Delta T = T_f - T_i = 32.8\ ^\circ C - 18.2\ ^\circ C = 14.6\ ^\circ C$$

STEP 3 **Write the heat equation.**

$$\text{Heat } (q) = m \times \Delta T \times SH$$

The heat equation must be rearranged to solve for mass (m), which is the heat divided by the temperature change and the specific heat.

$$\text{Mass} = \frac{\text{heat } (q)}{\Delta T \;\; SH}$$

STEP 4   **Substitute the given values into the equation and solve, making sure units cancel.**

$$\text{Mass} = \frac{655 \cancel{J}}{14.6 \cancel{°C} \;\; \dfrac{2.46 \cancel{J}}{g \;\cancel{°C}}}$$

$$\text{Mass} = 18.2 \text{ g}$$

STUDY CHECK

When 8.81 kJ are absorbed by a piece of iron, its temperature rises from 15 °C to 122 °C. What is the mass, in grams, if iron has a specific heat of 0.452 J/g °C?

Measuring Heat Changes

A **calorimeter** is used to measure the temperature change of water when a sample loses heat. A simple type of calorimeter can be set up using two Styrofoam coffee cups, a measured amount of water, and a thermometer. After measuring the initial temperature of the water, an object of known mass is heated and placed in the water. Eventually, the object and the water reach the same (or final) temperature. The heat lost by the object is equal to the heat gained by the water.

Heat ($-q$) lost by object = heat (q) gained by water

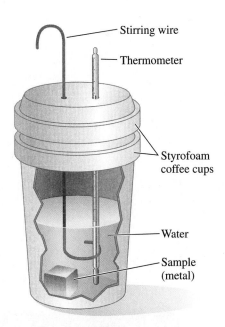

A coffee-cup calorimeter can be used to measure specific heat.

SAMPLE PROBLEM | 3.11

■ **Using a Calorimeter**

A 35.2-g sample of a metal heated to 100.0 °C is placed in a calorimeter containing 42.5 g of water at an initial temperature of 19.2 °C. If the final temperature of the metal and the water is 29.5 °C, what is the specific heat of the solid, assuming all the heat is transferred to the water?

SOLUTION

STEP 1   **List the given and needed data.**

Given

Unknown Metal	Water
Mass = 35.2 g	mass = 42.5 g
Initial temperature = 100.0 °C	initial temperature = 19.2 °C
Final temperature = 29.5 °C	final temperature = 29.5 °C
	SH of water = 4.184 J/g °C

Need SH of metal (J/g °C)

STEP 2   **Calculate the temperature changes.** Determine the temperature change for the solid and the water. The final temperature of the solid is the same as the final temperature of the water, which is 29.5 °C.

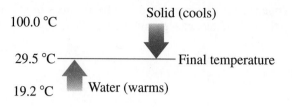

100.0 °C

Solid (cools)

29.5 °C —————————— Final temperature

19.2 °C ——— Water (warms)

$$\Delta T_{water} = T_f - T_i = 29.5\ °C - 19.2\ °C = 10.3\ °C$$

$$\Delta T_{solid} = T_f - T_i = 29.5\ °C - 100.0\ °C = -70.5\ °C$$

STEP 3 **Write the heat equation.**

$$q = m \times \Delta T \times SH$$

STEP 4 **Substitute the given values into the equations and solve.** We first calculate the heat gained by the water because we have all the data.

$$q_{water} = m \times \Delta T \times SH$$

$$q_{water} = 42.5\ \cancel{g} \times 10.3\ \cancel{°C} \times 4.184\ \frac{J}{\cancel{g}\ \cancel{°C}} = 1830\ J$$
$$\text{3 SFs}$$

The heat gained by the water is equal to the heat lost by the object. If water gained 1830 J, then the object lost 1830 J.

Heat gained by water ($q = 1830\ J$) = heat lost by object ($-q = -1830\ J$)

Now we rearrange the heat equation to solve for the specific heat of the metal.

$$q_{metal} = m \times \Delta T \times SH$$

$$SH_{metal} = \frac{q}{m \times \Delta T}$$

$$SH_{metal} = \frac{-1830\ J}{(35.2\ g)(-70.5\ °C)} = 0.737\ J/g\ °C$$

STUDY CHECK

A piece of granite weighing 250.0 g is heated to 100.0 °C and placed in a calorimeter containing 400.0 g of water. The temperature of the water increases from 20.0 °C to 28.5 °C. What is the specific heat of the granite (J/g °C), assuming all the heat is transferred to the water?

CHEMISTRY AND THE ENVIRONMENT

Carbon Dioxide and Global Warming

Earth's climate is a product of interactions between sunlight, the atmosphere, and the oceans. The Sun provides us with energy in the form of solar radiation. Some of this radiation is reflected back into space. The rest is absorbed by the clouds, atmospheric gases including carbon dioxide, and Earth's surface. For millions of years, concentrations of carbon dioxide (CO_2) have fluctuated. However, in the past 100 years, the amount of carbon dioxide (CO_2) gas in our atmosphere has increased significantly. From the years 1000 to 1800, the atmospheric carbon dioxide averaged 280 ppm. But since the beginning of the Industrial Revolution in 1800 up until 2005, the level of atmospheric carbon dioxide has risen from about 280 ppm to about 380 ppm, a 35% increase.

As the atmospheric CO_2 levels increase, more solar radiation is trapped by atmospheric gases, which raises the temperature at Earth's surface. Some scientists have estimated that if the carbon

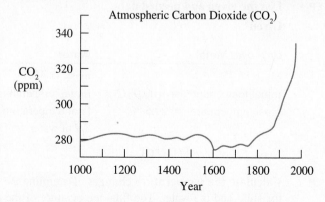

Atmospheric carbon dioxide levels are shown for the years from 1000 C.E. to 2000 C.E.

dioxide level doubles from its level before the Industrial Revolution, the average temperature globally could increase by 2.0 °C to 4.4 °C. Although this seems to be a small temperature change, it could have dramatic impact worldwide. Even now, glaciers and snow cover in much of the world have diminished. Ice sheets in Antarctica and Greenland are melting rapidly and breaking apart. Although no one knows for sure how rapidly the ice in the polar regions is melting, this accelerating change will contribute to a rise in sea level. In the twentieth century, the sea level rose 15 to 23 cm, and some scientists predict the sea level will rise 1 m in this century. Such an increase will have a major impact on coastal areas.

Until recently, carbon dioxide was maintained as algae in the oceans and trees in the forests utilized the carbon dioxide. However, the ability of these and other forms of plant life to absorb carbon diox-ide is not keeping up with the increase in carbon dioxide. Most scientists agree that the primary source of the increase of carbon dioxide is the burning of fossil fuels such as gasoline, coal, and natural gas. The cutting and burning of trees in the rain forests (deforestation) also reduces the amount of carbon dioxide removed from the atmosphere.

Worldwide efforts are being made to reduce the carbon dioxide produced by burning fossil fuels that heat our homes, run our cars, and provide energy for industries. Efforts are being made to explore alternative energy sources and to reduce the effects of deforestation. Meanwhile, we can reduce energy use in our homes by using appliances that are more energy efficient and replacing incandescent light bulbs with fluorescent lights. Such an effort worldwide will reduce the possible impact of global warming and at the same time save our fuel resources.

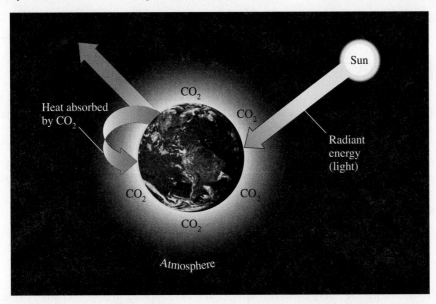

Heat from the Sun is trapped by the CO_2 layer in the atmosphere.

QUESTIONS AND PROBLEMS

Specific Heat

3.29 If the same amount of heat is supplied to samples of 10.0 g each of aluminum, iron, and copper all at 15.0 °C, which sample would reach the highest temperature (see Table 3.9)?

3.30 Substances A and B are the same mass and at the same initial temperature. When the same amount of heat is added to each, the final temperature of A is 75 °C and B is 35 °C. What does this tell you about the specific heats of A and B?

3.31 Calculate the specific heat for each of the following using Table 3.9:
   a. A 13.5-g sample of zinc heated from 24.2 °C to 83.6 °C that absorbs 312 J of heat
   b. A metal with a mass of 48.2 g that absorbs 345 J with a temperature change from 35.0 °C to 57.9 °C

3.32 Calculate the specific heat for each of the following using Table 3.9:
   a. A 18.5-sample of tin that absorbs 183 J of heat with a temperature increase from 35.0 °C to 78.6 °C
   b. A metal with a mass of 22.5 grams that absorbs 645 J with a temperature change from 36.2 °C to 92.0 °C

3.33 Calculate the energy, in joules and calories, for each of the following using Table 3.9:
   a. required to heat 25.0 g of water from 12.5 °C to 25.7 °C
   b. required to heat 38.0 g of copper (Cu) from 122 °C to 246 °C
   c. lost when 15.0 g of ethanol, C_2H_5OH, cools from 60.5 °C to −42.0 °C
   d. lost when 125 g of iron, Fe, cools from 118 °C to 55 °C

3.34 Calculate the energy, in joules and calories, for each of the following using Table 3.9:
   a. required to heat 5.25 g of water, H_2O, from 5.5 °C to 64.8 °C
   b. lost when 75.0 g of water, H_2O, cools from 86.4 °C to 2.1 °C
   c. required to heat 10.0 g of silver (Ag) from 112 °C to 275 °C
   d. lost when 18.0 g of gold (Au) cools from 224 °C to 118 °C

3.35 Calculate the mass, in grams, for each of the following using Table 3.9:
   a. a gold (Au) sample that absorbs 225 J to change its temperature from 15.0 °C to 47.0 °C
   b. an iron (Fe) object that loses 8.40 kJ when its temperature drops from 168.0 °C to 82.0 °C
   c. a sample of aluminum (Al) that absorbs 8.80 kJ when heated from 12.5 °C to 26.8 °C
   d. a sample of titanium (Ti) that loses 14 200 J when it cools from 185 °C to 42 °C

3.36 Calculate the mass, in grams, for each of the following using Table 3.9:

    a. a sample of water that absorbs 8250 J when its temperature rises from 18.4 °C to 92.6 °C

    b. a silver (Ag) sample that loses 3.22 kJ when its temperature drops from 145 °C to 24 °C

    c. a sample of aluminum (Al) that absorbs 1.65 kJ when its temperature rises from 65 °C to 187 °C

    d. an iron (Fe) bar that loses 2.52 kJ when its temperature drops from 252 °C to 75 °C

3.37 Calculate the rise in temperature (°C) for each of the following using Table 3.9:

    a. 20.0 g of iron (Fe) that absorbs 1580 J

    b. 150.0 g of water that absorbs 7.10 kJ

    c. 85.0 g of gold (Au) that absorbs 7680 J

    d. 50.0 g of copper (Cu) that absorbs 6.75 kJ

3.38 Calculate the decrease in temperature (°C) for each of the following using Table 3.9:

    a. 115 g of copper (Cu) that loses 2.45 kJ

    b. 22.0 g of silver (Ag) that loses 625 J

    c. 0.650 kg of liquid water that loses 5.48 kJ

    d. 35.0 g of silver (Ag) that loses 472 J

3.6 Energy and Nutrition

LEARNING GOAL

Use the energy values to calculate the kilocalories (kcal) or kilojoules (kJ) in a food.

TUTORIAL

Nutritional Energy

CASE STUDY

Calories from Hidden Sugar

The food we eat provides energy to do work in the body, which includes the growth and repair of cells. Carbohydrates are the primary fuel for the body, but if carbohydrate reserves are exhausted, fats and then proteins are used for energy.

For many years in the field of nutrition, the energy from food was measured as Calories or kilocalories. The nutritional unit **Calorie, Cal** (with an uppercase C), is the same as 1000 cal, or l kcal. Now the use of kilojoule (kJ) is becoming more prevalent. For example, a baked potato has an energy content of 120 Calories, which is 120 kcal or 500 kJ. A typical diet of 2000 Cal (kcal) is the same as an 8400 kJ diet.

Energy in Nutrition

    1 Cal = 1 kcal = 1000 cal

    1 Cal = 4.184 kJ = 4184 J

To determine the *energy value*, a sample of food is placed in a steel chamber filled with oxygen gas. A measured quantity of water surrounds the steel chamber. The food sample is ignited, releasing heat that increases the temperature of the surrounding water. From the mass of the food and water as well as the temperature increase, the energy value of the food is calculated. We will assume that the energy absorbed by the calorimeter is negligible.

Heat released from burning a food sample in a calorimeter is calculated from the temperature change of the water.

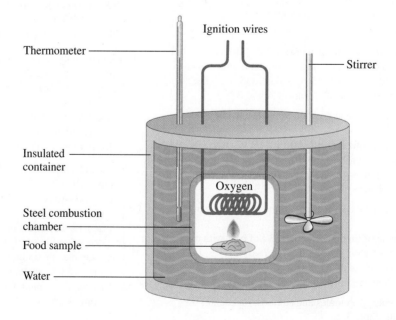

■ **Energy Values of Food**

A 1.8-oz serving of pasta provides 210 Cal. What is the energy value of pasta in Cal/g?

ANSWER

Using the equalities of 16 oz = 1 lb and 1 lb = 453.6 g, we can set up some calculations for the energy values of pasta.

$$\frac{210 \text{ Cal}}{1.8 \text{ oz}} \times \frac{16 \text{ oz}}{1 \text{ lb}} \times \frac{1 \text{ lb}}{453.6 \text{ g}} = 4.1 \text{ Cal/g}$$

SAMPLE PROBLEM | 3.12

■ **Calculating Food Energy Values**

A 2.3-g sample of butter, a fat, is placed in a calorimeter containing 1900 g of water at an initial temperature of 17 °C. After the complete combustion of the butter, the water has a temperature of 28 °C. What is the energy value (kcal/g and kJ/g) of butter? Assume that the energy absorbed by the calorimeter is negligible.

SOLUTION

STEP 1 Given mass = 1900 g of water; 2.3 g of butter

Initial temperature = 17 °C; final temperature = 28 °C

Need energy value in kcal/g and kJ/g

STEP 2 Calculate the temperature change. The temperature change ΔT is the difference between the two temperatures.

$$\Delta T = T_f - T_i = 28 \text{ °C} - 17 \text{ °C} = 11 \text{ °C}$$

STEP 3 Write the heat equation.

$$\text{Heat } (q) = m \times \Delta T \times SH$$

$$1900 \text{ g} \times 11 \text{ °C} \times \frac{1.00 \text{ cal}}{\text{g °C}} \times \frac{1 \text{ kcal}}{1000 \text{ cal}} = 21 \text{ kcal}$$

$$1900 \text{ g} \times 11 \text{ °C} \times \frac{4.184 \text{ J}}{\text{g °C}} \times \frac{1 \text{ kJ}}{1000 \text{ J}} = 87 \text{ kJ}$$

STEP 4 Substitute the energy in kcal and kJ into the expression for energy values for food. Because 2.3 g of fat provided the 21 kcal (87 kJ) of heat, the energy value of butter is calculated as follows:

$$\frac{21 \text{ kcal}}{2.3 \text{ g fat}} = 9.1 \text{ kcal/g of fat} \qquad \frac{87 \text{ kJ}}{2.3 \text{ g fat}} = 38 \text{ kJ/g of fat}$$

STUDY CHECK

A 4.5-g sample of the carbohydrate sucrose, table sugar, is placed in a calorimeter. The water in the container has a mass of 1500 g and an initial temperature of 15 °C. After all the sucrose is burned, the water temperature is 27 °C. What is the energy value, in kcal/g and kJ/g, for sucrose?

Energy Values for Foods

The **energy (caloric) values** of food are the kilocalories or kilojoules obtained from the complete combustion of 1 g of a carbohydrate, fat, or protein. Because the energy values for the metabolism of carbohydrates, fats, and proteins vary somewhat, the energy values determined using a calorimeter represent an average (see Table 3.10).

TABLE 3.10 Typical Energy (Caloric) Values for the Three Food Types

Food Type	kJ/g	kcal/g
Carbohydrate	17	4
Fat	38	9
Protein	17	4

The nutrition facts include the total Calories, Calories from fat, and total grams of carbohydrate.

TABLE 3.11 General Composition and Energy Content of Some Foods

Food	Carbohydrate (g)	Fat (g)	Protein (g)	Energy*
Banana, 1 medium	26	0	1	460 kJ (110 kcal)
Beef, ground, 3 oz	0	14	22	910 kJ (220 kcal)
Carrots, raw, 1 cup	11	0	1	200 kJ (50 kcal)
Chicken, no skin, 3 oz	0	3	20	460 kJ (110 kcal)
Egg, 1 large	0	6	6	330 kJ (80 kcal)
Milk, 4% fat, 1 cup	12	9	9	700 kJ (170 kcal)
Milk, nonfat, 1 cup	12	0	9	360 kJ (90 kcal)
Potato, baked	23	0	3	440 kJ (100 kcal)
Salmon, 3 oz	0	5	16	460 kJ (110 kcal)
Steak, 3 oz	0	27	19	1350 kJ (320 kcal)

*Energy values are rounded off to the tens place.

Using the energy values in Table 3.10, we can calculate the total energy of a food if the mass of each food type is known.

$$\text{Kilojoules} = \cancel{g} \times \frac{kJ}{\cancel{g}}$$

$$\text{Kilocalories} = \cancel{g} \times \frac{kcal}{\cancel{g}}$$

On packaged food, the energy content is listed in the Nutrition Facts label, usually in terms of the number of Calories for one serving. The general composition and energy content of some foods are given in Table 3.11.

SAMPLE PROBLEM | 3.13

■ **Energy Content for a Food**

What is the energy content (kJ and kcal) for a piece of chocolate cake that contains 34 g of carbohydrate, 10 g of fat, and 5 g of protein? Round off the answers of kJ and kcal to the tens place.

SOLUTION

Using the energy values for carbohydrate, fat, and protein (see Table 3.10), we can calculate the total number of kcal:

Food Type	Mass	Energy Values		Energy
Carbohydrate	$34 \cancel{g} \times$	$\dfrac{17 \text{ kJ (or 4 kcal)}}{1 \cancel{g}}$	$=$	580 kJ (or 140 kcal)
Fat	$10 \cancel{g} \times$	$\dfrac{38 \text{ kJ (or 9 kcal)}}{1 \cancel{g}}$	$=$	380 kJ (or 90 kcal)
Protein	$5 \cancel{g} \times$	$\dfrac{17 \text{ kJ (or 4 kcal)}}{1 \cancel{g}}$	$=$	90 kJ (or 20 kcal)
		Total energy content	$=$	1050 kJ (or 250 kcal)

STUDY CHECK

A 1-oz (28 g) serving of oat-bran hot cereal with half a cup of whole milk contains 22 g of carbohydrate, 7 g of fat, and 10 g of protein. If you eat two servings of the oat bran for breakfast, how many kilocalories will you obtain? Round off the kilocalories for each food type to the tens place.

Losing and Gaining Weight

The number of kilojoules or kilocalories needed in the daily diet of an adult depends on gender, age, and level of physical activity. Some typical levels of energy needs are given in Table 3.12.

TABLE 3.12 Typical Energy Requirements for Adults

Gender	Age	Moderately Active kcal (kJ)	Active kcal (kJ)
Female	19–30	2100 (8800)	2400 (10 000)
	31–50	2000 (8400)	2200 (9200)
Male	19–30	2700 (11 000)	3000 (13 000)
	31–50	2500 (10 500)	2900 (12 100)

A person gains weight when food intake exceeds energy output. The amount of food a person eats is regulated by the hunger center in the hypothalamus, which is located in the brain. Food intake is normally proportional to the nutrient stores in the body. If these nutrient stores are low, you feel hungry; if they are high, you do not feel like eating.

A person loses weight when food intake is less than energy output. Many diet products contain cellulose, which has no nutritive value but provides bulk and makes you feel full. Some diet drugs depress the hunger center and must be used with caution, because they excite the nervous system and elevate blood pressure. Because muscular exercise is an important way to expend energy, an increase in daily exercise aids weight loss. Table 3.13 lists some activities and the amount of energy they require.

TABLE 3.13 Energy Expended by a 70.0-kg (154-lb) Adult

Activity	Energy (kcal/h)	Energy (kJ/h)
Sleeping	60	250
Sitting	100	420
Walking	200	840
Swimming	500	2100
Running	750	3100

One hour of swimming uses 2100 kJ of energy.

QUESTIONS AND PROBLEMS

Energy and Nutrition

3.39 Using the following data, determine the kilojoules and kilocalories for each food burned in a calorimeter:
  a. one stalk of celery that heats 505 g of water from 25.2 °C to 35.7 °C
  b. a waffle that heats 4980 g of water from 20.6 °C to 62.4 °C

3.40 Using the following data, determine the kilojoules and kilocalories for each food burned in a calorimeter:
  a. one cup of popcorn that changes the temperature of 1250 g of water from 25.5 °C to 50.8 °C
  b. a sample of butter that produces energy to increase the temperature of 357 g of water from 22.7 °C to 38.8 °C

3.41 Using the energy values for food (see Table 3.10), determine each of the following (round off the answers to the tens place):
  a. the total kilojoules and kilocalories for 1 cup of orange juice that contains 26 g of carbohydrate, 2 g of protein, and no fat

  b. the grams of carbohydrate in one apple if the apple has no fat and no protein and provides 72 kcal of energy
  c. the total kilojoules and kilocalories in 1 tablespoon of vegetable oil, which contains 14 g of fat and no carbohydrate or protein
  d. the total kilojoules and kilocalories from a diet that consists of 68 g of carbohydrate, 150 g of protein, and 9.0 g of fat

3.42 Using the energy values for food (see Table 3.10), determine each of the following (round off the answers to the tens place):
  a. the total kilojoules and kilocalories in 2 tablespoons of crunchy peanut butter that contains 6 g of carbohydrate, 16 g of fat, and 7 g of protein
  b. the grams of protein in 1 cup of soup that provides 110 kcal with 6 g of fat and 9 g of carbohydrate
  c. the grams of sugar (carbohydrate) in one can of cola if it has 680 kilojoules and no fat and no protein
  d. one cup of clam chowder that contains 16 g of carbohydrate, 9 g of protein, and 12 g of fat

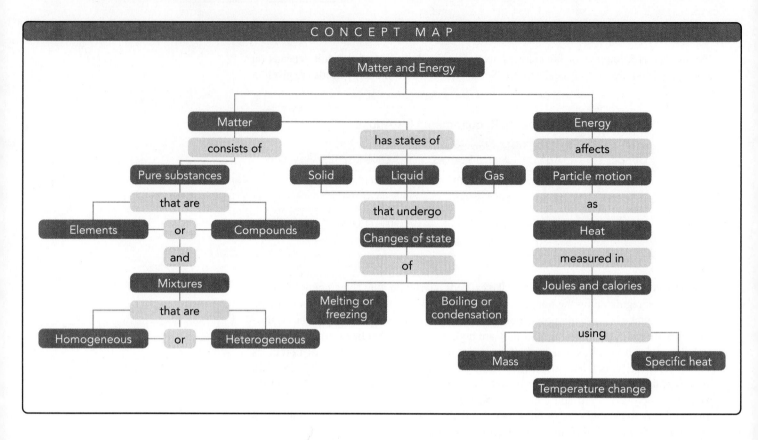

CHAPTER REVIEW

3.1 Classification of Matter

LEARNING GOAL: *Classify examples of matter as pure substances or mixtures.*

Matter is classified as pure substances or mixtures. Pure substances, which are elements or compounds, have fixed compositions. Mixtures have variable compositions, which are classified further as homogeneous or heterogeneous. The substances in mixtures can be separated using physical methods.

3.2 States and Properties of Matter

LEARNING GOAL: *Identify the states and the physical and chemical properties of matter.*

The three states of matter are solid, liquid, and gas. A physical property is a characteristic unique to a substance. A physical change occurs when physical properties change but not the composition of the substance. A chemical property indicates the ability of a substance to change into another substance. In a chemical change, at least one substance forms a new substance with new physical properties.

3.3 Temperature

LEARNING GOAL: *Given a temperature, calculate a corresponding value on another temperature scale.*

On the Celsius scale, there are 100 units between the freezing point of water (0 °C) and the boiling point (100 °C). On the Fahrenheit scale, there are 180 units between the freezing point of water (32 °F) and the boiling point (212 °F). A Fahrenheit temperature is related to

its Celsius temperature by the equation $T_F = 1.8\,T_C + 32$. The SI temperature of Kelvin is related to the Celsius temperature by the equation $T_K = T_C + 273$.

3.4 Energy

LEARNING GOAL: *Identify energy as kinetic or potential; convert between units of energy.*

Energy is the ability to do work. Kinetic energy is the energy of motion; potential energy is energy determined by position or composition. Common units of energy are the calorie (cal), kilocalorie (kcal), joule (J), and kilojoule (kJ).

3.5 Specific Heat

LEARNING GOAL: *Use specific heat to calculate heat loss or gain, temperature change, or mass of a sample.*

Specific heat is the amount of energy required to raise the temperature of exactly 1 g of a substance by exactly 1 °C. The heat lost or gained by a substance is determined by multiplying its mass, the temperature change, and its specific heat.

3.6 Energy and Nutrition

LEARNING GOAL: *Use the energy values to calculate the kilocalories (kcal) or kilojoules (kJ) in a food.*

The nutritional Calorie is the same amount of energy as 1 kcal or 1000 calories. The energy content of a food is the sum of kilocalories or kilojoules from carbohydrate, fat, and protein.

■ KEY TERMS

calorie (cal) The amount of heat energy that raises the temperature of exactly 1 g of water by exactly 1 °C.

chemical change A change during which the original substance is converted into a new substance that has a different composition and new chemical and physical properties.

chemical properties The properties that indicate the ability of a substance to change into a new substance.

compound A pure substance consisting of two or more elements, with a definite composition, that can be broken down into simpler substances only by chemical methods.

element A pure substance containing only one type of matter, which cannot be broken down by chemical methods.

energy The ability to do work.

energy (caloric) value The kilocalories obtained per gram of the three food types: carbohydrate, fat, and protein.

gas A state of matter that does not have a definite shape or volume.

heat The energy associated with the motion of particles that flows from a hotter to a colder object.

joule (J) The SI unit of heat energy; 4.184 J = 1 cal.

kinetic energy A type of energy that is required for actively doing work; energy of motion.

liquid A state of matter that takes the shape of the container but has a definite volume.

matter The material that makes up a substance and has mass and occupies space.

mixture The physical combination of two or more substances that does not change the identities of the mixed substances.

physical change A change in which the physical properties of a substance change without any change in its identity.

physical properties The properties that can be observed or measured without affecting the identity of a substance.

potential energy A type of energy related to position or composition of a substance.

pure substance A type of matter composed of elements and compounds that has a definite composition.

solid A state of matter that has its own shape and volume.

specific heat A quantity of heat that changes the temperature of exactly 1 g of a substance by exactly 1 °C.

states of matter Three forms of matter: solid, liquid, and gas.

work An activity that requires energy.

■ UNDERSTANDING THE CONCEPTS

3.43 Identify each of the following as an element, compound, or mixture:

a.
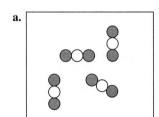

b.
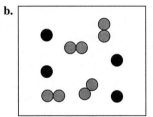

c.
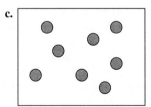

3.44 Which diagram illustrates a homogeneous mixture? Explain your choice. Which diagrams illustrate heterogeneous mixtures? Explain your choice.

a.   b.   c.

3.45 Classify each of the following as a homogeneous mixture or heterogeneous mixture:
a. lemon-flavored water   b. stuffed mushrooms
c. chicken noodle soup

3.46 Classify each of the following as a homogeneous mixture or heterogeneous mixture:
a. ketchup   b. hard-boiled egg
c. eye drops

3.47 Compost can be made at home from grass clippings, some kitchen scraps, and dry leaves. As microbes break down organic matter, heat is generated and the compost can reach a temperature of 155 °F, which kills most pathogens. What is this temperature in Celsius degrees?

Compost produced from decayed plant material is used to enrich the soil.

3.48 After a week, biochemical reactions in compost slow, and the temperature drops to 45 °C. The dark brown organic-rich mixture is ready for use in the garden. What is this temperature in Fahrenheit degrees? In kelvins?

3.49 Calculate the energy to heat three cubes (gold, aluminum, and silver) each with a volume of 10.0 cm^3 from 15 °C to 25 °C. Refer to Tables 2.9 and 3.9. What do you notice about the energy needed for each?

3.50 A 70.0-kg person had a quarter-pound cheeseburger, french fries, and a chocolate shake. Using Table 3.10 and Table 3.13, determine each of the following:
  a. the number of hours of sleep needed to "burn off" the kilocalories in this meal

b. the number of hours of running needed to "burn off" the kilocalories in this meal

Item	Protein (g)	Fat (g)	Carbohydrate (g)
Cheeseburger	31	29	34
French fries	3	11	26
Chocolate shake	11	9	60

ADDITIONAL QUESTIONS AND PROBLEMS

For instructor-assigned homework, go to
www.masteringchemistry.com.

3.51 Classify each of the following as an element, compound, or mixture:
  a. carbon in pencils
  b. carbon dioxide (CO_2) we exhale
  c. orange juice

3.52 Classify each of the following as an element, compound, or mixture:
  a. neon gas in lights
  b. a salad dressing of oil and vinegar
  c. sodium hypochlorite (NaOCl) in bleach

3.53 Classify each of the following mixtures as homogenous or heterogeneous:
  a. hot fudge sundae   **b.** herbal tea
  c. vegetable oil

3.54 Classify each of the following mixtures as homogenous or heterogeneous:
  a. water and sand   **b.** mustard
  c. blue ink

3.55 Identify each of the following as solid, liquid, or gas:
  a. vitamin tablets in a bottle   **b.** helium in a balloon
  c. milk in a bottle   **d.** the air you breathe
  e. charcoal briquettes on a barbecue

3.56 Identify each of the following as solid, liquid, or gas:
  a. popcorn in a bag   **b.** water in a garden hose
  c. a computer mouse   **d.** air in a tire
  e. hot tea in a teacup

3.57 Identify each of the following as a physical or chemical property:
  a. Gold is shiny.
  b. Gold melts at 1064 °C.
  c. Gold is a good conductor of electricity.
  d. When gold reacts with yellow sulfur, a black sulfide compound forms.

3.58 Identify each of the following as a physical or chemical property:
  a. A candle is 10 in. high and 2 in. in diameter.
  b. A candle burns.
  c. The wax of a candle softens on a hot day.
  d. A candle is blue.

3.59 Identify each of the following as a physical or chemical change:
  a. A plant grows a new leaf.
  b. Chocolate is melted for a dessert.
  c. Wood is chopped for the fireplace.
  d. Wood burns in a woodstove.

3.60 Identify each of the following as a physical or chemical change:
  a. Short hair grows until it is long.
  b. Carrots are grated for use in a salad.
  c. Malt undergoes fermentation to make beer.
  d. A copper pipe reacts with air and turns green.

3.61 Calculate each of the following temperatures in degrees Celsius and kelvins:
  a. The highest recorded temperature in the continental United States was 134 °F in Death Valley, California, on July 10, 1913.
  b. The lowest recorded temperature in the continental United States was −69.7 °F in Rodgers Pass, Montana, on January 20, 1954.

3.62 Calculate each of the following temperatures in kelvins and degrees Fahrenheit:
  a. The highest recorded temperature in the world was 58.0 °C in El Azizia, Libya, on September 13, 1922.
  b. The lowest recorded temperature in the world was −89.2 °C in Vostok, Antarctica, on July 21, 1983.

3.63 What is −15 °F in degrees Celsius and in kelvins?

3.64 The highest recorded body temperature that a person has survived is 46.5 °C. Calculate that temperature in degrees Fahrenheit and in kelvins.

3.65 If you want to lose 1 lb of "fat," which is 15% water, how many kilocalories do you need to expend?

3.66 Calculate the Cal (kcal) in 1 cup of whole milk that contains 12 g of carbohydrate, 9 g of fat, and 9 g of protein.

3.67 On a hot day, the beach sand gets hot but the water stays cool. Would you predict that the specific heat of sand is higher or lower than that of water? Explain.

3.68 A large bottle containing 883 g of water at 4 °C is removed from the refrigerator. How many kilojoules (kJ) are absorbed to warm the water to room temperature of 27 °C?

3.69 A hot-water bottle contains 725 g of water at 65 °C. If the water cools to body temperature (37 °C), how many kilojoules of heat could be transferred to sore muscles?

3.70 Copper is sometimes coated on the bottom surface of cooking pans. Copper has a specific heat of 0.385 J/g °C. A 25.0-g sample of copper heated to 85.0 °C is dropped into water at an initial temperature of 14.0 °C. If the final temperature reached by the copper–water mixture is 36.0 °C, how many grams of water are present?

3.71 A 25.0-g sample of a metal at 98.0 °C is placed in 50.0 g of water at 18.0 °C. If the final temperature of the alloy and water is 27.4 °C, what is the specific heat of the metal?

3.72 A 0.50-g sample of vegetable oil is placed in a calorimeter. When the sample is burned, 18.9 kJ is given off. What is the energy value (kcal/g) of the oil?

CHALLENGE QUESTIONS

3.73 When a 0.66-g sample of olive oil is burned in a calorimeter, the heat released increases the temperature of 370 g of water from 22.7 °C to 38.8 °C. What is the energy value of the olive oil in kJ/g and kcal/g?

3.74 When 1.0 g of gasoline burns, 11 500 kcal of energy are given off. If the density of gasoline is 0.74 g/mL, how many kilocalories of energy are obtained from 1.5 gal of gasoline?

3.75 A piece of copper metal (specific heat 0.385 J/g °C) at 86.0 °C is placed in 50.0 g of water at 16.0 °C. The metal and water come to the same temperature of 24.0 °C. What was the mass of the piece of copper?

3.76 Your friend has just eaten a slice of pizza, cola soft drink, and ice cream. What is the total kilocalories your friend obtained from this meal? How many hours will your friend need to swim to "burn off" the kilocalories in this meal if your friend has a mass of 70.0 kg (see Table 3.10 and Table 3.13)?

Item	Protein (g)	Fat (g)	Carbohydrate (g)
Pizza	13	10	29
Cola	0	0	51
Ice cream	8	28	44

3.77 A typical diet in the United States provides 15% of the calories from protein, 45% from carbohydrates, and the remainder from fats. Calculate the grams of protein, carbohydrate, and fat to be included each day in diets having the following calorie requirements:
a. 1200 kcal   **b.** 1900 kcal   **c.** 2600 kcal

3.78 A 125-g piece of metal is heated to 288 °C and dropped into 85.0 g of water at 26.0 °C. If the final temperature of the water and metal is 58.0 °C, what is the specific heat of the metal (J/g °C)?

3.79 Rearrange the heat equation to solve for each of the following:
a. the mass, in grams, of water that absorbs 8250 J when its temperature rises from 18.3 °C to 92.6 °C
b. the mass, in grams, of a gold sample that absorbs 225 J when the temperature rises from 15.0 °C to 47.0 °C
c. the rise in temperature when a 20.0-g sample of iron absorbs 1580 J
d. the specific heat of a metal when 8.50 g of the metal absorbs 28 cal and the temperature rises from 12 °C to 24 °C

ANSWERS

ANSWERS TO STUDY CHECKS

3.1 This is a heterogeneous mixture because it does not have a uniform composition.

3.2 **b** and **d** are chemical changes.

3.3 12 °F

3.4 39.8 °C

3.5 night, −260. °C; day, 410. °C

3.6 8.4 kcal

3.7 $SH = 1.23$ J/g °C

3.8 14.6 kJ

3.9 −360. J

3.10 182 g of iron

3.11 $SH = 0.80$ J/g °C

3.12 4.0 kcal/g of sucrose, 17 kJ/g of sucrose

3.13 380 kcal

ANSWERS TO SELECTED QUESTIONS AND PROBLEMS

3.1 **a.** pure substance **b.** mixture
c. pure substance **d.** pure substance
e. mixture

3.3 **a.** element **b.** compound
c. element **d.** compound
e. compound

3.5 **a.** heterogeneous **b.** homogeneous
c. homogeneous **d.** heterogeneous
e. heterogeneous

3.7 **a.** gas **b.** gas **c.** solid

3.9 **a.** physical **b.** chemical **c.** physical
d. chemical **e.** chemical

3.11 **a.** physical **b.** chemical **c.** physical
d. physical **e.** physical

3.13 **a.** chemical **b.** physical **c.** physical
d. chemical **e.** physical

3.15 In the United States, we still use the Fahrenheit temperature scale. In °F, normal body temperature is 98.6. On the Celsius scale, her temperature would be 37.7 °C, a mild fever.

3.17 **a.** 98.6 °F **b.** 18.5 °C **c.** 246 K
d. 335 K **e.** 46 °C **f.** 295 K

3.19 **a.** 41 °C
b. No. The temperature is equivalent to 39 °C.

3.21 When the car is at the top of the ramp, it has its maximum potential energy. As it descends, potential energy changes to kinetic energy. At the bottom, all the energy is kinetic.

3.23 **a.** potential **b.** kinetic **c.** potential
d. potential

3.25 **a.** 22 kJ **b.** 5300 cal **c.** 5.3 kcal

3.27 **a.** 3.5 kcal **b.** 99.2 cal **c.** 120 J
d. 1100 cal

3.29 Copper, which has the lowest specific heat, would reach the highest temperature.

3.31 **a.** 0.389 J/g °C **b.** 0.313 J/g °C

3.33 **a.** 1380 J; 330. cal **b.** 1810 J; 434 cal
c. −3780 J; −904 cal **d.** −3600 J; −850 cal

3.35 **a.** 54.5 g **b.** 216 g
c. 686 g **d.** 190. g

3.37 **a.** 175 °C **b.** 11.3 °C
c. 700. °C **d.** 351 °C

3.39 **a.** 22.2 kJ; 5.30 kcal **b.** 871 kJ; 208 kcal

3.41 **a.** 470 kJ; 110 kcal **b.** 18 g of carbohydrate
c. 530 kJ; 130 kcal **d.** 4050 kJ; 950 kcal

3.43 **a.** compound **b.** mixture **c.** element

3.45 **a.** homogeneous **b.** heterogeneous
c. heterogeneous

3.47 68.3 °C

3.49 gold 250 J or 59 cal; aluminum 240 J or 58 cal; silver 250 J or 59 cal

The heat needed for the three samples of metals is almost the same.

3.51 **a.** element **b.** compound **c.** mixture

3.53 **a.** heterogeneous **b.** homogeneous
c. homogeneous

3.55 **a.** Tablets are solid.
b. Helium in a balloon is a gas.
c. Milk is a liquid.
d. Air is a mixture of gases.
e. Charcoal is a solid.

3.57 **a.** Appearance is a physical property.
b. The melting point of gold is a physical property.
c. The ability of gold to conduct electricity is a physical property.
d. The ability of gold to form a new substance with sulfur is a chemical property.

3.59 **a.** Plant growth is a chemical change.
b. A change of state from solid to liquid is a physical change.
c. Chopping wood into smaller pieces is a physical change.
d. Burning wood, which forms new substances, is a chemical change.

3.61 **a.** 56.7 °C, 330. K **b.** −56.7 °C, 217 K

3.63 −26 °C, 247 K

3.65 3500 kcal

3.67 Water has a higher specific heat than sand, which means that a large amount of energy is required to cause a significant temperature change. Even a small amount of energy will cause a significant temperature change in the sand.

3.69 85 kJ

3.71 1.1 J/g °C

3.73 38 kJ/g; 9.1 kcal/g

3.75 70. g of copper

3.77 **a.** 45 g of protein, 140 g of carbohydrate, 53 g of fat
b. 71 g of protein, 210 g of carbohydrate, 84 g of fat
c. 98 g of protein, 290 g of carbohydrate, 120 g of fat

3.79 **a.** 26.5 g **b.** 54.5 g
c. 175 °C **d.** 0.27 cal/g °C

COMBINING IDEAS FROM CHAPTERS 1 TO 3

CI.1 Gold, one of the most sought-after metals in the world, has a density of 19.3 g/cm^3, a melting point of 1064 °C, and a specific heat of 0.129 J/g °C. A gold nugget found in Ruby, Alaska, in 1998 weighs 294.10 troy ounces.

Gold nuggets, also called native gold, can be found in streams and mines.

a. How many significant figures are in the measurement of the weight of the nugget?

b. If 1 troy ounce is 31.1035 g, what is the mass of the nugget in grams? In kilograms?

c. If the nugget were pure gold, what would its volume be in cm^3?

d. If the gold nugget were hammered into a foil with a thickness of 0.0035 in., what would be the area of the foil in m^2?

e. What is the melting point of gold in degrees Fahrenheit and kelvins?

f. How many kilojoules are required to raise the temperature of the nugget from 63 °F to 85 °F?

CI.2 The mileage for a motorcycle with a fuel-tank capacity of 22 L is 35 mi/gal. The density of gasoline is 0.74 g/mL.

a. How long can a trip, in kilometers, be made on a full tank of gasoline?

b. If the price of gasoline is $2.67/gal, what would be the cost of fuel for the trip?

c. If the average speed during the trip is 44 mi/h, how long will it take to reach the destination?

d. What is the mass, in grams, of the fuel in the tank?

e. When 1.0 g of gasoline burns, 47 kJ of energy is released. How many kilojoules are produced when the fuel in a full tank is burned?

CI.3 Answer the following questions for the water samples A and B shown in the diagrams:

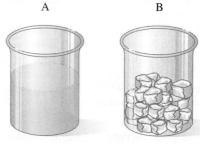

a. Which sample has its own shape?

b. When each sample is transferred to another container, what happens to the shape and volume of each?

c. Match the diagrams (1, 2, or 3) that represent the water particles with sample A and B. Give a reason for your choice.

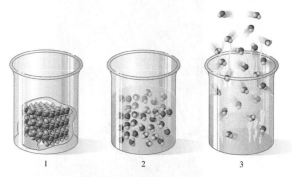

d. The state of matter indicated in diagram 1 is _____; in diagram 2 is _____; and in diagram 3 is _____.

e. When the water in sample A changes to sample B, it is a _____ change.

CI.4 The label of a lemon poppyseed energy bar lists the nutrition facts as 4 g of fat, 23 g of carbohydrate, and 10 g of protein and a mass of 68 g.

a. Using the energy values of carbohydrates, fats, and proteins (see Table 3.10), what are the kilocalories (Calories) for the lemon poppyseed bar? (Round off answers for each food type to the tens place.)

b. What are the kilojoules for the lemon poppyseed bar? (Round off answers for each food type to the tens place.)

c. If you obtain 160 kJ, how many grams of the lemon poppyseed bar did you eat?

d. If you are walking (840 kJ/h), how many minutes of walking will you need to expend the energy from two lemon poppyseed bars (see Table 3.13)?

CI.5 In a box of iron nails, there are 75 nails weighing 0.250 lb. The density of iron is 7.86 g/cm^3. The specific heat of iron is 0.452 J/g °C.

a. What is the volume, in cm^3, of the iron nails in the box?

b. If 30 nails are added to a graduated cylinder containing 17.6 mL of water, what is the new level of water in the cylinder?

c. How many joules must be added to the nails in the box to raise the temperature from 16 °C to 125 °C?

d. If all the iron nails at 55.0 °C are added to 325 g of water at 4.0 °C, what will be the final temperature (°C) of the nails and water?

CI.6 A hot tub with a surface area of 25 ft^2 is filled with water to a depth of 28 in. Hint: volume is calculated as area × height (A × h).

a. What is the volume, in liters, of water in the tub?

b. What is the mass, in kilograms, of water it contains?

c. How many kilojoules are needed to heat the water from 62 °F to 105 °F?

d. If the hot-tub heater provides 5900 kJ/min, how long, in hours, will it take to heat the water in the hot tub from 62 °F to 105 °F?

■ANSWERS

CI.1 a. 5 significant figures
b. 9147.5 g, 9.1475 kg
c. 474 cm^3
d. 5.3 m^2
e. 1947 °F, 1337 K
f. 14 kJ

CI.3 a. B
b. Sample A takes the shape of the new container, but its volume does not change. Sample B retains its own shape and its own volume.
c. Sample A is represented as particles in diagram 2. Sample B is represented as particles in diagram 1.
d. diagram 1, solid; diagram 2, liquid; diagram 3, gas
e. physical

CI.5 a. 14.4 cm^3
b. 23.4 mL
c. 5590 J
d. 5.8 °C

Atoms and Elements

4

"Many of my patients have diabetes, ulcers, hypertension, and cardiovascular problems," says Sylvia Lau, registered dietitian. "If a patient has diabetes, I discuss foods that raise blood sugar, such as fruit, milk, and starches. I talk about how dietary fat contributes to weight gain and complications from diabetes. For stroke patients, I suggest diets low in fat and cholesterol because high blood pressure increases the risk of another stroke."

   If a lab test shows low levels of iron, zinc, iodine, magnesium, or calcium, a dietitian discusses foods that provide those essential elements. For instance, Lau may recommend more beef for an iron deficiency, whole grain for zinc, leafy green vegetables for magnesium, dairy products for calcium, and iodized table salt and seafood for iodine.

Mastering**CHEMISTRY**

Visit **www.masteringchemistry.com**
for self-study materials and instructor-
assigned homework.

All matter is composed of *elements,* of which there are 117 different kinds. Of these, 88 elements occur naturally and make all the substances in our world. Many elements are already familiar to you. Perhaps you use aluminum in the form of foil or drink soft drinks from aluminum cans. You may have a ring or necklace made of gold, or silver, or perhaps platinum. If you play tennis or golf, then you may have noticed that your racket or clubs may be made from the elements titanium or carbon. In our bodies, calcium and phosphorus form the structure of bones and teeth, iron and copper are needed in the formation of red blood cells, and iodine is required for the proper functioning of the thyroid.

The correct amounts of certain elements are crucial to the proper growth and health of our bodies. Low levels of iron can lead to anemia, while low levels of iodine can cause hypothyroidism and goiter. Laboratory tests are used to confirm that elements such as iron, copper, zinc, or iodine are within normal ranges in our bodies.

4.1 Elements and Symbols

LEARNING GOAL

Given the name of an element, write its correct symbol; from the symbol, write the correct name.

Elements are pure substances from which all other things are built. As we discussed in Chapter 3, elements cannot be broken down into simpler substances. Over the centuries, elements have been named for planets, mythological figures, colors, minerals, geographic locations, and famous people. Some sources of names of elements are listed in Table 4.1. The names and symbols of all the elements are found on the inside cover of this text.

Chemical Symbols

Chemical symbols are one- or two-letter abbreviations for the names of the elements. Only the first letter of an element's symbol is capitalized. If the symbol has a second letter, it is lowercase so that we know when a different element is indicated. If two letters are capitalized, they represent the symbols of two different elements. For example, the element cobalt has the symbol Co. However, the two capital letters CO specify two elements, carbon (C) and oxygen (O).

One-Letter Symbols		Two-Letter Symbols	
C	carbon	Co	cobalt
S	sulfur	Si	silicon
N	nitrogen	Ne	neon
I	iodine	Ni	nickel

TUTORIAL

Element Names, Symbols, and Atomic Numbers

Although most of the symbols use letters from the current names, some are derived from their ancient names. For example, Na, the symbol for sodium, comes from the Latin word *natrium.* The symbol for iron, Fe, is derived from the Latin name *ferrum.* Table 4.2 lists the names and symbols of some common elements. Learning their names and symbols will greatly help your learning of chemistry.

TABLE 4.1 Some Elements and Their Names

Element	Source of Name
Uranium	The planet Uranus
Titanium	Titans (mythology)
Chlorine	*Chloros,* "greenish yellow" (Greek)
Iodine	*Ioeides,* "violet" (Greek)
Magnesium	Magnesia, a mineral
Californium	California
Curium	Marie and Pierre Curie

TABLE 4.2 Names and Symbols of Some Common Elements

Name[a]	Symbol	Name[a]	Symbol	Name[a]	Symbol
Aluminum	Al	Gold (*aurum*)	Au	Phosphorus	P
Argon	Ar	Helium	He	Platinum	Pt
Arsenic	As	Hydrogen	H	Potassium (*kalium*)	K
Barium	Ba	Iodine	I	Radium	Ra
Boron	B	Iron (*ferrum*)	Fe	Silicon	Si
Bromine	Br	Lead (*plumbum*)	Pb	Silver (*argentum*)	Ag
Cadmium	Cd	Lithium	Li	Sodium (*natrium*)	Na
Calcium	Ca	Magnesium	Mg	Strontium	Sr
Carbon	C	Manganese	Mn	Sulfur	S
Chlorine	Cl	Mercury (*hydrargyrum*)	Hg	Tin (*stannum*)	Sn
Chromium	Cr	Neon	Ne	Titanium	Ti
Cobalt	Co	Nickel	Ni	Uranium	U
Copper (*cuprum*)	Cu	Nitrogen	N	Zinc	Zn
Fluorine	F	Oxygen	O		

[a]Names given in parentheses are ancient Latin or Greek words from which the symbols are derived.

Aluminum

Carbon

Silver

Gold

Sulfur

CONCEPT CHECK 4.1

■ **Symbols of the Elements**

The symbol for carbon is C, and the symbol for sulfur is S. However, the symbol for cesium is Cs, not CS. Why?

ANSWER

When the symbol for an element has two letters, the first letter is capitalized, but the second letter is lowercase. If both letters are capitalized, such as in CS, two elements—carbon and sulfur—are indicated.

SAMPLE PROBLEM | **4.1**

■ **Writing Chemical Symbols**

What are the chemical symbols for the following elements?

a. nickel **b.** nitrogen
c. niobium **d.** neon

SOLUTION

a. Ni **b.** N **c.** Nb **d.** Ne

STUDY CHECK

What are the chemical symbols for the elements silicon, selenium, and silver?

CHEMISTRY AND HEALTH

Mercury

Mercury is a silvery, shiny transition element that is in Group 12 and Period 6 on the periodic table. It is the only metal that is a liquid at room temperature. Mercury can enter the body through inhaled mercury vapor, contact with the skin, or foods or water contaminated with mercury. In the body, mercury destroys proteins and disrupts cell function. Long-term exposure to mercury can damage the brain and kidneys, cause mental retardation, and decrease physical development. Blood, urine, and hair samples are used to test for mercury.

In both freshwater and saltwater, bacteria convert mercury into toxic methylmercury, which primarily attacks the central nervous system (CNS). Because fish absorb methylmercury, we are exposed to mercury by consuming mercury-contaminated fish. As levels of mercury ingested from fish became a concern, the Food and Drug Administration (FDA) set a maximum level of one part mercury per million parts seafood (1 ppm), which is the same as 1 μg of mercury in every gram of seafood. Fish higher in the food chain, such as swordfish and shark, can have such high levels of mercury that the Environmental Protection Agency (EPA) recommends they be consumed no more than once a week.

One of the worst incidents of mercury poisoning occurred in Minamata and Niigata, Japan, in 1950. At that time, the ocean was polluted with high levels of mercury from industrial wastes. Because fish were a major food in the diet, more than 2000 people were affected with mercury poisoning and died or developed neural damage. In the United States, between 1988 and 1997, industry decreased its use of mercury by 75% by banning mercury in paint and pesticides, reducing mercury in batteries, and regulating mercury in other products.

This mercury fountain, housed in glass, was designed by Alexander Calder for the 1937 World's Fair in Paris.

SAMPLE PROBLEM   4.2

■ Naming Chemical Elements

Give the name of the element that corresponds to each of the following chemical symbols:

a. Zn   **b.** K   **c.** H   **d.** Fe

SOLUTION

a. zinc   **b.** potassium   **c.** hydrogen   **d.** iron

STUDY CHECK

What are the names of the elements with the chemical symbols Mg, Al, and F?

QUESTIONS AND PROBLEMS

Elements and Symbols

4.1 Write the symbols for the following elements:
- **a.** copper
- **b.** platinum
- **c.** calcium
- **d.** manganese
- **e.** iron
- **f.** barium
- **g.** lead
- **h.** strontium

4.2 Write the symbols for the following elements:
- **a.** oxygen
- **b.** lithium
- **c.** uranium
- **d.** titanium
- **e.** hydrogen
- **f.** chromium
- **g.** tin
- **h.** gold

4.3 Write the name of the element for each symbol.
- **a.** C
- **b.** Cl
- **c.** I
- **d.** Hg
- **e.** Ag
- **f.** Ar
- **g.** B
- **h.** Ni

4.4 Write the name of the element for each symbol.
- **a.** He
- **b.** P
- **c.** Na
- **d.** As
- **e.** Ca
- **f.** Br
- **g.** Cd
- **h.** Si

4.5 What elements are in the following substances?
  a. table salt, NaCl
  b. plaster casts, $CaSO_4$
  c. Demerol, $C_{15}H_{22}ClNO_2$
  d. antacid, $CaCO_3$

4.6 What elements are in the following substances?
  a. water, H_2O
  b. baking soda, $NaHCO_3$
  c. lye, NaOH
  d. sugar, $C_{12}H_{22}O_{11}$

4.2 The Periodic Table

As more elements were discovered, it became necessary to organize them with some type of classification system. By the late 1800s, scientists recognized that certain elements looked alike and behaved much the same way. In 1872, a Russian chemist, Dmitri Mendeleev, arranged the 60 elements known at that time into groups with similar properties and placed them in order of increasing atomic masses. Today this arrangement, now with 117 elements, is known as the **periodic table** (see Figure 4.1).

LEARNING GOAL

Use the periodic table to identify the group and the period of an element and decide whether it is a metal, a metalloid, or a nonmetal.

Periodic Table of Elements

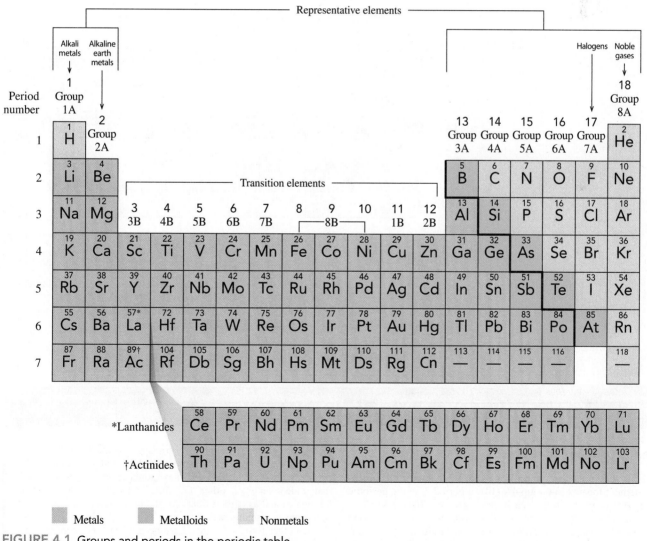

FIGURE 4.1 Groups and periods in the periodic table.

Q What is the symbol of the alkali metal in Period 3?

TUTORIAL

Elements and Symbols in the
Periodic Table

Periods and Groups

Each horizontal row in the periodic table is called a **period**. Each row or period is counted from the top of the table as Period 1 to Period 7. The first period contains only the elements hydrogen (H) and helium (He). The second period contains 8 elements: lithium (Li), beryllium (Be), boron (B), carbon (C), nitrogen (N), oxygen (O), fluorine (F), and neon (Ne). The third period also contains 8 elements, beginning with sodium (Na) and ending with argon (Ar). The fourth period, which begins with potassium (K), and the fifth period, which begins with rubidium (Rb), have 18 elements each. The sixth period, which begins with cesium (Cs), has 32 elements. The seventh period as of today contains the 31 remaining elements, although it could go up to 32 (see Figure 4.2).

Each vertical column on the periodic table contains a **group** (or family) of elements that have similar properties. At the top of each column is a number that is assigned to each group. The elements in the first two columns on the left of the periodic table and the last six columns on the right are called the **representative elements**. For many years, they have been given group numbers 1A–8A. In the center of the periodic table is a block of elements known as the **transition elements**, which are designated with the letter "B". A newer numbering system assigns group numbers of 1–18 going across the periodic table. Because both systems of assigning group numbers are currently in use, they are both indicated on the periodic table in this text and are included in our discussions of elements and group numbers. The lanthanides and actinides that are part of Periods 6 and 7 are placed at the bottom of the periodic table to allow it to fit on a page.

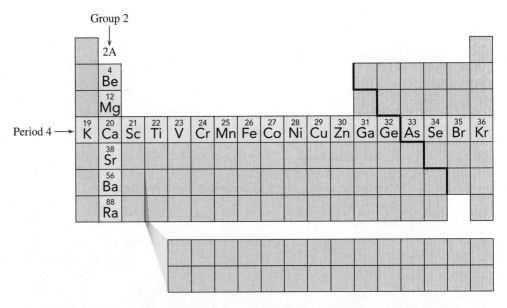

FIGURE 4.2 On the periodic table, each vertical column represents a group of elements, and each horizontal row of elements represents a period.

Q Are the elements Si, P, and S part of a group or a period?

CHEMISTRY AND INDUSTRY

Many Forms of Carbon

Carbon, which has the symbol C and atomic number 6, is a central element located in Group 4A (14) in Period 2 on the periodic table. However, its atoms can be arranged in different ways to give several different substances. Two forms of carbon—diamond and graphite—have been known since prehistoric times. In diamond, each carbon atom is attached to four other carbon atoms in a rigid structure. A diamond is transparent and harder than any other substance, whereas graphite is black and soft. In graphite, carbon atoms are arranged in hexagonal rings that form planes that slide over each other. Graphite is used as pencil lead, as lubricants, and as carbon fibers for the manufacture of lightweight golf clubs and tennis rackets.

Two other forms of carbon have been discovered more recently. In the form called Buckminsterfullerene, or buckyball, 60 carbon

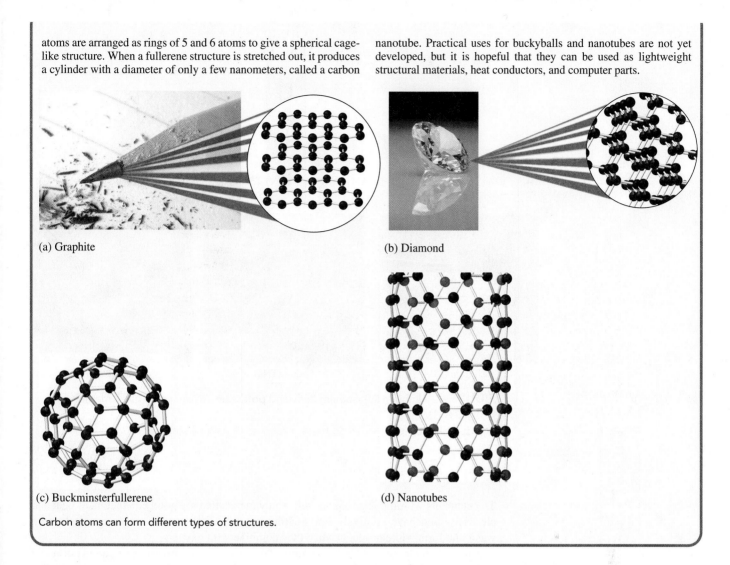

atoms are arranged as rings of 5 and 6 atoms to give a spherical cage-like structure. When a fullerene structure is stretched out, it produces a cylinder with a diameter of only a few nanometers, called a carbon nanotube. Practical uses for buckyballs and nanotubes are not yet developed, but it is hopeful that they can be used as lightweight structural materials, heat conductors, and computer parts.

(a) Graphite

(b) Diamond

(c) Buckminsterfullerene

(d) Nanotubes

Carbon atoms can form different types of structures.

Classification of Groups

Several groups in the periodic table have special names (see Figure 4.3). Group 1A (1) elements—lithium (Li), sodium (Na), potassium (K), rubidium (Rb), cesium (Cs), and francium (Fr)—are a family of elements known as the **alkali metals** (see Figure 4.4).

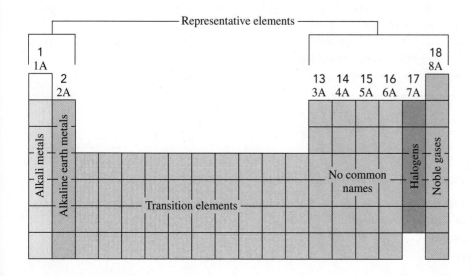

FIGURE 4.3 Certain groups on the periodic table have common names.

Q What is the common name for the group of elements that includes helium and argon?

Group
1A (1)

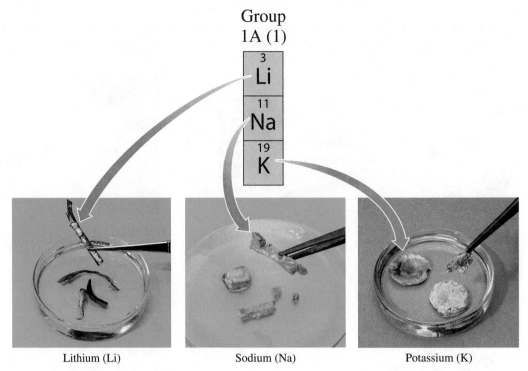

FIGURE 4.4 Lithium (Li), sodium (Na), and potassium (K) are some alkali metals from Group 1A (1).

Q What physical properties do these alkali metals have in common?

Group
7A (17)

The elements within this group are soft, shiny metals that are good conductors of heat and electricity and have relatively low melting points. Alkali metals react vigorously with water and form white products when they combine with oxygen.

Although hydrogen (H) is at the top of Group 1A (1), hydrogen is not an alkali metal and has very different properties from the rest of the elements in this group. Thus, hydrogen is not included in the alkali metals. In some periodic tables, H is placed at the top of Group 7A (17).

Group 2A (2) elements—beryllium (Be), magnesium (Mg), calcium (Ca), strontium (Sr), barium (Ba), and radium (Ra)—are called the **alkaline earth metals**. They are shiny metals, like those in Group 1A, but they are not as reactive.

The **halogens** are found on the right side of the periodic table in Group 7A (17). They include the elements fluorine (F), chlorine (Cl), bromine (Br), iodine (I), and astatine (At), as shown in Figure 4.5. The halogens, especially fluorine and chlorine, are highly reactive and form compounds with most of the elements.

Group 8A (18) contains the **noble gases**—helium (He), neon (Ne), argon (Ar), krypton (Kr), xenon (Xe), and radon (Rn). They are quite unreactive and are seldom found in combination with other elements.

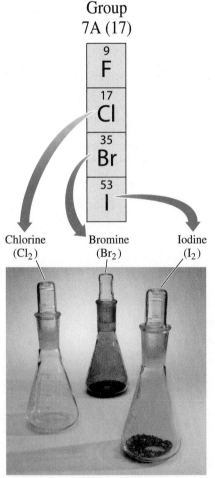

FIGURE 4.5 Chlorine (Cl_2), bromine (Br_2), and iodine (I_2) are halogens from Group 7A (17).

Q What elements are in the halogen group?

SAMPLE PROBLEM | 4.3

■ **Group and Period Numbers of Some Elements**

Give the period and group for each of the following elements and identify as a representative or a transition element:

a. iodine   **b.** manganese

c. barium   **d.** gold

SOLUTION

a. Iodine (I), Period 5, Group 7A (17), is a representative element.
b. Manganese (Mn), Period 4, Group 7B (7), is a transition element.
c. Barium (Ba), Period 6, Group 2A (2), is a representative element.
d. Gold (Au), Period 6, Group 1B (11), is a transition element.

STUDY CHECK

Strontium is an element that gives a brilliant red color to fireworks.

a. In what group is strontium found?
b. What is the name of this chemical family?
c. In what period is strontium found?
d. For the same group, what element is in Period 3?
e. What alkali metal, halogen, and noble gas are in the same period as strontium?

Strontium provides the red color in fireworks.

Metals, Metalloids, and Nonmetals

Another feature of the periodic table is the heavy zigzag line that separates the elements into the *metals* and the *nonmetals*. The metals are those elements on the left of the line *except for hydrogen*, and the nonmetals are the elements on the right (see Figure 4.6).

In general, most **metals** are shiny solids. They can be shaped into wires (ductile) or hammered into a flat sheet (malleable). Metals are good conductors of heat and electricity. They usually melt at higher temperatures than nonmetals. All of the metals are solids at room temperature, except for mercury (Hg), which is a liquid. Some typical metals are sodium (Na), magnesium (Mg), copper (Cu), gold (Au), silver (Ag), iron (Fe), and tin (Sn).

Nonmetals are not especially shiny, malleable, or ductile, and they are often poor conductors of heat and electricity. They typically have low melting points and low densities. You may have heard of nonmetals such as hydrogen (H), carbon (C), nitrogen (N), oxygen (O), chlorine (Cl), and sulfur (S).

Except for aluminum, the elements located along the heavy zigzag line are **metalloids**: B, Si, Ge, As, Sb, Te, Po, and At. Metalloids are elements that exhibit some properties that are typical of metals and other properties that are characteristic of

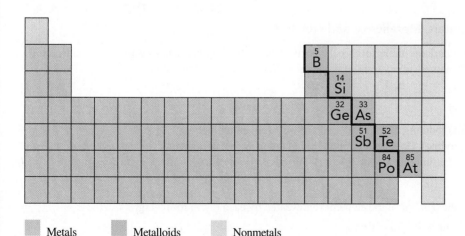

Metals ▪   Metalloids ▪   Nonmetals ▪

FIGURE 4.6 Along the heavy zigzag line on the periodic table that separates the metals and nonmetals are metalloids that exhibit characteristics of both metals and nonmetals.

Q On which side of the heavy zigzag line are the nonmetals located?

A silver cup is shiny, antimony is a blue-gray solid, and sulfur is a dull, yellow color.

TABLE 4.3 Some Characteristics of a Metal, a Metalloid, and a Nonmetal

Silver (Ag)	Antimony (Sb)	Sulfur (S)
Metal	Metalloid	Nonmetal
Shiny	Blue-gray, shiny	Dull, yellow
Extremely ductile	Brittle	Brittle
Can be hammered into sheets (malleable)	Shatters when hammered	Shatters when hammered
Good conductor of heat and electricity	Poor conductor of heat and electricity	Poor conductor of heat and electricity
Used in coins, jewelry, tableware	Used to harden lead, color glass and plastics	Used in gunpowder, rubber, fungicides
Density 10.5 g/mL	Density 6.7 g/mL	Density 2.1 g/mL
Melting point 962 °C	Melting point 630 °C	Melting point 113 °C

the nonmetals. For example, they are better conductors of heat and electricity than the nonmetals, but not as good as the metals. The metalloids are semiconductors because they can be easily modified to function as conductors or insulators. Table 4.3 compares some characteristics of silver, a metal, with those of antimony, a metalloid, and sulfur, a nonmetal.

CONCEPT CHECK 4.2

■ Groups and Periods on the Periodic Table

Consider the elements aluminum, silicon, and phosphorus.

a. In what group and period are they found?
b. Identify each as a metal, metalloid, or nonmetal.

ANSWER

a. They are all found in Period 3. Aluminum is in Group 3A (13), silicon is in Group 4A (14), and phosphorus is in Group 5A (15).
b. Aluminum is a metal, silicon is a metalloid, and phosphorus is a nonmetal.

SAMPLE PROBLEM | 4.4

■ Metals, Metalloids, and Nonmetals

Use a periodic table to classify each of the following elements by its group and period, group names (if any), and if it is a metal, metalloid, and nonmetal:
a. Na   **b.** I   **c.** Sb

SOLUTION

a. Na (sodium), Group 1A (1), Period 3, is an alkali metal.
b. I (iodine), Group 7A (17), Period 5, halogen, is a nonmetal.
c. Sb (antimony), Group 5A (15), Period 5, is a metalloid.

STUDY CHECK

Give the symbol of the element represented by each of the following:

a. Group 5A (15), Period 4
b. a noble gas in Period 6
c. a metalloid in Period 2

CHEMISTRY AND HEALTH

Elements Essential to Health

Of the 117 elements, only about 20 are essential for the well-being and survival of the human body. Of those, four elements—oxygen, carbon, hydrogen, and nitrogen—located in Period 1 and Period 2 on the periodic table, make up 96% of our body mass. Most of the food in our daily diet provides these elements to maintain a healthy body. These major elements are found in carbohydrates such as blood glucose ($C_6H_{12}O_6$) and large molecules of the lipids (fats) that typically use 18-carbon fatty acids such as stearic acid ($C_{18}H_{36}O_2$). These elements also make up long carbon chains in proteins of our muscles, enzymes, and hormones such as vasopressin formula: ($C_{46}H_{65}N_{15}O_{12}S_2$), as well as the nucleic acids of DNA and RNA. Most of the hydrogen and oxygen is found in water, which makes up 55–60% of our body mass.

The macrominerals—Ca, P, K, Cl, S, Na, and Mg—are located in Period 3 and Period 4 of the periodic table. They are involved as positive or negative ions in the formation of bones and teeth, maintenance of heart and blood vessels, muscle contractions, nerve impulses, acid–base balance of body fluids, and regulation of cellular metabolism. The macrominerals are present in lower amounts than the major elements, so smaller amounts are required in our daily diets.

The other essential elements, called microminerals or trace elements, are mostly transition elements in Period 4, along with Mo and I in Period 5. They are present in the human body in very small amounts, usually less than 100 mg. Over the years, the detection of such small amounts has improved to where researchers can identify more of the roles of trace elements. Some trace elements, such as arsenic, chromium, and selenium, are toxic at high levels in the body but are still required by the body in very small amounts. Other elements, such as tin and nickel, are thought to be essential, but their metabolic role has not yet been determined. Some examples and the amounts present in a 60-kg person are listed in Table 4.4.

TABLE 4.4 Typical Amounts of Essential Elements in a 60-kg Adult

Element	Quantity	Function
Major Elements		
Oxygen (O)	39 kg	Building block of biomolecules and water (H_2O)
Carbon (C)	11 kg	Building block of organic and biomolecules
Hydrogen (H)	6 kg	Component of biomolecules, water (H_2O), and pH of body fluids, stomach acid (HCl)
Nitrogen (N)	1.5 kg	Component of proteins and nucleic acids
Macrominerals		
Calcium (Ca)	1000 g	Bone and teeth, muscle contraction, nerve impulses
Phosphorus (P)	600 g	Bone and teeth, nucleic acids, ATP
Potassium (K)	120 g	Most abundant positive ion (K^+) in cells, muscle contraction, nerve impulses
Chlorine (Cl)	100 g	Most abundant negative ion (Cl^-) in fluids outside cells, stomach acid (HCl)
Sulfur (S)	86 g	Proteins, liver, vitamin B_1, insulin
Sodium (Na)	60 g	Most abundant positive ion (Na^+) in fluids outside cells, water balance, muscle contraction, nerve impulses
Magnesium (Mg)	36 g	Bone, required for metabolic reactions
Microminerals (trace elements)		
Iron (Fe)	3600 mg	Component of oxygen carrier hemoglobin
Silicon (Si)	3000 mg	Growth and maintenance of bone and teeth, tendons and ligaments, hair and skin
Zinc (Zn)	2000 mg	Metabolic reactions in cells, DNA synthesis, growth of bone, teeth, connective tissue, immune system
Copper (Cu)	240 mg	Blood vessels, blood pressure, immune system
Manganese (Mn)	60 mg	Bone growth, blood clotting, necessary for metabolic reactions
Iodine (I)	20 mg	Proper thyroid function
Molybdenum (Mo)	12 mg	Needed to process Fe and N from diets
Arsenic (As)	3 mg	Growth and reproduction
Chromium (Cr)	3 mg	Maintenance of blood sugar levels, synthesis of biomolecules
Cobalt (Co)	3 mg	Vitamin B_{12}, red blood cells
Selenium (Se)	2 mg	Immune system, health of heart and pancreas
Vanadium (V)	2 mg	Formation of bone and teeth, energy from food

	1 Group 1A																	18 Group 8A
1	1 H	2 Group 2A											13 Group 3A	14 Group 4A	15 Group 5A	16 Group 6A	17 Group 7A	2 He
2	3 Li	4 Be											5 B	6 C	7 N	8 O	9 F	10 Ne
3	11 Na	12 Mg	3 3B	4 4B	5 5B	6 6B	7 7B	8	9 8B	10	11 1B	12 2B	13 Al	14 Si	15 P	16 S	17 Cl	18 Ar
4	19 K	20 Ca	21 Sc	22 Ti	23 V	24 Cr	25 Mn	26 Fe	27 Co	28 Ni	29 Cu	30 Zn	31 Ga	32 Ge	33 As	34 Se	35 Br	36 Kr
5	37 Rb	38 Sr	39 Y	40 Zr	41 Nb	42 Mo	43 Tc	44 Ru	45 Rh	46 Pd	47 Ag	48 Cd	49 In	50 Sn	51 Sb	52 Te	53 I	54 Xe
6	55 Cs	56 Ba	57* La	72 Hf	73 Ta	74 W	75 Re	76 Os	77 Ir	78 Pt	79 Au	80 Hg	81 Tl	82 Pb	83 Bi	84 Po	85 At	86 Rn
7	87 Fr	88 Ra	89† Ac	104 Rf	105 Db	106 Sg	107 Bh	108 Hs	109 Mt	110 Ds	111 Rg	112 —	113 —	114 —	115 —	116 —		118 —

☐ Major elements in human body   ☐ Macrominerals   ☐ Microminerals (trace elements)

QUESTIONS AND PROBLEMS

The Periodic Table

4.7 Identify the group or period number described by each of the following statements:
a. contains the elements C, N, and O
b. begins with helium
c. the alkali metals
d. ends with neon

4.8 Identify the group or period number described by each of the following statements:
a. contains Na, K, and Rb
b. the row that begins with Li
c. the noble gases
d. contains F, Cl, Br, and I

4.9 Classify each of the following as an alkali metal, alkaline earth metal, transition element, halogen, or noble gas:
a. Ca   **b.** Fe   **c.** Xe
d. K   **e.** Cl

4.10 Classify each of the following as an alkali metal, alkaline earth metal, transition element, halogen, or noble gas:
a. Ne   **b.** Mg   **c.** Cu
d. Br   **e.** Ba

4.11 Give the symbol of the element described by each of the following:
a. Group 4A (14), Period 2
b. a noble gas in Period 1
c. an alkali metal in Period 3
d. Group 2A (2), Period 4
e. Group 3A (13), Period 3

4.12 Give the symbol of the element described by each of the following:
a. an alkaline earth metal in Period 2
b. Group 5A (15), Period 3
c. a noble gas in Period 4
d. a halogen in Period 5
e. Group 4A (14), Period 4

4.13 Is each of the following elements a metal, metalloid, or nonmetal?
a. calcium
b. sulfur
c. a shiny element
d. an element that is a gas at room temperature
e. located in Group 8A (18)
f. bromine
g. boron
h. silver

4.14 Is each of the following elements a metal, metalloid, or nonmetal?
a. located in Group 2A (2)
b. a good conductor of electricity
c. chlorine
d. arsenic
e. an element that is not shiny
f. oxygen
g. nitrogen
h. tin

4.3 The Atom

All the elements listed on the periodic table are made up of atoms. An **atom** is the smallest particle of an element that retains the characteristics of that element. You have probably seen the element aluminum. Imagine that you are tearing a piece of aluminum foil into smaller and smaller pieces. Now imagine that you have a piece so small that you cannot tear it down further. Then you would have a single atom of aluminum.

LEARNING GOAL

Describe the electrical charge and location in an atom for a proton, a neutron, and an electron.

SELF STUDY ACTIVITY
Atoms and Isotopes

TUTORIAL
The Anatomy of Atoms

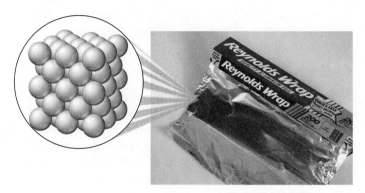

Aluminum foil consists of atoms of aluminum.

The concept of the atom is relatively recent. Although the Greek philosophers in 500 B.C.E. reasoned that everything must contain minute particles they called *atomos,* the idea of atoms did not become a scientific theory until 1808. Then John Dalton (1766–1844) developed an atomic theory that proposed that atoms were responsible for the combinations of elements found in compounds.

Dalton's Atomic Theory

1. All matter is made up of tiny particles called atoms.

2. All atoms of a given element are identical to one another and different from atoms of other elements.

3. Atoms of two or more different elements combine to form compounds. A particular compound is always made up of the same kinds of atoms and the same number of each kind of atom.

4. A chemical reaction involves the rearrangement, separation, or combination of atoms. Atoms are never created or destroyed during a chemical reaction.

Dalton's atomic theory formed the basis of current atomic theory, although we have modified some of Dalton's statements. We now know that atoms of the same element are not completely identical to each other and consist of even smaller particles. However, an atom is still the smallest particle that retains the properties of an element.

Although atoms are the building blocks of everything we see around us, we cannot see an atom or even a billion atoms with the naked eye. However, when billions and billions of atoms are packed together, the characteristics of each atom are added to those of the next until we can see the characteristics we associate with the element. For example, a small piece of the shiny silvery-colored element we call nickel consists of many, many nickel atoms. A special kind of microscope called a *scanning tunneling microscope* (STM) produces images of individual atoms (see Figure 4.7).

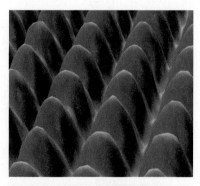

FIGURE 4.7 Images of nickel atoms are produced when nickel is magnified millions of times by a scanning tunneling microscope (STM).

Q Why is a microscope with extremely high magnification needed to see atoms?

FIGURE 4.8 Like charges repel, and unlike charges attract.

Q Why are the electrons attracted to the protons in the nucleus of an atom?

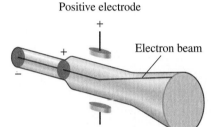

Positive electrode

Electron beam

Cathode ray tube

Negatively charged cathode rays (electrons) are attracted to the positive electrode.

TUTORIAL
Atomic Structure and Properties of Subatomic Particles

Electrical Charges in an Atom

By the end of the 1800s, experiments with electricity showed that atoms were not solid spheres but were composed of even smaller bits of matter, called **subatomic particles**, three of which are the *proton, neutron, and electron.* Two of these subatomic particles were discovered because they have electrical charges.

An electrical charge can be positive or negative. Experiments show that like charges repel, or push away from each other. When you brush your hair on a dry day, electrical charges that are alike build up on the brush and in your hair. As a result, your hair flies away from the brush. Opposite or unlike charges attract. The crackle of clothes taken from the clothes dryer indicates the presence of electrical charges. The clinginess of the clothing results from the attraction of opposite, unlike charges, as shown in Figure 4.8.

Structure of the Atom

In 1897, J. J. Thomson, an English physicist, applied electricity to a glass tube, which produced streams of small particles called *cathode rays*. Because these rays were attracted to a positively charged electrode, Thomson realized that the particles in the rays must be negatively charged. In further experiments, these particles called **electrons** were found to be much smaller than the atom and to have an extremely small mass. Because atoms are neutral, scientists soon discovered that atoms contained positively charged particles called **protons** that were much heavier than the electrons.

Thomson proposed a "plum-pudding" model for the atom in which the electrons and protons were randomly distributed through the atom. In 1911, Ernest Rutherford worked with Thomson to test this model. In Rutherford's experiment, positively charged particles were aimed at a thin sheet of gold foil (see Figure 4.9). If the Thomson model were correct, the particles would travel in straight paths through the gold foil. Rutherford was greatly surprised to find that some of the particles were deflected slightly as they passed through the gold foil, and a few particles were deflected so much that they went back in the opposite direction.

According to Rutherford, it was as though he had shot a cannonball at a piece of tissue paper and it bounced back at him. Rutherford realized that the protons must be contained in a small, positively charged region at the center of the atom, which he called the **nucleus**. He proposed that the electrons in the atom occupy the space surrounding the nucleus through which most of the particles traveled undisturbed. Only the particles that came near this dense, positive center were deflected. If an atom were the size of a football stadium, the nucleus would be about the size of a golf ball placed in the center of the field.

Scientists knew that the nucleus was heavier than the mass of the protons, so they looked for another subatomic particle. Eventually, they discovered that the nucleus also contained a particle that is neutral, which they called a **neutron**. Thus, the masses of the protons and neutrons in the nucleus determine its mass (see Figure 4.10).

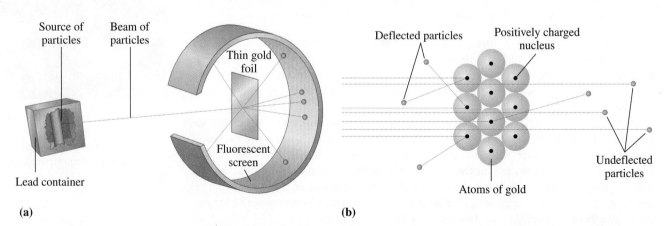

(a) **(b)**

FIGURE 4.9 **(a)** Positive particles are aimed at a piece of gold foil. **(b)** Particles that come close to the atomic nuclei are deflected from their straight path.

Q Why are some particles deflected while most pass through the gold foil undeflected?

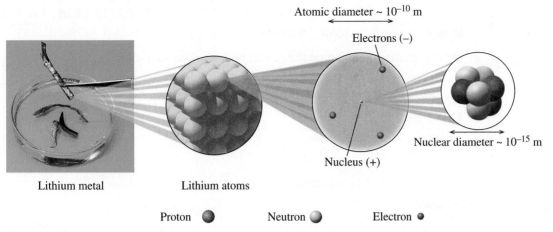

Proton ●   Neutron ●   Electron ●

FIGURE 4.10 In an atom, the protons (positive charge) and neutrons (neutral) that make up almost all the mass of the atom are packed into the tiny volume of the nucleus. The rapidly moving electrons (negative charge) surround the nucleus and account for the large volume of the atom.

Q Why do scientists consider the atom to be mostly empty space?

Mass of the Atom

All of the subatomic particles are extremely small compared with the things you see around you. One proton has a mass of 1.673×10^{-24} g, and the neutron is about the same. The mass of the electron is 9.110×10^{-28} g, which is much less than either a proton or neutron. Because the masses of subatomic particles are so small, chemists use a unit called an **atomic mass unit (amu)**. An amu is defined as one-twelfth of the mass of the carbon atom with 6 protons and 6 neutrons, a standard with which the mass of every other atom is compared. In biology, the atomic mass unit is called a *dalton* in honor of John Dalton. On the amu scale, the proton and neutron each have a mass of about 1 amu. By comparison, the mass of a proton or neutron is nearly 2000 times greater than that of an electron. Because the electron mass is so small, it is usually ignored in atomic mass calculations. Table 4.5 summarizes some information about the subatomic particles in an atom.

Thomson proposed the "plum-pudding" model of the atom.

TABLE 4.5 Subatomic Particles in the Atom

Particle	Symbol	Relative Charge	Mass (g)	Mass (amu)	Location in Atom
Proton	p or p^+	1+	1.673×10^{-24}	1.007	Nucleus
Neutron	n or n^0	0	1.675×10^{-24}	1.008	Nucleus
Electron	e^-	1−	9.110×10^{-28}	0.000 55	Outside nucleus

CONCEPT CHECK 4.3

■ **Identifying Subatomic Particles**

Is each of the following statements true or false?

a. Protons are heavier than electrons.
b. Protons are attracted to neutrons.
c. Electrons are so small that they have no electrical charge.
d. The nucleus contains all the protons and neutrons of an atom.

ANSWER

a. True
b. False; protons are attracted to electrons.
c. False; electrons have a 1− charge.
d. True

SAMPLE PROBLEM | 4.5

■ Identifying Subatomic Particles

Identify the subatomic particle that has the following characteristics:

a. no charge
b. a mass of 0.000 55 amu
c. a mass about the same as a neutron

SOLUTION

a. neutron   **b.** electron   **c.** proton

STUDY CHECK

Give the symbol, electrical charge, and location of a proton in an atom.

QUESTIONS AND PROBLEMS

The Atom

4.15 Identify each description as either a proton, neutron, or electron.
  a. has the smallest mass
  b. carries a 1+ charge
  c. is found outside the nucleus
  d. is electrically neutral

4.16 Identify each description as either a proton, neutron, or electron.
  a. has a mass about the same as a proton
  b. is found in the nucleus
  c. is attracted to the protons
  d. has a 1− charge

4.17 What did Rutherford determine about the structure of the atom from his gold-foil experiment?

4.18 Why does the nucleus in every atom have a positive charge?

4.19 Is each of the following statements true or false?
  a. A proton and an electron have opposite charges.
  b. The nucleus contains most of the mass of an atom.
  c. Electrons repel each other.
  d. A proton is attracted to a neutron.

4.20 Is each of the following statements true or false?
  a. A proton is attracted to an electron.
  b. A neutron has twice the mass of a proton.
  c. Neutrons repel each other.
  d. Electrons and neutrons have opposite charges.

4.21 On a dry day, your hair flies away when you brush it. How would you explain this?

4.22 Sometimes clothes cling together when removed from a dryer. What kinds of charges are on the clothes?

4.4 Atomic Number and Mass Number

LEARNING GOAL

Given the atomic number and the mass number of an atom, state the number of protons, neutrons, and electrons.

TUTORIAL
Atomic Number and Mass Number

All of the atoms of the same element always have the same number of protons. This feature distinguishes atoms of one element from atoms of all the other elements.

Atomic Number

An **atomic number**, which is equal to the number of protons in the nucleus of an atom, is used to identify and define each element.

Atomic number = number of protons in an atom

On the inside front cover of this text is a periodic table, which gives all of the elements in order of increasing atomic number. The atomic number is the whole number that appears above the symbol for each element. For example, a hydrogen atom, with atomic number 1, has one proton; a lithium atom, with atomic number 3, has three protons; an atom of carbon, with atomic number 6, has six protons; and gold, with atomic number 79, has 79 protons; and so forth.

An atom is electrically neutral. That means that the number of protons in an atom is equal to the number of electrons. This electrical balance gives an atom an overall charge of zero. Thus, in every atom, the atomic number also gives the number of electrons.

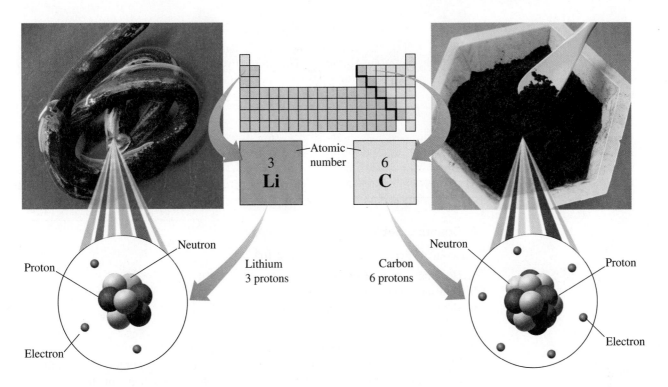

All atoms of lithium (left) contain three protons, and all atoms of carbon (right) contain six protons.

SAMPLE PROBLEM | 4.6

■ **Using the Atomic Number to Find the Number of Protons and Electrons**

Using the periodic table, state the atomic number, number of protons, and number of electrons for an atom of each of the following elements:

a. nitrogen   **b.** magnesium   **c.** bromine

SOLUTION

a. atomic number 7; seven protons and seven electrons
b. atomic number 12; 12 protons and 12 electrons
c. atomic number 35; 35 protons and 35 electrons

STUDY CHECK

Consider an atom that has 79 electrons.

a. How many protons are in its nucleus?
b. What is its atomic number?
c. What is its name, and what is its symbol?

Mass Number

We now know that the protons and neutrons determine the mass of the nucleus. For any atom, the **mass number** is the sum of the number of protons and neutrons in the nucleus.

Mass number = number of protons + number of neutrons

For example, an atom of oxygen that contains eight protons and eight neutrons has a mass number of 16. An atom of iron that contains 26 protons and 30 neutrons has a mass number of 56. Table 4.6 illustrates the relationship between atomic number, mass number, and the number of protons, neutrons, and electrons in some atoms of different elements.

TABLE 4.6 Composition of Some Atoms of Different Elements

Element	Symbol	Atomic Number	Mass Number	Number of Protons	Number of Neutrons	Number of Electrons
Hydrogen	H	1	1	1	0	1
Nitrogen	N	7	14	7	7	7
Chlorine	Cl	17	37	17	20	17
Iron	Fe	26	56	26	30	26
Gold	Au	79	197	79	118	79

CONCEPT CHECK 4.4

■ **Counting Subatomic Particles in Atoms**

An atom of silver has a mass number of 109.

a. How many protons are in the nucleus?
b. How many neutrons are in the nucleus?
c. How many electrons are in the atom?

ANSWER

a. Silver (Ag) in Period 5 with atomic number 47 has 47 protons.
b. The number of neutrons is calculated by subtracting the number of protons from the mass number. $109 - 47 = 62$ neutrons for Ag with a mass number of 109
c. An atom is neutral, which means that the number of electrons is equal to the number of protons. An atom of silver with 47 protons has 47 electrons.

SAMPLE PROBLEM | **4.7**

■ **Calculating Numbers of Protons, Neutrons, and Electrons**

For an atom of zinc that has a mass number of 68, determine the following:

a. the number of protons   **b.** the number of neutrons
c. the number of electrons

SOLUTION

a. On the periodic table, the atomic number of zinc is 30. A zinc atom has 30 protons.
b. The number of neutrons in this atom is found by subtracting the atomic number from the mass number. The number of neutrons is 38.

$$\text{Mass number} - \text{atomic number} = \text{number of neutrons}$$
$$68 \quad - \quad 30 \quad = \quad 38$$

c. Because an atom is neutral, the number of electrons is equal to the number of protons. A zinc atom has 30 electrons.

STUDY CHECK

How many neutrons are in the nucleus of a bromine atom that has a mass number of 80?

QUESTIONS AND PROBLEMS

Atomic Number and Mass Number

4.23 Indicate whether you would you use atomic number, mass number, or both to obtain the following:
    a. number of protons in an atom
    b. number of neutrons in an atom
    c. number of particles in the nucleus
    d. number of electrons in a neutral atom

4.24 What do you know about the subatomic particles from each of the following?
    a. atomic number
    b. mass number
    c. mass number − atomic number
    d. mass number + atomic number

4.25 Write the name and symbol of the element with each of the following atomic numbers:

  a. 3   **b.** 9
  c. 20   **d.** 30
  e. 10   **f.** 14
  g. 53   **h.** 8

4.26 Write the name and symbol of the element with each of the following atomic numbers:

  a. 1   **b.** 11
  c. 19   **d.** 82
  e. 35   **f.** 47
  g. 15   **h.** 2

4.27 How many protons and electrons are there in a neutral atom of each of the following elements?

  a. argon   **b.** zinc
  c. iodine   **d.** potassium

4.28 How many protons and electrons are there in a neutral atom of each of the following elements?

  a. carbon   **b.** fluorine
  c. calcium   **d.** sulfur

4.29 Complete the following table for a neutral atom of each element:

Name of Element	Symbol	Atomic Number	Mass Number	Number of Protons	Number of Neutrons	Number of Electrons
	Al		27			
		12			12	
Potassium					20	
				16	15	
			56			26

4.30 Complete the following table for a neutral atom of each element:

Name of Element	Symbol	Atomic Number	Mass Number	Number of Protons	Number of Neutrons	Number of Electrons
	N		15			
Calcium			42			
				38	50	
		14			16	
		56	138			

4.5 Isotopes and Atomic Mass

We have seen that all atoms of the same element have the same number of protons and electrons. However, the atoms of any one element are not completely identical, because they can have different numbers of neutrons.

Atoms and Isotopes

Isotopes are atoms of the same element that have different numbers of neutrons. For example, all atoms of the element magnesium (Mg) have 12 protons. However, some naturally occurring magnesium atoms have 12 neutrons, others have 13 neutrons, and still others have 14 neutrons. The differences in numbers of neutrons for these magnesium atoms cause their mass numbers to be different but not their chemical behavior. The three isotopes of magnesium have the same atomic number but different mass numbers.

On the periodic table, the atomic number appears above the element symbol. To distinguish between the different isotopes of an element, we use an **atomic symbol**, which has the mass number in the upper left corner and the atomic number in the lower left corner.

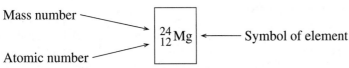

Atomic symbol for an isotope of magnesium

An isotope may be referred to by its name or symbol followed by the mass number, such as magnesium-24 or Mg-24. Magnesium has three naturally occurring isotopes, as shown in Table 4.7.

LEARNING GOAL

Give the number of protons, electrons, and neutrons in an isotope of an element; calculate the atomic mass of an element.

SELF STUDY ACTIVITY
Atoms and Isotopes

TUTORIAL
Isotopes

TABLE 4.7 Isotopes of Magnesium

Atomic symbol	$^{24}_{12}Mg$	$^{25}_{12}Mg$	$^{26}_{12}Mg$
Number of protons	12	12	12
Number of electrons	12	12	12
Mass number	**24**	**25**	**26**
Number of neutrons	**12**	**13**	**14**
Mass of isotope (amu)	**23.99**	**24.99**	**25.98**
% abundance	78.70%	10.13%	11.17%

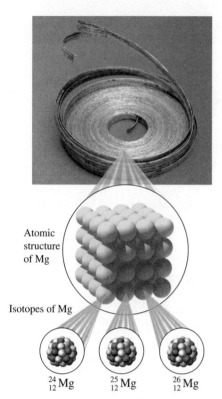

Atomic structure of Mg

Isotopes of Mg

$^{24}_{12}Mg$ $^{25}_{12}Mg$ $^{26}_{12}Mg$

The nuclei of three naturally occurring magnesium isotopes have different numbers of neutrons.

SAMPLE PROBLEM | 4.8

■ Identifying Protons and Neutrons in Isotopes

State the number of protons and neutrons in each of the following isotopes of neon (Ne):

a. $^{20}_{10}Ne$ **b.** $^{21}_{10}Ne$ **c.** $^{22}_{10}Ne$

SOLUTION

The atomic number of Ne is 10, which means that the nucleus of each isotope has 10 protons. The number of neutrons is found by subtracting the atomic number (10) from each of the mass numbers.

a. 10 protons; 10 neutrons $(20 - 10)$
b. 10 protons; 11 neutrons $(21 - 10)$
c. 10 protons; 12 neutrons $(22 - 10)$

STUDY CHECK

Write a symbol for each of the following isotopes:

a. a nitrogen atom with eight neutrons
b. an atom with 20 protons and 22 neutrons
c. an atom with mass number 27 and 14 neutrons

Atomic Mass

In laboratory work, a scientist uses samples that contain many atoms of an element. Among those atoms are all the various isotopes with their different masses. To obtain a convenient mass to work with, chemists use the mass of an "average atom" of each element. This average atom has an **atomic mass**, which is the *weighted average* of the masses of all of the naturally occurring isotopes of that element. On the periodic table, the atomic mass is given below the symbol of each element.

Most elements consist of several isotopes, which is one reason that the atomic masses on the periodic table are seldom whole numbers. For example, a sample of chlorine atoms consists of two isotopes, $^{35}_{17}Cl$ and $^{37}_{17}Cl$. With an atomic mass of 35.45 amu, there is a higher percentage of $^{35}_{17}Cl$ atoms than of $^{37}_{17}Cl$ atoms. In fact, there are about three atoms of $^{35}_{17}Cl$ for every atom of $^{37}_{17}Cl$ in a sample of chlorine atoms.

Calculating Atomic Mass

TUTORIAL
Atomic Mass Calculations

To calculate the atomic mass of an element, the percentage of each isotope and the mass of each isotope must be determined experimentally. For example, a sample of chlorine atoms consists of 75.76% of $^{35}_{17}Cl$ atoms and 24.24% of $^{37}_{17}Cl$ atoms. The atomic mass is calculated using the percentage of each isotope and its mass: $^{35}_{17}Cl$ has a mass of 34.97 amu, and $^{37}_{17}Cl$ has a mass of 36.97 amu.

$$\text{Atomic mass of Cl} = \text{mass } ^{35}_{17}Cl \times \frac{^{35}_{17}Cl\%}{100\%} + \text{mass } ^{37}_{17}Cl \times \frac{^{37}_{17}Cl\%}{100\%}$$

amu from $^{35}_{17}Cl$ amu from $^{37}_{17}Cl$

Isotope	Mass (amu)	×	Abundance (%)	=	Contribution to Average Cl Atom
$^{35}_{17}Cl$	34.97	×	$\dfrac{75.76}{100}$	=	26.49 amu
$^{37}_{17}Cl$	36.97	×	$\dfrac{24.24}{100}$	=	8.962 amu
			Atomic mass of Cl	=	35.45 amu

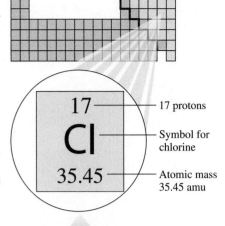

17 ——— 17 protons
Cl ——— Symbol for chlorine
35.45 ——— Atomic mass 35.45 amu

The atomic mass of 35.45 amu is the weighted average mass of a sample of Cl atoms, although no individual Cl atom actually has this mass.

Table 4.8 lists the naturally occurring isotopes of selected elements and their atomic masses.

TABLE 4.8 The Atomic Mass of Some Elements

Element	Isotopes	Atomic Mass (Weighted Average)
Lithium	$^{6}_{3}Li$, $^{7}_{3}Li$	6.941 amu
Carbon	$^{12}_{6}C$, $^{13}_{6}C$, $^{14}_{6}C$	12.01 amu
Oxygen	$^{16}_{8}O$, $^{17}_{8}O$, $^{18}_{8}O$	16.00 amu
Fluorine	$^{19}_{9}F$	19.00 amu
Sulfur	$^{32}_{16}S$, $^{33}_{16}S$, $^{34}_{16}S$, $^{36}_{16}S$	32.07 amu
Copper	$^{63}_{29}Cu$, $^{65}_{29}Cu$	63.55 amu

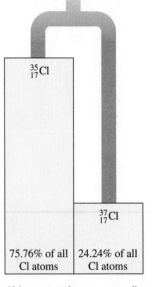

$^{35}_{17}Cl$

$^{37}_{17}Cl$

75.76% of all Cl atoms   24.24% of all Cl atoms

Chlorine, with two naturally occurring isotopes, has an atomic mass of 35.45.

CONCEPT CHECK 4.5

■ **Average Atomic Mass**

Carbon consists of three naturally occurring isotopes: $^{12}_{6}C$, $^{13}_{6}C$, and $^{14}_{6}C$. Using the atomic mass on the periodic table, which isotope of carbon is the most prevalent?

ANSWER

Using the periodic table, we find that the atomic mass for all the naturally occurring isotopes of carbon is 12.01 amu. Thus, the isotope $^{12}_{6}C$ must be the most prevalent isotope in a carbon sample.

SAMPLE PROBLEM   4.9

■ **Calculating Atomic Mass**

Using Table 4.7, calculate the atomic mass for magnesium.

SOLUTION

$^{24}_{12}Mg$	23.99	×	$\dfrac{78.70}{100}$	=	18.88 amu
$^{25}_{12}Mg$	24.99	×	$\dfrac{10.13}{100}$	=	2.531 amu
$^{26}_{12}Mg$	25.98	×	$\dfrac{11.17}{100}$	=	2.902 amu
	Atomic mass of Mg			=	24.31 amu

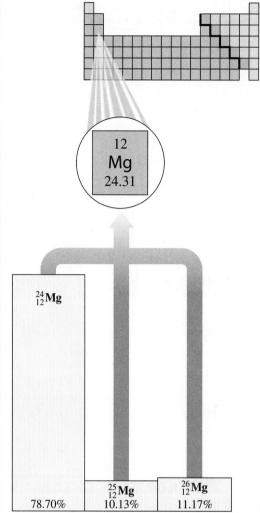

Magnesium, with three naturally occurring isotopes, has an atomic mass of 24.31.

STUDY CHECK

There are two naturally occurring isotopes of boron. The isotope $^{10}_{5}B$ has a mass of 10.01 amu with an abundance of 19.80%, and $^{11}_{5}B$ has a mass of 11.01 amu with an abundance of 80.20%. What is the atomic mass of boron?

QUESTIONS AND PROBLEMS

Isotopes and Atomic Mass

4.31 What are the number of protons, neutrons, and electrons in the following isotopes?
    a. $^{27}_{13}Al$ **b.** $^{52}_{24}Cr$ **c.** $^{34}_{16}S$ **d.** $^{81}_{35}Br$

4.32 What are the number of protons, neutrons, and electrons in the following isotopes?
    a. $^{2}_{1}H$ **b.** $^{14}_{7}N$ **c.** $^{26}_{14}Si$ **d.** $^{70}_{30}Zn$

4.33 Write the atomic symbols for isotopes with the following characteristics:
    a. 15 protons and 16 neutrons
    b. 35 protons and 45 neutrons
    c. 50 electrons and 72 neutrons
    d. a chlorine atom with 18 neutrons
    e. a mercury atom with 122 neutrons

4.34 Write the atomic symbols for isotopes with the following characteristics:
    a. an oxygen atom with 10 neutrons
    b. four protons and five neutrons
    c. 25 electrons and 28 neutrons
    d. a mass number of 24 and 13 neutrons
    e. a nickel atom with 32 neutrons

4.35 There are three naturally occurring isotopes of argon, with mass numbers 36, 38, and 40.
    a. Write the atomic symbol for each of these atoms.
    b. How are these isotopes alike?
    c. How are they different?
    d. Why is the atomic mass of argon listed on the periodic table not a whole number?
    e. Which isotope is the most abundant in a sample of argon?

4.36 There are four naturally occurring isotopes of strontium, with mass numbers 84, 86, 87, and 88.
  a. Write the atomic symbol for each of these atoms.
  b. How are these isotopes alike?
  c. How are they different?
  d. Why is the atomic mass of strontium listed on the periodic table not a whole number?
  e. Which isotope is the most abundant in a sample of strontium?

4.37 What is the difference between the mass of an isotope and the atomic mass of an element?

4.38 What is the difference between the mass number and atomic mass of the element?

4.39 Copper consists of two isotopes, $^{63}_{29}Cu$ and $^{65}_{29}Cu$. If the atomic mass for copper on the periodic table is 63.55, are there more atoms of $^{63}_{29}Cu$ or $^{65}_{29}Cu$ in a sample of copper?

4.40 A fluorine sample consists of only one type of atom, $^{19}_{9}F$, which has a mass of 19.00 amu. How would the mass of a $^{19}_{9}F$ atom compare to the atomic mass listed on the periodic table?

4.41 There are three naturally occurring isotopes of neon: $^{20}_{10}Ne$, $^{21}_{10}Ne$, and $^{22}_{10}Ne$. Use the atomic mass of neon listed on the periodic table to identify the most abundant isotope.

4.42 Zinc consists of five naturally occurring isotopes: $^{64}_{30}Zn$, $^{66}_{30}Zn$, $^{67}_{30}Zn$, $^{68}_{30}Zn$, and $^{70}_{30}Zn$. None of these isotopes has the atomic mass of 65.41 listed for zinc on the periodic table. Explain.

4.43 Two isotopes of gallium are naturally occurring, with $^{69}_{31}Ga$ at 60.11% (68.93 amu) and $^{71}_{31}Ga$ at 39.89% (70.92 amu). What is the atomic mass of gallium?

4.44 Two isotopes of copper are naturally occurring, with $^{63}_{29}Cu$ at 69.09% (62.93 amu) and $^{65}_{29}Cu$ at 30.91% (64.93 amu). What is the atomic mass of copper?

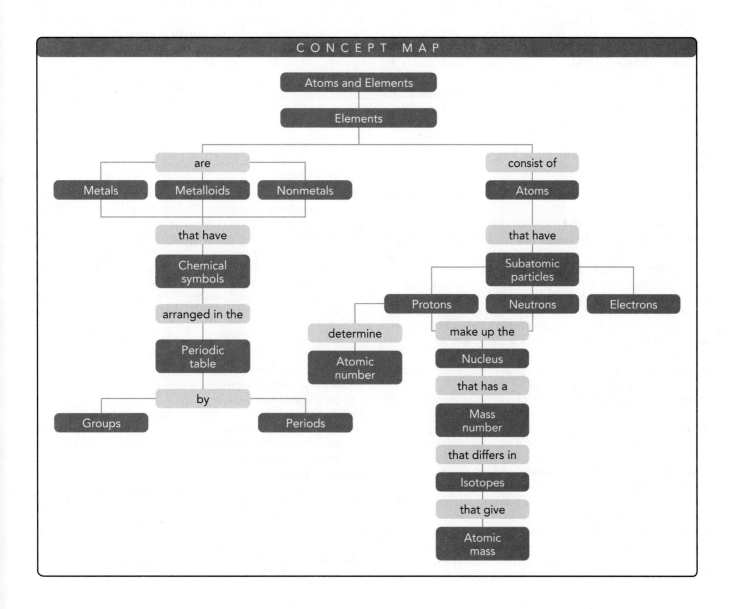

CONCEPT MAP

CHAPTER REVIEW

4.1 Elements and Symbols

LEARNING GOAL: *Given the name of an element, write its correct symbol; from the symbol, write the correct name.*

Elements are the primary substances of matter. Chemical symbols are one- or two-letter abbreviations of the names of the elements.

4.2 The Periodic Table

LEARNING GOAL: *Use the periodic table to identify the group and the period of an element and decide whether it is a metal, a metalloid, or a nonmetal.*

The periodic table is an arrangement of the elements by increasing atomic number. A vertical column on the periodic table containing elements with similar properties is called a group. A horizontal row is called a period. Elements in Group 1A (1) are called the alkali metals; Group 2A (2), the alkaline earth metals; Group 7A (17), the halogens; and Group 8A (18), the noble gases. On the periodic table, metals are located on the left of the heavy zigzag line, and nonmetals are to the right of the heavy zigzag line. Except for aluminum, elements located along the heavy zigzag line are called metalloids.

4.3 The Atom

LEARNING GOAL: *Describe the electrical charge and location in an atom for a proton, a neutron, and an electron.*

An atom is the smallest particle that retains the characteristics of an element. Atoms are composed of three types of subatomic particles. Protons have a positive charge ($+$), electrons carry a negative charge ($-$), and neutrons are electrically neutral. The protons and neutrons are found in the tiny, dense nucleus. Electrons are located outside the nucleus.

4.4 Atomic Number and Mass Number

LEARNING GOAL: *Given the atomic number and the mass number of an atom, state the number of protons, neutrons, and electrons.*

The atomic number gives the number of protons in all the atoms of the same element. In a neutral atom, there is an equal number of protons and electrons. The mass number is the total number of protons and neutrons in an atom.

4.5 Isotopes and Atomic Mass

LEARNING GOAL: *Give the number of protons, electrons, and neutrons in an isotope of an element; calculate the atomic mass of an element.*

Atoms that have the same number of protons but different numbers of neutrons are called isotopes. The atomic mass of an element is the weighted average mass of all the isotopes in a naturally occurring sample of that element.

KEY TERMS

alkali metals Elements of Group 1A (1) except hydrogen; these are soft, shiny metals.

alkaline earth metals Group 2A (2) elements.

atom The smallest particle of an element that retains the characteristics of the element.

atomic mass The weighted average mass of all the naturally occurring isotopes of an element.

atomic mass unit (amu) A small mass unit used to describe the mass of very small particles such as atoms and subatomic particles; 1 amu is equal to one-twelfth the mass of a carbon-12 atom.

atomic number A number that is equal to the number of protons in an atom.

atomic symbol An abbreviation used to indicate the mass number and atomic number of an isotope.

chemical symbol An abbreviation that represents the name of an element.

electron A negatively charged subatomic particle having a very small mass that is usually ignored in mass calculations; its symbol is e^-.

group A vertical column in the periodic table that contains elements having similar physical and chemical properties.

halogen Group 7A (17) elements of fluorine, chlorine, bromine, iodine, and astatine.

isotope An atom that differs only in mass number from another atom of the same element. Isotopes have the same atomic number (number of protons) but different numbers of neutrons.

mass number The total number of neutrons and protons in the nucleus of an atom.

metal An element that is shiny, malleable, ductile, and a good conductor of heat and electricity. The metals are located to the left of the heavy zigzag line in the periodic table.

metalloid Elements with properties of both metals and nonmetals located along the heavy zigzag line on the periodic table.

neutron A neutral subatomic particle having a mass of 1 amu and found in the nucleus of an atom; its symbol is n or n^0.

noble gas An element in Group 8A (18) of the periodic table, generally unreactive and seldom found in combination with other elements.

nonmetal An element that is not shiny and is a poor conductor of heat and electricity. The nonmetals are located to the right of the heavy zigzag line in the periodic table.

nucleus The compact, very dense center of an atom, containing the protons and neutrons of the atom.

period A horizontal row of elements in the periodic table.

periodic table An arrangement of elements by increasing atomic number such that elements having similar chemical behavior are grouped in vertical columns.

proton A positively charged subatomic particle having a mass of 1 amu and found in the nucleus of an atom; its symbol is p or p^+.

representative element An element found in Groups 1A (1) through 8A (18) excluding B groups (3–12) of the periodic table.

subatomic particle A particle within an atom; protons, neutrons, and electrons are subatomic particles.

transition element An element located between Groups 2A (2) and 3A (13) on the periodic table.

UNDERSTANDING THE CONCEPTS

4.45 According to Dalton's atomic theory, are each of the following true or false?
  a. Atoms of an element are identical to atoms of other elements.
  b. Every element is made of atoms.
  c. Atoms of two different elements combine to form compounds.
  d. In a chemical reaction, some atoms disappear and new atoms appear.

4.46 Use Rutherford's gold-foil experiment to answer each of the following:
  a. What did Rutherford expect to happen when he aimed particles at the gold foil?
  b. How did the results differ from what he expected?
  c. How did he use the results to propose a new model of the atom?

4.47 Use the subatomic particles (1–3) to define each of the following:
  1. protons   **2.** neutrons   **3.** electrons

  a. atomic mass
  b. atomic number
  c. positive charge
  d. negative charge
  e. mass number – atomic number

4.48 Use the subatomic particles (1–3) to define each of the following:
  1. protons   **2.** neutrons   **3.** electrons

  a. mass number
  b. surround the nucleus
  c. nucleus
  d. charge of 0
  e. equal to number of electrons

4.49 Consider the following atoms in which X represents the chemical symbol of the element:

$$^{16}_{8}X \qquad ^{16}_{9}X \qquad ^{18}_{10}X \qquad ^{17}_{8}X \qquad ^{18}_{8}X$$

  a. What atoms have the same number of protons?
  b. Which atoms are isotopes? Of what element?
  c. Which atoms have the same mass number?
  d. What atoms have the same number of neutrons?

4.50 For each of the following, write the symbol and name for X and the number of protons and neutrons. Which are isotopes of each other?
  a. $^{37}_{17}X$ **b.** $^{116}_{49}X$ **c.** $^{116}_{50}X$
  d. $^{124}_{50}X$ **e.** $^{116}_{48}X$

4.51 Which of the following pairs of atoms have the same number of protons? Neutrons? Electrons?
  a. $^{37}_{17}Cl$, $^{38}_{18}Ar$ **b.** $^{36}_{16}S$, $^{34}_{16}S$
  c. $^{79}_{34}Se$, $^{81}_{35}Br$ **d.** $^{40}_{18}Ar$, $^{39}_{17}Cl$

4.52 Complete the following table for the three naturally occurring isotopes of silicon, the major component in computer chips.

	Isotope		
	$^{28}_{14}Si$	$^{29}_{14}Si$	$^{30}_{14}Si$
Number of protons			
Number of neutrons			
Number of electrons			
Atomic number			
Mass number			

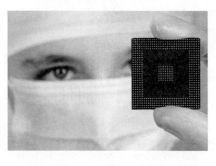

Computer chips consist primarily of the element silicon.

4.53 For each representation of a nucleus, write the atomic symbol and determine which ones are isotopes.

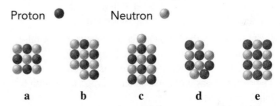

Proton ●   Neutron ◐

  a   b   c   d   e

4.54 Identify the element represented by each nucleus in Problem 4.53 as a metal, metalloid, or nonmetal.

4.55 Provide the following information:
  a. the atomic number and symbol of the lightest alkali metal
  b. the atomic number and symbol of the heaviest noble gas
  c. the atomic mass and symbol of the alkaline earth metal in Period 3
  d. the atomic mass and symbol of the halogen with the fewest electrons

4.56 Provide the following information:
  a. the atomic number and symbol of the heaviest metalloid in Group 4A (14)
  b. the atomic number and symbol of the only metal in Group 5A (15)
  c. the atomic mass and symbol of the alkali metal in Period 4
  d. the atomic mass and symbol of the heaviest halogen

ADDITIONAL QUESTIONS AND PROBLEMS

For instructor-assigned homework, go to
www.masteringchemistry.com.

4.57 Why is Co the symbol for cobalt, not CO?

4.58 Which of the following is correct? Write the correct symbol if needed.
- **a.** copper, Cp
- **b.** silicon, SI
- **c.** iron, Fe
- **d.** fluorine, Fl
- **e.** potassium, P
- **f.** sodium, Na
- **g.** gold, Au
- **h.** lead, PB

4.59 Give the symbol and name of the element found in the following group and period on the periodic table:
- **a.** Group 2A (2), Period 3
- **b.** Group 7A (17), Period 4
- **c.** Group 3A (13), Period 3
- **d.** Group 6A (16), Period 2

4.60 Give the group and period number for the following elements:
- **a.** potassium
- **b.** phosphorus
- **c.** carbon
- **d.** neon

4.61 Write the names of two elements that are in the following groups:
- **a.** halogens
- **b.** noble gases
- **c.** alkali metals
- **d.** alkaline earth metals

4.62 The following are trace elements that have been found to be crucial to the biochemical and physiological processes in the body. Indicate whether each is a metal or nonmetal.
- **a.** zinc
- **b.** cobalt
- **c.** manganese
- **d.** iodine
- **e.** copper
- **f.** selenium

4.63 Indicate if each of the following statements is true or false:
- **a.** The proton is a negatively charged particle.
- **b.** The neutron is 2000 times as heavy as a proton.
- **c.** The atomic mass unit is based on a carbon atom with 6 protons and 6 neutrons.
- **d.** The nucleus is the largest part of the atom.
- **e.** The electrons are located outside the nucleus.

4.64 Indicate if each of the following statements is true or false:
- **a.** The neutron is electrically neutral.
- **b.** Most of the mass of an atom is due to the protons and neutrons.
- **c.** The charge of an electron is equal, but opposite, to the charge of a neutron.
- **d.** The proton and the electron have about the same mass.
- **e.** The mass number is the number of protons.

4.65 Complete the following statements:
- **a.** The atomic number gives the number of _____ in the nucleus.
- **b.** In an atom, the number of electrons is equal to the number of _____.
- **c.** Sodium and potassium are examples of elements called _____.

4.66 Complete the following statements:
- **a.** The number of protons and neutrons in an atom is also the _____ number.
- **b.** The elements in Group 7A (17) are called the _____.
- **c.** Elements that are shiny and conduct heat are called _____.

4.67 Write the name and symbol of the element with the following atomic number:
- **a.** 28
- **b.** 56
- **c.** 88
- **d.** 33
- **e.** 50
- **f.** 55
- **g.** 79
- **h.** 80

4.68 Write the name and symbol of the element with the following atomic number:
- **a.** 10
- **b.** 22
- **c.** 48
- **d.** 26
- **e.** 54
- **f.** 78
- **g.** 83
- **h.** 92

4.69 Give the number of protons and electrons in neutral atoms of each of the following:
- **a.** Mn
- **b.** phosphorus
- **c.** Sr
- **d.** Co
- **e.** uranium

4.70 Give the number of protons and electrons in neutral atoms of each of the following:
- **a.** chromium
- **b.** Cs
- **c.** copper
- **d.** chlorine
- **e.** Cd

4.71 For the following atoms, determine the number of protons, neutrons, and electrons:
- **a.** $^{107}_{47}Ag$
- **b.** $^{98}_{43}Tc$
- **c.** $^{208}_{82}Pb$
- **d.** $^{222}_{86}Rn$
- **e.** $^{136}_{54}Xe$

4.72 For the following atoms, give the number of protons, neutrons, and electrons:
- **a.** $^{22}_{10}Ne$
- **b.** $^{127}_{53}I$
- **c.** $^{75}_{35}Br$
- **d.** $^{133}_{55}Cs$
- **e.** $^{195}_{78}Pt$

4.73 Write the atomic symbol and mass number for each of the following:
- **a.** an atom with four protons and five neutrons
- **b.** an atom with 12 protons and 14 neutrons
- **c.** a calcium atom with a mass number of 46
- **d.** an atom with 30 electrons and 40 neutrons

4.74 Write the atomic symbol and mass number for each of the following:
- **a.** an aluminum atom with 14 neutrons
- **b.** an atom with atomic number 26 and 32 neutrons
- **c.** a strontium atom with 50 neutrons
- **d.** an atom with a mass number of 72 and atomic number 33

4.75 Complete the following table:

Name	Atomic Symbol	Number of Protons	Number of Neutrons	Number of Electrons
	$^{34}_{16}S$			
		28	34	
Magnesium			14	
	$^{220}_{86}Rn$			

4.76 Complete the following table:

Name	Atomic Symbol	Number of Protons	Number of Neutrons	Number of Electrons
Potassium			22	
	$^{51}_{23}V$			
		48	64	
Barium			82	

4.77 The most abundant isotope of gold is Au-197.
- **a.** How many protons, neutrons, and electrons are in this isotope?
- **b.** What is the atomic symbol of another isotope of gold with 116 neutrons?
- **c.** What is the atomic symbol of an atom with atomic number of 78 and 116 neutrons?

4.78 Cadmium, atomic number 48, consists of eight naturally occurring isotopes. Do you expect any of the isotopes to have the atomic mass listed on the periodic table for cadmium? Explain.

4.79 The most abundant isotope of lead is $^{208}_{82}$Pb.
 a. How many protons, neutrons, and electrons are in the atom?
 b. What is the atomic symbol of an isotope of lead with 132 neutrons?
 c. What is the name and atomic symbol of an isotope with the same mass number as in part **b**, and 131 neutrons?

4.80 The most abundant isotope of silver is $^{107}_{47}$Ag.
 a. How many protons, neutrons, and electrons are in the atom?
 b. What is the atomic symbol of an isotope of silver with 62 neutrons?
 c. What is the name and atomic symbol of an isotope with the same mass number as in part **b**, and 61 neutrons?

CHALLENGE QUESTIONS

4.81 Lead consists of four naturally occurring isotopes. Calculate the atomic mass of lead.

Isotope	Mass (amu)	Abundance (%)
$^{204}_{82}$Pb	203.97	1.40
$^{206}_{82}$Pb	205.97	24.10
$^{207}_{82}$Pb	206.98	22.10
$^{208}_{82}$Pb	207.98	52.40

4.82 Indium (In), with an atomic mass of 114.8 amu, consists of two naturally occurring isotopes, $^{113}_{49}$In and $^{115}_{49}$In. If 4.30% of indium is $^{113}_{49}$In, which has a mass of 112.90 amu, what is the mass of the $^{115}_{49}$In?

4.83 Silicon has three isotopes that occur in nature: $^{28}_{14}$Si (27.977 amu) has an abundance of 92.23%, $^{29}_{14}$Si (28.976 amu) has a 4.68% abundance, and $^{30}_{14}$Si (29.974 amu) has a 3.09% abundance. What is the atomic mass of silicon?

4.84 Antimony (Sb), which has an atomic weight of 121.75 amu, has two naturally occurring isotopes: Sb-121 and Sb-123. If a sample of antimony is 42.70% Sb-123, which has a mass of 122.90 amu, what is the mass of Sb-121?

4.85 If the diameter of a sodium atom is 3.14×10^{-8} cm, how many sodium atoms would fit along a line exactly 1 in. long?

4.86 A lead atom has a mass of 3.4×10^{-22} g. How many lead atoms are in a cube of lead that has a volume of 2.00 cm^3 if the density of lead is 11.3 g/cm^3?

ANSWERS

ANSWERS TO STUDY CHECKS

4.1 Si, Se, and Ag

4.2 magnesium, aluminum, and fluorine

4.3 **a.** Strontium is in Group 2A (2).
 b. This is the alkaline earth metals.
 c. Strontium is in Period 5.
 d. Magnesium, Mg
 e. Alkali metal, Rb; halogen, I; noble gas, Xe

4.4 **a.** As **b.** Rn **c.** B

4.5 A proton, symbol p or p^+ with a charge of 1+, is found in the nucleus of an atom.

4.6 **a.** 79 protons **b.** 79 **c.** gold, Au

4.7 45 neutrons

4.8 **a.** $^{15}_{7}$N **b.** $^{42}_{20}$Ca **c.** $^{27}_{13}$Al

4.9 10.81 amu

ANSWERS TO SELECTED QUESTIONS AND PROBLEMS

4.1 **a.** Cu **b.** Pt **c.** Ca
 d. Mn **e.** Fe **f.** Ba
 g. Pb **h.** Sr

4.3 **a.** carbon **b.** chlorine **c.** iodine
 d. mercury **e.** silver **f.** argon
 g. boron **h.** nickel

4.5 **a.** sodium, chlorine
 b. calcium, sulfur, oxygen
 c. carbon, hydrogen, chlorine, nitrogen, oxygen
 d. calcium, carbon, oxygen

4.7 **a.** Period 2 **b.** Group 8A (18)
 c. Group 1A (1) **d.** Period 2

4.9 **a.** alkaline earth metal **b.** transition element
 c. noble gas **d.** alkali metal
 e. halogen

4.11 **a.** C **b.** He **c.** Na
 d. Ca **e.** Al

4.13 **a.** metal **b.** nonmetal **c.** metal
 d. nonmetal **e.** nonmetal **f.** nonmetal
 g. metalloid **h.** metal

4.15 **a.** electron **b.** proton
 c. electron **d.** neutron

4.17 Rutherford determined that an atom contains a small, compact nucleus that is positively charged.

4.19 Answers **a**, **b**, and **c** are true, but **d** is false.

4.21 In the process of brushing hair, strands of hair become charged with like charges that repel each other.

4.23 **a.** atomic number **b.** both
 c. mass number **d.** atomic number

4.25 **a.** lithium, Li **b.** fluorine, F
 c. calcium, Ca **d.** zinc, Zn
 e. neon, Ne **f.** silicon, Si
 g. iodine, I **h.** oxygen, O

4.27 **a.** 18 **b.** 30 **c.** 53 **d.** 19

4.29 See Table 4.13.

4.31 **a.** 13 protons, 14 neutrons, 13 electrons
 b. 24 protons, 28 neutrons, 24 electrons
 c. 16 protons, 18 neutrons, 16 electrons
 d. 35 protons, 46 neutrons, 35 electrons

4.33 **a.** $^{31}_{15}$P **b.** $^{80}_{35}$Br **c.** $^{122}_{50}$Sn
 d. $^{35}_{17}$Cl **e.** $^{202}_{80}$Hg

TABLE 4.13

Name of Element	Symbol	Atomic Number	Mass Number	Number of Protons	Number of Neutrons	Number of Electrons
Aluminum	Al	13	27	13	14	13
Magnesium	Mg	12	24	12	12	12
Potassium	K	19	39	19	20	19
Sulfur	S	16	31	16	15	16
Iron	Fe	26	56	26	30	26

4.35 a. $^{36}_{18}Ar$ $^{38}_{18}Ar$ $^{40}_{18}Ar$
b. They all have the same number of protons and electrons.
c. They have different numbers of neutrons, which gives them different mass numbers.
d. The atomic mass of Ar listed on the periodic table is the weighted average atomic mass of all the naturally occurring isotopes.
e. The isotope Ar-40 is most abundant because its mass is closest to the atomic mass on the periodic table.

4.37 The mass of an isotope is the mass of an individual atom. The atomic mass is the weighted average of all the naturally occurring isotopes of that element.

4.39 Since the atomic mass of copper is closer to 63 amu, there are more atoms of $^{63}_{29}Cu$.

4.41 Since the atomic mass of neon is 20.18 amu, the most abundant isotope is $^{20}_{10}Ne$.

4.43 69.72 amu

4.45 a. false   **b.** true   **c.** true
d. false

4.47 a. 1 and 2   **b.** 1   **c.** 1
d. 3   **e.** 2

4.49 a. $^{16}_{8}X$, $^{17}_{8}X$, $^{18}_{8}X$ all have eight protons.
b. $^{16}_{8}X$, $^{17}_{8}X$, $^{18}_{8}X$ all are isotopes of oxygen.
c. $^{16}_{8}X$ and $^{16}_{9}X$ have mass numbers of 16, whereas $^{18}_{8}X$ and $^{18}_{10}X$ have mass numbers of 18.
d. $^{16}_{8}X$ and $^{18}_{10}X$ both have eight neutrons.

4.51 a. Both have 20 neutrons.
b. Both have 16 protons and 16 electrons.
c. Not the same
d. Both have 22 neutrons.

4.53 a. $^{9}_{4}Be$ **b.** $^{11}_{5}B$ **c.** $^{13}_{6}C$
d. $^{10}_{5}B$ **e.** $^{12}_{6}C$

b and **d** are isotopes of boron; **c** and **e** are isotopes of carbon.

4.55 a. 3, Li   **b.** 86, Rn
c. 24.31 amu, Mg   **d.** 19.00 amu, F

4.57 The first letter of a symbol is a capital, but a second letter is lowercase. The symbol Co is for cobalt, but the symbols in CO are for carbon and oxygen.

4.59 a. Mg, magnesium   **b.** Br, bromine
c. Al, aluminum   **d.** O, oxygen

4.61 a. any two elements in Group 7A (17), such as fluorine, chlorine, bromine, or iodine
b. any two elements in Group 8A (18), such as helium, neon, argon, krypton, xenon, or radon

c. any two elements in Group 1A (1), such as lithium, sodium, and potassium, except hydrogen
d. any two elements in Group 2A (2), such as magnesium, calcium, and barium

4.63 a. false   **b.** false   **c.** true
d. false   **e.** true

4.65 a. protons   **b.** protons
c. alkali metals

4.67 a. nickel, Ni   **b.** barium, Ba
c. radium, Ra   **d.** arsenic, As
e. tin, Sn   **f.** cesium, Cs
g. gold, Au   **h.** mercury, Hg

4.69 a. 25 protons, 25 electrons   **b.** 15 protons, 15 electrons
c. 38 protons, 38 electrons   **d.** 27 protons, 27 electrons
e. 92 protons, 92 electrons

4.71 a. 47 protons, 60 neutrons, 47 electrons
b. 43 protons, 55 neutrons, 43 electrons
c. 82 protons, 126 neutrons, 82 electrons
d. 86 protons, 136 neutrons, 86 electrons
e. 54 protons, 82 neutrons, 54 electrons

4.73 a. $^{9}_{4}Be$ **b.** $^{26}_{12}Mg$ **c.** $^{46}_{20}Ca$
d. $^{70}_{30}Zn$

4.75

Name	Atomic Symbol	Number of Protons	Number of Neutrons	Number of Electrons
Sulfur	$^{34}_{16}S$	16	18	16
Nickel	$^{62}_{28}Ni$	28	34	28
Magnesium	$^{26}_{12}Mg$	12	14	12
Radon	$^{220}_{86}Rn$	86	134	86

4.77 a. 79 protons, 118 neutrons, 79 electrons
b. $^{195}_{79}Au$
c. $^{194}_{78}Pt$

4.79 a. 82 protons, 126 neutrons, 82 electrons
b. $^{214}_{82}Pb$
c. $^{214}_{83}Bi$, bismuth

4.81 207.2 amu

4.83 28.09 amu

4.85 8.09×10^{7} sodium atoms

Electronic Structure and Periodic Trends

5

Visit **www.masteringchemistry.com** for self-study materials and instructor-assigned homework.

"The unique qualities of semiconducting metals make it possible for us to create sophisticated electronic circuits," says Tysen Streib, Global Product Manager, Applied Materials. "Elements from groups 3A (13), 4A (14), and 5A (15) of the periodic table often make good semiconductors because they readily form covalently bonded crystals. When small amounts of impurities are added, free-flowing electrons can travel through the crystal with very little interference. Without these covalent bonds and loosely bound electrons, we wouldn't have any of the microchips that we use in computers, cell phones, and thousands of other devices."

Materials scientists study the chemical properties of materials to find new uses for them in products such as cars, bridges, and clothing. They also develop materials that can be used as superconductors or in integrated-circuit chips and fuel cells. Chemistry is important in materials science because it provides information about structure and composition.

A rainbow forms when light passes through water droplets.

W hen sunlight passes through a prism or a drop of water, the light bends at different angles and separates into many colors. These are the colors we see in a rainbow that forms when sunlight passes through raindrops. Similar colors are also seen when certain elements are heated. You see this effect during a fireworks display. Lithium and strontium make red colors in fireworks, sodium gives yellow, magnesium is used for white, barium makes a green color, copper is blue, and mixtures of strontium and copper make purple.

In this chapter, we will discuss light, which is a tool scientists use to probe the behavior of electrons in atoms. Then we will look at how electrons in atoms are arranged in energy levels and how the behavior of the elements determines their position in the periodic table. Electron arrangement determines the similarity of atoms in a group and how atoms in each group of elements react to form new substances.

After Rutherford concluded that the atom contained a nucleus and a large volume occupied only by electrons, scientists wondered how electrons move about the nucleus. Eventually, they proposed that electrons had different energies and must be arranged in specific energy levels within atoms. Further work by physicists and chemists revealed that electrons behaved both as particles and light waves. Eventually, the modern view of the atom emerged.

5.1 Electromagnetic Radiation

LEARNING GOAL

Compare the wavelength of radiation with its energy.

The wavelength is the distance between adjacent wave peaks.

When we listen to a radio, use a microwave oven, turn on a light, see the colors of a rainbow, or have an X-ray, we are experiencing various forms of **electromagnetic radiation**. All of these types of electromagnetic radiation, including light, consist of energy particles that move as waves of energy.

Wavelength and Frequency

You are probably familiar with the action of waves in the ocean. If you were at a beach, you might notice that the water in each wave rises and falls as the wave comes in to shore. The highest point on the wave is called a peak or a crest. On a calm day, there might be long distances between peaks. However, if there were a storm with a lot of energy, the peaks would be much closer together.

The waves of electromagnetic radiation have peaks, too. The **wavelength** (symbol λ, lambda) is the distance from a peak in one wave to the peak in the next wave (see Figure 5.1). In some types of radiation, the peaks are far apart, while in others, they are close together.

(a) **(b)**

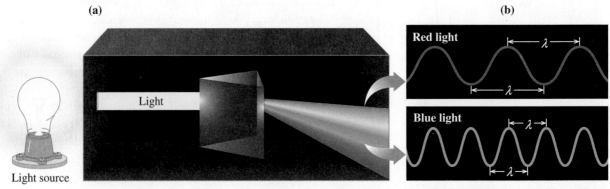

FIGURE 5.1 **(a)** Light passing through a prism is separated into a spectrum of colors we see in a rainbow. **(b)** The wavelength (λ) is the distance from a point in one wave to the same point in the next. Waves with different wavelengths have different frequencies.

Q How does the wavelength of red light compare to that of blue light?

The **frequency** (symbol ν, nu) is the number of waves that pass a certain point in 1 s. All electromagnetic radiation travels at the speed of light (c), 3.00×10^8 m/s, which is equal to the product of the wavelength and frequency. The speed of light is about a million times faster than the speed of sound, which is the reason we see lightning before we hear thunder during an electrical storm.

Speed of light (c) $= 3.00 \times 10^8$ m/s $=$ wavelength (λ) $\times$ frequency (ν)

Electromagnetic Spectrum

The **electromagnetic spectrum** is an arrangement of electromagnetic radiation in order of decreasing wavelengths or increasing frequencies (see Figure 5.2). At one end are the *radio waves* with long wavelengths that are used for AM and FM radio bands, cellular phones, and TV signals. The wavelength of a typical AM radio wave can be as long as a football field. *Microwaves* have shorter wavelengths and higher frequencies than radio waves. Microwaves are used in radar. *Infrared radiation* (IR) is responsible for the heat we feel from sunlight and the heat of infrared lamps used to warm food in restaurants. When we change the volume or the station on a TV set, we use a remote to send infrared impulses to the infrared receiver in the TV. However, a remote must be pointed at a TV set because infrared requires "line-of-sight" positioning and is limited to one-to-one communication. Bluetooth and other wireless technology, which do not require "line-of-sight," use radiation with a lower energy than infrared to connect many electronic devices, including mobile and cordless phones and laptops.

Visible light with wavelengths from 700 to 400 nm is the only light our eyes can detect. Red light has the longest wavelength at 700 nm; orange is about 600 nm; green is about 500 nm; and violet at 400 nm has the shortest wavelength of visible light. We see objects as different colors because the objects reflect only certain wavelengths, which are absorbed by our eyes. We see white light from a lightbulb because all the wavelengths of visible light are mixed together.

Ultraviolet (UV) light has shorter wavelengths and higher frequencies than violet light of the visible range. UV light in sunlight can cause serious sunburn, which may lead to skin cancer. While some UV light from the Sun is blocked by the ozone layer, the cosmetic industry has developed sunscreens to prevent the absorption of UV light by the skin. *X-rays* and *gamma rays* have shorter wavelengths than ultraviolet light, which means they have some of the highest frequencies. X-rays can pass through soft substances but not metals or bone, which is why they are used to scan luggage at airports and to see images of the bones and teeth in the body. Gamma rays are produced by radioactive atoms and in nuclear processes in the Sun and stars. Gamma rays are dangerous because they kill cells in the body, which is the reason gamma rays are used in the treatment of tumors and cancers.

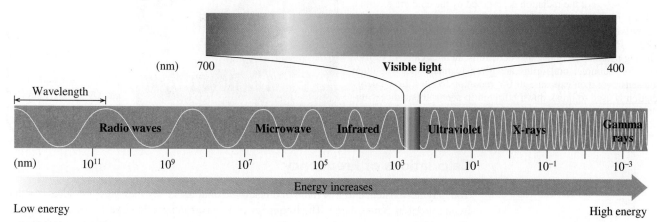

FIGURE 5.2 The electromagnetic spectrum shows the arrangement of wavelengths of electromagnetic radiation. The visible portion consists of wavelengths of 700 nm to 400 nm.

Q How does the wavelength of ultraviolet light compare to that of a microwave?

CONCEPT CHECK 5.1

■ **The Electromagnetic Spectrum**

1. Arrange the following in order of decreasing wavelengths: X-rays, ultraviolet light, FM radio waves, and microwaves.
2. Visible light contains colors from red to violet.
   a. What color has the shortest wavelength?
   b. What color has the lowest frequency?

ANSWER

1. The electromagnetic radiation with the longest wavelength is FM radio waves, then microwaves, followed by ultraviolet light and then X-rays, which have the shortest wavelengths.
2. The shortest wavelengths are in the blue-violet range of the visible light; the longest wavelengths are in the red range.
   a. Violet light has the shortest wavelength.
   b. Red light has the longest wavelength. Because frequency is inversely related to wavelength, red light would also have the lowest frequency.

CHEMISTRY AND HEALTH

Biological Reactions to UV Light

Our everyday life depends on sunlight, but exposure to sunlight can have damaging effects on living cells, and too much exposure can even cause their death. The light energy, especially ultraviolet (UV), excites electrons and may lead to unwanted chemical reactions. The list of damaging effects of sunlight includes sunburn; wrinkling and premature aging of the skin; changes in the DNA of the cells, which can lead to skin cancers; inflammation of the eyes; and perhaps cataracts. Some drugs, like the acne medications Accutane and Retin-A, as well as antibiotics, diuretics, sulfonamides, and estrogen, make the skin extremely sensitive to light.

High-energy radiation is the most damaging biologically. Most of the radiation in this range is absorbed in the epidermis of the skin. The degree to which radiation is absorbed depends on the thickness of the epidermis, the hydration of the skin, the amount of pigments and proteins of the skin, and the arrangement of the blood vessels. In light-skinned people, 85–90% of the radiation is absorbed by the epidermis, with the rest reaching the lower dermis layer. In dark-skinned people, 90–95% of the radiation is absorbed by the epidermis, with a smaller percentage reaching the dermis.

However, medicine does take advantage of the beneficial effect of sunlight. Phototherapy can be used to treat certain skin conditions, including psoriasis, eczema, and dermatitis. In the treatment of psoriasis, for example, oral drugs are given to make the skin more photosensitive; then exposure to UV radiation follows. Low-energy radiation is used to break down bilirubin in neonatal jaundice. Sunlight is also a factor in stimulating the immune system.

In cutaneous T-cell lymphoma, an abnormal increase in T cells causes painful ulceration of the skin. The skin is treated by photophoresis, in which the patient receives a photosensitive chemical, and then blood is removed from the body and exposed to ultraviolet light. The blood is returned to the patient, and the treated T cells stimulate the immune system to respond to the cancer cells.

In a disorder called Seasonal Affective Disorder, or S.A.D., people experience mood swings and depression during the winter. Some research suggests that S.A.D. is result of a decrease in serotonin, or an increase in melatonin, when there are fewer hours of sunlight. One treatment for S.A.D. is therapy using bright light provided by a lamp called a light box. A daily exposure to intense light for 30–60 minutes seems to reduce symptoms of S.A.D.

Blood is irradiated with UV light to treat T-cell lymphoma.

Calculations of Frequency

You may have listened to FM radio and heard the announcer say they were broadcasting at a frequency of 91.5 megahertz. The frequency of an electromagnetic wave is measured in hertz (Hz), which is equal to cycles per s. Thus, one megahertz (MHz) is equal to 1×10^6 Hz.

$$91.5 \; \cancel{\text{MHz}} \times \frac{1 \times 10^6 \; \text{Hz}}{1 \; \cancel{\text{MHz}}} = 91.5 \times 10^6 \; \text{Hz}$$

SAMPLE PROBLEM | 5.1

■ Frequency

A student uses a microwave oven to make popcorn. If the microwaves have a frequency of 2.5×10^9 Hz, what is their frequency in megahertz (MHz)?

SOLUTION

STEP 1   **Given** $\nu = 2.5 \times 10^9$ Hz   **Need** MHz

STEP 2   **Plan**

$$\text{Hz} \xrightarrow{\text{Metric factor}} \text{MHz}$$

STEP 3   **Equalities/Conversion Factors**

$$1 \text{ MHz} = 1 \times 10^6 \text{ Hz}$$

$$\frac{1 \times 10^6 \text{ Hz}}{1 \text{ MHz}} \quad \text{and} \quad \frac{1 \text{ MHz}}{1 \times 10^6 \text{ Hz}}$$

STEP 4   **Set Up Problem**

$$2.5 \times 10^9 \text{ Hz} \times \frac{1 \text{ MHz}}{1 \times 10^6 \text{ Hz}} = 2.5 \times 10^3 \text{ MHz or 2500 MHz}$$

STUDY CHECK

A shortwave radio station broadcasts at 6100 kHz. What is this frequency in megahertz?

A microwave oven uses microwave radiation to heat food.

QUESTIONS AND PROBLEMS

Electromagnetic Radiation

5.1 What is meant by the wavelength of UV light?

5.2 How are the wavelength and frequency of light related?

5.3 What is the difference between "white" light and blue or red light?

5.4 Why can we use X-rays, but not radio waves or microwaves, to give an image of bones and teeth?

5.5 An AM radio station broadcasts news at 650 kHz. What is the frequency of this AM radio wave in hertz?

5.6 An FM radio station broadcasts news at 98.0 MHz. What is the frequency, in hertz, of this radio wave?

5.7 If orange light has a wavelength of 6.3×10^{-5} cm, what is its wavelength in meters and nanometers?

5.8 A wavelength of 850 nm is used for fiber-optic transmission. What is its wavelength in meters?

5.9 Which type of electromagnetic radiation, ultraviolet light, microwaves, or X-rays, has the longest wavelengths?

5.10 Of radio waves, infrared light, and UV light, which has the shortest wavelengths?

5.11 Place the following types of electromagnetic radiation in order of increasing wavelengths: the blue color in a rainbow, X-rays, microwaves from an oven, infrared radiation from a heat lamp.

5.12 Place the following types of electromagnetic radiation in order of decreasing frequencies: AM music station, UV radiation from the Sun, police radar.

5.2 Atomic Spectra and Energy Levels

When the white light from the Sun or a lightbulb is passed through a prism, it produces a continuous spectrum, like a rainbow. Perhaps you have seen this happen when sunlight goes through a prism or through raindrops. When atoms of elements are heated, they also produce light. At night, you may have seen the yellow color of sodium streetlamps or the red color of neon lights.

LEARNING GOAL

Explain how atomic spectra correlate with the energy levels in atoms.

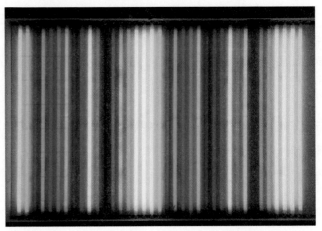

Colors are produced when electricity excites electrons in noble gases.

A CD or DVD is read when light from a laser is reflected from the pits on the surface.

Photons

Any light emitted in an atomic spectrum is a stream of small particles called **photons**. Each photon is a packet of energy also known as a *quantum*. All photons travel at the speed of light as a wave of energy. We say that photons have both particle and wave characteristics. The energy of a photon is directly proportional to its frequency. Thus, high-frequency photons have high energy and short wavelengths, whereas low-frequency photons have low energy and long wavelengths.

Photons play an important role in our modern world, particularly in the use of lasers, which use a narrow range of wavelengths. For example, lasers use photons of a single frequency to read pits on compact discs (CDs) and digital versatile discs (DVDs) or to scan bar codes on labels when we buy groceries. A CD is read by a laser with a wavelength of 780 nm. The newer DVDs are read by a laser with a wavelength of 640 nm. The shorter wavelength for DVDs allows a smaller pit size on the disc, which means that the disc has a greater storage capacity. In hospitals, high-energy photons are used in light treatments to reach tumors deep within the tissues without damaging the surrounding tissue.

> ### CONCEPT CHECK 5.2
>
> ■ **Energy, Frequency, and Wavelengths of Photons**
>
> Compare the energy, frequency, and wavelength of photons of gamma radiation and green visible light.
>
> ANSWER
>
> The energy of a photon is directly related to its frequency and inversely related to its wavelength. Thus, a photon of gamma radiation has higher energy and higher frequency but a shorter wavelength than a photon of green visible light.

Atomic Spectra

When the light emitted from heated elements is passed through a prism, it does not produce a continuous spectrum. Instead, an **atomic spectrum** is produced that consists of lines of different colors separated by dark areas (see Figure 5.3). This separation of colors indicates that only certain wavelengths of light are produced when an element is heated, which gives each element a unique atomic spectrum.

Electron Energy Levels

Scientists associate the lines in atomic spectra with changes in the energies of the electrons. In an atom, each electron has a fixed or specific energy known as its

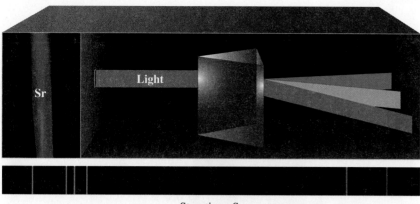

Strontium, Sr

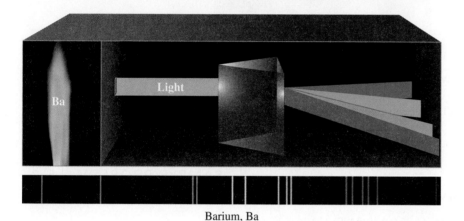

Barium, Ba

FIGURE 5.3 An atomic spectrum unique to an element is produced when light emitted by heating the element passes through a prism, which separates the light into colored lines.

Q Why don't the elements form a continuous spectrum as seen with white light?

energy level. The energy levels are assigned values called **principal quantum numbers** (n), which are positive integers ($n = 1$, $n = 2$, ...). Generally, electrons in the lower energy levels are closer to the nucleus, while electrons in the higher energy levels are farther away. The energy of an electron is *quantized,* which means that the energy of an electron can only have specific energy values, but cannot have values between them.

Principal Quantum Number (n)

$1 < 2 < 3 < 4 < 5 < 6 < 7$

Energy of electrons increases ⟶

As an analogy, we can think of the energy levels of an atom as similar to the shelves in a bookcase. The lowest energy level is the first shelf; the second energy level would be the second shelf. If we have a stack of books on the floor, it takes less energy to put them on the bottom shelf first, and then the second shelf, and so on. However, we could never get any book to stay in the space between any of the shelves. Unlike standard bookcases, however, there is a big difference in the energy of the first and second energy levels, and thereafter the energy levels are closer.

Changes in Energy Levels

When the electrons in an atom occupy the lowest energy levels, they are in their *ground state.* By absorbing the amount of energy equal to the difference in energy levels, an electron is raised to a higher energy level. An electron loses energy when it falls to a lower energy level and emits electromagnetic radiation equal to the energy level difference (see Figure 5.4). If the electromagnetic radiation emitted has a wavelength in the visible range, we see a color.

TUTORIAL
Energy Levels

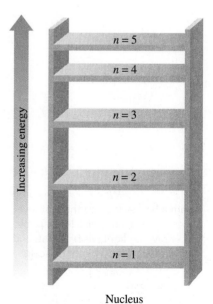

An electron can have only the energy of one of the energy levels in an atom.

FIGURE 5.4 Electrons absorb a specific amount of energy to move to a higher energy level. When electrons lose energy, photons with specific energies are emitted.

Q How does the energy of a photon of green light compare to the energy of a photon of red light?

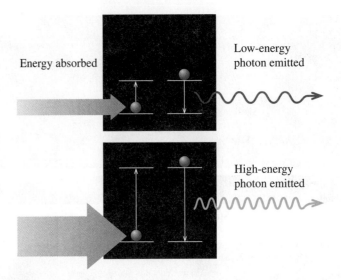

Energy absorbed

Low-energy photon emitted

High-energy photon emitted

Ultraviolet light is produced when electrons drop from higher energy levels to the first ($n = 1$) energy level. A series of colored lines of visible light can be produced when electrons fall from higher energy levels to the second ($n = 2$) level. Photons in the infrared are produced when electrons drop to the third ($n = 3$) energy level (see Figure 5.5).

SAMPLE PROBLEM | 5.2

■ Change in Energy Levels

a. How does an electron move to a higher energy level?
b. When an electron drops to a lower energy level, how is energy lost?

SOLUTION

a. An electron moves to a higher energy level when it absorbs an amount of energy equal to the difference in energy levels.
b. Energy, equal to the difference in energy levels, is emitted as a photon when an electron drops to a lower-energy level.

STUDY CHECK

Why did scientists propose that electrons occupy specific energy levels in an atom?

FIGURE 5.5 When electrons drop from a higher level to the first level, second level, and third level, photons of ultraviolet light, visible light, and infrared are emitted (not to scale).

Q Why is a different photon of light emitted when an electron drops from energy level 5 to level 3 than when an electron drops from energy level 4 to level 2?

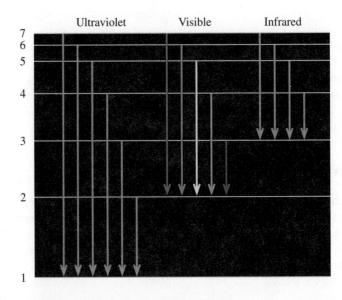

Ultraviolet   Visible   Infrared

CHEMISTRY AND THE ENVIRONMENT

Energy-Saving Fluorescent Bulbs

A compact fluorescent light (CFL) is a type of fluorescent bulb that is replacing the standard lightbulb we use in our homes and workplaces. Compared to a standard lightbulb, the CFL has a longer life and uses less electricity. Within about 20 days of use, the fluorescent bulb saves enough money in electricity costs to pay for its higher initial cost.

A standard incandescent lightbulb has a thin tungsten filament inside a sealed glass bulb. When the light is switched on, electricity flows through this filament, and electrical energy is converted to heat energy. When the filament reaches a temperature around 2300 °C, we see white light. We say that the lightbulb is incandescent.

A fluorescent bulb produces light in a different way. When the switch is turned on, electrons move between two electrodes and collide with mercury atoms in a gas mixture of mercury and argon inside the bulb. When the electrons in the mercury atoms absorb energy from the collisions, they are raised to higher energy levels. As electrons fall to lower energy levels, energy in the ultraviolet range is emitted. This ultraviolet light strikes the phosphor coating inside the tube, and fluorescence occurs as visible light is emitted.

The production of light in a fluorescent bulb is more efficient than in an incandescent lightbulb. A 75-watt incandescent bulb can be replaced by a 20-watt fluorescent bulb that gives the same amount of light, providing a 70% reduction in electricity costs. A typical lightbulb lasts for one to two months, whereas a compact fluorescent lightbulb lasts from one to two years.

A compact fluorescent light (CFL) uses up to 70% less energy.

QUESTIONS AND PROBLEMS

Atomic Spectra and Energy Levels

5.13 What feature of an atomic spectrum indicates that the energy emitted by heating an element is not continuous?

5.14 How can we explain the distinct lines that appear in an atomic spectrum?

5.15 Electrons can jump to higher energy levels when they _____ (absorb/emit) a photon.

5.16 Electrons drop to lower energy levels when they _____ (absorb/emit) a photon.

5.17 An electron drop to what energy level is likely to emit a photon in the infrared region?

5.18 An electron drop to what energy level is likely to emit a photon of red light?

5.19 Identify the type of photon in each pair with the greater energy.
a. green light or yellow light
b. red light or blue light

5.20 Identify the type of photon in each pair with the greater energy.
a. orange light or violet light
b. infrared light or ultraviolet light

5.3 Sublevels and Orbitals

There is a limit to the number of electrons allowed in each energy level. Only a few electrons can occupy the lower energy levels, while more electrons can be accommodated in higher energy levels. The maximum number of electrons allowed in any energy level is calculated using the formula $2n^2$ (two times the square of the principal quantum number). Table 5.1 shows the maximum number of electrons allowed in the first four energy levels.

LEARNING GOAL

Describe the sublevels and orbitals in atoms.

TABLE 5.1 Maximum Number of Electrons Allowed in Energy Levels 1–4

Energy Level (n)	1	2	3	4
$2n^2$	$2(1)^2$	$2(2)^2$	$2(3)^2$	$2(4)^2$
Maximum Number of Electrons	2	8	18	32

FIGURE 5.6 The number of sublevels in an energy level is the same as the principal quantum number, n.

Q How many sublevels are in energy level n = 5?

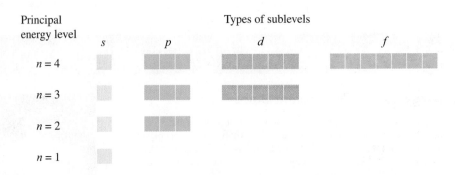

Sublevels

Within each energy level, there are one or more **sublevels** that contain electrons with identical energy. The sublevels are identified by the letters s, p, d, and f. The number of sublevels within an energy level is equal to the principal quantum number, n. The first energy level (n = 1) has only one sublevel, 1s. The second energy level (n = 2) has two sublevels, 2s and 2p. The third energy level (n = 3) has three sublevels, 3s, 3p, and 3d. The fourth energy level (n = 4) has four sublevels, 4s, 4p, 4d, and 4f (see Figure 5.6). Energy levels from n = 5 and higher also have the same number of sublevels as the value of n, but only s, p, d, and f sublevels are needed to hold the electrons in atoms of the elements known today.

Within each energy level, the s sublevel has the lowest energy. If there are additional sublevels, the p sublevel has the next lowest energy, then the d sublevel, and finally the f sublevel.

Order of Increasing Energy of Sublevels in an Energy Level

$$s < p < d < f$$

Lowest ⟶ Highest
energy   energy

CONCEPT CHECK 5.3

■ **Energy Levels and Sublevels**

Complete the following table with the energy level and sublevels:

Energy Level (n)	Sublevels			
	s	p	d	f
			4d	
1				
	2s			
		3p		

ANSWER

The number of sublevels is equal to the principal quantum number, n. The available sublevels begin with the lowest energy s, followed by p, d, and f. For n = 1, there is one sublevel s; n = 2, two sublevels 2s and 2p; n = 3, three sublevels 3s, 3p, and 3d; n = 4, four sublevels 4s, 4p, 4d, and 4f.

Energy Level (n)	Sublevels			
	s	p	d	f
4	4s	4p	4d	4f
1	1s			
2	2s	2p		
3	3s	3p	3d	

Orbitals

There is no way to know the exact location of an electron in an atom. Instead, scientists describe the location of an electron in terms of probability. A region in an atom where there is the highest probability (about 90%) of finding an electron is called an **orbital**. Suppose you could draw an imaginary circle with a 100-m radius around your chemistry classroom. There is a high probability of finding you within that area when your chemistry class is in session. But once in a while, you may be found outside that circle because you were sick or your car did not start. The shapes of orbitals represent the three-dimensional volumes in which electrons have the highest probability of being found.

Shapes of Orbitals

Each sublevel within an energy level is composed of the same type of orbitals. There is an *s* orbital for each *s* sublevel; *p* orbitals for each *p* sublevel; *d* orbitals for each *d* sublevel; and *f* orbitals for each *f* sublevel.

Each type of orbital has a unique shape. In an *s* orbital, an electron is most likely found in a region with a spherical shape. Imagine that you take a picture of the location of an electron in an *s* orbital every second. Suppose that after one hour, you put all these pictures together. The result, called a probability density, might look like Figure 5.7. For convenience, we represent this density as a spherical volume called a 1*s* orbital.

While the shape of every *s* orbital is spherical, there is an increase in the size of the *s* orbitals in higher energy levels (see Figure 5.8).

A *p* sublevel consists of three *p* orbitals, each of which has two lobes. The three *p* orbitals are arranged in three perpendicular directions, along the *x, y,* and *z* axes around the nucleus (see Figure 5.9). At higher energy levels, the shape of *p* orbitals is the same, but the volume increases.

Now we can look at energy level $n = 2$, which consists of *s* and *p* sublevels. The total available orbitals are one *s* orbital and three *p* orbitals.

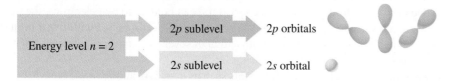

Energy level $n = 2$ consists of one 2s orbital and three 2p orbitals.

Energy level $n = 3$ consists of three sublevels *s, p,* and *d*. The *d* sublevels contain five *d* orbitals. Energy level $n = 4$ consists of four sublevels *s, p, d,* and *f*. The *f* sublevels contain seven *f* orbitals. The shapes of *d* orbitals and *f* orbitals are complex and we have not included them in this text.

FIGURE 5.7 An *s* orbital is a sphere that represents the region of highest probability of finding an *s* electron around the nucleus of an atom.

Q Is the probability high or low of finding an *s* electron outside an *s* orbital?

FIGURE 5.8 All *s* orbitals have the same shape, but the volume increases when they contain electrons at higher energy levels.

Q How would you compare the energy of an electron in the 1s, 2s, and 3s orbitals?

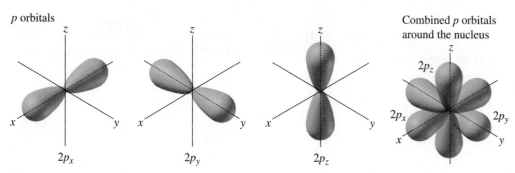

FIGURE 5.9 Each of the *p* orbitals has a dumbbell shape and is aligned along a different axis.

Q What are some similarities and differences of the *p* orbitals in energy level $n = 3$?

Electron spinning counterclockwise   Electron spinning clockwise

Opposite spins of electrons in an orbital

An orbital can hold up to two electrons with opposite spins.

SAMPLE PROBLEM | 5.3

■ **Sublevels and Orbitals**

Indicate the type and number of orbitals in each of the following energy levels or sublevels:

a. $3p$ sublevel   **b.** $n = 2$

c. $n = 3$ **d.** $4d$ sublevel

SOLUTION

a. The $3p$ sublevel contains three $3p$ orbitals.
b. The $n = 2$ energy level consists of $2s$ (one) and $2p$ (three) orbitals.
c. The $n = 3$ energy level consists of $3s$ (one), $3p$ (three), and $3d$ (five) orbitals.
d. The $4d$ sublevel contains five $4d$ orbitals.

STUDY CHECK

What is similar and what is different for $1s$, $2s$, and $3s$ orbitals?

Orbital Capacity and Electron Spin

The *Pauli exclusion principle* states that an orbital can hold up to a maximum of two electrons. According to a useful model for electron behavior, an electron is seen as spinning on its axis, which generates a magnetic field. When two electrons are in the same orbital, they will repel each other unless their magnetic fields cancel. This happens only when the two electrons spin in opposite directions. We can represent the spins of the electrons in the same orbital with one arrow pointing up and the other pointing down.

Number of Electrons in Sublevels

There is a maximum number of electrons that can occupy each sublevel. An s sublevel holds one or two electrons. Because each p orbital can hold up to two electrons, the three p orbitals in a p sublevel can accommodate six electrons. A d sublevel with five d orbitals can hold a maximum of 10 electrons. With a total of seven f orbitals, an f sublevel can hold up to 14 electrons.

As mentioned earlier, higher energy levels such as $n = 5, 6$, and 7, would have 5, 6, and 7 sublevels, but those beyond sublevel f are not utilized by the electrons in the atoms of elements known today. The total number of electrons in all the sublevels adds up to give the electrons allowed in an energy level. The number of sublevels, number of orbitals, and the maximum number of electrons for energy levels 1–4 are shown in Table 5.2.

TABLE 5.2 Electron Capacity in Sublevels for Energy Levels 1–4

Energy Level (n)	Number of Sublevels	Type of Sublevels	Number of Orbitals	Maximum Number of Electrons	Total Electrons ($2n^2$)
4	4	$4f$	7	14	32
		$4d$	5	10	
		$4p$	3	6	
		$4s$	1	2	
3	3	$3d$	5	10	18
		$3p$	3	6	
		$3s$	1	2	
2	2	$2p$	3	6	8
		$2s$	1	2	
1	1	$1s$	1	2	2

QUESTIONS AND PROBLEMS

Sublevels and Orbitals

5.21 Describe the shape of the following orbitals:
    a. $1s$ **b.** $2p$ **c.** $5s$

5.22 Describe the shape of the following orbitals:
    a. $3p$ **b.** $6s$ **c.** $4p$

5.23 What is similar about the following?
    a. $1s$ and $2s$ orbitals   **b.** $3s$ and $3p$ sublevels
    c. $3p$ and $4p$ sublevels   **d.** three $3p$ orbitals

5.24 What is similar about the following?
    a. $5s$ and $6s$ orbitals   **b.** $3p$ and $4p$ orbitals
    c. $3s$ and $4s$ sublevels   **d.** $2s$ and $2p$ orbitals

5.25 Indicate the number of each in the following:
    a. orbitals in the $3d$ sublevel
    b. sublevels in the $n = 1$ energy level

c. orbitals in the $6s$ sublevel
d. orbitals in the $n = 3$ energy level

5.26 Indicate the number of each in the following:
    a. orbitals in $n = 2$ energy level
    b. sublevels in $n = 4$ energy level
    c. orbitals in the $5f$ sublevel
    d. orbitals in the $6p$ sublevel

5.27 Indicate the maximum number of electrons in the following:
    a. $2p$ orbital   **b.** $3p$ sublevel
    c. energy level $n = 4$ **d.** $5d$ sublevel

5.28 Indicate the maximum number of electrons in the following:
    a. $3s$ sublevel   **b.** $4p$ orbital
    c. energy level $n = 3$ **d.** $4f$ sublevel

5.4 Drawing Orbital Diagrams and Writing Electron Configurations

LEARNING GOAL

Draw the orbital diagram and write the electron configuration for an element.

We can now look at how electrons are arranged in the orbitals within an atom. In an **orbital diagram**, boxes (or circles) represent the orbitals. We see from the energy diagram (Figure 5.10) that the electrons in the $1s$ orbital have a lower energy level than in the

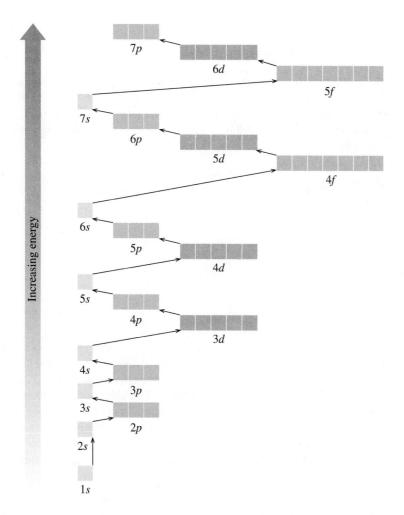

FIGURE 5.10 The sublevels fill in order of increasing energy beginning with 1s.

Q Why do $3d$ orbitals fill after the $4s$ orbital?

2s orbital. Thus, the first two electrons in an atom will go into the 1s orbital. Because the 1s sublevel can hold only two electrons, the next two electrons will go into the 2s orbital, which has the next lowest energy on the energy diagram.

With few exceptions (which will be noted later in this chapter), the "building" of electrons continues to the next lowest energy sublevel that is available until all the electrons are placed. For example, the atomic number of carbon is 6, which means that a carbon atom has six electrons. The first two electrons go into the 1s orbital; the next two electrons go into the 2s orbital. In the orbital diagram, the two electrons in the 1s and 2s orbitals are shown with opposite spins, the first arrow is up and the second down. The last two electrons in carbon begin to fill the 2p sublevel, the next lowest energy sublevel. However, there are three 2p orbitals of equal energy. Because the negatively charged electrons repel each other, they are placed in different 2p orbitals.

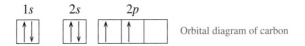

Orbital diagram of carbon

Electron Configurations

Chemists use a notation called the **electron configuration** to indicate the placement of the electrons of an atom in order of increasing energy. For example, the electron configuration for carbon is written by starting with the orbitals of the lowest energy sublevel first followed by the orbitals of the next lower energy sublevel.

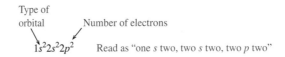

Period 1 Hydrogen and Helium

We will begin drawing orbital diagrams and writing electron configurations with the elements H and He in Period 1. The 1s orbital (which is the 1s sublevel) is written first because it has the lowest energy. Hydrogen has one electron in the 1s sublevel; helium has two. In the orbital diagram, the electrons for helium are shown with opposite spins.

Atomic Number	Element	Orbital Diagram	Electron Configuration
		1s	
1	H	↑	$1s^1$
2	He	↑↓	$1s^2$

Period 2 Lithium to Neon

Period 2 begins with lithium, which has three electrons. The first two electrons fill the 1s orbital, while the third electron goes into the 2s orbital, the sublevel with the next lowest energy. In beryllium, another electron is added to complete the 2s orbital. The next six electrons are now placed in the 2p orbitals. The electrons are added one at a time from boron to nitrogen, which gives three half-filled 2p orbitals. Because orbitals in the same sublevel are equal in energy, there is less repulsion when electrons are placed in separate orbitals. From oxygen to neon, the remaining electrons are paired up using opposite spins until the 2p sublevel is complete. In writing the electron configurations for the elements in Period 2, the 1s is written first followed by the 2s and then 2p.

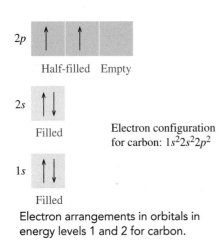

Electron arrangements in orbitals in energy levels 1 and 2 for carbon.

TUTORIAL

Electron Configurations

An electron configuration can also be written in an *abbreviated electron configuration*. The electron configuration of the preceding noble gas is replaced by writing its symbol inside square brackets. For example, the electron configuration for lithium, $1s^2 2s^1$, can be abbreviated as $[He]2s^1$, where $[He]$ replaces $1s^2$.

Atomic Number	Element	Orbital Diagram	Electron Configuration	Abbreviated Electron Configuration
3	Li	$1s$ $2s$	$1s^2 2s^1$	$[He]2s^1$
4	Be		$1s^2 2s^2$	$[He]2s^2$
5	B	$2p$	$1s^2 2s^2 2p^1$	$[He]2s^2 2p^1$
6	C		$1s^2 2s^2 2p^2$	$[He]2s^2 2p^2$
		Unpaired electrons		
7	N		$1s^2 2s^2 2p^3$	$[He]2s^2 2p^3$
8	O		$1s^2 2s^2 2p^4$	$[He]2s^2 2p^4$
9	F		$1s^2 2s^2 2p^5$	$[He]2s^2 2p^5$
10	Ne		$1s^2 2s^2 2p^6$	$[He]2s^2 2p^6$

CONCEPT CHECK 5.4

■ Orbital Diagrams and Electron Configurations

Draw or write each of the following for a nitrogen atom:

a. orbital diagram
b. electron configuration
c. abbreviated electron configuration

ANSWER

On the periodic table, nitrogen has atomic number 7, which means it has seven electrons.

a. For the orbital diagram, we draw boxes to represent the $1s$, $2s$, and $2p$ orbitals.

$1s$ $2s$ $2p$

First, we place a pair of electrons with opposite spins in both the $1s$ and $2s$ orbitals. Then we place the three remaining electrons in separate $2p$ orbitals with parallel spins.

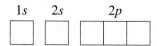

Orbital diagram for nitrogen (N)

b. The electron configuration for nitrogen is written to show the electrons in the same order of increasing energy as drawn for the orbital diagrams.

$1s^2 2s^2 2p^3$ Electron configuration for nitrogen (N)

c. The abbreviated electron configuration for nitrogen is written by substituting the symbol of $[He]$, the noble gas that precedes Period 2, for $1s^2$ in the electron configuration.

$[He]2s^2 2p^3$ Abbreviated electron configuration for nitrogen (N)

Period 3 Sodium to Argon

In Period 3, electrons enter the orbitals of the $3s$ and $3p$ sublevels, but not the $3d$ sublevel. We notice that the elements sodium to argon, which are directly below the elements lithium to neon in Period 2, have a similar pattern of filling their s and p orbitals. In sodium and magnesium, one and two electrons go into the $3s$ orbital. The remaining electrons for aluminum, silicon, and phosphorus go into different $3p$ orbitals so that phosphorus has a half-filled $3p$ sublevel. We can draw the complete orbital diagram for phosphorus as follows:

For elements in Period 3 through Period 7, we abbreviate the orbital diagram by using the symbol from the preceding noble gas, followed by the orbital boxes for the electrons in the last filled period.

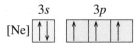

The remaining electrons in sulfur, chlorine, and argon are paired up (opposite spins) with the electrons already in the $3p$ orbitals. For the abbreviated electron configurations of Period 3, the symbol [Ne] replaces $1s^2 2s^2 2p^6$.

Atomic Number	Element	Orbital Diagram (3s and 3p orbitals only)	Electron Configuration	Abbreviated Electron Configuration
11	Na	[Ne] ↑ ☐☐☐	$1s^2 2s^2 2p^6 3s^1$	$[Ne]3s^1$
12	Mg	[Ne] ↑↓ ☐☐☐	$1s^2 2s^2 2p^6 3s^2$	$[Ne]3s^2$
13	Al	[Ne] ↑↓ ↑☐☐	$1s^2 2s^2 2p^6 3s^2 3p^1$	$[Ne]3s^2 3p^1$
14	Si	[Ne] ↑↓ ↑↑☐	$1s^2 2s^2 2p^6 3s^2 3p^2$	$[Ne]3s^2 3p^2$
15	P	[Ne] ↑↓ ↑↑↑	$1s^2 2s^2 2p^6 3s^2 3p^3$	$[Ne]3s^2 3p^3$
16	S	[Ne] ↑↓ ↑↓↑↑	$1s^2 2s^2 2p^6 3s^2 3p^4$	$[Ne]3s^2 3p^4$
17	Cl	[Ne] ↑↓ ↑↓↑↓↑	$1s^2 2s^2 2p^6 3s^2 3p^5$	$[Ne]3s^2 3p^5$
18	Ar	[Ne] ↑↓ ↑↓↑↓↑↓	$1s^2 2s^2 2p^6 3s^2 3p^6$	$[Ne]3s^2 3p^6$

SAMPLE PROBLEM | 5.4

■ Orbital Diagrams and Electron Configurations

For the element silicon, draw or write each of the following:

a. orbital diagram
b. abbreviated orbital diagram
c. electron configuration
d. abbreviated electron configuration

SOLUTION

a. Silicon in Period 3 has atomic number 14, which tells us that it has 14 electrons. In the orbital diagram, we draw boxes to represent the orbitals from $1s$ to $3p$.

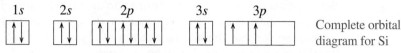

Starting with the $1s$ orbital, add paired electrons with opposite spins up to $3s$. Then place the last two electrons in separate $3p$ orbitals with parallel spins.

Complete orbital diagram for Si

b. For silicon, the preceding noble gas is neon. In the abbreviated orbital diagram, the orbitals $1s$, $2s$, and $2p$ are replaced by [Ne].

Abbreviated orbital diagram for Si

c. The electron configuration gives the electrons that fill the orbitals in order of increasing energy. The first 10 electrons in silicon would complete the orbitals in Periods 1 and 2: $1s^2$, $2s^2$, and $2p^6$. In Period 3, two electrons go into the $3s$ orbital, and the remaining two electrons go into the $3p$ orbitals.

Si $1s^2 2s^2 2p^6 3s^2 3p^2$ Electron configuration for Si

d. For silicon, the preceding noble gas is neon. For the abbreviated electron configuration of silicon, we replace $1s^2 2s^2 2p^6$ with [Ne].

[Ne] $3s^2 3p^2$ Abbreviated electron configuration for Si

STUDY CHECK

Write the complete and abbreviated electron configurations for sulfur.

QUESTIONS AND PROBLEMS

Drawing Orbital Diagrams and Writing Electron Configurations

5.29 Compare the terms electron configuration and abbreviated electron configuration.

5.30 Compare the terms orbital diagram and electron configuration.

5.31 Draw the abbreviated orbital diagram for each of the following:
   a. boron   **b.** aluminum
   c. phosphorus   **d.** argon

5.32 Draw the abbreviated orbital diagram for each of the following:
   a. fluorine   **b.** potassium
   c. magnesium   **d.** barium

5.33 Write a complete electron configuration for each of the following:
   a. nitrogen   **b.** sodium
   c. bromine   **d.** nickel

5.34 Write a complete electron configuration for each of the following:
   a. carbon   **b.** iron
   c. oxygen   **d.** arsenic

5.35 Write an abbreviated electron configuration for each of the following:
   a. calcium   **b.** strontium
   c. gallium   **d.** zinc

5.36 Write an abbreviated electron configuration for each of the following:
   a. lead   **b.** cadmium
   c. antimony   **d.** iodine

5.37 Give the symbol of the element with each of the following electron configurations:
   a. $1s^2 2s^1$ **b.** $1s^2 2s^2 2p^6 3s^2 3p^4$
   c. [Ne]$3s^2 3p^2$ **d.** [He]$2s^2 2p^5$

5.38 Give the symbol of the element with each of the following electron configurations:
   a. $1s^2 2s^2 2p^4$ **b.** [Ne]$3s^2$
   c. $1s^2 2s^2 2p^6 3s^2 3p^6$ **d.** [Ne]$3s^2 3p^1$

5.39 Give the symbol of the element that meets the following conditions:
   a. has three electrons in energy level $n = 3$
   b. has two $2p$ electrons
   c. completes the $3p$ sublevel
   d. completes the $2s$ sublevel

5.40 Give the symbol of the element that meets the following conditions:
   a. has five electrons in the $3p$ sublevel
   b. has three $2p$ electrons
   c. completes the $3s$ sublevel
   d. has one electron in the $3s$ sublevel

5.5 Electron Configurations and the Periodic Table

Until now, we have written electron configurations using the energy-level diagram. As configurations involve more sublevels, this becomes tedious. However, the electron configurations of the elements are related to their position in the periodic table. Different sections or blocks within the periodic table correspond to the *s*, *p*, *d*, and *f* sublevels (see Figure 5.11). Therefore, we can "build" the electron configurations of atoms by reading the periodic table in order of increasing atomic number.

Blocks on the Periodic Table

1. The *s* **block** includes hydrogen and helium as well as the elements in Group 1A (1) and Group 2A (2). This means that the final one or two electrons in the elements of the *s* block are located in *s* sublevels. The period number indicates the particular *s* sublevel that is filling: 1*s*, 2*s*, and so on.

2. The *p* **block** consists of the elements in Group 3A (13) to Group 8A (18). There are six *p* block elements in each period because each *p* sublevel can hold as many as six electrons. The period number indicates the particular *p* sublevel that is filling: 2*p*, 3*p*, and so on.

3. The *d* **block**, containing the transition elements, first appears after calcium (atomic number 20). There are 10 elements in the *d* block because the five *d* orbitals in each *d* sublevel can hold up to 10 electrons. The particular *d* sublevel is one less ($n - 1$) than the period number. For example, in Period 4, the first *d* block is the 3*d* sublevel. In Period 5, the second *d* block is the 4*d* sublevel.

4. The *f* **block** includes all the elements in the two rows at the bottom of the periodic table. There are 14 elements in each *f* block because an *f* sublevel can hold up to 14 electrons. Elements that have atomic numbers higher than 57 (La) have electrons in the 4*f* block. The particular *f* sublevel is two less ($n - 2$) than the period number. For example, in Period 6, the first *f* block is the 4*f* sublevel. In Period 7, the second *f* block is the 5*f* sublevel.

Writing Electron Configurations Using Sublevel Blocks

Now we can write electron configurations using the sublevel blocks on the periodic table. As before, each configuration begins at H. But now we move across the table from left to right, writing down each sublevel block we come to until we reach the element for which we are writing an electron configuration. For example, let's write the electron configuration for chlorine (atomic number 17) from the sublevel blocks on the periodic table.

FIGURE 5.11 Electron configuration follows the order of sublevels on the periodic table.

Q If neon is in the Group 8A (18), Period 2, how many electrons are in the 1*s*, 2*s*, and 2*p* sublevels of neon?

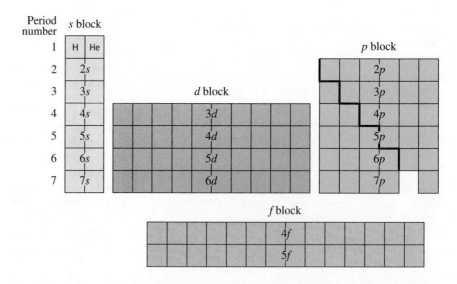

To write the electron configuration for chlorine (atomic number 17) from the sublevel blocks, we can use the following steps:

STEP 1   Locate the element on the periodic table.
Chlorine (atomic number 17) is in Group 7 (17) and Period 3.

STEP 2   Write the filled sublevels in order.

Period		Sublevel Blocks Filled
1	$1s$ sublevel (H $\longrightarrow$ He)	$1s^2$
2	$2s$ sublevel (Li $\longrightarrow$ Be)	$2s^2$
	$2p$ sublevel (B $\longrightarrow$ Ne)	$2p^6$
3	$3s$ sublevel (Na $\longrightarrow$ Mg)	$3s^2$

STEP 3   Count the number of electrons in the final block up to the given element and complete the sublevel notation.
Because chlorine is the fifth element in the $3p$ block, there are five electrons in the $3p$ sublevel.

Period		Last Sublevel Block
3	$3p$ sublevel (Al $\longrightarrow$ Cl)	$3p^5$

The electron configuration is written with the sequence of filled sublevel blocks and the last sublevel block, which gives

$1s^22s^22p^63s^23p^5$ $\longleftarrow$ Final sublevel block where Cl appears

Period 4

In Period 4, we see that the $4s$ block comes before the $3d$ block because the $4s$ orbital has a lower energy than the $3d$ orbitals. We see this occur again in Period 5 when the $5s$ sublevel fills before the $4d$ sublevel, and again in Period 6 when the $6s$ fills before the $5d$.

At the beginning of Period 4, the last one and two electrons in potassium (19) and calcium (20) go into the $4s$ orbital. In scandium, the electron following the filled $4s$ orbital goes into the $3d$ block. The $3d$ block continues to fill until it is complete with 10 electrons at zinc (30). Once the $3d$ block is complete, the next six electrons, gallium to krypton, go into the $4p$ block.

Atomic Number	Element	Electron Configuration	Abbreviated Electron Configuration
4s Block			
19	K	$1s^22s^22p^63s^23p^64s^1$	$[Ar]4s^1$
20	Ca	$1s^22s^22p^63s^23p^64s^2$	$[Ar]4s^2$
3d Block			
21	Sc	$1s^22s^22p^63s^23p^64s^23d^1$	$[Ar]4s^23d^1$
22	Ti	$1s^22s^22p^63s^23p^64s^23d^2$	$[Ar]4s^23d^2$
23	V	$1s^22s^22p^63s^23p^64s^23d^3$	$[Ar]4s^23d^3$
24	Cr*	$1s^22s^22p^63s^23p^64s^13d^5$	$[Ar]4s^13d^5$ (half-filled d sublevel is stable)
25	Mn	$1s^22s^22p^63s^23p^64s^23d^5$	$[Ar]4s^23d^5$
26	Fe	$1s^22s^22p^63s^23p^64s^23d^6$	$[Ar]4s^23d^6$
27	Co	$1s^22s^22p^63s^23p^64s^23d^7$	$[Ar]4s^23d^7$
28	Ni	$1s^22s^22p^63s^23p^64s^23d^8$	$[Ar]4s^23d^8$
29	Cu*	$1s^22s^22p^63s^23p^64s^13d^{10}$	$[Ar]4s^13d^{10}$ (filled d sublevel is stable)
30	Zn	$1s^22s^22p^63s^23p^64s^23d^{10}$	$[Ar]4s^23d^{10}$

(continued on next page)

Atomic Number	Element	Electron Configuration	Abbreviated Electron Configuration
4p Block			
31	Ga	$1s^2 2s^2 2p^6 3s^2 3p^6 4s^2 3d^{10} 4p^1$	$[Ar]4s^2 3d^{10} 4p^1$
32	Ge	$1s^2 2s^2 2p^6 3s^2 3p^6 4s^2 3d^{10} 4p^2$	$[Ar]4s^2 3d^{10} 4p^2$
33	As	$1s^2 2s^2 2p^6 3s^2 3p^6 4s^2 3d^{10} 4p^3$	$[Ar]4s^2 3d^{10} 4p^3$
34	Se	$1s^2 2s^2 2p^6 3s^2 3p^6 4s^2 3d^{10} 4p^4$	$[Ar]4s^2 3d^{10} 4p^4$
35	Br	$1s^2 2s^2 2p^6 3s^2 3p^6 4s^2 3d^{10} 4p^5$	$[Ar]4s^2 3d^{10} 4p^5$
36	Kr	$1s^2 2s^2 2p^6 3s^2 3p^6 4s^2 3d^{10} 4p^6$	$[Ar]4s^2 3d^{10} 4p^6$

*Exceptions to the order of filling.

Some Exceptions in Sublevel Block Order

Within the filling of the $3d$ sublevel, exceptions occur for chromium and copper. In Cr and Cu, the $3d$ sublevel is close to being a half-filled or filled sublevel, which is particularly stable. Thus, chromium has only one electron in the $4s$ and 5 electrons in the $3d$ sublevel to give the added stability of a half-filled d sublevel. This is shown in the abbreviated orbital diagram for chromium that follows:

A similar exception occurs when copper achieves a completely filled $3d$ sublevel by placing only 1 electron in the $4s$ sublevel and using 10 electrons to complete the $3d$ sublevel. This is shown in the abbreviated orbital diagram for copper that follows:

After the $4s$ and $3d$ sublevels are completed, the $4p$ sublevel fills as expected from gallium to krypton, the noble gas that completes Period 4. There are also exceptions in filling for the higher d and f electron sublevels, some caused by the added stability of half-filled shells and others where the cause is not known.

SAMPLE PROBLEM | 5.5

■ **Using Sublevel Blocks to Write Electron Configurations**

Use the sublevel blocks on the periodic table to write the electron configuration for selenium.

SOLUTION

STEP 1 **Locate the element on the periodic table.** Selenium is in Period 4 and Group 6A (16), which is in a p block.

STEP 2 **Write the filled sublevels in order going across each period.** Beginning with $1s$, go across the periodic table writing each filled sublevel block as follows:

Period 1	$1s^2$
Period 2	$2s^2$
	$2p^6$
Period 3	$3s^2$
	$3p^6$
Period 4	$4s^2$
	$3d^{10}$

Guide to Writing Electron Configurations with Sublevel Blocks

STEP 1
Locate the element on the periodic table.

STEP 2
Write the filled sublevels in order going across each period.

STEP 3
Complete the configuration by counting the electrons in the unfilled block.

STEP 3 **Complete the configuration by counting the electrons in the unfilled block.** There are four electrons in the $4p$ sublevel for Se, which complete the electron configuration for Se: $1s^2 2s^2 2p^6 3s^2 3p^6 4s^2 3d^{10} 4p^4$.

STUDY CHECK

Use the sublevel blocks on the periodic table to write the electron configuration for tin.

QUESTIONS AND PROBLEMS

Electron Configurations and the Periodic Table

5.41 Write a complete electron configuration for an atom of each of the following:
 a. arsenic   **b.** iron
 c. palladium   **d.** iodine

5.42 Write a complete electron configuration for an atom of each of the following:
 a. calcium   **b.** cobalt
 c. gallium   **d.** cadmium

5.43 Write an abbreviated electron configuration for an atom of each of the following:
 a. titanium   **b.** strontium
 c. barium   **d.** lead

5.44 Write an abbreviated electron configuration for an atom of each of the following:
 a. nickel   **b.** arsenic
 c. tin   **d.** antimony

5.45 Give the symbol of the element with each of the following electron configurations:
 a. $1s^2 2s^2 2p^6 3s^2 3p^3$
 b. $1s^2 2s^2 2p^6 3s^2 3p^6 4s^2 3d^7$
 c. $[Ar]4s^2 3d^{10}$
 d. $[Ar]4s^2 3d^{10} 4p^5$

5.46 Give the symbol of the element with each of the following electron configurations:
 a. $1s^2 2s^2 2p^6 3s^2 3p^6 4s^2 3d^8$ **b.** $[Kr]5s^2 4d^4$
 c. $1s^2 2s^2 2p^6 3s^2 3p^6 4s^2 3d^{10} 4p^2$ **d.** $[Xe]6s^2 4f^{14} 5d^{10} 6p^3$

5.47 Give the symbol of the element that meets the following conditions:
 a. has three electrons in energy level $n = 4$
 b. has three $2p$ electrons
 c. completes the $5p$ sublevel
 d. has two electrons in the $4d$ sublevel

5.48 Give the symbol of the element that meets the following conditions:
 a. has five electrons in energy level $n = 3$
 b. has one electron in the $6p$ sublevel
 c. completes the $7s$ sublevel
 d. has four $5p$ electrons

5.49 Give the number of electrons in the indicated orbitals for the following:
 a. $3d$ in zinc   **b.** $2p$ in sodium
 c. $4p$ in arsenic   **d.** $5s$ in rubidium

5.50 Give the number of electrons in the indicated orbitals for the following:
 a. $3d$ in manganese **b.** $5p$ in antimony
 c. $6p$ in lead   **d.** $3s$ in magnesium

5.6 Periodic Trends of the Elements

The electron configurations of atoms are an important factor in the physical and chemical behavior of the elements and in the properties of the compounds that they form. In this section, we will look at the *valence electrons* in atoms, the trends in the sizes of atoms, and *ionization energy*. Going across a period, there is a pattern of regular change in these properties from one group to the next. Known as *periodic properties,* each property increases or decreases across a period, and then the trend is repeated in each successive period. We can use the seasonal changes in temperatures as an analogy for periodic properties. In the winter, temperatures are usually cold and become warmer in the spring. By summer, the outdoor temperatures are hot but begin to cool in the fall. By winter, we expect cold temperatures again as the pattern of decreasing and increasing temperatures repeats for another year.

Group Number and Valence Electrons

The chemical properties of representative elements are mostly due to the **valence electrons**, which are the electrons in the outermost energy levels. These valence electrons occupy the s and p sublevels with the highest quantum number n. The group numbers indicate the number of valence (outer) electrons for the elements in each vertical column. For

LEARNING GOAL

Use the electron configurations of elements to explain periodic trends.

TUTORIAL
Patterns in the Periodic Table

TUTORIAL
Electron Configurations and the Periodic Table

TABLE 5.3 Valence Electrons for Representative Elements in Periods 1–4

1A (1)	2A (2)	3A (13)	4A (14)	5A (15)	6A (16)	7A (17)	8A (18)
1 H $1s^1$							2 He $1s^2$
3 Li $2s^1$	4 Be $2s^2$	5 B $2s^2 2p^1$	6 C $2s^2 2p^2$	7 N $2s^2 2p^3$	8 O $2s^2 2p^4$	9 F $2s^2 2p^5$	10 Ne $2s^2 2p^6$
11 Na $3s^1$	12 Mg $3s^2$	13 Al $3s^2 3p^1$	14 Si $3s^2 3p^2$	15 P $3s^2 3p^3$	16 S $3s^2 3p^4$	17 Cl $3s^2 3p^5$	18 Ar $3s^2 3p^6$
19 K $4s^1$	20 Ca $4s^2$	31 Ga $4s^2 4p^1$	32 Ge $4s^2 4p^2$	33 As $4s^2 4p^3$	34 Se $4s^2 4p^4$	35 Br $4s^2 4p^5$	36 Kr $4s^2 4p^6$

example, the elements in Group 1A (1), such as lithium, sodium, and potassium, all have one electron in an s orbital. Looking at the sublevel block, we can represent the valence electron in the alkali metals of Group 1A (1) as ns^1. All the elements in Group 2A (2), the alkaline earth metals, have two valence electrons, ns^2. The halogens in Group 7A (17) have seven valence electrons, $ns^2 np^5$.

We can see the repetition of the outermost s and p electrons for the representative elements for Periods 1 to 4 in Table 5.3. Helium is included in Group 8A (18) because it is a noble gas, but it has only two electrons in its complete energy level.

SAMPLE PROBLEM | 5.6

■ Using Group Numbers

Using the periodic table, write the group number, the period, and the electron configuration of the valence electrons for the following:

a. calcium   **b.** selenium   **c.** lead

SOLUTION

a. Calcium is in Group 2A (2), Period 4. It has a valence electron configuration of $4s^2$.
b. Selenium is in Group 6A (16), Period 4. It has a valence electron configuration of $4s^2 4p^4$.
c. Lead is in Group 4A (14), Period 6. It has a valence electron configuration of $6s^2 6p^2$. The electrons in the $5d$ and $4f$ sublevels are not valence electrons.

STUDY CHECK

What are the group numbers, the periods, and the valence electron configurations for sulfur and strontium?

Electron-Dot Symbols

An **electron-dot symbol** is a convenient way to represent the valence electrons, which are shown as dots placed on the sides, top, or bottom of the symbol for the element. One to four valence electrons are arranged as single dots. When there are five to eight electrons, one or more electrons are paired. Any of the following would be an acceptable electron-dot symbol for magnesium, which has two valence electrons:

Possible Electron-Dot Symbols for the Two Valence Electrons in Magnesium

Ṁg·   Ṁg·   ·Ṁg   ·Mg·   Ṃg·   ·Ṃg

TABLE 5.4 Electron-Dot Symbols for Selected Elements in Periods 1–4

	Group Number							
	1A (1)	2A (2)	3A (13)	4A (14)	5A (15)	6A (16)	7A (17)	8A (18)
Number of Valence Electrons	1	2	3	4	5	6	7	8
Electron-Dot Symbols	H·							He:
	Li·	Be·	·B·	·C·	·N·	·O:	·F:	:Ne:
	Na·	Mg·	·Al·	·Si·	·P·	·S:	·Cl:	:Ar:
	K·	Ca·	·Ga·	·Ge·	·As·	·Se:	·Br:	:Kr:

Atoms of magnesium

SAMPLE PROBLEM | 5.7

■ Writing Electron-Dot Symbols

Write the electron-dot symbol for each of the following elements:

a. bromine   **b.** aluminum

SOLUTION

a. Because the group number for bromine is 7A (17), bromine has seven valence electrons.

·B̈r:

b. Aluminum, in Group 3A (13), has three valence electrons.

·Al·

STUDY CHECK

What is the electron-dot symbol for phosphorus?

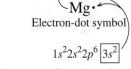

Mg·

Electron-dot symbol

$1s^2 2s^2 2p^6 \boxed{3s^2}$

Electron configuration of magnesium

Electron-dot symbols for selected elements are given in Table 5.4.

Atomic Radius

Although there are no fixed boundaries to atoms, we can use the probability of finding electrons to determine the size of an atom. Then, the **atomic radius** is the distance from the nucleus to the energy level that contains the valence (outermost) electrons. Each increasing energy level contains valence electrons that will usually be farther away from the nucleus. Thus, the valence electrons for Ar, which are in the $n = 3$ energy level, would be farther from the nucleus than the valence electrons for He in the $n = 1$ energy level (see Figure 5.12).

If we look at a group of representative elements, we see that the atomic radius within a group increases from the top to the bottom. For example, the one valence electron of the alkali metals is found for Li in the $n = 2$ energy level; for Na in the $n = 3$ energy level; for K in the $n = 4$ energy level; and for Rb in the $n = 5$ energy level.

The atomic radius of representative elements is affected by the attractive forces of the protons in the nucleus on the electrons in the outermost level. Going from left to right in a period, additional protons increase the positive charge of the nucleus. The increase in nuclear attraction pulls all of the electrons closer to the nucleus. As a result, the atomic radius decreases going across a period.

The atomic radii of the transition elements within a period change only slightly because electrons add to d orbitals rather than to the outermost energy level. Because the increase in nuclear charge is canceled by an increase in d electrons, the attraction of the valence electrons by the nucleus remains about the same. Because there is little change in the nuclear attraction for the valence electrons, the atomic radius remains constant.

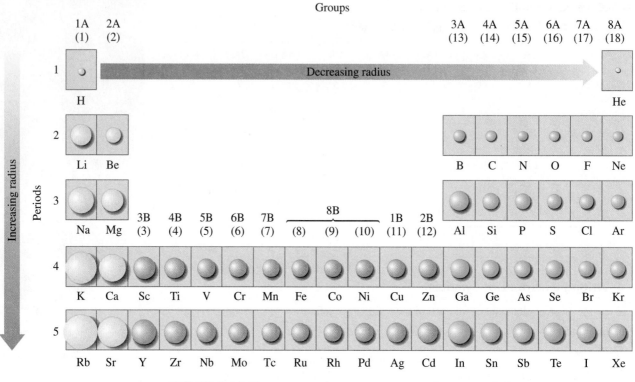

FIGURE 5.12 The atomic radius increases going down a group but decreases going from left to right across a period.

Q Why does the atomic radius increase going down a group?

CONCEPT CHECK 5.5

■ **Atomic Radius**

Why is the radius of a phosphorus atom larger than the radius of a nitrogen atom but smaller than the radius of a silicon atom?

ANSWER

The radius of a phosphorus atom is larger than the radius of a nitrogen atom because phosphorus has valence electrons in a higher energy level, which is farther from the nucleus. A phosphorus atom has one more proton than a silicon atom; therefore, the nucleus in phosphorus has a stronger attraction for the valence electrons, which decreases its radius compared to a silicon atom.

Ionization Energy

Electrons are held in atoms by their attraction to the nucleus. Therefore, energy is required to remove an electron from an atom. The **ionization energy** is the energy needed to remove the least tightly bound electron from an atom in the gaseous (g) state. When an electron is removed from a neutral atom, a cation with a $1+$ charge is formed.

$$\text{Na}(g) + \text{energy (ionization)} \longrightarrow \text{Na}^+ + e^-$$

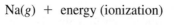

The ionization energy generally decreases going down a group. Less energy is needed to remove an electron as nuclear attraction for electrons farther from the nucleus decreases (see Figure 5.13). Going across a period from left to right, the ionization energy generally increases. As the attraction for outermost electrons increases across a period, more energy is required to remove an electron. In general, the ionization energy is low for the metals and high for the nonmetals. In addition to these general trends, there are specific differences in ionization energies that can be explained by the special stability of a filled or a half-filled sublevel.

In Period 1, the valence electrons are close to the nucleus and strongly held. H and He have high ionization energies because a large amount of energy is required to remove an electron. The ionization energy for He is the highest of any element because He has a full, stable 1s sublevel, which is disrupted by removing an electron. The high ionization energies of the noble gases indicate that their electron configurations are especially stable (see Figure 5.14).

In Period 2, there is a decrease in ionization energy from Group 2A (2) to Group 3A (13). A 2p electron, which is in a higher sublevel, is farther from the nucleus and more easily removed than a 2s electron, and removing this electron results in a stable, filled sublevel. In Groups 4A (14) and 5A (15), the increased charge of the nucleus attracts the 2p electrons more strongly, which requires a greater ionization energy. Then there is a slight decrease in ionization energy for Group 6A (16) compared to Group 5A (15). For Group 6A (16), the removal of a 2p electron requires less energy because it results in a stable, half-filled 2p sublevel. Ionization energy increases for both Group 7A (17) and Group 8A (18) due to increased nuclear charge. Disrupting the stable ns^2np^6 electron configuration of the noble gases requires the highest ionization energies of all the groups.

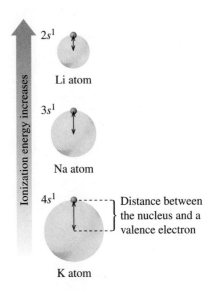

FIGURE 5.13 As the distance to the nucleus in Li, Na, and K atoms increases, the attraction to the nucleus decreases and less energy is required to remove the valence electron.

Q Why would Cs have a lower ionization energy than K?

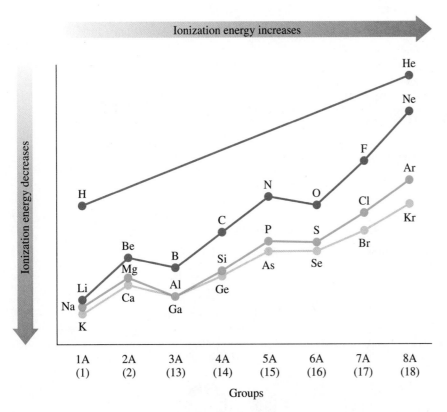

FIGURE 5.14 Ionization energies for the representative elements tend to decrease going down a group and increase going across a period.

Q Why is the ionization energy for Li less than that for O?

SAMPLE PROBLEM   5.8

■ Ionization Energy

Indicate the element in each pair that has the higher ionization energy and explain your choice.

a. K or Na   **b.** Mg or Cl   **c.** F or N

SOLUTION

a. Na. In Na, an electron is removed from an orbital closer to the nucleus.
b. Cl. The increased nuclear charge increases the attraction for the outermost electrons for elements in the same period.
c. F. The increased nuclear charge of F requires a higher ionization energy compared to N.

STUDY CHECK

Arrange Sn, Sr, and I in order of increasing ionization energy.

Sizes of Atoms and Their Ions

In Figure 5.15, the sizes of ions formed by atoms of metals and nonmetals are compared. We see that positive ions in Group 1A (1) are much smaller than the corresponding metal atoms. This occurs because metal atoms lose all of their electrons from the outermost energy level. If we look at the second and third energy levels for sodium, we see that the electron in the third energy level is lost to form the sodium ion, (Na^+), which has an octet in the second energy level.

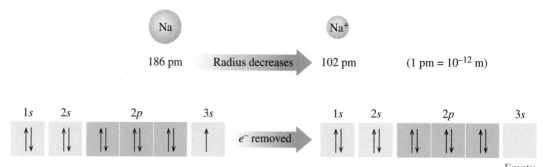

The positive ions are smaller than their corresponding atoms.

FIGURE 5.15 The positive ions of metals are about half the size of the corresponding metal atoms. The negative ions of nonmetals are about twice the size of the corresponding nonmetal atoms.

Q How would you compare the size of positive ions going down a group?

Group 1A (1)

Atoms   Ions

Li   Li⁺
Na   Na⁺
K   K⁺
Rb   Rb⁺
Cs   Cs⁺

Group 7A (17)

Atoms   Ions

F   F⁻
Cl   Cl⁻
Br   Br⁻
I   I⁻

When a nonmetal atom adds electrons, its size increases because of the repulsion between electrons. For example, a fluoride ion is larger than a fluorine atom because a valence electron is added to the second energy level, which completes an octet.

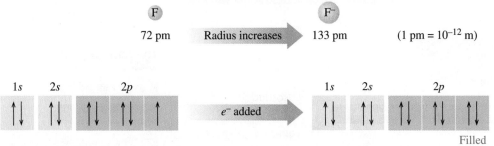

The negative ions are larger than their corresponding atoms.

QUESTIONS AND PROBLEMS

Periodic Trends of the Elements

5.51 What does the group number of an element indicate about its electron configuration?

5.52 What is similar and what is different about the valence electrons of the elements in a group?

5.53 Write the group number using both A/B notation and 1–18 numbering of elements that have the following outer electron configuration:
  a. $2s^2$ **b.** $3s^2 3p^3$
  c. $4s^2 3d^5$ **d.** $5s^2 4d^{10} 5p^4$

5.54 Write the group number using both A/B notation and 1–18 numbering of elements that have the following outer electron configuration:
  a. $4s^2 4p^5$ **b.** $4s^1$
  c. $4s^2 3d^8$ **d.** $5s^2 4d^{10} 5p^2$

5.55 Write the valence electron configuration for each of the following (example $ns^2 np^4$):
  a. alkali metals   **b.** Group 4A
  c. Group 13   **d.** Group 5A

5.56 Write the valence electron configuration for each of the following (example $ns^2 np^4$):
  a. halogens   **b.** Group 6A
  c. Group 10   **d.** alkaline earth metals

5.57 Indicate the number of valence (outermost) electrons in each of the following:
  a. aluminum   **b.** Group 5A
  c. nickel   **d.** F, Cl, Br, and I

5.58 Indicate the number of valence (outermost) electrons in each of the following:
  a. Li, Na, K, Rb, and Cs   **b.** zinc and cadmium
  c. C, Si, Ge, Sn, and Pb   **d.** Group 8A

5.59 Give the group number and write the electron-dot symbol for each of the following elements:
  a. sulfur   **b.** nitrogen
  c. calcium   **d.** sodium
  e. gallium

5.60 Give the group number and write the electron-dot symbol for each of the following elements:
  a. carbon   **b.** oxygen
  c. argon   **d.** lithium
  e. chlorine

5.61 Place the elements in each set in order of decreasing atomic radius.
  a. Mg, Al, Si   **b.** Cl, Br, I
  c. I, Sb, Sr   **d.** P, Si, Na

5.62 Place the elements in each set in order of decreasing atomic radius.
  a. Cl, S, P   **b.** Ge, Si, C
  c. Ba, Ca, Sr   **d.** O, S, Se

5.63 Select the larger atom in each pair.
  a. Na or O   **b.** Na or Rb
  c. Na or Mg   **d.** Na or Cl

5.64 Select the larger atom in each pair.
  a. S or Cl   **b.** S or O
  c. S or Se   **d.** S or Al

5.65 Arrange each set of elements in order of increasing ionization energy.
  a. F, Cl, Br   **b.** Na, Cl, Al
  c. Na, K, Cs   **d.** As, Sb, Sn

5.66 Arrange each set of elements in order of increasing ionization energy.
  a. O, N, C   **b.** S, P, Cl
  c. As, P, N   **d.** Al, Si, P

5.67 Select the element in each pair with the higher ionization energy.
  a. Br or I   **b.** Mg or Al
  c. S or P   **d.** I or Xe

5.68 Select the element in each pair with the higher ionization energy.
  a. O or Ne   **b.** K or Br
  c. Ca or Ba   **d.** N or O

5.69 Why is a potassium ion smaller than a potassium atom?

5.70 Why is a bromide ion larger than a bromine atom?

5.71 Which is larger in each of the following?
  a. Na or Na^+ **b.** Cl or Cl^-
  c. S or S^{2-}

5.72 Which is smaller in each of the following?
  a. I or I^- **b.** Ca or Ca^{2+}
  c. Rb or Rb^+

CONCEPT MAP

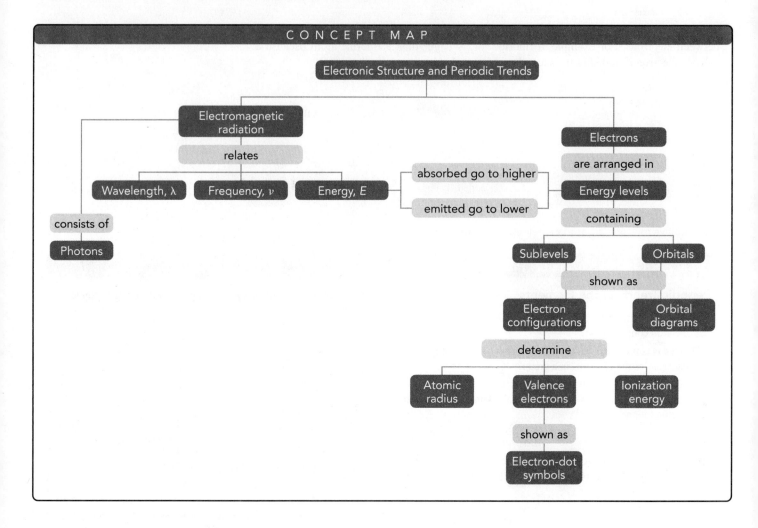

CHAPTER REVIEW

5.1 Electromagnetic Radiation

LEARNING GOAL: *Compare the wavelength of radiation with its energy.*
Electromagnetic radiation such as radio waves and visible light is energy that travels at the speed of light. Each particular type of radiation has a specific wavelength and frequency. A wavelength is the distance between a peak on one wave and a peak on the next wave. Frequency is the number of wave peaks that pass a point in 1 second. Long-wavelength radiation has low frequencies, while short-wavelength radiation has high frequencies. Radiation with a high frequency has high energy.

5.2 Atomic Spectra and Energy Levels

LEARNING GOAL: *Explain how atomic spectra correlate with the energy levels in atoms.*
The atomic spectra of elements are related to the specific energy levels occupied by electrons. Light consists of photons, which are particles of a specific energy or *quanta*. When an electron absorbs a photon of a particular energy, it attains a higher energy level. When an electron drops to a lower energy level, a photon of a particular energy is emitted. Each element has it own unique atomic spectrum.

5.3 Sublevels and Orbitals

LEARNING GOAL: *Describe the sublevels and orbitals in atoms.*
An orbital is a region around the nucleus where an electron with a specific energy is most likely to be found. Each orbital holds a maximum of two electrons, which must have opposite spins. In each energy level (*n*), electrons occupy orbitals of identical energy within sublevels. An *s* sublevel contains one *s* orbital, a *p* sublevel contains three *p* orbitals, a *d* sublevel contains five *d* orbitals, and an *f* sublevel contains seven *f* orbitals. Each type of orbital has a unique shape.

5.4 Drawing Orbital Diagrams and Writing Electron Configurations

LEARNING GOAL: *Draw the orbital diagram and write the electron configuration for an element.*
Within a sublevel, electrons enter orbitals in the same energy level one at a time until all the orbitals are half-filled. Additional electrons enter with opposite spins until the orbitals in that sublevel are filled with two electrons each. The electrons in an atom can be depicted in an orbital diagram, which shows the orbitals that are occupied by paired and unpaired electrons. The electron configuration shows the number of electrons in each sublevel. An abbreviated electron configuration places the symbol of a noble gas in brackets to represent the filled sublevels.

5.5 Electron Configurations and the Periodic Table

LEARNING GOAL: *Write the electron configuration for an atom using the sublevel blocks on the periodic table.*
The periodic table consists of *s*, *p*, *d*, and *f* sublevel blocks. An electron configuration can be written following the order of the sublevel blocks

in the periodic table. Beginning with $1s$, an electron configuration is obtained by writing the sublevel blocks in order going across the periodic table until the element is reached.

5.6 Periodic Trends of the Elements
LEARNING GOAL: *Use the electron configurations of elements to explain periodic trends.*

The properties of elements are related to the valence electrons of the atoms. With only a few minor exceptions, each group of elements has the same arrangement of valence electrons differing only in the energy level. The radius of an atom increases going down a group and decreases going across a period. The energy required to remove a valence electron is the ionization energy, which generally decreases going down a group and generally increases going across a period.

◼ KEY TERMS

atomic radius The distance of the outermost electrons from the nucleus.

atomic spectrum A series of lines specific for each element produced by photons emitted by electrons dropping to lower energy levels.

***d* block** The 10 elements in Groups 3B (3) to 2B (12) in which electrons fill the five *d* orbitals.

electromagnetic radiation Forms of energy such as visible light, microwaves, radio waves, infrared, ultraviolet light, and X-rays that travel as waves at the speed of light.

electromagnetic spectrum The arrangement of types of radiation from long wavelengths to short wavelengths.

electron-dot symbol The representation of an atom that shows valence electrons as dots around the symbol of the element.

electron configuration A list of the number of electrons in each sublevel within an atom, arranged by increasing energy.

***f* block** The 14 elements in the rows at the bottom of the periodic table in which electrons fill the seven *4f* and *5f* orbitals.

frequency The number of times the peaks of a wave pass a point in 1 second.

ionization energy The energy needed to remove the least tightly bound electron from the outermost energy level of an atom.

orbital The region around the nucleus of an atom where electrons of certain energy are most likely to be found: *s* orbitals are spherical; *p* orbitals have two lobes.

orbital diagram A diagram that shows the distribution of electrons in the orbitals of the energy levels.

***p* block** The elements in Groups 3A (13) to 8A (18) in which electrons fill the *p* orbitals.

photon The smallest particle of light.

principal quantum number (*n*) The numbers ($n = 1, n = 2, \ldots$) assigned to energy levels.

***s* block** The elements in Groups 1A (1) and 2A (2) in which electrons fill the *s* orbitals.

sublevel A group of orbitals of equal energy within energy levels. The number of sublevels in each energy level is the same as the principal quantum number (*n*).

valence electrons The electrons in the outermost energy levels.

wavelength The distance between the peaks of two adjacent waves.

◼ UNDERSTANDING THE CONCEPTS

Use the following diagram for Problems 5.73 and 5.74.

A.
B.
C.

5.73 Select diagram A, B, or C that
  a. has the longest wavelength
  b. has the shortest wavelength
  c. has the highest frequency
  d. has the lowest frequency

5.74 Select diagram A, B, or C that
  a. has the highest energy
  b. has the lowest energy
  c. would represent blue light
  d. would represent red light
  e. would represent green light

5.75 Match the following with an *s* or *p* orbital:

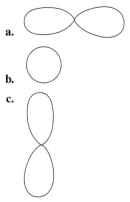

5.76 Match the following with *s* or *p* orbitals:
  a. two lobes   **b.** spherical shape
  c. found in $n = 2$ **d.** found in $n = 3$

5.77 Indicate if the following orbital diagrams are possible or not and explain. If possible, indicate the element it represents.

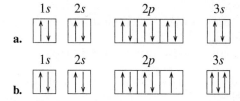

5.78 Indicate if the following abbreviated orbital diagrams are possible or not and explain. When possible, indicate the element it represents.

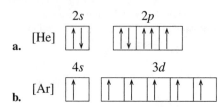

a. [He]

b. [Ar]

5.79 Match the spheres represented with atoms of Li, Na, K, and Rb.

A.   **B.**   **C.**   **D.**

5.80 Match the spheres represented with atoms of K, Ge, Ca, and Kr.

A.   **B.**   **C.**   **D.**

ADDITIONAL QUESTIONS AND PROBLEMS

For instructor-assigned homework, go to
www.masteringchemistry.com.

5.81 What is the difference between a continuous spectrum and atomic spectra?

5.82 Why does a neon sign give off red light?

5.83 What is the Pauli exclusion principle?

5.84 Why would there be five unpaired electrons in a *d* sublevel, but no paired electrons?

5.85 How are each of the following similar and how are they different?
a. 2*p* and 3*p* orbitals
b. 3*s* and 3*p* orbitals
c. the orbitals in the 4*p* sublevel

5.86 Indicate the number of unpaired electrons in each of the following:
a. chromium   **b.** an element in Group 16
c. an element in Group 3A   **d.** F, Cl, Br, and I

5.87 Which of the following orbitals are possible in an atom: 4*p*, 2*d*, 3*f*, and 5*f*?

5.88 Which of the following orbitals are possible in an atom: 1*p*, 4*f*, 6*s*, and 4*d*?

5.89 **a.** What electron sublevel starts to fill after completion of the 3*s* sublevel?
b. What electron sublevel starts to fill after completion of the 4*p* sublevel?
c. What electron sublevel starts to fill after completion of the 3*d* sublevel?
d. What electron sublevel starts to fill after completion of the 3*p* sublevel?

5.90 **a.** What electron sublevel starts to fill after completion of the 5*s* sublevel?
b. What electron sublevel starts to fill after completion of the 4*d* sublevel?
c. What electron sublevel starts to fill after completion of the 4*f* sublevel?
d. What electron sublevel starts to fill after completion of the 5*p* sublevel?

5.91 **a.** How many 3*d* electrons are in Fe?
b. How many 5*p* electrons are in Ba?
c. How many 4*d* electrons are in I?
d. How many 6*s* electrons are in Ba?

5.92 **a.** How many 4*s* electrons are in Zn?
b. How many 4*p* electrons are in Br?
c. How many 6*p* electrons are in Bi?
d. How many 5*s* electrons are in Cd?

5.93 What do the elements Ca, Sr, and Ba have in common in terms of their electron configuration? Where are they located in the periodic table?

5.94 What do the elements O, S, and Se have in common in terms of their electron configuration? Where are they located in the periodic table?

5.95 Consider three elements with the following abbreviated electron configurations:
X = [Ar]4s^2 Y = [Ne]3$s^2$3p^4
Z = [Ar]4$s^2$3d^{10}4p^4
a. Identify each element as a metal, metalloid, or nonmetal.
b. Which element has the largest atomic radius?
c. Which element has the highest ionization energy?
d. Which element has the smallest atomic radius?

5.96 Consider three elements with the following abbreviated electron configurations:
X = [Ar]$4s^2 3d^5$ Y = [Ar]$4s^2 3d^{10} 4p^1$
Z = [Ar]$4s^2 3d^{10} 4p^6$
  a. Identify each element as a metal, metalloid, or nonmetal.
  b. Which element has the smallest atomic radius?
  c. Which element has the highest ionization energy?
  d. Which element has a half-filled sublevel?

5.97 Name the element that corresponds to each of the following:
  a. $1s^2 2s^2 2p^6 3s^2 3p^3$
  b. alkali metal with the smallest atomic radius
  c. [Kr] $5s^2 4d^{10}$
  d. Group 5A element with the highest ionization energy
  e. Period 3 element with the largest atomic radius

5.98 Name the element that corresponds to each of the following:
  a. $1s^2 2s^2 2p^6 3s^2 3p^6 4s^1 3d^5$
  b. [Xe] $6s^2 4f^{14} 5d^{10} 6p^5$
  c. halogen with the highest ionization energy
  d. Group 5A element with the smallest ionization energy
  e. Period 4 element with the smallest atomic radius

5.99 An oxide ion, O^{2-}, is about twice the size of an oxygen atom. How would you explain this?

5.100 An aluminum ion, Al^{3+}, is only one-third the size of an aluminum atom. How would you explain this?

5.101 For each pair, select the smaller.
  a. Sr or Sr^{2+} **b.** Se or Se^{2-}
  c. Br^- or I^-

5.102 For each pair, select the smaller.
  a. Li^+ or Li   **b.** Br or Br^-
  c. Ca^{2+} or Mg^{2+}

5.103 Why is the ionization energy of Ca higher than K but lower than Mg?

5.104 Why is the ionization energy of Cl lower than F but higher than S?

5.105 The ionization energy generally increases going from left to right across a period. Why do Group 3A elements have a lower ionization energy than Group 2A elements?

5.106 The ionization energy generally increases going from left to right across a period. Why do Group 6A elements have a lower ionization energy than Group 5A elements?

5.107 Of the elements Na, P, Cl, and F, which
  a. is a metal?
  b. has the largest atomic radius?
  c. has the highest ionization energy?
  d. loses an electron most easily?
  e. is found in Group 7A, Period 3?

5.108 Of the elements K, Ca, Br, Kr, which
  a. is a noble gas?
  b. has the smallest atomic radius?
  c. has the lowest ionization energy?
  d. requires the most energy to remove an electron?
  e. is found in Group 2A, Period 4?

5.109 Write the abbreviated electron configuration and group number for each of the following elements:
  a. Si   **b.** Se
  c. Mn   **d.** Sb

5.110 Write the abbreviated electron configuration and group number for each of the following elements:
  a. Zn   **b.** Rh
  c. Tc   **d.** Pb

5.111 Give the symbol of the element that has the
  a. smallest atomic radius in Group 6A
  b. smallest atomic radius in Period 3
  c. highest ionization energy in Group 13
  d. lowest ionization energy in Period 3
  e. abbreviated electron configuration [Kr]$5s^2 4d^6$

5.112 Give the symbol of the element that has the
  a. largest atomic radius in Period 5
  b. largest atomic radius in Group 3A
  c. highest ionization energy in Group 18
  d. lowest ionization energy in Period 2
  e. abbreviated electron configuration [Kr]$5s^2 4d^{10} 5p^2$

CHALLENGE QUESTIONS

5.113 Compare the speed, wavelengths, and frequencies of ultraviolet light and microwaves.

5.114 Radio waves, which travel at the speed of light, are used to communicate with satellites in space. If a message from flight control requires 170 s to reach a satellite, how far, in kilometers, is the satellite from Earth?

5.115 How do scientists explain the colored lines observed in the spectra of heated atoms?

5.116 Even though H has only one electron, there are many lines in the atomic spectrum of H. Explain.

5.117 What is meant by an energy level, a sublevel, and an orbital?

5.118 In some periodic tables, H is placed in Group 1A (1). In other periodic tables, H is placed in Group 7A (17). Why?

5.119 Compare F, S, and Cl in terms of atomic radius and ionization energy.

Global Positioning Satellites (GPS) in orbit transmit microwaves to receivers for navigation and mapmaking.

ANSWERS

ANSWERS TO STUDY CHECKS

5.1 6.1 MHz

5.2 Because the spectra of elements consisted of only discrete, separated lines, scientists concluded that electrons occupied only certain energy levels in the atom.

5.3 The $1s$, $2s$, and $3s$ orbitals are all spherical, but they increase in volume because the electron is most likely to be found farther from the nucleus for higher energy levels.

5.4 $1s^2 2s^2 2p^6 3s^2 3p^4$ Complete electron configuration for sulfur (S)
[Ne] $3s^2 3p^4$ Abbreviated electron configuration for sulfur (S)

5.5 Tin has the electron configuration:
$1s^2 2s^2 2p^6 3s^2 3p^6 4s^2 3d^{10} 4p^6 5s^2 4d^{10} 5p^2$

5.6 Sulfur in Group 6A (16), Period 3, and has a $3s^2 3p^4$ valence electron configuration. Strontium in Group 2A (2), Period 5, and has a $5s^2$ valence electron configuration.

5.7 $\cdot \ddot{P} \cdot$

5.8 Ionization energy increases going across a period: Sr is lowest, Sn is higher, and I is the highest of this group.

ANSWERS TO SELECTED QUESTIONS AND PROBLEMS

5.1 The wavelength of UV light is the distance between crests of the wave.

5.3 White light has all the colors of the spectrum, including red and blue light.

5.5 6.5×10^5 Hz

5.7 6.3×10^{-7} m; 630 nm

5.9 Microwaves have a longer wavelength than ultraviolet light or X-rays.

5.11 From shortest to longest wavelength: X-rays, blue light, infrared, microwaves.

5.13 Atomic spectra consist of a series of lines separated by dark sections, indicating that the energy emitted by the elements is not continuous.

5.15 absorb

5.17 A photon in the infrared region of the spectrum is emitted when an excited electron drops to the third energy level.

5.19 The photon with greater energy is
a. green light   **b.** blue light

5.21 **a.** spherical   **b.** two lobes
c. spherical

5.23 **a.** Both are spherical.
b. Both are part of the third energy level.
c. Both contain three p orbitals.
d. All have two lobes and belong in the third energy level.

5.25 **a.** There are five orbitals in the $3d$ sublevel.
b. There is one sublevel in the $n = 1$ energy level.
c. There is one orbital in the $6s$ sublevel.
d. There are nine orbitals in the $n = 3$ energy level.

5.27 **a.** There is a maximum of two electrons in a $2p$ orbital.
b. There is a maximum of six electrons in the $3p$ sublevel.
c. There is a maximum of 32 electrons in the $n = 4$ energy level.
d. There is a maximum of 10 electrons in the $5d$ sublevel.

5.29 The electron configuration shows the number of electrons in each sublevel of an atom. The abbreviated electron configuration uses the symbol of the preceding noble gas to show completed sublevels.

5.31

5.33 **a.** N $1s^2 2s^2 2p^3$
b. Na $1s^2 2s^2 2p^6 3s^1$
c. Br $1s^2 2s^2 2p^6 3s^2 3p^6 4s^2 3d^{10} 4p^5$
d. Ni $1s^2 2s^2 2p^6 3s^2 3p^6 4s^2 3d^8$

5.35 **a.** Ca [Ar]$4s^2$ **b.** Sr [Kr]$5s^2$
c. Ga [Ar]$4s^2 3d^{10} 4p^1$ **d.** Zn [Ar]$4s^2 3d^{10}$

5.37 **a.** Li   **b.** S
c. Si   **d.** F

5.39 **a.** Al   **b.** C
c. Ar   **d.** Be

5.41 **a.** As $1s^2 2s^2 2p^6 3s^2 3p^6 4s^2 3d^{10} 4p^3$
b. Fe $1s^2 2s^2 2p^6 3s^2 3p^6 4s^2 3d^6$
c. Pd $1s^2 2s^2 2p^6 3s^2 3p^6 4s^2 3d^{10} 4p^6 5s^2 4d^8$
d. I $1s^2 2s^2 2p^6 3s^2 3p^6 4s^2 3d^{10} 4p^6 5s^2 4d^{10} 5p^5$

5.43 **a.** Ti [Ar]$4s^2 3d^2$
b. Sr [Kr]$5s^2$
c. Ba [Xe]$6s^2$
d. Pb [Xe]$6s^2 4f^{14} 5d^{10} 6p^2$

5.45 **a.** P   **b.** Co
c. Zn   **d.** Br

5.47 **a.** Ga   **b.** N
c. Xe   **d.** Zr

5.49 **a.** 10   **b.** 6
c. 3   **d.** 1

5.51 The group numbers 1A–8A or the ones digit in 1, 2, and 13–18 indicate 1–8 valence electrons. The group numbers 3–12 give the electrons in the s and d sublevel. The Group B indicates that the d sublevel is filling.

5.53 **a.** 2A (2)   **b.** 5A (15)
c. 7B (7)   **d.** 6A (16)

5.55 **a.** ns^1 **b.** $ns^2 np^2$
c. $ns^2 np^1$ **d.** $ns^2 np^3$

5.57 **a.** 3   **b.** 5
c. 2   **d.** 7

5.59 **a.** Sulfur is in Group 6A (16); $\cdot \ddot{S} :$
b. Nitrogen is in Group 5A (15); $\cdot \ddot{N} \cdot$
c. Calcium is in Group 2A (2); $\cdot Ca \cdot$
d. Sodium is in Group 1A (1); Na$\cdot$
e. Gallium is in Group 3A (13); $\cdot \ddot{Ga} \cdot$

5.61 **a.** Mg, Al, Si
b. I, Br, Cl
c. Sr, Sb, I
d. Na, Si, P

5.63 **a.** Na
b. Rb
c. Na
d. Na

5.65 **a.** Br, Cl, F
b. Na, Al, Cl
c. Cs, K, Na
d. Sn, Sb, As

5.67 **a.** Br
b. Mg
c. P
d. Xe

5.69 When a potassium ion forms, it loses the only electron in its outermost energy level and is smaller than a potassium atom.

5.71 **a.** Na
b. Cl^-
c. S^{2-}

5.73 **a.** C has the longest wavelength.
b. A has the shortest wavelength.
c. A has the highest frequency.
d. C has the lowest frequency.

5.75 **a.** p
b. s
c. p

5.77 **a.** This is possible. This element is magnesium.
b. Not possible. The 2p sublevel would fill before the 3s, and only two electrons are allowed in an s orbital.

5.79 Li is D, Na is A, K is C, and Rb is B.

5.81 A continuous spectrum from white light contains wavelengths of all energies. Atomic spectra are line spectra in which a series of lines corresponds to energy emitted when electrons drop from a higher energy level to a lower level.

5.83 The Pauli exclusion principle states that two electrons in the same orbital must have opposite spins.

5.85 **a.** A 2p and a 3p orbital have the same shape with two lobes; each p orbital can hold up to two electrons with opposite spins. However, the 3p orbital is larger because a 3p electron has a higher energy level and is more likely to be found farther from the nucleus.
b. A 3s orbital and a 3p orbital are found in the same energy level, $n = 3$, and each can hold up to two electrons with opposite spins. However, the shapes of a 3s orbital and a 3p orbital are different.
c. The orbitals in the 4p sublevel all have the same energy level and shape. However, there are three 4p orbitals directed along the $x, y,$ and z axes around the nucleus.

5.87 A 4p orbital is possible because $n = 4$ has four sublevels, including a p sublevel. A 2d orbital is not possible, because $n = 2$ has only s and p sublevels. There are no 3f orbitals, because only $s, p,$ and d sublevels are allowed for $n = 3$. A 5f sublevel is possible in $n = 5$ because five sublevels are allowed.

5.89 **a.** 3p
b. 5s
c. 4p
d. 4s

5.91 **a.** 6
b. 6
c. 10
d. 2

5.93 Ca, Sr, and Ba all have two valence electrons, ns^2, which place them in Group 2A (2).

5.95 **a.** X is a metal; Y and Z are nonmetals.
b. X has the largest atomic radius.
c. Y has the highest ionization energy.
d. Y has the smallest atomic radius.

5.97 **a.** phosphorus
b. lithium (H is a nonmetal)
c. cadmium
d. nitrogen
e. sodium

5.99 When two electrons are added to the outermost electron level in an oxygen atom, there is an increase in electron repulsion that pushes the electrons apart and increases the size of the oxide ion compared to the oxygen atom.

5.101 **a.** Sr^{2+}
b. Se
c. Br^-

5.103 Calcium has a greater number of protons than K. The least tightly bound electron in Ca is farther from the nucleus than in Mg and needs less energy to remove.

5.105 In Group 3A (13), ns^2np^1, the p electron is farther from the nucleus and easier to remove.

5.107 **a.** Na
b. Na
c. F
d. Na
e. Cl

5.109 **a.** [Ne]$3s^23p^2$; Group 4A (14)
b. [Ar]$4s^23d^{10}4p^4$; Group 6A (16)
c. [Ar]$4s^23d^5$; Group 7B (7)
d. [Kr]$5s^24d^{10}5p^3$; Group 5A (15)

5.111 **a.** O
b. Ar
c. B
d. Na
e. Ru

5.113 Ultraviolet light and microwaves both travel at the same speed: 3.0×10^8 m/s. The wavelength of ultraviolet light is shorter than the wavelength of microwaves, and the frequency of ultraviolet light is higher than the frequency of microwaves.

5.115 The series of lines separated by dark sections in spectra indicate that the energy emitted by the elements is not continuous and that electrons are moving between discrete energy levels.

5.117 The energy level contains all the electrons with similar energy. A sublevel contains electrons with the same energy, while an orbital is the region around the nucleus where electrons of a certain energy are most likely to be found.

5.119 S has a larger atomic radius than Cl; Cl is larger than F: S > Cl > F. F has a higher ionization energy than Cl; Cl has a higher ionization energy than S: F > Cl > S.

6

Inorganic and Organic Compounds: Names and Formulas

"Dentures replace natural teeth that are extracted due to cavities, bad gums, or trauma," says Dr. Irene Hilton, dentist, La Clinica de La Raza in Oakland, California. *"I make an impression of teeth using alginate, which is a polysaccharide extracted from seaweed. I mix the compound with water and place the gel-like material in the patient's mouth, where it becomes a hard cementlike substance. I fill this mold with gypsum ($CaSO_4$) and water, which form a solid to which I add teeth made of plastic or porcelain. When I get a good match to the patient's own teeth, I prepare a preliminary wax denture. This is placed in the patient's mouth to check the bite and adjust the position of the replacement teeth. Then a permanent denture is made using a hard plastic polymer (methyl methacrylate)."*

In nature, atoms of most of the elements on the periodic table are found in combination with other atoms. Only the atoms of the noble gases—He, Ne, Ar, Kr, Xe, and Rn—do not typically combine in nature with other atoms. As discussed in Chapter 3, a *compound* is a pure substance, composed of two or more elements, with a definite composition. A compound forms when atoms of one element combine with one or more atoms of a different element.

In a typical ionic compound, one or more electrons are transferred from the atoms of metals to atoms of nonmetals. The attraction that results is called an ionic bond. We use many ionic compounds every day. When we cook, we use ionic compounds such as salt ($NaCl$) and baking soda ($NaHCO_3$). Milk of magnesia ($Mg(OH)_2$) or calcium carbonate ($CaCO_3$) may be taken to settle an upset stomach. In a mineral supplement, iron may be present as iron(II) sulfate ($FeSO_4$). Certain sunscreens contain zinc oxide (ZnO), and the tin(II) fluoride (SnF_2) in toothpaste provides fluoride to help prevent tooth decay.

The structures of ionic crystals result in the beautiful facets seen in gems. Sapphires and rubies are made of aluminum oxide (Al_2O_3). Impurities of chromium make rubies red, and iron and titanium make sapphires blue.

Small amounts of metals cause the different colors of gemstones.

In compounds of nonmetals, covalent bonding occurs by atoms sharing one or more valence electrons. Covalent compounds consist of molecules, which are discrete groups of atoms. A molecule of oxygen gas (O_2) consists of two oxygen atoms; a molecule of water (H_2O) consists of two atoms of hydrogen and one atom of oxygen. There are many more covalent compounds than there are ionic compounds. Covalent compounds consisting of mostly carbon and hydrogen are known as organic compounds. Familiar examples include propane (C_3H_8), butane (C_4H_{10}), and ethyl alcohol (C_2H_6O).

Your food contains much bigger organic molecules, known as biomolecules, such as the polysaccharide starch, which contains many covalent bonds between atoms of carbon, hydrogen, and oxygen. Carbohydrates break down in digestion to provide us with glucose ($C_6H_{12}O_6$) for energy. When you have iced tea, perhaps you add sugar (sucrose), which is an organic covalent compound ($C_{12}H_{22}O_{11}$). Other organic covalent compounds include the antibiotic amoxicillin ($C_{16}H_{19}N_3O_5S$) and the antidepressant Prozac ($C_{17}H_{18}F_3NO$).

6.1 Octet Rule and Ions

Most of the elements, except the noble gases, combine to form compounds. The noble gases are so stable that they form compounds only under extreme conditions. One explanation for the stability of noble gases is that they have eight valence electrons, known as an octet. The exception is helium, which is stable with two electrons that fill its first energy level.

Compounds are the result of the formation of chemical bonds between two or more different elements. Ionic bonds occur when metal atoms lose valence electrons and atoms of nonmetals gain valence electrons. For example, sodium atoms lose electrons and chlorine atoms gain electrons to form the ionic compound $NaCl$. Covalent bonds form when atoms of nonmetals share valence electrons. In the covalent compound NCl_3, atoms of nitrogen and chlorine share electrons.

In the formation of either an *ionic bond* or a *covalent bond*, atoms lose, gain, or share valence electrons to acquire an octet of eight valence electrons. A few elements achieve the stability of helium with two valence electrons. This tendency for atoms to attain a noble gas electron configuration is known as the **octet rule** and provides a key to our understanding of the ways in which atoms bond and form compounds.

LEARNING GOAL

Using the octet rule, write the symbols of the single ions for the representative elements.

TUTORIAL
Octet Rule and Ions

CHEMISTRY AND INDUSTRY

Some Uses for Noble Gases

Noble gases may be used when an unreactive substance is required. Scuba divers normally use a pressurized mixture of nitrogen and oxygen gases for breathing under water. However, when the air mixture is used at depths where pressure is high, the nitrogen gas is absorbed into the blood, where it can cause mental disorientation. To avoid this problem, a breathing mixture of oxygen and helium may be substituted (see *Breathing Mixtures for Scuba* in Chapter 3). The diver still obtains the necessary oxygen, but the unreactive helium that dissolves in the blood does not cause mental disorientation. However, its lower density does change the vibrations of the vocal cords, and the diver will sound like Donald Duck.

Helium is also used to fill blimps and balloons. When dirigibles were first designed, they were filled with hydrogen, the lightest gas. However, when they encountered any type of spark or heating source, the dirigibles exploded violently because of the extreme reactivity of

hydrogen gas with oxygen present in the air. Today blimps are filled with unreactive helium gas, which presents no danger of explosion. Lighting bulbs are generally filled with a noble gas such as neon, because the filament will burn out when oxygen is present.

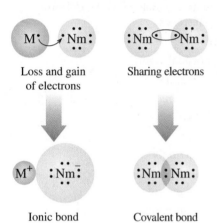

Loss and gain of electrons

Sharing electrons

Ionic bond

Covalent bond

M is a metal
Nm is a nonmetal

Positive Ions

Ions, which have electrical charges, form when atoms lose or gain electrons to form octets. Because the ionization energies of metals of Groups 1A (1), 2A (2), and 3A (13) are low, these metal atoms readily lose their valence electrons to nonmetals. In doing so, they acquire the electron configuration of a noble gas (usually 8 valence electrons) and form ions with positive charges. For example, when a sodium atom loses its single valence electron, the remaining electrons have the electron configuration of the noble gas neon. By losing an electron, sodium has 10 electrons instead of 11. Because there are still 11 protons in its nucleus, the atom is no longer neutral. It has become a sodium ion with an electrical charge, called an **ionic charge**, of 1+. In the symbol for the sodium ion, the ionic charge is written as + in the upper right-hand corner as Na^+. For 1+, the 1 is understood and not written.

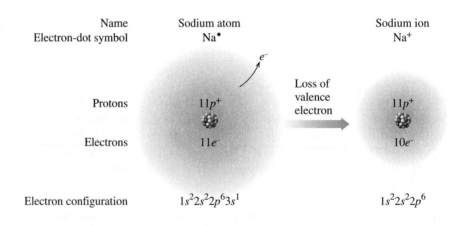

Name	Sodium atom	Sodium ion
Electron-dot symbol	$Na^{\bullet}$	Na^+
Protons	$11p^+$	$11p^+$
		Loss of valence electron
Electrons	$11e^-$	$10e^-$
Electron configuration	$1s^2 2s^2 2p^6 3s^1$	$1s^2 2s^2 2p^6$

Metals in ionic compounds lose their valence electrons to form positively charged ions called **cations** (pronounced *cat'-i-ons*). Magnesium, a metal in Group 2A (2), attains a noble gas configuration like neon by losing two valence electrons to form a positive ion with a 2+ ionic charge. A metal ion is named by its element name. Thus, Mg^{2+} is named the *magnesium* ion.

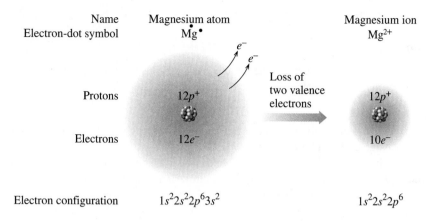

Negative Ions

In the last chapter, we learned that the ionization energies of nonmetals of Groups 5A (15), 6A (16), and 7A (17) are high. Therefore, in an ionic compound, a non-metal gains one or more electrons to become an ion with a negative charge. For example, when an atom of chlorine in Group 7A (17) with seven valence electrons gains one electron, it attains an octet to give it the electron configuration of argon. This *monatomic* ion, formed from a single atom, is called a *chloride* ion (Cl^-) and has a 1− charge. A negatively charged monatomic ion, also called an **anion** (pronounced *an'-i-on*), is named by using the first syllable of its name followed by *ide*.

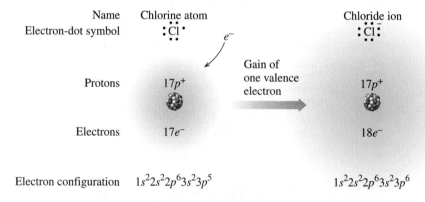

Table 6.1 lists the names of some important metal and nonmetal ions.

TABLE 6.1 Formulas and Names of Some Common Monatomic Ions

Group Number	Formula of Ion	Name of Ion	Group Number	Formula of Ion	Name of Ion
	Metals			Nonmetals	
1A (1)	Li^+	Lithium	5A (15)	N^{3-}	Nitride
	Na^+	Sodium		P^{3-}	Phosphide
	K^+	Potassium	6A (16)	O^{2-}	Oxide
2A (2)	Mg^{2+}	Magnesium		S^{2-}	Sulfide
	Ca^{2+}	Calcium	7A (17)	F^-	Fluoride
	Ba^{2+}	Barium		Cl^-	Chloride
3A (13)	Al^{3+}	Aluminum		Br^-	Bromide
				I^-	Iodide

CONCEPT CHECK 6.1

■ **Ions**

a. Write the symbol and name of the ion that has 7 protons and 10 electrons.
b. State the number of protons and electrons in a calcium ion, Ca^{2+}.

ANSWER

a. The element with 7 protons is nitrogen. An ion of nitrogen with 10 electrons has an ionic charge of 3− and is called a *nitride* ion, N^{3-}.
b. In a calcium ion, Ca^{2+}, there are 20 protons. The ionic charge of 2+ indicates a loss of 2 electrons, which gives a total of 18 electrons in the calcium ion.

Ionic Charges from Group Numbers

For the representative elements, we can use group numbers on the periodic table to determine the ionic charges. The elements in Groups 1A (1), 2A (2), and 3A (13) lose 1, 2, or 3 electrons, respectively. Group 1A (1) metals form ions with 1+ charges, Group 2A (2) metals form ions with 2+ charges, and Group 3A (13) metals form ions with 3+ charges.

The nonmetals from Groups 5A (15), 6A (16), and 7A (17) gain 3, 2, or 1 electron, respectively. Group 5A (15) nonmetals form ions with 3− charges, Group 6A (16) nonmetals form ions with 2− charges, and Group 7A (17) nonmetals form monatomic ions with 1− charges. The elements of Group 4A (14) do not typically form ions. Table 6.2 lists the ionic charges for some common ions of representative elements.

TABLE 6.2 Examples of Monatomic Ions and Their Nearest Noble Gases

Noble Gases		Metals Lose Valence Electrons			Nonmetals Gain Valence Electrons				Noble Gases
		1A (1)	2A (2)	3A (13)	5A (15)	6A (16)	7A (17)		
He	⇐	Li^+							
Ne	⇐	Na^+	Mg^{2+}	Al^{3+}	N^{3-}	O^{2-}	F^-	⇒	Ne
Ar	⇐	K^+	Ca^{2+}		P^{3-}	S^{2-}	Cl^-	⇒	Ar
Kr	⇐	Rb^+	Sr^{2+}				Br^-	⇒	Kr
Xe	⇐	Cs^+	Ba^{2+}				I^-	⇒	Xe

SAMPLE PROBLEM | 6.1

■ **Writing Ions**

Consider the elements aluminum and oxygen.

a. Identify each as a metal or a nonmetal.
b. State the number of valence electrons for each.
c. State the number of electrons that must be lost or gained for each to achieve an octet.
d. Write the symbol and name of each resulting ion, including its ionic charge.

SOLUTION

Aluminum	Oxygen
a. metal	nonmetal
b. three valence electrons	six valence electrons
c. loses 3 e^-	gains 2 e^-
d. Al^{3+}, aluminum ion	O^{2-}, oxide ion

STUDY CHECK

What are the symbols for the ions formed by potassium and sulfur?

CHEMISTRY AND HEALTH

Some Important Ions in the Body

Several ions in body fluids have important physiological and metabolic functions. Some of them are listed in Table 6.3.

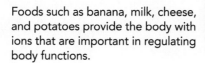
Foods such as banana, milk, cheese, and potatoes provide the body with ions that are important in regulating body functions.

TABLE 6.3 Ions in the Body

Ion	Occurrence	Function	Source	Result of Too Little	Result of Too Much
Na^+	Principal cation outside the cell	Regulation and control of body fluids	Salt	Hyponatremia, anxiety, diarrhea, circulatory failure, decrease in body fluid	Hypernatremia, little urine, thirst, edema
K^+	Principal cation inside the cell	Regulation of body fluids and cellular functions	Bananas, orange juice, milk, prunes, potatoes	Hypokalemia (hypopotassemia), lethargy, muscle weakness, failure of neurological impulses	Hyperkalemia (hyperpotassemia), irritability, nausea, little urine, cardiac arrest
Ca^{2+}	Cation outside the cell; 90% of calcium in the body in bone as $Ca_3(PO_4)_2$ or $CaCO_3$	Major cation of bone; needed for muscle contraction	Milk, yogurt, cheese, greens, spinach	Hypocalcemia, tingling fingertips, muscle cramps, osteoporosis	Hypercalcemia, relaxed muscles, kidney stones, deep bone pain
Mg^{2+}	Cation outside the cell; 70% of magnesium in the body in bone structure	Essential for certain enzymes, muscles, nerve control	Widely distributed (part of chlorophyll of all green plants), nuts, whole grains	Disorientation, hypertension, tremors, slow pulse	Drowsiness
Cl^-	Principal anion outside the cell	Gastric juice, regulation of body fluids	Salt	Same as for Na^+	Same as for Na^+

QUESTIONS AND PROBLEMS

Octet Rule and Ions

6.1 **a.** How does the octet rule explain the formation of a sodium ion?
b. Group 1A (1) and Group 2A (2) elements are found in many compounds but Group 8A (18) elements are not. Why?

6.2 **a.** How does the octet rule explain the formation of a chloride ion?
b. Group 7A (17) elements are found in many compounds, but Group 8A (18) elements are not. Why?

6.3 State the number of electrons that must be lost by atoms of each of the following elements to acquire a noble gas electron configuration:
a. Li   **b.** Mg   **c.** Al
d. Cs   **e.** Ba

6.4 State the number of electrons that must be gained by atoms of each of the following elements to acquire a noble gas electron configuration:
a. Cl   **b.** S   **c.** N
d. I   **e.** P

6.5 What noble gas has the same electron configuration as each of the following ions?
a. Li^+ **b.** Mg^{2+} **c.** K^+
d. O^{2-} **e.** Br^-

6.6 What noble gas has the same electron configuration as each of the following ions?
a. Na^+ **b.** Sr^{2+} **c.** S^{2-}
d. Al^{3+} **e.** I^-

6.7 State the number of electrons lost or gained when the following elements form ions:
a. Sr   **b.** P   **c.** Group 7A (17)
d. Na   **e.** Ga

6.8 State the number of electrons lost or gained when the following elements form ions:
a. O   **b.** Group 2A (2)   **c.** F
d. Rb   **e.** N

6.9 Write the symbols of the ions with the following number of protons and electrons:
a. 3 protons, 2 electrons   **b.** 9 protons, 10 electrons
c. 12 protons, 10 electrons   **d.** 26 protons, 23 electrons
e. 30 protons, 28 electrons

6.10 How many protons and electrons are in the following ions?
a. O^{2-} **b.** K^+ **c.** Br^-
d. S^{2-} **e.** Sr^{2+}

6.2 Ionic Compounds

LEARNING GOAL

Using charge balance, write the correct formula for an ionic compound.

TUTORIAL

Ionic Compounds

Ionic compounds consist of positive and negative ions. The ions are held together by strong electrical attractions between the opposite charges called **ionic bonds**.

Properties of Ionic Compounds

The physical and chemical properties of an ionic compound such as NaCl are very different from those of the original elements. For example, the original elements of NaCl were sodium, which is a soft, shiny metal, and chlorine, which is a yellow-green poisonous gas. However, when they react and form positive and negative ions, they produce NaCl, which is ordinary table salt, a hard, white, crystalline substance that is important in our diet.

While positive ions repel and negative ions repel, the ions of an ionic compound are arranged to allow many positive and negative attractions. Thus, in a crystal of NaCl, the larger Cl^- ions are packed together in a three-dimensional structure, while the smaller Na^+ ions occupy the holes between the Cl^- ions (see Figure 6.1). In such a structure, every Na^+ ion is surrounded by six Cl^- ions, and every Cl^- ion is surrounded by six Na^+ ions, which provides many strong electrostatic attractions between oppositely charged ions. These strong attractions between positive and negative ions account for the high melting points of ionic compounds, often more than 500 °C. For example, the melting point of NaCl is 801 °C. At room temperature, ionic compounds are solids.

Charge Balance in Ionic Compounds

The **formula** of an ionic compound indicates the number and kinds of ions that make up the ionic compound. The sum of the ionic charges in the formula is always zero, which means that the total amount of positive charge is equal to the total amount of negative charge. For example, the formula NaCl indicates that the compound consists of one

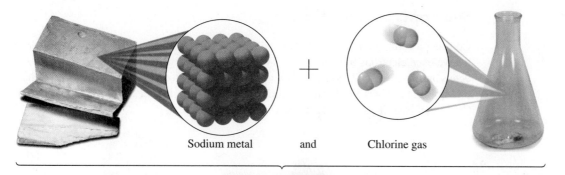

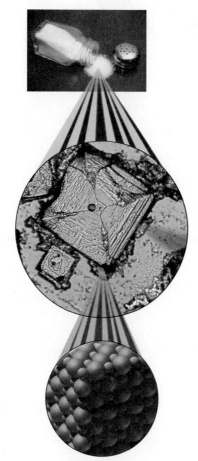

Sodium chloride

FIGURE 6.1 The elements sodium and chlorine react to form the ionic compound sodium chloride, the compound that makes up table salt. The magnification of NaCl crystals shows the arrangement of Na^+ ions and Cl^- ions in a crystal of NaCl.

Q What is the type of bonding between Na^+ and Cl^- ions in NaCl?

sodium ion, Na^+, for every chloride ion, Cl^-. Although all the ions have charges, the ionic charges are not shown in the formula of an ionic compound.

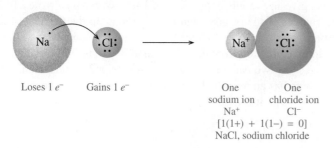

Subscripts in Formulas

Consider a compound of magnesium and chlorine. To achieve an octet, a Mg atom loses its two valence electrons to form Mg^{2+}. Two Cl atoms each gain 1 electron to complete their octets and form two Cl^- ions. The two Cl^- ions are needed to balance the positive charge of Mg^{2+}. This gives the formula $MgCl_2$, magnesium chloride, in which the subscript 2 shows that two Cl^- ions are required for charge balance.

Loses 2 e^- Each gains 1 e^-

One
magnesium ion
Mg^{2+}

Two
chloride ions
$2Cl^-$

$[1(2+) + 2(1-) = 0]$
$MgCl_2$, magnesium chloride

CONCEPT CHECK 6.2

■ Diagramming an Ionic Compound

Diagram the formation of the ionic compound aluminum fluoride, AlF_3.

ANSWER

In their electron-dot symbols, aluminum has three valence electrons and fluorine has seven. The aluminum loses its three valence electrons, and each fluorine atom gains one electron, to give ions with noble gas configurations in the ionic compound AlF_3.

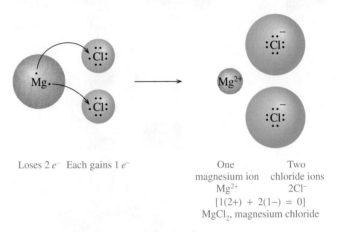

Loses
3 e^-

Each
gains
1 e^-

One
aluminum ion
Al^{3+}

Three
fluoride ions
$3F^-$

$[1(3+) + 3(1-) = 0]$

AlF_3, aluminum fluoride

Writing Ionic Formulas from Ionic Charges

The subscripts in the formula of an ionic compound represent the number of positive and negative ions that give an overall charge of zero. Thus, we can now write a formula directly from the ionic charges of the positive and negative ions. In the formula of an ionic compound, the cation is written first, followed by the anion. Suppose we wish to write the formula of the ionic compound containing Na^+ and S^{2-} ions. To balance the ionic charge of the S^{2-} ion, we show two Na^+ ions by using a subscript 2 in the formula. This gives the formula Na_2S, which has an overall charge of zero. When there is no subscript for a symbol such as the S in Na_2S, it is assumed to be 1.

(MC)

TUTORIAL
Writing Ionic Formulas

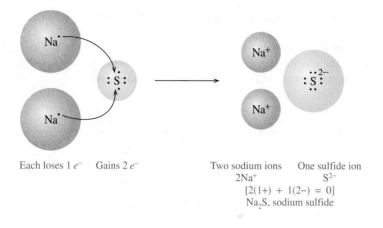

Each loses 1 e^- Gains 2 e^-

Two sodium ions   One sulfide ion
$2Na^+$ S^{2-}
$[2(1+) + 1(2-) = 0]$
Na_2S, sodium sulfide

CONCEPT CHECK 6.3

■ **Writing Formulas from Ionic Charges**

Determine the ionic charges and write the formula for the ionic compound formed when lithium and nitrogen react.

ANSWER

Lithium in Group 1A (1) forms Li^+; nitrogen in Group 5A (15) forms N^{3-}. The charge of $3-$ for N^{3-} is balanced by three Li^+ ions. Writing the positive ion first gives the formula Li_3N.

QUESTIONS AND PROBLEMS

Ionic Compounds

6.11 Which of the following pairs of elements are likely to form an ionic compound?
  a. lithium and chlorine   **b.** oxygen and bromine
  c. potassium and oxygen   **d.** sodium and neon
  e. sodium and magnesium   **f.** nitrogen and fluorine

6.12 Which of the following pairs of elements are likely to form an ionic compound?
  a. helium and oxygen   **b.** magnesium and chlorine
  c. chlorine and bromine   **d.** potassium and sulfur
  e. sodium and potassium   **f.** nitrogen and iodine

6.13 Using electron-dot symbols, diagram the formation of the following:
  a. KF   **b.** $BaCl_2$ **c.** Na_3N

6.14 Using electron-dot symbols, diagram the formation of the following:
  a. MgS   **b.** $GaCl_3$ **c.** Li_2O

6.15 Write the correct ionic formula for compounds formed between the following:
  a. Na^+ and O^{2-} **b.** Al^{3+} and Br^-
  c. Ba^{2+} and N^{3-} **d.** Mg^{2+} and F^-
  e. Al^{3+} and S^{2-}

6.16 Write the correct ionic formula for compounds formed between the following:
  a. Al^{3+} and Cl^- **b.** Ca^{2+} and S^{2-}
  c. Li^+ and S^{2-} **d.** Rb^+ and P^{3-}
  e. Cs^+ and I^-

6.17 Determine the ions and write the correct formula for ionic compounds formed by the following:
  a. sodium and sulfur   **b.** potassium and nitrogen
  c. aluminum and iodine   **d.** gallium and oxygen

6.18 Determine the ions and write the correct formula for ionic compounds formed by the following:
  a. calcium and chlorine   **b.** barium and bromine
  c. sodium and phosphorus   **d.** magnesium and oxygen

6.3 Naming and Writing Ionic Formulas

As we determined in Section 6.1, the name of a metal ion is the same as its elemental name. The name of a nonmetal ion is obtained by using the first syllable of its elemental name followed by *ide*. The name of the compound includes a space between the names of the metal and nonmetal.

LEARNING GOAL

Given the formula of an ionic compound, write the correct name; given the name of an ionic compound, write the correct formula.

Iodized salt contains KI to prevent iodine deficiency.

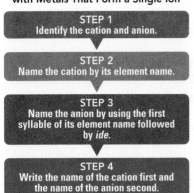

Guide to Naming Ionic Compounds with Metals That Form a Single Ion

STEP 1
Identify the cation and anion.

STEP 2
Name the cation by its element name.

STEP 3
Name the anion by using the first syllable of its element name followed by *ide*.

STEP 4
Write the name of the cation first and the name of the anion second.

Naming Ionic Compounds Containing Two Elements

In the name of an ionic compound made up of two elements, the metal ion is given first, followed by the name of the nonmetal ion. Subscripts are never mentioned; they are understood because of the charge balance of the ions in the compound.

Compound	Metal Ion	Nonmetal Ion	Name
KI	K^+ Potassium	I^- Iodide	Potassium iodide
$MgBr_2$	Mg^{2+} Magnesium	Br^- Bromide	Magnesium bromide
Al_2O_3	Al^{3+} Aluminum	O^{2-} Oxide	Aluminum oxide

SAMPLE PROBLEM | 6.2

■ Naming Ionic Compounds

Write the name of the ionic compound Mg_3N_2.

SOLUTION

STEP 1 Identify the cation and anion. The cation from Group 2A (2) is Mg^{2+} and the anion from Group 5A (15) is N^{3-}.

STEP 2 Name the cation by its element name. The cation Mg^{2+} is magnesium.

STEP 3 Name the anion by using the first syllable of its element name followed by *ide*. The anion N^{3-} is nitride.

STEP 4 Write the name of the cation first and the name of the anion second. Mg_3N_2 is named magnesium nitride.

STUDY CHECK

Name the compound Ga_2S_3.

Metals with Variable Charges

The transition metals typically form two or more kinds of positive ions because they lose their outer electrons as well as electrons from a lower energy level. For example, in some ionic compounds, iron is in the Fe^{2+} form; in other compounds, it takes the Fe^{3+} form. Copper also forms two different ions: Cu^+ is present in some compounds and Cu^{2+} in others. When a metal can form two or more ions, it is not possible to predict the ionic charge from the group number. We say that it has a *variable charge*.

When different ions are possible for a metal, a naming system is needed to identify the particular cation in a compound. To do this, a Roman numeral that is equal to the ionic charge is placed in parentheses immediately after the elemental name of the metal. For iron, Fe^{2+} is named iron(II), and Fe^{3+} is named iron(III). Table 6.4 lists the ions of some common metals that produce more than one ion.

Figure 6.2 shows some common ions and their location on the periodic table. Typically, the transition elements form more than one positive ion except for zinc (Zn^{2+}), cadmium (Cd^{2+}), and silver (Ag^+), which have fixed charges and form only one ion. Thus, the elemental names of zinc, cadmium, and silver are sufficient when naming their cations in ionic compounds. Metals in Groups 4A (14) and 5A (15) also form

TABLE 6.4 Some Metals That Form More Than One Positive Ion

Element	Possible Ions	Name of Ion
Chromium	Cr^{2+}	Chromium(II)
	Cr^{3+}	Chromium(III)
Cobalt	Co^{2+}	Cobalt(II)
	Co^{3+}	Cobalt(III)
Copper	Cu^{+}	Copper(I)
	Cu^{2+}	Copper(II)
Gold	Au^{+}	Gold(I)
	Au^{3+}	Gold(III)
Iron	Fe^{2+}	Iron(II)
	Fe^{3+}	Iron(III)
Lead	Pb^{2+}	Lead(II)
	Pb^{4+}	Lead(IV)
Manganese	Mn^{2+}	Manganese(II)
	Mn^{3+}	Manganese(III)
Mercury	Hg_2^{2+}	Mercury(I)*
	Hg^{2+}	Mercury(II)
Nickel	Ni^{2+}	Nickel(II)
	Ni^{3+}	Nickel(III)
Tin	Sn^{2+}	Tin(II)
	Sn^{4+}	Tin(IV)

*Mercury(I) ions form pairs with a 2+ charge

more than one positive ion. For example, lead and tin in Group 4A (14) form cations with charges of 2+ and 4+.

When a Roman numeral in parentheses is needed, its value is the ionic charge of the metal ion in the formula. For example, we use charge balance to calculate the charge of the copper cation in the formula $CuCl_2$. Two chloride ions each have a 1− charge, which gives

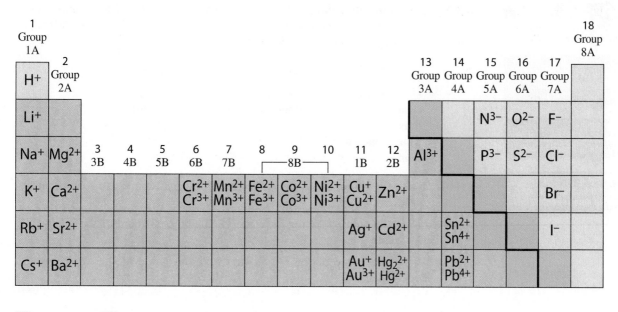

Metals   Metalloids   Nonmetals

FIGURE 6.2 On the periodic table, positive ions are produced from metals and negative ions are produced from nonmetals.

Q What are the typical ions produced by calcium, copper, and oxygen?

TABLE 6.5 Some Ionic Compounds of Metals That Form Two Kinds of Positive Ions

Compound	Systematic Name
$FeCl_2$	Iron(II) chloride
Fe_2O_3	Iron(III) oxide
Cu_3P	Copper(I) phosphide
$CuBr_2$	Copper(II) bromide
$SnCl_2$	Tin(II) chloride
PbS_2	Lead(IV) sulfide

a total negative charge of 2−. To balance the 2− charge, the copper ion must have a positive charge of 2+, which is a Cu^{2+} ion:

$$CuCl_2$$

Cu charge $+ 2Cl^-$ charge	= 0	
(?) $+ 2(1-)$	= 0	
(2+) $+ 2-$	= 0	

To indicate the copper ion Cu^{2+}, we place (II) immediately after *copper* when naming the compound: copper(II) chloride

Table 6.5 lists names of some ionic compounds in which the metals form more than one positive ion.

SAMPLE PROBLEM | 6.3

■ **Naming Ionic Compounds with Variable Charge Metal Ions**

Antifouling paint contains Cu_2O, which prevents the growth of barnacles and algae on the bottom of boats. What is the name of Cu_2O?

SOLUTION

Guide to Naming Ionic Compounds with Variable Charge Metals

> **STEP 1**
> Determine the charge of the cation from the anion.

> **STEP 2**
> Name the cation by its element name and use a Roman numeral in parentheses for the charge.

> **STEP 3**
> Name the anion by using the first syllable of its element name followed by *ide*.

> **STEP 4**
> Write the name of the cation first and the name of the anion second.

STEP 1 **Determine the charge of the cation from the anion.** The nonmetal O in Group 6A (16) forms the O^{2-} ion. Because there are two Cu ions to balance the O^{2-}, the charge of each Cu ion is 1+.

	Metal	Nonmetal
Elements	Copper	Oxygen
Groups	Transition elements	6A (16)
Ions	Cu?	O^{2-}
Charge balance	2(1+) $+$	(2−) = 0
Ions	Cu^+	O^{2-}

STEP 2 **Name the cation by its element name and use a Roman numeral in parentheses for the charge.** copper(I)

STEP 3 **Name the anion by using the first syllable of its element name followed by *ide*.** oxide

STEP 4 **Write the name of the cation first and the name of the anion second.** copper(I) oxide

STUDY CHECK

Write the name of the compound whose formula is $AuCl_3$.

Writing Formulas from the Name of an Ionic Compound

The formula of an ionic compound is written from the first part of the name that describes the metal ion and the second part that specifies the nonmetal. Subscripts are added as needed to balance the charge. The steps for writing a formula from the name of an ionic compound are shown in Sample Problem 6.4.

TUTORIAL
Writing Ionic Formulas

SAMPLE PROBLEM | **6.4**

■ **Writing Formulas of Ionic Compounds**

Write the formula of aluminum sulfide.

SOLUTION

STEP 1 Identify the cation and anion.

	Cation	Anion
Ions	Aluminum	Sulfide
Groups	3A (13)	6A (16)
Symbols	Al^{3+}	S^{2-}

STEP 2 Balance the charges. Two Al^{3+} ions (6+) are needed to balance the charges of three S^{2-} ions (6−).

$$Al^{3+} \qquad S^{2-}$$
$$Al^{3+} \qquad S^{2-}$$
$$\qquad \quad S^{2-}$$
$$\overline{2(3+) \; + \; 3(2-) = 0}$$

Use as subscripts in formula

STEP 3 Write the formula, cation first, using the subscripts from the charge balance.

$$Al_2S_3$$

STUDY CHECK

Write the ions and the formula for each of the following ionic compounds:

a. magnesium bromide
b. lithium oxide

Guide to Writing Formulas from the Name of an Ionic Compound

> **STEP 1**
> Identify the cation and anion.

> **STEP 2**
> Balance the charges.

> **STEP 3**
> Write the formula, cation first, using the subscripts from the charge balance.

SAMPLE PROBLEM | **6.5**

■ **Writing Formulas of Ionic Compounds**

Write the formula for iron(III) chloride.

SOLUTION

STEP 1 Identify the cation and anion. The Roman numeral (III) indicates that the charge of the iron ion is 3+, Fe^{3+}.

	Cation	Anion
Ions	Iron(III)	Chloride
Groups	Transition elements	7A (17)
Symbols	Fe^{3+}	Cl^-

STEP 2 Balance the charges.

$$Fe^{3+} \quad Cl^-$$
$$Cl^-$$
$$\underline{\qquad\qquad Cl^- \qquad\qquad}$$
$$\mathbf{1}(3+) + \mathbf{3}(1-) = 0$$

Becomes a subscript in the formula

STEP 3 Write the formula, cation first, using the subscripts from the charge balance.

$$FeCl_3$$

STUDY CHECK

The pigment chrome oxide green contains chromium(III) oxide. Write the correct formula for chromium(III) oxide.

QUESTIONS AND PROBLEMS

Naming and Writing Ionic Formulas

6.19 Write the symbol for the ion of each of the following:
  a. chlorine   **b.** potassium
  c. oxygen   **d.** aluminum

6.20 Write the symbol for the ion of each of the following:
  a. fluorine   **b.** strontium
  c. sodium   **d.** lithium

6.21 What is the name of each of the following ions?
  a. Li^+ **b.** S^{2-}
  c. Ca^{2+} **d.** N^{3-}

6.22 What is the name of each of the following ions?
  a. Mg^{2+} **b.** Ba^{2+}
  c. I^- **d.** Br^-

6.23 Write names for the following ionic compounds:
  a. Al_2O_3 **b.** $CaCl_2$
  c. Na_2O **d.** Mg_3P_2
  e. KI   **f.** BaF_2

6.24 Write names for the following ionic compounds:
  a. $MgCl_2$ **b.** K_3P
  c. Li_2S **d.** CsF
  e. MgO   **f.** $SrBr_2$

6.25 Why is a Roman numeral placed after the name of most transition element ions?

6.26 The compound $CaCl_2$ is named calcium chloride; the compound $CuCl_2$ is named copper(II) chloride. Explain why a Roman numeral is used in one name but not in the other.

6.27 Write the name for each of the following (include the Roman numeral when necessary):
  a. Fe^{2+} **b.** Cu^{2+}
  c. Zn^{2+} **d.** Pb^{4+}
  e. Cr^{3+} **f.** Mn^{2+}

6.28 Write the name for each of the following (include the Roman numeral when necessary):
  a. Ag^+ **b.** Cu^+ **c.** Fe^{3+}
  d. Sn^{2+} **e.** Au^{3+} **f.** Ni^{2+}

6.29 Write the name for each of the following:
  a. $SnCl_2$ **b.** FeO
  c. Cu_2S **d.** CuS
  e. $CdBr_2$ **f.** $HgCl_2$

6.30 Write the name for each of the following:
  a. Ag_3P **b.** PbS
  c. SnO_2 **d.** $MnCl_3$
  e. Cr_2O_3 **f.** CoS

6.31 Give the symbol of the cation in each of the following:
  a. $AuCl_3$ **b.** Fe_2O_3
  c. PbI_4 **d.** $SnCl_2$

6.32 Give the symbol of the cation in each of the following:
  a. $FeCl_2$ **b.** CrO
  c. Mn_2S_3 **d.** AlP

6.33 Write the formula for each of the following:
  a. magnesium chloride   **b.** sodium sulfide
  c. copper(I) oxide   **d.** zinc phosphide
  e. gold(III) nitride

6.34 Write the formula for each of the following:
  a. nickel(III) oxide   **b.** barium fluoride
  c. tin(IV) chloride   **d.** silver sulfide
  e. copper(II) iodide

6.35 Write the formula for each of the following:
  a. cobalt(III) chloride   **b.** lead(IV) oxide
  c. silver iodide   **d.** calcium nitride
  e. copper(I) phosphide   **f.** chromium(II) chloride

6.36 Write the formula for each of the following:
  a. zinc bromide   **b.** iron(III) sulfide
  c. manganese(IV) oxide   **d.** chromium(III) iodide
  e. lithium nitride   **f.** gold(I) oxide

6.4 Polyatomic Ions

A **polyatomic ion** is a group of two or more atoms that has an ionic charge. Most poly-atomic ions consist of a nonmetal such as phosphorus, sulfur, carbon, or nitrogen bonded to oxygen atoms. These oxygen-containing polyatomic ions have an ionic charge of $1-$, $2-$, or $3-$. Only one common polyatomic ion, NH_4^+, is positively charged. Some common polyatomic ions are shown in Figure 6.3.

Plaster molding
$CaSO_4$

Fertilizer
NH_4NO_3

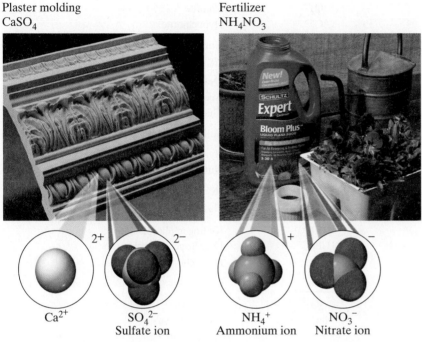

FIGURE 6.3 Many products contain polyatomic ions, which are groups of atoms that carry an ionic charge.

Q What is the charge of a sulfate ion?

Naming Polyatomic Ions

The names of the most common negatively charged polyatomic ions end in *ate*. When a related ion has one less oxygen atom, the *ite* ending is used for its name. The hydroxide ion (OH^-) and cyanide ion (CN^-) are exceptions to this naming pattern. By recognizing these endings, you can identify a polyatomic ion in the name of a compound.

There is no easy way to learn polyatomic ions. By memorizing the formula and the name of the common ion shown in boxes in Table 6.6, you can derive the related ions. Note that the *ate* and *ite* ions of a particular nonmetal have the same ionic charge. For example, the sulfate ion is SO_4^{2-}. We write the formula of the sulfite ion also with a $2-$ charge, but with one less oxygen atom, SO_3^{2-}. Phosphate and phosphite ions each have $3-$ charges; nitrate and nitrite have $1-$ charges; and chlorate and chlorite (and other halogens) have $1-$ charges.

The formula of hydrogen carbonate, or *bicarbonate,* can be written by placing a hydrogen cation (H^+) in front of the polyatomic ion formula for carbonate (CO_3^{2-}), and decreasing the charge from $2-$ to $1-$ to give HCO_3^-.

$$CO_3^{2-} + H^+ = HCO_3^-$$

The elements in Group 7A (17) form more than two types of polyatomic anions. Prefixes are added to the names, and the ending is changed to distinguish among these ions. The prefix *per* is used for the polyatomic ion that has one more oxygen atom than the *ate*

TABLE 6.6 Names and Formulas of Some Common Polyatomic Ions

Nonmetal	Formula of Ion[a]	Name of Ion
Hydrogen	OH^-	Hydroxide
Nitrogen	NH_4^+	Ammonium
	$\boxed{NO_3^-}$	**Nitrate**
	NO_2^-	Nitrite
Chlorine	ClO_4^-	Perchlorate
	$\boxed{ClO_3^-}$	**Chlorate**
	ClO_2^-	Chlorite
	ClO^-	Hypochlorite
Carbon	$\boxed{CO_3^{2-}}$	**Carbonate**
	HCO_3^-	Hydrogen carbonate (or bicarbonate)
	CN^-	Cyanide
	$C_2H_3O_2^-$	Acetate
	SCN^-	Thiocyanate
Sulfur	$\boxed{SO_4^{2-}}$	**Sulfate**
	HSO_4^-	Hydrogen sulfate (or bisulfate)
	SO_3^{2-}	Sulfite
	HSO_3^-	Hydrogen sulfite (or bisulfite)
Phosphorus	$\boxed{PO_4^{3-}}$	**Phosphate**
	HPO_4^{2-}	Hydrogen phosphate
	$H_2PO_4^-$	Dihydrogen phosphate
	PO_3^{3-}	Phosphite
Chromium	$\boxed{CrO_4^{2-}}$	**Chromate**
	$Cr_2O_7^{2-}$	Dichromate
Manganese	MnO_4^-	Permanganate

[a]Boxed formulas are the most common polyatomic ion for that element.

form of the polyatomic ion. The ending *ate* is changed to *ite* for the ion with one less oxygen. The prefix *hypo* is used for the polyatomic ion that has one less oxygen than in the *ite* form. We can see this in the polyatomic ions of chlorine combined with oxygen. Note that all the polyatomic ions of chlorine have the same ionic charge.

ClO_4^-	*per*chlor*ate* ion	one more O than common (*ate*) ion
ClO_3^-	chlor*ate* ion	most common ion
ClO_2^-	chlor*ite* ion	one less O than common (*ate*) ion
ClO^-	*hypo*chlor*ite* ion	one less O than *ite* ion

Writing Formulas for Compounds Containing Polyatomic Ions

No polyatomic ion exists by itself. Like any ion, a polyatomic ion must be associated with ions of opposite charge. The bonding between polyatomic ions and other ions is one of electrical attraction. For example, the compound sodium sulfate consists of sodium ions (Na^+) and sulfate ions (SO_4^{2-}) held together by ionic bonds.

To write correct formulas for compounds containing polyatomic ions, we follow the same rules of charge balance that we used for writing the formulas of simple ionic compounds. The total negative and positive charges must equal zero. For example, consider the formula for a compound containing calcium ions and carbonate ions. The ions are written as

$$Ca^{2+} \qquad CO_3^{2-}$$

Calcium ion   Carbonate ion

$$(2+) + (2-) = 0$$

Because one ion of each balances the charge, the formula is written as

$CaCO_3$
Calcium carbonate

When more than one polyatomic ion is needed for charge balance, parentheses are used to enclose the formula of the ion. A subscript is written outside the closing parenthesis to indicate the number needed for charge balance. Consider the formula for magnesium nitrate. The ions in this compound are the magnesium ion and the nitrate ion, a polyatomic ion.

Mg^{2+} NO_3^-
Magnesium ion   Nitrate ion

To balance the 2+ charge of magnesium, two nitrate ions are needed. In the formula, parentheses are placed around the nitrate ion, as follows:

NO_3^-
Mg^{2+}
NO_3^-
$1(2+) + 2(1-) = 0$

Magnesium nitrate

$Mg(NO_3)_2$

Parentheses enclose the formula of the nitrate ion

Subscript outside the parentheses indicates the use of two nitrate ions

---

SAMPLE PROBLEM | 6.6

■ **Writing Formulas with Polyatomic Ions**

Write the formula of aluminum bicarbonate.

SOLUTION

STEP 1 **Identify the cation and anion.** The cation is aluminum, Al^{3+}, and the anion is bicarbonate, HCO_3^-.

Cation	Anion
Al^{3+}	HCO_3^-

STEP 2 **Balance the charges.**

Al^{3+} HCO_3^-
                  HCO_3^-
                  HCO_3^-
$1(3+) + 3(1-) = 0$

Becomes a subscript in the formula

STEP 3 **Write the formula, cation first, using the subscripts from the charge balance.** The formula for the compound is written by enclosing the formula of the bicarbonate ion, HCO_3^-, in parentheses and writing the subscript 3 outside the right parenthesis.

$Al(HCO_3)_3$

STUDY CHECK

Write the formula for a compound containing ammonium ions and phosphate ions.

CONCEPT CHECK 6.4

■ **Polyatomic Ions in Bones and Teeth**

Bones and teeth contain a mineral substance called hydroxyapatite, $Ca_{10}(PO_4)_6(OH)_2$, a solid formed from calcium ions, phosphate ions, and hydroxide ions. What polyatomic ions are contained in the mineral substance of bone and teeth?

ANSWER

The polyatomic ions, which contain atoms of two or more elements, are phosphate (PO_4^{3-}) ions and hydroxide (OH^-) ions.

Naming Compounds Containing Polyatomic Ions

When naming ionic compounds containing polyatomic ions, we write the positive ion, usually a metal, first, and then we write the name of the polyatomic ion. It is important that you learn to recognize the polyatomic ion in the formula and name it correctly. As with other ionic compounds, no prefixes are used.

Na_2SO_4	$FePO_4$	$Al_2(CO_3)_3$
Na_2 $\boxed{SO_4}$	Fe $\boxed{PO_4}$	Al_2 ($\boxed{CO_3}$)$_3$
Sodium sulfate	Iron(III) phosphate	Aluminum carbonate

Table 6.7 lists the formulas and names of some ionic compounds that include polyatomic ions and also gives their uses in medicine and industry.

TABLE 6.7 Some Compounds That Contain Polyatomic Ions

Formula	Name	Use
$BaSO_4$	Barium sulfate	Contrast medium for X-rays
$CaCO_3$	Calcium carbonate	Antacid, calcium supplement
$CaSO_3$	Calcium sulfite	Preservative in cider and fruit juices
$CaSO_4$	Calcium sulfate	Plaster casts
$AgNO_3$	Silver nitrate	Topical anti-infective
$NaHCO_3$	Sodium bicarbonate	Antacid
$Zn_3(PO_4)_2$	Zinc phosphate	Dental cements
$FePO_4$	Iron(III) phosphate	Food and bread enrichment
K_2CO_3	Potassium carbonate	Alkalizer, diuretic
$Al_2(SO_4)_3$	Aluminum sulfate	Antiperspirant, anti-infective
$AlPO_4$	Aluminum phosphate	Antacid
$MgSO_4$	Magnesium sulfate	Cathartic, Epsom salts

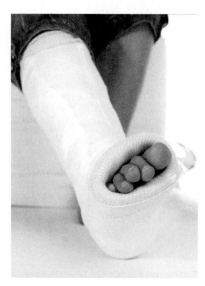

A plaster cast made of $CaSO_4$ immobilizes a broken leg.

SAMPLE PROBLEM 6.7

■ Naming Compounds Containing Polyatomic Ions

Name the following ionic compounds:
a. $Cu(NO_2)_2$
b. $KClO_3$
c. $Mn(OH)_2$

SOLUTION

We can name compounds with polyatomic ions by separating the compound into a cation and an anion, which is usually the polyatomic ion.

	STEP 1		STEP 2	STEP 3	STEP 4
Formula	Cation	Anion	Name of Cation	Name of Anion	Name of Compound
a. $Cu(NO_2)_2$	Cu^{2+}	NO_2^-	Copper(II) ion	Nitrite ion	Copper(II) nitrite
b. $KClO_3$	K^+	ClO_3^-	Potassium ion	Chlorate ion	Potassium chlorate
c. $Mn(OH)_2$	Mn^{2+}	OH^-	Manganese(II) ion	Hydroxide ion	Manganese(II) hydroxide

STUDY CHECK

What is the name of $Ca_3(PO_4)_2$, which is used to replenish calcium?

Guide to Naming Ionic Compounds with Polyatomic Ions

STEP 1
Identify the cation and polyatomic ion (anion).

STEP 2
Name the cation using a Roman numeral, if needed.

STEP 3
Name the polyatomic ion usually ending with *ite* or *ate*.

STEP 4
Write the name of the compound, cation first and the polyatomic ion second.

Summary of Naming Ionic Compounds

Throughout this chapter, we have examined strategies for naming ionic compounds. Now we can summarize the rules, as illustrated in Figure 6.4. In general, ionic compounds having two elements are named by stating the first element, followed by the second element with an *ide* ending. For ionic compounds, it is necessary to determine whether the metal can form more than one positive ion; if so, a Roman numeral following the name of the metal indicates the particular ionic charge. Ionic compounds having three or more elements include some type of polyatomic ion. They are named by ionic rules but usually have an *ate* or *ite* ending when the polyatomic ion has a negative charge.

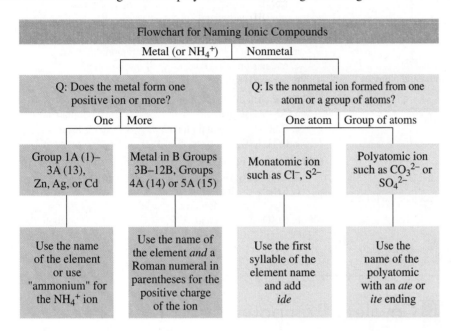

FIGURE 6.4 A flowchart for naming ionic compounds.

Q Why are the names of some metal ions followed by a Roman numeral in the name of a compound?

SAMPLE PROBLEM | 6.8

■ Naming Ionic Compounds

Name the following compounds:

a. Na_3P **b.** $CuSO_4$ **c.** $Cr(ClO)_3$

SOLUTION

Formula	STEP 1 Cation Anion		STEP 2 Name of Cation	STEP 3 Name of Anion	STEP 4 Name of Compound
a. Na_3P	Na^+	P^{3-}	Sodium ion	Phosphide ion	Sodium phosphide
b. $CuSO_4$	Cu^{2+}	SO_4^{2-}	Copper(II) ion	Sulfate ion	Copper(II) sulfate
c. $Cr(ClO)_3$	Cr^{3+}	ClO^-	Chromium(III) ion	Hypochlorite ion	Chromium(III) hypochlorite

STUDY CHECK

What is the name of $Fe(NO_3)_2$?

QUESTIONS AND PROBLEMS

Polyatomic Ions

6.37 Write the formulas including the charge for the following polyatomic ions:
    a. bicarbonate   **b.** ammonium   **c.** phosphate
    d. hydrogen sulfate   **e.** hypochlorite

6.38 Write the formulas including the charge for the following polyatomic ions:
    a. nitrite   **b.** sulfite   **c.** hydroxide
    d. phosphite   **e.** acetate

6.39 Name the following polyatomic ions:
   a. SO_4^{2-} **b.** CO_3^{2-} **c.** PO_4^{3-}
   d. NO_3^- **e.** ClO_4^-

6.40 Name the following polyatomic ions:
   a. OH^- **b.** HSO_3^- **c.** CN^-
   d. NO_2^- **e.** CrO_4^{2-}

6.41 Complete the following table with the formula and name of the compound:

	NO_2^-	CO_3^{2-}	HSO_4^-	PO_4^{3-}
Li^+				
Cu^{2+}				
Ba^{2+}				

6.42 Complete the following table with the formula and name of the compound:

	NO_3^-	HCO_3^-	SO_3^{2-}	HPO_4^{2-}
NH_4^+				
Al^{3+}				
Pb^{4+}				

6.43 Write the formula for the polyatomic ion in each of the following and name each compound:
   a. Na_2CO_3 **b.** NH_4Cl
   c. Na_3PO_3 **d.** $Mn(NO_2)_2$
   e. $FeSO_3$ **f.** $KC_2H_3O_2$

6.44 Write the formula for the polyatomic ion in each of the following and name each compound:
   a. KOH **b.** $NaNO_3$
   c. Au_2CO_3 **d.** $NaHCO_3$
   e. $BaSO_4$ **f.** $Ca(ClO)_2$

Questions 6.45 to 6.48 are a review of ionic compounds with either a monatomic anion or a polyatomic ion.

6.45 Write the correct formula for the following:
   a. barium hydroxide   **b.** sodium sulfate
   c. iron(II) nitrate   **d.** zinc phosphate
   e. iron(III) carbonate

6.46 Write the correct formula for the following:
   a. aluminum chlorate   **b.** ammonium oxide
   c. magnesium bicarbonate **d.** sodium nitrite
   e. copper(I) sulfate

6.47 Name each of the following compounds:
   a. $Al_2(SO_4)_3$ antiperspirant   **b.** $CaCO_3$ antacid
   c. Cr_2O_3 green pigment   **d.** Na_3PO_4 laxative
   e. $(NH_4)_2SO_4$ fertilizer   **f.** Fe_2O_3 pigment

6.48 Name each of the following compounds:
   a. $Co_3(PO_4)_2$ violet pigment   **b.** $Mg_3(PO_4)_2$ antacid
   c. $FeSO_4$ iron supplement   **d.** $MgSO_4$ Epsom salts
                in vitamins
   e. Cu_2O fungicide   **f.** SnF_2 tooth decay
                                  preventative

6.5 Covalent Compounds and Their Names

LEARNING GOAL

Given the formula of a covalent compound, write its correct name; given the name of a covalent compound, write its formula.

SELF STUDY ACTIVITY
Covalent Bonds

TUTORIAL
Writing Electron-Dot Formulas

A **covalent compound** forms when atoms of two nonmetals share electrons. Because of the high ionization energies of the nonmetals, electrons are not transferred between atoms but are shared to achieve stability. When atoms share electrons, the bond is a **covalent bond**. When two or more atoms share electrons, they form a **molecule**.

Formation of a Hydrogen Molecule

The simplest covalent molecule is hydrogen gas, H_2. When two hydrogen atoms are far apart, there is no attraction between them. As the H atoms move closer, the positive charge of each nucleus attracts the electron of the other atom. This attraction, which is greater than the repulsion between the valence electrons, pulls the atoms closer until they share a pair of valence electrons (see Figure 6.5). In this covalent bond, the shared electrons provide each H atom with the noble gas configuration of He. Thus, the H_2 molecule is more stable than two individual H atoms.

Electron-Dot Formulas of Covalent Molecules

The valence electrons in covalent molecules are shown using an electron-dot formula, or Lewis structure. The shared electrons, or **bonding pairs**, are shown as two dots or a single line between atoms. The nonbonding pairs of electrons, or **lone pairs**, are placed on the outside. For example, a fluorine molecule (F_2) consists of two fluorine atoms, Group 7A (17), each with seven valence electrons. Both F atoms achieve octets by sharing their unpaired valence electrons. In the F_2 molecule, each F atom has the noble gas configuration of neon.

A lone pair → A bonding pair →

$$:\ddot{F}\cdot + \cdot\ddot{F}: \longrightarrow :\ddot{F}:\ddot{F}: \qquad :\ddot{F} — \ddot{F}: \qquad = \qquad F_2$$

Electrons to share   A shared pair of electrons   A covalent bond   A fluorine molecule

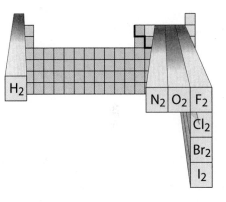

FIGURE 6.5 A covalent bond forms as H atoms move close together to share electrons.

Q What determines the attraction between two H atoms?

Increasing energy

H·

·H Far apart; no attractions

H· ·H Attractions pull atoms closer

H:H

H_2 molecule

Distance between nuclei decreases ⟶

Hydrogen (H_2) and fluorine (F_2) are examples of nonmetals whose natural state is diatomic; that is, they contain two like atoms. The elements that exist as diatomic molecules are listed in Table 6.8.

Sharing Electrons Between Atoms of Different Elements

The number of electrons that a nonmetal atom shares and the number of covalent bonds it forms are usually equal to the number of electrons it needs to acquire a noble gas configuration. For example, the element carbon combines with hydrogen to form many covalent compounds known as *organic compounds*. A carbon atom has four valence electrons and needs to acquire four more electrons for an octet. Thus, carbon shares four valence electrons with four hydrogen atoms, each sharing one electron, which forms four covalent bonds. The compound, CH_4, called methane, is a component of natural gas. The electron-dot formula for the methane molecule is written with the carbon atom in the center and the hydrogen atoms on each of the sides. Table 6.9 gives the formulas and three-dimensional models of some covalent molecules.

TABLE 6.8 Elements That Exist as Diatomic, Covalent Molecules

Element	Diatomic Molecule	Name
H	H_2	Hydrogen
N	N_2	Nitrogen
O	O_2	Oxygen
F	F_2	Fluorine
Cl	Cl_2	Chlorine
Br	Br_2	Bromine
I	I_2	Iodine

TABLE 6.9 Electron-Dot Formulas for Some Covalent Compounds

CH_4	NH_3	H_2O

Formulas Using Electron Dots

H
··
H:C:H
··
H

H:N:H
··
H

··
:O:H
··
H

Formulas Using Bonds and Electron Dots

H
|
H—C—H
|
H

H—N̈—H
|
H

:Ö—H
|
H

Molecular Models

Methane molecule Ammonia molecule Water molecule

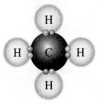

Methane, CH_4

TUTORIAL

Covalent Molecules and the Octet Rule

TABLE 6.10 Prefixes Used in Naming Covalent Compounds

1 mono	6 hexa
2 di	7 hepta
3 tri	8 octa
4 tetra	9 nona
5 penta	10 deca

Names and Formulas of Covalent Compounds

When naming a covalent compound, the first nonmetal in the formula is named by its elemental name; the second nonmetal is named by using the first syllable of its elemental name with the ending changed to *ide*. Subscripts that indicate two or more atoms of an element are expressed as prefixes placed in front of each name. Table 6.10 lists prefixes used in naming covalent compounds. The names of covalent compounds need prefixes because several different compounds can form from the same two nonmetals. For example, carbon and oxygen can form two different compounds, carbon monoxide (CO) and carbon dioxide (CO_2). When the vowels *o* and *o* or *a* and *o* appear together, the first vowel is omitted, as in carbon monoxide. With the exception of carbon monoxide, the prefix *mono* is omitted, as in NO, nitrogen oxide. Table 6.11 lists some common covalent compounds.

TABLE 6.11 Some Common Covalent Compounds

Formula	Name	Commercial Uses
CS_2	Carbon disulfide	Manufacture of rayon
CO_2	Carbon dioxide	Carbonation of beverages, fire extinguishers, propellant in aerosols, dry ice
NO	Nitrogen oxide	Stabilizer
N_2O	Dinitrogen oxide	Inhaled anesthetic, "laughing gas"
SO_2	Sulfur dioxide	Preserving fruits, vegetables; disinfectant in wineries; bleaching textiles
SO_3	Sulfur trioxide	Manufacture of explosives
SF_6	Sulfur hexafluoride	Electrical circuits (insulation)

CONCEPT CHECK 6.5

■ Naming Covalent Compounds

Why does the name of the covalent compound BrCl, bromine chloride, not include a prefix, but the name of SCl_2, sulfur dichloride, does?

ANSWER

When a formula has one atom of each element, no prefix (mono) is needed, as in the name bromine chloride. In SCl_2, two atoms of chlorine are indicated by the prefix *di*, sulfur dichloride.

SAMPLE PROBLEM | **6.9**

■ Naming Covalent Compounds

Name the covalent compound P_4O_6.

SOLUTION

STEP 1 Name the first nonmetal by its element name. In P_4O_6, the first nonmetal (P) is phosphorus.

STEP 2 Name the second nonmetal by using the first syllable of its element name followed by *ide*. The second nonmetal (O) is named oxide.

STEP 3 Add prefixes to indicate the number of atoms (subscripts). Because there are four P atoms, we use the prefix *tetra* to write *tetraphosphorus*. The six oxygen atoms use the prefix *hexa*, which gives the name *hexoxide*. When the vowels *a* and *o* appear together, as in *hexa + oxide*, the ending (*a*) of the prefix is dropped.

P_4O_6 *tetra*phosphorus *hex*oxide

STUDY CHECK

Write the name of each of the following compounds:

a. $SiBr_4$ **b.** Br_2O

Guide to Naming Covalent Compounds with Two Nonmetals

STEP 1
Name the first nonmetal by its element name.

STEP 2
Name the second nonmetal by using the first syllable of its element name followed by *ide*.

STEP 3
Add prefixes to indicate the number of atoms (subscripts).

Writing Formulas from the Names of Covalent Compounds

In the name of a covalent compound, the names of two nonmetals are given, along with prefixes for the number of atoms of each. To write the formula from the name, we use the symbol for each element and a subscript for any prefix.

SAMPLE PROBLEM | 6.10

■ **Writing Formulas for Covalent Compounds**

Write the formula for diboron trioxide.

SOLUTION

STEP 1 **Write the symbols in order of the elements in the name.** The first nonmetal is boron, and the second nonmetal is oxygen.

B O

STEP 2 **Write any prefixes as subscripts.** The prefix *di* in *di*boron indicates that there are two atoms of boron, shown as a subscript 2 in the formula. The prefix *tri* in *tri*oxide indicates that there are three atoms of oxygen, shown as a subscript 3 in the formula.

B_2O_3

STUDY CHECK

What is the formula of iodine pentafluoride?

Guide to Writing Formulas for Covalent Compounds

> STEP 1
> Write the symbols in the order of the elements in the name.

> STEP 2
> Write any prefixes as subscripts.

Summary of Naming Inorganic Compounds

Up to now, we have examined strategies for naming ionic and covalent compounds. In general, compounds having two elements are named by stating the first element, followed by the second element with an *ide* ending. If the first element is a metal, the compound is usually ionic; if the first element is a nonmetal, the compound is usually covalent. For ionic compounds, it is necessary to determine whether the metal can form more than one type of positive ion; if so, a Roman numeral following the name of the metal indicates the particular ionic charge. One exception is the ammonium ion, NH_4^+, which is also written first as a positively charged polyatomic ion. In naming covalent compounds having two elements, prefixes are necessary to indicate the number of atoms of each nonmetal, as shown in that particular formula. Ionic compounds having three or more elements include some type of polyatomic ion. They are named by ionic rules but have an *ate* or *ite* ending when the polyatomic ion has a negative charge.

Organic compounds of C and H, such as CH_4 and C_2H_6, use a different system of naming that we will discuss in the next section.

(CONCEPT CHECK 6.6)

■ **Naming Ionic and Covalent Compounds**

Identify each of the following compounds as ionic or covalent and give its name:

a. Ca_3N_2 **b.** Cu_3PO_4 **c.** SO_3

ANSWER

a. Ca_3N_2, consisting of a metal and nonmetal, is an ionic compound. As a representative metal, Ca forms a single ion, Ca^{2+}, *calcium*. The nonmetal forms a negative ion, N^{3-}, named *nitride*. The compound is named calcium nitride.

b. Cu_3PO_4 is an ionic compound with a polyatomic ion. As a transition element, Cu forms more than one ion. In this formula, the positive ion is Cu^+, named *copper*(I) because 3(1+) balances the 3− charge on PO_4^{3-}, named *phosphate*. The compound is named copper(I) phosphate.

c. SO_3 consists of two nonmetals, which indicates that it is named as a covalent compound. The first element, S, is *sulfur* (no prefix is needed). The second element, O, *oxide*, has a subscript 3, which requires a prefix *tri* in the name. The compound is named sulfur trioxide.

QUESTIONS AND PROBLEMS

Covalent Compounds and Their Names

6.49 What elements on the periodic table are most likely to form covalent compounds?

6.50 How does the bond that forms between Na and Cl differ from a bond that forms between N and Cl?

6.51 State the number of valence electrons, bonding pairs, and lone pairs in each of the following electron-dot formulas:

   a. H:H   **b.** H:B̈r: **c.** :B̈r:B̈r:

6.52 State the number of valence electrons, bonding pairs, and lone pairs in each of the following electron-dot formulas:

              H
   a. H:Ö: **b.** H:N̈:H   **c.** :B̈r:Ö:B̈r:
              H

6.53 Name each of the following:

   a. PBr_3 **b.** CBr_4
   c. SiO_2 **d.** HF
   e. NI_3

6.54 Name each of the following:

   a. CS_2 **b.** P_2O_5
   c. Cl_2O **d.** PCl_3
   e. CO

6.55 Name each of the following:

   a. N_2O_3 **b.** NCl_3
   c. $SiBr_4$ **d.** PCl_5
   e. N_2S_3

6.56 Name each of the following:

   a. SiF_4 **b.** IBr_3
   c. CO_2 **d.** SO_2
   e. N_2O

6.57 Write the formula for each of the following:

   a. carbon tetrachloride   **b.** carbon monoxide
   c. phosphorus trichloride   **d.** dinitrogen tetroxide

6.58 Write the formula for each of the following:

   a. sulfur dioxide   **b.** silicon tetrachloride
   c. iodine trichloride   **d.** dinitrogen oxide

6.59 Write the formula for each of the following:

   a. oxygen difluoride   **b.** boron trifluoride
   c. dinitrogen trioxide   **d.** sulfur hexafluoride

6.60 Write the formula for each of the following:

   a. sulfur dibromide   **b.** carbon disulfide
   c. tetraphosphorus hexoxide   **d.** dinitrogen pentoxide

6.61 Name each of the following ionic or covalent compounds:

   a. $AlCl_3$ **b.** B_2O_3
   c. N_2O_4 **d.** $Sn(NO_3)_2$
   e. $Cu(ClO_2)_2$

6.62 Name each of the following ionic or covalent compounds:

   a. N_2 **b.** $Mg(BrO)_2$
   c. SiI_4 **d.** $NiSO_4$
   e. Fe_2S_3

6.6 Organic Compounds

LEARNING GOAL

Identify properties characteristic of organic and inorganic compounds.

SELF STUDY ACTIVITY

Introduction to Organic Molecules

Vegetable oil, an organic compound, is not soluble in water.

Organic chemistry is the chemistry of covalent compounds that contain carbon and hydrogen. The element carbon has a special role in chemistry because it bonds with other carbon atoms to give a vast array of molecules. The variety of molecules is so great that we find organic compounds in many common products we use, such as gasoline, medicine, shampoos, plastic bottles, and perfumes. The food we eat is composed of different organic compounds that supply us with fuel for energy and the carbon atoms needed to build and repair the cells of our bodies.

Although many organic compounds occur in nature, chemists have synthesized even more. The cotton, wool, or silk in your clothes contains naturally occurring organic compounds, whereas materials such as polyester, nylon, or plastic have been synthesized. Learning about the structures and reactions of organic compounds will provide you with a foundation for understanding the more complex molecules of biochemistry.

Organic chemistry is the study of carbon compounds. **Organic compounds** always contain carbon and hydrogen, and sometimes oxygen, sulfur, nitrogen, phosphorus, or a halogen. The formulas of organic compounds are written with carbon first, followed by hydrogen and then any other elements. Organic compounds typically have low melting and boiling points, are not soluble in water, and are less dense than water. For example, vegetable oil, which is a mixture of organic compounds, does not dissolve in water but floats on top. Many organic compounds undergo combustion and burn vigorously in air.

By contrast, inorganic compounds containing elements other than carbon and hydrogen are ionic, with high melting and boiling points. Inorganic compounds that are ionic are usually soluble in water and most do not burn in air. Table 6.12 contrasts some of the properties associated with organic and inorganic compounds, such as propane, C_3H_8, and sodium chloride, NaCl.

TABLE 6.12 Some Properties of Organic and Inorganic Compounds

Property	Organic	Example: C_3H_8	Inorganic	Example: NaCl
Elements	C and H, sometimes O, S, N, P or Cl (F, Br, I)	C and H	Most metals and nonmetals	Na and Cl
Bonding	Mostly covalent	Covalent (4 bonds to each C)	Ionic or covalent (inorganic)	Ionic
Melting point	Usually low	$-188\ °C$	Usually high	$801\ °C$
Boiling point	Usually low	$-42\ °C$	Usually high	$1413\ °C$
Flammability	High	Burns in air	Low	Does not burn
Solubility in water	Not usually soluble	No	Most are soluble	Yes

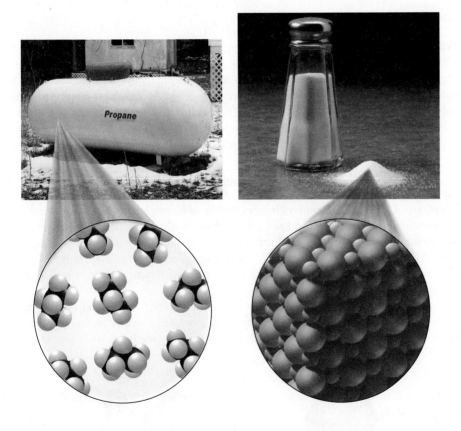

Propane, C_3H_8, is an organic compound, whereas sodium chloride, NaCl, is an inorganic compound.

CONCEPT CHECK 6.7

■ Properties of Organic Compounds

Identify the following as properties typical of organic or inorganic compounds:

a. not soluble in water
b. high melting point
c. burns in air

ANSWER

a. Many organic compounds are not soluble in water.
b. Inorganic compounds are most likely to have high melting points.
c. Organic compounds are most likely to be flammable.

FIGURE 6.6 Three-dimensional representations of methane, CH_4: **(a)** tetrahedron, **(b)** ball-and-stick model, **(c)** space-filling model, **(d)** expanded structural formula.

Q Why does methane have a tetrahedral shape and not a flat shape?

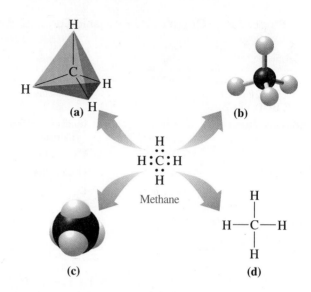

Bonding in Organic Compounds

Hydrocarbons, as the name suggests, are organic compounds that consist of only carbon and hydrogen. In the simplest hydrocarbon, methane (CH_4), the carbon atom forms an octet by sharing four valence electrons with four hydrogen atoms. In the electron-dot formula, each shared pair of electrons represents a single covalent bond. In organic molecules, every carbon atom has four bonds. A hydrocarbon is referred to as a *saturated hydrocarbon* when all of the bonds in the molecules are single bonds. An **expanded structural formula** is written when we show the bonds between all of the atoms.

The Tetrahedral Structure of Carbon

In methane, CH_4, the covalent bonds from carbon to hydrogen are directed to the corners of a tetrahedron. This gives the lowest amount of repulsion between the H atoms. This three-dimensional structure of methane can be illustrated as a ball-and-stick model or a space-filling model (see Figure 6.6). The expanded structural formula is a two-dimensional representation in which the bonds from carbon to each hydrogen atom are shown.

In ethane, C_2H_6, each carbon atom is bonded to another carbon and three hydrogen atoms. As in methane, each carbon retains the tetrahedral shape. In the ball-and-stick model of ethane, two tetrahedra are attached to each other (see Figure 6.7).

FIGURE 6.7 Three-dimensional representations of ethane, C_2H_6: **(a)** tetrahedral shape of each carbon, **(b)** ball-and-stick model, **(c)** space-filling model, **(d)** expanded structural formula.

Q How is the tetrahedral shape maintained in a molecule with two carbon atoms?

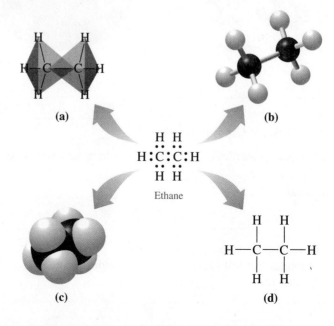

QUESTIONS AND PROBLEMS

Organic Compounds

6.63 Identify the following as formulas of organic or inorganic compounds:

a. KCl
b. C_4H_{10}
c. C_2H_6O
d. H_2SO_4
e. $CaCl_2$
f. C_2H_5Cl

6.64 Identify the following as formulas of organic or inorganic compounds:

a. $C_6H_{12}O_6$
b. Na_2SO_4
c. I_2
d. C_4H_9Br
e. $C_{10}H_{22}$
f. CH_4

6.65 Identify the following properties as most typical of organic or inorganic compounds:

a. soluble in water
b. low boiling point
c. burns in air
d. high melting point

6.66 Identify the following properties as most typical of organic or inorganic compounds:

a. contains Na
b. is a gas at room temperature
c. contains covalent bonds
d. produces ions in water

6.67 Match the following physical and chemical properties with the compounds ethane, C_2H_6, or sodium bromide, NaBr:

a. boils at −89 °C
b. burns vigorously
c. is a solid at 250 °C
d. dissolves in water

6.68 Match the following physical and chemical properties with the compounds cyclohexane, C_6H_{12}, or calcium nitrate, $Ca(NO_3)_2$:

a. melts at 500 °C
b. insoluble in water
c. does not burn in air
d. is a liquid at room temperature

6.69 How are the hydrogen atoms arranged in space in methane?

6.70 In a propane molecule with three carbon atoms, what is the geometry around each carbon atom?

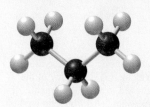

Propane

6.7 Names and Formulas of Hydrocarbons: Alkanes

More than 90% of the known compounds in the world are organic compounds. The large number of carbon compounds is possible because carbon atoms can form long, stable chains. To help us study organic compounds, we organize them into groups that have similar structures and chemical properties.

The **alkanes** are a class of hydrocarbons in which carbon and hydrogen atoms are connected by covalent bonds. One of the most common uses of alkanes is as fuels. Methane, used in gas heaters and gas cooktops, is an alkane with one carbon atom. The alkanes ethane, propane, and butane contain two, three, and four carbon atoms, respectively, connected in a row or a *continuous* chain. As we can see, all the names for alkanes end in *ane*. These names are part of the **IUPAC** (International Union of Pure and Applied Chemistry) **system** used by chemists to name organic compounds. Alkanes with five or more carbon atoms in a chain are named using the prefixes *pent* (5), *hex* (6), *hept* (7), *oct* (8), *non* (9), and *dec* (10) (see Table 6.13). Figure 6.8 is a summary of the rules for naming ionic and covalent compounds.

LEARNING GOAL

Write the IUPAC name and draw the condensed structural formula for an alkane.

TUTORIAL
IUPAC Naming of Alkanes

TABLE 6.13 IUPAC Names of the First 10 Alkanes

Number of Carbon Atoms	Prefix	Name	Molecular Formula	Condensed Structural Formula
1	Meth	Methane	CH_4	CH_4
2	Eth	Ethane	C_2H_6	CH_3-CH_3
3	Prop	Propane	C_3H_8	$CH_3-CH_2-CH_3$
4	But	Butane	C_4H_{10}	$CH_3-CH_2-CH_2-CH_3$
5	Pent	Pentane	C_5H_{12}	$CH_3-CH_2-CH_2-CH_2-CH_3$
6	Hex	Hexane	C_6H_{14}	$CH_3-CH_2-CH_2-CH_2-CH_2-CH_3$
7	Hept	Heptane	C_7H_{16}	$CH_3-CH_2-CH_2-CH_2-CH_2-CH_2-CH_3$
8	Oct	Octane	C_8H_{18}	$CH_3-CH_2-CH_2-CH_2-CH_2-CH_2-CH_2-CH_3$
9	Non	Nonane	C_9H_{20}	$CH_3-CH_2-CH_2-CH_2-CH_2-CH_2-CH_2-CH_2-CH_3$
10	Dec	Decane	$C_{10}H_{22}$	$CH_3-CH_2-CH_2-CH_2-CH_2-CH_2-CH_2-CH_2-CH_2-CH_3$

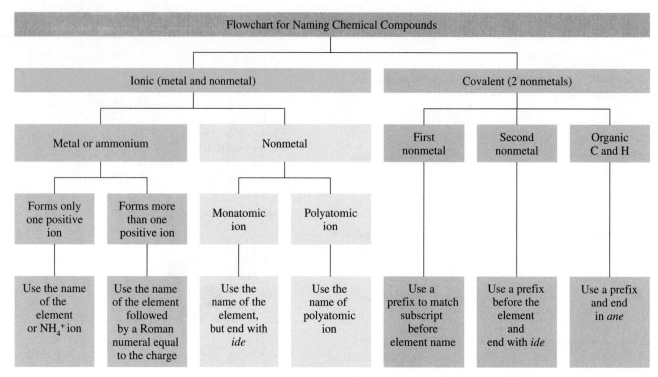

FIGURE 6.8 A flowchart for naming inorganic ionic compounds, inorganic covalent compounds, and organic compounds.

Q Why is sulfur hexafluoride an inorganic compound, but hexane is an organic compound?

SAMPLE PROBLEM | 6.11

■ **Naming Alkanes**

Give the IUPAC name for each of the following:
a. $CH_3—CH_2—CH_3$ **b.** C_6H_{14}

SOLUTION

a. A chain with three carbon atoms is propane.
b. An alkane with six carbon atoms is hexane.

STUDY CHECK

What is the IUPAC name of C_8H_{18}?

Condensed Structural Formulas

In a **condensed structural formula**, each carbon atom and its attached hydrogen atoms are written as a group. A subscript indicates the number of hydrogen atoms bonded to each carbon atom.

$$H—\overset{\overset{\displaystyle H}{|}}{\underset{\underset{\displaystyle H}{|}}{C}}— \quad = \quad CH_3— \qquad —\overset{\overset{\displaystyle H}{|}}{\underset{\underset{\displaystyle H}{|}}{C}}— \quad = \quad —CH_2—$$

Expanded   Condensed   Expanded   Condensed

By contrast, the molecular formula gives the total number of each kind of atom but does not indicate the arrangement of the atoms in the molecule. When a molecule consists of a chain of three or more carbon atoms, the carbon atoms do not lie in a straight line. The tetrahedral shape of carbon arranges the carbon bonds in a zigzag pattern, which is seen in the ball-and-stick model of hexane.

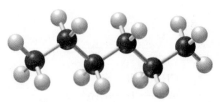

A ball-and-stick model of hexane.

CONCEPT CHECK 6.8

■ **Drawing Expanded and Condensed Structural Formulas for Alkanes**

A molecule of butane, C_4H_{10}, has four carbon atoms in a row. What are its expanded and condensed structural formulas?

ANSWER

In the expanded structural formula, four carbon atoms are connected to each other and to hydrogen atoms to give each carbon atom a total of four bonds. In the condensed structural formula, each carbon atom and its attached hydrogen atoms are written as CH_3— or —CH_2—.

Expanded structural formula

$CH_3—CH_2—CH_2—CH_3$ Condensed structural formula

CHEMISTRY AND INDUSTRY

Crude Oil

Crude oil, or petroleum, contains a wide variety of hydrocarbons. At an oil refinery, the components in crude oil are separated by fractional distillation, a process that removes groups or fractions of hydrocarbons by continually heating the mixture to higher temperatures (see Table 6.14). Fractions containing alkanes with longer carbon chains require higher temperatures before they reach their boiling temperature and form gases. The gases are removed and passed through a distillation column where they cool and condense back to liquids. The major use of crude oil is to obtain gasoline. To increase the production of gasoline, heating oils are broken down using specialized catalysts to give the lower-weight alkanes.

A refinery converts crude oil into gasoline, heating oil, and other organic products.

TABLE 6.14 Typical Alkane Mixtures Obtained by Distillation of Crude Oil

Distillation Temperatures (°C)	Number of Carbon Atoms	Product
Below 30	1–4	Natural gas
30–200	5–12	Gasoline
200–250	12–16	Kerosene, jet fuel
250–350	16–18	Diesel fuel, heating oil
350–450	18–25	Lubricating oil
Nonvolatile residue	Over 25	Asphalt, tar

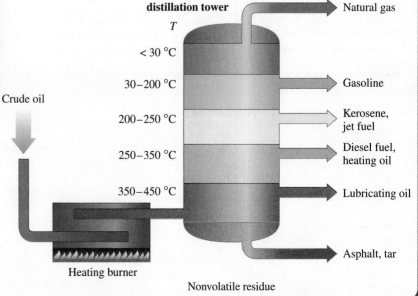

CAREER FOCUS

Geologist

"Chemistry underpins geology," says Vic Abadie, consulting geologist. "I am a self-employed geologist consulting in exploration for petroleum and natural gas. Predicting the occurrence of an oil reservoir depends in part on understanding chemical reactions of minerals. This is because over geologic time, such reactions create or destroy pore spaces that host crude oil or gas in a reservoir rock formation. Chemical analysis can match oil in known reservoir formations with distant formations that generated oil and from which oil migrated into the reservoirs in the geologic past. This can help identify target areas to explore for new reservoirs.

"I evaluate proposals to drill for undiscovered oil and gas. I recommend that my clients invest in proposed wells that my analysis suggests have strong geologic and economic merit. This is a commercial application of the scientific method: The proposal to drill is the hypothesis, and the drill bit tests it. A successful well validates the hypothesis and generates oil or gas production and revenue for my clients and me. I do this and other consulting for private and corporate clients. The risk is high, the work is exciting, and my time is flexible."

Some Uses of Alkanes

The first four alkanes—methane, ethane, propane, and butane—are gases at room temperature and are widely used as heating fuels.

Alkanes having 5–8 carbon atoms (pentane, hexane, heptane, and octane) are liquids at room temperature. They are highly volatile, which makes them useful in fuels such as gasoline. Liquid alkanes with 9–17 carbon atoms have higher boiling points and are found in kerosene, diesel, and jet fuels. Motor oil is a mixture of liquid hydrocarbons and is used to lubricate the internal components of engines. Mineral oil is a mixture of liquid hydrocarbons and is used as a laxative and a lubricant. Alkanes with 18 or more carbon atoms are waxy solids at room temperature. Known as paraffins, these compounds are used in waxy coatings for fruits and vegetables to retain moisture, inhibit mold growth, and enhance appearance. Petrolatum, or Vaseline, is a mixture of liquid hydrocarbons with low boiling points that are encapsulated in solid hydrocarbons. It is used in ointments and cosmetics and as a lubricant.

The solid alkanes that make up waxy coatings on fruits and vegetables help to retain moisture, inhibit mold, and enhance appearance.

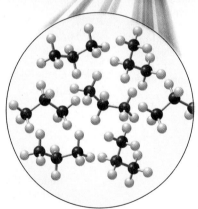

Propane is used in a mini torch to solder copper pipes.

QUESTIONS AND PROBLEMS

Names and Formulas of Hydrocarbons: Alkanes

6.71 Give the IUPAC name for each of the following alkanes:
   a. $CH_3-CH_2-CH_2-CH_2-CH_3$
   b. CH_3-CH_3
   c. $CH_3-CH_2-CH_2-CH_2-CH_2-CH_3$

6.72 Give the IUPAC name for each of the following alkanes:
   a. CH_4
   b. $CH_3-CH_2-CH_2-CH_3$
   c. $CH_3-CH_2-CH_3$

6.73 Draw the condensed structural formula for each of the following:
   a. methane   **b.** ethane   **c.** pentane

6.74 Draw the condensed structural formula for each of the following:
   a. propane   **b.** hexane   **c.** octane

6.75 Heptane, C_7H_{16}, has a density of 0.68 g/mL.
   a. What is the condensed structural formula of heptane?
   b. Is it a solid, liquid, or gas at room temperature?
   c. Is it soluble in water?
   d. Will it float or sink in water?

6.76 Nonane, C_9H_{20}, has a density of 0.79 g/mL.
   a. What is the condensed structural formula of nonane?
   b. Is it a solid, liquid, or gas at room temperature?
   c. Is it soluble in water?
   d. Will it float or sink in water?

CHEMISTRY AND HEALTH

Fatty Acids in Lipids, Soaps, and Cell Membranes

Many of the things we learned in this chapter about inorganic and organic compounds can also be applied to compounds found in living systems. The same elements, carbon and hydrogen, that make up organic compounds are also the elements found in the molecules of life, an area of chemistry known as biochemistry. For example, a family of biomolecules, known as *lipids*, is soluble in organic solvents but not in water.

One type of lipid, called a fatty acid, is combined with glycerol to form fats, including the ever present "body fat." The structure of a fatty acid consists of a long chain of carbon atoms with an ionic group called carboxylate at one end. An example of a fatty acid is palmitic acid, which is a 16-carbon acid found in palm oil and many fats.

$$CH_3 - (CH_2)_{14} - COO^-$$

Because a fatty acid contains mostly carbon and hydrogen, the carbon of a fatty acid has similar properties to an alkane, which makes that portion of the fatty-acid chain insoluble in water. However, the ionic group at one end of a fatty acid has an ionic charge that makes that end soluble in water.

Soaps are typically prepared from fats such as coconut oil. We say that soap is the salt of a long-chain fatty acid because it has a long hydrocarbon-like chain with a carboxylate negative end combined with a positive sodium or potassium ion. Soaps have dual properties because the parts of soap have different solubilities.

The sodium or potassium carboxylate end is ionic and very soluble in water ("hydrophilic") but not in oils or grease. However, the long carbon chain is not soluble in water ("hydrophobic") but does dissolve in substances such as oil or grease. When soap is used, the hydrocarbon tails of the soap molecules dissolve in the fats and oils that accompany dirt. The soap molecules and the oil or grease form clusters called *micelles*, in which the water-loving salt ends of the soap molecules extend outside where they can dissolve in water. As a result, small globules of oil and fat coated with soap molecules are pulled into the water and rinsed away.

In the body, lipids are found in cell membranes, fat-soluble vitamins, and steroid hormones. The membrane of a cell separates the contents of a cell from the external aqueous environment. A cell

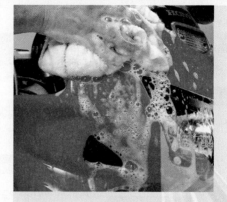

The hydrocarbon tails dissolve in grease and oil to form micelles, which are pulled by the ionic heads into rinse water.

membrane consists of a double row of lipids arranged like a sandwich called a *bilayer*. The hydrocarbon tails form the center of the bilayer away from the aqueous environment, and the ionic heads form the outside and inside surfaces of the membrane, which are in contact with external and internal fluids.

$$CH_3CH_2CH_2CH_2CH_2CH_2CH_2CH_2CH_2CH_2CH_2CH_2CH_2CH_2CH_2CH_2CH_2 - \overset{\overset{\displaystyle O}{\|}}{C} - O^- Na^+$$

Hydrocarbon tail
(Hydrophobic)

Ionic head
(Hydrophilic)

This "soap" is the salt of the long-chain stearic acid.

A cell membrane is composed of two layers of lipids.

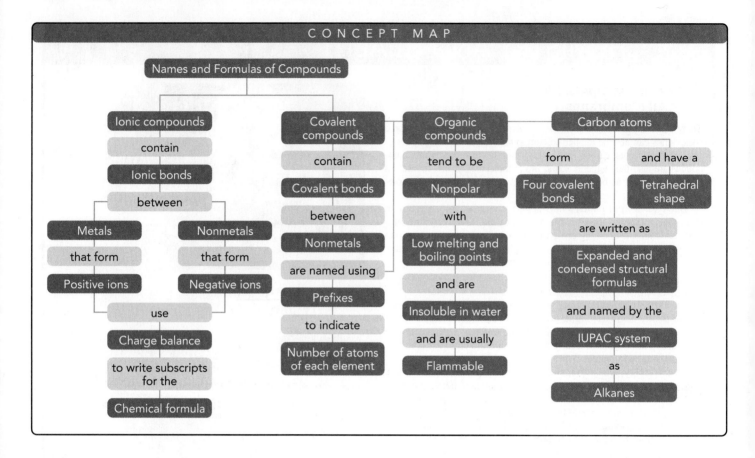

CHAPTER REVIEW

6.1 Octet Rule and Ions

LEARNING GOAL: *Using the octet rule, write the symbols of the single ions for the representative elements.*

The stability of the noble gases is associated with eight valence electrons; helium needs two electrons for stability. Atoms of elements in Groups 1A–7A (1, 2, 13–17) achieve stability by losing, gaining, or sharing their valence electrons in the formation of compounds. Metals of the representative elements form octets by losing valence electrons to form positively charged ions (cations): Group 1A (1), 1+, Group 2A (2), 2+, and Group 3A (13), 3+. When reacting with metals, nonmetals gain electrons to form octets and form negatively charged ions (anions): Groups 5A (15), 3−, 6A (16), 2−, and 7A (17), 1−.

6.2 Ionic Compounds

LEARNING GOAL: *Using charge balance, write the correct formula for an ionic compound.*

The total positive and negative ionic charge is balanced in the formula of an ionic compound. Charge balance is achieved by using subscripts after each symbol so that the overall charge is zero.

6.3 Naming and Writing Ionic Formulas

LEARNING GOAL: *Given the formula of an ionic compound, write the correct name; given the name of an ionic compound, write the correct formula.*

In naming ionic compounds, the positive ion is given first, followed by the name of the negative ion. The names of ionic compounds containing two elements end with *ide*. Except for Ag, Cd, and Zn, transition elements form cations with two or more ionic charges. The charge of the cation is determined from the total negative charge in the formula and included as a Roman numeral immediately following the name of the metal.

6.4 Polyatomic Ions

LEARNING GOAL: *Write the name and formula of a compound containing a polyatomic ion.*

A polyatomic ion is a group of atoms with an electrical charge; for example, the carbonate ion has the formula CO_3^{2-}. Most polyatomic ions contain a nonmetal and one or more oxygen atoms. The ammonium ion, NH_4^+, is a positive polyatomic ion.

6.5 Covalent Compounds and Their Names

LEARNING GOAL: *Given the formula of a covalent compound, write its correct name; given the name of a covalent compound, write its formula.*

Two nonmetals can form two or more different covalent compounds. In a covalent bond, atoms of nonmetals share valence electrons. In most covalent compounds, the atoms achieve a noble gas electron configuration. In the names of covalent compounds, prefixes are used to indicate the subscripts in the formulas. The ending of the second nonmetal is changed to *ide*.

6.6 Organic Compounds

LEARNING GOAL: *Identify properties characteristic of organic and inorganic compounds.*

Organic compounds are covalent compounds that contain carbon and hydrogen. Typically, they have low melting points and low boiling points, are not very soluble in water, and burn vigorously in air. In contrast, typical inorganic compounds are ionic compounds, have high melting and boiling points, are usually soluble in water, and do not burn in air. In organic compounds, carbon atoms share four valence electrons to form four covalent bonds. In the simplest organic molecule, methane, CH_4, the four hydrogen atoms bonded to the carbon atom are in the corners of a tetrahedron.

6.7 Names and Formulas of Hydrocarbons: Alkanes

LEARNING GOAL: *Write the IUPAC name and draw the condensed structural formula for an alkane.*

Alkanes are hydrocarbons that have only C—C single bonds. In the expanded structural formula, a separate line is drawn for every bonded atom. A condensed structural formula depicts each carbon atom and the number of attached hydrogen atoms. The IUPAC name of an alkane indicates the number of carbon atoms and ends in *ane*.

◼ KEY TERMS

alkanes Hydrocarbons containing only single bonds between carbon atoms.

anion A negatively charged ion.

bonding pair A pair of electrons shared between two atoms.

cation A positively charged ion.

condensed structural formula A structural formula that shows the arrangement of the carbon atoms in a molecule but groups each carbon atom with its bonded hydrogen atoms.

covalent bond A sharing of valence electrons by atoms.

covalent compound A combination of atoms in which noble gas configurations are attained by electron sharing.

expanded structural formula A type of structural formula that shows the arrangement of the atoms by drawing each bond between carbon atoms or between carbon and hydrogen.

formula The group of symbols representing the elements in a compound with subscripts for the number of each.

hydrocarbons Organic compounds consisting of only carbon and hydrogen.

ion An atom or group of atoms having an electrical charge because of a loss or gain of electrons.

ionic bond The attraction between oppositely charged ions.

ionic charge The difference between the number of protons (positive) and the number of electrons (negative), written in the upper right corner of the symbol for the ion.

ionic compound A compound of positive and negative ions held together by ionic bonds.

IUPAC system A system for naming organic compounds determined by the International Union of Pure and Applied Chemistry.

lone pair Electrons in a molecule that are not shared in a bond, but complete the octet for an element.

molecule The smallest unit of two or more atoms held together by covalent bonds.

octet rule Representative elements react with other elements to produce a noble gas configuration with eight valence electrons.

organic compounds Compounds of carbon that have covalent bonds that are mostly nonpolar with properties that include low melting and boiling points, insolubility in water, and flammability.

polyatomic ion A group of covalently bonded atoms that has an overall electrical charge.

◼ UNDERSTANDING THE CONCEPTS

6.77 **a.** How does the octet rule explain the formation of a calcium ion?
b. What noble gas has the same electron configuration as the calcium ion?
c. Why are Group 1A (1) and Group 2A (2) elements found in many compounds but not Group 8A (18) elements?

6.78 **a.** How does the octet rule explain the formation of a sulfide ion?
b. What noble gas has the same electron configuration as the sulfide ion?
c. Why are Group 6A (16) elements found in many compounds but not Group 8A (18) elements?

6.79 Identify each of the following atoms or ions:

$15p^+$ $16n$ $18e^-$ $8p^+$ $8n$ $8e^-$ $30p^+$ $35n$ $28e^-$ $26p^+$ $28n$ $23e^-$

$\quad$ **A** $\qquad$ **B** $\qquad$ **C** $\qquad$ **D**

6.80 Identify each of the following atoms or ions:

$2e^-$ $3p^+$ $4n$ $0e^-$ $1p^+$ $3e^-$ $3p^+$ $4n$

$\quad$ **A** $\qquad$ **B** $\qquad$ **C**

$10e^-$ $7p^+$ $8n$ $1e^-$ $1p^+$ $2n$

$\quad$ **D** $\qquad$ **E**

6.81 Consider the following electron-dot formulas for elements X and Y:

$$X\cdot \quad \cdot \ddot{Y}\cdot$$

a. What are the group numbers of X and Y?
b. Will a compound of X and Y be ionic or covalent?
c. What ions would be formed by X and Y?
d. What would be the formula of a compound of X and Y?
e. What would be the formula of a compound of X and chlorine?
f. What would be the formula of a compound of Y and chlorine?

6.82 Identify each of the following atoms or ions:
a. 35 protons, 45 neutrons, and 36 electrons
b. 47 protons, 60 neutrons, and 46 electrons
c. 50 protons, 68 neutrons, and 46 electrons
d. 15 protons, 16 neutrons, and 15 electrons
e. 82 protons, 126 neutrons, and 82 electrons
f. 34 protons, 46 neutrons, and 36 electrons

6.83 Write the formulas and names of the ionic compounds for the elements indicated by the period and electron-dot symbols in the following table:

Period	Electron-Dot Symbols	Formula of Compound	Name of Compound
2	$X\cdot$ and $\cdot\ddot{Y}\cdot$		
4	$\dot{X}\cdot$ and $\cdot\ddot{Y}\cdot$		
4	$\cdot\dot{X}\cdot$ and $\cdot\ddot{Y}\cdot$		

6.84 Write the formulas and names of the ionic compounds for the elements indicated by the period and electron-dot symbols in the following table:

Period	Electron-Dot Symbols	Formula of Compound	Name of Compound
3	$\dot{\text{X}}\cdot$ and $\cdot\ddot{\text{Y}}\cdot$		
3	$\cdot\dot{\text{X}}\cdot$ and $\ddot{\text{Y}}\cdot$		
5	$\dot{\text{X}}\cdot$ and $\cdot\ddot{\text{Y}}\colon$		

6.85 Write the ions, formulas, and names of the ionic compounds using the electron configuration of the elements.

Electron Configurations		Symbols of Ions		Formula of Compound	Name of Compound
Metal	**Nonmetal**	**Cation**	**Anion**		
$1s^2 2s^2 2p^6 3s^2$	$1s^2 2s^2 2p^3$				
$1s^2 2s^2 2p^6 3s^2 3p^6 4s^1$	$1s^2 2s^2 2p^4$				
$1s^2 2s^2 2p^6 3s^2 3p^1$	$1s^2 2s^2 2p^5$				

6.86 Write the ions, formulas, and names of the ionic compounds using the electron configuration of the elements.

Electron Configurations		Symbols of Ions		Formula of Compound	Name of Compound
Metal	**Nonmetal**	**Cation**	**Anion**		
$1s^2 2s^1$	$1s^2 2s^2 2p^6 3s^2 3p^4$				
$1s^2 2s^2 2p^6 3s^2 3p^6 4s^2$	$1s^2 2s^2 2p^6 3s^2 3p^3$				
$1s^2 2s^2 2p^6 3s^1$	$1s^2 2s^2 2p^6 3s^2 3p^5$				

6.87 Match the following physical and chemical properties with potassium chloride, KCl, used in salt substitutes, or butane, C_4H_{10}, used in lighters:

a. melts at $-138\ °C$ **b.** burns vigorously in air
c. melts at $770\ °C$ **d.** contains ionic bonds
e. is a gas at room temperature

6.88 Match the following physical and chemical properties with octane, C_8H_{18}, found in gasoline, or magnesium sulfate, $MgSO_4$, also called Epsom salts:
a. contains only covalent bonds
b. melts at 1124 °C
c. is insoluble in water
d. is a liquid at room temperature
e. contains ions

6.89 Identify each of the following compounds as inorganic (ionic or covalent) or organic (covalent) and give the name:
a. C_7H_{16} **b.** NO_2
c. K_3PO_4 **d.** C_2H_6

6.90 Identify each of the following compounds as inorganic (ionic or covalent) or organic (covalent) and give the name:
a. $Cr_2(SO_4)_3$ **b.** K_2S
c. PCl_5 **d.** $C_{10}H_{22}$

ADDITIONAL QUESTIONS AND PROBLEMS

For instructor-assigned homework, go to
www.masteringchemistry.com.

6.91 What noble gas has the same electron configuration as each of the following ions?
a. N^{3-} b. Sr^{2+} c. I^- d. O^{2-} e. Cs^+

6.92 What noble gas has the same electron configuration as each of the following ions?
a. Al^{3+} b. Br^- c. Ca^{2+} d. Na^+ e. S^{2-}

6.93 Consider an ion with the symbol X^{2+} formed from a representative element.
a. What is the group number of the element?
b. What is the electron-dot symbol of the element?
c. If X is in Period 2, what is the element?
d. What is the formula of the compound formed from X and the nitride ion?

6.94 Consider the following electron-dot symbols of representative elements X and Y:

$\cdot X \cdot$ $\cdot \ddot{Y} \cdot$

a. What are the group numbers of X and Y?
b. What ions would be formed by X and Y?
c. What would be the formula of a compound of X and Y?
d. What would be the formula of a compound of X and chlorine?

6.95 One of the ions of tin is tin(IV).
a. What is the symbol for this ion?
b. How many protons and electrons are in the ion?
c. What is the formula of tin(IV) oxide?
d. What is the formula of tin(IV) phosphate?

6.96 One of the ions of gold is gold(III).
a. What is the symbol for this ion?
b. How many protons and electrons are in the ion?
c. What is the formula of gold(III) sulfate?
d. What is the formula of gold(III) chloride?

6.97 Identify the group number in the periodic table of X, a representative element, in each of the following ionic compounds:
a. XCl_3 b. Al_2X_3 c. XCO_3

6.98 Identify the group number in the periodic table of X, a representative element, in each of the following ionic compounds:
a. X_2O_3 b. X_2SO_3 c. Na_3X

6.99 Name the following ionic compounds:
a. $FeCl_3$ b. $Ca_3(PO_4)_2$ c. $Al_2(CO_3)_3$ d. $PbCl_4$
e. $MgCO_3$ f. $SnSO_4$ g. CuS

6.100 Name the following ionic compounds:
a. $CaSO_4$ b. $Ba(NO_3)_2$ c. MnS d. $LiClO_4$
e. $CrPO_3$ f. Na_2HPO_4 g. $CaCl_2$

6.101 Write the formula for the following ionic compounds:
a. copper(I) nitride
b. potassium hydrogen sulfite
c. lead(IV) sulfide
d. gold(III) carbonate
e. zinc perchlorate

6.102 Write the formula for the following ionic compounds:
a. iron(III) nitrate   b. copper(II) hydrogen carbonate
c. tin(IV) sulfite   d. barium dihydrogen phosphate
e. cadmium hypochlorite

6.103 Write the formula of the following ionic compounds:
a. nickel(III) chloride   b. lead(IV) oxide
c. silver bromide   d. calcium nitride
e. copper(I) phosphide   f. chromium(II) sulfide

6.104 Write the formula of the following ionic compounds:
a. tin(II) oxide   b. iron(III) chloride
c. cobalt(II) sulfide   d. chromium(III) iodide
e. lithium nitride   f. gold(I) oxide

6.105 Write the name for each of the following:
a. MgO b. $Cr(HCO_3)_3$ c. $Mn_2(CrO_4)_3$

6.106 Write the name for each of the following:
a. Cu_2S b. $Fe_3(PO_4)_2$ c. $Ca(ClO)_2$

6.107 Name each of the following covalent compounds:
a. NCl_3 b. SCl_2 c. N_2O
d. F_2 e. PF_5 f. P_2O_5

6.108 Name each of the following covalent compounds:
a. CBr_4 b. SF_6 c. Br_2
d. N_2O_4 e. SO_2 f. CS_2

6.109 Give the formula for each of the following:
a. iodine heptafluoride
b. boron trifluoride
c. hydrogen iodide
d. sulfur dichloride

6.110 Give the formula for each of the following:
a. diphosphorus pentasulfide
b. carbon tetraiodide
c. sulfur trioxide
d. dinitrogen difluoride

6.111 Classify each of the following as ionic or covalent, and give its name:
a. $CoCl_3$ b. Na_2SO_4
c. N_2O_3 d. I_2
e. ICl_3 f. CF_4

6.112 Classify each of the following as ionic or covalent, and give its name:
a. $Al_2(SO_3)_3$ b. SO c. H_2
d. Mg_3N_2 e. SiS_2 f. $CrPO_4$

6.113 Write the formula for each of the following:
a. tin(II) carbonate   b. lithium phosphide
c. silicon tetrachloride   d. chromium(II) sulfide
e. carbon dioxide   f. calcium bromide
g. pentane

6.114 Write the formula for each of the following:
a. sodium carbonate   b. nitrogen dioxide
c. aluminum nitrate   d. cobalt(II) nitride
e. potassium phosphate   f. manganese(III) fluoride
g. nonane

CHALLENGE QUESTIONS

6.115 Why are only the valence electrons of the representative elements involved in the formation of positive and negative ions?

6.116 How does the octet rule determine the loss or gain of electrons by representative elements?

6.117 How are the ions of Group 2A (2) and Group 6A (16) different? How are they similar?

6.118 Identify the errors in the following formulas or names. Write a correct formula or name.
   a. $Ca(NO_3)_2$ is calcium dinitrate.
   b. Copper(II) oxide has the formula Cu_2O.
   c. Potassium carbonate has the formula $(K)_2CO_3$.
   d. Na_2S is sodium sulfate.
   e. Silver sulfate has the formula $Ag_3(SO_4)$.

6.119 Indicate the type of compound (ionic or covalent) and complete the table:

Formula of Compound	Type of Compound	Name of Compound
$FeSO_4$		
		Silicon dioxide
		Ammonium nitrate
$Al_2(SO_4)_3$		
		Cobalt(III) sulfide

6.120 Give the symbol of the noble gas that has the same electron configuration as
   a. Cl^- **b.** Ba^{2+} **c.** Se^{2-}

6.121 Classify each of the following compounds as inorganic (ionic or covalent) or organic (covalent) and name each:
   a. Li_2O **b.** $CH_3—CH_2—CH_3$ **c.** CF_4
   d. CH_4 **e.** MgF_2 **f.** C_4H_{10}

6.122 Classify each of the following compounds as inorganic (ionic or covalent) or organic (covalent) and name each:
   a. $FeCl_2$ **b.** C_7H_{16} **c.** N_2
   d. $Ca_3(PO_4)_2$ **e.** PCl_3 **f.** $Al(NO_3)_3$

6.123 When sodium reacts with sulfur, an ionic compound forms. If a sample of this compound contains 4.8×10^{22} sodium ions, how many sulfide ions does it contain?

6.124 When magnesium reacts with nitrogen, an ionic compound forms. If a sample of this compound contains 1.6×10^{23} magnesium ions, how many nitride ions does it contain?

ANSWERS

ANSWERS TO STUDY CHECKS

6.1 K^+ and S^{2-}

6.2 gallium sulfide

6.3 gold(III) chloride

6.4 **a.** Mg^{2+}, Br^- $MgBr_2$ **b.** Li^+, O^{2-} Li_2O

6.5 Cr_2O_3

6.6 $(NH_4)_3PO_4$

6.7 calcium phosphate

6.8 iron(II) nitrate

6.9 **a.** silicon tetrabromide   **b.** dibromine oxide

6.10 IF_5

6.11 octane

ANSWERS TO SELECTED QUESTIONS AND PROBLEMS

6.1 **a.** When a sodium atom loses its valence electron, its second energy level has a complete octet.
   b. Group 1A (1) and 2A (2) elements can lose one or two electrons to attain a noble gas configuration. Group 8A (18) elements already have an octet of valence electrons (two for helium), so they do not lose or gain electrons and are not normally found in compounds.

6.3 **a.** 1   **b.** 2   **c.** 3
   d. 1   **e.** 2

6.5 **a.** helium   **b.** neon   **c.** argon
   d. neon   **e.** krypton

6.7 **a.** lose 2 e^- **b.** gain 3 e^- **c.** gain 1 e^-
   d. lose 1 e^- **e.** lose 3 e^-

6.9 **a.** Li^+ **b.** F^- **c.** Mg^{2+}
   d. Fe^{3+} **e.** Zn^{2+}

6.11 **a** and **c**

6.13 **a.** $K \cdot$ $\cdot \ddot{\underset{..}{F}} : \longrightarrow K^+ [: \ddot{\underset{..}{F}} :]^- \longrightarrow KF$

   b. $\cdot Ba \cdot$ $\cdot \ddot{\underset{..}{Cl}} : \longrightarrow Ba^{2+} [: \ddot{\underset{..}{Cl}} :]^- \longrightarrow BaCl_2$
   $\cdot \ddot{\underset{..}{Cl}} :$ $[: \ddot{\underset{..}{Cl}} :]^-$

   c. $Na \cdot$ Na^+
   $Na \cdot \ddot{N} \cdot \longrightarrow Na^+ [: \ddot{\underset{..}{N}} :]^{3-} \longrightarrow Na_3N$
   $Na \cdot$ Na^+

6.15 **a.** Na_2O **b.** $AlBr_3$ **c.** Ba_3N_2
   d. MgF_2 **e.** Al_2S_3

6.17 **a.** Na^+, S^{2-} Na_2S **b.** K^+, N^{3-} K_3N
   c. Al^{3+}, I^- AlI_3 **d.** Ga^{3+}, O^{2-} Ga_2O_3

6.19 **a.** Cl^- **b.** K^+ **c.** O^{2-} **d.** Al^{3+}

6.21 **a.** lithium   **b.** sulfide   **c.** calcium   **d.** nitride

6.23 **a.** aluminum oxide   **b.** calcium chloride
   c. sodium oxide   **d.** magnesium phosphide
   e. potassium iodide   **f.** barium fluoride

6.25 Most of the transition elements form more than one positive ion. The specific ion is indicated in a name by writing a Roman numeral that is the same as the ionic charge. For example, iron forms Fe^{2+} and Fe^{3+} ions, which are named iron(II) and iron(III).

6.27 **a.** iron(II)   **b.** copper(II)   **c.** zinc
   d. lead(IV)   **e.** chromium(III)   **f.** manganese(II)

6.29 a. tin(II) chloride **b.** iron(II) oxide
c. copper(I) sulfide **d.** copper(II) sulfide
e. cadmium bromide **f.** mercury(II) chloride

6.31 a. Au^{3+} **b.** Fe^{3+} **c.** Pb^{4+} **d.** Sn^{2+}

6.33 a. $MgCl_2$ **b.** Na_2S **c.** Cu_2O
d. Zn_3P_2 **e.** AuN

6.35 a. $CoCl_3$ **b.** PbO_2 **c.** AgI
d. Ca_3N_2 **e.** Cu_3P **f.** $CrCl_2$

6.37 a. HCO_3^- **b.** NH_4^+ **c.** PO_4^{3-}
d. HSO_4^- **e.** ClO^-

6.39 a. sulfate **b.** carbonate **c.** phosphate **d.** nitrate
e. perchlorate

6.41

	NO_2^-	CO_3^{2-}	HSO_4^-	PO_4^{3-}
Li^+	$LiNO_2$ Lithium nitrite	Li_2CO_3 Lithium carbonate	$LiHSO_4$ Lithium hydrogen sulfate	Li_3PO_4 Lithium phosphate
Cu^{2+}	$Cu(NO_2)_2$ Copper(II) nitrite	$CuCO_3$ Copper(II) carbonate	$Cu(HSO_4)_2$ Copper(II) hydrogen sulfate	$Cu_3(PO_4)_2$ Copper(II) phosphate
Ba^{2+}	$Ba(NO_2)_2$ Barium nitrite	$BaCO_3$ Barium carbonate	$Ba(HSO_4)_2$ Barium hydrogen sulfate	$Ba_3(PO_4)_2$ Barium phosphate

6.43 a. CO_3^{2-}, sodium carbonate **b.** NH_4^+, ammonium chloride
c. PO_3^{3-}, sodium phosphite **d.** NO_2^-, manganese(II) nitrite
e. SO_3^{2-}, iron(II) sulfite **f.** $C_2H_3O_2^-$, potassium acetate

6.45 a. $Ba(OH)_2$ **b.** Na_2SO_4 **c.** $Fe(NO_3)_2$
d. $Zn_3(PO_4)_2$ **e.** $Fe_2(CO_3)_3$

6.47 a. aluminum sulfate **b.** calcium carbonate
c. chromium(III) oxide **d.** sodium phosphate
e. ammonium sulfate **f.** iron(III) oxide

6.49 The nonmetallic elements are most likely to form covalent bonds.

6.51 a. 2 valence electrons, 1 bonding pair and 0 lone pairs
b. 8 valence electrons, 1 bonding pair and 3 lone pairs
c. 14 valence electrons, 1 bonding pair and 6 lone pairs

6.53 a. phosphorus tribromide **b.** carbon tetrabromide
c. silicon dioxide **d.** hydrogen fluoride
e. nitrogen triiodide

6.55 a. dinitrogen trioxide **b.** nitrogen trichloride
c. silicon tetrabromide **d.** phosphorus pentachloride
e. dinitrogen trisulfide

6.57 a. CCl_4 **b.** CO **c.** PCl_3 **d.** N_2O_4

6.59 a. OF_2 **b.** BF_3 **c.** N_2O_3 **d.** SF_6

6.61 a. aluminum chloride **b.** diboron trioxide
c. dinitrogen tetroxide **d.** tin(II) nitrate
e. copper(II) chlorite

6.63 a. inorganic **b.** organic **c.** organic
d. inorganic **e.** inorganic **f.** organic

6.65 a. inorganic **b.** organic **c.** organic **d.** inorganic

6.67 a. ethane **b.** ethane **c.** NaBr **d.** NaBr

6.69 The four bonds from carbon to hydrogen in CH_4 are as far apart as possible, which means that the hydrogen atoms are at the corners of a tetrahedron.

6.71 a. pentane **b.** ethane **c.** hexane

6.73 a. CH_4
b. CH_3-CH_3
c. $CH_3-CH_2-CH_2-CH_2-CH_3$

6.75 a. $CH_3-CH_2-CH_2-CH_2-CH_2-CH_2-CH_3$
b. liquid
c. no
d. float

6.77 a. By losing two valence electrons from the fourth energy level, calcium achieves an octet in the third energy level.
b. Ar
c. Group 1A (1) and 2A (2) elements acquire octets by losing electrons when they form compounds. Group 8A (18) elements are stable with octets (or two electrons for helium).

6.79 a. P^{3-} ion **b.** O atom **c.** Zn^{2+} ion **d.** Fe^{3+} ion

6.81 a. X = 1A (1); Y = 6A (16) **b.** ionic **c.** X^+, Y^{2-}
d. X_2Y **e.** XCl **f.** YCl_2

6.83

Period	Electron-Dot Symbols	Formula of Compound	Name of Compound
2	X· and ·Ÿ·	Li_3N	Lithium nitride
4	X· and :Ÿ·	$CaBr_2$	Calcium bromide
4	·X· and :Ÿ·	Ga_2Se_3	Gallium selenide

6.85

Electron Configurations		Symbols of Ions			
Metal	Nonmetal	Cation	Anion	Formula of Compound	Name of Compound
$1s^2 2s^2 2p^6 3s^2$	$1s^2 2s^2 2p^3$	Mg^{2+}	N^{3-}	Mg_3N_2	Magnesium nitride
$1s^2 2s^2 2p^6 3s^2 3p^6 4s^1$	$1s^2 2s^2 2p^4$	K^+	O^{2-}	K_2O	Potassium oxide
$1s^2 2s^2 2p^6 3s^2 3p^1$	$1s^2 2s^2 2p^5$	Al^{3+}	F^-	AlF_3	Aluminum fluoride

6.87 a. butane   **b.** butane
c. potassium chloride   **d.** potassium chloride
e. butane

6.89 a. organic, covalent, heptane
b. inorganic, covalent, nitrogen dioxide
c. inorganic, ionic, potassium phosphate
d. organic, covalent, ethane

6.91 a. Ne   **b.** Kr   **c.** Xe
d. Ne   **e.** Xe

6.93 a. 2A (2)   **b.** · X ·   **c.** Be   **d.** X_3N_2

6.95 a. Sn^{4+} **b.** 50 protons, 46 electrons
c. SnO_2 **d.** $Sn_3(PO_4)_4$

6.97 a. 3A (13)   **b.** 6A (16)   **c.** 2A (2)

6.99 a. iron(III) chloride   **b.** calcium phosphate
c. aluminum carbonate   **d.** lead(IV) chloride
e. magnesium carbonate   **f.** tin(II) sulfate
g. copper(II) sulfide

6.101 a. Cu_3N **b.** $KHSO_3$ **c.** PbS_2
d. $Au_2(CO_3)_3$ **e.** $Zn(ClO_4)_2$

6.103 a. $NiCl_3$ **b.** PbO_2 **c.** AgBr
d. Ca_3N_2 **e.** Cu_3P **f.** CrS

6.105 a. magnesium oxide
b. $Cr(HCO_3)_3$ is chromium(III) hydrogen carbonate or chromium(III) bicarbonate.
c. manganese(III) chromate

6.107 a. nitrogen trichloride   **b.** sulfur dichloride
c. dinitrogen oxide   **d.** fluorine
e. phosphorus pentafluoride
f. diphosphorus pentoxide

6.109 a. IF_7 **b.** BF_3 **c.** HI   **d.** SCl_2

6.111 a. ionic, cobalt(III) chloride
b. ionic, sodium sulfate
c. covalent, dinitrogen trioxide
d. covalent, iodine
e. covalent, iodine trichloride
f. covalent, carbon tetrafluoride

6.113 a. $SnCO_3$ **b.** Li_3P **c.** $SiCl_4$ **d.** CrS
e. CO_2 **f.** $CaBr_2$
g. $CH_3-CH_2-CH_2-CH_2-CH_3$, C_5H_{12}

6.115 The valence electrons are the electrons in the highest energy levels lost or gained in the formation of ionic compounds.

6.117 Elements in Group 2A (2) will lose two electrons to attain an octet; elements in Group 6A (16) will gain two electrons to attain an octet. Both either gain or lose two electrons.

6.119

Formula of Compound	Type of Compound	Name of Compound
$FeSO_4$	Ionic	Iron(II) sulfate
SiO_2	Covalent	Silicon dioxide
NH_4NO_3	Ionic	Ammonium nitrate
$Al_2(SO_4)_3$	Ionic	Aluminum sulfate
Co_2S_3	Ionic	Cobalt(III) sulfide

6.121 a. inorganic (ionic), lithium oxide
b. organic (covalent), propane
c. inorganic (covalent), carbon tetrafluoride
d. organic (covalent), methane
e. inorganic (ionic), magnesium fluoride
f. organic (covalent), butane

6.123 2.4×10^{22} S^{2-} ions

Chemical Quantities

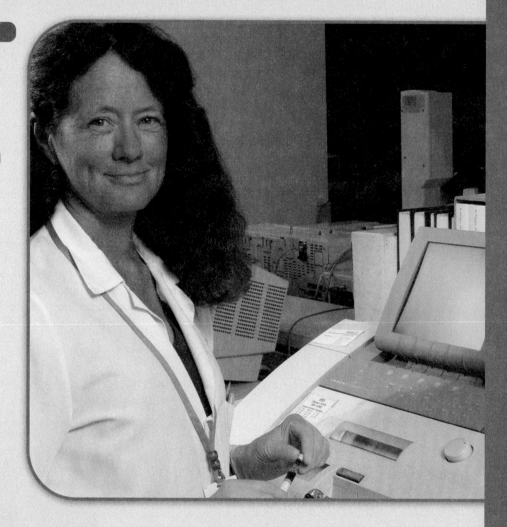

"In a stat lab, we are sent blood samples of patients in emergency situations," says Audrey Trautwein, clinical laboratory technician, Stat Lab, Santa Clara Valley Medical Center. *"We may need to assess the status of a trauma patient in ER or a patient who is in surgery. For example, an acidic blood pH diminishes cardiac function and affects the actions of certain drugs. In a stat situation, it is critical that we obtain our results fast. This is done using a blood gas analyzer. As I put a blood sample into the analyzer, a small probe draws out a measured volume, which is tested simultaneously for pH, P_{O_2}, and P_{CO_2} as well as electrolytes, glucose, and hemoglobin. In about one minute, we have our test results, which are sent to the doctor's computer."*

Visit **www.masteringchemistry.com** for self-study materials and instructor-assigned homework.

In chemistry, we calculate and measure the amounts of substances to use in the lab. Actually, measuring the amount of a substance is something you do every day. When you cook, you measure out the proper amounts of ingredients so you don't have too much of one and too little of another. At the gas station, you pump a certain amount of fuel into your gas tank. If you paint the walls of a room, you measure the area and purchase the amount of paint that will cover the walls. If you take a pain reliever, you read the label to see how many tablets of aspirin or ibuprofen are needed. You read the nutrition label on food packaging to determine the quantities of carbohydrate, fat, sodium, iron, or zinc.

Each substance that we use in the chemistry lab is measured so it functions properly in an experiment. The formula of a substance tells us the number and kinds of atoms it has, which we then use to determine the mass of the substance to use in an experiment. Similarly, when we know the percentage and mass of the elements in a substance, we can determine its formula.

7.1 The Mole

LEARNING GOAL

Use Avogadro's number to determine the number of particles in a given number of moles.

TUTORIAL

Using Avogadro's Number

There are special terms we use to represent specific numbers of items. At the grocery store, you buy eggs by the dozen. In an office-supply store, you may buy pencils by the gross and paper by the ream. In a restaurant, the beverage manager orders soda and water by the case. Each of these terms—*dozen, gross, ream,* and *case*—counts the number of items present. For example, when you buy a dozen eggs, you know you will get 12 eggs in the carton. In chemistry, tiny particles such as atoms, molecules, and ions are counted by the **mole** (abbreviated *mol* in calculations), a unit that contains 6.022×10^{23} items. This very large number is called **Avogadro's number** after Amedeo Avogadro, an Italian physicist. It looks like this when written with four significant figures:

Avogadro's Number

$$602\ 200\ 000\ 000\ 000\ 000\ 000\ 000 = 6.022 \times 10^{23}$$

Collections of items include dozen, gross, and mole.

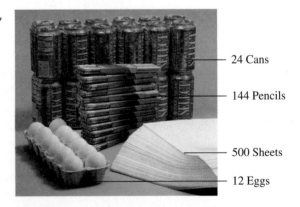

24 Cans

144 Pencils

500 Sheets

12 Eggs

One mole of any element always contains Avogadro's number of atoms. For example, 1 mol of carbon contains 6.022×10^{23} carbon atoms; 1 mol of aluminum contains 6.022×10^{23} aluminum atoms; 1 mol of sulfur contains 6.022×10^{23} sulfur atoms.

1 mol of an element $= 6.022 \times 10^{23}$ atoms of that element

Avogadro's number indicates that 1 mol of a compound contains 6.022×10^{23} of the particular type of particles that make up that compound. One mol of a covalent compound contains Avogadro's number of molecules. For example, 1 mol of CO_2 contains 6.022×10^{23} molecules of CO_2. One mol of an ionic compound contains Avogadro's number of **formula units**, which are the groups of ions represented by the formula of an ionic compound. For example, 1 mol of NaCl contains 6.022×10^{23} formula units of NaCl (Na^+, Cl^-). Table 7.1 gives examples of the number of particles in some 1-mol quantities.

TABLE 7.1 Number of Particles in 1-Mol Samples

Substance	Number and Type of Particles
1 mol of Al	6.022×10^{23} atoms of Al
1 mol of Fe	6.022×10^{23} atoms of Fe
1 mol of H_2O	6.022×10^{23} molecules of H_2O
1 mol of vitamin C ($C_6H_8O_6$)	6.022×10^{23} molecules of vitamin C
1 mol of NaCl	6.022×10^{23} formula units of NaCl

We can use Avogadro's number as a conversion factor to convert between the moles of a substance and the number of particles it contains.

$$\frac{6.022 \times 10^{23} \text{ particles}}{1 \text{ mol}} \quad \text{and} \quad \frac{1 \text{ mol}}{6.022 \times 10^{23} \text{ particles}}$$

For example, we use Avogadro's number to convert 4.00 mol of iron to atoms of iron.

$$4.00 \text{ mol Fe} \times \frac{6.022 \times 10^{23} \text{ atoms Fe}}{1 \text{ mol Fe}} = 2.41 \times 10^{24} \text{ atoms of Fe}$$

Avogadro's number as a conversion factor

We can also use Avogadro's number to convert 3.01×10^{24} molecules of CO_2 to moles of CO_2.

$$3.01 \times 10^{24} \text{ molecules } CO_2 \times \frac{1 \text{ mol } CO_2 \text{ molecules}}{6.022 \times 10^{23} \text{ molecules } CO_2} = 5.00 \text{ mol of } CO_2 \text{ molecules}$$

Avogadro's number as a conversion factor

In calculations that convert between moles and particles, the number of moles will be a small number compared to the number of atoms or molecules, which will be a large number.

CONCEPT CHECK 7.1

■ **Moles and Particles**

Explain why 0.20 mol of aluminum is a small number, but the number of atoms in 0.20 mol is a large number: 1.2×10^{23} atoms of aluminum.

ANSWER

The term mole, abbreviated *mol*, is used as a collection term that represents 6.022×10^{23} particles. Because atoms are submicroscopic particles, a large number of atoms are in 1 mol of aluminum.

SAMPLE PROBLEM | 7.1

■ **Calculating the Number of Molecules in a Mole**

Methane, CH_4, is the major component of natural gas. How many molecules of methane are present in 1.75 mol of methane?

SOLUTION

STEP 1 **Given** 1.75 mol of CH_4 **Need** molecules of CH_4

STEP 2 **Plan** moles → Avogadro's number → molecules

STEP 3 **Equalities/Conversion Factors**

$$1 \text{ mol of } CH_4 = 6.022 \times 10^{23} \text{ molecules of } CH_4$$

$$\frac{6.022 \times 10^{23} \text{ molecules } CH_4}{1 \text{ mol } CH_4} \quad \text{and} \quad \frac{1 \text{ mol } CH_4}{6.022 \times 10^{23} \text{ molecules } CH_4}$$

Guide to Calculating the Atoms or Molecules of a Substance

STEP 1
Determine the given number of moles.

STEP 2
Write a plan to convert moles to atoms or molecules.

STEP 3
Use Avogadro's number to write conversion factors.

STEP 4
Set up problem to convert given moles to atoms or molecules.

STEP 4 Set Up Problem Calculate the number of CH_4 molecules.

$$1.75 \text{ mol CH}_4 \times \frac{6.022 \times 10^{23} \text{ molecules CH}_4}{1 \text{ mol CH}_4} = 1.05 \times 10^{24} \text{ molecules of CH}_4$$

STUDY CHECK

How many moles of water, H_2O, contain 2.60×10^{23} molecules of water?

TUTORIAL

Moles and the Chemical Formula

TUTORIAL

Conversions Involving Moles

Moles of Elements in a Formula

We have seen that the subscripts in a chemical formula indicate the number of atoms of each type of element in a compound. For example, one molecule of NH_3 consists of 1 N atom and 3 H atoms. If we have 6.022×10^{23} molecules (1 mol) of NH_3, it would contain 6.022×10^{23} atoms (1 mol) of N and $3 \times 6.022 \times 10^{23}$ atoms (3 mol) of H. Thus, each subscript in a formula also refers to the moles of each kind of atom in 1 mol of a compound. For example, the subscripts in the NH_3 formula specify that 1 mol of NH_3 molecules contains 1 mol of N atoms and 3 mol of H atoms.

The formula subscript specifies the

	NH_3	
Atoms in 1 molecule	1 atom of N	3 atoms of H
Moles of each element in 1 mol	1 mol of N	3 mol of H

Aspirin, $C_9H_8O_4$, is a drug used to reduce pain and inflammation in the body. In the aspirin formula, there are 9 C atoms, 8 H atoms, and 4 O atoms. The subscripts also state the number of moles of each element in 1 mol of aspirin: 9 mol of C, 8 mol of H, and 4 mol of O.

The formula subscript specifies the

	$C_9H_8O_4$	
Atoms in 1 molecule	9 atoms of C	8 atoms of H	4 atoms of O
Moles of each element in 1 mol	9 mol of C	8 mol of H	4 mol of O

The subscripts in a formula are helpful when we need to determine the amount of any of the elements in the formula. For the aspirin formula, $C_9H_8O_4$, we can write the following sets of conversion factors for each of the elements in 1 mol of aspirin:

$$\frac{9 \text{ mol C}}{1 \text{ mol C}_9\text{H}_8\text{O}_4} \quad \frac{8 \text{ mol H}}{1 \text{ mol C}_9\text{H}_8\text{O}_4} \quad \frac{4 \text{ mol O}}{1 \text{ mol C}_9\text{H}_8\text{O}_4}$$

$$\frac{1 \text{ mol C}_9\text{H}_8\text{O}_4}{9 \text{ mol C}} \quad \frac{1 \text{ mol C}_9\text{H}_8\text{O}_4}{8 \text{ mol H}} \quad \frac{1 \text{ mol C}_9\text{H}_8\text{O}_4}{4 \text{ mol O}}$$

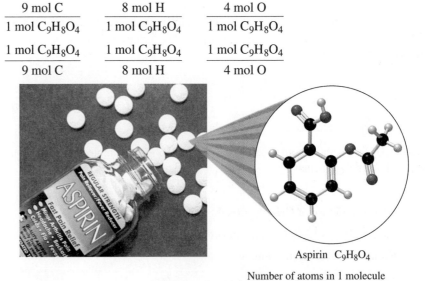

Aspirin $C_9H_8O_4$

Number of atoms in 1 molecule

Carbon (C)   Hydrogen (H)   Oxygen (O)

CONCEPT CHECK 7.2

■ **Using Subscripts of a Formula**

Indicate the moles of each element in 1 mol of each of the following:

a. $C_8H_9NO_2$, acetaminophen used in Tylenol
b. $Zn(C_2H_3O_2)_2$, zinc dietary supplement

ANSWER

a. The subscripts in the formula indicate that there are 8 mol of C, 9 mol of H, 1 mol of N, and 2 mol of O in 1 mol of acetaminophen.
b. In the formula, the subscript 1 is understood for Zn as 1 mol of Zn. When the subscripts inside the parentheses ($C_2H_3O_2$) are multiplied by the subscript 2 outside the parentheses, there is a total of 4 mol of C, 6 mol of H, and 4 mol of O in 1 mol of $Zn(C_2H_3O_2)_2$.

SAMPLE PROBLEM   7.2

■ **Calculating the Moles of an Element in a Compound**

How many moles of carbon are present in 1.50 mol of aspirin, $C_9H_8O_4$?

SOLUTION

STEP 1 Given 1.50 mol of $C_9H_8O_4$ **Need** moles of C

STEP 2 Plan moles of $C_9H_8O_4$ Subscript   moles of C

STEP 3 Equalities/Conversion Factors

$$1 \text{ mol of } C_9H_8O_4 = 9 \text{ mol of C}$$
$$\frac{9 \text{ mol C}}{1 \text{ mol } C_9H_8O_4} \quad \text{and} \quad \frac{1 \text{ mol } C_9H_8O_4}{9 \text{ mol C}}$$

STEP 4 Set Up Problem

$$1.50 \text{ mol } \cancel{C_9H_8O_4} \times \frac{9 \text{ mol C}}{1 \text{ mol } \cancel{C_9H_8O_4}} = 13.5 \text{ mol of C}$$

STUDY CHECK

How many moles of aspirin, $C_9H_8O_4$, contain 0.480 mol of O?

SAMPLE PROBLEM   7.3

■ **Particles of an Element in a Compound**

How many Na^+ ions are in 3.00 mol of Na_2O?

SOLUTION

STEP 1 Given 3.00 mol of Na_2O **Need** number of Na^+ ions

STEP 2 Plan Use the subscript for Na in the formula to convert from moles of Na_2O to moles of Na^+ ions and use Avogadro's number to calculate the number of Na^+ ions.

moles of Na_2O Subscript   moles of Na^+ ions   Avogadro's number   number of Na^+ ions

STEP 3 Equalities/Conversion Factors In Na_2O, an ionic compound, there are 2 mol of Na^+ ions in 1 mol of Na_2O.

$$1 \text{ mol of } Na_2O = 2 \text{ mol of } Na^+ \text{ ions}$$
$$\frac{2 \text{ mol } Na^+ \text{ ions}}{1 \text{ mol } Na_2O} \quad \text{and} \quad \frac{1 \text{ mol } Na_2O}{2 \text{ mol } Na^+ \text{ ions}}$$

$$1 \text{ mol of } Na^+ \text{ ions} = 6.022 \times 10^{23} \, Na^+ \text{ ions}$$
$$\frac{6.022 \times 10^{23} \, Na^+ \text{ ions}}{1 \text{ mol } Na^+ \text{ ions}} \quad \text{and} \quad \frac{1 \text{ mol } Na^+ \text{ ions}}{6.022 \times 10^{23} \, Na^+ \text{ ions}}$$

STEP 4 Set Up Problem

$$3.00 \text{ mol } Na_2O \times \frac{2 \text{ mol } Na^+ \text{ ions}}{1 \text{ mol } Na_2O} \times \frac{6.022 \times 10^{23} \, Na^+ \text{ ions}}{1 \text{ mol } Na^+ \text{ ions}} = 3.61 \times 10^{24} \, Na^+ \text{ ions}$$

STUDY CHECK

How many SO_4^{2-} ions are in 2.50 mol of $Fe_2(SO_4)_3$, a compound used in water and sewage treatment?

QUESTIONS AND PROBLEMS

The Mole

7.1 Calculate each of the following:
   a. number of Ag atoms in 0.200 mol of Ag
   b. number of C_3H_8O molecules in 0.750 mol of C_3H_8O
   c. moles of Au in 2.88×10^{23} atoms of Au

7.2 Calculate each of the following:
   a. number of Ni atoms in 3.4 mol of Ni
   b. number of $Mg(OH)_2$ formula units in 1.20 mol of $Mg(OH)_2$
   c. moles of Zn in 5.6×10^{24} atoms of Zn

7.3 Quinine, $C_{20}H_{24}N_2O_2$, is a component of tonic water and bitter lemon.
   a. How many moles of hydrogen are in 1.0 mol of quinine?
   b. How many moles of carbon are in 5.0 mol of quinine?
   c. How many moles of nitrogen are in 0.020 mol of quinine?

7.4 Aluminum sulfate, $Al_2(SO_4)_3$, is used in some antiperspirants.
   a. How many moles of sulfur are present in 3.0 mol of $Al_2(SO_4)_3$?
   b. How many moles of aluminum ions are present in 0.40 mol of $Al_2(SO_4)_3$?

   c. How many moles of sulfate ions (SO_4^{2-}) are present in 1.5 mol of $Al_2(SO_4)_3$?

7.5 Calculate each of the following:
   a. number of C atoms in 0.500 mol of C
   b. number of SO_2 molecules in 1.28 mol of SO_2
   c. moles of Fe in 5.22×10^{22} atoms of Fe
   d. moles of C_2H_5OH in 8.50×10^{24} molecules of C_2H_5OH

7.6 Calculate each of the following:
   a. number of Li atoms in 4.5 mol of Li
   b. number of CO_2 molecules in 0.0180 mol of CO_2
   c. moles of Cu in 7.8×10^{21} atoms of Cu
   d. moles of C_2H_6 in 3.754×10^{23} molecules of C_2H_6

7.7 Calculate each of the following in 2.00 mol of H_3PO_4:
   a. moles of H   b. moles of O
   c. atoms of P   d. atoms of O

7.8 Calculate each of the following in 0.185 mol of dipropyl ether, $(C_3H_7)_2O$:
   a. moles of C   b. moles of O
   c. atoms of H   d. atoms of C

7.2 Molar Mass

LEARNING GOAL

Given the chemical formula of a substance, calculate its molar mass.

SELF STUDY ACTIVITY

Stoichiometry

A single atom or molecule is much too small to weigh, even on the most accurate balance. In fact, it takes a huge number of atoms or molecules to make enough of a substance for you to see. An amount of water that contains Avogadro's number of water molecules is only a few sips. However, in the laboratory, we can use a balance to weigh out Avogadro's number of particles for 1 mol of substance.

For any element, the quantity called **molar mass** is the quantity in grams that equals the atomic mass of that element. For example, carbon has an atomic mass of 12.01 on the periodic table. This means 1 mol of carbon atoms has a mass of 12.01 g. Thus, to obtain 1 mol of carbon atoms, we would need to weigh out 12.01 g of carbon. We use the periodic table to determine the molar mass of any element by stating its atomic mass in grams. Thus, sulfur has a molar mass of 32.07 g and silver has a molar mass of 107.9 g.

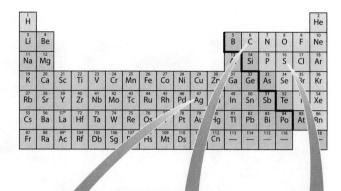

6.022 × 10²³ atoms of C

1 mol of C atoms

12.01 g of C atoms

47
Ag
107.9

1 mol of
silver has
a mass
of 107.9 g

6
C
12.01

1 mol of
carbon has
a mass
of 12.01 g

16
S
32.07

1 mol of
sulfur has
a mass
of 32.07 g

Molar Mass of a Compound

To determine the molar mass of a compound, multiply the molar mass of each element by its subscript in the formula and add the results. For example, the molar mass of sulfur trioxide, SO_3, is obtained by adding the molar masses of 1 mol of sulfur and 3 mol of oxygen.

CONCEPT CHECK 7.3

■ Calculating Molar Mass

Acid rain forms when sulfur trioxide, released during volcanic eruptions and the burning of sulfur-containing fuels such as coal, is converted to sulfuric acid in the atmosphere. What is the molar mass of SO_3?

ANSWER

From the periodic table, we can obtain the molar masses of sulfur and oxygen.

$$\frac{32.07 \text{ g S}}{1 \text{ mol S}} \qquad \frac{16.00 \text{ g O}}{1 \text{ mol O}}$$

With a subscript of one understood for S, and a subscript of three for O, we know that there are 1 mol of S and 3 mol of O in 1 mol of compound. The number of grams for each element is obtained by multiplying each molar mass by its subscript.

Grams of S for 1 mol of S

$$1 \text{ mol S} \times \frac{32.07 \text{ g S}}{1 \text{ mol S}} = 32.07 \text{ g of S}$$

Grams of O for 3 mol of O

$$3 \text{ mol O} \times \frac{16.00 \text{ g O}}{1 \text{ mol O}} = 48.00 \text{ g of O}$$

The molar mass of the compound is obtained by adding the grams of each element.

1 mol of S	=	32.07 g of S
3 mol of O	=	+ 48.00 g of O
Molar mass of SO_3 =		80.07 g of SO_3

1-Mol Quantities

| S | Fe | NaCl | $K_2Cr_2O_7$ | $C_{12}H_{22}O_{11}$ |

TABLE 7.2 The Molar Mass of Selected Elements and Compounds

Substance	Molar Mass
1 mol of C	12.01 g
1 mol of Na	22.99 g
1 mol of Fe	55.85 g
1 mol of NaF	41.99 g
1 mol of $C_6H_{12}O_6$ (glucose)	180.16 g
1 mol of $C_8H_{10}N_4O_2$ (caffeine)	194.20 g

FIGURE 7.1 One-mol samples: sulfur, S (32.07 g); iron, Fe (55.85 g); salt, NaCl (58.44 g); potassium dichromate, $K_2Cr_2O_7$ (294.2 g); and sugar, sucrose, $C_{12}H_{22}O_{11}$ (342.3 g).

Q How is the molar mass for $K_2Cr_2O_7$ obtained?

Figure 7.1 shows some 1-mol quantities of substances. Table 7.2 lists the molar mass for several 1-mol samples.

SAMPLE PROBLEM | **7.4**

■ Calculating the Molar Mass of a Compound

Find the molar mass of lithium carbonate, Li_2CO_3, used to produce red color in fireworks.

SOLUTION

STEP 1 **Using the periodic table, obtain the molar masses of lithium, carbon, and oxygen.**

$$\frac{6.941 \text{ g Li}}{1 \text{ mol Li}} \qquad \frac{12.01 \text{ g C}}{1 \text{ mol C}} \qquad \frac{16.00 \text{ g O}}{1 \text{ mol O}}$$

Guide to Calculating Molar Mass

STEP 1 Obtain the molar mass of each element.

STEP 2 Multiply each molar mass by the number of moles (subscript) in the formula.

STEP 3 Calculate the molar mass by adding the masses of the elements.

STEP 2 **Obtain the mass of each element in the formula by multiplying each molar mass by its number of moles (subscript) in the formula.**

Grams from 2 mol of Li

$$2 \text{ mol Li} \times \frac{6.941 \text{ g Li}}{1 \text{ mol Li}} = 13.88 \text{ g of Li}$$

Grams from 1 mol of C

$$1 \text{ mol C} \times \frac{12.01 \text{ g C}}{1 \text{ mol C}} = 12.01 \text{ g of C}$$

Grams from 3 mol of O

$$3 \text{ mol O} \times \frac{16.00 \text{ g O}}{1 \text{ mol O}} = 48.00 \text{ g of O}$$

STEP 3 **Calculate the molar mass by adding the masses of the elements.** Obtain the molar mass of Li_2CO_3 by adding the masses of 2 mol of Li, 1 mol of C, and 3 mol of O.

2 mol of Li	= 13.88 g of Li
1 mol of C	= 12.01 g of C
3 mol of O	= 48.00 g of O
Molar mass of Li_2CO_3	= 73.89 g of Li_2CO_3

Lithium carbonate produces a red color in fireworks.

STUDY CHECK

Calculate the molar mass of salicylic acid, $C_7H_6O_3$, which is used to make aspirin.

QUESTIONS AND PROBLEMS

Molar Mass

7.9 Calculate the molar mass for each of the following:
  a. NaCl (table salt)
  b. Fe_2O_3 (rust)
  c. $C_{19}H_{20}FNO_3$ (Paxil, an antidepressant)
  d. $Al_2(SO_4)_3$ (antiperspirant)
  e. $KC_4H_5O_6$ (cream of tartar)
  f. $C_{16}H_{19}N_3O_5S$ (amoxicillin, an antibiotic)

7.10 Calculate the molar mass for each of the following:
  a. $FeSO_4$ (iron supplement)
  b. Al_2O_3 (absorbent and abrasive)
  c. $C_7H_5NO_3S$ (saccharin)

d. C_3H_8O (rubbing alcohol)
e. $(NH_4)_2CO_3$ (baking powder)
f. $Zn(C_2H_3O_2)_2$ (zinc dietary supplement)

7.11 Calculate the molar mass for each of the following:
  a. Cl_2 **b.** $C_3H_6O_3$
  c. $Mg_3(PO_4)_2$ **d.** AlF_3
  e. $C_2H_4Cl_2$ **f.** SnF_2

7.12 Calculate the molar mass for each of the following:
  a. O_2 **b.** KH_2PO_4
  c. $Fe(ClO_4)_3$ **d.** $C_4H_8O_4$
  e. $Ga_2(CO_3)_3$ **f.** $KBrO_4$

7.3 Calculations Using Molar Mass

The molar mass of an element or compound is one of the most useful conversion factors in chemistry. Molar mass is used to change from moles of a substance to grams, or from grams to moles. To do these calculations, we use the molar mass as a conversion factor. For example, 1 mol of magnesium has a mass of 24.31 g. To express molar mass as an equality, we can write

  1 mol of Mg = 24.31 g of Mg

From this equality, two conversion factors can be written:

$$\frac{24.31 \text{ g Mg}}{1 \text{ mol Mg}} \quad \text{and} \quad \frac{1 \text{ mol Mg}}{24.31 \text{ g Mg}}$$

Conversion factors are written for compounds in the same way. For example, the molar mass of the compound H_2O is $(2 \times 1.008) + (1 \times 16.00) = 18.02$ g.

  1 mol of H_2O = 18.02 g of H_2O

The conversion factors from the molar mass of H_2O are written:

$$\frac{18.02 \text{ g } H_2O}{1 \text{ mol } H_2O} \quad \text{and} \quad \frac{1 \text{ mol } H_2O}{18.02 \text{ g } H_2O}$$

We can now change from moles to grams, or grams to moles, using the conversion factors derived from the molar mass. (Remember, you must determine the molar mass of the substance first.)

LEARNING GOAL

Given the number of moles of a substance, calculate the mass in grams; given the mass, calculate the number of moles.

SELF STUDY ACTIVITY
Stoichiometry

Silver metal is used to make jewelry.

SAMPLE PROBLEM | **7.5**

■ Converting Moles of an Element to Grams

Silver metal is used in the manufacture of tableware, mirrors, jewelry, and dental alloys. If the design for a piece of jewelry requires 0.750 mol of silver, how many grams of silver are needed?

SOLUTION

STEP 1 Given 0.750 mol of Ag   **Need** grams of Ag

STEP 2 Plan moles of Ag   Molar mass factor   grams of Ag

STEP 3 Equalities/Conversion Factors

$$1 \text{ mol of Ag} = 107.9 \text{ g of Ag}$$
$$\frac{107.9 \text{ g Ag}}{1 \text{ mol Ag}} \quad \text{and} \quad \frac{1 \text{ mol Ag}}{107.9 \text{ g Ag}}$$

Guide to Calculating the Moles (or Grams) of a Substance from Grams (or Moles)

STEP 1
Determine the given number of moles (or grams).

STEP 2
Write a plan to convert moles to grams (or grams to moles).

STEP 3
Determine the molar mass and write conversion factors.

STEP 4
Set up problem to convert given moles to grams (or grams to moles).

STEP 4 **Set Up Problem** Calculate the grams of silver using the molar mass factor that cancels "mol Ag".

$$0.750 \text{ mol Ag} \times \frac{107.9 \text{ g Ag}}{1 \text{ mol Ag}} = 80.9 \text{ g of Ag}$$

STUDY CHECK

Calculate the number of grams of gold (Au) present in 0.124 mol of gold.

CONCEPT CHECK 7.4

■ Comparing Grams, Moles, and Atoms

Compounds called chlorofluorocarbons (CFCs) were used for many years in aerosol cans and in refrigerators. Eventually, it was determined that, in the stratosphere, CFCs interacted with sunlight, which has destroyed part of the ozone layer. The CFC Freon-13 has the formula of $CClF_3$.

a. What element has the most atoms in 1 mol of Freon-13?
b. Are there more moles of Cl or F in 1 mol of Freon-13?
c. Are there more grams of Cl or F in 1 mol of Freon-13?
d. Is the number of moles of carbon larger or smaller than the number of atoms of carbon?

ANSWER

a. In one molecule of $CClF_3$, there are 1 atom of C, 1 atom of Cl, and 3 atoms of F. Thus, in 1 mol of $CClF_3$, there would be 3 mol of F, or $3 \times 6.022 \times 10^{23}$ atoms of F.
b. From the formula $CClF_3$, there are 1 mol Cl and 3 mol of F. Thus, there are more moles of F than Cl in 1 mol of Freon-13.
c. To determine the greater number of grams, we multiply the molar mass of each by the number of moles (subscripts). One mol of Cl $\times$ 35.45 g/mol $=$ 35.45 g of Cl, and 3 mol of F $\times$ 19.00 g/mol $=$ 57.00 g of F. Thus, there are more grams of fluorine than chlorine in the compound.
d. Because there is Avogadro's number of atoms in 1 mol, the number of moles of carbon is smaller than the number of atoms of carbon.

TUTORIAL

Converting Between Grams and Moles

Table salt is NaCl.

SAMPLE PROBLEM | 7.6

■ Converting Mass of a Compound to Moles

A box of salt contains 737 g of NaCl. How many moles of NaCl are present in the box?

SOLUTION

STEP 1 **Given** 737 g of NaCl   **Need** moles of NaCl

STEP 2 **Plan** grams of NaCl   Molar mass factor   moles of NaCl

STEP 3 **Equalities/Conversion Factors**

$$1 \text{ mol of NaCl} = 58.44 \text{ g of NaCl}$$
$$\frac{58.44 \text{ g NaCl}}{1 \text{ mol NaCl}} \text{ and } \frac{1 \text{ mol NaCl}}{58.44 \text{ g NaCl}}$$

STEP 4 **Set Up Problem** Calculate the moles of NaCl using the molar mass.

$$737 \text{ g NaCl} \times \frac{1 \text{ mol NaCl}}{58.44 \text{ g NaCl}} = 12.6 \text{ mol of NaCl}$$

One gel cap of an antacid contains 311 mg of $CaCO_3$ and 232 mg of $MgCO_3$. In a recommended dosage of two gel caps, how many moles each of $CaCO_3$ and $MgCO_3$ are present?

We can now combine the calculations from the previous problems to convert the mass in grams of a compound to the number of molecules. In these calculations, such as shown in Sample Problem 7.7, we see the central role of moles in the conversion of mass to particles.

SAMPLE PROBLEM   7.7

■ **Converting Grams to Particles**

A 10.00-lb bag of table sugar contains 4536 g of sucrose, $C_{12}H_{22}O_{11}$. How many molecules of sugar are present?

SOLUTION

STEP 1   Given 4536 g of sucrose   **Need** moles of sucrose

STEP 2   Plan Convert the grams of sugar to moles of sugar using the molar mass of $C_{12}H_{22}O_{11}$. Then use Avogadro's number to calculate the number of molecules of sugar.

| g of sugar | Molar mass factor | moles | Avogadro's number | molecules |

STEP 3   Equalities/Conversion Factors The molar mass of $C_{12}H_{22}O_{11}$ is the sum of the masses of 12 mol of C + 22 mol of H + 11 mol of O:

$$(144.1 \text{ g/mol}) + (22.18 \text{ g/mol}) + (176.0 \text{ g/mol}) = 342.3 \text{ g/mol}$$

$$1 \text{ mol of } C_{12}H_{22}O_{11} = 342.3 \text{ g of } C_{12}H_{22}O_{11}$$

$$\frac{342.3 \text{ g } C_{12}H_{22}O_{11}}{1 \text{ mol } C_{12}H_{22}O_{11}} \quad \text{and} \quad \frac{1 \text{ mol } C_{12}H_{22}O_{11}}{342.3 \text{ g } C_{12}H_{22}O_{11}}$$

$$1 \text{ mol of } C_{12}H_{22}O_{11} = 6.022 \times 10^{23} \text{ molecules of } C_{12}H_{22}O_{11}$$

$$\frac{6.022 \times 10^{23} \text{ molecules}}{1 \text{ mol } C_{12}H_{22}O_{11}} \quad \text{and} \quad \frac{1 \text{ mol } C_{12}H_{22}O_{11}}{6.022 \times 10^{23} \text{ molecules}}$$

STEP 4   Set Up Problem Using the molar mass as a conversion factor, convert the grams of sucrose to moles of sucrose and then use Avogadro's number to convert moles of sucrose to molecules.

$$4536 \text{ g } C_{12}H_{22}O_{11} \times \frac{1 \text{ mol } C_{12}H_{22}O_{11}}{342.3 \text{ g } C_{12}H_{22}O_{11}} \times \frac{6.022 \times 10^{23} \text{ molecules}}{1 \text{ mol } C_{12}H_{22}O_{11}}$$

$$= 7.980 \times 10^{24} \text{ molecules of sucrose}$$

In the body, caffeine, which has a formula of $C_8H_{10}N_4O_2$, acts as a stimulant and diuretic. How many moles of nitrogen are in 2.50 g of caffeine?

Table sugar is sucrose, $C_{12}H_{22}O_{11}$.

Guide to Converting Grams to Particles

| STEP 1 |
| Determine the given number of grams. |

| STEP 2 |
| Write a plan that converts grams to particles. |

| STEP 3 |
| Write the conversion factors for molar mass and Avogadro's number. |

| STEP 4 |
| Set up the problem to convert grams to moles to particles. |

Figure 7.2 gives a summary of all the calculations in this section by showing the connections between the moles, mass in grams, and number of molecules or formula units, and the moles and atoms of each element in that compound.

FIGURE 7.2 The moles of a compound are related to its mass in grams by molar mass, to the number of molecules (or formula units) by Avogadro's number, and to the moles of each element by the subscripts in the formula.

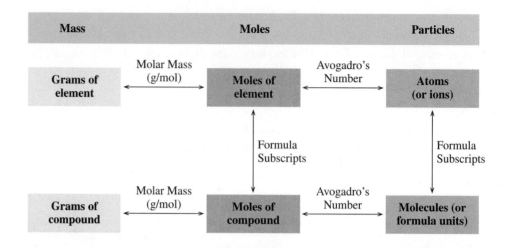

Mass	Moles	Particles
Grams of element ←→ Molar Mass (g/mol) →	**Moles of element** ←→ Avogadro's Number →	**Atoms (or ions)**
	Formula Subscripts ↕	Formula Subscripts ↕
Grams of compound ←→ Molar Mass (g/mol) →	**Moles of compound** ←→ Avogadro's Number →	**Molecules (or formula units)**

Q What steps are needed to calculate the number of H atoms in 5.00 g of CH_4?

QUESTIONS AND PROBLEMS

Calculations Using Molar Mass

7.13 Calculate the mass, in grams, for each of the following:
  a. 1.50 mol of Na   **b.** 2.80 mol of Ca
  c. 0.125 mol of CO_2 **d.** 0.0485 mol Na_2CO_3
  e. 7.14×10^2 mol of PCl_3

7.14 Calculate the mass, in grams, for each of the following:
  a. 5.12 mol of Al   **b.** 0.75 mol of Cu
  c. 3.52 mol of $MgBr_2$ **d.** 0.145 mol of C_2H_6O
  e. 2.08 mol of $(NH_4)_2SO_4$

7.15 Calculate the mass, in grams, in 0.150 mol of each of the following:
  a. Ne   **b.** I_2
  c. Na_2O **d.** $Ca(NO_3)_2$
  e. C_6H_{14}

7.16 Calculate the mass, in grams, in 2.28 mol of each of the following:
  a. N_2 **b.** SO_3
  c. $C_3H_6O_3$ **d.** $Mg(HCO_3)_2$
  e. SF_6

7.17 Calculate the number of moles in each of the following:
  a. 82.0 g of Ag   **b.** 0.288 g of C
  c. 15.0 g of ammonia, NH_3 **d.** 7.25 g of propane, C_3H_8
  e. 245 g of Fe_2O_3

7.18 Calculate the number of moles in each of the following:
  a. 85.2 g of Ni   **b.** 144 g of K
  c. 6.4 g of H_2O **d.** 308 g of $BaSO_4$
  e. 252.8 g of fructose, $C_6H_{12}O_6$

7.19 Calculate the number of moles in 25.0 g of each of the following:
  a. He   **b.** O_2
  c. $Al(OH)_3$ **d.** Ga_2S_3
  e. C_4H_{10}, butane

7.20 Calculate the number of moles in 4.00 g of each of the following:
  a. Au   **b.** SnO_2
  c. $Cr(OH)_3$ **d.** Ca_3N_2
  e. $C_6H_8O_6$, vitamin C

7.21 Calculate the number of atoms of C in each of the following:
  a. 25.0 g of C   **b.** 0.688 mol of CO_2
  c. 275 g of C_3H_8 **d.** 1.84 mol of C_2H_6O
  e. 7.5×10^{24} molecules of CH_4

7.22 Calculate the number of atoms of N in each of the following:
  a. 0.755 mol of N_2
  b. 0.82 g of $NaNO_3$
  c. 40.0 g of N_2O
  d. 6.24×10^{23} molecules of NH_3
  e. 1.4×10^{22} molecules of N_2O_4

7.23 Propane gas, C_3H_8, a hydrocarbon, is used as a fuel for many barbecues.
  a. How many grams of the compound are in 1.50 mol of propane?
  b. How many moles of the compound are in 34.0 g of propane?
  c. How many grams of carbon are in 34.0 g of propane?
  d. How many atoms of hydrogen are in 0.254 g of propane?

7.24 Allyl sulfide, $(C_3H_5)_2S$, is the substance that gives garlic, onions, and leeks their characteristic odor.
  a. How many moles of sulfur are in 23.2 g of $(C_3H_5)_2S$?
  b. How many atoms of hydrogen are in 0.75 mol of $(C_3H_5)_2S$?
  c. How many grams of carbon are in 4.20×10^{23} molecules of $(C_3H_5)_2S$?
  d. How many atoms of carbon are in 15.0 g of $(C_3H_5)_2S$?

7.4 Percent Composition and Empirical Formulas

LEARNING GOAL

Given the formula of a compound, calculate the percent composition; from the percent composition, determine the empirical formula of a compound.

We have seen that the atoms of the elements present in a compound are combined in a definite proportion. Now we can also say that the molar mass of any compound contains a definite proportion by mass of its elements. Using molar mass, we can determine the **percent composition** of a compound, which is the percent by mass of each element present.

$$\text{Mass percent of each element in a compound} = \frac{\text{mass of each element}}{\text{molar mass of compound}} \times 100\%$$

TUTORIAL
Mass Percent

For example, let us calculate the percent composition of dinitrogen oxide, N_2O, "laughing gas," which is used as an anesthetic for surgery and in dentistry.

STEP 1   **Determine the total mass of each element in the molar mass.**

$$2 \text{ mol N} \times \frac{14.01 \text{ g N}}{1 \text{ mol N}} = 28.02 \text{ g of N}$$

$$1 \text{ mol O} \times \frac{16.00 \text{ g O}}{1 \text{ mol O}} = 16.00 \text{ g of O}$$

$$\text{Molar mass of } N_2O \quad = 44.02 \text{ g of } N_2O$$

STEP 2   **Calculate the percent composition of each element by multiplying the mass ratio by 100%.**

$$\text{Mass \% N} = \frac{28.02 \text{ g N}}{44.02 \text{ g } N_2O} \times 100\% = 63.65\% \text{ N}$$

$$\text{Mass \% O} = \frac{16.00 \text{ g O}}{44.02 \text{ g } N_2O} \times 100\% = 36.35\% \text{ O}$$

Check that the total percent is equal or very close to 100%.

$$63.65\% \text{ N} + 36.35\% \text{ O} = 100.00\%$$

SAMPLE PROBLEM   7.8

■ **Calculating Percent Composition**

The odor of pears is due to the organic compound called propyl acetate, which has a formula of $C_5H_{10}O_2$. What is its percent composition by mass?

SOLUTION

STEP 1   **Determine the total mass of each element in the molar mass of a formula.**

$$5 \text{ mol C} \times \frac{12.01 \text{ g C}}{1 \text{ mol C}} = 60.05 \text{ g of C}$$

$$10 \text{ mol H} \times \frac{1.008 \text{ g H}}{1 \text{ mol H}} = 10.08 \text{ g of H}$$

$$2 \text{ mol O} \times \frac{16.00 \text{ g O}}{1 \text{ mol O}} = 32.00 \text{ g of O}$$

$$\text{Molar mass of } C_5H_{10}O_2 = 102.13 \text{ g of } C_5H_{10}O_2$$

STEP 2   **Divide the total mass of each element by the molar mass and multiply by 100%.**

$$\text{Mass \% C} = \frac{60.05 \text{ g C}}{102.13 \text{ g } C_5H_{10}O_2} \times 100\% = 58.80\% \text{ C}$$

$$\text{Mass \% H} = \frac{10.08 \text{ g H}}{102.13 \text{ g } C_5H_{10}O_2} \times 100\% = 9.870\% \text{ H}$$

$$\text{Mass \% O} = \frac{32.00 \text{ g O}}{102.13 \text{ g } C_5H_{10}O_2} \times 100\% = 31.33\% \text{ O}$$

Check that the total percent is equal to 100%.

$$58.80\% \text{ C} + 9.870\% \text{ H} + 31.33\% \text{ O} = 100.00\%$$

STUDY CHECK

The organic compound ethylene glycol, $C_2H_6O_2$, used as automobile antifreeze, is a sweet-tasting liquid, which is toxic to humans and animals. What is the percent composition by mass of ethylene glycol?

Guide to Calculating Percent Composition

STEP 1
Determine the total mass of each element in the molar mass of a formula.

STEP 2
Divide the total mass of each element by the molar mass and multiply by 100%.

CHEMISTRY AND THE ENVIRONMENT

Fertilizers

Every year in the spring, homeowners and farmers add fertilizers to the soil to produce greener lawns and larger crops. Plants require several nutrients, but the major ones are nitrogen, phosphorus, and potassium. Nitrogen promotes green growth, phosphorus promotes strong root development for strong plants and abundant flowers, and potassium helps plants defend against diseases and weather extremes. The numbers on a package of fertilizer give the percentages each of N, P, and K by mass. For example, the set of numbers 30–3–4 describes a fertilizer that contains 30% N, 3% P, and 4% K.

Nitrogen (N) 30%
Phosphorus (P) 3%
Potassium (K) 4%

The label on a bag of fertilizer states the percentages of N, P, and K.

The major nutrient, nitrogen, is present in huge quantities as N_2 in the atmosphere, but plants cannot utilize nitrogen in this form. Bacteria in the soil convert atmospheric N_2 to usable forms by nitrogen fixation. To provide additional nitrogen to plants, several types of nitrogen-containing chemicals, including ammonia, nitrates, and ammonium compounds, are added to soil. The nitrates are absorbed directly, but ammonia and ammonium salts are first converted to nitrates by the soil bacteria.

The percent nitrogen depends on the type of nitrogen compound used in the fertilizer. The percent nitrogen by mass in each type is calculated using percent composition.

Type of Fertilizer		Percent Nitrogen by Mass
NH_3	$\dfrac{14.01 \text{ g N}}{17.03 \text{ g NH}_3}$	$\times\ 100\% = 82.27\%$ N
NH_4NO_3	$\dfrac{28.02 \text{ g N}}{80.05 \text{ g NH}_4NO_3}$	$\times\ 100\% = 35.00\%$ N
$(NH_4)_2SO_4$	$\dfrac{28.02 \text{ g N}}{132.15 \text{ g (NH}_4)_2SO_4}$	$\times\ 100\% = 21.20\%$ N
$(NH_4)_2HPO_4$	$\dfrac{28.02 \text{ g N}}{132.06 \text{ g (NH}_4)_2HPO_4}$	$\times\ 100\% = 21.22\%$ N

The choice of a fertilizer depends on its use and convenience. A fertilizer can be prepared as crystals or a powder, in a liquid solution, or as a gas such as ammonia. The ammonia and ammonium fertilizers are water soluble and quick-acting. Other forms may be made to slow-release by enclosing water-soluble ammonium salts in a thin plastic coating. The most commonly used fertilizer is NH_4NO_3 because it is easy to apply and has a high percent of N by mass.

TABLE 7.3 Examples of Molecular and Empirical Formulas

Name	Molecular (actual formula)	Empirical (simplest formula)
Acetylene	C_2H_2	CH
Benzene	C_6H_6	CH
Ammonia	NH_3	NH_3
Hydrazine	N_2H_4	NH_2
Ribose	$C_5H_{10}O_5$	CH_2O
Glucose	$C_6H_{12}O_6$	CH_2O

Empirical Formulas

Up to now, the formulas you have seen have been **molecular formulas**, which are the actual or true formulas of compounds. If we write a formula that represents the lowest whole-number ratio of the atoms in a compound, it is called the simplest, or **empirical formula**. For example, the compound benzene, with molecular formula C_6H_6, has the empirical formula CH. Some molecular formulas and their empirical formulas are shown in Table 7.3.

The empirical formula of a compound is determined by converting the number of grams of each element to moles and finding the lowest whole-number ratio to use as subscripts, as shown in Sample Problem 7.9.

Water-treatment plants use chemicals to purify sewage.

SAMPLE PROBLEM | **7.9**

■ **Calculating an Empirical Formula**

An iron compound is used to purify water in water-treatment plants. If laboratory analysis shows that a sample of the iron compound contains 6.87 g of iron and 13.1 g of chlorine, what is the empirical formula of the iron compound?

SOLUTION

STEP 1 Calculate the moles of each element.

$$6.87 \ g\text{-Fe} \times \frac{1 \ mol \ Fe}{55.85 \ g\text{-Fe}} = 0.123 \ mol \ of \ Fe$$

$$13.1 \ g\text{-Cl} \times \frac{1 \ mol \ Cl}{35.45 \ g\text{-Cl}} = 0.370 \ mol \ of \ Cl$$

STEP 2 Divide by the smaller number of moles. In this problem, the 0.123 mol of Fe is the smaller amount.

$$\frac{0.123 \ mol \ Fe}{0.123} = 1.00 \ mol \ of \ Fe$$

$$\frac{0.370 \ mol \ Cl}{0.123} = 3.01 \ mol \ of \ Cl$$

STEP 3 Use the lowest whole-number ratio of moles as subscripts. The relationship of mol of Fe to mol of Cl is 1 to 3, which we obtain by rounding off 3.01 to 3.

$$Fe_{1.00}Cl_{3.01} \longrightarrow Fe_1Cl_3, \ \text{written as } FeCl_3 \qquad \text{Empirical formula}$$

STUDY CHECK

Phosphine is a highly toxic compound used for pest and rodent control. If a sample of phosphine contains 0.456 g of P and 0.0440 g of H, what is its empirical formula?

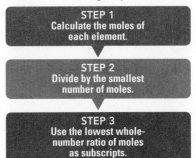

Guide to Calculating Empirical Formula

STEP 1
Calculate the moles of each element.

STEP 2
Divide by the smallest number of moles.

STEP 3
Use the lowest whole-number ratio of moles as subscripts.

Often, the relative amounts of the elements are given as the percent composition of a compound. The percent composition is true for any amount of compound. Thus, in any sample of methane, CH_4, there is always 74.9% C and 25.1% H. However, we can use the percent composition to determine the grams of C and H in any size sample of methane. In a small sample, there would be fewer grams of each element than in a large sample. For convenience, we will convert the percent composition to grams of each element in 100. g of the compound, as shown in Sample Problem 7.10.

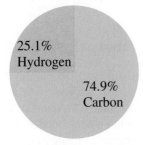

25.1%
Hydrogen

74.9%
Carbon

Percent composition of methane, CH_4

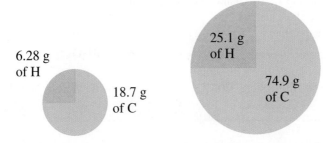

6.28 g
of H

18.7 g
of C

25.1 g
of H

74.9 g
of C

Composition, in grams, for 25.0 g of methane, CH_4

Composition, in grams, for 100. g of methane, CH_4

The percent composition for any amount of methane is always the same.

Sulfate of potash supplies sulfur and potassium.

SAMPLE PROBLEM | 7.10

■ Calculating an Empirical Formula from Percent Composition

The nonstick coating Teflon used on cooking pans consists of a carbon and fluorine polymer that is made from a compound called tetrafluoroethene. What is the empirical formula of tetrafluoroethene if it has 24.0% C and 76.0% F?

SOLUTION

STEP 1   **Calculate the moles of each element.** We must first change the percentages of the elements to grams by choosing a sample size of exactly 100 g. Then the numerical values remain the same, but we have grams instead of the % signs. In exactly 100 g of this compound, there are 24.0 g of C and 76.0 g of F. Then the number of moles for each element is calculated.

$$24.0 \text{ g C} \times \frac{1 \text{ mol C}}{12.01 \text{ g C}} = 2.00 \text{ mol of C}$$

$$76.0 \text{ g F} \times \frac{1 \text{ mol F}}{19.00 \text{ g F}} = 4.00 \text{ mol of F}$$

STEP 2   **Divide each of the calculated mole values by the smaller number of moles.**

$$\frac{2.00 \text{ mol C}}{2.00} = 1.00 \text{ mol of C}$$

$$\frac{4.00 \text{ mol F}}{2.00} = 2.00 \text{ mol of F}$$

STEP 3   **Use the lowest whole-number ratio of moles as subscripts.**

$$C_{1.00}F_{2.00} = CF_2$$

STUDY CHECK

Sulfate of potash is the common name of a compound used in fertilizers to supply potassium and sulfur. What is the empirical formula of this compound if it has a percent composition of 44.9% K, 18.4% S, and 36.7% O?

Converting Decimal Numbers to Whole Numbers

Sometimes the result of dividing by the smallest number of moles gives a decimal instead of a whole number. Decimal values that are very close to whole numbers can be rounded off. For example, 2.04 rounds off to 2 and 6.98 rounds off to 7. However, a decimal that is greater than 0.1 or less than 0.9 should not be rounded off. Instead, we multiply by a small integer until we obtain a whole number. Some multipliers that are typically used are listed in Table 7.4.

Let us suppose the numbers of moles we obtain give subscripts in the ratio of $C_{1.00}H_{2.33}O_{0.99}$. While 0.99 rounds off to 1, we cannot round off 2.33. If we multiply 2.33 × 2, we obtain 4.66, which is still not a whole number. If we multiply 2.33 by 3, the answer is 6.99, which rounds off to 7. To complete the empirical formula, all the other subscripts must be multiplied by 3.

$$C_{(1.00 \times 3)}H_{(2.33 \times 3)}O_{(0.99 \times 3)} = C_{3.00}H_{6.99}O_{2.97} \longrightarrow C_3H_7O_3$$

TABLE 7.4 Some Multipliers That Convert Decimals to Whole Numbers

Decimal	Multiply by	Example		Whole Number
0.20	5	1.20 × 5	=	6
0.25	4	2.25 × 4	=	9
0.33	3	1.33 × 3	=	4
0.50	2	2.50 × 2	=	5
0.67	3	1.67 × 3	=	5

SAMPLE PROBLEM 7.11

■ Calculating an Empirical Formula Using Multipliers

Ascorbic acid (vitamin C), found in citrus fruits and vegetables, is important in metabolic reactions in the body, in the synthesis of collagen, and to prevent scurvy. If ascorbic acid contains carbon (40.9%), hydrogen (4.58%), and oxygen (54.5%), what is the empirical formula of ascorbic acid?

Citrus fruits are a good source of vitamin C.

SOLUTION

STEP 1 Calculate the moles for each element. From the percent composition, we can write that exactly 100 g of ascorbic acid contains 40.9% C or 40.9 g of C, 4.58% H or 4.58 g of H, and 54.5% O or 54.5 g of O.

$$C = 40.9 \text{ g} \qquad H = 4.58 \text{ g} \qquad O = 54.5 \text{ g}$$

Now we convert the grams of each element to moles.

$$40.9 \text{ g C} \times \frac{1 \text{ mol C}}{12.01 \text{ g C}} = 3.41 \text{ mol of C}$$

$$4.58 \text{ g H} \times \frac{1 \text{ mol H}}{1.008 \text{ g H}} = 4.54 \text{ mol of H}$$

$$54.5 \text{ g O} \times \frac{1 \text{ mol O}}{16.00 \text{ g O}} = 3.41 \text{ mol of O}$$

STEP 2 Divide by the smallest number of moles. For this problem, the smallest number of moles is 3.41.

$$\frac{3.41 \text{ mol C}}{3.41} = 1.00 \text{ mol of C}$$

$$\frac{4.54 \text{ mol H}}{3.41} = 1.33 \text{ mol of H}$$

$$\frac{3.41 \text{ mol O}}{3.41} = 1.00 \text{ mol of O}$$

STEP 3 Use the lowest whole-number ratio of moles as subscripts. As calculated thus far, the ratio of moles gives the formula

$$C_{1.00}H_{1.33}O_{1.00}$$

However, the subscript for H is not close to a whole number, which means we cannot round it off. If we multiply each of the subscripts by 3, we obtain a subscript for H that is close enough to 4 to round off. Thus, we see that the empirical formula of ascorbic acid is $C_3H_4O_3$.

$$C_{(1.00 \times 3)}H_{(1.33 \times 3)}O_{(1.00 \times 3)} = C_{3.00}H_{3.99}O_{3.00} \longrightarrow C_3H_4O_3$$

STUDY CHECK

An organic compound called glyoxylic acid is used by plants and bacteria to convert fats into glucose. What is the empirical formula of glyoxylic acid if it contains 32.5% C, 2.70% H, and 64.8% O?

QUESTIONS AND PROBLEMS

Percent Composition and Empirical Formulas

7.25 Calculate the percent composition by mass of each of the following compounds:
   a. MgF_2, magnesium fluoride
   b. $Ca(OH)_2$, calcium hydroxide
   c. $C_4H_8O_4$, erythrose, a carbohydrate
   d. $(NH_4)_3PO_4$, ammonium phosphate
   e. $C_{17}H_{19}NO_3$, morphine

7.26 Calculate the percent composition by mass of each of the following compounds:
   a. $CaCl_2$, calcium chloride
   b. $Na_2Cr_2O_7$, sodium dichromate
   c. $C_2H_3Cl_3$, trichloroethane, a cleaning solvent
   d. $Ca_3(PO_4)_2$, calcium phosphate, found in bone and teeth
   e. $C_{18}H_{36}O_2$, stearic acid, a fatty acid

7.27 Calculate the percent by mass of N in each of the following compounds:
  a. N_2O_5, dinitrogen pentoxide
  b. NH_4NO_3, ammonium nitrate, fertilizer
  c. $C_2H_8N_2$, dimethylhydrazine, rocket fuel
  d. $C_9H_{15}N_5O$, Rogaine, stimulates hair growth
  e. $C_{14}H_{22}N_2O$, Lidocaine, local anesthetic

7.28 Calculate the percent by mass of S in each of the following compounds:
  a. Na_2SO_4, sodium sulfate
  b. Al_2S_3, aluminum sulfide
  c. SO_3, sulfur trioxide
  d. C_2H_6SO, dimethylsulfoxide, topical anti-inflammatory
  e. $C_{10}H_{10}N_4O_2S$, sulfadiazine, antibacterial

7.29 Calculate the empirical formula for each of the following substances:
  a. 3.57 g of N and 2.04 g of O
  b. 7.00 g of C and 1.75 g of H
  c. 0.175 g of H, 2.44 g of N, and 8.38 g of O
  d. 2.06 g of Ca, 2.66 g of Cr, and 3.28 g of O

7.30 Calculate the empirical formula for each of the following substances:
  a. 2.90 g of Ag and 0.430 g of S
  b. 2.22 g of Na and 0.774 g of O

  c. 2.11 g of Na, 0.0900 g of H, 2.94 g of S, and 5.86 g of O
  d. 5.52 g of K, 1.45 g of P, and 3.00 g of O

7.31 In an experiment, 2.51 g of sulfur combines with fluorine to give 11.44 g of a fluoride compound. What is the empirical formula of the compound?

7.32 In an experiment, 1.26 g of iron combines with oxygen to give a compound that has a mass of 1.80 g. What is the empirical formula of the compound?

7.33 Calculate the empirical formula for each of the following substances:
  a. 70.9% K and 29.1% S
  b. 55.0% Ga and 45.0% F
  c. 31.0% B and 69.0% O
  d. 18.8% Li, 16.3% C, and 64.9% O
  e. 51.7% C, 6.95% H, and 41.3% O

7.34 Calculate the empirical formula for each of the following substances:
  a. 55.5% Ca and 44.5% S
  b. 78.3% Ba and 21.7% F
  c. 76.0% Zn and 24.0% P
  d. 29.1% Na, 40.6% S, and 30.3% O
  e. 19.8% C, 2.20% H, and 78.0% Cl

7.5 Molecular Formulas

LEARNING GOAL

Determine the molecular formula of a substance from the empirical formula and molar mass.

TUTORIAL
Empirical and Molecular Formulas

TUTORIAL
Formula Mass

An empirical formula represents the lowest whole-number ratio of atoms in a compound. However, empirical formulas do not necessarily represent the actual number of atoms in a molecule. A molecular formula is related to the empirical formula by a small integer such as 1, 2, or 3.

Molecular formula = small integer × empirical formula

For example, in Table 7.5, we see several different compounds that have the same empirical formula, CH_2O. The molecular formulas are related to the empirical formulas by small whole numbers (integers). The same relationship is true for the molar mass and empirical formula mass. The molar mass of each of the different compounds is related to the mass of the empirical formula (30.03 g) by the same small integer.

Molecular formula	Small integer	Molar mass
Empirical formula		Empirical formula mass

Molar mass = small integer × empirical formula mass

TABLE 7.5 Comparing the Molar Mass of Some Compounds with the Empirical Formula of CH_2O

Compound	Empirical Formula	Molecular Formula	Molar Mass (g)	Integer × Empirical Formula	Integer × Empirical Formula Mass
Acetaldehyde	CH_2O	CH_2O	30.03	$(CH_2O)_1$	1 × 30.03
Acetic acid	CH_2O	$C_2H_4O_2$	60.06	$(CH_2O)_2$	2 × 30.03
Lactic acid	CH_2O	$C_3H_6O_3$	90.09	$(CH_2O)_3$	3 × 30.03
Erythrose	CH_2O	$C_4H_8O_4$	120.12	$(CH_2O)_4$	4 × 30.03
Ribose	CH_2O	$C_5H_{10}O_5$	150.15	$(CH_2O)_5$	5 × 30.03

■ **Empirical and Molecular Formulas**

A compound has an empirical formula of CH_2O. Indicate if each of the following would be a possible molecular formula. Explain your answer.

a. CH_2O

b. C_2H_4O

c. $C_2H_4O_2$

d. $C_5H_{10}O_5$

e. $C_6H_{12}O$

ANSWER

a. The empirical formula and molecular formula have a ratio of 1:1, which gives an empirical formula that is the same as the molecular formula.

b. Because only the C and H are twice their moles in the empirical formula, this cannot be a corresponding molecular formula.

c. This can be a possible molecular formula because the C, H, and O are twice the moles of the empirical formula.

d. This can be a possible molecular formula because the C, H, and O are five times the moles of the empirical formula.

e. Because only the C and H are six times their moles in the empirical formula, this cannot be a corresponding molecular formula.

Relating Empirical and Molecular Formulas

Once we determine the empirical formula, we can calculate its mass in grams. If we are given the molar mass of the compound, we can calculate the value of the small integer.

$$\text{Small integer} = \frac{\text{molar mass of compound}}{\text{empirical formula mass}}$$

For example, when the molar mass of ribose is divided by the empirical formula mass, the integer is 5.

$$\text{Small integer} = \frac{\text{molar mass of ribose}}{\text{empirical formula mass of ribose}} = \frac{150.15 \text{ g}}{30.03 \text{ g}} = 5$$

Multiplying the subscripts in the empirical formula (CH_2O) by 5 gives the molecular formula of ribose, $C_5H_{10}O_5$.

$$5 \times \text{empirical formula } (CH_2O) = \text{molecular formula } (C_5H_{10}O_5)$$

Calculating a Molecular Formula

Earlier, in Sample Problem 7.11, we worked out the empirical formula of ascorbic acid (vitamin C) to be $C_3H_4O_3$. If we are given a molar mass of 176.12 g for ascorbic acid, we can determine its molecular formula as follows:

STEP 1   **Calculate the mass of the empirical formula.** The mass of the empirical formula $C_3H_4O_3$ is obtained in the same way as molar mass.

Empirical formula   = 3 mol of C + 4 mol of H + 3 mol of O

Empirical formula mass = (3 × 12.01 g) + (4 × 1.008 g) + (3 × 16.00 g)

= 88.06 g

STEP 2   **Divide the molar mass by the empirical formula mass to obtain a small integer.**

$$\text{Small integer} = \frac{\text{molar mass of ascorbic acid}}{\text{empirical formula mass of } C_3H_4O_3} = \frac{176.12 \text{ g}}{88.06 \text{ g}} = 2$$

STEP 3   **Multiply the empirical formula by the small integer to obtain the molecular formula.** Multiplying all the subscripts in the empirical formula of ascorbic acid by 2 gives its molecular formula.

$$C_{(3 \times 2)}H_{(4 \times 2)}O_{(3 \times 2)} = C_6H_8O_6$$

Brightly colored dishes are made of melamine.

Guide to Calculating a Molecular Formula from an Empirical Formula

| STEP 1 |
| Calculate the empirical formula mass. |

| STEP 2 |
| Divide the molar mass by the empirical formula mass to obtain a small integer. |

| STEP 3 |
| Multiply the empirical formula by the small integer to obtain the molecular formula. |

SAMPLE PROBLEM | **7.12**

■ **Determination of a Molecular Formula**

Melamine, which is used to make plastic items such as dishes and toys, contains 28.57% carbon, 4.80% hydrogen, and 66.64% nitrogen. If the molar mass is about 126 g, what is the molecular formula of melamine?

SOLUTION

STEP 1   Calculate the empirical formula mass. Write each percentage as grams in exactly 100 g of melamine and determine the number of moles of each.

$$C = 28.57 \text{ g} \qquad H = 4.80 \text{ g} \qquad N = 66.64 \text{ g}$$

$$28.57 \text{ g C} \times \frac{1 \text{ mol C}}{12.01 \text{ g C}} = 2.38 \text{ mol of C}$$

$$4.80 \text{ g H} \times \frac{1 \text{ mol H}}{1.008 \text{ g H}} = 4.76 \text{ mol of H}$$

$$66.64 \text{ g N} \times \frac{1 \text{ mol N}}{14.01 \text{ g N}} = 4.76 \text{ mol of N}$$

Divide the moles of each element by the smallest number of moles, 2.38, to obtain the subscripts of each element in the formula.

$$\frac{2.38 \text{ mol C}}{2.38} = 1.00 \text{ mol of C}$$

$$\frac{4.76 \text{ mol H}}{2.38} = 2.00 \text{ mol of H}$$

$$\frac{4.76 \text{ mol N}}{2.38} = 2.00 \text{ mol of N}$$

Using these values as subscripts, $C_{1.00}H_{2.00}N_{2.00}$, we write the empirical formula of melamine as CH_2N_2.

$$C_{1.00}H_{2.00}N_{2.00} = CH_2N_2$$

Now we calculate the molar mass of this empirical formula as follows:

Empirical formula = 1 mol of C + 2 mol of H + 2 mol of N

Empirical formula mass = $(1 \times 12.01) + (2 \times 1.008) + (2 \times 14.01) = 42.05 \text{ g}$

STEP 2   Divide the molar mass by the empirical formula mass to obtain a small integer.

$$\text{Small integer} = \frac{\text{molar mass of melamine}}{\text{empirical formula mass of } CH_2N_2} = \frac{126 \text{ g}}{42.05 \text{ g}} = 3$$

STEP 3   Multiply the empirical formula by the small integer to obtain the molecular formula. Because the molar mass was three times the empirical formula mass, the subscripts in the empirical formula are multiplied by 3 to give the molecular formula.

$$C_{(1 \times 3)} H_{(2 \times 3)} N_{(2 \times 3)} = C_3H_6N_6$$

STUDY CHECK

The insecticide lindane has a percent composition of 24.78% C, 2.08% H, and 73.14% Cl. If its molar mass is about 291 g/mol, what is the molecular formula?

QUESTIONS AND PROBLEMS

Molecular Formulas

7.35 Write the empirical formula of each of the following substances:
   a. H_2O_2, peroxide
   b. $C_{18}H_{12}$, chrysene, used in the manufacture of dyes
   c. $C_{10}H_{16}O_2$, chrysanthemic acid, in pyrethrum flowers
   d. $C_9H_{18}N_6$, altretamine, an anticancer medication
   e. $C_2H_4N_2O_2$, oxamide, a fertilizer

7.36 Write the empirical formula of each of the following substances:
   a. $C_6H_6O_3$, pyrogallol, a developer in photography
   b. $C_6H_{12}O_6$, galactose, a carbohydrate
   c. $C_8H_6O_4$, terephthalic acid, used in the manufacture of plastic bottles
   d. C_6Cl_6, hexachlorobenzene, a fungicide
   e. $C_{24}H_{16}O_{12}$, laccaic acid, a crimson dye

7.37 The carbohydrate fructose found in honey and fruits has an empirical formula of CH_2O. If the molar mass of fructose is about 180 g, what is its molecular formula?

7.38 Caffeine has an empirical formula of $C_4H_5N_2O$. If it has a molar mass of about 194 g, what is the molecular formula of caffeine?

Coffee beans are a source of caffeine.

7.39 Benzene and acetylene have the same empirical formula, CH. However, benzene has a molar mass of about 78 g, and acetylene has a molar mass of about 26 g. What are the molecular formulas of benzene and acetylene?

7.40 Glyoxyl, used in textiles; maleic acid, used to retard oxidation of fats and oils; and acontic acid, a plasticizer, all have the same empirical formula, CHO. However, glyoxyl has a molar mass of about 58 g, maleic acid about 117 g, and acontic acid about 174 g. What are the molecular formulas of glyoxyl, maleic acid, and acontic acid?

7.41 Mevalonic acid is involved in the biosynthesis of cholesterol. Mevalonic acid is 48.64% C, 8.16% H, and 43.20% O. If mevalonic acid has a molar mass of about 148 g, what is its molecular formula?

7.42 Chloral hydrate, a sedative, contains 14.52% C, 1.83% H, 64.30% Cl, and 19.35% O. If it has a molar mass of about 165 g, what is the molecular formula of chloral hydrate?

7.43 Vanillic acid contains 57.14% C, 4.80% H, and 38.06% O and has a molar mass of about 168 g. What is the molecular formula of vanillic acid?

7.44 Lactic acid, the substance that builds up in muscles during exercise, has a molar mass of about 90 g and has a composition of 40.0% C, 6.71% H, and the rest is oxygen. What is the molecular formula of lactic acid?

7.45 A sample of nicotine, a poisonous compound found in tobacco leaves, is 74.0% C, 8.7% H, and the remainder is nitrogen. If the molar mass of nicotine is about 162 g, what is the molecular formula of nicotine?

7.46 Adenine is a nitrogen-containing compound found in DNA and RNA. If adenine has a composition of 44.5% C, 3.70% H, and 51.8% N, and a molar mass of about 135 g, what is the molecular formula of adenine?

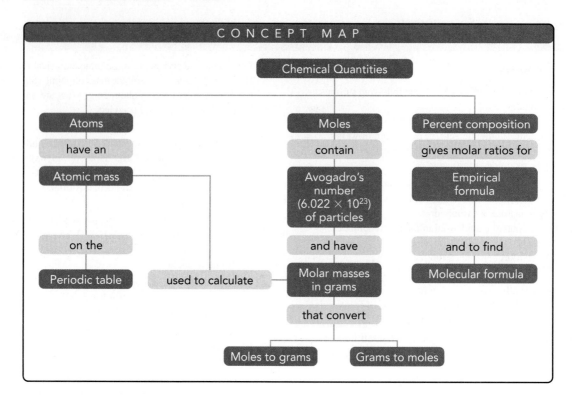

CHAPTER REVIEW

7.1 The Mole
LEARNING GOAL: *Use Avogadro's number to determine the number of particles in a given number of moles.*
One mole of an element contains 6.022×10^{23} atoms; a mole of a compound contains 6.022×10^{23} molecules or formula units.

7.2 Molar Mass
LEARNING GOAL: *Given the chemical formula of a substance, calculate its molar mass.*
The molar mass of a substance is the mass in grams equal numerically to its atomic mass, or the sum of the atomic masses, which have been multiplied by their subscripts in a formula.

7.3 Calculations Using Molar Mass
LEARNING GOAL: *Given the number of moles of a substance, calculate the mass in grams; given the mass, calculate the number of moles.*
The molar mass is the mass in grams of 1 mol of an element or a compound. It becomes an important conversion factor when it is used to change a quantity in grams to moles, or to change a given number of moles to grams.

7.4 Percent Composition and Empirical Formulas
LEARNING GOAL: *Given the formula of a compound, calculate the percent composition; from the percent composition, determine the empirical formula of a compound.*
The percent composition of a compound is obtained by dividing the mass in grams of each element in a compound by the molar mass of that compound. The empirical formula can be calculated by determining the mole ratio from the grams of the elements present in a sample of a compound or the percent composition.

7.5 Molecular Formulas
LEARNING GOAL: *Determine the molecular formula of a substance from the empirical formula and molar mass.*
A molecular formula is equal to, or a multiple of, the empirical formula. The molar mass, which must be known, is divided by the mass of the empirical formula to obtain the multiple.

KEY TERMS

Avogadro's number The number of items in a mole, equal to 6.022×10^{23}.

empirical formula The simplest or smallest whole-number ratio of the atoms in a formula.

formula unit The group of ions represented by the formula of an ionic compound.

molar mass The mass in grams of 1 mol of an element is equal numerically to its atomic mass. The molar mass of a compound is equal to the sum of the masses of the elements in the formula.

mole A group of atoms, molecules, or formula units that contains 6.022×10^{23} of these items.

molecular formula The actual formula that gives the number of atoms of each type of element in the compound.

percent composition The percent by mass of the elements in a formula.

UNDERSTANDING THE CONCEPTS

7.47 A dandruff shampoo contains dipyrithione, $C_{10}H_8N_2O_2S_2$, which acts as antibacterial and antifungal agent.

Dandruff shampoo contains dipyrithione.

a. What is the empirical formula of dipyrithione?
b. What is the molar mass of dipyrithione?
c. What is the percent composition of dipyrithione?
d. How many atoms of C are in 25.0 g of dipyrithione?
e. How many moles of dipyrithione contain 8.2×10^{24} atoms of nitrogen?

7.48 Ibuprofen, an anti-inflammatory, has the formula $C_{13}H_{18}O_2$.

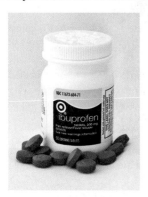

Ibuprofen is an anti-inflammatory.

a. What is the percent by mass of oxygen in ibuprofen?
b. How many atoms of carbon are in 0.425 g of ibuprofen?
c. How many grams of hydrogen are in 3.75×10^{22} molecules of ibuprofen?
d. How many atoms of hydrogen are in 0.245 g of ibuprofen?

7.49 Using the following models of the molecules, determine (black = C, white = H, yellow = S, green = Cl)

1.

2.
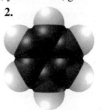

a. molecular formula
b. empirical formula
c. molar mass
d. percent composition

7.50 Using the following models of the molecules, determine (black = C, white = H, yellow = S, red = O)

1.

2.

a. molecular formula
b. empirical formula
c. molar mass
d. percent composition

ADDITIONAL QUESTIONS AND PROBLEMS

For instructor-assigned homework, go to
www.masteringchemistry.com.

7.51 Calculate the molar mass of each of the following:
    a. $ZnSO_4$, zinc sulfate, zinc supplement
    b. $Ca(IO_3)_2$, calcium iodate, iodine source in table salt
    c. $C_5H_8NNaO_4$, monosodium glutamate, flavor enhancer
    d. $C_6H_{12}O_2$, isoamyl formate, used to make artificial fruit syrups

7.52 Calculate the molar mass of each of the following:
    a. $MgCO_3$, magnesium carbonate, an antacid
    b. $Au(OH)_3$, gold(III) hydroxide, used in gold plating
    c. $C_{18}H_{34}O_2$, oleic acid, from olive oil
    d. $C_{21}H_{26}O_5$, prednisone, anti-inflammatory

7.53 Calculate the percent composition for each of the following compounds:
    a. K_2CrO_4 **b.** $Al(HCO_3)_3$
    c. $C_6H_{12}O_6$

7.54 Calculate the percent by mass of P in each of the following compounds:
    a. P_4O_{10} **b.** Mg_3P_2 **c.** $Ca_3(PO_4)_2$

7.55 During heavy exercise and workouts, lactic acid, $C_3H_6O_3$, accumulates in the muscles where it can cause pain and soreness.

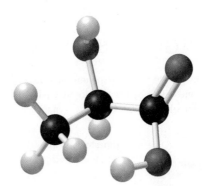

a. What is the percent by mass of O in lactic acid?
b. How many atoms of carbon are in 125 g of lactic acid?
c. How many grams of lactic acid contain 3.50 g of hydrogen?
d. What is the empirical formula of lactic acid?

7.56 Ammonium sulfate, $(NH_4)_2SO_4$, is used in fertilizers.
  a. What is the percent by mass of N in $(NH_4)_2SO_4$?
  b. How many atoms of hydrogen are in 0.75 mol of $(NH_4)_2SO_4$?
  c. How many grams of oxygen are in 4.50×10^{23} formula units of $(NH_4)_2SO_4$?
  d. What mass of $(NH_4)_2SO_4$ contains 2.50 g of sulfur?

7.57 Aspirin, $C_9H_8O_4$, is used to reduce inflammation and reduce fever.
  a. What is the percent composition by mass for aspirin?
  b. How many moles of aspirin contain 5.0×10^{24} atoms of carbon?
  c. How many atoms of oxygen are in 7.50 g of aspirin?
  d. How many molecules of aspirin contain 2.50 g of hydrogen?

7.58 Rolaids is an antacid used for an upset stomach. One tablet contains 550 mg of calcium carbonate, $CaCO_3$, and 110 mg of magnesium hydroxide, $Mg(OH)_2$. One package contains 12 tablets.
  a. How many moles of calcium carbonate are present in one package?
  b. How many calcium ions, Ca^{2+}, would you obtain from two tablets?
  c. How many atoms of magnesium would you ingest if you take three tablets?
  d. How many hydroxide ions are in one tablet?

7.59 A mixture contains 0.250 mol of Mn_2O_3 and 20.0 g of MnO_2.
  a. How many atoms of oxygen are present in the mixture?
  b. How many grams of manganese are in the mixture?

7.60 A mixture contains 4.00×10^{23} molecules of PCl_3 and 0.250 mol of PCl_5.
  a. How many grams of chlorine are present in the mixture?
  b. How many moles of phosphorus are in the mixture?

7.61 Write the empirical formula for each of the following compounds:
  a. $C_5H_5N_5$, adenine, a nitrogen compound in RNA and DNA
  b. FeC_2O_4, iron(II) oxalate, a photographic developer
  c. $C_{16}H_{16}N_4$, stilbamidine, an antibiotic for animals
  d. $C_6H_{14}N_2O_2$, lysine, an amino acid needed for growth

7.62 Write the empirical formula for each of the following compounds:
  a. N_2H_4, hydrazine, rocket fuel
  b. $C_{10}H_{10}O_5$, opianic acid, used to synthesize a drug to treat tuberculosis
  c. $CrCl_3$, chromium(III) chloride, used in chrome plating
  d. $C_{16}H_{16}N_2O_2$, lysergic acid, a controlled substance from ergot

7.63 Calculate the empirical formula of each compound that contains
  a. 2.20 g of S and 7.81 g of F
  b. 6.35 g of Ag, 0.825 g of N, and 2.83 g of O
  c. 89.2 g of Au and 10.9 g of O

7.64 Calculate the empirical formula of each compound from the percent composition.
  a. 61.0% Sn and 39.0% F
  b. 25.9% N and 74.1% O
  c. 22.1% Al, 25.4% P, and 52.5% O

7.65 Oleic acid, a component of olive oil, is 76.54% C, 12.13% H, and 11.33% O. The experimental value of the molar mass is about 282 g.
  a. What is the molecular formula of oleic acid?
  b. If oleic acid has a density of 0.895 g/mL, how many molecules of oleic acid are in 3.00 mL of oleic acid?

7.66 Iron pyrite, commonly known as "fool's gold," is 46.5% Fe and 53.5% S.

Iron pyrite is commonly known as "fool's gold".

  a. If a crystal of iron pyrite contains 4.58 g of iron, what is the mass, in grams, of the crystal?
  b. If the empirical formula and the molecular formula are the same, what is the formula of the compound?
  c. How many moles of iron are in the crystal?

7.67 Succinic acid is 40.7% C, 5.12% H, and 54.2% O. If it has a molar mass of about 118 g, what are the empirical and molecular formulas?

7.68 A compound is 70.6% Hg, 12.5% Cl, and 16.9% O. If it has a molar mass of about 568 g, what are the empirical and molecular formulas?

7.69 A sample of a compound contains 1.65×10^{23} atoms of C, 0.552 g of H, and 4.39 g of O. If 1 mol of the compound contains 4 mol of O, what is the molecular formula and molar mass of the compound?

7.70 What is the molecular formula of a compound if 0.500 mol of the compound contains 0.500 mol of Sr, 1.81×10^{24} atoms of O, and 35.5 g of Cl?

■ CHALLENGE QUESTIONS

7.71 A toothpaste contains 0.240% by mass sodium fluoride used to prevent tooth decay and 0.30% by mass triclosan, $C_{12}H_7Cl_3O_2$, a preservative and antigingivitis agent. One tube contains 119 g of toothpaste.

Toothpaste contains NaF to prevent tooth decay.

  a. How many moles of NaF are in the tube of toothpaste?
  b. How many fluoride ions, F^-, are in the tube of toothpaste?
  c. How many grams of sodium ion, Na^+, are in 1.50 g of toothpaste?
  d. How many molecules of triclosan are in the tube of toothpaste?
  e. What is the percent composition by mass of triclosan?

7.72 Sorbic acid, an inhibitor of mold in cheese, has a percent composition of 64.27% C, 7.19% H, and 28.54% O. If sorbic acid has a molar mass of about 112 g, what is its molecular formula?

Cheese contains sorbic acid, which prevents the growth of mold.

7.73 Iron(III) chromate, a yellow powder used as a pigment in paints, contains 24.3% Fe, 33.9% Cr, and 41.8% O. If it has a molar mass of about 460 g, what are the empirical and molecular formulas?

Iron(III) chromate is a yellow pigment used in paints.

7.74 A gold bar is 3.11 cm long, 1.48 cm wide, and 0.758 cm thick.
  a. If gold has a density of 19.3 g/mL, what is the mass of the gold bar?
  b. How many atoms of gold are in the bar?
  c. When the same mass of gold combines with oxygen, the oxide product has a mass of 75.6 g. How many moles of oxygen atoms are combined with the gold?
  d. What is the molecular formula of the oxide product if it is the same as the empirical formula?

A gold bar consists of gold atoms.

ANSWERS

ANSWERS TO STUDY CHECKS

7.1 0.432 mol of H_2O

7.2 0.120 mol of aspirin

7.3 4.52×10^{24} SO_4^{2-} ions

7.4 138.12 g

7.5 24.4 g of Au

7.6 6.21×10^{-3} mol of $CaCO_3$; 5.50×10^{-3} mol of $MgCO_3$

7.7 0.0515 mol of N

7.8 %C = 38.70%, %H = 9.744%, %O = 51.55%

7.9 PH_3

7.10 K_2SO_4

7.11 $C_2H_2O_3$

7.12 $C_6H_6Cl_6$

ANSWERS TO SELECTED QUESTIONS AND PROBLEMS

7.1 **a.** 1.20×10^{23} atoms of Ag
  b. 4.52×10^{23} molecules of C_3H_8O
  c. 0.478 mol of Au

7.3 **a.** 24 mol of H
  b. 1.0×10^2 mol of C
  c. 0.040 mol of N

7.5 **a.** 3.01×10^{23} atoms of C
  b. 7.71×10^{23} molecules of SO_2
  c. 0.0867 mol of Fe
  d. 14.1 mol of C_2H_5OH

7.7 **a.** 6.00 mol of H
  b. 8.00 mol of O
  c. 1.20×10^{24} atoms of P
  d. 4.82×10^{24} atoms of O

7.9 **a.** 58.44 g   **b.** 159.7 g
     c. 329.4 g   **d.** 342.2 g
     e. 188.18 g   **f.** 365.5 g

7.11 **a.** 70.90 g   **b.** 90.08 g
     c. 262.9 g   **d.** 83.98 g
     e. 98.95 g   **f.** 156.7 g

7.13 **a.** 34.5 g   **b.** 112 g
     c. 5.50 g   **d.** 5.14 g
     e. 9.80×10^4 g

7.15 **a.** 3.03 g   **b.** 38.1 g
     c. 9.30 g   **d.** 24.6 g
     e. 12.9 g

7.17 **a.** 0.760 mol of Ag   **b.** 0.0240 mol of C
     c. 0.881 mol of NH_3 **d.** 0.164 mol of C_3H_8
     e. 1.53 mol of Fe_2O_3

7.19 **a.** 6.25 mol of He   **b.** 0.781 mol of O_2
     c. 0.321 mol of $Al(OH)_3$ **d.** 0.106 mol of Ga_2S_3
     e. 0.430 mol of C_4H_{10}

7.21 **a.** 1.25×10^{24} atoms of C   **b.** 4.14×10^{23} atoms of C
     c. 1.13×10^{25} atoms of C   **d.** 2.22×10^{24} atoms of C
     e. 7.5×10^{24} atoms of C

7.23 **a.** 66.1 g of propane
     b. 0.771 mol of propane
     c. 27.8 g of C
     d. 2.78×10^{22} atoms of H

7.25 **a.** 39.01% Mg; 60.99% F
     b. 54.09% Ca; 43.18% O; 2.72% H
     c. 40.00% C; 6.71% H; 53.29% O
     d. 28.19% N; 8.12% H; 20.77% P; 42.92% O
     e. 71.55% C; 6.71% H; 4.91% N; 16.82% O

7.27 **a.** 25.94% N   **b.** 35.00% N
     c. 46.62% N   **d.** 33.47% N
     e. 11.96% N

7.29 **a.** N_2O **b.** CH_3
     c. HNO_3 **d.** $CaCrO_4$

7.31 SF_6

7.33 **a.** K_2S **b.** GaF_3
     c. B_2O_3 **d.** Li_2CO_3
     e. $C_5H_8O_3$

7.35 **a.** HO   **b.** C_3H_2
     c. C_5H_8O **d.** $C_3H_6N_2$
     e. CH_2NO

7.37 $C_6H_{12}O_6$

7.39 benzene C_6H_6; acetylene C_2H_2

7.41 $C_6H_{12}O_4$

7.43 $C_8H_8O_4$

7.45 $C_{10}H_{14}N_2$

7.47 **a.** C_5H_4NOS
     b. 252.3 g
     c. 47.60% C, 3.20% H, 11.11% N, 12.68% O, 25.42% S
     d. 5.97×10^{23} atoms of C
     e. 6.8 mol of dipyrithione

7.49 **1. a.** S_2Cl_2 **b.** SCl
     c. 135.04 g   **d.** 47.50% S, 52.50% Cl
     2. a. C_6H_6 **b.** CH
     c. 78.11 g   **d.** 92.25% C, 7.74% H

7.51 **a.** 161.48 g   **b.** 389.9 g
     c. 169.11 g   **d.** 116.16 g

7.53 **a.** 40.27% K; 26.78% Cr; 32.96% O
     b. 12.85% Al; 1.44% H; 17.16% C; 68.57% O
     c. 40.00% C; 6.72% H; 53.29% O

7.55 **a.** 53.29% O
     b. 2.51×10^{24} atoms of C
     c. 52.1 g of lactic acid
     d. CH_2O

7.57 **a.** 59.99% C, 4.48% H, 35.52% O
     b. 0.923 mol of aspirin
     c. 1.00×10^{23} atoms of O
     d. 1.87×10^{23} molecules of aspirin

7.59 **a.** 7.29×10^{23} atoms of O   **b.** 40.1 g of Mn

7.61 **a.** CHN   **b.** FeC_2O_4
     c. C_4H_4N **d.** C_3H_7NO

7.63 **a.** SF_6 **b.** $AgNO_3$
     c. Au_2O_3

7.65 **a.** $C_{18}H_{34}O_2$
     b. 5.72×10^{21} molecules of oleic acid

7.67 The empirical formula is $C_2H_3O_2$; the molecular formula is $C_4H_6O_4$.

7.69 The molecular formula is $C_4H_8O_4$; the molar mass is 120.10 g.

7.71 **a.** 0.00680 mol of NaF
     b. 4.10×10^{21} F^- ions
     c. 0.00197 g of Na^+ ions
     d. 7.4×10^{20} molecules of triclosan
     e. 49.76% C, 2.44% H, 36.74% Cl, 11.05% O

7.73 The empirical formula is $Fe_2Cr_3O_{12}$. The molecular formula is $Fe_2Cr_3O_{12}$.

COMBINING IDEAS FROM CHAPTERS 4 TO 7

CI.7 For parts **a** to **f**, consider the loss of electrons by atoms of the element X, and a gain of electrons by atoms of the element Y, if X is in Group 2A (2), Period 3, and Y is in Group 7A (17), Period 3.

$$X \qquad Y \qquad Y$$

  a. Which spheres represent a metal? How do you know?
  b. Which spheres represent a nonmetal? How do you know?
  c. What are the ionic charges of X and Y?
  d. Write the electron configurations of the atoms X and Y.
  e. Write the electron configurations of the ions of X and Y.
  f. Write the actual formula and name of the ionic compound indicated by the ions.

CI.8 A bracelet of sterling silver marked 925 contains 92.5% silver by mass and 7.5% other metals. It has a volume of 25.6 cm^3 and a density of 10.2 g/cm^3.

Sterling silver is 92.5% silver by mass.

  a. What is the mass, in kilograms, of the sterling silver bracelet?
  b. How many atoms of silver are in the bracelet?
  c. Determine the number of protons and neutrons in each of the two stable isotopes of silver:

$$^{107}_{47}Ag \qquad ^{109}_{47}Ag$$

  d. When silver combines with oxygen, a compound forms that contains 93.10% Ag by mass. What is the name and molecular formula of the oxide product if it is the same as the empirical formula?

CI.9 Oxalic acid, an organic compound found in plants and vegetables such as rhubarb, has a percent composition of 26.7% C, 2.24% H, and 71.1% O. Oxalic acid can interfere with respiration and cause kidney or bladder stones. Rhubarb leaves contain about 0.5% by mass of oxalic acid. If a large quantity of rhubarb leaves is ingested, the oxalic acid can be toxic. The lethal dose (LD$_{50}$) in rats for oxalic acid is 375 mg/kg.

Rhubarb leaves are a plant source of oxalic acid.

  a. What is the empirical formula of oxalic acid?
  b. If oxalic acid has a molar mass of about 90 g, what is its molecular formula?
  c. Using the LD$_{50}$, how many grams of oxalic acid would be toxic for a 160-lb person?
  d. How many kilograms of rhubarb leaves would the person in part **c** need to eat to reach the toxic level of oxalic acid?

CI.10 The active ingredient in Tums has the following percent composition: 40.0% Ca, 12.0% C, and 48.0% O.

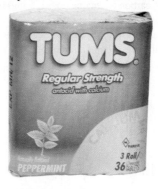

The active ingredient in Tums is used to neutralize excess stomach acid.

  a. If the empirical and molecular formulas are the same, what is its molecular formula?
  b. What is the name of the ingredient?
  c. If one Tums tablet contains 500. mg of this ingredient and a person takes two tablets a day, how many calcium ions does this person obtain from the Tums tablets?

CI.11 Tamiflu (Oseltamivir), C$_{16}$H$_{28}$N$_2$O$_4$, is an antiviral drug that is used to treat influenza. The preparation of Tamiflu begins with the extraction of shikimic acid from the seedpods of the Chinese spice, star anise. From 2.6 g of star anise, 0.13 g of shikimic acid can be obtained and used to produce one capsule containing 75 mg of Tamiflu. The usual adult dosage for treatment of influenza is 75 mg of Tamiflu twice daily for 5 days.

Shikimic acid

The spice called star anise is a plant source of shikimic acid.

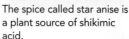

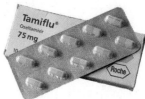

a. What is the empirical formula of Tamiflu?

b. What is the percent composition of Tamiflu?

c. What is the molecular formula of shikimic acid? (Black spheres are carbon, white spheres are hydrogen, and red spheres are oxygen.)

d. How many moles of shikimic acid are contained in 1.3 g of shikimic acid?

e. How many capsules containing 75 mg of Tamiflu could be produced from 155 g of star anise?

f. How many grams of carbon are in one dose (75 mg) of Tamiflu?

g. How many kilograms of Tamiflu would be needed to treat all the people in a city with a population of 500 000 people?

CI.12 The compound butyric acid gives rancid butter its characteristic odor.

 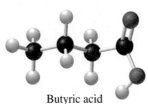

Butyric acid

a. If black spheres are carbon atoms, white spheres are hydrogen atoms, and red spheres are oxygen atoms, what is the molecular formula?

b. What is the empirical formula of butyric acid?

c. What is the percent composition of butyric acid?

d. How many grams of carbon are in 0.850 g of butyric acid?

e. How many grams of butyric acid contain 3.28×10^{23} oxygen atoms?

f. Butyric acid has a density of 0.959 g/mL at 20 °C. How many moles of butyric acid are contained in 0.565 mL of butyric acid?

■ANSWERS

CI.7 **a.** X is a metal. When an atom of a metal loses electrons, the size of the ion is smaller than the corresponding atom.

b. Y is a nonmetal. When an atom of a nonmetal gains electrons, the size of the ion is larger than the corresponding atom.

c. X^{2+}, Y^-

d. $X = 1s^2 2s^2 2p^6 3s^2$ $Y = 1s^2 2s^2 2p^6 3s^2 3p^5$

e. $X^{2+} = 1s^2 2s^2 2p^6$ $Y^- = 1s^2 2s^2 2p^6 3s^2 3p^6$

f. $MgCl_2$, magnesium chloride

CI.9 **a.** CHO_2

b. $C_2H_2O_4$

c. 27 g of oxalic acid

d. 5 kg of rhubarb

CI.11 **a.** $C_8H_{14}NO_2$

b. 61.51% C, 9.03% H, 8.97% N, 20.49% O

c. $C_7H_{10}O_5$

d. 7.5×10^{-3} mol of shikimic acid

e. 59 capsules

f. 0.046 g of C

g. 4×10^2 kg

Chemical Reactions of Inorganic and Organic Compounds

8

Visit **www.masteringchemistry.com** for self-study materials and instructor-assigned homework.

"We use mass spectrometry to analyze and confirm the presence of drugs," says Valli Vairavan, clinical lab technologist–Mass Spectrometry, Santa Clara Valley Medical Center. *"A mass spectrometer separates and identifies compounds including drugs by mass. When we screen a urine sample, we look for metabolites, which are the products of drugs that have metabolized in the body. If the presence of one or more drugs such as heroin and cocaine is indicated, we confirm it by using mass spectrometry."*

Drugs or their metabolites are detected in urine 24–48 hours after use. Cocaine metabolizes to benzoylecgonine and hydroxycocaine, morphine to morphine-3-glucuronide, and heroin to acetylmorphine. Amphetamines and methamphetamines are detected unchanged.

The fuel in our cars burns with oxygen to provide energy to make the cars move and run the air conditioner. When we cook our food or bleach our hair, chemical reactions take place. In our bodies, chemical reactions convert food into molecules that build muscles and move them. In the leaves of trees and plants, carbon dioxide and water are converted into carbohydrates.

Some chemical reactions are simple, whereas others are quite complex. However, they can all be written with equations used to describe chemical reactions. In every chemical reaction, the atoms in the reacting substances, called *reactants*, are rearranged to give new substances called *products*.

In this chapter, we will see how equations are written and how we can determine the amount of reactant or product involved. We do the same thing at home when we use a recipe to make bread or cookies. At the automotive repair shop, a mechanic does essentially the same thing by adjusting the fuel system of an engine to allow for the correct amounts of fuel and oxygen. In the hospital, a respiratory therapist evaluates the levels of CO_2 and O_2 in the blood.

LEARNING GOAL

Identify a balanced chemical equation and determine the number of atoms in the reactants and products.

TABLE 8.1 Types of Visible Evidence of a Chemical Reaction

1. Change in the color
2. Formation of a gas (bubbles)
3. Formation of a solid (precipitate)
4. Heat (or a flame) produced or heat absorbed

8.1 Equations for Chemical Reactions

As we discussed in Chapter 3, Section 3.2, a *chemical change* occurs when a substance is converted into one or more new substances. For example, when silver tarnishes, the shiny silver metal (Ag) reacts with sulfur (S) to become the dull, black substance we call tarnish (Ag_2S) (see Figure 8.1).

A **chemical reaction** always involves chemical change because atoms of the reacting substances form new combinations with new properties. For example, a chemical reaction takes place when an antacid tablet is dropped into a glass of water. The tablet fizzes and bubbles as $NaHCO_3$ and citric acid ($C_6H_8O_7$) in the tablet react to form carbon dioxide (CO_2) gas (see Figure 8.2). During a chemical change, new properties become visible, which are an indication that a chemical reaction has taken place (see Table 8.1).

A chemical change: the tarnishing of silver

FIGURE 8.1 A chemical change produces new substances with new properties.

Q Why is the formation of tarnish a chemical change?

FIGURE 8.2 A chemical reaction of an antacid ($NaHCO_3$) tablet in water produces bubbles of carbon dioxide (CO_2) and water (H_2O).

Q What is the evidence for chemical change in this chemical reaction?

CO_2

$NaHCO_3$

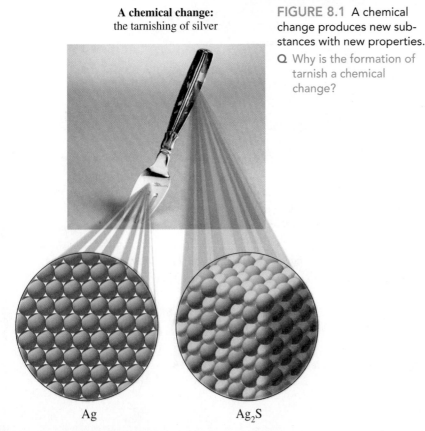

Ag Ag_2S

CONCEPT CHECK 8.1

■ **Evidence of a Chemical Reaction**

Indicate the visible evidence of a chemical reaction in each of the following:

a. burning propane fuel in a barbecue
b. using peroxide to change the color of hair

ANSWER
a. The production of heat during burning of propane fuel is evidence of a chemical reaction.
b. The change in hair color is evidence of a chemical reaction.

When you build a model airplane, prepare a new recipe, or mix a medication formulation, you follow a set of directions. These directions tell you what materials to use and the products you will obtain. In chemistry, a **chemical equation** tells us the materials we need and the products that will form in a chemical reaction.

SELF STUDY ACTIVITY
What Is Chemistry?

Writing a Chemical Equation

Suppose you work in a bicycle shop assembling wheels and frames into bicycles. You could represent this process by a simple equation:

SELF STUDY ACTIVITY
Chemical Reactions and Equations

Equation: 2 Wheels + 1 Frame ⟶ 1 Bicycle

Reactants   Product

When you burn charcoal in a grill, the carbon in the charcoal combines with oxygen to form carbon dioxide. We can represent this reaction by the following chemical equation:

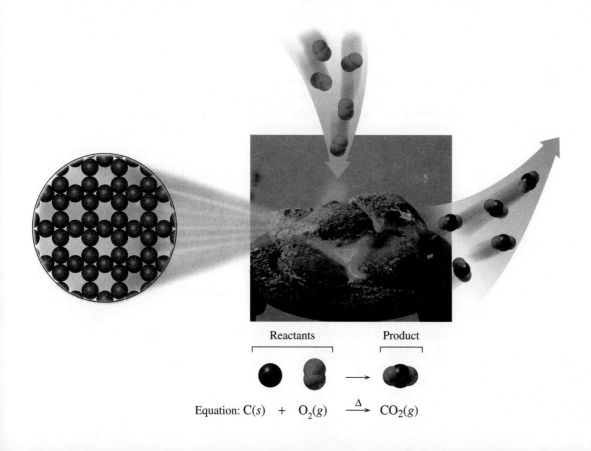

Reactants   Product

Equation: $C(s)$ + $O_2(g)$ $\xrightarrow{\Delta}$ $CO_2(g)$

TABLE 8.2 Some Symbols Used in Writing Equations

Symbol	Meaning
+	Separates two or more formulas
$\longrightarrow$	React(s) to form products
Δ	Reactants are heated
(s)	Solid
(l)	Liquid
(g)	Gas
(aq)	Aqueous

In an equation, the formulas of the **reactants** are written on the left of the arrow and the formulas of the **products** on the right. When there are two or more formulas on the same side, they are separated by plus (+) signs. The delta sign (Δ) indicates that heat was used to start the reaction.

Generally, each formula in an equation is followed by an abbreviation, in parentheses, that gives the physical state of the substance: solid (s), liquid (l), or gas (g). If a substance is dissolved in water, it is in an aqueous (aq) solution. Table 8.2 summarizes some of the symbols used in equations.

Identifying a Balanced Chemical Equation

When a chemical reaction takes place, the bonds between the atoms of the reactants are broken and new bonds are formed to give the products. All atoms are conserved, which means that atoms cannot be gained, lost, or changed into other types of atoms during a chemical reaction. Every chemical reaction is written as a **balanced equation**, which shows the same number of atoms for each element in the reactants as in the products. For example, the chemical equation we wrote above for burning carbon is balanced because there is one carbon atom and two oxygen atoms in both the reactants and the products:

$$C(s) + O_2(g) \longrightarrow CO_2(g)$$

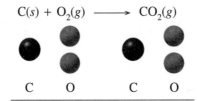

C	O	C	O

Reactant atoms = Product atoms

Now consider the reaction in which hydrogen reacts with oxygen to form water. The formulas of the reactants and products are written as follows:

$$H_2(g) + O_2(g) \longrightarrow H_2O(g)$$

When we add up the atoms of each element on each side, we find that the equation is *not balanced*. There are two oxygen atoms to the left of the arrow, but only one to the right. To balance this equation, we place whole numbers called **coefficients** in front of the formulas. If we write a coefficient of 2 in front of the H_2O formula, it represents two molecules of water. Because the coefficient multiplies all the atoms in H_2O, there are now four hydrogen atoms in the products. To obtain four atoms of hydrogen in the reactants, we must write a coefficient of 2 in front of the formula H_2. *Do not change subscripts*; this alters the chemical identity of a reactant or product. Now the number of hydrogen atoms and oxygen atoms is the same in the reactants as in the products. The equation is *balanced*. It is important to use only coefficients to balance any chemical equation.

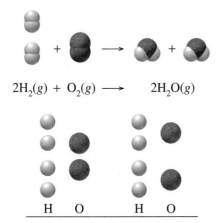

$$2H_2(g) + O_2(g) \longrightarrow 2H_2O(g)$$

H	O	H	O

Reactant atoms = Product atoms

SAMPLE PROBLEM | 8.1

■ **Chemical Equations**

Hydrogen and nitrogen react to form ammonia, NH_3.

$$3H_2(g) + N_2(g) \longrightarrow 2NH_3(g)$$
Ammonia

a. What are the coefficients in the equation?
b. How many atoms of each element are in the reactants and products of the equation?

SOLUTION

a. The coefficients are three (3) in front of H_2; one (1), which is understood, in front of N_2; and two (2) in front of NH_3.
b. In the reactants, there are six hydrogen atoms and two nitrogen atoms. In the product, there are also six hydrogen atoms and two nitrogen atoms.

STUDY CHECK

When the hydrocarbon ethane (C_2H_6) burns in oxygen, the products are carbon dioxide and water. The balanced equation is as follows:

$$2C_2H_6(g) + 7O_2(g) \longrightarrow 4CO_2(g) + 6H_2O(g)$$

State the total number of atoms of each element on each side of the equation.

QUESTIONS AND PROBLEMS

Equations for Chemical Reactions

8.1 State the number of atoms of each element on the reactant and on the product sides of the following equations:
  a. $2NO(g) + O_2(g) \longrightarrow 2NO_2(g)$
  b. $5C(s) + 2SO_2(g) \longrightarrow CS_2(g) + 4CO(g)$
  c. $2C_2H_2(g) + 5O_2(g) \longrightarrow 4CO_2(g) + 2H_2O(g)$
  d. $N_2H_4(g) + 2H_2O_2(g) \longrightarrow N_2(g) + 4H_2O(g)$

8.2 State the number of atoms of each element on the reactant and on the product sides of the following equations:
  a. $CH_4(g) + 2O_2(g) \longrightarrow CO_2(g) + 2H_2O(g)$
  b. $4P(s) + 5O_2(g) \longrightarrow P_4O_{10}(s)$
  c. $4NH_3(g) + 6NO(g) \longrightarrow 5N_2(g) + 6H_2O(g)$
  d. $6CO_2(g) + 6H_2O(l) \longrightarrow C_6H_{12}O_6(aq) + 6O_2(g)$
  Glucose

8.3 Determine whether each of the following equations is balanced or not balanced:
  a. $S(s) + O_2(g) \longrightarrow SO_3(g)$
  b. $2Al(s) + 3Cl_2(g) \longrightarrow 2AlCl_3(s)$
  c. $2NaOH(aq) + H_2SO_4(aq) \longrightarrow Na_2SO_4(aq) + H_2O(l)$
  d. $C_3H_8(g) + 5O_2(g) \longrightarrow 3CO_2(g) + 4H_2O(g)$

8.4 Determine whether each of the following equations is balanced or not balanced:
  a. $PCl_3(l) + Cl_2(g) \longrightarrow PCl_5(s)$
  b. $CO(g) + 2H_2(g) \longrightarrow CH_3OH(l)$
  c. $2KClO_3(s) \longrightarrow 2KCl(s) + O_2(g)$
  d. $Mg(s) + N_2(g) \longrightarrow Mg_3N_2(s)$

8.5 All of the following are balanced equations. State the number of atoms of each element in the reactants and in the products.
  a. $2Na(s) + Cl_2(g) \longrightarrow 2NaCl(s)$
  b. $PCl_3(l) + 3H_2(g) \longrightarrow PH_3(g) + 3HCl(g)$
  c. $P_4O_{10}(s) + 6H_2O(l) \longrightarrow 4H_3PO_4(aq)$

8.6 All of the following are balanced equations. State the number of atoms of each element in the reactants and in the products.
  a. $2N_2(g) + 3O_2(g) \longrightarrow 2N_2O_3(g)$
  b. $Al_2O_3(s) + 6HCl(aq) \longrightarrow 2AlCl_3(aq) + 3H_2O(l)$
  c. $C_5H_{12}(l) + 8O_2(g) \longrightarrow 5CO_2(g) + 6H_2O(g)$

8.2 Balancing a Chemical Equation

We can now show the process of balancing a chemical equation for the reaction of the alkane methane, CH_4, with oxygen to produce carbon dioxide and water. This is the reaction that occurs in the flame of a laboratory burner or a gas cooktop.

STEP 1 **Write an equation using the correct formulas for the reactants and products.**

$$CH_4(g) + O_2(g) \longrightarrow CO_2(g) + H_2O(g)$$

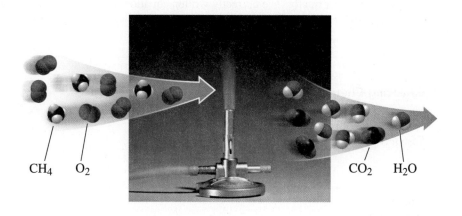

CH$_4$ O$_2$ CO$_2$ H$_2$O

STEP 2   **Count the atoms of each element in the reactants and products.**   In the initial unbalanced equation, a coefficient of 1 is understood and not usually written. When we compare the atoms on the reactant side and the product side, we see that there are more hydrogen atoms in the reactants and more oxygen atoms in the products.

$$CH_4(g) + O_2(g) \longrightarrow CO_2(g) + H_2O(g)$$

1 C atom	1 C atom	Balanced
4 H atoms	2 H atoms	Not balanced
2 O atoms	3 O atoms	Not balanced

STEP 3   **Use coefficients to balance each element.**   We will start by balancing the 4 H in CH_4 because it has the most atoms. By placing a 2 in front of the formula for water, a total of 4 H atoms in the products is obtained.

$$CH_4(g) + O_2(g) \longrightarrow CO_2(g) + 2H_2O(g)$$

We can balance the 4 O atoms on the product side by placing a 2 in front of the formula O_2.

$$CH_4(g) + 2O_2(g) \longrightarrow CO_2(g) + 2H_2O(g)$$

STEP 4   **Check the final equation to confirm that it is balanced.**   In the final equation, the numbers of atoms of carbon, hydrogen, and oxygen are the same in the reactants and the products.

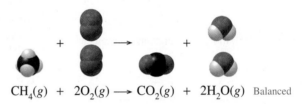

$$CH_4(g) + 2O_2(g) \longrightarrow CO_2(g) + 2H_2O(g) \quad \text{Balanced}$$

Reactants	Products
1 C atom	1 C atom
4 H atoms	4 H atoms
4 O atoms	4 O atoms

In a balanced equation, the coefficients must be the lowest set of whole numbers. Suppose you had added coefficients to the equation and obtained the following:

$$2CH_4(g) + 4O_2(g) \longrightarrow 2CO_2(g) + 4H_2O(g) \quad \text{Incorrect}$$

Although there are equal numbers of atoms on both sides of the equation, this is not written correctly. The correctly written equation is obtained by dividing all the coefficients by 2.

CONCEPT CHECK 8.2

Balancing Chemical Equations

Indicate the number of each in the following equation:

$$Fe_2S_3(s) + 6HCl(aq) \longrightarrow 2FeCl_3(aq) + 3H_2S(g)$$

	Reactants	Products
Atoms of Cl		
Atoms of S		
Atoms of Fe		
Molecules of H_2S		

ANSWER The total number of atoms in each formula is obtained by multiplying by its coefficient. The number of molecules is obtained from the coefficient.

	Reactants	Products
Atoms of Cl	6	6
Atoms of S	3	3
Atoms of Fe	2	2
Molecules of H_2S	0	3

SAMPLE PROBLEM 8.2

■ Balancing Chemical Equations

Balance the following equation:

$$Na_3PO_4(aq) + MgCl_2(aq) \longrightarrow Mg_3(PO_4)_2(s) + NaCl(aq)$$

SOLUTION

STEP 1 **Write the equation using the correct formulas.**

$$Na_3PO_4(aq) + MgCl_2(aq) \longrightarrow Mg_3(PO_4)_2(s) + NaCl(aq)$$

STEP 2 **Count the atoms or ions in the reactants and the products.**   When we compare the number of ions in the reactants and products, we find that the equation is not balanced. In this equation, we can balance the phosphate ion as a group of atoms because it appears on both sides of the equation.

**Guide to Balancing a
Chemical Equation**

STEP 1
Write an equation using the
correct formulas in the
reactants and products.

STEP 2
Count the atoms of each element
in reactants and products.

STEP 3
Use coefficients to balance
each element.

STEP 4
Check the final equation
for balance.

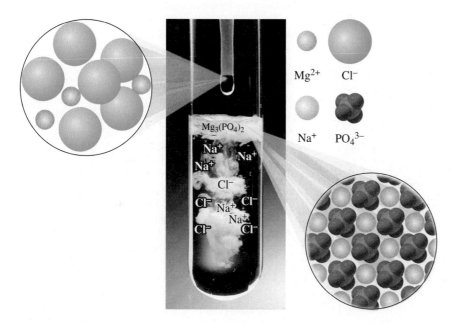

Reactants	Products	
$Na_3PO_4(aq) + MgCl_2(aq) \rightarrow Mg_3(PO_4)_2(s) + NaCl(aq)$		
$3\ Na^+$	$1\ Na^+$	Not balanced
$1\ PO_4^{3-}$	$2\ PO_4^{3-}$	Not balanced
$1\ Mg^{2+}$	$3\ Mg^{2+}$	Not balanced
$2\ Cl^-$	$1\ Cl^-$	Not balanced

STEP 3 **Use coefficients to balance each element.**   We typically begin with the formula that has the highest subscript values, $Mg_3(PO_4)_2$ in this case. A 3 in front of $MgCl_2$ balances Mg, and a 2 in front of Na_3PO_4 balances the phosphate ion (PO_4^{3-}).

$$2Na_3PO_4(aq) + 3MgCl_2(aq) \longrightarrow Mg_3(PO_4)_2(s) + NaCl(aq)$$

Looking again at each of the ions in the reactants and products, we see that the sodium and chloride ions are not yet equal. A 6 in front of the NaCl balances the equation.

$$2Na_3PO_4(aq) + 3MgCl_2(aq) \longrightarrow Mg_3(PO_4)_2(s) + 6NaCl(aq)$$

STEP 4 **Check the final equation for balance.** A check of the total number of ions confirms that the equation is balanced. A coefficient of 1 is understood and not usually written.

Reactants	Products
6 Na^+	6 Na^+
2 PO_4^{3-}	2 PO_4^{3-}
3 Mg^{2+}	3 Mg^{2+}
6 Cl^-	6 Cl^-

$$2Na_3PO_4(aq) + 3MgCl_2(aq) \longrightarrow Mg_3(PO_4)_2(s) + 6NaCl(aq) \qquad \text{Balanced}$$

STUDY CHECK

Balance the following equation:

$$Fe(s) + O_2(g) \longrightarrow Fe_3O_4(s)$$

QUESTIONS AND PROBLEMS

Balancing a Chemical Equation

8.7 Balance the following equations:
   a. $N_2(g) + O_2(g) \longrightarrow NO(g)$
   b. $HgO(s) \longrightarrow Hg(l) + O_2(g)$
   c. $Fe(s) + O_2(g) \longrightarrow Fe_2O_3(s)$
   d. $Na(s) + Cl_2(g) \longrightarrow NaCl(s)$
   e. $Cu_2O(s) + O_2(g) \longrightarrow CuO(s)$

8.8 Balance the following equations:
   a. $Ca(s) + Br_2(l) \longrightarrow CaBr_2(s)$
   b. $P_4(s) + O_2(g) \longrightarrow P_4O_{10}(s)$
   c. $C_4H_8(g) + O_2(g) \longrightarrow CO_2(g) + H_2O(g)$
   d. $Sb_2S_3(s) + HCl(aq) \longrightarrow SbCl_3(s) + H_2S(g)$
   e. $Fe_2O_3(s) + C(s) \longrightarrow Fe(s) + CO(g)$

8.9 Balance the following equations:
   a. $Mg(s) + AgNO_3(aq) \longrightarrow Mg(NO_3)_2(aq) + Ag(s)$
   b. $CuCO_3(s) \longrightarrow CuO(s) + CO_2(g)$
   c. $Al(s) + CuSO_4(aq) \longrightarrow Cu(s) + Al_2(SO_4)_3(aq)$
   d. $Pb(NO_3)_2(aq) + NaCl(aq) \longrightarrow PbCl_2(s) + NaNO_3(aq)$
   e. $Al(s) + HCl(aq) \longrightarrow AlCl_3(aq) + H_2(g)$

8.10 Balance the following equations:
   a. $Zn(s) + H_2SO_4(aq) \longrightarrow ZnSO_4(aq) + H_2(g)$
   b. $Al(s) + H_2SO_4(aq) \longrightarrow Al_2(SO_4)_3(aq) + H_2(g)$
   c. $K_2SO_4(aq) + BaCl_2(aq) \longrightarrow BaSO_4(s) + KCl(aq)$
   d. $CaCO_3(s) \longrightarrow CaO(s) + CO_2(g)$
   e. $Al_2(SO_4)_3(aq) + KOH(aq) \longrightarrow Al(OH)_3(s) + K_2SO_4(aq)$

8.11 Balance the following equations:
   a. $Fe_2O_3(s) + CO(g) \longrightarrow Fe(s) + CO_2(g)$
   b. $Li_3N(s) \longrightarrow Li(s) + N_2(g)$
   c. $Al(s) + HBr(aq) \longrightarrow AlBr_3(aq) + H_2(g)$
   d. $Ba(OH)_2(aq) + Na_3PO_4(aq) \longrightarrow$
   $$Ba_3(PO_4)_2(s) + NaOH(aq)$$
   e. $As_4S_6(s) + O_2(g) \longrightarrow As_4O_6(s) + SO_2(g)$

8.12 Balance the following equations:
   a. $K(s) + H_2O(l) \longrightarrow KOH(aq) + H_2(g)$
   b. $Cr(s) + S_8(s) \longrightarrow Cr_2S_3(s)$
   c. $BCl_3(s) + H_2O(l) \longrightarrow H_3BO_3(aq) + HCl(aq)$
   d. $Fe(OH)_3(s) + H_2SO_4(aq) \longrightarrow Fe_2(SO_4)_3(aq) + H_2O(l)$
   e. $BaCl_2(aq) + Na_3PO_4(aq) \longrightarrow Ba_3(PO_4)_2(s) + NaCl(aq)$

8.13 Write a balanced equation using the correct formulas and include conditions (s, l, g, or aq) for each of the following reactions:
   a. Lithium metal reacts with liquid water to form hydrogen gas and aqueous lithium hydroxide.
   b. Solid phosphorus reacts with chlorine gas to form solid phosphorus pentachloride.
   c. Solid iron(II) oxide reacts with carbon monoxide gas to form solid iron and carbon dioxide gas.
   d. Liquid pentene (C_5H_{10}) burns in oxygen gas to form carbon dioxide gas and water vapor.
   e. Hydrogen sulfide gas and solid iron(III) chloride react to form solid iron(III) sulfide and hydrogen chloride gas.

8.14 Write a balanced equation using the correct formulas and include conditions (s, l, g, or aq) for each of the following reactions:
   a. Solid calcium carbonate decomposes to produce solid calcium oxide and carbon dioxide gas.
   b. Nitrogen oxide gas reacts with carbon monoxide gas to produce nitrogen gas and carbon dioxide gas.
   c. Iron metal reacts with solid sulfur to produce solid iron(III) sulfide.
   d. Solid calcium reacts with nitrogen gas to produce solid calcium nitride.
   e. In the *Apollo* lunar module, hydrazine gas, N_2H_4, reacts with dinitrogen tetroxide gas to produce gaseous nitrogen and water vapor.

8.3 Reaction Types

A great number of reactions occur in nature, in biological systems, and in the laboratory. However, there are some general patterns among all reactions that help us classify reactions. Some reactions may fit into more than one reaction type.

Combination Reactions

In a **combination reaction**, two or more elements or compounds bond to form one product. For example, sulfur and oxygen combine to form the product sulfur dioxide.

$$S(s) + O_2(g) \longrightarrow SO_2(g)$$

In Figure 8.3, the elements magnesium and oxygen combine to form a single product, magnesium oxide.

$$2Mg(s) + O_2(g) \longrightarrow 2MgO(s)$$

In other examples of combination reactions, elements or compounds combine to form a single product.

$$N_2(g) + 3H_2(g) \longrightarrow 2NH_3(g) \quad \text{Ammonia}$$
$$Cu(s) + S(s) \longrightarrow CuS(s)$$
$$MgO(s) + CO_2(g) \longrightarrow MgCO_3(s)$$

LEARNING GOAL

Identify a reaction as a combination, decomposition, single replacement, or double replacement.

Two or more reactants	combine to yield	a single product
A + B	$\longrightarrow$	A B

SELF STUDY ACTIVITY

Chemical Reactions and Equations

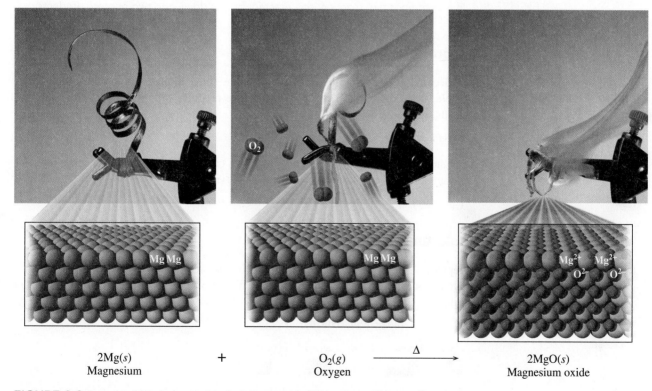

$$\underset{\text{Magnesium}}{2Mg(s)} \quad + \quad \underset{\text{Oxygen}}{O_2(g)} \quad \xrightarrow{\Delta} \quad \underset{\text{Magnesium oxide}}{2MgO(s)}$$

FIGURE 8.3 In a combination reaction, two or more substances combine to form one substance as product.

Q What happens to the atoms of the reactants in a combination reaction?

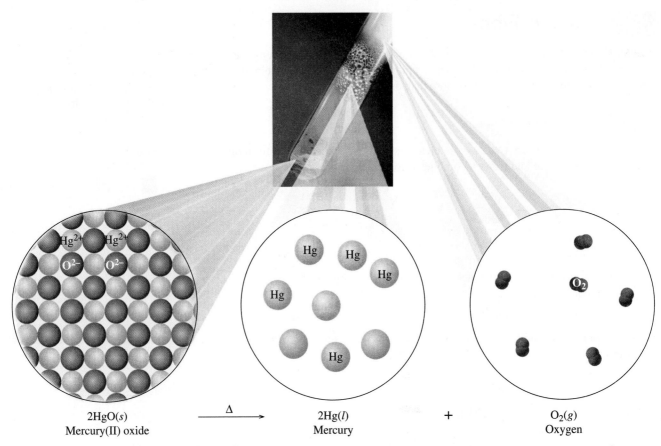

$$2HgO(s) \xrightarrow{\Delta} 2Hg(l) + O_2(g)$$
Mercury(II) oxide   Mercury   Oxygen

FIGURE 8.4 In a decomposition reaction, one reactant breaks down into two or more products.
Q How do the differences in the reactant and products classify this as a decomposition reaction?

Decomposition Reactions

A   splits   two or more
reactant   into   products

| A B | → | A | + | B |

In a **decomposition reaction**, a single reactant splits into two or more products. For example, when mercury(II) oxide is heated, the products are the elements mercury and oxygen (see Figure 8.4).

$$2HgO(s) \xrightarrow{\Delta} 2Hg(l) + O_2(g)$$

In another example of a decomposition reaction, calcium carbonate breaks apart into simpler compounds of calcium oxide and carbon dioxide.

$$CaCO_3(s) \xrightarrow{\Delta} CaO(s) + CO_2(g)$$

Replacement Reactions

In replacement reactions, elements in compounds are replaced by other elements. In a **single replacement reaction**, an uncombined element takes the place of an element in a compound.

Single replacement

One element   replaces   another element

| A | + | B C | → | A C | + | B |

In the single replacement reaction shown in Figure 8.5, zinc replaces hydrogen in hydrochloric acid, HCl(*aq*).

$$Zn(s) + 2HCl(aq) \longrightarrow ZnCl_2(aq) + H_2(g)$$

In a single replacement reaction, chlorine replaces bromine in the compound potassium bromide.

$$Cl_2(g) + 2KBr(s) \longrightarrow 2KCl(s) + Br_2(l)$$

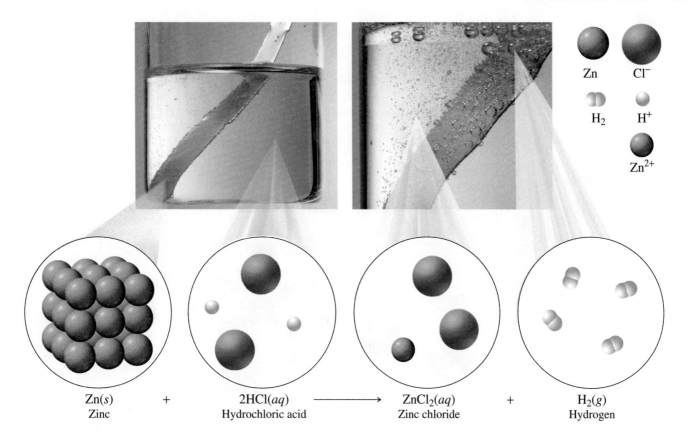

$$\underset{\text{Zinc}}{\text{Zn}(s)} \quad + \quad \underset{\text{Hydrochloric acid}}{2\text{HCl}(aq)} \quad \longrightarrow \quad \underset{\text{Zinc chloride}}{\text{ZnCl}_2(aq)} \quad + \quad \underset{\text{Hydrogen}}{\text{H}_2(g)}$$

FIGURE 8.5 In a single replacement reaction, an atom or ion replaces an atom or ion in a compound.
Q What changes in the formulas of the reactants identify this equation as a single replacement reaction?

In a **double replacement reaction**, the positive ions in the reacting compounds switch places.

Double replacement

Two elements   replace each other

A B + C D ⟶ A D + C B

In the reaction shown in Figure 8.6, barium ions change places with sodium ions in the reactants to form sodium chloride and a white solid precipitate of barium sulfate.

$$\text{Na}_2\text{SO}_4(aq) + \text{BaCl}_2(aq) \longrightarrow \text{BaSO}_4(s) + 2\text{NaCl}(aq)$$

When sodium hydroxide and hydrochloric acid (HCl) react, sodium and hydrogen ions switch places, forming sodium chloride and water.

$$\text{NaOH}(aq) + \text{HCl}(aq) \longrightarrow \text{NaCl}(aq) + \text{HOH}(l)$$

Table 8.3 summarizes the reaction types and gives examples.

TABLE 8.3 Summary of Reaction Types

Reaction Type	Example
Combination	
A + B ⟶ AB	$\text{Ca}(s) + \text{Cl}_2(g) \longrightarrow \text{CaCl}_2(s)$
Decomposition	
AB ⟶ A + B	$\text{Fe}_2\text{S}_3(s) \longrightarrow 2\text{Fe}(s) + 3\text{S}(s)$
Single Replacement	
A + BC ⟶ AC + B	$\text{Cu}(s) + 2\text{AgNO}_3(aq) \longrightarrow \text{Cu(NO}_3)_2(aq) + 2\text{Ag}(s)$
Double Replacement	
AB + CD ⟶ AD + CB	$\text{BaCl}_2(aq) + \text{K}_2\text{SO}_4(aq) \longrightarrow \text{BaSO}_4(s) + 2\text{KCl}(aq)$

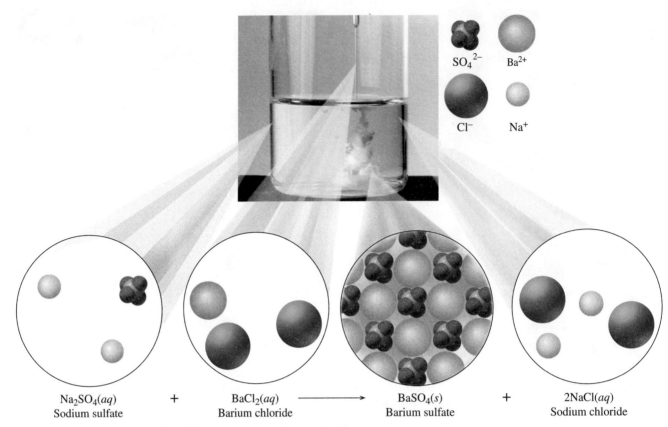

$$\underset{\substack{\text{Sodium sulfate}}}{Na_2SO_4(aq)} \quad + \quad \underset{\substack{\text{Barium chloride}}}{BaCl_2(aq)} \quad \longrightarrow \quad \underset{\substack{\text{Barium sulfate}}}{BaSO_4(s)} \quad + \quad \underset{\substack{\text{Sodium chloride}}}{2NaCl(aq)}$$

FIGURE 8.6 In a double replacement reaction, the positive ions in the reactants replace each other.

Q How do the changes in the formulas of the reactants identify this equation as a double replacement reaction?

SAMPLE PROBLEM | 8.3

■ Identifying Reactions and Predicting Products

1. Classify the following reactions as combination, decomposition, single replacement, or double replacement:
   a. $2H_2O_2(aq) \longrightarrow 2H_2O(l) + O_2(g)$
   b. $2K_3PO_4(aq) + 3CuCl_2(aq) \longrightarrow Cu_3(PO_4)_2(s) + 6KCl(aq)$

2. Predict the product(s) for each of the following and balance the equation:
   a. Single replacement: $Al(s) + CuCl_2(aq) \longrightarrow$ _____ + _____
   b. Combination: $K(s) + Cl_2(g) \longrightarrow$ _____

SOLUTION

1. a. When one reactant breaks down into two products, the reaction is a *decomposition*.
   b. The positive ions K^+ and Cu^{2+} from the reactants have exchanged places to form two products, which makes this a *double replacement* reaction.

2. a. To complete this single replacement reaction, the Cu in the $CuCl_2$ compound is replaced by Al.

   $$2Al(s) + 3CuCl_2(aq) \longrightarrow 2AlCl_3(aq) + 3Cu(s)$$

   b. To complete this combination reaction, K and Cl form KCl.

   $$2K(s) + Cl_2(g) \longrightarrow 2KCl(s)$$

STUDY CHECK

Nitrogen gas and oxygen gas react to form nitrogen dioxide gas. Write the balanced equation and identify the reaction type.

QUESTIONS AND PROBLEMS

Reaction Types

8.15 a. Why would the following be a decomposition reaction?

$$2Al_2O_3(s) \xrightarrow{\Delta} 4Al(s) + 3O_2(g)$$

b. Why would the following be a single replacement reaction?

$$Br_2(g) + BaI_2(s) \longrightarrow BaBr_2(s) + I_2(g)$$

8.16 a. Why would the following be a combination reaction?

$$H_2(g) + Br_2(g) \longrightarrow 2HBr(g)$$

b. Why would the following be a double replacement reaction?

$$AgNO_3(aq) + NaCl(aq) \longrightarrow AgCl(s) + NaNO_3(aq)$$

8.17 Classify each of the following reactions as a combination, decomposition, single replacement, or double replacement:

a. $4Fe(s) + 3O_2(g) \longrightarrow 2Fe_2O_3(s)$

b. $Mg(s) + 2AgNO_3(aq) \longrightarrow Mg(NO_3)_2(aq) + 2Ag(s)$

c. $CuCO_3(s) \longrightarrow CuO(s) + CO_2(g)$

d. $NaOH(aq) + HCl(aq) \longrightarrow NaCl(aq) + H_2O(l)$

e. $ZnCO_3(s) \longrightarrow CO_2(g) + ZnO(s)$

f. $Al_2(SO_4)_3(aq) + 6KOH(aq) \longrightarrow$
$$2Al(OH)_3(s) + 3K_2SO_4(aq)$$

g. $Pb(s) + O_2(g) \longrightarrow PbO_2(s)$

8.18 Classify each of the following reactions as a combination, decomposition, single replacement, or double replacement:

a. $CuO(s) + 2HCl(aq) \longrightarrow CuCl_2(aq) + H_2O(l)$

b. $2Al(s) + 3Br_2(g) \longrightarrow 2AlBr_3(s)$

c. $Pb(NO_3)_2(aq) + 2NaCl(aq) \longrightarrow PbCl_2(s) + 2NaNO_3(aq)$

d. $2Mg(s) + O_2(g) \longrightarrow 2MgO(s)$

e. $Fe_2O_3(s) + 3C(s) \longrightarrow 2Fe(s) + 3CO(g)$

f. $C_6H_{12}O_6(aq) \longrightarrow 2C_2H_6O(aq) + 2CO_2(g)$

g. $BaCl_2(aq) + K_2CO_3(aq) \longrightarrow BaCO_3(s) + 2KCl(aq)$

8.19 Try your hand at predicting the products that would result from the following types of reactions. Balance each equation you write.

a. Combination: $Mg(s) + Cl_2(g) \longrightarrow$ _____

b. Decomposition: $HBr(g) \longrightarrow$ _____ + _____

c. Single replacement: $Mg(s) + Zn(NO_3)_2(aq) \longrightarrow$
_____ + _____

d. Double replacement: $K_2S(aq) + Pb(NO_3)_2(aq) \longrightarrow$
_____ + _____

8.20 Try your hand at predicting the products that would result from the reactions of the following. Balance each equation you write.

a. Combination: $Ca(s) + S(s) \longrightarrow$ _____

b. Decomposition: $PbO_2(s) \longrightarrow$ _____ + _____

c. Single replacement: $KI(s) + Cl_2(g) \longrightarrow$
_____ + _____

d. Double replacement: $CuCl_2(aq) + Na_2S(aq) \longrightarrow$
_____ + _____

CHEMISTRY AND HEALTH

Smog and Health Concerns

There are two types of smog. One, photochemical smog, requires sunlight to initiate reactions that produce pollutants such as nitrogen oxides and ozone. The other type of smog, called industrial or London smog, occurs in areas where coal containing sulfur is burned and the unwanted product sulfur dioxide is emitted.

Photochemical smog is most prevalent in cities where people are dependent on cars for transportation. On a typical day in Los Angeles, for example, nitrogen oxide (NO) emissions from car exhausts increase as traffic increases on the roads. The nitrogen oxide is formed when N_2 and O_2 react at high temperatures in car and truck engines.

$$N_2(g) + O_2(g) \xrightarrow{Heat} 2NO(g)$$

Then, NO reacts with oxygen in the air to produce NO_2, a reddish brown gas that is irritating to the eyes and damaging to the respiratory tract.

$$2NO(g) + O_2(g) \longrightarrow 2NO_2(g)$$

The reddish-brown color of smog is due to nitrogen dioxide.

When NO_2 molecules are exposed to sunlight, they are converted to NO molecules and oxygen atoms.

$$NO_2(g) \xrightarrow{Sunlight} NO(g) + O(g)$$
$$\text{Oxygen atoms}$$

Oxygen atoms are so reactive that they combine with oxygen molecules in the atmosphere, forming ozone.

$$O(g) + O_2(g) \longrightarrow O_3(g)$$
$$\text{Ozone}$$

In the upper atmosphere (the stratosphere), ozone is beneficial because it protects us from harmful ultraviolet radiation that comes from the Sun. However, in the lower atmosphere, ozone irritates the eyes and respiratory tract, which causes coughing and decreased lung function. Ozone also causes fabric to deteriorate, rubber to crack, and it damages trees and crops.

Industrial smog is produced when sulfur is converted to sulfur dioxide during the burning of coal or other high sulfur fuels.

$$S(s) + O_2(g) \longrightarrow SO_2(g)$$

The SO_2 is damaging to plants and is corrosive to metals such as steel. The SO_2 is also damaging to humans and can cause lung impairment and respiratory difficulties. In the air, SO_2 reacts with oxygen to form SO_3, which can combine with water to form sulfuric acid. When rain falls, it absorbs the sulfuric acid, which makes acid rain.

$$2SO_2(g) + O_2(g) \longrightarrow 2SO_3(g)$$
$$SO_3(g) + H_2O(l) \longrightarrow H_2SO_4(aq)$$
$$\text{Sulfuric acid}$$

The presence of sulfuric acid in rivers and lakes causes an increase in the acidity of the water, reducing the ability of animals and plants to survive.

8.4 Functional Groups in Organic Compounds

In Chapter 6, we discussed the hydrocarbons called alkanes, which are organic compounds composed of carbon and hydrogen. Now we can look at other organic compounds in which carbon atoms may also bond with the nonmetals oxygen and nitrogen. Table 8.4 lists the number of covalent bonds formed by elements found in organic compounds. In a typical organic compound, carbon forms four covalent bonds, hydrogen forms one covalent bond, nitrogen forms three covalent bonds, and oxygen forms two covalent bonds.

TABLE 8.4 Covalent Bonds for Elements in Organic Compounds

Element	Group	Covalent Bonds	Structure of Atoms
H	1A (1)	1	H—
C	4A (14)	4	—C—
N	5A (15)	3	—N—
O	6A (16)	2	—O—

There are millions of organic compounds, and new ones are synthesized every day. We organize the organic compounds according to specific groups of atoms called **functional groups**. Compounds that have the same functional groups have similar properties. By identifying functional groups, we can classify organic compounds according to their structure and predict some of their common reactions.

Alkenes, Alkynes, and Aromatic Compounds

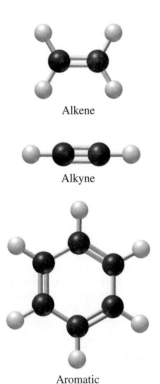

Alkene

Alkyne

Aromatic

The hydrocarbon family also includes alkenes, alkynes, and aromatic compounds. An **alkene** contains at least one double bond, which forms when two adjacent carbon atoms share two pairs of valence electrons. The simplest alkene, ethene (or ethylene), is $CH_2 = CH_2$. For an alkene, the ending of the name of the corresponding alkane is changed from *ane* to *ene*. To give carbon four bonds, each carbon atom (black) in the double bond of ethene is attached to two hydrogen atoms (white).

Ethene is an important plant hormone involved in promoting the ripening of fruit. Commercially grown fruit, such as avocados, bananas, and tomatoes, are often picked before they are ripe. Before the fruit is brought to market, it is exposed to ethene to accelerate the ripening process. Ethene also accelerates the breakdown of cellulose in plants, which causes flowers to wilt and leaves to fall from trees.

You may recognize the name ethylene in the name polyethylene, a synthetic polymer used for plastic bottles, film, and insulation material. Many of the small alkenes are used to make long-chain polymers such as polypropylene in clothing and carpets, and polystyrene used for plastic coffee cups.

In an **alkyne**, a triple bond forms when two carbon atoms share three pairs of valence electrons. The simplest alkyne is called ethyne, $HC \equiv CH$, commonly known as acetylene. In ethyne, each carbon atom in the triple bond is attached to one hydrogen atom. Alkenes and alkynes are known as *unsaturated hydrocarbons* because they contain double or triple bonds.

In 1825, Michael Faraday isolated a hydrocarbon called *benzene*, which had the formula C_6H_6. Because compounds containing benzene often had fragrant odors, the family became known as **aromatic compounds**. A benzene molecule consists of a ring of six carbon atoms with one hydrogen atom attached to each carbon. Later, scientists proposed that the carbon atoms be arranged in a ring with alternating single and double bonds.

Today, the benzene structure is typically represented as a hexagon with a circle in the center:

Fruit is picked early and ripened by exposure to ethene.

Alkene   Alkyne   Aromatic

<hr>

CONCEPT CHECK 8.3

■ Identifying Alkanes, Alkenes, and Alkynes

Classify each of the following condensed structural formulas as an alkane, alkene, or alkyne:

a. $CH_3 — C \equiv C — CH_3$
b. $CH_3 — CH_2 — CH_3$
c. $CH_3 — CH_2 — CH_2 — CH = CH_2$

ANSWER
a. This is an alkyne because it has a condensed structural formula with a triple bond.
b. This is an alkane because it has a condensed structural formula with only single bonds between carbon atoms.
c. This is an alkene because it has a condensed structural formula with a double bond.

Alcohols and Ethers

In an **alcohol**, a *hydroxyl* (—OH) *group* replaces a hydrogen atom in a hydrocarbon. The oxygen (O) atom is shown in red in the ball-and-stick models.

$$CH_3 — CH_2 — \textbf{OH}$$
Alcohol

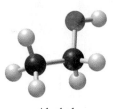

Alcohol

Ethanol (ethyl alcohol), $CH_3 — CH_2 — OH$, has been known since prehistoric times as an intoxicating product formed by the fermentation of the sugars and starches in grains, grapes, and other fruits. Ethanol is the alcohol found in alcoholic beverages.

$$C_6H_{12}O_6 \xrightarrow{\text{Fermentation}} 2CH_3 — CH_2 — OH + 2CO_2$$

Ethanol is also used as a solvent for perfumes, varnishes, and some medicines, such as tincture of iodine. Recent interest in alternative fuels has led to increased production of ethanol by the fermentation of sugars from grains such as corn, wheat, and rice. "Gasohol" is a fuel comprised of ethanol and gasoline.

An **ether** contains an oxygen atom (red) that is attached by single bonds to two carbon atoms. The oxygen atom also has two unshared pairs of electrons, but they are not shown in the condensed structural formulas.

$$CH_3 — \textbf{O} — CH_3$$
Ether

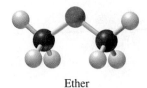

Ether

The term *ether* has been associated with anesthesia because diethyl ether was the most widely used anesthetic for more than a hundred years.

$$CH_3 — CH_2 — O — CH_2 — CH_3$$
Diethyl ether

Although diethyl ether was easy to administer, ether is very volatile and highly flammable. A small spark in the operating room could cause an explosion. Since the 1950s, anesthetics such as isoflurane and halothane have been developed that are not as flammable.

CONCEPT CHECK 8.4

■ **Identifying Functional Groups**

Describe the differences between the functional groups found in alcohols and ethers.

ANSWER In alcohols, the functional group is a hydroxyl group (—OH) attached to a carbon atom. The functional group in ethers is an oxygen atom bonded to two carbon atoms (—O—).

TUTORIAL
Aldehydes and Ketones

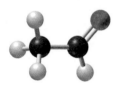

Aldehyde

Ketone

Aldehydes and Ketones

Aldehydes and ketones contain a *carbonyl group* (C=O), which is a carbon with a double bond to oxygen. In an **aldehyde**, the carbon atom of the carbonyl group is bonded to a hydrogen atom. Only the simplest aldehyde (CH_2O) has a carbonyl group attached to two hydrogen atoms. In a **ketone**, the carbonyl group is bonded to two other carbon atoms.

$$CH_3 - \overset{\overset{\textstyle O}{\|}}{C} - H \qquad CH_3 - \overset{\overset{\textstyle O}{\|}}{C} - CH_3$$
Aldehyde   Ketone

Ketones and aldehydes are in many items we use or eat each day. For example, they are used in the food industry to produce flavorings such as vanilla, cinnamon, and spearmint. When we buy a small bottle of liquid flavoring, the aldehyde or ketone is dissolved in alcohol because the compounds are not very soluble in water.

Formaldehyde, HCHO, the simplest aldehyde, is a colorless gas with a pungent odor. Industrially, it is a reactant in the synthesis of polymers used to make fabrics, insulation materials, carpeting, pressed wood products such as plywood, and plastics for kitchen counters. An aqueous solution called formalin, which contains 40 percent formaldehyde, is used as a germicide and to preserve biological specimens. Exposure to formaldehyde fumes can irritate the eyes, nose, and upper respiratory tract and cause skin rashes, headaches, dizziness, and general fatigue.

The simplest ketone, known as *acetone* or propanone (dimethyl ketone), is a colorless liquid with a mild odor that has wide use as a solvent in cleaning fluids, paint and nail polish removers, and rubber cement. Acetone is extremely flammable, and care must be taken when using acetone. In the body, acetone may be produced in uncontrolled diabetes, fasting, and high-protein diets when large amounts of fats are metabolized for energy.

Carbohydrates, such as table sugar, are made of carbon, hydrogen, and oxygen. The structures of the simple sugars known as *monosaccharides* consist of a chain typically of six carbon atoms attached to hydroxyl groups (—OH), and a carbonyl group as an aldehyde (—CHO) on the first carbon, or as a ketone (C=O) on the second carbon atom. The three common monosaccharides—glucose, galactose, and fructose—have the same molecular formula, $C_6H_{12}O_6$, but have different structural formulas.

The most common monosaccharide—glucose, also known as blood sugar—is an aldehyde. It is a building block of the disaccharides lactose and sucrose, and the polysaccharides such as starch, cellulose, and glycogen. Fructose, the sweetest carbohydrate, is a ketone. Both glucose and fructose are found in honey.

$$H_3C - \overset{\overset{\textstyle O}{\|}}{C} - CH_3$$
Acetone

CONCEPT CHECK 8.5

Galactose is a monosaccharide in milk and milk products.

Galactose

a. What is the molecular formula of galactose?
b. What functional groups are present in galactose?

ANSWER
a. The total number of C atoms is 6, of H atoms is 12, and of O atoms is 6. The molecular formula is $C_6H_{12}O_6$.
b. The functional groups present in galactose are an aldehyde group on carbon 1, and five hydroxyl groups on carbons 2–6.

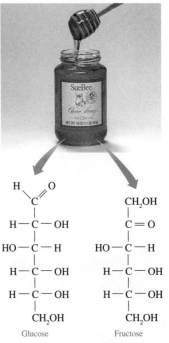

Glucose   Fructose

Honey is a mixture of fructose and glucose.

(MC)

SELF STUDY ACTIVITY
Carboxylic Acids

Carboxylic Acids and Esters

In a **carboxylic acid**, the functional group is the *carboxyl group*, which is a combination of the *carb*onyl and hydr*oxyl* groups.

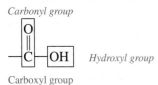

In all but the smallest carboxylic acid, the carbon of the carboxyl group is attached to another carbon atom:

$$CH_3-\overset{O}{\overset{\|}{C}}-O-H \quad \text{or} \quad CH_3-COOH \quad \text{or} \quad CH_3-CO_2H$$

Carboxylic acid

Carboxylic acids are common in nature. Formic acid, HCOOH, is injected under the skin from red ant and bee stings. Acetic acid, CH_3-COOH, is formed when the ethanol in wine and apple cider reacts with oxygen in the air. A solution of acetic acid and water is the vinegar used in food preparation and salad dressings.

An **ester** is similar to a carboxylic acid, except the oxygen is attached to a carbon and not to hydrogen.

$$CH_3-\overset{O}{\overset{\|}{C}}-O-CH_3 \quad \text{or} \quad CH_3-COO-CH_3 \quad \text{or} \quad CH_3-CO_2-CH_3$$

Ester

Many of the fragrances of perfumes and flowers and the flavors of fruits are due to esters. Small esters are volatile, so we can smell them, and soluble in water, so we can taste them. Esters are responsible for the odor and flavor of oranges, bananas, pears, pineapples, and strawberries. For example, the ester named ethyl butanoate produces the smell and flavor of pineapples.

Carboxylic acid

The sour taste of vinegar is due to acetic acid.

Ester

Esters such as ethyl butanoate provide the odor and flavor of many fruits such as pineapples.

$$CH_3-CH_2-CH_2-\overset{\overset{\displaystyle O}{\|}}{C}-O-CH_2-CH_3$$

Ethyl butanoate
(ethyl butyrate)

CHEMISTRY AND HEALTH

Fatty Acids

The fats and oils in our diets contain fatty acids, which are long chains of carbon atoms with a carboxylic acid group at one end. An example is lauric acid, a 12-carbon acid found in coconut oil, which has a structure that can be written in several forms.

Writing Formulas for Lauric Acid

$$CH_3-(CH_2)_{10}-\overset{\overset{\displaystyle O}{\|}}{C}-OH \qquad CH_3-(CH_2)_{10}-COOH$$

$$CH_3-CH_2-CH_2-CH_2-CH_2-CH_2-CH_2-CH_2-CH_2-CH_2-CH_2-\overset{\overset{\displaystyle O}{}}{\underset{\displaystyle OH}{C}}$$

Condensed structural formula

Saturated fatty acids such as lauric acid contain only single bonds between carbons. Saturated fatty acids are generally solids at room temperature. *Unsaturated fatty acids* have one or more double bonds in the carbon chain and are typically liquid oils at room temperature. Table 8.5 lists some common fatty acids.

Omega-3 Fatty Acids in Fish Oils

Because unsaturated fatty acids are now recognized as beneficial to health, American diets have changed to include more unsaturated fatty acids and less saturated fatty acids. This change is a response to research that indicates that atherosclerosis and heart disease are associated with high levels of fats in the diet. However, the Inuit peoples of Alaska have a diet with high levels of unsaturated fats as well as high levels of blood cholesterol, but they have a very low occur-

rence of heart disease. The fats in the Inuit diet are primarily unsaturated fats from fish rather than from land animals.

The fatty acids in vegetable oils are omega-6, in which the first double bond occurs at carbon 6 from the CH_3- end. However, the fatty acids in fish oils are mostly the omega-3 type, in which the first double bond occurs at the third carbon from the CH_3- end.

In atherosclerosis and heart disease, cholesterol forms plaques that adhere to the walls of the blood vessels. Blood pressure rises as blood has to squeeze through a smaller opening in the blood vessel. As more plaque forms, there is also a possibility of blood clots blocking the blood vessels and causing a heart attack. Omega-3 fatty acids lower the tendency of blood clots blocking the blood vessels. However, high levels of omega-3 fatty acids can increase bleeding if the ability of the platelets to form blood clots is reduced too much. It appears that a diet that includes fish such as salmon, tuna, and herring can provide higher amounts of the omega-3 fatty acids, which lessen the possibility of developing heart disease.

Cold-water fish are a source of omega-3 fatty acids.

Omega-6 Fatty Acids

Linoleic acid
$$\underset{1}{CH_3}-(CH_2)_4-\underset{6}{CH}=CH-CH_2-CH=CH-(CH_2)_7-COOH$$

Arachidonic acid
$$\underset{1}{CH_3}-(CH_2)_4-\underset{6}{(CH}=CH-CH_2)_4-(CH_2)_2-COOH$$

Omega-3 Fatty Acids

Linolenic acid
$$\underset{1}{CH_3}-CH_2-\underset{3}{(CH}=CH-CH_2)_3-(CH_2)_6-COOH$$

Eicosapentaenoic acid (EPA)
$$\underset{1}{CH_3}-CH_2-\underset{3}{(CH}=CH-CH_2)_5-(CH_2)_2-COOH$$

Docosahexaenoic acid (DHA)
$$\underset{1}{CH_3}-CH_2-\underset{3}{(CH}=CH-CH_2)_6-CH_2-COOH$$

TABLE 8.5 Structures and Sources of Some Common Fatty Acids

Name	Carbon Atoms	Double Bonds	Condensed Structural Formula	Sources
Saturated				
Palmitic acid	16	0	$CH_3-(CH_2)_{14}-COOH$	Palm
Stearic acid	18	0	$CH_3-(CH_2)_{16}-COOH$	Animal fat
Unsaturated				
Oleic acid	18	1	$CH_3-(CH_2)_7-CH=CH-(CH_2)_7-COOH$	Olives, corn
Linoleic acid	18	2	$CH_3-(CH_2)_4-CH=CH-CH_2-CH=CH-(CH_2)_7-COOH$	Soybean, safflower, sunflower

SAMPLE PROBLEM   8.4

■ Classifying Organic Compounds

Classify the following organic compounds according to their functional groups:

a. $CH_3-CH_2-CH_2-OH$ **b.** $CH_3-CH=CH-CH_3$

c. $CH_3-CH_2-CH_2-COOH$

SOLUTION

a. When the functional group $-OH$ is bonded to a carbon atom, the compound is an alcohol.

b. An alkene contains one or more double bonds between carbon atoms.

c. The functional group $-COOH$ indicates that this compound is a carboxylic acid.

STUDY CHECK

Classify the following organic compound according to its functional group:

$CH_3-CH_2-O-CH_3$

Amines and Amides

Amines are derivatives of ammonia, NH_3, in which carbon groups replace one, two, or three of the hydrogen atoms. From Table 8.4, we know that a nitrogen atom forms three bonds.

$$NH_3 \qquad CH_3-NH_2 \qquad CH_3-\underset{\underset{CH_3}{|}}{N}H \qquad CH_3-\underset{\underset{CH_3}{|}}{N}-CH_3$$

Ammonia   Examples of amines

Amine

One characteristic of fish is their odor, which is due to amines. Amines, such as putrescine and cadaverine, are produced when proteins decay and have a particularly pungent and offensive odor.

$$H_2N-CH_2-CH_2-CH_2-CH_2-NH_2$$
Putrescine

$$H_2N-CH_2-CH_2-CH_2-CH_2-CH_2-NH_2$$
Cadaverine

In an **amide**, the hydroxyl group of a carboxylic acid is replaced by a nitrogen group.

$$CH_3-\overset{\overset{\displaystyle O}{\|}}{C}-NH_2$$
Amide

The simplest natural amide is urea, an end product of protein metabolism in the body. The kidneys remove urea from the blood and provide for its excretion in urine. Urea is also used as a component of fertilizer to increase nitrogen in the soil.

$$H_2N-\overset{\overset{\displaystyle O}{\|}}{C}-NH_2$$
Urea

CASE STUDY
Death by Chocolate

Amide

CHEMISTRY AND HEALTH

Amines in Health and Medicine

In response to allergic reactions or injury to cells, the body increases the production of an amine known as histamine, which causes blood vessels to dilate and redness and swelling to occur. Administering an antihistamine helps block the effects of histamine.

In the body, hormones called biogenic amines carry messages between the central nervous system and nerve cells. Epinephrine (adrenaline) and norepinephrine (noradrenaline) are released by the adrenal medulla in "fight or flight" situations to raise the blood glucose level and move the blood to the muscles. Used in remedies for colds, hay fever, and asthma, Benzedrine, Neo-Synephrine (phenylephrine), and norepinephrine contract the capillaries in the mucous membranes of the respiratory passages. Sometimes Benzedrine is taken internally to combat the desire to sleep, but it has side effects. Parkinson's disease is a result of a deficiency in another biogenic amine called dopamine.

Epinephrine (adrenaline)

Produced synthetically, amphetamines (known as "uppers") are stimulants of the central nervous system much like epinephrine, but they also increase cardiovascular activity and depress the appetite. They are sometimes used to bring about weight loss, but they can cause chemical dependency. Methedrine is used to treat depression and in the illegal form is known as "crank" or "crystal meth." The prefix *meth* means that there is one more methyl group CH_3— on the nitrogen atom.

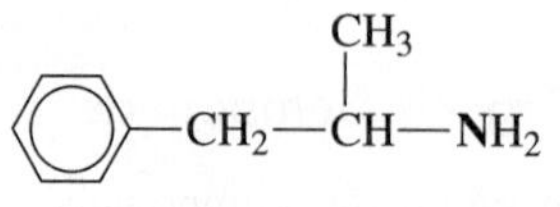

Benzedrine (amphetamine)

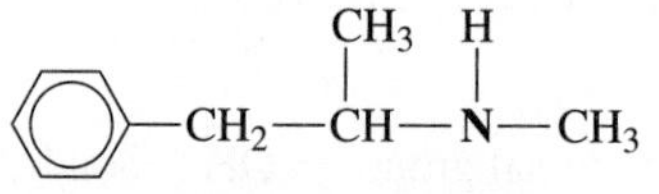

Methamphetamine (methedrine)

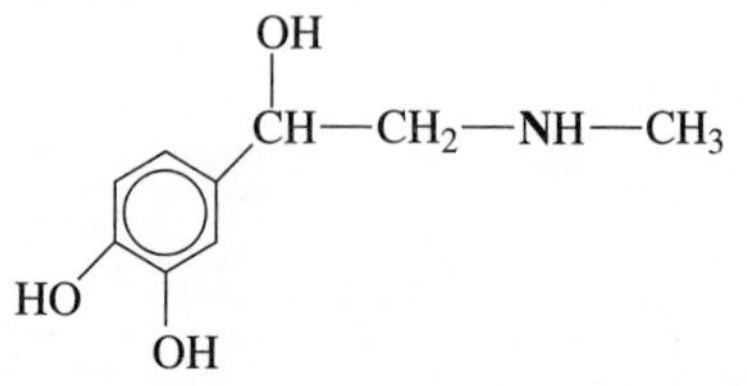

Many artificial sweeteners contain amides.

Synthetic amides are used as substitutes for sugar and aspirin. Aspartame, which is marketed as NutraSweet, is used in a large number of sugar-free products. It is a noncarbohydrate sweetener made of aspartic acid and a methyl ester of phenylalanine, which is 180 times sweeter than sucrose (table sugar). Aspartame does have some caloric value, but it is so sweet that only a small amount is needed. However, one breakdown product, phenylalanine, poses a danger to anyone who cannot metabolize it properly, a condition called phenylketonuria (PKU).

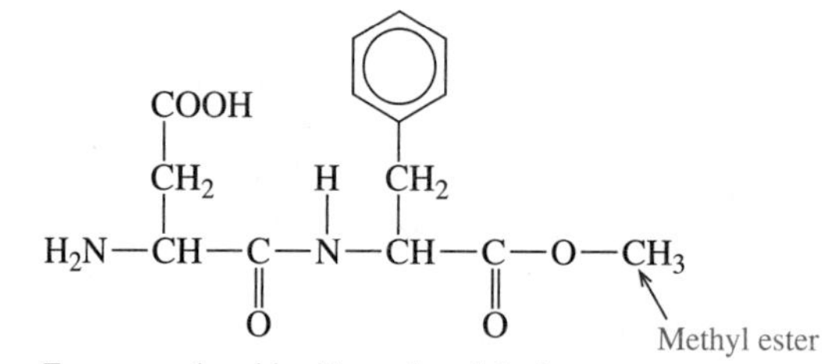

From aspartic acid   From phenylalanine
Aspartame (NutraSweet)

Amino Acids

Amino acids, which are the molecular building blocks of proteins, contain an amino group ($—NH_2$) and a carboxylic acid group ($—COOH$) bonded to a central carbon atom. The unique characteristics of amino acids are due to different side chains (R). In biological systems, the amino and carboxylic acids groups are present in ionic forms.

General Structure of an α-Amino Acid

The structures, names, and three-letter abbreviations of some common amino acids found in proteins are listed in Table 8.6.

TABLE 8.6 Some Typical Amino Acids in Proteins

Glycine (Gly) Alanine (Ala) Valine (Val) Phenylalanine (Phe)

Serine (Ser) Threonine (Thr) Cysteine (Cys)

Peptides

The linking of two or more amino acids forms a *peptide*. A *peptide bond* is an amide bond that forms when the —COO⁻ group of one amino acid reacts with the $H_3\overset{+}{N}$— group of the next amino acid.

Amino acid 1 Amino acid 2 Dipeptide

For example, the combination of the amino acids glycine and alanine produces an amide bond in the dipeptide Gly–Ala. For convenience, the order of amino acids in the peptide is written as the sequence of three-letter abbreviations, starting with the amino acid with the free —NH₃⁺ and ending with the amino acid that contains the free —COO⁻.

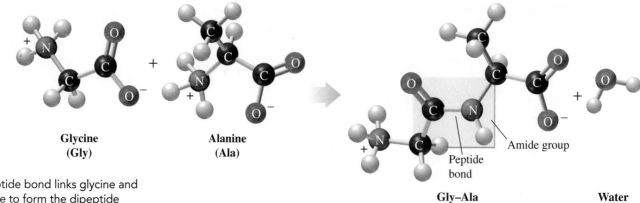

Glycine (Gly)   **Alanine (Ala)**   **Gly–Ala**   **Water**

Amide group

Peptide bond

A peptide bond links glycine and alanine to form the dipeptide Gly–Ala.

CONCEPT CHECK 8.6

Writing Dipeptide Structures

Draw the condensed structural formula for the dipeptide Val–Ser.

ANSWER Valine is joined to serine by a peptide bond; valine has a free $-NH_3^+$, and serine has a free $-COO^-$.

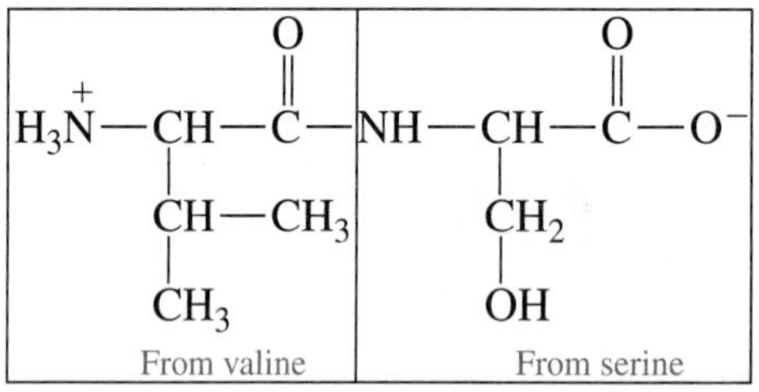

$$\overset{+}{H_3N}-CH-\overset{\overset{\displaystyle O}{\|}}{C}-NH-CH-\overset{\overset{\displaystyle O}{\|}}{C}-O^-$$

From valine | From serine

Val–Ser

A list of the common functional groups in organic compounds is shown in Table 8.7.

TABLE 8.7 Classification of Organic Compounds

Class	Functional Group	Example
Alkene	$>\!C\!=\!C\!<$	$H_2C\!=\!CH_2$
Alkyne	$-C\!\equiv\!C-$	$HC\!\equiv\!CH$
Alcohol	$-OH$	CH_3-CH_2-OH
Ether	$-O-$	CH_3-O-CH_3
Aldehyde	$-\overset{\overset{\displaystyle O}{\|}}{C}-H$	$CH_3-\overset{\overset{\displaystyle O}{\|}}{C}-H$
Ketone	$-\overset{\overset{\displaystyle O}{\|}}{C}-$	$CH_3-\overset{\overset{\displaystyle O}{\|}}{C}-CH_3$
Carboxylic acid	$-\overset{\overset{\displaystyle O}{\|}}{C}-O-H$	$CH_3-\overset{\overset{\displaystyle O}{\|}}{C}-O-H$
Ester	$-\overset{\overset{\displaystyle O}{\|}}{C}-O-$	$CH_3-\overset{\overset{\displaystyle O}{\|}}{C}-O-CH_3$
Amine	$-N-$	CH_3-NH_2
Amide	$-\overset{\overset{\displaystyle O}{\|}}{C}-N-$	$CH_3-\overset{\overset{\displaystyle O}{\|}}{C}-NH_2$

CONCEPT CHECK 8.7

■ **Organic Compounds**

Classify the following organic compounds according to their functional groups:

a. $CH_3—CH_2—NH—CH_3$
b. $HC≡C—CH_3$
c. $CH_3—COOH$

ANSWER

a. amine
b. alkyne
c. carboxylic acid

SAMPLE PROBLEM | 8.5

■ **Identifying Functional Groups**

Classify the following organic compounds according to their functional groups:

a. $CH_3—CH_2—O—CH_3$
b. $CH_3—CH=CH—CH_2—CH_3$
c. $CH_3—CH_2—COOH$
d. $CH_3—CH_2—CH_2—CH_2—OH$

SOLUTION

a. ether   **b.** alkene
c. carboxylic acid **d.** alcohol

STUDY CHECK

How does a carboxylic acid differ from an ester?

QUESTIONS AND PROBLEMS

Functional Groups in Organic Compounds

8.21 Identify the class of compounds that contains each of the following functional groups:
  a. hydroxyl group attached to a carbon chain
  b. carbon–carbon double bond
  c. carbonyl group attached to a hydrogen atom
  d. carboxyl group attached to two carbon atoms

8.22 Identify the class of compounds that contains each of the following functional groups:
  a. a nitrogen atom attached to one or more carbon atoms
  b. carboxyl group
  c. oxygen atom bonded to two carbon atoms
  d. a carbonyl group between two carbon atoms

8.23 Classify the following molecules according to their functional groups. The possibilities are alcohol, ether, ketone, carboxylic acid, or amine.
  a. $CH_3—CH_2—O—CH_2—CH_3$

  b. $CH_3—\overset{\displaystyle OH}{\underset{|}{CH}}—CH_3$

  c. $CH_3—\overset{\displaystyle O}{\overset{\|}{C}}—CH_2—CH_3$

  d. $CH_3—CH_2—CH_2—CH_2—CH_2—COOH$

  e. $CH_3—CH_2—NH_2$

8.24 Classify the following molecules according to their functional groups. The possibilities are alkene, aldehyde, carboxylic acid, ester, or amine.

  a. $CH_3—\overset{\displaystyle O}{\overset{\|}{C}}—O—CH_2—CH_3$

  b. $CH_3—\overset{\displaystyle CH_3}{\underset{|}{N}}—CH_3$

  c. $CH_3—CH_2—CH_2—\overset{\displaystyle O}{\overset{\|}{C}}—H$

  d. $CH_3—CH_2—COOH$

  e. $CH_3—CH=CH—CH_2—CH_2—CH_3$

8.5 Reactions of Organic Compounds

The most common type of reaction of organic compounds is the *combustion* reaction, which is typical of hydrocarbons used as fuels. The burning of gasoline in the engine of a car and in a snow blower are examples of combustion reactions.

Combustion Reactions

In a **combustion reaction**, an organic compound reacts with oxygen in the air to produce carbon dioxide, water, and energy usually as heat. For example, methane (CH_4) in natural gas undergoes combustion with oxygen, producing a flame and heat. The combustion of methane is used to cook our food on a gas cooktop and to heat our homes. We can write the balanced chemical equation for the combustion of methane as:

$$CH_4(g) + 2O_2(g) \xrightarrow{\Delta} CO_2(g) + 2H_2O(g) + \text{heat}$$

Propane is the fuel used in portable heaters and gas barbecues. Gasoline, a mixture of liquid hydrocarbons, is the fuel that powers our cars, lawn mowers, and snow blowers. As alkanes, they all undergo combustion. The balanced equation for the combustion of propane (C_3H_8) is:

$$C_3H_8(g) + 5O_2(g) \xrightarrow{\Delta} 3CO_2(g) + 4H_2O(g) + \text{energy}$$

Acetylene, $H—C\equiv C—H$ or C_2H_2, is used in oxyacetylene gas welding to fuse metals. During the combustion of acetylene, a temperature of about 3300 °C is obtained. The balanced equation for the combustion of acetylene is written:

$$2C_2H_2(g) + 5O_2(g) \xrightarrow{\Delta} 4CO_2(g) + 2H_2O(g) + \text{energy}$$

The propane from the torch undergoes combustion, which provides energy to solder metals.

Combustion reactions also occur in the cells of the body in order to metabolize food, which provides energy for the activities we want to do. We absorb oxygen (O_2) from the air to burn glucose ($C_6H_{12}O_6$) from our food, and eventually our cells produce CO_2, H_2O, and energy:

$$C_6H_{12}O_6(aq) + 6O_2(g) \longrightarrow 6CO_2(g) + 6H_2O(l) + \text{energy}$$

A mixture of acetylene and oxygen undergoes combustion during the welding of metals.

SAMPLE PROBLEM | 8.6

■ **Combustion**

Write a balanced equation for the complete combustion of pentane.

SOLUTION

For a combustion reaction, the initial reactants are pentane (C_5H_{12}) and oxygen, and the products are carbon dioxide and water:

$$C_5H_{12}(g) + O_2(g) \xrightarrow{\Delta} CO_2(g) + H_2O(g) + energy$$

The C and H atoms in pentane are balanced with coefficients to give $5CO_2$ and $6H_2O$:

$$C_5H_{12}(g) + O_2(g) \xrightarrow{\Delta} 5CO_2(g) + 6H_2O(g) + energy$$

The total of 16 O atoms in the products is balanced with $8O_2$. The balanced equation for the complete combustion of pentane is written:

$$C_5H_{12}(g) + 8O_2(g) \xrightarrow{\Delta} 5CO_2(g) + 6H_2O(g) + energy$$

STUDY CHECK

Ethene, used to ripen fruit, has the formula C_2H_4. Write a balanced equation for the complete combustion of ethene.

CHEMISTRY AND HEALTH

Incomplete Combustion: Toxicity of Carbon Monoxide

When a propane heater, fireplace, or woodstove is used in a closed room, there must be adequate ventilation. If the supply of oxygen is limited, *incomplete combustion* from burning gas, oil, or wood produces carbon monoxide. The incomplete combustion of methane in natural gas is written:

$$2CH_4(g) + 3O_2(g) \xrightarrow{\Delta} 2CO(g) + 4H_2O(g) + heat$$

Limited oxygen supply   Carbon monoxide

Carbon monoxide (CO) is a colorless, odorless, poisonous gas. When inhaled, CO passes into the bloodstream, where it attaches to hemoglobin. When CO binds to the hemoglobin, it reduces the amount of oxygen (O_2) reaching the organs and cells. As a result, a healthy person can experience a reduction in exercise capability, visual perception, and manual dexterity.

When the amount of hemoglobin bound to CO (COHb) is about 10 percent, a person may experience shortness of breath, mild headache, and drowsiness. Heavy smokers can have as high as 9 percent COHb in their blood. When as much as 30 percent of the hemoglobin is bound to CO, a person may experience more severe symptoms, including dizziness, mental confusion, severe headache, and nausea. If 50 percent or more of the hemoglobin is bound to CO, a person could become unconscious and die if not treated immediately with oxygen.

Combustion of Alcohols

We have seen that hydrocarbons undergo combustion in the presence of oxygen. Alcohols burn with oxygen, too. For example, in a restaurant, a dessert may be prepared by pouring a liqueur on fruit or ice cream and lighting it. The balanced equation for the combustion of the ethanol in the liquor is written as:

$$CH_3{-}CH_2{-}OH(g) + 3O_2(g) \xrightarrow{\Delta} 2CO_2(g) + 3H_2O(g) + energy$$

Hydrogenation: An Addition Reaction

The most characteristic reaction of alkenes and alkynes is the *addition* of atoms to the double or triple bond. Addition reactions occur because double and triple bonds are easily broken, which provides electrons to form new single bonds. In **hydrogenation**, atoms of hydrogen from H_2 add to the carbons in a double or triple bond to form alkanes. A catalyst

A dessert is heated by a combustion reaction.

TUTORIAL

Addition Reactions

CHEMISTRY AND INDUSTRY

Hydrogenation of Unsaturated Fats

Vegetable oils such as corn oil or safflower oil contain a high proportion of unsaturated fatty acids with double bonds. Commercially, the process of hydrogenation is used to convert many, but not all, of the double bonds to single bonds, which produces more saturated solid fats, such as margarine.

In commercial hydrogenation, the addition of hydrogen is stopped before all the double bonds in a vegetable oil become completely saturated. Complete hydrogenation gives a very brittle product, whereas the partial hydrogenation of a vegetable oil changes it to a soft, semisolid fat. As the fat becomes more saturated, the substances become more solid at room temperature. By control-

ling the amount of hydrogen, manufacturers can produce the various types of products on the market today, such as soft margarines, solid stick margarines, and solid shortenings. Although these products now contain more saturated fatty acids than the original oils, they contain no cholesterol, unlike similar products from animal sources, such as butter and lard. For example, when oleic acid, a typical unsaturated fatty acid in olive oil, is hydrogenated, it is converted to stearic acid, a saturated fatty acid.

$$CH_3-(CH_2)_7-CH=CH-(CH_2)_7-COOH + H_2 \xrightarrow{Pt}$$
$$CH_3-(CH_2)_{16}-COOH$$

Oleic acid (unsaturated) → Stearic acid (saturated)

In nature, the more prevalent structure of the double bonds in unsaturated fatty acids is the *cis* structure, which means that the hydrogen atoms are on one side of the double bond and the carbon chains attached to the double bond are on the other side. During the process of hydrogenation of vegetable oils, some of the cis double bonds open, but then form *trans* double bonds, which have the hydrogen atoms and carbon chains on opposite sides. If the label on a product states that the oils have been "partially" or "fully hydrogenated," that product will also contain trans fatty acids.

The concern about trans fatty acids is that they behave like saturated fatty acids in the body. Several studies reported that trans fatty acids raise the levels of LDL-cholesterol, low-density lipoproteins containing cholesterol that can accumulate in the arteries. More research is still needed to determine the overall impact of trans fatty acids present in fats in our diets. Since 2006, food labels have given the grams of trans fat per serving.

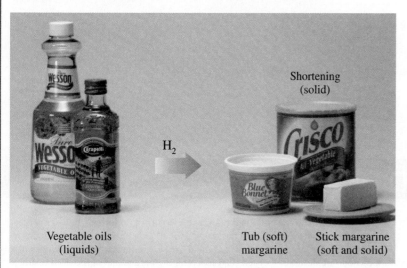

Hydrogenation converts vegetable oils to solid products.

such as platinum (Pt), nickel (Ni), or palladium (Pd) is added to speed up the reaction. Below is an example of the hydrogenation of an alkene:

$$CH_3-CH=CH-CH_3 \; + \; H-H \; \xrightarrow{Pt} \; CH_3-\overset{\overset{\displaystyle H}{|}}{CH}-\overset{\overset{\displaystyle H}{|}}{CH}-CH_3$$

2-Butene → Butane

When an alkyne undergoes hydrogenation, two molecules of hydrogen are required to form the alkane product.

$$CH_3-C\equiv C-CH_3 \; + \; 2H-H \; \xrightarrow{Pt} \; CH_3-\overset{\overset{\displaystyle H}{|}}{\underset{\underset{\displaystyle H}{|}}{C}}-\overset{\overset{\displaystyle H}{|}}{\underset{\underset{\displaystyle H}{|}}{C}}-CH_3$$

2-Butyne → Butane

SAMPLE PROBLEM | 8.7

■ Writing Equations for Hydrogenation

Draw the condensed structural formula for the product of the following hydrogenation reactions:

a. $CH_3-CH=CH_2 + H_2 \xrightarrow{Ni}$ **b.** $HC\equiv CH + 2H_2 \xrightarrow{Ni}$

SOLUTION

In a hydrogenation reaction, hydrogen adds to a double or triple bond to give an alkane.

a. $CH_3—CH_2—CH_3$ **b.** $H_3C—CH_3$

STUDY CHECK

Draw the condensed structural formula of the alkane formed by the hydrogenation of
$H—C{\equiv}C—CH_2—CH_3$.

QUESTIONS AND PROBLEMS

Reactions of Organic Compounds

8.25 Write a balanced equation for the complete combustion of each of the following:
    a. propane   **b.** octane
    c. $CH_3—CH_2—O—CH_3$ **d.** C_6H_{12}

8.26 Write a balanced equation for the complete combustion of each of the following:
    a. ethane   **b.** heptane
    c. $CH_3—CH_2—CH_2—CH_2—OH$ **d.** C_3H_6O

8.27 Draw the condensed structural formula of the product in each of the following:

    a. $CH_3—CH_2—CH_2—CH{=}CH_2 + H_2 \xrightarrow{Pt}$

    b. $CH_3—CH{=}CH—CH_3 + H_2 \xrightarrow{Ni}$

    c. $CH_3—CH_2—CH_2—CH_2—CH{=}CH_2 + H_2 \xrightarrow{Pt}$

8.28 Draw the condensed structural formula of the product in each of the following:

    a. $CH_3—CH_2—CH{=}CH_2 + H_2 \xrightarrow{Pt}$

    b. $CH_3—\overset{\overset{\displaystyle CH_3}{|}}{C}{=}CH—CH_2—CH_3 + H_2 \xrightarrow{Pt}$

    c. $CH_3—\overset{\overset{\displaystyle CH_3}{|}}{C}H—C{\equiv}CH + 2H_2 \xrightarrow{Pt}$

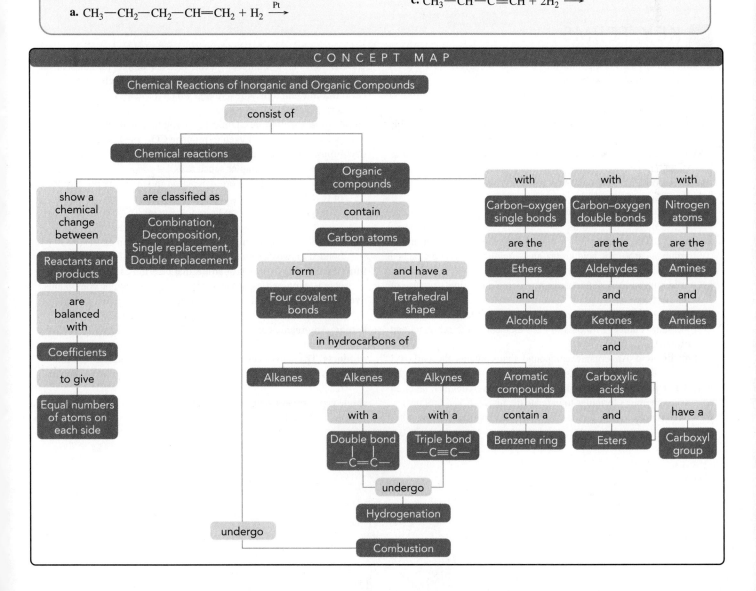

CONCEPT MAP

■ CHAPTER REVIEW

8.1 Equations for Chemical Reactions

LEARNING GOAL: *Identify a balanced chemical equation and determine the number of atoms in the reactants and products.*

A chemical change occurs when the atoms of the initial substances rearrange to form new substances. A chemical equation shows the formulas of the substances that react on the left of the reaction arrow and the products that form on the right side of the reaction arrow.

8.2 Balancing a Chemical Equation

LEARNING GOAL: *Write a balanced chemical equation from the formulas of the reactants and products for a reaction.*

An equation is balanced by writing the smallest whole numbers (coefficients) in front of formulas to equalize the atoms of each element in the reactants and the products.

8.3 Reaction Types

LEARNING GOAL: *Identify a reaction as a combination, decomposition, single replacement, or double replacement.*

Many chemical reactions are organized by reaction type: combination, decomposition, single replacement, or double replacement.

8.4 Functional Groups in Organic Compounds

LEARNING GOAL: *Classify organic molecules according to their functional groups.*

An organic molecule contains a characteristic group of atoms called a functional group that determines the molecule's family and chemical reactivity. Functional groups are used to classify organic compounds, act as reactive sites in the molecule, and provide a system of naming for organic compounds. Some common functional groups include the hydroxyl group in alcohols, the carbonyl group in aldehydes and ketones, and a nitrogen atom in amines.

8.5 Reactions of Organic Compounds

LEARNING GOAL: *Draw the condensed structural formulas of the products of reactions of organic compounds.*

Combustion is a characteristic reaction of alkanes and other organic compounds. When a substance undergoes combustion, it reacts with oxygen to produce carbon dioxide and water. In hydrogenation, hydrogen atoms are attached to the carbon atoms in a double or triple bond, which converts an alkene or an alkyne to an alkane.

■ KEY TERMS

alcohol An organic compound that contains the hydroxyl ($-OH$) group bonded to a carbon atom.

aldehydes A class of organic compounds that contains a carbonyl group ($C=O$) bonded to at least one hydrogen atom.

alkene A type of hydrocarbon that contains carbon–carbon double bonds ($C=C$).

alkyne A hydrocarbon that contains carbon–carbon triple bonds ($C\equiv C$).

amide An organic compound in which a carbon containing the carbonyl group is attached to a nitrogen atom.

amine An organic compound that contains a nitrogen atom bonded to one or more carbon atoms.

amino acid The building block of proteins, consisting of an amino group, a carboxylic acid group, and a unique side group attached to the alpha carbon.

aromatic compounds Compounds that contain the ring structure of benzene.

balanced equation The final form of a chemical reaction that shows the same number of atoms of each element in the reactants and products.

carboxylic acid An organic compound that contains the carboxyl functional group ($-COOH$).

chemical equation A shorthand way to represent a chemical reaction using chemical formulas to indicate the reactants and products.

chemical reaction The process by which a chemical change takes place.

coefficients Whole numbers placed in front of the formulas in an equation to balance the number of atoms of each element.

combination reaction A reaction in which reactants combine to form a single product.

combustion reaction A chemical reaction in which an organic compound reacts with oxygen to produce CO_2, H_2O, and energy.

decomposition reaction A reaction in which a single reactant splits into two or more simpler substances.

double replacement reaction A reaction in which the positive ions of two different reactants exchange places.

ester An organic compound that contains a carboxyl group ($-COO-$) with an oxygen atom bonded to carbon.

ether An organic compound that contains an oxygen atom ($-O-$) bonded to two carbon atoms.

functional group A group of atoms that determines the physical and chemical properties and naming of a class of organic compounds.

hydrogenation The addition of hydrogen (H_2) to the double bond of alkenes or the triple bond of alkynes to yield alkanes.

ketone An organic compound in which a carbonyl group ($C=O$) is bonded to two carbon atoms.

products The substances formed as a result of a chemical reaction.

reactants The initial substances that undergo change in a chemical reaction.

single replacement reaction A reaction in which an element replaces a different element in a compound.

■ SUMMARY OF REACTIONS

Combination

$$S(s) + O_2(g) \longrightarrow SO_2(g)$$

Decomposition

$$2HgO(s) \longrightarrow 2Hg(l) + O_2(g)$$

Single Replacement

$$Zn(s) + 2HCl(aq) \longrightarrow ZnCl_2(aq) + H_2(g)$$

Double Replacement

$$AgNO_3(aq) + HCl(aq) \longrightarrow AgCl(s) + HNO_3(aq)$$

Combustion

$$C_3H_8(g) + 5O_2(g) \xrightarrow{\Delta} 3CO_2(g) + 4H_2O(g) + energy$$

Hydrogenation

$$CH_3-CH=CH_2(g) + H_2(g) \xrightarrow{Pt} CH_3-CH_2-CH_3(g)$$

■ UNDERSTANDING THE CONCEPTS

8.29 If red spheres represent oxygen atoms and blue spheres represent nitrogen atoms:

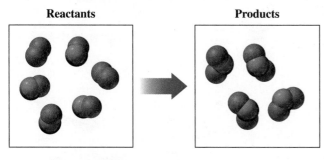

a. Write a balanced equation for the reaction.
b. Indicate the type of reaction as combination, decomposition, single replacement, or double replacement.

8.30 If purple spheres represent iodine atoms and white spheres represent hydrogen atoms:

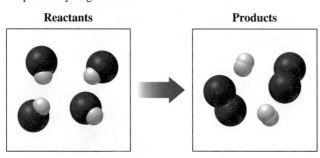

a. Write a balanced equation for the reaction.
b. Indicate the type of reaction as combination, decomposition, single replacement, or double replacement.

8.31 If blue spheres represent nitrogen atoms and purple spheres represent iodine atoms:

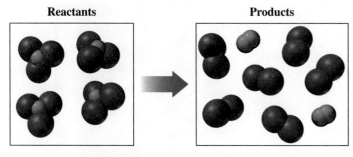

a. Write a balanced equation for the reaction.
b. Indicate the type of reaction as combination, decomposition, single replacement, or double replacement.

8.32 If green spheres represent chlorine atoms, yellow-green spheres represent fluorine atoms, and white spheres represent hydrogen atoms:

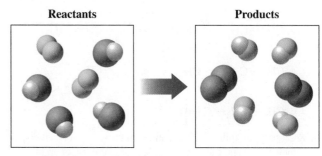

a. Write a balanced equation for the reaction.
b. Indicate the type of reaction as combination, decomposition, single replacement, or double replacement.

8.33 If green spheres represent chlorine atoms and red spheres represent oxygen atoms:

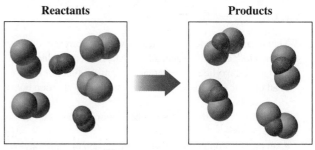

a. Write a balanced equation for the reaction.
b. Indicate the type of reaction as decomposition, combination, single replacement, or double replacement.

8.34 If blue spheres represent nitrogen atoms and purple spheres represent iodine atoms:

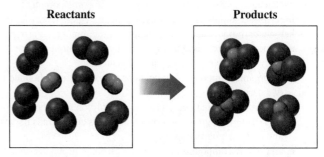

a. Write a balanced equation for the reaction.
b. Indicate the type of reaction as decomposition, combination, single replacement, or double replacement.

8.35 Balance each of the following by adding coefficients, and identify the type of reaction for each:

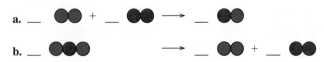

8.36 Balance each of the following by adding coefficients, and identify the type of reaction for each:

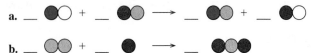

8.37 Classify the following as alkenes, alkynes, alcohols, ethers, aldehydes, ketones, carboxylic acids, esters, amines, or amides:

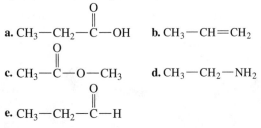

a. $CH_3—CH_2—\overset{\overset{O}{\|}}{C}—OH$ b. $CH_3—CH=CH_2$

c. $CH_3—\overset{\overset{O}{\|}}{C}—O—CH_3$ d. $CH_3—CH_2—NH_2$

e. $CH_3—CH_2—\overset{\overset{O}{\|}}{C}—H$

8.38 Classify the following as alkenes, alkynes, alcohols, ethers, aldehydes, ketones, carboxylic acids, esters, amines, or amides:

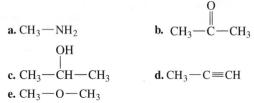

a. $CH_3—NH_2$ b. $CH_3—\overset{\overset{O}{\|}}{C}—CH_3$

c. $CH_3—\overset{\overset{OH}{|}}{CH}—CH_3$ d. $CH_3—C≡CH$

e. $CH_3—O—CH_3$

8.39 Match each of the following terms with the corresponding description: alkane, alkene, alkyne, alcohol, ether, aldehyde, ketone, carboxylic acid, ester, amine, amide, functional group
a. an organic compound that contains a hydroxyl group bonded to a carbon
b. a hydrocarbon that contains one or more carbon–carbon double bonds
c. an organic compound in which the carbon of a carbonyl group is bonded to a hydrogen
d. a hydrocarbon that contains only carbon–carbon single bonds
e. an organic compound in which the carbon of a carbonyl group is bonded to a hydroxyl group

8.40 Match each of the following terms with the corresponding description: alkane, alkene, alkyne, alcohol, ether, aldehyde, ketone, carboxylic acid, ester, amine, amide, functional group
a. an organic compound in which the hydrogen atom of a carboxyl group is replaced by a carbon atom
b. an organic compound that contains an oxygen atom bonded to two carbon atoms
c. a hydrocarbon that contains a carbon–carbon triple bond
d. a characteristic group of atoms that makes compounds behave and react in a particular way
e. an organic compound in which the carbonyl group is bonded to two carbon atoms

8.41 Identify the functional groups in each of the following:

a.   almonds

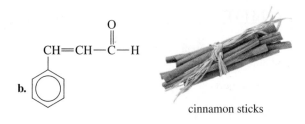

b.   cinnamon sticks

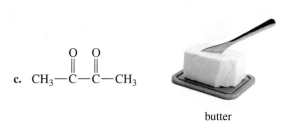

c. $CH_3—\overset{\overset{O}{\|}}{C}—\overset{\overset{O}{\|}}{C}—CH_3$ butter

8.42 Identify the functional groups in each of the following:
a. BHA, an antioxidant used as a preservative in foods such as baked goods, butter, meats, and snack foods

b. vanillin, a flavoring, obtained from the seeds of the vanilla bean

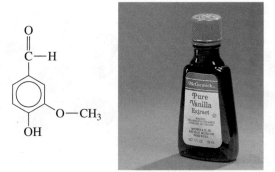

8.43 The sweetener aspartame is made from two amino acids: aspartic acid and phenylalanine. Identify the functional groups in aspartame.

8.44 Some aspirin substitutes contain phenacetin to reduce fever. Identify the functional groups in phenacetin.

ADDITIONAL QUESTIONS AND PROBLEMS

For instructor-assigned homework, go to
www.masteringchemistry.com.

8.45 Identify the type of reaction for each of the following as combination, decomposition, single replacement, double replacement, or combustion:

a. A metal and a nonmetal element form an ionic compound.

b. A hydrocarbon reacts with oxygen.

c. Two compounds react to produce two new compounds.

d. Heating calcium carbonate produces calcium oxide and carbon dioxide.

e. Zinc replaces copper in $Cu(NO_3)_2$.

8.46 Identify the type of reaction for each of the following as combination, decomposition, single replacement, double replacement, or combustion:

a. A compound breaks apart into its elements.

b. An element replaces the ion in a compound.

c. Copper and bromine form copper(II) bromide.

d. Iron(II) sulfite breaks down to iron(II) oxide and sulfur dioxide.

e. Silver ion from $AgNO_3(aq)$ forms a solid with bromide ion from $KBr(aq)$.

8.47 Balance each of the following unbalanced equations and identify the type of reaction:

a. $NH_3(g) + HCl(g) \longrightarrow NH_4Cl(s)$

b. $Fe_3O_4(s) + H_2(g) \longrightarrow Fe(s) + H_2O(g)$

c. $Sb(s) + Cl_2(g) \longrightarrow SbCl_3(s)$

d. $NI_3(s) \longrightarrow N_2(g) + I_2(g)$

e. $KBr(aq) + Cl_2(aq) \longrightarrow KCl(aq) + Br_2(l)$

f. $Fe(s) + H_2SO_4(aq) \longrightarrow Fe_2(SO_4)_3(aq) + H_2(g)$

g. $Al_2(SO_4)_3(aq) + NaOH(aq) \longrightarrow$
$$Na_2SO_4(aq) + Al(OH)_3(s)$$

8.48 Balance each of the following unbalanced equations and identify the type of reaction:

a. $Si_3N_4(s) \longrightarrow Si(s) + N_2(g)$

b. $Mg(s) + N_2(g) \longrightarrow Mg_3N_2(s)$

c. $Al(s) + H_2SO_4(aq) \longrightarrow Al_2(SO_4)_3(aq) + H_2(g)$

d. $Mg(s) + H_3PO_4(aq) \longrightarrow Mg_3(PO_4)_2(s) + H_2(g)$

e. $Cr_2O_3(s) + H_2(g) \longrightarrow Cr(s) + H_2O(g)$

f. $Al(s) + Cl_2(g) \longrightarrow AlCl_3(s)$

g. $MgCl_2(aq) + AgNO_3(aq) \longrightarrow Mg(NO_3)_2(aq) + AgCl(s)$

8.49 Predict the products and write a balanced equation for each of the following:

a. Single replacement: $Zn(s) + HCl(aq) \longrightarrow$
$$\underline{\hspace{1.5cm}} + \underline{\hspace{1.5cm}}$$

b. Decomposition: $BaCO_3(s) \xrightarrow{\Delta} \underline{\hspace{1.5cm}} + \underline{\hspace{1.5cm}}$

c. Double replacement: $NaOH(aq) + HCl(aq) \longrightarrow$
$$\underline{\hspace{1.5cm}} + \underline{\hspace{1.5cm}}$$

d. Combination: $Al(s) + F_2(g) \longrightarrow \underline{\hspace{1.5cm}}$

8.50 Predict the products and write a balanced equation for each of the following:

a. Decomposition: $NaCl(s) \xrightarrow{Electricity} \underline{\hspace{1.5cm}} + \underline{\hspace{1.5cm}}$

b. Combination: $Ca(s) + Br_2(g) \longrightarrow \underline{\hspace{1.5cm}}$

c. Combination: $SO_2(g) + O_2(g) \longrightarrow \underline{\hspace{1.5cm}}$

d. Double replacement: $NiCl_2(aq) + NaOH(aq) \longrightarrow$
$$\underline{\hspace{1.5cm}} + \underline{\hspace{1.5cm}}$$

8.51 Write a balanced equation for each of the following reactions and identify the type of reaction:

a. Sodium metal reacts with oxygen gas to form solid sodium oxide.

b. Aqueous sodium chloride and aqueous silver nitrate react to form solid silver chloride and aqueous sodium nitrate.

8.52 Write a balanced equation for each of the following reactions and identify the type of reaction:

a. Solid potassium chlorate is heated to form solid potassium chloride and oxygen gas.

b. Carbon monoxide gas and oxygen gas combine to form carbon dioxide gas.

CHALLENGE QUESTIONS

8.53 Write a balanced equation for each of the following reaction descriptions and identify each type of reaction:
  a. An aqueous solution of lead(II) nitrate is mixed with aqueous sodium phosphate to produce solid lead(II) phosphate and aqueous sodium nitrate.
  b. Gallium metal heated in oxygen gas forms solid gallium(III) oxide.
  c. When solid sodium nitrate is heated, solid sodium nitrite and oxygen gas are produced.

8.54 Write a balanced equation for each of the following reaction descriptions and identify each type of reaction:
  a. Solid bismuth(III) oxide and solid carbon react to form bismuth metal and carbon monoxide gas.
  b. Solid sodium chlorate is heated to form solid sodium chloride and oxygen gas.
  c. Butane gas (C_4H_{10}) reacts with oxygen gas to form two gaseous products: carbon dioxide and water.

8.55 In the following diagram, blue spheres are the element X and yellow spheres are the element Y:

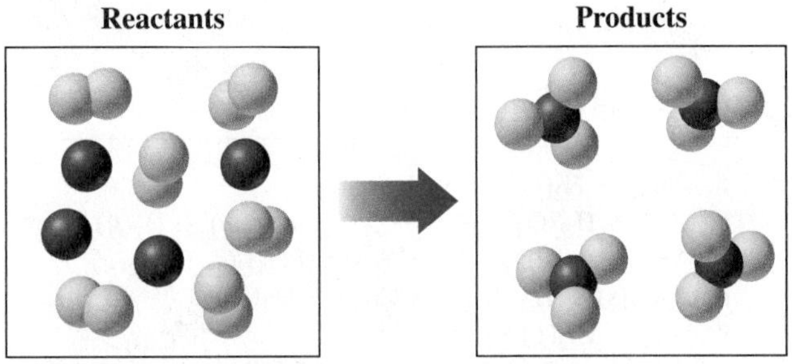

Reactants	Products

  a. Write the formulas of the reactants and the products.
  b. Write the balanced equation for the reaction.
  c. Identify the type of reaction.

8.56 In the following diagram, red spheres are the element A, white spheres are the element B, and green spheres are the element C:

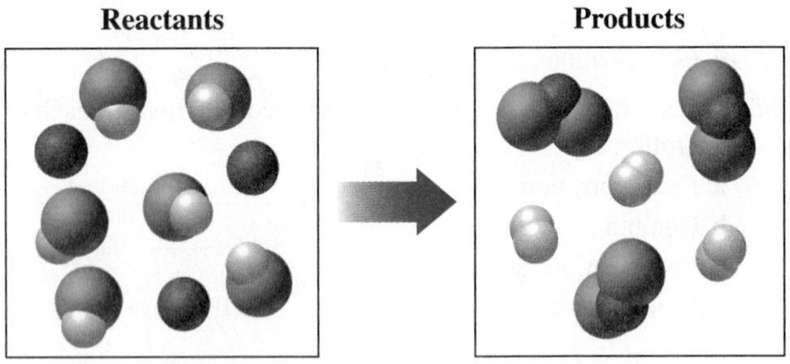

Reactants	Products

  a. Write the formulas of the reactants and the products.
  b. Write the balanced equation for the reaction.
  c. Identify the type of reaction.

8.57 When a compound containing C and H is burned in oxygen, the reaction produces 0.40 mol of CO_2 and 0.60 mol of H_2O.
  a. Determine the number of moles of carbon and moles of hydrogen in the compound.

  b. Determine the empirical formula of the compound.
  c. If one molecule of the compound contains 6 H atoms, what is the molecular formula?
  d. Write the balanced equation for the combustion reaction.

8.58 A copper wire with a mass of 4.32 g reacts with sulfur to form 5.41 g of a copper sulfide compound.
  a. Determine the number of moles of copper and sulfur in the compound.
  b. Determine the empirical formula of the product.
  c. If the actual formula of the sulfide compound contains 1 S, what is the molecular formula of the compound?
  d. Write the balanced equation for the combination reaction.

8.59 Toradol is used in dentistry to relieve pain. Name the functional groups in this molecule.

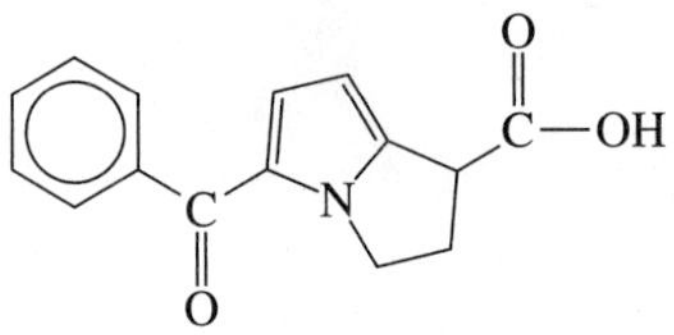

8.60 Voltaren is indicated for acute and chronic treatment of the symptoms of rheumatoid arthritis. Name the functional groups in this molecule.

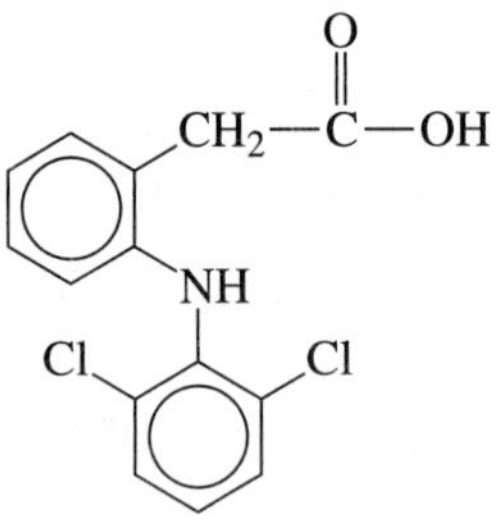

8.61 Identify the functional group in each of the following:
  a. $CH_3-CH_2-CH_2-OH$
  b. $CH_3-CH_2-CH_2-NH-CH_3$

  c. $CH_3-CH_2-\overset{\overset{\displaystyle O}{\|}}{C}-CH_2-CH_3$

  d. $CH_3-\overset{\overset{\displaystyle CH_3}{|}}{CH}-CH=CH-CH_3$

  e. $CH_3-CH_2-\overset{\overset{\displaystyle O}{\|}}{C}-OH$

8.62 Complete and balance each of the following reactions:

  a. $C_5H_{12} + O_2 \xrightarrow{\Delta}$
  b. $CH_3-C{\equiv}C-CH_3 + 2H_2 \xrightarrow{Ni}$
  c. $CH_3-CH_2-CH_2-CH_2-CH=CH_2 + O_2 \xrightarrow{\Delta}$

ANSWERS

ANSWERS TO STUDY CHECKS

8.1 There are 4 carbon atoms, 12 hydrogen atoms, and 14 oxygen atoms on the reactant side of the equation and on the product side.

8.2 $3Fe(s) + 2O_2(g) \longrightarrow Fe_3O_4(s)$

8.3 $N_2(g) + 2O_2(g) \longrightarrow 2NO_2(g)$ combination reaction

8.4 ether

8.5 A carboxylic acid contains a carboxyl group attached to H, while an ester contains a carboxyl group attached to C.

8.6 $C_2H_4(g) + 3O_2(g) \xrightarrow{\Delta} 2CO_2(g) + 2H_2O(g) + energy$

8.7 $CH_3-CH_2-CH_2-CH_3$

ANSWERS TO SELECTED QUESTIONS AND PROBLEMS

8.1 **a.** Reactant side: 2 N atoms, 4 O atoms
    Product side: 2 N atoms, 4 O atoms
   b. Reactant side: 5 C atoms, 2 S atoms, 4 O atoms
    Product side: 5 C atoms, 2 S atoms, 4 O atoms
   c. Reactant side: 4 C atoms, 4 H atoms, 10 O atoms
    Product side: 4 C atoms, 4 H atoms, 10 O atoms
   d. Reactant side: 2 N atoms, 8 H atoms, 4 O atoms
    Product side: 2 N atoms, 8 H atoms, 4 O atoms

8.3 **a.** not balanced   **b.** balanced
   c. not balanced   **d.** balanced

8.5 **a.** 2 Na atoms, 2 Cl atoms
   b. 1 P atom, 3 Cl atoms, 6 H atoms
   c. 4 P atoms, 16 O atoms, 12 H atoms

8.7 **a.** $N_2(g) + O_2(g) \longrightarrow 2NO(g)$
   b. $2HgO(s) \longrightarrow 2Hg(l) + O_2(g)$
   c. $4Fe(s) + 3O_2(g) \longrightarrow 2Fe_2O_3(s)$
   d. $2Na(s) + Cl_2(g) \longrightarrow 2NaCl(s)$
   e. $2Cu_2O(s) + O_2(g) \longrightarrow 4CuO(s)$

8.9 **a.** $Mg(s) + 2AgNO_3(aq) \longrightarrow Mg(NO_3)_2(aq) + 2Ag(s)$
   b. $CuCO_3(s) \longrightarrow CuO(s) + CO_2(g)$
   c. $2Al(s) + 3CuSO_4(aq) \longrightarrow 3Cu(s) + Al_2(SO_4)_3(aq)$
   d. $Pb(NO_3)_2(aq) + 2NaCl(aq) \longrightarrow PbCl_2(s) + 2NaNO_3(aq)$
   e. $2Al(s) + 6HCl(aq) \longrightarrow 2AlCl_3(aq) + 3H_2(g)$

8.11 **a.** $Fe_2O_3(s) + 3CO(g) \longrightarrow 2Fe(s) + 3CO_2(g)$
   b. $2Li_3N(s) \longrightarrow 6Li(s) + N_2(g)$
   c. $2Al(s) + 6HBr(aq) \longrightarrow 2AlBr_3(aq) + 3H_2(g)$
   d. $3Ba(OH)_2(aq) + 2Na_3PO_4(aq) \longrightarrow$
                $Ba_3(PO_4)_2(s) + 6NaOH(aq)$
   e. $As_4S_6(s) + 9O_2(g) \longrightarrow As_4O_6(s) + 6SO_2(g)$

8.13 **a.** $2Li(s) + 2H_2O(l) \longrightarrow H_2(g) + 2LiOH(aq)$
   b. $2P(s) + 5Cl_2(g) \longrightarrow 2PCl_5(s)$
   c. $FeO(s) + CO(g) \longrightarrow Fe(s) + CO_2(g)$
   d. $2C_5H_{10}(l) + 15O_2(g) \longrightarrow 10CO_2(g) + 10H_2O(g)$
   e. $3H_2S(g) + 2FeCl_3(s) \longrightarrow Fe_2S_3(s) + 6HCl(g)$

8.15 **a.** A single reactant splits into two simpler substances (elements).
   b. One element in the reacting compound is replaced by the other reactant.

8.17 **a.** combination   **b.** single replacement
   c. decomposition   **d.** double replacement
   e. decomposition   **f.** double replacement
   g. combination

8.19 **a.** $Mg(s) + Cl_2(g) \longrightarrow MgCl_2(s)$
   b. $2HBr(g) \longrightarrow H_2(g) + Br_2(g)$
   c. $Mg(s) + Zn(NO_3)_2(aq) \longrightarrow Mg(NO_3)_2(aq) + Zn(s)$
   d. $K_2S(aq) + Pb(NO_3)_2(aq) \longrightarrow PbS(s) + 2KNO_3(aq)$

8.21 **a.** alcohol   **b.** alkene
   c. aldehyde   **d.** ester

8.23 **a.** ether   **b.** alcohol
   c. ketone   **d.** carboxylic acid
   e. amine

8.25 **a.** $C_3H_8(g) + 5O_2(g) \xrightarrow{\Delta} 3CO_2(g) + 4H_2O(g) + energy$

b. $2C_8H_{18}(l) + 25O_2(g) \xrightarrow{\Delta} 16CO_2(g) + 18H_2O(g) + energy$

c. $2C_3H_8O(l) + 9O_2(g) \xrightarrow{\Delta} 6CO_2(g) + 8H_2O(g) + energy$

d. $C_6H_{12}(l) + 9O_2(g) \xrightarrow{\Delta} 6CO_2(g) + 6H_2O(g) + energy$

8.27 **a.** $CH_3-CH_2-CH_2-CH_2-CH_3$
   b. $CH_3-CH_2-CH_2-CH_3$
   c. $CH_3-CH_2-CH_2-CH_2-CH_2-CH_3$

8.29 **a.** $2NO(g) + O_2(g) \longrightarrow 2NO_2(g)$
   b. combination

8.31 **a.** $2NI_3(s) \longrightarrow N_2(g) + 3I_2(g)$
   b. decomposition

8.33 **a.** $2Cl_2(g) + O_2(g) \longrightarrow 2OCl_2(g)$
   b. combination

8.35 **a.** a. 1,1,2 combination reaction
   b. 2,2,1 decomposition reaction

8.37 **a.** carboxylic acid   **b.** alkene   **c.** ester
   d. amine   **e.** aldehyde

8.39 **a.** alcohol   **b.** alkene   **c.** aldehyde
   d. alkane   **e.** carboxylic acid

8.41 **a.** aromatic, aldehyde   **b.** aromatic, aldehyde, alkene
   c. ketone

8.43 carboxylic acid, aromatic, amine, amide, ester

8.45 **a.** combination   **b.** combustion
   c. double replacement   **d.** decomposition
   e. single replacement

8.47 **a.** $NH_3(g) + HCl(g) \longrightarrow NH_4Cl(s)$ combination
   b. $Fe_3O_4(s) + 4H_2(g) \longrightarrow$
             $3Fe(s) + 4H_2O(g)$ single replacement
   c. $2Sb(s) + 3Cl_2(g) \longrightarrow 2SbCl_3(s)$ combination
   d. $2NI_3(s) \longrightarrow N_2(g) + 3I_2(g)$ decomposition
   e. $2KBr(aq) + Cl_2(aq) \longrightarrow$
             $2KCl(aq) + Br_2(l)$ single replacement
   f. $2Fe(s) + 3H_2SO_4(aq) \longrightarrow$
          $Fe_2(SO_4)_3(aq) + 3H_2(g)$ single replacement
   g. $Al_2(SO_4)_3(aq) + 6NaOH(aq) \longrightarrow$
        $3Na_2SO_4(aq) + 2Al(OH)_3(s)$ double replacement

8.49 **a.** $Zn(s) + 2HCl(aq) \longrightarrow ZnCl_2(aq) + H_2(g)$
   b. $BaCO_3(s) \xrightarrow{\Delta} BaO(s) + CO_2(g)$
   c. $NaOH(aq) + HCl(aq) \longrightarrow NaCl(aq) + H_2O(l)$
   d. $2Al(s) + 3F_2(g) \longrightarrow 2AlF_3(s)$

8.51 **a.** $4Na(s) + O_2(g) \longrightarrow 2Na_2O(s)$ combination
   b. $NaCl(aq) + AgNO_3(aq) \longrightarrow$
          $AgCl(s) + NaNO_3(aq)$ double replacement

8.53 **a.** $3Pb(NO_3)_2(aq) + 2Na_3PO_4(aq) \longrightarrow$
         $Pb_3(PO_4)_2(s) + 6NaNO_3(aq)$ double replacement
   b. $4Ga(s) + 3O_2(g) \xrightarrow{\Delta} 2Ga_2O_3(s)$ combination
   c. $2NaNO_3(s) \xrightarrow{\Delta} 2NaNO_2(s) + O_2(g)$ decomposition

8.55 **a.** Reactants: X and Y_2; Products: XY_3
   b. $2X + 3Y_2 \longrightarrow 2XY_3$
   c. combination

8.57 **a.** 0.40 mol of carbon and 1.2 mol of H, hydrogen
   b. The ratio is 0.40 mol of C : 1.2 mol of H, which gives the empirical formula CH_3.
   c. C_2H_6
   d. $2C_2H_6(g) + 7O_2(g) \xrightarrow{\Delta} 4CO_2(g) + 6H_2O(g) + energy$

8.59 aromatic, alkene, ketone, amine, carboxylic acid

8.61 **a.** alcohol   **b.** amine   **c.** ketone
   d. alkene   **e.** carboxylic acid

9 Chemical Quantities in Reactions

"In our food science laboratory, I develop a variety of food products, from cake donuts to energy beverages," says Anne Cristofano, senior food technologist at Mattson & Company. *"When I started the donut project, I researched the ingredients and then weighed them out in the lab. I added water to make a batter and cooked the donuts in a fryer. The batter and the oil temperature make a big difference. If I don't get the right taste or texture, I adjust the ingredients, such as sugar and flour, or adjust the temperature."* A food technologist studies the physical and chemical properties of food and develops scientific ways to process and preserve it for extended shelf life. The food products are tested for texture, color, and flavor. The results of these tests help improve the quality and safety of food.

Visit **www.masteringchemistry.com** for self-study help and instructor-assigned homework.

When we know the balanced chemical equation for a reaction, we can determine the mole and mass relationships between the reactants and products. Then we use molar masses to calculate the quantities of substances used or produced in a particular reaction. We do much the same thing at home when we use a recipe to make a cake or add the right quantity of water to make soup. In the manufacturing of chemical compounds, side reactions decrease the percent of product obtained. From the actual amount of product, we can determine the percent yield for a reaction. Knowing how to determine the quantitative results of a chemical reaction is essential to chemists, engineers, pharmacists, respiratory therapists, and other scientists and health professionals.

We will also look at energy changes in a balanced chemical equation. A reaction is exothermic if it produces heat along with the other products. A reaction is endothermic when it requires heat for the reaction. In the body, chemical reactions require energy when we need to produce large molecules such as protein and glycogen from smaller molecules. When chemical reactions in the body break down large molecules, energy is produced, which is stored in high-energy compounds.

9.1 Mole Relationships in Chemical Equations

In Chapter 8, we saw that equations are balanced in terms of the numbers of each type of atom in the reactants and products. However, when experiments are done in the laboratory or a medication is prepared in the pharmacy, samples contain so many atoms and molecules that it is usually impractical to count them individually. What can be measured conveniently is mass, using a balance. Because mass is related to the number of particles through the molar mass, measuring the mass is equivalent to counting the number of particles or moles.

Conservation of Mass

In any chemical reaction, the total amount of matter in the reactants is equal to the total amount of matter in the products. If all of the reactants were weighed, they would have a total mass equal to the total mass of the products. This is known as the **law of conservation of mass**, which says that there is no change in the total mass of the substances reacting in a balanced chemical reaction. Thus, no material is lost or gained as original substances are changed to new substances.

For example, tarnish forms when silver reacts with sulfur to form silver sulfide.

$$2Ag(s) + S(s) \longrightarrow Ag_2S(s)$$

In this reaction, the number of silver atoms that reacts is two times the number of sulfur atoms. When 200 silver atoms react, 100 sulfur atoms are required. However, many more atoms would actually be present in this reaction. If we are dealing with molar amounts, then the coefficients in the equation can be interpreted in terms of moles. Thus, 2 mol of silver react with each 1 mol of sulfur. Since the molar mass of each can be determined, the quantities of Ag and S can also be stated in terms of mass in grams of each. Thus, 215.8 g of Ag and 32.1 g of S react to form 247.9 g of Ag_2S. The total mass of the reactants (247.9 g) is equal to the 247.9 g of product. The various ways in which a chemical equation can be interpreted are seen in Table 9.1.

LEARNING GOAL

Given a quantity in moles of reactant or product, use a mole–mole factor from the balanced equation to calculate the number of moles of another substance in the reaction.

TUTORIAL
Stoichiometry

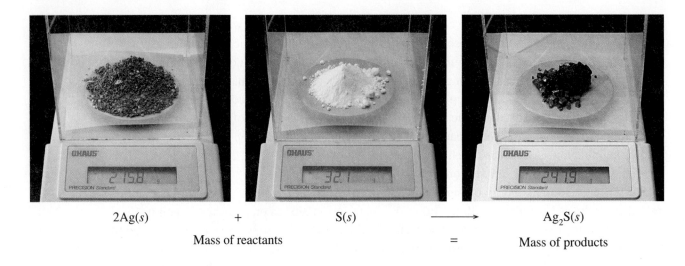

$2Ag(s)$ + $S(s)$ $\longrightarrow$ $Ag_2S(s)$

Mass of reactants   =   Mass of products

TABLE 9.1 Information Available from a Balanced Equation

	Reactants		Products
Equation	**2Ag(s)**	**+ S(s)**	$\longrightarrow$ **Ag$_2$S(s)**
Atoms	2 Ag atoms	+ 1 S atom	$\longrightarrow$ 1 Ag$_2$S formula unit
	200 Ag atoms	+ 100 S atoms	$\longrightarrow$ 100 Ag$_2$S formula units
Avogadro's number of atoms	$2(6.022 \times 10^{23})$ Ag atoms	$+ 1(6.022 \times 10^{23})$ S atoms	$\longrightarrow 1(6.022 \times 10^{23})$ Ag$_2$S formula units
Moles	2 mol of Ag	+ 1 mol of S	$\longrightarrow$ 1 mol of Ag$_2$S
Mass (g)	2(107.9 g) of Ag	+ 1(32.1 g) of S	$\longrightarrow$ 1(247.9 g) of Ag$_2$S
Total mass (g)	247.9 g		$\longrightarrow$ 247.9 g

TUTORIAL
Law of Conservation of Mass

CONCEPT CHECK 9.1

■ **Conservation of Mass**

The combustion of the alkane CH_4 with oxygen produces carbon dioxide and water. Calculate the total mass of reactants and products for the following equation when 1 mol of CH_4 reacts:

$$CH_4(g) + 2O_2(g) \longrightarrow CO_2(g) + 2H_2O(g)$$

ANSWER

Interpreting the coefficients in the equation as the number of moles of each substance and multiplying by the respective molar masses gives the total mass of reactants and products. The quantities of moles are exact because the coefficients in the balanced equation are exact.

	Reactants	Products
Equation	$CH_4(g) + 2O_2(g)$	$\longrightarrow CO_2(g) + 2H_2O(g)$
Moles	1 mol of CH_4 + 2 mol of O_2	$\longrightarrow$ 1 mol of CO_2 + 2 mol of H_2O
Mass	16.04 g of CH_4 + 64.00 g of O_2	$\longrightarrow$ 44.01 g of CO_2 + 36.03 g of H_2O
Total mass	**80.04 g of reactants**	**=** **80.04 g of products**

Mole–Mole Factors in an Equation

When iron reacts with sulfur, the product is iron(III) sulfide.

$$2Fe(s) + 3S(s) \longrightarrow Fe_2S_3(s)$$

Because the equation is balanced, we know the proportions of iron and sulfur in the reaction. For this reaction, we see that when 2 mol of iron react, 3 mol of sulfur will also react, forming 1 mol of iron(III) sulfide. Actually, any amount of iron or sulfur may be used, but the *ratio* of iron reacting with sulfur will be the same. From the coefficients, we can write **mole–mole factors** between reactants and between reactants and products. The coefficients used in the mole–mole factors are exact numbers; they do not limit the number of significant figures.

Iron (Fe)   Sulfur (S)   Iron(III) sulfide (Fe_2S_3)
$2Fe(s)$ $+$ $3S(s)$ $\longrightarrow$ $Fe_2S_3(s)$

Fe and S: $\dfrac{2 \text{ mol Fe}}{3 \text{ mol S}}$ and $\dfrac{3 \text{ mol S}}{2 \text{ mol Fe}}$

Fe and Fe_2S_3: $\dfrac{2 \text{ mol Fe}}{1 \text{ mol } Fe_2S_3}$ and $\dfrac{1 \text{ mol } Fe_2S_3}{2 \text{ mol Fe}}$

S and Fe_2S_3: $\dfrac{3 \text{ mol S}}{1 \text{ mol } Fe_2S_3}$ and $\dfrac{1 \text{ mol } Fe_2S_3}{3 \text{ mol S}}$

CONCEPT CHECK 9.2

■ Writing Mole–Mole Factors

Consider the following balanced equation:

$$4Na(s) + O_2(g) \longrightarrow 2Na_2O(s)$$

Write the mole–mole factors for

a. Na and O_2 **b.** Na and Na_2O

ANSWER

a. The mole–mole factors for Na and O_2 use the coefficient of Na to write 4 mol of Na, and the coefficient of 1 (understood) to write 1 mol of O_2. Two mole–mole factors can be written:

4 mol of Na = 1 mol of O_2

$\dfrac{4 \text{ mol Na}}{1 \text{ mol } O_2}$ and $\dfrac{1 \text{ mol } O_2}{4 \text{ mol Na}}$

b. The mole–mole factors for Na and Na_2O use the coefficient of Na to write 4 mol of Na, and the coefficient of Na_2O to write 2 mol of Na_2O. Two mole–mole factors can be written:

$$4 \text{ mol of Na } = 2 \text{ mol of } Na_2O$$

$$\frac{4 \text{ mol Na}}{2 \text{ mol } Na_2O} \quad \text{and} \quad \frac{2 \text{ mol } Na_2O}{4 \text{ mol Na}}$$

Using Mole–Mole Factors in Calculations

Whenever you use a recipe, adjust an engine for the proper mixture of fuel and air, or prepare medicines in a pharmaceutical laboratory, you need to know the proper amounts of reactants to use and how much of the product will form. Earlier, we wrote all the possible conversion factors that can be obtained from the balanced equation $2Fe(s) + 3S(s) \longrightarrow Fe_2S_3(s)$. Now we will show how mole–mole factors are used in chemical calculations.

SAMPLE PROBLEM | 9.1

■ Calculating Moles of a Reactant

In the reaction of iron and sulfur, how many moles of sulfur are needed to react with 1.42 mol of iron?

$$2Fe(s) + 3S(s) \longrightarrow Fe_2S_3(s)$$

SOLUTION

Guide to Using Mole–Mole Factors

STEP 1
Write the given and needed number of moles.

STEP 2
Write a plan to convert the given to the needed moles.

STEP 3
Use coefficients to write relationships and mole–mole factors.

STEP 4
Set up problem using the mole–mole factor that cancels given moles.

STEP 1 Write the given and needed number of moles.   In this problem, we need to find the number of moles of S that react with 1.42 mol of Fe.
Given   1.42 mol of Fe   **Need**   moles of S

STEP 2 Write a plan to convert the given to the needed moles.

$$\text{moles of Fe} \quad \boxed{\begin{array}{c} \text{Mole–mole} \\ \text{factor} \end{array}} \quad \text{moles of S}$$

STEP 3 Use coefficients to write relationships and mole–mole factors.   Use coefficients to write the mole–mole factors for the given and needed substances.

$$2 \text{ mol of Fe } = 3 \text{ mol of S}$$

$$\frac{2 \text{ mol Fe}}{3 \text{ mol S}} \quad \text{and} \quad \frac{3 \text{ mol S}}{2 \text{ mol Fe}}$$

STEP 4 Set up problem using the mole–mole factor that cancels given moles.   Use a mole–mole factor to cancel the given moles and provide needed moles.

$$1.42 \text{ mol Fe} \times \frac{3 \text{ mol S}}{2 \text{ mol Fe}} = 2.13 \text{ mol of S}$$

The answer is given with three significant figures because the given quantity of 1.42 mol of Fe has three SFs. The coefficients used in the mole–mole factor are exact.

STUDY CHECK

Using the equation in Sample Problem 9.1, calculate the number of moles of iron needed to react with 2.75 mol of sulfur.

SAMPLE PROBLEM | 9.2

■ **Calculating Moles of a Product**

In a combustion reaction, propane (C_3H_8) reacts with oxygen. How many moles of CO_2 can be produced when 2.25 mol of C_3H_8 react?

$$C_3H_8(g) + 5O_2(g) \longrightarrow 3CO_2(g) + 4H_2O(g)$$

(MC)

TUTORIAL

Moles of Reactants and Products

SOLUTION

STEP 1 Write the given and needed moles.   In this problem, the given is 2.25 mol of C_3H_8. We need to find the number of moles of CO_2 that can be produced.
Given   2.25 mol of C_3H_8 **Need**   moles of CO_2

STEP 2 Write a plan to convert the given to the needed moles.

moles of C_3H_8 **Mole–mole factor**   moles of CO_2

STEP 3 Use coefficients to write relationships and mole–mole factors.   Use coefficients to write the mole–mole factors for the given and needed substances.

$$1 \text{ mol of } C_3H_8 = 3 \text{ mol of } CO_2$$

$$\frac{1 \text{ mol } C_3H_8}{3 \text{ mol } CO_2} \quad \text{and} \quad \frac{3 \text{ mol } CO_2}{1 \text{ mol } C_3H_8}$$

STEP 4 Set up problem using the mole–mole factor that cancels given moles.   Use a mole–mole factor to cancel the given moles and provide needed moles.

$$2.25 \text{ mol } C_3H_8 \times \frac{3 \text{ mol } CO_2}{1 \text{ mol } C_3H_8} = 6.75 \text{ mol of } CO_2$$

The answer is given with three significant figures because the given quantity of 2.25 mol of C_3H_8 has three SFs. The coefficients used in the mole–mole factor are exact.

STUDY CHECK

Using the equation in Sample Problem 9.2, calculate the moles of oxygen that must react to produce 0.756 mol of water.

QUESTIONS AND PROBLEMS

Mole Relationships in Chemical Equations

9.1 Describe each of the following equations in terms of (1) number of particles and (2) number of moles:
a. $2SO_2(g) + O_2(g) \longrightarrow 2SO_3(g)$
b. $4P(s) + 5O_2(g) \longrightarrow 2P_2O_5(s)$

9.2 Describe each of the following equations in terms of (1) number of particles and (2) number of moles:
a. $2Al(s) + 3Cl_2(g) \longrightarrow 2AlCl_3(s)$
b. $4HCl(g) + O_2(g) \longrightarrow 2Cl_2(g) + 2H_2O(g)$

9.3 Calculate the total masses of the reactants and the products in each of the equations of Problem 9.1.

9.4 Calculate the total masses of the reactants and the products in each of the equations of Problem 9.2.

9.5 Write all of the mole–mole factors for the equations listed in Problem 9.1.

9.6 Write all of the mole–mole factors for the equations listed in Problem 9.2.

9.7 The reaction of hydrogen with oxygen produces water.

$$2H_2(g) + O_2(g) \longrightarrow 2H_2O(g)$$

a. How many moles of O_2 are required to react with 2.0 mol of H_2?
b. How many moles of H_2 are needed to react with 5.0 mol of O_2?
c. How many moles of H_2O form when 2.5 mol of O_2 reacts?

9.8 Ammonia is produced by the reaction of hydrogen and nitrogen.

$$N_2(g) + 3H_2(g) \longrightarrow 2NH_3(g)$$
Ammonia

a. How many moles of H_2 are needed to react with 1.0 mol of N_2?
b. How many moles of N_2 reacted if 0.60 mol of NH_3 is produced?
c. How many moles of NH_3 are produced when 1.4 mol of H_2 reacts?

9.9 Carbon disulfide and carbon monoxide are produced when carbon is heated with sulfur dioxide.

$$5C(s) + 2SO_2(g) \longrightarrow CS_2(l) + 4CO(g)$$

a. How many moles of C are needed to react with 0.500 mol of SO_2?
b. How many moles of CO are produced when 1.2 mol of C reacts?
c. How many moles of SO_2 are required to produce 0.50 mol of CS_2?
d. How many moles of CS_2 are produced when 2.5 mol of C reacts?

9.10 In the acetylene torch, acetylene gas (C_2H_2) burns in oxygen to produce carbon dioxide and water.

$$2C_2H_2(g) + 5O_2(g) \longrightarrow 4CO_2(g) + 2H_2O(g)$$

a. How many moles of O_2 are needed to react with 2.00 mol of C_2H_2?
b. How many moles of CO_2 are produced when 3.5 mol of C_2H_2 reacts?
c. How many moles of C_2H_2 are required to produce 0.50 mol of H_2O?
d. How many moles of CO_2 are produced from 0.100 mol of O_2?

9.2 Mass Calculations for Reactions

LEARNING GOAL

Given the mass in grams of a substance in a reaction, calculate the mass in grams of another substance in the reaction.

When you perform a chemistry experiment in the laboratory, you use a laboratory balance to measure out a certain mass of the reactant. From the mass in grams, you can determine the number of moles of reactant. By using mole–mole factors, you can predict the moles of product that can be produced. Then the molar mass of the product is used to convert the moles back into mass in grams. The process in Concept Check 9.3 can be used to set up and solve problems that involve calculations of quantities for substances in chemical reactions.

MC

TUTORIAL
Masses of Reactants and Products

CONCEPT CHECK 9.3

■ **Mass of Product from Mass of Reactant**

In the engines of cars and trucks, nitrogen and oxygen from the air react at high temperature to produce nitrogen oxide, a component of smog. Complete the following to help answer the question: Calculate the grams of NO produced when 2.15 mol of O_2 reacts.

$$N_2(g) + O_2(g) \longrightarrow 2NO(g)$$

a. What mole–mole equality will be needed in the calculation?
b. What are the mole–mole factors from the equality you wrote in **a**?
c. What equality and molar mass factor will be needed to complete the calculation?

ANSWER

a. The equality for the mole–mole relationship between O_2 and NO uses the coefficients from the balanced equation:

$$1 \text{ mol of } O_2 = 2 \text{ mol of NO}$$

b. From the equality, two mole–mole factors can be written:

$$\frac{2 \text{ mol NO}}{1 \text{ mol } O_2} \quad \text{and} \quad \frac{1 \text{ mol } O_2}{2 \text{ mol NO}}$$

c. To convert the number of moles of NO to grams of NO, we need to write the equality and molar mass factors for NO:

$$1 \text{ mol of NO} = 30.01 \text{ g of NO}$$

$$\frac{30.01 \text{ g NO}}{1 \text{ mol NO}} \quad \text{and} \quad \frac{1 \text{ mol NO}}{30.01 \text{ g NO}}$$

SAMPLE PROBLEM | 9.3

■ Mass of Product from Mass of Reactant

$$2C_2H_2(g) + 5O_2(g) \longrightarrow 4CO_2(g) + 2H_2O(g)$$

How many grams of carbon dioxide are produced when 54.6 g of C_2H_2 is burned?

SOLUTION

STEP 1   Use molar mass to convert grams of given to moles.
   Given   54.6 g of C_2H_2 **Need**   grams of CO_2

$$1 \text{ mol of } C_2H_2 = 26.04 \text{ g of } C_2H_2$$

$$\frac{26.04 \text{ g } C_2H_2}{1 \text{ mol } C_2H_2} \quad \text{and} \quad \frac{1 \text{ mol } C_2H_2}{26.04 \text{ g } C_2H_2}$$

$$54.6 \text{ g } \cancel{C_2H_2} \times \frac{1 \text{ mol } C_2H_2}{26.04 \text{ g } \cancel{C_2H_2}} = 2.10 \text{ mol of } C_2H_2$$

STEP 2   Write a mole–mole factor from the coefficients in the equation.

$$2 \text{ mol of } C_2H_2 = 4 \text{ mol of } CO_2$$

$$\frac{2 \text{ mol } C_2H_2}{4 \text{ mol } CO_2} \quad \text{and} \quad \frac{4 \text{ mol } CO_2}{2 \text{ mol } C_2H_2}$$

STEP 3   Convert moles of given to moles of needed substance using the mole–mole factor.   Select the appropriate mole–mole factor that will cancel the moles of C_2H_2 and give the moles of CO_2.

$$2.10 \text{ } \cancel{\text{mol } C_2H_2} \times \frac{4 \text{ mol } CO_2}{2 \text{ } \cancel{\text{mol } C_2H_2}} = 4.20 \text{ mol of } CO_2$$

STEP 4   Convert moles of needed substance to grams using molar mass.

$$1 \text{ mol of } CO_2 = 44.01 \text{ g of } CO_2$$

$$\frac{44.01 \text{ g } CO_2}{1 \text{ mol } CO_2} \quad \text{and} \quad \frac{1 \text{ mol } CO_2}{44.01 \text{ g } CO_2}$$

$$4.20 \text{ } \cancel{\text{mol } CO_2} \times \frac{44.01 \text{ g } CO_2}{1 \text{ } \cancel{\text{mol } CO_2}} = 185 \text{ g of } CO_2$$

The solution is also obtained by placing the conversion factors in sequence:

$$54.6 \text{ g } \cancel{C_2H_2} \times \frac{1 \text{ } \cancel{\text{mol } C_2H_2}}{26.04 \text{ g } \cancel{C_2H_2}} \times \frac{4 \text{ } \cancel{\text{mol } CO_2}}{2 \text{ } \cancel{\text{mol } C_2H_2}} \times \frac{44.01 \text{ g } CO_2}{1 \text{ } \cancel{\text{mol } CO_2}} = 185 \text{ g of } CO_2$$

STUDY CHECK

Using the equation in Sample Problem 9.3, calculate the grams of CO_2 that can be produced when 25.0 g of O_2 reacts.

In a combustion reaction, acetylene (C_2H_2) burns with oxygen to produce carbon dioxide and water.

Guide to Calculating the Masses of Reactants and Products for a Chemical Reaction

> **STEP 1**
> Use molar mass to convert grams of given to moles (if necessary).

> **STEP 2**
> Write a mole–mole factor from the coefficients in the equation.

> **STEP 3**
> Convert moles of given to moles of needed substance using mole–mole factor.

> **STEP 4**
> Convert moles of needed substance to grams using molar mass.

SAMPLE PROBLEM | 9.4

■ Calculating the Mass of Reactant

The alkane heptane, C_7H_{16}, is designated as the zero point in the octane rating of gasoline. Heptane is an undesirable compound in gasoline because it burns rapidly and causes engine knocking. How many grams of O_2 are required to react with 22.50 g of C_7H_{16}?

$$C_7H_{16}(g) + 11O_2(g) \longrightarrow 7CO_2(g) + 8H_2O(g)$$

SOLUTION

STEP 1   Use molar mass to convert grams of given to moles.

Given   22.50 g of C_7H_{16} **Need**   grams of O_2

$$1 \text{ mol of } C_7H_{16} = 100.20 \text{ g of } C_7H_{16}$$

$$\frac{100.20 \text{ g } C_7H_{16}}{1 \text{ mol } C_7H_{16}} \quad \text{and} \quad \frac{1 \text{ mol } C_7H_{16}}{100.20 \text{ g } C_7H_{16}}$$

$$22.50 \text{ g } C_7H_{16} \times \frac{1 \text{ mol } C_7H_{16}}{100.20 \text{ g } C_7H_{16}} = 0.2246 \text{ mol of } C_7H_{16}$$

STEP 2   Write a mole–mole factor from the coefficients in the equation.

$$1 \text{ mol of } C_7H_{16} = 11 \text{ mol of } O_2$$

$$\frac{1 \text{ mol } C_7H_{16}}{11 \text{ mol } O_2} \quad \text{and} \quad \frac{11 \text{ mol } O_2}{1 \text{ mol } C_7H_{16}}$$

STEP 3   Convert moles of given to moles of needed substance using the mole–mole factor.   Select the appropriate mole–mole factor that converts the number of moles of C_7H_{16} to the number of moles of O_2.

$$0.2246 \text{ mol } C_7H_{16} \times \frac{11 \text{ mol } O}{1 \text{ mol } C_7H_{16}} = 2.471 \text{ mol of } O_2$$

STEP 4   Convert moles of needed substance to grams using molar mass.

$$1 \text{ mol of } O_2 = 32.00 \text{ g of } O_2$$

$$\frac{32.00 \text{ g } O_2}{1 \text{ mol } O_2} \quad \text{and} \quad \frac{1 \text{ mol } O_2}{32.00 \text{ g } O_2}$$

$$2.471 \text{ mol } O_2 \times \frac{32.00 \text{ g } O_2}{1 \text{ mol } O_2} = 79.07 \text{ g of } O_2$$

The solution can also be obtained by placing the conversion factors in sequence.

$$22.50 \text{ g } C_7H_{16} \times \frac{1 \text{ mol } C_7H_{16}}{100.20 \text{ g } C_7H_{16}} \times \frac{11 \text{ mol } O_2}{1 \text{ mol } C_7H_{16}} \times \frac{32.00 \text{ g } O_2}{1 \text{ mol } O_2} = 79.04 \text{ g of } O_2$$

When a calculation is set up as a sequence, the calculations lead to a final numerical answer without writing out the intermediate values. This accounts for the slight differences in the final digit due to rounding off intermediate values.

STUDY CHECK

Using the equation in Sample Problem 9.4, calculate the grams of C_7H_{16} that are needed to produce 15.0 g of H_2O.

QUESTIONS AND PROBLEMS

Mass Calculations for Reactions

9.11 Sodium reacts with oxygen to produce sodium oxide.

$$4Na(s) + O_2(g) \longrightarrow 2Na_2O\ (s)$$

a. How many grams of Na_2O are produced when 2.50 mol of Na react?

b. If you have 18.0 g of Na, how many grams of O_2 are required for reaction?

c. How many grams of O_2 are needed in a reaction that produces 75.0 g of Na_2O?

9.12 Nitrogen gas reacts with hydrogen gas to produce ammonia by the following equation:

$$N_2(g) + 3H_2(g) \longrightarrow 2NH_3(g)$$

a. If you have 1.80 mol of H_2, how many grams of NH_3 can be produced?

b. How many grams of H_2 are needed to react with 2.80 g of N_2?

c. How many grams of NH_3 can be produced from 12.0 g of H_2?

9.13 Ammonia and oxygen react to form nitrogen and water.

$$4NH_3(g) + 3O_2(g) \longrightarrow 2N_2(g) + 6H_2O(g)$$
Ammonia

   a. How many grams of O_2 are needed to react with 8.00 mol of NH_3?

   b. How many grams of N_2 can be produced when 6.50 g of O_2 reacts?

   c. How many grams of water are formed from the reaction of 34.0 g of NH_3?

9.14 Iron(III) oxide reacts with carbon to give iron and carbon monoxide.

$$Fe_2O_3(s) + 3C(s) \longrightarrow 2Fe(s) + 3CO(g)$$

   a. How many grams of C are required to react with 2.50 mol of Fe_2O_3?

   b. How many grams of CO are produced when 36.0 g of C reacts?

   c. How many grams of Fe can be produced when 6.00 g of Fe_2O_3 reacts?

9.15 Nitrogen dioxide and water react to produce nitric acid, HNO_3, and nitrogen oxide.

$$3NO_2(g) + H_2O(l) \longrightarrow 2HNO_3(aq) + NO(g)$$

   a. How many grams of H_2O are required to react with 28.0 g of NO_2?

   b. How many grams of NO are obtained from 15.8 g of NO_2?

   c. How many grams of HNO_3 are produced from 8.25 g of NO_2?

9.16 Calcium cyanamide, $CaCN_2$, reacts with water to form calcium carbonate and ammonia.

$$CaCN_2(s) + 3H_2O(l) \longrightarrow CaCO_3(s) + 2NH_3(g)$$

   a. How many grams of water are needed to react with 75.0 g of $CaCN_2$?

   b. How many grams of NH_3 are produced from 5.24 g of $CaCN_2$?

   c. How many grams of $CaCO_3$ form if 155 g of water reacts?

9.17 When the ore lead(II) sulfide burns in oxygen, the products are lead(II) oxide and sulfur dioxide gas.

   a. Write the balanced equation for the reaction.

   b. How many grams of oxygen are required to react with 0.125 mol of lead(II) sulfide?

   c. How many grams of sulfur dioxide can be produced when 65.0 g of lead(II) sulfide reacts?

   d. How many grams of lead(II) sulfide are used to produce 128 g of lead(II) oxide?

9.18 When the gases dihydrogen sulfide and oxygen react, they form the gases sulfur dioxide and water vapor.

   a. Write the balanced equation for the reaction.

   b. How many grams of oxygen are required to react with 2.50 g of dihydrogen sulfide?

   c. How many grams of sulfur dioxide can be produced when 38.5 g of oxygen reacts?

   d. How many grams of oxygen are required to produce 55.8 g of water vapor?

9.3 Limiting Reactants

When you make peanut butter sandwiches for lunch, you need 2 slices of bread and 1 tablespoon of peanut butter for each sandwich. As an equation, we could write:

2 slices of bread + 1 tablespoon peanut butter $\longrightarrow$ 1 peanut butter sandwich

If you have 8 slices of bread and a full jar of peanut butter, you will run out of bread after you make 4 peanut butter sandwiches. You cannot make any more sandwiches once the bread is used up, even though there is a lot of peanut butter left in the jar. The number of slices of bread has limited the number of sandwiches you can make. On a different day, you might have 8 slices of bread but only a tablespoon of peanut butter left in the peanut butter jar. You will run out of peanut butter after you make just 1 peanut butter sandwich with 6 slices of bread left over. The small amount of peanut butter available has limited the number of sandwiches you can make.

LEARNING GOAL

Identify a limiting reactant when given the quantities of two reactants; calculate the amount of product formed from the limiting reactant.

TUTORIAL

Limiting Reactant and Yield: Mole Calculations

 +   +

8 slices of bread + 1 jar of peanut butter make 4 peanut butter sandwiches + peanut butter left over

 +  +

8 slices of bread + 1 tablespoon of peanut butter make 1 peanut butter sandwich + 6 slices of bread left over

In a similar way, the availability of reactants in a chemical reaction can limit the amount of product that forms. In many reactions, the reactants are not combined in quantities that allow each to be used up at exactly the same time. Then one reactant, called the **limiting reactant**, is used up before the other. The other reactant, called the **excess reactant**, is left over.

TUTORIAL

What Will Run Out First?

Bread	Peanut Butter	Sandwiches	Limiting Reactant	Excess Reactant
1 Loaf (20 slices)	1 Tablespoon	1	Peanut butter	Bread
4 Slices	1 Full jar	2	Bread	Peanut butter
8 Slices	1 Full jar	4	Bread	Peanut butter

CONCEPT CHECK 9.4

■ Limiting Reactants

For a picnic that you are planning, you have 10 spoons, 8 forks, and 6 knives. If each person you invite requires 1 spoon, 1 fork, and 1 knife, how many people can you ask to your picnic?

ANSWER

The relationship of utensils required by each person can be written:

1 person = 1 spoon, 1 fork, and 1 knife

The maximum number of people for each utensil can be determined as follows:

$$10 \text{ spoons} \times \frac{1 \text{ person}}{1 \text{ spoon}} = 10 \text{ people}$$

$$8 \text{ forks} \times \frac{1 \text{ person}}{1 \text{ fork}} = 8 \text{ people}$$

$$6 \text{ knives} \times \frac{1 \text{ person}}{1 \text{ knife}} = 6 \text{ people (smallest number of people)}$$

The limiting utensil is six knives, which means that six people (including yourself) can be at your picnic.

Calculating Moles of Product from a Limiting Reactant

Consider the reaction in which hydrogen and chlorine form hydrogen chloride:

$$H_2(g) + Cl_2(g) \longrightarrow 2HCl(g)$$

Suppose the reaction mixture contains 2 mol of H_2 and 5 mol of Cl_2. From the equation, we see that 1 mol of hydrogen reacts with 1 mol of chlorine to produce 2 mol of hydrogen chloride. Now we need to calculate the amount of product that is possible from each of the reactants. We are looking for the limiting reactant, which is the one that produces the smaller amount of product.

The mole–mole factors from the equation are written as:

$$2 \text{ mol of HCl} = 1 \text{ mol of } H_2 \qquad 2 \text{ mol of HCl} = 1 \text{ mol of } Cl_2$$

$$\frac{2 \text{ mol HCl}}{1 \text{ mol } H_2} \text{ and } \frac{1 \text{ mol } H_2}{2 \text{ mol HCl}} \qquad \frac{2 \text{ mol HCl}}{1 \text{ mol } Cl_2} \text{ and } \frac{1 \text{ mol } Cl_2}{2 \text{ mol HCl}}$$

Moles of HCl from H_2

$$2 \text{ mol } H_2 \times \frac{2 \text{ mol HCl}}{1 \text{ mol } H_2} = 4 \text{ mol of HCl (smaller amount of product)}$$

Moles of HCl from Cl_2

$$5 \text{ mol } Cl_2 \times \frac{2 \text{ mol HCl}}{1 \text{ mol } Cl_2} = 10 \text{ mol of HCl (not possible)}$$

In this reaction mixture, H_2 is the limiting reactant. When 2 mol of H_2 are used up, the reaction stops. At that point the reaction is complete. The excess reactant, 3 mol of Cl_2, is left over and cannot react. We can show the changes in each reactant and the product as follows:

	Reactants		Product
Equation	H_2 +	Cl_2 $\longrightarrow$	$2HCl$
Initial moles	2 mol	5 mol	0
Moles used/formed	−2 mol	−2 mol	+4 mol
Moles left	0 (2 − 2)	3 mol (5 − 2)	4 mol (0 + 4)
Identify as	Limiting reactant	Excess reactant	Product possible

---

CONCEPT CHECK 9.5

■ Moles of Product from Limiting Reactant

Consider the reaction for the synthesis of methanol (CH_3OH):

$$CO(g) + 2H_2(g) \longrightarrow CH_3OH(g)$$

In the laboratory, 3.00 mol of CO and 5.00 mol of H_2 are combined.

a. What mole–mole equalities will be needed in the calculation?
b. What are the mole–mole factors from the equalities you wrote in **a**?
c. What is the number of moles of CH_3OH produced from each reactant?
d. What is the limiting reactant for the reaction?

ANSWER

a. Two equalities are needed: one for the mole–mole relationship between CO and CH_3OH and another for the mole–mole relationship between H_2 and CH_3OH using the coefficients from the balanced equation.

$$1 \text{ mol of CO} = 1 \text{ mol of } CH_3OH \qquad 2 \text{ mol of } H_2 = 1 \text{ mol of } CH_3OH$$

b. From each equality, two mole–mole factors can be written:

$$\frac{1 \text{ mol } CH_3OH}{1 \text{ mol CO}} \text{ and } \frac{1 \text{ mol CO}}{1 \text{ mol } CH_3OH} \qquad \frac{1 \text{ mol } CH_3OH}{2 \text{ mol } H_2} \text{ and } \frac{2 \text{ mol } H_2}{1 \text{ mol } CH_3OH}$$

c. Using separate calculations, calculate the moles of CH_3OH that are possible from each of the reactants.

$$3.00 \text{ mol CO} \times \frac{1 \text{ mol } CH_3OH}{1 \text{ mol CO}} = 3.00 \text{ mol of } CH_3OH$$

$$5.00 \text{ mol } H_2 \times \frac{1 \text{ mol } CH_3OH}{2 \text{ mol } H_2} = 2.50 \text{ mol of } CH_3OH$$

d. The smaller amount, which is 2.50 mol of CH_3OH, is the maximum number of moles of methanol that can be produced. Because this smaller quantity is produced from H_2, the limiting reactant is H_2.

Calculating Mass of Product from a Limiting Reactant

The quantities of the reactants can also be given in grams. The calculations to identify the limiting reactant are the same as before, but the grams of each reactant must first be converted to moles. Once the limiting reactant is determined, the moles of possible product can be converted to grams using molar mass. This calculation is shown in Sample Problem 9.5.

MC

TUTORIAL
Limiting Reactant and Yield: Mass Calculations

A ceramic brake disc in a sports car withstands temperature of 1400 °C.

Guide to Calculating Product from a Limiting Reactant

STEP 1
Use molar mass to convert the grams of each reactant to moles.

STEP 2
Write mole–mole factors using the coefficients in the equation.

STEP 3
Calculate moles of product from each reactant and determine the limiting reactant.

STEP 4
Determine the moles of product or calculate grams of product using molar mass.

SAMPLE PROBLEM | 9.5

■ **Mass of Product from a Limiting Reactant**

When silicon dioxide (sand) and carbon are heated, the products are silicon carbide, SiC, and carbon monoxide. Silicon carbide is a ceramic material that tolerates extreme temperatures and is used as an abrasive and in the brake discs of sports cars. How many grams of CO are formed from 70.0 g of SiO_2 and 50.0 g of C?

$$SiO_2(s) + 3C(s) \xrightarrow{\text{Heat}} SiC(s) + 2CO(g)$$

SOLUTION

After we convert the grams of each reactant to moles, we can calculate the moles of CO that are possible from each reactant. Using the smaller number of moles of CO, we calculate the grams of CO from the limiting reactant using molar mass.

STEP 1 **Use molar mass to convert grams of each reactant to moles.**
Given 70.0 g of SiO_2 and 50.0 g of C
Need grams of CO

1 mol of SiO_2 = 60.09 g of SiO_2	1 mol of C = 12.01 g of C
$\dfrac{1 \text{ mol } SiO_2}{60.09 \text{ g } SiO_2}$ and $\dfrac{60.09 \text{ g } SiO_2}{1 \text{ mol } SiO_2}$	$\dfrac{1 \text{ mol C}}{12.01 \text{ g C}}$ and $\dfrac{12.01 \text{ g C}}{1 \text{ mol C}}$

$$70.0 \text{ g } SiO_2 \times \frac{1 \text{ mol } SiO_2}{60.09 \text{ g } SiO_2} = 1.16 \text{ mol of } SiO_2 \quad 50.0 \text{ g C} \times \frac{1 \text{ mol C}}{12.01 \text{ g C}} = 4.16 \text{ mol of C}$$

STEP 2 **Write mole–mole factors using the coefficients in the equation.**

2 mol of CO = 1 mol of SiO_2	3 mol of C = 2 mol of CO
$\dfrac{2 \text{ mol CO}}{1 \text{ mol } SiO_2}$ and $\dfrac{1 \text{ mol } SiO_2}{2 \text{ mol CO}}$	$\dfrac{2 \text{ mol CO}}{3 \text{ mol C}}$ and $\dfrac{3 \text{ mol C}}{2 \text{ mol CO}}$

STEP 3 **Calculate moles of product from each reactant and determine the limiting reactant.**

$$1.16 \text{ mol } SiO_2 \times \frac{2 \text{ mol CO}}{1 \text{ mol } SiO_2} = 2.32 \text{ mol of CO (smaller number of moles)}$$

$$4.16 \text{ mol C} \times \frac{2 \text{ mol CO}}{3 \text{ mol C}} = 2.77 \text{ mol of CO}$$

The smaller amount of 2.32 mol of CO is the most CO that can be produced. Thus, SiO_2 is the limiting reactant.

STEP 4 **Determine grams of product using molar mass.**

1 mol of CO = 28.01 g of CO
$\dfrac{1 \text{ mol CO}}{28.01 \text{ g CO}}$ and $\dfrac{28.01 \text{ g CO}}{1 \text{ mol CO}}$

The grams of product CO are calculated by converting the smaller number of moles of CO to grams using molar mass.

$$2.32 \text{ mol CO} \times \frac{28.01 \text{ g CO}}{1 \text{ mol CO}} = 65.0 \text{ g of CO}$$

STUDY CHECK

Hydrogen sulfide burns with oxygen to give sulfur dioxide and water. How many grams of sulfur dioxide are formed from the reaction of 0.250 mol of H_2S and 0.300 mol of O_2?

$$2H_2S(g) + 3O_2(g) \longrightarrow 2SO_2(g) + 2H_2O(g)$$

QUESTIONS AND PROBLEMS

Limiting Reactants

9.19 A taxi company has 10 taxis.
- **a.** On a certain day, only 8 taxi drivers show up for work. How many taxis can be used to pick up passengers?
- **b.** On another day, 10 taxi drivers show up for work but 3 taxis are in the repair shop. How many taxis can be driven?

9.20 A clock maker has 15 clock faces. Each clock requires 1 face and 2 hands.
- **a.** If the clock maker also has 42 hands, how many clocks can be produced?
- **b.** If the clock maker has only 8 hands, how many clocks can be produced?

9.21 Nitrogen and hydrogen react to form ammonia.

$$N_2(g) + 3H_2(g) \longrightarrow 2NH_3(g)$$

Determine the limiting reactant in each of the following mixtures of reactants:
- **a.** 3.0 mol of N_2 and 5.0 mol of H_2
- **b.** 8.0 mol of N_2 and 4.0 mol of H_2
- **c.** 3.0 mol of N_2 and 12.0 mol of H_2

9.22 Iron and oxygen react to form iron(III) oxide.

$$4Fe(s) + 3O_2(g) \longrightarrow 2Fe_2O_3(s)$$

Determine the limiting reactant in each of the following mixtures of reactants:
- **a.** 2.0 mol of Fe and 6.0 mol of O_2
- **b.** 5.0 mol of Fe and 4.0 mol of O_2
- **c.** 16.0 mol of Fe and 20.0 mol of O_2

9.23 For each of the following reactions, calculate the moles of indicated product produced when 2.00 mol of each reactant is used:
- **a.** $2SO_2(g) + O_2(g) \longrightarrow 2SO_3(g)$ (SO_3)
- **b.** $3Fe(s) + 4H_2O(l) \longrightarrow Fe_3O_4(s) + 4H_2(g)$ (Fe_3O_4)
- **c.** $C_7H_{16}(g) + 11O_2(g) \longrightarrow$
$$7CO_2(g) + 8H_2O(g)   (CO_2)$$

9.24 For each of the following reactions, calculate the moles of indicated product produced when 3.00 mol of each reactant is used:
- **a.** $4Li(s) + O_2(g) \longrightarrow 2Li_2O(s)$ (Li_2O)
- **b.** $Fe_2O_3(s) + 3H_2(g) \longrightarrow 2Fe(s) + 3H_2O(l)$ (Fe)
- **c.** $Al_2S_3(s) + 6H_2O(l) \longrightarrow$
$$2Al(OH)_3(aq) + 3H_2S(g)   (H_2S)$$

9.25 For each of the following reactions, calculate the grams of indicated product produced when 20.0 g of each reactant is used:
- **a.** $2Al(s) + 3Cl_2(g) \longrightarrow 2AlCl_3(s)$ $(AlCl_3)$
- **b.** $4NH_3(g) + 5O_2(g) \longrightarrow 4NO(g) + 6H_2O(g)$ (H_2O)
- **c.** $CS_2(g) + 3O_2(g) \longrightarrow CO_2(g) + 2SO_2(g)$ (SO_2)

9.26 For each of the following reactions, calculate the grams of indicated product produced when 20.0 g of each reactant is used:
- **a.** $4Al(s) + 3O_2(g) \longrightarrow 2Al_2O_3(s)$ (Al_2O_3)
- **b.** $3NO_2(g) + H_2O(l) \longrightarrow$
$$2HNO_3(aq) + NO(g)   (HNO_3)$$
- **c.** $C_2H_5OH(l) + 3O_2(g) \longrightarrow 2CO_2(g) + 3H_2O(g)$ (H_2O)

9.4 Percent Yield

In all of our calculations up to now, we have calculated the amount of product as the maximum quantity possible, or 100%. In each problem, we assumed that all of the reactants were changed completely to product. While this would be an ideal situation, it does not usually happen. As we run a reaction and transfer products from one container to another, some product is usually lost. In the lab as well as commercially, the starting materials may not be completely pure, and side reactions may use some of the reactants to give unwanted products. Thus, 100% of the desired product is not actually obtained.

When we run a chemical reaction in the laboratory, we measure out specific quantities of the reactants and place them in a reaction flask. We calculate the **theoretical yield** for the reaction, which is the amount of product (100%) we would expect if all the reactants are converted to desired product. When the reaction ends, we collect and measure the mass of the product, which is the **actual yield** for the product. Because some product is usually lost, the actual yield is less than the theoretical yield. Using the actual yield and the theoretical yield for a product, we can calculate the **percent yield**.

$$\text{Percent yield (\%)} = \frac{\text{Actual yield}}{\text{Theoretical yield}} \times 100\%$$

LEARNING GOAL

Given the actual quantity of product, determine the percent yield for a reaction.

■ **Calculating Percent Yield**

For your chemistry class party, you have prepared cookie dough from a recipe that makes 5 dozen cookies. You place dough for 12 cookies on a baking sheet and place it in the oven. But then the phone rings and you run to answer it. While you are talking, the cookies on the baking sheet burn and you have to throw them out. You proceed to prepare four more baking sheets with 12 cookies each. If the rest of the cookies are edible, what is the percent yield of cookies you provide for the chemistry party?

ANSWER

The theoretical yield of cookies is 5 dozen or 60 cookies, which is the maximum or 100% of the possible number of cookies. The actual yield is 48 edible cookies, which is 60 cookies minus the 12 cookies that burned. The percent yield is the ratio of 48 edible cookies divided by the theoretical yield of 60 cookies that were possible multiplied by 100%.

Theoretical yield: 60 cookies possible
Actual yield: 48 cookies to eat
Percent yield: $\dfrac{48 \text{ cookies (actual)}}{60 \text{ cookies (theoretical)}} \times 100\% = 80\%$

SAMPLE PROBLEM | 9.6

■ **Calculating Percent Yield**

On a space shuttle, LiOH is used to absorb exhaled CO_2 from breathing air to form $LiHCO_3$.

$$LiOH(s) + CO_2(g) \longrightarrow LiHCO_3(s)$$

What is the percent yield of the reaction if 50.0 g of LiOH gives 72.8 g of $LiHCO_3$?

SOLUTION

STEP 1   Write the given and needed quantities.
   Given 50.0 g of LiOH and 72.8 g of $LiHCO_3$ (actually produced)
   Need percent yield of $LiHCO_3$

STEP 2   Write a plan to calculate the theoretical yield and the percent yield.
   Calculation of theoretical yield:

grams of LiOH → Molar mass → moles of LiOH → Mole–mole factor → moles of $LiHCO_3$ → Molar mass → grams of $LiHCO_3$

   Calculation of percent yield:
   $\dfrac{\text{Actual yield}}{\text{Theoretical yield}} \times 100\%$

STEP 3   Write the molar mass for the reactant and the mole–mole factors from the balanced equation.

1 mol of LiOH = 23.95 g of LiOH

$\dfrac{1 \text{ mol LiOH}}{23.95 \text{ g LiOH}}$ and $\dfrac{23.95 \text{ g LiOH}}{1 \text{ mol LiOH}}$

1 mol of $LiHCO_3$ = 1 mol of LiOH

$\dfrac{1 \text{ mol LiHCO}_3}{1 \text{ mol LiOH}}$ and $\dfrac{1 \text{ mol LiOH}}{1 \text{ mol LiHCO}_3}$

1 mol of $LiHCO_3$ = 67.96 g of $LiHCO_3$

$\dfrac{67.96 \text{ g LiHCO}_3}{1 \text{ mol LiHCO}_3}$ and $\dfrac{1 \text{ mol LiHCO}_3}{67.96 \text{ g LiHCO}_3}$

On a space shuttle, the LiOH in the canisters removes CO_2 from the air.

Guide to Calculations for Percent Yield

STEP 1
Write the given and needed quantities.

STEP 2
Write a plan to calculate the theoretical yield and the percent yield.

STEP 3
Write the molar mass for the reactant and the mole–mole factors from the balanced equation.

STEP 4
Calculate the percent yield ratio by dividing the actual yield (given) by the theoretical yield × 100%.

STEP 4 **Calculate the percent yield ratio by dividing the actual yield (given) by the theoretical yield x 100%.** Calculation of theoretical yield:

$$50.0 \text{ g LiOH} \times \frac{1 \text{ mol LiOH}}{23.95 \text{ g LiOH}} \times \frac{1 \text{ mol LiHCO}_3}{1 \text{ mol LiOH}} \times \frac{67.96 \text{ g LiHCO}_3}{1 \text{ mol LiHCO}_3} = 142 \text{ g of LiHCO}_3$$

Calculation of percent yield:

$$\frac{\text{Actual yield (given)}}{\text{Theoretical yield (calculated)}} \times 100\% = \frac{72.8 \text{ g LiHCO}_3}{142 \text{ g LiHCO}_3} \times 100\% = 51.3\%$$

A percent yield of 51.3% means that 72.8 g of the theoretical amount of 142 g of $LiHCO_3$ was actually produced by the reaction.

STUDY CHECK

For the reaction in Sample Problem 9.6, what is the percent yield if 8.00 g of CO_2 produces 10.5 g of $LiHCO_3$?

QUESTIONS AND PROBLEMS

Percent Yield

9.27 Carbon disulfide is produced by the reaction of carbon and sulfur dioxide.

$$5C(s) + 2SO_2(g) \longrightarrow CS_2(g) + 4CO(g)$$

a. What is the percent yield for the reaction if 40.0 g of carbon produces 36.0 g of carbon disulfide?
b. What is the percent yield for the reaction if 32.0 g of sulfur dioxide produces 12.0 g of carbon disulfide?

9.28 Iron(III) oxide reacts with carbon monoxide to produce iron and carbon dioxide.

$$Fe_2O_3(s) + 3CO(g) \longrightarrow 2Fe(s) + 3CO_2(g)$$

a. What is the percent yield for the reaction if 65.0 g of iron(III) oxide produces 15.0 g of iron?
b. What is the percent yield for the reaction if 75.0 g of carbon monoxide produces 85.0 g of carbon dioxide?

9.29 Aluminum reacts with oxygen to produce aluminum oxide.

$$4Al(s) + 3O_2(g) \longrightarrow 2Al_2O_3(s)$$

The reaction of 50.0 g of aluminum and sufficient oxygen has a 75.0% yield. How many grams of aluminum oxide are produced?

9.30 Propane (C_3H_8) burns in oxygen to produce carbon dioxide and water.

$$C_3H_8(g) + 5O_2(g) \longrightarrow 3CO_2(g) + 4H_2O(g)$$

Calculate the mass of CO_2 that can be produced if the reaction of 45.0 g of propane and sufficient oxygen has a 60.0% yield.

9.31 When 30.0 g of carbon is heated with silicon dioxide, 28.2 g of carbon monoxide is produced. What is the percent yield of this reaction?

$$SiO_2(s) + 3C(s) \longrightarrow SiC(s) + 2CO(g)$$

9.32 Calcium and nitrogen react to form calcium nitride.

$$3Ca(s) + N_2(g) \longrightarrow Ca_3N_2(s)$$

If 56.6 g of calcium is mixed with nitrogen gas and 32.4 g of calcium nitride is produced, what is the percent yield of the reaction?

9.5 Energy in Chemical Reactions

Almost every chemical reaction involves a loss or gain of energy. To discuss energy change for a reaction, we refer to the *system*, which includes all of the reactants and products we are looking at. The *surroundings* are all the things that contain and interact with the system, such as the reaction flask and the air in the room.

Heat of Reaction (Enthalpy Change)

The **heat of reaction** is the amount of heat absorbed or released during a reaction that takes place at constant pressure. A change of energy occurs as reactants interact, bonds break apart, and products form. We determine a heat of reaction or *enthalpy change*, symbol ΔH, as the difference in the energy of the products and the reactants.

$$\Delta H = H_{\text{products}} - H_{\text{reactants}}$$

LEARNING GOAL

Given the heat of reaction (enthalpy change), calculate the loss or gain of heat for an exothermic or endothermic reaction.

TUTORIAL

Heat of Reaction

In an **endothermic reaction** (*endo* means "within"), the energy of the products is greater than that of the reactants. In these reactions, heat flows out of the surroundings into the system, where it is used to convert the reactants to products. For an endothermic reaction, the heat of reaction can be written as one of the reactants. It can also be written as a ΔH value with a positive sign (+). Let us look at the equation for the endothermic reaction in which 570 kJ of heat is needed to convert two moles of carbon dioxide to two moles of carbon monoxide and one mole of oxygen.

Endothermic, Heat Flows In

Heat Is a Reactant

$$2CO_2(g) + 570 \text{ kJ} \longrightarrow 2CO(g) + O_2(g)$$
$$2CO_2(g) \longrightarrow 2CO(g) + O_2(g)$$

$\Delta H = +570 \text{ kJ}$
Positive sign

Heat Flow for Endothermic Reactions

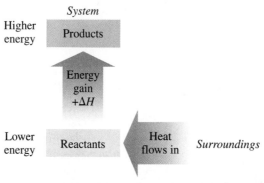

In an **exothermic reaction** (*exo* means "out"), the energy of the reactants is greater than that of the products. In these reactions, heat flows out of a system into the surroundings. In these equations, the heat of reaction can be written as one of the products or as ΔH with a negative sign (−). For example, in the thermite reaction, the reaction of aluminum and iron(III) oxide produces a great amount of heat. The amount of heat produced during the thermite reaction is so immense that the products reach temperatures of 2500 °C, forming liquids before they cool to solids. This reaction has been used to cut or weld railroad tracks.

Exothermic, Heat Flows Out

Heat Is a Product

$$2Al(s) + Fe_2O_3(s) \longrightarrow 2Fe(s) + Al_2O_3(s) + 850 \text{ kJ}$$
$$2Al(s) + Fe_2O_3(s) \longrightarrow 2Fe(s) + Al_2O_3(s)$$

$\Delta H = -850 \text{ kJ}$
Negative sign

Heat Flow for Exothermic Reactions

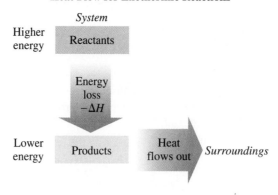

Aluminum and iron(III) oxide produce heat in an exothermic reaction.

Reaction	Energy Change	Heat in the Equation	Sign of ΔH
Endothermic	Heat absorbed	Reactant side	Positive sign (+)
Exothermic	Heat released	Product side	Negative sign (−)

CONCEPT CHECK 9.7

■ **Exothermic and Endothermic Reactions**

In the reaction of 1 mol of carbon with oxygen gas, the energy of the carbon dioxide produced is 393 kJ less than that of the reactants.

a. Is the reaction exothermic or endothermic?
b. Write the balanced chemical equation including the heat of the reaction.
c. What is the value (in kJ) for the heat of reaction?

ANSWER

a. When the energy of the products is less than that of the reactants, the reaction gives off heat, which means that it is exothermic.
b. In an exothermic reaction, the heat is written as a product.

$$C(s) + O_2(g) \longrightarrow CO_2(g) + 393 \text{ kJ}$$

c. The heat of reaction for an exothermic reaction has a negative sign: $\Delta H = -393$ kJ

Calculations of Heat in Reactions

The value of ΔH refers to the heat change for the number of moles (kJ/mol) of each substance in the balanced equation for the reaction. Consider the following decomposition reaction:

$$2H_2O(l) \longrightarrow 2H_2(g) + O_2(g) \quad \Delta H = +572 \text{ kJ}$$
$$2H_2O(l) + 572 \text{ kJ} \longrightarrow 2H_2(g) + O_2(g)$$

For this reaction, 572 kJ are absorbed by 2 mol of H_2O to produce 2 mol of H_2 and one mol of O_2. We can write heat conversion factors for each substance in this reaction as follows:

$$\frac{572 \text{ kJ}}{2 \text{ mol } H_2O} \qquad \frac{572 \text{ kJ}}{2 \text{ mol } H_2} \qquad \frac{572 \text{ kJ}}{1 \text{ mol } O_2}$$

Suppose in this reaction that 9.00 g of H_2O undergoes reaction. We can calculate the heat absorbed as

$$9.00 \text{ g } H_2O \times \frac{1 \text{ mol } H_2O}{18.02 \text{ g } H_2O} \times \frac{572 \text{ kJ}}{2 \text{ mol } H_2O} = 143 \text{ kJ absorbed}$$

SAMPLE PROBLEM | 9.7

■ **Calculating Heat in a Reaction**

The heat of reaction for the formation of ammonia from hydrogen and nitrogen has a $\Delta H = -92.2$ kJ.

$$N_2(g) + 3H_2(g) \longrightarrow 2NH_3(g) \quad \Delta H = -92.2 \text{ kJ}$$

How much heat, in kilojoules, is released when 50.0 g of ammonia forms?

SOLUTION

STEP 1 **List given and needed data for the equation.**
Given 50.0 g of NH_3
Need heat in kilojoules (kJ) to form NH_3

STEP 2 **Write a plan using heat of reaction and any molar mass needed.** Use conversion factors to convert the grams of NH_3 to moles of NH_3 and to heat in kilojoules.

grams of NH_3 → Molar mass → moles of NH_3 → Heat of reaction → kilojoules

Guide to Calculations Using Heat of Reaction (ΔH)

STEP 1
List given and needed data for the equation.

STEP 2
Write a plan using heat of reaction and any molar mass needed.

STEP 3
Write the conversion factors including heat of reaction.

STEP 4
Set up the problem.

STEP 3 Write the conversion factors including heat of reaction.

$$1 \text{ mol of } NH_3 = 17.03 \text{ g of } NH_3$$

$$\frac{17.03 \text{ g } NH_3}{1 \text{ mol } NH_3} \quad \text{and} \quad \frac{1 \text{ mol } NH_3}{17.03 \text{ g } NH_3}$$

$$2 \text{ mol of } NH_3 = 92.2 \text{ kJ}$$

$$\frac{92.2 \text{ kJ}}{2 \text{ mol } NH_3} \quad \text{and} \quad \frac{2 \text{ mol } NH_3}{92.2 \text{ kJ}}$$

STEP 4 Set up the problem.

$$50.0 \text{ g } NH_3 \times \frac{1 \text{ mol } NH_3}{17.03 \text{ g } NH_3} \times \frac{92.2 \text{ kJ}}{2 \text{ mol } NH_3} = 135 \text{ kJ}$$

STUDY CHECK

Mercury(II) oxide decomposes to mercury and oxygen.

$$2HgO(s) \longrightarrow 2Hg(l) + O_2(g) \qquad \Delta H = 182 \text{ kJ}$$

a. Is the reaction exothermic or endothermic?
b. How many kJ are needed if 25.0 g of mercury(II) oxide reacts?

CHEMISTRY AND HEALTH

Hot Packs and Cold Packs

In a hospital, at a first-aid station, or at an athletic event, a *cold pack* may be used to reduce swelling from an injury, remove heat from inflammation, or decrease capillary size to lessen the effect of hemorrhaging. Inside the plastic container of a cold pack, there is a compartment containing solid ammonium nitrate (NH_4NO_3) that is separated from a compartment containing water. The pack is activated when it is hit or squeezed hard enough to break the walls between the compartments and cause the ammonium nitrate to mix with the water (shown as H_2O over the reaction arrow). In an endothermic process, each gram of NH_4NO_3 that dissolves absorbs 330 J of heat from the water. The temperature drops and the pack becomes cold and ready to use.

Endothermic Reaction in a Cold Pack

$$26 \text{ kJ} + NH_4NO_3(s) \xrightarrow{H_2O} NH_4NO_3(aq)$$

Hot packs are used to relax muscles, lessen aches and cramps, and increase circulation by expanding capillary size. Constructed in the

same way as cold packs, a hot pack may contain the salt $CaCl_2$. The dissolving of the salt in water is exothermic and releases 670 J/g of salt. The temperature rises and the pack becomes hot and ready to use.

Exothermic Reaction in a Hot Pack

$$CaCl_2(s) \xrightarrow{H_2O} CaCl_2(aq) + 75 \text{ kJ}$$

Cold packs use an endothermic reaction.

QUESTIONS AND PROBLEMS

Energy in Chemical Reactions

9.33 In an exothermic reaction, is the energy of the products less or greater than that of the reactants?

9.34 In an endothermic reaction, is the energy of the products less or greater than that of the reactants?

9.35 Classify the following as exothermic or endothermic reactions:
   a. 550 kJ is released.
   b. The energy level of the products is higher than that of the reactants.
   c. The metabolism of glucose in the body provides energy.

9.36 Classify the following as exothermic or endothermic reactions:
   a. The energy level of the products is lower than that of the reactants.

   b. In the body, the synthesis of proteins requires energy.
   c. 125 kJ is absorbed.

9.37 Classify the following as exothermic or endothermic reactions and give ΔH for each:
   a. gas burning in a Bunsen burner:

$$CH_4(g) + 2O_2(g) \longrightarrow CO_2(g) + 2H_2O(g) + 890 \text{ kJ}$$

   b. dehydrating limestone:

$$Ca(OH)_2(s) + 65.3 \text{ kJ} \longrightarrow CaO(s) + H_2O(l)$$

   c. formation of aluminum oxide and iron from aluminum and iron(III) oxide:

$$2Al(s) + Fe_2O_3(s) \longrightarrow Al_2O_3(s) + 2Fe(s) + 850 \text{ kJ}$$

9.38 Classify the following as exothermic or endothermic reactions and give ΔH for each:

a. combustion of propane:

$$C_3H_8(g) + 5O_2(g) \longrightarrow 3CO_2(g) + 4H_2O(g) + 2220 \text{ kJ}$$

b. formation of "table" salt:

$$2Na(s) + Cl_2(g) \longrightarrow 2NaCl(s) + 819 \text{ kJ}$$

c. decomposition of phosphorus pentachloride:

$$PCl_5(g) + 67 \text{ kJ} \longrightarrow PCl_3(g) + Cl_2(g)$$

9.39 The equation for the formation of silicon tetrachloride from silicon and chlorine is

$$Si(s) + 2Cl_2(g) \longrightarrow SiCl_4(g) \qquad \Delta H = -657 \text{ kJ}$$

How many kilojoules are released when 125 g of Cl_2 reacts with silicon?

9.40 Methanol (CH_3OH), which is used as a cooking fuel, undergoes combustion to produce carbon dioxide and water.

$$2CH_3OH(l) + 3O_2(g) \longrightarrow 2CO_2(g) + 4H_2O(l) \quad \Delta H = -726 \text{ kJ}$$

How many kilojoules are released when 75.0 g of methanol is burned?

9.6 Energy in the Body

In our cells, there is a variety of "high-energy" compounds. One of the most important of these is adenosine triphosphate, or ATP. The energy released from the foods we eat is stored in the high-energy phosphate bonds of ATP. The high-energy **ATP** molecule is composed of adenine, ribose, and three phosphate groups. The corners in the adenine and ribose structures represent carbon atoms.

LEARNING GOAL

Describe the role of ATP in providing energy for the cells of the body.

SELF STUDY ACTIVITY
ATP and Energy

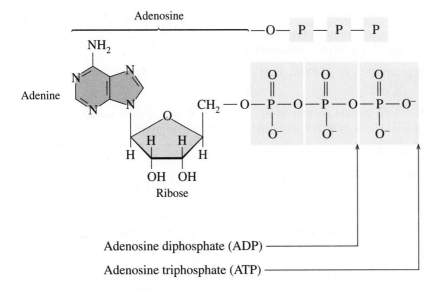

ATP (adenosine triphosphate) consists of adenine, ribose, and three phosphate groups.

In the cells of the body, large molecules are degraded to small molecules in reactions that give off energy for ATP synthesis. At the same time, the synthesis of larger molecules in the cells requires ATP energy. When ATP reacts with water in the cells, energy is released along with adenosine diphosphate (abbreviated as **ADP**) and a HPO_4^{2-}, called inorganic phosphate (abbreviated as P_i). The loss of a phosphate group releases 7.3 kcal/mol (31 kJ/mol) of ATP. The equation for this exothermic reaction can be written as follows:

$$ATP \longrightarrow ADP + P_i + 7.3 \text{ kcal (31 kJ)}$$

Every time we contract our muscles, move substances across cellular membranes, or send nerve signals, we use energy from the breakdown of ATP. In a cell that is doing work, as many as 2 million ATP molecules may be utilized every second. The total amount of ATP used in one day can be very large, even though only about 1 g of ATP is present in our bodies at any given time.

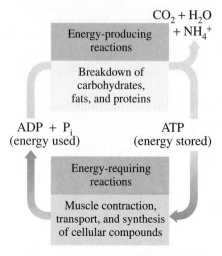

The energy-storage molecule, ATP, links energy-producing reactions with energy-requiring reactions in the cells.

TUTORIAL

ATP: Energy Storage

When we eat, the energy released from metabolic reactions of carbohydrates, lipids, and proteins is used to generate more ATP in our cells. In an endothermic reaction, 7.3 kcal/mol (31 kJ/mol) is required to convert ADP and P_i to energy-rich ATP.

$$ADP + P_i + 7.3 \text{ kcal } (31 \text{ kJ}) \longrightarrow ATP$$

ATP Energy Drives Energy-Requiring Reactions

ATP is needed in our cells because it can be linked with reactions that require energy. For example, when we digest carbohydrates, the glucose that is obtained must add a phosphate group before it can be utilized by the body. However, adding a phosphate group is not a spontaneous reaction and requires 3.3 kcal/mol (14 kJ/mol). Thus, the addition of phosphate occurs only when it is linked with the release of energy from the breakdown of ATP to ADP and P_i. This combination of reactions utilizes the energy from ATP to push, or "drive," the nonspontaneous, energy-requiring reaction.

ATP	$\longrightarrow$	$ADP + P_i + 7.3$ kcal (31 kJ)	Provides energy
Glucose + P_i + 3.3 kcal (14 kJ)	$\longrightarrow$	glucose-6-phosphate	Requires energy
ATP + glucose	$\longrightarrow$	ADP + glucose-6-phosphate + 4.0 kcal (17 kJ)	

The linking of a reaction that requires energy with a reaction that supplies energy is a very important concept in biochemistry. Many of the reactions essential to a cell cannot proceed by themselves, but they can be made to proceed by pairing them with a reaction that releases energy. Similar kinds of combined reactions are also used to transmit nerve impulses, transport substances across membranes to higher concentrations, and to contract muscles.

CONCEPT CHECK 9.8

■ **High-Energy Phosphate Compounds**

Describe the components of ATP and ADP.

ANSWER

ATP and ADP both contain adenine and ribose. In ATP, adenosine is attached to three phosphate groups, but in ADP, it is attached to two phosphate groups.

SAMPLE PROBLEM   9.8

■ **ATP**

Write an equation for the breakdown of ATP, including the energy values.

SOLUTION

The equation for the breakdown of ATP produces ADP, P_i, and energy.

$$ATP \longrightarrow ADP + P_i + 7.3 \text{ kcal } (31 \text{ kJ})$$

STUDY CHECK

Write the equation for the formation of ATP from ADP and P_i, including the energy values.

Energy Fuels in the Cells of the Body

Glucose is the primary fuel used by the body to provide energy for the synthesis of ATP. Glucose, $C_6H_{12}O_6$, is the product when the carbohydrates in our diets undergo digestion and metabolism. Lipids and proteins in our diets also play an important role in metabolism and energy production, but only when glucose is depleted. The digestion of lipids produces fatty acids, and the digestion of proteins produces amino acids.

CHEMISTRY AND HEALTH

ATP Energy and Ca²⁺ Needed to Contract Muscles

Our muscles consist of thousands of parallel fibers. Within these muscle fibers are fibrils composed of two proteins called filaments. Arranged in alternating rows, the thick filaments of myosin overlap the thin filaments containing actin. During a muscle contraction, the thin filaments slide inward over the thick filaments, causing a shortening of the muscle fibers.

Calcium ion, Ca^{2+}, and ATP play an important role in muscle contraction. An increase in the Ca^{2+} concentration in the muscle fibers causes the filaments to slide, while a decrease stops the process. In a relaxed muscle, the Ca^{2+} concentration is low. However, when a nerve impulse reaches the muscle, the calcium channels open, allowing the Ca^{2+} ions to flow into the fluid surrounding the filaments. The muscle contracts as myosin binds to actin and pulls the filaments inward. The energy for the contraction is provided by the breakdown of ATP to ADP + P_i.

Muscle contraction continues as long as both ATP and Ca^{2+} levels are high around the filaments. When the nerve impulse ends, the calcium channels close. Energy from ATP moves Ca^{2+} out of the filaments, which causes the muscle to relax. In rigor mortis, the Ca^{2+} ion concentration remains high within the muscle fibers, causing a continued state of rigidity. After approximately 24 hours, the cells deteriorate, causing a decrease in Ca^{2+} and relaxing the muscles.

Muscle contraction uses the energy from the breakdown of ATP.

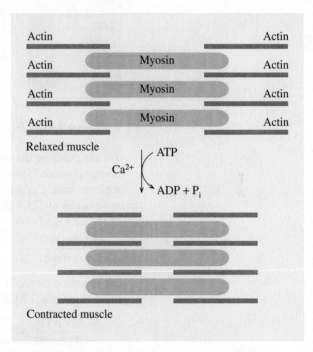

Muscles contract when myosin binds to actin.

When our diet supplies more glucose than we need for immediate energy use, the excess is stored as glycogen (a long chain of glucose molecules) in the liver and muscles. When enough glycogen is stored, excess glucose is converted to fats.

Percent Yield for ATP Production from Glucose

In a nutrition laboratory, a calorimeter is used to measure the energy from the combustion of glucose. In a calorimeter, the combustion of 1 mol of glucose produces 680 kcal (2800 kJ).

$$C_6H_{12}O_6(aq) + 6O_2(g) \longrightarrow 6CO_2(g) + H_2O(g) + 680 \text{ kcal (2800 kJ)}$$

In the body, the complete combustion of glucose provides the energy for 36 ATP.

$$C_6H_{12}O_6 + 6O_2 + 36 \text{ ADP} + 36 \text{ P}_i \longrightarrow 6CO_2 + 6H_2O + 36 \text{ ATP}$$

We can use the energy from the breakdown of ATP (7.3 kcal/mol or 31 kJ/mol) to calculate the total energy in kcal when 1 mol of glucose undergoes combustion in the body.

$$\frac{36 \text{ mol ATP}}{1 \text{ mol glucose}} \times \frac{7.3 \text{ kcal (or 31 kJ)}}{1 \text{ mol ATP}} = 260 \text{ kcal/mol of glucose (or 1100 kJ/mol)}$$

We can use our discussion in Section 9.5 about percent yield to calculate the percent yield, also called "efficiency," of energy production by the body. The energy produced from

1 mol of glucose in a calorimeter can be considered as the theoretical yield (kcal), and the ATP energy produced in the body would be the actual yield (kcal). By comparing the energy produced by the combustion of glucose in a calorimeter to the energy produced in our cells, we see that the cells of the body are about 38% efficient in converting the total available chemical energy in glucose to ATP. The rest of the energy from glucose is lost as heat.

$$\frac{260 \text{ kcal (cells of the body)}}{680 \text{ kcal (calorimeter)}} \times 100\% = 38\%$$

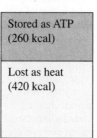

Calorimeter

Energy produced by 1 mol of glucose (680 kcal)

Cells

Stored as ATP (260 kcal)

Lost as heat (420 kcal)

Energy from Fatty Acids (Lipids)

We can compare the energy from the utilization by the body of 1 mol of a fatty acid to the energy produced from 1 mol of glucose. For example, in the body, 1 mol of the fatty acid myristic acid, $C_{14}H_{28}O_2$, produces sufficient energy to make 112 mol of ATP. Using the molar mass of 228.4 g/mol for myristic acid, we can calculate the ATP produced per gram of the fatty acid.

$$\frac{112 \text{ mol ATP}}{1 \text{ mol myristic acid}} \times \frac{1 \text{ mol myristic acid}}{228.4 \text{ g myristic acid}} = 0.490 \text{ mol of ATP/g of myristic acid (fat)}$$

We already know that 1 mol of glucose in the body generates 36 mol of ATP. Using the molar mass of glucose, 180.20 g/mol, we can calculate the ATP produced per gram of glucose.

$$\frac{36 \text{ mol ATP}}{1 \text{ mol glucose}} \times \frac{1 \text{ mol glucose}}{180.20 \text{ g glucose}} = 0.200 \text{ mol of ATP/g of glucose}$$

From these calculations, we see that 1 g of fat produces more than twice the ATP than does 1 g of glucose. This means that we obtain more than twice the number of calories from 1 g of fat (9 kcal/g or 38 kJ/g) than we do from 1 g of carbohydrate (4 kcal/g or 17 kJ/g).

QUESTIONS AND PROBLEMS

Energy in the Body

9.41 What is the full name of ATP?

9.42 What is the full name of ADP?

9.43 What is the primary fuel for energy in the body?

9.44 What are the functional groups in the ATP molecule?

9.45 Why would the breakdown of ATP be an exothermic reaction?

9.46 Why would the formation of ATP be an endothermic reaction?

9.47 Why is ATP considered an energy-rich compound?

9.48 What is meant when we say that the energy from the breakdown of ATP is used to "drive" a reaction?

9.49 Why are the fats in the adipose tissues of the body considered the major form of stored energy?

9.50 When does the body utilize amino acids for energy needs?

CHEMISTRY AND HEALTH

Stored Fat and Obesity

The storage of fat is an important survival feature in the lives of many animals. In hibernating animals, large amounts of stored fat provide the energy for the entire hibernation period, which could be several months. In camels, large amounts of food are stored in the camel's hump, which is actually a huge fat deposit. When food resources are low, the camel can survive months without food or water by utilizing the fat reserves in the hump. Migratory birds preparing to fly long distances also store large amounts of fat. Whales are kept warm by a layer of body fat called "blubber" (which can be as thick as 2 ft) under their skin. Blubber also provides energy when whales must survive long periods without food. Penguins also have blubber, which protects them from the cold and provides energy when they are incubating their eggs.

Humans also have the capability to store large amounts of fat, although they do not hibernate or usually have to survive for long periods without food. When humans survived on sparse diets that were mostly vegetarian, about 20% of the dietary calories were from fat. Today, a typical diet includes more dairy products and foods with high fat levels, and as much as 60% of the calories are from fat. The U.S. Public Health Service now estimates that in the United States, more than one-third of adults are obese. Obesity is defined as a body weight that is more than 20 percent over an ideal weight. Obesity is a major factor in health problems such as diabetes, heart disease, high blood pressure, stroke, and gallstones, as well as some cancers and forms of arthritis.

At one time, scientists thought that obesity was simply a problem of eating too much. In 1995, scientists discovered that a hormone called *leptin* is produced in fat cells. When fat cells are full, high levels of leptin signal the brain to limit the intake of food. When fat stores are low, leptin production decreases, which signals the brain to increase food intake. Some obese persons have high levels of leptin, although the leptin does not cause them to decrease how much they eat.

The causes of obesity have become a major research field. Scientists are studying differences in the rate of leptin production, degrees of resistance to leptin, and possible combinations of these factors. After a person has dieted and lost weight, the leptin level drops. This decrease in leptin may cause an increase in hunger, slow metabolism, and increased food intake, which starts the weight-gain cycle all over again. Currently, studies are being made to assess the safety of leptin therapy following weight loss.

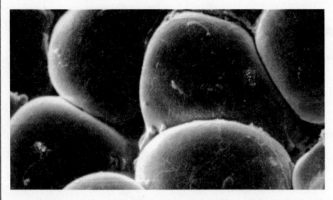

The fat cells that make up adipose tissue are capable of storing unlimited quantities of fat.

Animals store fat for insulation and as an energy reserve.

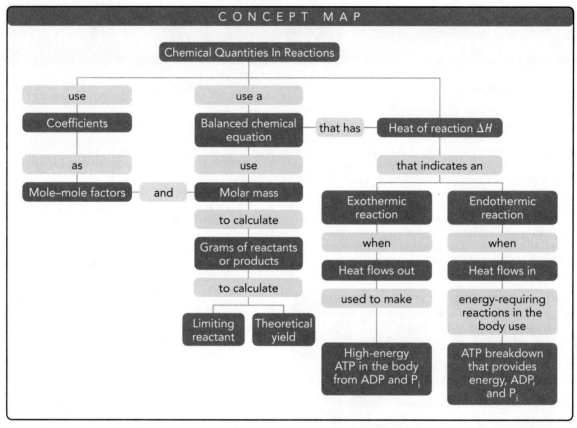

CHAPTER REVIEW

9.1 Mole Relationships in Chemical Equations

LEARNING GOAL: *Given a quantity in moles of reactant or product, use a mole–mole factor from the balanced equation to calculate the number of moles of another substance in the reaction.*

In a balanced equation, the total mass of the reactants is equal to the total mass of the products. The coefficients in an equation describing the relationship between the moles of any two components are used to write mole–mole factors. When the number of moles for one substance is known, a mole–mole factor is used to find the moles of a different substance in the reaction.

9.2 Mass Calculations for Reactions

LEARNING GOAL: *Given the mass in grams of a substance in a reaction, calculate the mass in grams of another substance in the reaction.*

In calculations using equations, the molar masses of the substances and their mole–mole factors are used to change the number of grams of one substance to the corresponding grams of a different substance.

9.3 Limiting Reactants

LEARNING GOAL: *Identify a limiting reactant when given the quantities of two reactants; calculate the amount of product formed from the limiting reactant.*

A limiting reactant is the reactant that produces the smaller amount of product while the other reactant is left over. When the masses of two reactants are given, the mass of a product is calculated from the limiting reactant.

9.4 Percent Yield

LEARNING GOAL: *Given the actual quantity of product, determine the percent yield for a reaction.*

The percent yield of a reaction indicates the percent of product actually produced during a reaction. The percent yield is calculated by dividing the actual yield in grams of a product by the theoretical yield in grams.

9.5 Energy in Chemical Reactions

LEARNING GOAL: *Given the heat of reaction (enthalpy change), calculate the loss or gain of heat for an exothermic or endothermic reaction.*

In chemical reactions, the heat of reaction (ΔH) is the energy difference between the reactants and the products. In an exothermic reaction, the energy of the products is lower than that of the reactants. Heat is released, and ΔH is negative. In an endothermic reaction, the energy of the products is higher than that of the reactants. Heat is absorbed, and the ΔH is positive.

9.6 Energy in the Body

LEARNING GOAL: *Describe the role of ATP in providing energy for the cells of the body.*

Metabolism includes reactions that break down large molecules into smaller ones with a release of energy, and reactions that require energy to build larger molecules from smaller ones. Energy obtained from breaking down large molecules is stored primarily in adenosine triphosphate (ATP), a high-energy compound that can provide energy for energy-requiring reactions.

SUMMARY OF KEY REACTIONS

Formation of ATP

$$ADP + P_i + 7.3 \text{ kcal (31 kJ)} \longrightarrow ATP$$

Breakdown of ATP

$$ATP \longrightarrow ADP + P_i + 7.3 \text{ kcal (31 kJ)}$$

Complete Combustion of Glucose

$$C_6H_{12}O_6 + 6O_2 + 36\,ADP + 36\,P_i \longrightarrow 6CO_2 + 6H_2O + 36\,ATP$$

KEY TERMS

actual yield The actual amount of product produced by a reaction.

ADP Adenosine diphosphate, formed by the breakdown of ATP with water; consists of adenine, ribose, and two phosphate groups.

ATP Adenosine triphosphate, a high-energy compound that stores energy in the cells; consists of adenine, ribose, and three phosphate groups.

endothermic reaction A reaction wherein the energy of the products is greater than that of the reactants.

excess reactant The reactant that remains when the limiting reactant is used up in a reaction.

exothermic reaction A reaction wherein the energy of the reactants is greater than that of the products.

heat of reaction The heat (symbol ΔH) absorbed or released when a reaction takes place at constant pressure.

law of conservation of mass In a chemical reaction, the total mass of the reactants is equal to the total mass of the products; matter is neither lost nor gained.

limiting reactant The reactant used up during a chemical reaction, which limits the amount of product that can form.

mole–mole factor A conversion factor that relates the number of moles of two compounds in an equation derived from their coefficients.

percent yield The ratio of the actual yield of a reaction to the theoretical yield possible for the reaction.

theoretical yield The maximum amount of product that a reaction can produce from a given amount of reactant.

UNDERSTANDING THE CONCEPTS

9.51 If red spheres represent oxygen atoms and blue spheres represent nitrogen atoms,

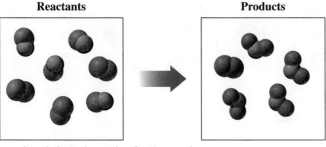

a. write a balanced equation for the reaction.
b. identify the limiting reactant.

9.52 If purple spheres represent iodine atoms and white spheres represent hydrogen atoms,

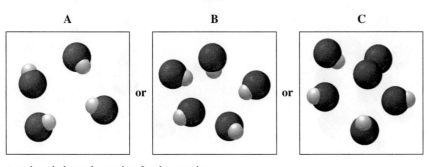

a. write a balanced equation for the reaction.
b. identify the diagram that shows the products.

9.53 If blue spheres represent nitrogen atoms and white spheres represent hydrogen atoms,

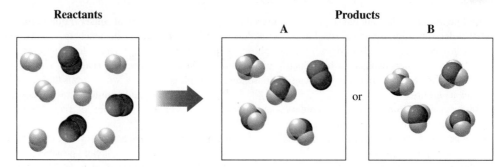

Reactants   **Products**
    A   B

or

    a. write a balanced equation for the reaction.
    b. identify the diagram that shows the products.

9.54 If green spheres represent chlorine atoms, gray-green spheres represent fluorine atoms, and white spheres represent hydrogen atoms,

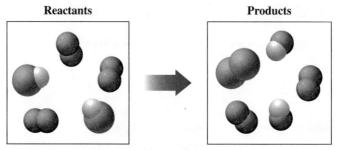

Reactants   **Products**

    a. write a balanced equation for the reaction.
    b. identify the limiting reactant.

9.55 If blue spheres represent nitrogen atoms and purple spheres represent iodine atoms,

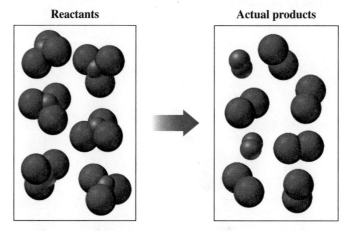

Reactants   **Actual products**

    a. write a balanced equation for the reaction.
    b. from the diagram of the actual products that result, calculate the percent yield for the reaction.

9.56 If green spheres represent chlorine atoms and red spheres represent oxygen atoms,

Reactants   **Actual products**

    a. write a balanced equation for the reaction.
    b. identify the limiting reactant.
    c. from the diagram of the actual products that result, calculate the percent yield for the reaction.

ADDITIONAL QUESTIONS AND PROBLEMS

For instructor-assigned homework, go to
www.masteringchemistry.com.

9.57 At a winery, glucose ($C_6H_{12}O_6$) in grapes undergoes fermentation to produce ethanol (C_2H_6O) and carbon dioxide.

$$C_6H_{12}O_6(aq) \longrightarrow 2C_2H_6O(aq) + 2CO_2(g)$$
Glucose   Ethanol

Glucose in grapes ferments to produce ethanol.

a. How many moles of glucose are required to form 124 g of ethanol?
b. How many grams of ethanol would be formed from the reaction of 0.240 kg of glucose?

9.58 Gasohol is a fuel containing ethanol (C_2H_6O) that burns in oxygen (O_2) to give carbon dioxide and water.
a. Write the balanced chemical equation for the reaction.
b. How many moles of O_2 are needed to completely react with 4.0 mol of C_2H_6O?
c. If a car produces 88 g of CO_2, how many grams of O_2 are used up in the reaction?
d. If you add 125 g of C_2H_6O to your gasoline, how many grams of CO_2 and H_2O can be produced?

9.59 The *unbalanced* reaction for ammonia and fluorine is the following:

$$NH_3(g) + F_2(g) \longrightarrow N_2F_4(g) + HF(g)$$

a. Write the balanced equation for the reaction.
b. How many moles of each reactant are needed to produce 4.00 mol of HF?
c. How many grams of F_2 are required to react with 1.50 mol of NH_3?
d. How many grams of N_2F_4 can be produced when 3.40 g of NH_3 reacts?

9.60 Propane gas, C_3H_8, reacts with oxygen to produce water and carbon dioxide. Propane has a density of 2.02 g/L at room temperature.

$$C_3H_8(g) + 5O_2(g) \longrightarrow 3CO_2(g) + 4H_2O(l)$$

a. How many moles of water form when 5.00 L of propane gas completely reacts?
b. How many grams of CO_2 are produced from 18.5 g of oxygen gas and excess propane?
c. How many grams of H_2O can be produced from the reaction of 8.50×10^{22} molecules of propane gas?

9.61 When a mixture of 12.8 g of Na and 10.2 g of Cl_2 reacts, what is the mass of NaCl that is produced?

$$2Na(s) + Cl_2(g) \longrightarrow 2NaCl(s)$$

9.62 If a mixture of 35.8 g of CH_4 and 75.5 g of S reacts, how many grams of H_2S are produced?

$$CH_4(g) + 4S(g) \longrightarrow CS_2(g) + 2H_2S(g)$$

9.63 Pentane gas, C_5H_{12}, reacts with oxygen to produce carbon dioxide and water:

$$C_5H_{12}(g) + 8O_2(g) \longrightarrow 5CO_2(g) + 6H_2O(g)$$

a. How many grams of pentane must react to produce 4.0 mol of water?
b. How many grams of CO_2 are produced from 32.0 g of oxygen and excess pentane?
c. How many grams of CO_2 are formed if 44.5 g of C_5H_{12} is mixed with 108 g of O_2?

9.64 When nitrogen dioxide (NO_2) from car exhaust combines with water in the air, it forms nitric acid (HNO_3), which causes acid rain, and nitrogen oxide.

$$3NO_2(g) + H_2O(l) \longrightarrow 2HNO_3(aq) + NO(g)$$

a. How many molecules of NO_2 are needed to react with 0.250 mol of H_2O?
b. How many grams of HNO_3 are produced when 60.0 g of NO_2 completely reacts?
c. How many grams of HNO_3 can be produced if 225 g of NO_2 is mixed with 55.2 g of H_2O?

9.65 Acetylene gas, C_2H_2, undergoes hydrogenation to produce ethane C_2H_6:

$$C_2H_2(g) + 2H_2(g) \xrightarrow{\text{Pt}} C_2H_6(g)$$

When 28.0 g of acetylene reacts with hydrogen, 24.5 g of ethane is produced. What is the percent yield for the reaction?

9.66 When 50.0 g of iron(III) oxide reacts with carbon monoxide, 32.8 g of iron is produced. What is the percent yield of the reaction?

$$Fe_2O_3(s) + 3CO(g) \longrightarrow 2Fe(s) + 3CO_2(g)$$

9.67 A reaction of nitrogen and sufficient hydrogen produced 30.0 g of ammonia, which is a 65.0% yield for the reaction. How many grams of nitrogen reacted?

$$N_2(g) + 3H_2(g) \longrightarrow 2NH_3(g)$$

9.68 Consider the *unbalanced* equation for the decomposition of potassium chlorate:

$$KClO_3(s) \longrightarrow KCl(s) + O_2(g)$$

a. Write the *balanced* equation for the reaction.
b. When 46.0 g of $KClO_3$ is heated, 12.1 g of O_2 is formed. How many grams of KCl is also formed?
c. What is the percent yield of O_2 for the reaction?

9.69 Acetylene, C_2H_2, used in welders' torches, burns according to the following equation:

$$2C_2H_2(g) + 5O_2(g) \longrightarrow 4CO_2(g) + 2H_2O(g)$$

a. How many molecules of oxygen are needed to react with 22.0 g of acetylene?
b. How many grams of carbon dioxide could be produced from the complete reaction of the acetylene in part **a**?
c. If the reaction in part **a** produces 64.0 g of CO_2, what is the percent yield for the reaction?

9.70 Sodium and nitrogen combine to form sodium nitride.

$$6Na(s) + N_2(g) \longrightarrow 2Na_3N(s)$$

a. If 80.0 g of sodium is mixed with 20.0 g of nitrogen gas, what mass of sodium nitride forms?
b. If the reaction in part **a** has a percent yield of 75.0%, how much sodium nitride is actually produced?

9.71 The formation of 2 mol of nitrogen oxide, NO, from $N_2(g)$ and $O_2(g)$, requires 90.2 kJ of heat.

$$N_2(g) + O_2(g) \longrightarrow 2NO(g) \qquad \Delta H = 90.2 \text{ kJ}$$

a. How many kJ are required to form 3.00 g of NO?
b. What is the complete equation (including heat) for the decomposition of NO?
c. How many kJ are released when 5.00 g of NO decomposes to N_2 and O_2?

9.72 The formation of rust (Fe_2O_3) from solid iron and oxygen gas releases 1.7×10^3 kJ.

$$4Fe(s) + 3O_2(g) \longrightarrow 2Fe_2O_3(s) \quad \Delta H = -1.7 \times 10^3 \text{ kJ}$$

a. How many kJ are released when 2.00 g of Fe reacts?
b. How many grams of rust form when 150 kcal are released?

9.73 Each of the following is a reaction that occurs in the cells of the body. Identify which is exothermic and endothermic.
a. Succinyl-CoA + $H_2O \longrightarrow$ succinate + CoA + 8.9 kcal
b. GDP + P_i + 34 kJ $\longrightarrow$ GTP + H_2O

9.74 Each of the following is a reaction that is needed by the cells of the body. Identify which is exothermic and endothermic.
a. Phosphocreatine + $H_2O \longrightarrow$ creatine + P_i + 10.2 kcal
b. Fructose-6-phosphate + P_i + 16 kJ $\longrightarrow$
fructose-1,6-bisphosphate

9.75 Each of the following is a reaction that is needed by the cells of the body. Why could reaction (1) be used to drive reaction (2)?
1. Phosphoenolpyruvate $\longrightarrow$
pyruvate + P_i + 14.8 kcal (61.9 kJ)
2. ADP + P_i + 7.3 kcal (31 kJ) $\longrightarrow$ ATP

9.76 Each of the following is a reaction that is needed by the cells of the body. Why could reaction (1) be used to drive reaction (2)?
1. Phosphoenolpyruvate $\longrightarrow$
pyruvate + P_i + 14.8 kcal (61.9 kJ)
2. Fructose + P_i + 3.8 kcal (16 kJ) $\longrightarrow$ fructose-6-phosphate

9.77 One mol of the fatty acid palmitic acid ($C_{16}H_{32}O_2$) produces 129 mol of ATP. How many kcal are produced from 1.0 mol of palmitic acid?

9.78 One mol of the fatty acid lauric acid ($C_{12}H_{24}O_2$) produces 95 mol of ATP. How many kcal are produced from 1.0 mol of lauric acid?

CHALLENGE QUESTIONS

9.79 Carbon monoxide reacts with hydrogen to form the alcohol methanol, CH_3OH:

$$CO(g) + 2H_2(g) \longrightarrow CH_3OH(l)$$

Suppose you mix 50.0 g of CO and 10.0 g of H_2.
a. What is the limiting reactant?
b. What is the excess reactant?
c. How many grams of methanol can be produced?
d. How many grams of the excess reactant are left over?

9.80 The combustion of the alkyne propyne, $CH_3-C\equiv CH$, releases heat when it burns according to the following equation:

$$C_3H_4(g) + 4O_2(g) \longrightarrow 3CO_2(g) + 2H_2O(g)$$

a. How many moles of water are produced from the complete reaction of 64.0 g of oxygen?
b. How many moles of oxygen are needed to react completely with 2.25×10^{24} molecules of propyne?
c. How many grams of carbon dioxide are produced from the complete reaction of 78.0 g of propyne?
d. If the reaction in part **c** produces 186 g of CO_2, what is the percent yield for the reaction?

9.81 Consider the following *unbalanced* equation:

$$Cr(s) + O_2(g) \longrightarrow Cr_2O_3(s)$$

a. Write the *balanced* equation.
b. Identify the type of reaction.
c. How many moles of oxygen react with 4.50 mol of Cr?
d. How many grams of chromium(III) oxide are produced when 24.8 g of chromium reacts?
e. When 0.500 mol of chromium reacts with 8.00 g of oxygen, how many grams of chromium(III) oxide can form?
f. If 74.0 g of chromium and 62.0 g of oxygen undergo a reaction that has a 70.0% yield, what mass of chromium(III) oxide forms?

9.82 When peroxide (H_2O_2) is used in rocket fuels, it produces water, oxygen, and heat:

$$2H_2O_2(l) \longrightarrow 2H_2O(l) + O_2(g)$$

a. If the reaction of 2.00 g of H_2O_2 releases 5.76 kJ, what is the heat of reaction?
b. How many kilojoules are released when 275 g of peroxide reacts?

9.83 For the reaction:

$$2S(s) + 3O_2(g) \longrightarrow 2SO_3(g) \quad \Delta H = -790 \text{ kJ}$$

a. Is the reaction endothermic or exothermic?
b. How many kJ are released when 1.5 mol of S reacts?
c. How many kJ are released when 125 g of SO_3 is formed?
d. What is the ΔH, in kilojoules, for the following reaction?

$$2SO_3(g) \longrightarrow 2S(s) + 3O_2(g)$$

e. Is the reaction in part **d** endothermic or exothermic?

9.84 A camel's hump contains 14 kg of fat.
a. Using the value of 0.490 mol of ATP per gram of fat, how many moles of ATP could be produced by the fat in the camel's hump?
b. If the breakdown of ATP releases 7.3 kcal/mol, how many kcal are produced by the utilization of the fat?

9.85 In the body, there may be as many as 2 million (2×10^6) ATP molecules broken down in one cell every second. Researchers estimate that the human body has about 10^{13} cells.
a. How much energy, in kcal, could be produced by the cells in the body in 24 hours?
b. If ATP has a molar mass of about 507 g/mol, how many grams of ATP are broken down in 24 hours?

ANSWERS

ANSWERS TO STUDY CHECKS

9.1 1.83 mol of Fe

9.2 0.945 mol of O_2

9.3 27.5 g of CO_2

9.4 10.4 g of C_7H_{16}

9.5 12.8 g of SO_2

9.6 84.7%

9.7 a. endothermic

b. 10.5 kJ

9.8 $ADP + P_i + 7.3 \text{ kcal (31 kJ)} \longrightarrow ATP$

ANSWERS TO SELECTED QUESTIONS AND PROBLEMS

9.1 a. (1) Two molecules of sulfur dioxide gas react with one molecule of oxygen gas to produce two molecules of sulfur trioxide gas.
(2) Two mol of sulfur dioxide gas react with 1 mol of oxygen gas to produce 2 mol of sulfur trioxide gas.
b. (1) Four atoms of solid phosphorus react with five molecules of oxygen gas to produce two molecules of solid diphosphorus pentoxide.
(2) Four mol of solid phosphorus react with 5 mol of oxygen gas to produce 2 mol of solid diphosphorus pentoxide.

9.3 a. 160.14 g of reactants = 160.14 g of products
b. 283.88 g of reactants = 283.88 g of products

9.5 a. $\dfrac{2 \text{ mol } SO_2}{1 \text{ mol } O_2}$ and $\dfrac{1 \text{ mol } O_2}{2 \text{ mol } SO_2}$

$\dfrac{2 \text{ mol } SO_2}{2 \text{ mol } SO_3}$ and $\dfrac{2 \text{ mol } SO_3}{2 \text{ mol } SO_2}$

$\dfrac{2 \text{ mol } SO_3}{1 \text{ mol } O_2}$ and $\dfrac{1 \text{ mol } O_2}{2 \text{ mol } SO_3}$

b. $\dfrac{4 \text{ mol } P}{5 \text{ mol } O_2}$ and $\dfrac{5 \text{ mol } O_2}{4 \text{ mol } P}$

$\dfrac{4 \text{ mol } P}{2 \text{ mol } P_2O_5}$ and $\dfrac{2 \text{ mol } P_2O_5}{4 \text{ mol } P}$

$\dfrac{5 \text{ mol } O_2}{2 \text{ mol } P_2O_5}$ and $\dfrac{2 \text{ mol } P_2O_5}{5 \text{ mol } O_2}$

9.7 a. 1.0 mol of O_2 **b.** 10. mol of H_2
c. 5.0 mol of H_2O

9.9 a. 1.25 mol of C **b.** 0.96 mol of CO
c. 1.0 mol of SO_2 **d.** 0.50 mol of CS_2

9.11 a. 77.5 g of Na_2O **b.** 6.26 g of O_2
c. 19.4 g of O_2

9.13 a. 192 g of O_2 **b.** 3.79 g of N_2
c. 54.0 g of H_2O

9.15 a. 3.66 g of H_2O **b.** 3.44 g of NO
c. 7.53 g of HNO_3

9.17 a. $2PbS(s) + 3O_2(g) \longrightarrow 2PbO(s) + 2SO_2(g)$
b. 6.00 g of O_2
c. 17.4 g of SO_2
d. 137 g of PbS

9.19 a. Eight taxis can be used to pick up passengers.
b. Seven taxis can be driven.

9.21 a. 5.0 mol of H_2 **b.** 4.0 mol of H_2
c. 3.0 mol of N_2

9.23 a. 2.00 mol of SO_3 **b.** 0.500 mol of Fe_3O_4
c. 1.27 mol of CO_2

9.25 a. 25.1 g of $AlCl_3$ **b.** 13.5 g of H_2O
c. 26.7 g of SO_2

9.27 a. 71.0% **b.** 63.2%

9.29 70.9 g of Al_2O_3

9.31 60.5%

9.33 a. In exothermic reactions, the energy of the products is less than that of the reactants.

9.35 a. exothermic
b. endothermic
c. exothermic

9.37 a. Heat is released, which makes the reaction exothermic with $\Delta H = -890$ kJ.
b. Heat is absorbed, which makes the reaction endothermic with $\Delta H = 65.3$ kJ.
c. Heat is released, which makes the reaction exothermic with $\Delta H = -850$ kJ.

9.39 579 kJ

9.41 ATP

9.43 glucose

9.45 The ΔH for the breakdown of ATP is -7.3 kcal/mol (-31 kJ/mol), which makes the reaction exothermic.

9.47 When ATP is reacted, enough energy is released to drive energy-requiring reactions in the cell.

9.49 Fats can be stored in unlimited quantities in the body, so they are the major form of stored energy in the body.

9.51 a. $2NO + O_2 \longrightarrow 2NO_2$
b. NO is the limiting reactant.

9.53 a. $N_2 + 3H_2 \longrightarrow 2NH_3$ **b.** A

9.55 a. $2NI_3 \longrightarrow N_2 + 3I_2$ **b.** 67% yield

9.57 a. 1.35 mol of glucose **b.** 123 g of ethanol

9.59 a. $2NH_3(g) + 5F_2(g) \longrightarrow N_2F_4(g) + 6HF(g)$
b. 1.33 mol of NH_3 and 3.33 mol of F_2
c. 143 g of F_2
d. 10.4 g of N_2F_4

9.61 16.8 g of NaCl

9.63 a. 48 g of C_5H_{12} **b.** 27.5 g of CO_2
c. 92.8 g of CO_2

9.65 75.9%

9.67 38.0 g of N_2 reacted.

9.69 a. 1.27×10^{24} molecules of O_2
    b. 74.4 g of CO_2
    c. 86.0%

9.71 a. 4.51 kJ
    b. $2\,NO(g) \longrightarrow N_2(g) + O_2(g) + 90.2$ kJ
    c. 7.51 kJ

9.73 a. exothermic   **b.** endothermic

9.75 Reaction (1) releases a large amount of energy, which provides the energy required for reaction (2).

9.77 940 kcal

9.79 a. CO is the limiting reactant.
    b. H_2 is the excess reactant.

    c. 57.4 g of methanol is formed.
    d. 2.8 g of H_2 is left over.

9.81 a. $4Cr(s) + 3O_2(g) \longrightarrow 2Cr_2O_3(s)$
    b. This is a combination reaction.
    c. 3.38 mol of oxygen
    d. 36.2 g of chromium(III) oxide
    e. 25.4 g of chromium(III) oxide
    f. 75.7 g of chromium(III) oxide

9.83 a. exothermic   **b.** 590 kJ is released
    c. 620 kJ is released   **d.** $\Delta H = +790$ kJ
    e. endothermic

9.85 a. 21 kcal   **b.** 1500 g of ATP

Structures of Solids and Liquids

10

"The pharmacy is one of the many factors in the final integration of chemistry and medicine in patient care," says Dorothea Lorimer, pharmacist, Kaiser Hospital. *"If someone is allergic to a particular drug, I have to find out if a new drug has similar structural features. For instance, some people are allergic to sulfur. If there is sulfur in the new medication, there is a chance it will cause a reaction."*

A prescription indicates a specific amount of a medication. At the pharmacy, the chemical name, formula, and quantity in milligrams or micrograms are checked. Then the prescribed number of capsules is prepared and placed in a container. If it is a liquid, a specific volume is measured and poured into a bottle for liquid prescriptions.

Visit **www.masteringchemistry.com** for self-study materials and instructor-assigned homework.

In Chapter 6, we looked at the formation of ionic and covalent compounds. Now we can investigate more complex chemical bonds and how they contribute to the structure of a molecule or a polyatomic ion. Once we have this basis of understanding, we can look at the differences in properties of compounds and their reactivity.

Electron-dot formulas can be used to diagram the sharing of valence electrons in molecules and polyatomic ions. The presence of multiple bonds can be identified, and possible resonance structures can be drawn. From the electron-dot formulas, we can predict the three-dimensional shapes and polarities of molecules. Then we examine how the different attractive forces between the particles of ions and molecules influence the physical properties of substances, such as melting and boiling point. Finally, we review the discussion of physical states of solids, liquids, and gases from Chapter 3 and describe the energy involved in changes of state.

10.1 Electron-Dot Formulas

LEARNING GOAL

Draw the electron-dot formulas for covalent compounds or polyatomic ions with multiple bonds and show resonance structures.

In Chapter 5, we used electron-dot symbols for atoms in which dots representing the valence electrons were placed around the symbol of an element. In Chapter 6, we drew the electron-dot formulas for simple covalent compounds to show the sharing of valence electrons and the nonbonding electron pairs.

Electron-dot symbols for atoms of the representative elements in Periods 1 to 4 are shown in Table 10.1. For elements with one to four valence electrons, the dots are placed one at a time on the sides, top, and bottom of the atomic symbol. For elements with more than four valence electrons, each added dot is paired with one of the first four dots.

TABLE 10.1 Valence Electrons of Some Elements and Their Electron-Dot Symbols

	Group Number							
	1A (1)	2A (2)	3A (13)	4A (14)	5A (15)	6A (16)	7A (17)	8A (18)
Number of Valence Electrons	1	2	3	4	5	6	7	8
Valence Electron Configuration	ns^1	ns^2	ns^2np^1	ns^2np^2	ns^2np^3	ns^2np^4	ns^2np^5	ns^2np^6
Electron-Dot Symbols	H·							He: $(1s^2)$
	Li·	Be·	·B·	·C·	·N̈·	·Ö·	·F̈:	:Ne:
	Na·	Mg·	·Al·	·Si·	·P̈·	·S̈:	·C̈l:	:Är:
	K·	Ca·	·Ga·	·Ge·	·Äs·	·S̈e:	·B̈r:	:Kr:

In this chapter, we will develop a method that enables us to draw electron-dot formulas for more complex covalent compounds and polyatomic ions.

Drawing Electron-Dot Formulas

When we draw an electron-dot formula for a molecule or polyatomic ion, we show the sequence of atoms, the bonding pairs of electrons shared between atoms, and the nonbonding or *lone pairs* of electrons.

TUTORIAL

Writing Electron-Dot Structures

CONCEPT CHECK 10.1

■ Lone Pairs and Bonding Pairs

State the number of valence electrons, bonding pairs, and nonbonding (lone) electron pairs in each of the following electron-dot formulas:

$$
\text{a. } H\!:\!\overset{\displaystyle H}{\underset{\displaystyle \cdot\cdot}{\overset{\cdot\cdot}{O}}}\!: \qquad \text{b. } \overset{\displaystyle :\overset{\cdot\cdot}{Br}:}{:\overset{\cdot\cdot}{Br}\!:\!\underset{\cdot\cdot}{\overset{\cdot\cdot}{O}}\!:}
$$

ANSWER

a. There is a total of eight valence electrons: each H has one valence electron, and the O atom has six valence electrons. There are two bonding pairs: one between each H atom and the central O atom. The O atom also has two lone pairs.

b. There is a total of 20 valence electrons: each Br has seven valence electrons, and the O atom has six valence electrons. There are two bonding pairs: one between each Br atom and the central O atom. The O atom also has two lone pairs; each Br atom has three lone pairs.

SAMPLE PROBLEM | 10.1

■ Drawing Electron-Dot Formulas

Draw the electron-dot formula for PCl_3, phosphorus trichloride, used commercially to prepare insecticides and flame retardants.

SOLUTION

STEP 1 Determine the arrangement of atoms. In PCl_3, the central atom is P.

Cl P Cl

Cl

STEP 2 Determine the total number of valence electrons. We can use the group numbers to determine the number of valence electrons for each of the atoms in the molecule.

Element	Group	Atoms	Valence Electrons	=	Total
P	5A (15)	1 P	$\times\, 5\,e^-$	=	$5\,e^-$
Cl	7A (17)	3 Cl	$\times\, 7\,e^-$	=	$21\,e^-$
	Total valence electrons for PCl_3			=	$26\,e^-$

STEP 3 Attach each bonded atom to the central atom with a pair of electrons. A pair of bonding electrons is placed between each Cl atom and the central P atom. Each bonding pair can be represented by a bond line.

$$
\underset{\displaystyle Cl}{Cl\!:\!\underset{\cdot\cdot}{P}\!:\!Cl} \quad \text{or} \quad \underset{\displaystyle |}{\underset{Cl}{Cl-P-Cl}}
$$

STEP 4 Place the remaining electrons using single or multiple bonds to complete octets. A total of six electrons $(3 \times 2\,e^-)$ is used to form three single bonds between the central P atom and the Cl atoms. Of the remaining 20 valence electrons, 18 electrons are drawn as lone pairs to complete the octets of the Cl atoms.

26 valence e^- − 6 bonding e^- = 20 e^- remaining

$$
\underset{\displaystyle :\overset{\cdot\cdot}{Cl}:}{:\overset{\cdot\cdot}{Cl}\!:\!\underset{\cdot\cdot}{P}\!:\!\overset{\cdot\cdot}{Cl}:} \quad \text{or} \quad \underset{\displaystyle :\overset{\cdot\cdot}{Cl}:}{:\overset{\cdot\cdot}{Cl}-P-\overset{\cdot\cdot}{Cl}:}
$$

Guide to Drawing Electron-Dot Formulas

> **STEP 1**
> Determine the arrangement of atoms.

> **STEP 2**
> Determine the total number of valence electrons.

> **STEP 3**
> Attach each bonded atom to the central atom with a pair of electrons.

> **STEP 4**
> Place the remaining electrons using single or multiple bonds to complete octets (two for H, six for B).

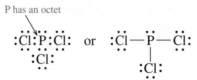

The ball-and-stick model of PCl_3 consists of P and Cl atoms.

The two remaining electrons are drawn on the P atom, which completes its octet. All of the 26 available valence electrons, 3 bonding pairs and 10 lone pairs, have been used to draw complete octets for all the atoms.

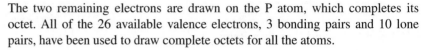

P has an octet

$$:\ddot{C}l:\overset{..}{P}:\ddot{C}l: \quad \text{or} \quad :\ddot{C}l-\overset{|}{P}-\ddot{C}l:$$
$$:\ddot{C}l: \qquad\qquad :\ddot{C}l:$$

STUDY CHECK

Draw the electron-dot formula for Cl_2O.

Exceptions to the Octet Rule

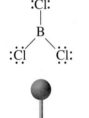

In BCl_3, the central B atom is bonded to three Cl atoms.

While the octet rule is useful for bonding in many compounds, there are exceptions. We have already seen that a hydrogen (H_2) molecule requires just two electrons or a single bond. Usually the nonmetals form octets. However, in BCl_3, the B atom has only three valence electrons to share. Boron compounds typically have six valence electrons on the central B atoms and form just three bonds. While we will generally see compounds of P, S, Cl, Br, and I with octets, they can form molecules in which they share more of their valence electrons. This expands their valence electrons to 10, 12, or even 14 electrons. For example, we have seen that the P atom in PCl_3 has an octet, but in PCl_5, the P atom has five bonds with 10 valence electrons. In H_2S, the S atom has an octet, but in SF_6, there are six bonds to sulfur with 12 valence electrons.

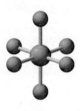

In SF_6, the central S atom is bonded to six F atoms.

SAMPLE PROBLEM | 10.2

■ Drawing Electron-Dot Formulas for Polyatomic Ions

Sodium chlorite, $NaClO_2$, is an ionic compound that contains the chlorite ion ClO_2^-. It is used to bleach textiles, pulp, and paper. Draw the electron-dot formula for the chlorite ion.

SOLUTION

STEP 1 **Determine the arrangement of atoms.** For the polyatomic ion ClO_2^-, the central atom is Cl. The atoms and electrons of a polyatomic ion are placed in brackets, and the charge is written outside to the upper right.

$$\left[\text{O} \quad \text{Cl} \quad \text{O} \right]^-$$

STEP 2 **Determine the total number of valence electrons.** We can use the group numbers to determine the number of valence electrons for each of the atoms in the molecule. Because the ion has a negative charge, one more electron is added to the valence electrons.

Element	Group	Atoms	Valence Electrons	=	Total
O	6A (16)	2 O	$\times 6\,e^-$	=	$12\,e^-$
Cl	7A (17)	1 Cl	$\times 7\,e^-$	=	$7\,e^-$
Ionic charge (negative) add		$1\,e^-$		=	$1\,e^-$
	Total valence electrons for ClO_2^-			=	$20\,e^-$

STEP 3 **Attach each bonded atom to the central atom with a pair of electrons.** A pair of bonding electrons is placed between each O atom and the central Cl atom. Each bonding pair is also represented by a line, which indicates a single bond.

$$[O:Cl:O]^- \quad \text{or} \quad [O-Cl-O]^-$$

In ClO_2^-, the central Cl atom is bonded to two O atoms.

STEP 4 **Place the remaining electrons using single or multiple bonds to complete octets.** A total of four electrons is used to bond the O atoms to the central Cl atom, which leaves 16 valence electrons.

$$20 \text{ valence } e^- - 4 \text{ bonding } e^- = 16 \ e^- \text{ remaining}$$

Of these, 12 electrons are drawn as lone pairs to complete the octets of the O atoms.

$$\left[\ddot{\underset{..}{O}} : Cl : \ddot{\underset{..}{O}} \right]^- \quad \text{or} \quad \left[\ddot{\underset{..}{O}} - Cl - \ddot{\underset{..}{O}} \right]^-$$

The remaining four electrons are placed as two lone pairs on the central Cl atom.

$$\left[\ddot{\underset{..}{O}} : \ddot{\underset{..}{Cl}} : \ddot{\underset{..}{O}} \right]^- \quad \text{or} \quad \left[\ddot{\underset{..}{O}} - \ddot{\underset{..}{Cl}} - \ddot{\underset{..}{O}} \right]^-$$

When we check the electron-dot formula, we see that all 20 available valence electrons—two bonding pairs and eight lone pairs—have been used to draw complete octets for all the atoms.

STUDY CHECK

Draw the electron-dot formula for the polyatomic ion, amide or amino ion, NH_2^-.

Sodium chlorite is used in the processing and bleaching of pulp from wood fibers and recycled cardboard.

Multiple Covalent Bonds and Resonance

Up to now, we have looked at covalent bonding in molecules or polyatomic ions that have only single bonds. However, some covalent compounds, such as alkenes and alkynes, contain atoms that share two or three pairs of electrons. A **double bond** occurs when two pairs of electrons are shared between two atoms; in a **triple bond**, three pairs of electrons are shared. Atoms of carbon, oxygen, nitrogen, and sulfur are most likely to form multiple bonds. Atoms of hydrogen and the halogens do not form double or triple bonds.

Double and triple bonds form when the number of valence electrons is not adequate to draw octets for all the atoms in the molecule. Then one or more lone pairs from the atoms attached to the central atom are shared with the central atom.

(CONCEPT CHECK 10.2)

■ Multiple Bonds in Covalent Molecules

The covalent molecule N_2 contains a triple bond. Describe how the atoms of N achieve octets by forming a triple bond.

ANSWER

Because nitrogen is in Group 5A (15), each N atom has five valence electrons. An octet for each N atom cannot be achieved by sharing only one or two pairs of electrons. However, each N atom achieves an octet by sharing three bonding pairs of electrons, which is called a triple bond.

Octets

$$\cdot \ddot{N} \cdot \ \cdot \ddot{N} \cdot \longrightarrow \ :N \vdots\vdots N: \quad :N \equiv N: \ N_2$$

Three shared   Triple bond   Nitrogen
pairs   molecule

SAMPLE PROBLEM | 10.3

■ **Drawing Electron-Dot Formulas with Multiple Bonds**

Draw the electron-dot formula for carbon dioxide, CO_2, in which the central atom is C.

SOLUTION

STEP 1 **Determine the arrangement of atoms.**

O C O

STEP 2 **Determine the total number of valence electrons.**

Element	Group	Atoms	Valence Electrons	=	Total
O	6A (16)	2 O	$\times\ 6\ e^-$	=	$12\ e^-$
C	4A (14)	1 C	$\times\ 4\ e^-$	=	$4\ e^-$
		Total valence electrons for CO_2		=	$16\ e^-$

STEP 3 **Attach each bonded atom to the central atom with a pair of electrons.** A pair of bonding electrons (single bond) is drawn between each O atom and the central C atom.

O:C:O or O—C—O

STEP 4 **Place the remaining electrons using single or multiple bonds to complete octets.** Four electrons are used to attach the O atoms to the central C atom, which leaves 12 valence electrons.

16 valence e^- − 4 bonding e^- = 12 e^- remaining

We can use the 12 remaining electrons as lone pairs to complete the octets of the O atoms but not the octet of the central C atom.

:Ö:C:Ö: or :Ö—C—Ö:

An octet for the C atom must be achieved by sharing lone pairs from each O atom. When atoms share two bonding pairs of electrons, it is called a double bond.

Lone pairs converted to bonding pairs

:Ö:C:Ö: or :Ö——C——Ö:

Double bonds   Double bonds

:Ö::C::Ö: or :O=C=O:

Molecule of carbon dioxide

STUDY CHECK

Draw the electron-dot formula for HCN (atoms arranged as H C N).

Resonance Structures

When a molecule or polyatomic ion contains multiple bonds, it may be possible to draw more than one electron-dot formula. Suppose we begin to draw the electron-dot formula for ozone, O_3, a component in the stratosphere that protects us from the ultraviolet rays of the Sun.

STEP 1 **Determine the arrangement of atoms.** In O_3, one of the O atoms is the central atom.

O O O

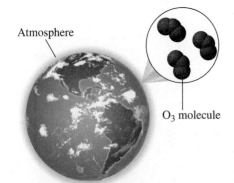

Atmosphere

O_3 molecule

Ozone, O_3, is a component in the stratosphere that protects us from the ultraviolet rays of the Sun.

STEP 2 **Determine the total number of valence electrons.** We can use the group numbers to determine the number of valence electrons.

Element	Group	Atoms	Valence Electrons	=	Total
O	6A (16)	3 O	$\times$ 6 e^-	=	18 e^-

STEP 3 **Attach each bonded atom to the central atom with a pair of electrons.** Using four of the available valence electrons, we draw a bonding pair between the O atoms on the end and the central O atom.

O—O—O

STEP 4 **Place the remaining electrons using single or multiple bonds to complete octets.** We can use the remaining 14 electrons to complete the octets on the end O atoms, but not for the central O atom.

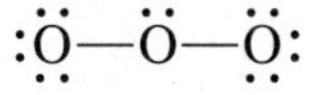

:Ö—Ö—Ö:

To achieve an octet for the central O atom, one lone pair from an end O atom is shared. But which one should be used? One possibility is to form a double bond on the left, and the other possibility is to form a double bond on the right.

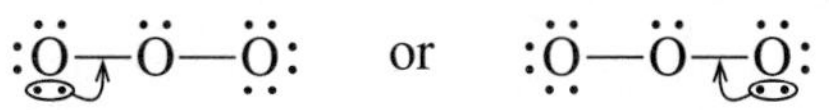

:Ö—Ö—Ö: or   :Ö—Ö—Ö:

When two or more electron-dot formulas can be written, they are called **resonance structures**, which are shown with a double-headed arrow.

:Ö=Ö—Ö: ⟷ :Ö—Ö=Ö:

Resonance structures

Experiments show that the actual bond lengths are equivalent to a molecule with a "one-and-a-half" bond between the central O atom and each outside O atom. In the actual ozone molecule, the electrons are spread equally over all the O atoms. When we draw resonance structures of molecules or polyatomic ions, the true structure is really an average of those structures.

SAMPLE PROBLEM | 10.4

■ **Drawing Resonance Structures**

Sulfur dioxide is produced naturally from volcanic activity and the burning of sulfur-containing coal. Once in the atmosphere, the SO_2 is converted to SO_3, which combines with water forming sulfuric acid, H_2SO_4, a component of acid rain. Draw two resonance structures for sulfur dioxide.

SOLUTION

STEP 1 **Determine the arrangement of atoms.** In SO_2, the S atom is the central atom.

O S O

STEP 2 **Determine the total number of valence electrons.** We can use the group numbers to determine the number of valence electrons.

Element	Group	Atoms	Valence Electrons	=	Total
S	6A (16)	1 S	$\times$ 6 e^-	=	6 e^-
O	6A (16)	2 O	$\times$ 6 e^-	=	12 e^-
		Total valence electrons for SO_2		=	18 e^-

STEP 3 **Attach each bonded atom to the central atom with a pair of electrons.** We will use a single line to represent each pair of bonding electrons.

O—S—O

The ball-and-stick model of SO_2, consists of S (yellow) and O (red) atoms.

STEP 4 **Place the remaining electrons using single or multiple bonds to complete octets.** After four electrons are used to form single bonds between the S atom and the O atoms, the remaining 14 electrons are drawn as lone pairs to complete the octets of the O atoms but not the S atom.

$$:\ddot{O}—\ddot{S}—\ddot{O}:$$

To complete the octet for S, one lone pair from one of the O atoms is shared to form a double bond. Because the lone pair that is shared can come from either O atom, two resonance structures can be drawn.

$$:\ddot{O}—\ddot{S}=O: \longleftrightarrow :O=\ddot{S}—\ddot{O}:$$

STUDY CHECK

Draw three resonance structures for SO_3.

CONCEPT CHECK 10.3

■ **Resonance Structures**

Explain why SO_2 has resonance structures but SCl_2 does not.

ANSWER

In the electron-dot formula of SCl_2, the sharing of the valence electrons of each chlorine atom completes the octet of the sulfur atom. However, in SO_2, the central sulfur atom must form a double bond with one of the oxygen atoms. Thus, when two or more electron-dot formulas can be drawn, the molecule has two or more resonance structures.

Table 10.2 summarizes this method of drawing electron-dot formulas for several molecules and ions.

TABLE 10.2 Using Valence Electrons to Draw Electron-Dot Formulas

Molecule or Polyatomic Ion	Total Valence Electrons	Form Single Bonds to Attach Atoms (electrons used)	Electrons Remaining	Completed Octets (or H:)
Cl_2	$2(7) = 14$	$Cl—Cl$ ($2\,e^-$)	$14 - 2 = 12$	$:\ddot{Cl}—\ddot{Cl}:$
HCl	$1 + 7 = 8$	$H—Cl$ ($2\,e^-$)	$8 - 2 = 6$	$H—\ddot{Cl}:$
H_2O	$2(1) + 6 = 8$	$H—O—H$ ($4\,e^-$)	$8 - 4 = 4$	$H—\ddot{O}—H$
PCl_3	$5 + 3(7) = 26$	$Cl—\underset{\underset{Cl}{\vert}}{P}—Cl$ ($6\,e^-$)	$26 - 6 = 20$	$:\ddot{Cl}—\underset{\underset{:\ddot{Cl}:}{\vert}}{P}—\ddot{Cl}:$
ClO_3^-	$7 + 3(6) + 1 = 26$	$\left[O—\underset{\underset{O}{\vert}}{Cl}—O\right]^-$ ($6\,e^-$)	$26 - 6 = 20$	$\left[:\ddot{O}—\underset{\underset{:\ddot{O}:}{\vert}}{Cl}—\ddot{O}:\right]^-$
NO_2^-	$5 + 2(6) + 1 = 18$	$[O—N—O]^-$ ($4\,e^-$)	$18 - 4 = 14$	$\left[:\ddot{O}—\ddot{N}=\ddot{O}:\right]^-$ $\updownarrow$ $\left[:\ddot{O}=\ddot{N}—\ddot{O}:\right]^-$

QUESTIONS AND PROBLEMS

Electron-Dot Formulas

10.1 Determine the total number of valence electrons in each of the following:
a. H_2S **b.** I_2
c. CCl_4 **d.** OH^-

10.2 Determine the total number of valence electrons in each of the following:
a. SBr_2 **b.** NBr_3
c. CH_3OH **d.** NH_4^+

10.3 Draw the electron-dot formula for each of the following molecules or ions:
a. HF   **b.** SF_2
c. NBr_3 **d.** BH_4^-
e. CH_3OH (methyl alcohol) H C O H
                                               H
f. N_2H_4 (hydrazine) H N N H
                                 H H

10.4 Draw the electron-dot formula for each of the following molecules or ions:
a. H_2O **b.** CCl_4
c. H_3O^+ **d.** SiF_4
                                                   H H
e. CF_2Cl_2 **f.** C_2H_6 H C C H
                                                   H H

10.5 When is it necessary to draw a multiple bond in an electron-dot formula?

10.6 If the available valence electrons for a molecule or polyatomic ion does not complete all of the octets in an electron-dot formula, what should you do?

10.7 What is resonance?

10.8 When does a covalent compound have resonance?

10.9 Draw the electron-dot formula for each of the following molecules or ions:
a. CO
b. H_2CCH_2 (ethene)
c. H_2CO (C is the central atom)

10.10 Draw the electron-dot formula for each of the following molecules or ions:
a. HCCH (ethyne)   **b.** CS_2 (C is the central atom)
c. NO^+

10.11 Draw resonance structures for each of the following molecules or ions:
a. $ClNO_2$ (N is the central atom)
b. OCN^-

10.12 Draw resonance structures for each of the following molecules or ions:
a. HCO_2^- (C is the central atom)
b. N_2O (N N O)

10.2 Shapes of Molecules and Ions (VSEPR Theory)

Using the information in the previous section about electron-dot formulas, we can predict the three-dimensional shapes of many molecules or polyatomic ions. The shape is important in our understanding of how molecules interact with enzymes or certain antibiotics or produce our sense of taste and smell.

The three-dimensional shape is determined by drawing an electron-dot formula and identifying the number of electron groups (one or more electron pairs) around the central atom. In the **valence shell electron-pair repulsion (VSEPR) theory**, the electron groups, which can be lone pairs, single, or multiple bonds, are arranged as far apart as possible around a central atom to minimize the repulsion between their negative charges. The number of electron groups is determined by drawing the electron-dot formula of a molecule or polyatomic ion. Then the specific shape is based on the number of atoms attached to the electron groups of the central atom.

LEARNING GOAL

Predict the three-dimensional structure of a molecule or polyatomic ion and classify it as polar or nonpolar.

TUTORIAL
Molecular Shapes

Two Electron Groups

In $BeCl_2$, two chlorine atoms are bonded to a central beryllium atom. Because an atom of Be has a strong attraction for valence electrons, it forms a covalent rather than ionic compound. With only two electron groups (two electron pairs) around the central atom, the electron-dot formula of $BeCl_2$ is one of the exceptions to the octet rule. The best arrangement of two electron groups for minimal repulsion is to place them on opposite sides of the Be atom. This gives a **linear** shape and a bond angle of 180° to the $BeCl_2$ molecule.

:C̈l—Be—C̈l:

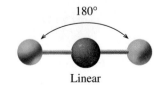

180°

Linear

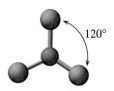

180°

Linear

Another example of a linear molecule is CO_2. In predicting shapes, we count a *double* or *triple* bond (two or three electron pairs) as one electron group. Thus, a multiple bond is counted as a single electron group in determining electron repulsion. In the electron-dot structure of CO_2, two electron groups (two double bonds) are placed on opposite sides of the central C atom, which gives a bond angle of 180°. The shape of the CO_2 molecule is *linear*.

$$:\overset{..}{O}\!\!=\!\!C\!\!=\!\!\overset{..}{O}:$$

Three Electron Groups

In the electron-dot formula of BF_3, the central atom B is attached to three fluorine atoms by three electron groups (another exception to the octet rule). When three electron groups are as far apart as possible, the bond angles are 120° and the shape is **trigonal planar**. The BF_3 molecule is flat with all the atoms in the same plane and bond angles of 120°.

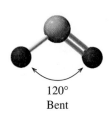

120°

Trigonal planar

$$:\overset{..}{\underset{..}{F}}:$$
$$:\overset{..}{\underset{..}{F}}\!\!-\!\!B\!\!-\!\!\overset{..}{\underset{..}{F}}:$$

Electron-dot formula

$$:\overset{..}{\underset{..}{F}}:$$
$$B$$
$$:\overset{..}{\underset{..}{F}}:\quad :\overset{..}{\underset{..}{F}}:$$

Electron arrangement

In the electron-dot formula for SO_2, there are three electron groups around the sulfur atom: a single-bonded O atom, a double-bonded O atom, and a lone pair of electrons. Minimal repulsion between the three electron groups is achieved when they form a trigonal planar arrangement. However, the shape of the SO_2 molecule is only determined by the two oxygen atoms bonded to the central atom. Therefore, with two O atoms bonded to the central S atom, the shape of the SO_2 molecule is **bent**. In the SO_2 molecule, the bond angle is slightly decreased by the lone pair but is close to 120°.

120°
Bent

$$:\overset{..}{\underset{..}{O}}\!\!-\!\!\overset{..}{S}\!\!=\!\!\overset{..}{\underset{..}{O}}:$$

Electron-dot formula

$$\overset{..}{S}$$
$$:\overset{..}{\underset{..}{O}}:\quad\overset{..}{\underset{..}{O}}:$$

Electron arrangement

Four Electron Groups

Up to now, the molecular shapes we have discussed have all been planar. However, when there are four electron groups around a central atom, the minimum repulsion is obtained by placing the four electron groups at the corners of a three-dimensional tetrahedron. A regular tetrahedron has four sides that are equilateral triangles. For molecules with four electron groups, the central atom is located in the center of a tetrahedron.

As we have seen in Chapter 6, Section 6.6, the alkane methane, CH_4, consists of a carbon central atom bonded to four hydrogen atoms. When the electron-dot formula of CH_4 is drawn, it may appear to be planar, with 90° bond angles. However, VSEPR theory indicates that the best arrangement to minimize the repulsion between four electron groups is **tetrahedral**, which places the four electron groups in CH_4 at the corners of a tetrahedron. With four H atoms (and no lone pairs) bonded to the central C, the shape of the CH_4 molecule is also *tetrahedral*, with bond angles of 109.5°.

Tetrahedron

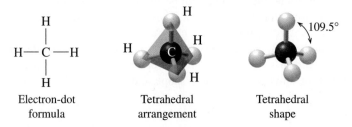

$$H\!\!-\!\!\overset{\displaystyle H}{\underset{\displaystyle H}{C}}\!\!-\!\!H$$

Electron-dot
formula

Tetrahedral
arrangement

109.5°

Tetrahedral
shape

Now we will look at molecules that have four electron groups but only two or three attached atoms. For example, in ammonia, NH_3, the three bonded hydrogen atoms and one lone electron pair occupy the corners of a tetrahedron. Because only three electrons groups are attached to hydrogen atoms, the shape of the NH_3 molecule is **trigonal pyramidal**. In the NH_3 molecule, the bond angles are decreased slightly, to about 107°, by the negatively charged lone pair.

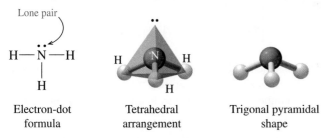

| Electron-dot formula | Tetrahedral arrangement | Trigonal pyramidal shape |

In the electron-dot formula of water, H_2O, there are two bonded hydrogen atoms and two lone electron pairs. According to the VSEPR theory, four electron groups have a tetrahedral arrangement around the central oxygen atom. A molecule with four electron groups but only two bonded atoms has a bent shape. In the H_2O molecule, the bond angle is decreased by the negatively charged lone electron pairs to about 105°. Table 10.3 gives the molecular shapes for molecules with two, three, and four bonded atoms.

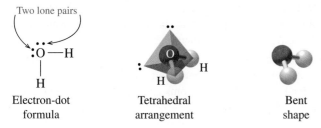

| Electron-dot formula | Tetrahedral arrangement | Bent shape |

TABLE 10.3 Molecular Shapes for a Central Atom with Two, Three, and Four Bonded Atoms

Electron Groups	Electron Arrangement	Bonded Atoms	Lone Pairs	Bond Angle	Molecular Shape	Example	
2	Linear	2	0	180°	Linear	$BeCl_2$	
3	Trigonal planar	3	0	120°	Trigonal planar	BF_3	
3	Trigonal planar	2	1	<120°	Bent	SO_2	
4	Tetrahedral	4	0	109.5°	Tetrahedral	CH_4	
4	Tetrahedral	3	1	<109.5°	Trigonal pyramidal	NH_3	
4	Tetrahedral	2	2	<109.5°	Bent	H_2O	

CONCEPT CHECK 10.4

■ Shapes of Molecules

Determine the shape of a PH_3 molecule.

ANSWER

To determine the shape of a molecule, we draw the electron-dot formula. According to VSEPR theory, the total number of electron groups around the central atom determines the electron geometry, and the number of atoms attached to the central atom determines the shape.

STEP 1   **Draw the electron-dot formula.** The P atom has five valence electrons, and each H atom has one valence electron, which gives a total of eight valence electrons for the electron-dot formula:

$$H-\overset{\displaystyle ..}{P}-H$$
$$|$$
$$H$$

STEP 2   **Arrange the electron groups around the central atom to minimize repulsion.** There are four electron groups around the central P atom, which gives the electron groups a tetrahedral arrangement.

STEP 3   **Use the atoms bonded to the central atom to determine the molecular shape.** The shape of a molecule is determined by the number of bonded atoms. For PH_3, which has three bonded atoms and one lone pair, the shape is *trigonal pyramidal*.

SAMPLE PROBLEM   10.5

■ Predicting Shapes

Use VSEPR theory to predict the shape of the following molecules or polyatomic ions:

a. H_2Se **b.** NO_3^-

SOLUTION

a. H_2Se

STEP 1   **Draw the electron-dot formula.** The Se atom has six valence electrons, and each H atom has one valence electron, which gives a total of eight valence electrons for the electron-dot formula:

$$:\overset{\displaystyle ..}{Se}-H$$
$$|$$
$$H$$

STEP 2   **Arrange the electron groups around the central atom to minimize repulsion.** In the electron-dot formula for H_2Se, there are four electron groups around Se. To minimize repulsion, the electron geometry would have a tetrahedral arrangement.

STEP 3   **Use the atoms bonded to the central atom to determine the molecular shape.** Because H_2Se has two bonded atoms along with two lone pairs, H_2Se has a *bent* shape.

b. NO_3^-

STEP 1   **Draw the electron-dot formula.** For the polyatomic ion, NO_3^-, the N atom with five electrons, each O atom with six electrons, and one negative charge gives a total of 24 valence electrons. Because the double bond can be drawn to each of three O atoms, there are three possible resonance structures.

$$\left[:\overset{..}{\underset{..}{O}}-N=\overset{..}{\underset{..}{O}}:\right]^- \longleftrightarrow \left[:\overset{..}{\underset{..}{O}}-N-\overset{..}{\underset{..}{O}}:\right]^- \longleftrightarrow \left[:\overset{..}{\underset{..}{O}}=N-\overset{..}{\underset{..}{O}}:\right]^-$$
$$\underset{\displaystyle :\overset{..}{\underset{..}{O}}:}{} \qquad \underset{\displaystyle :\overset{..}{\underset{..}{O}}:}{} \qquad \underset{\displaystyle :\overset{..}{\underset{..}{O}}:}{}$$

Guide to Predicting Molecular Shape (VSEPR Theory)

STEP 1
Draw the electron-dot formula.

STEP 2
Arrange the electron groups around the central atom to minimize repulsion.

STEP 3
Use the atoms bonded to the central atom to determine the molecular shape.

STEP 2 **Arrange the electron groups around the central atom to minimize repulsion.** In the electron-dot formula, the central atom, N, has three electron groups. To minimize repulsion, three electron groups give a trigonal planar arrangement.

STEP 3 **Use the atoms bonded to the central atom to determine the molecular shape.** With three bonded atoms but no lone pairs on the central N atom, the polyatomic ion, NO_3^-, has a *trigonal planar* shape.

STUDY CHECK

Use VSEPR theory to predict the shape of ClO_2^-.

QUESTIONS AND PROBLEMS

Shapes of Molecules and Ions (VSEPR Theory)

10.13 Predict the shape of a molecule with each of the following:
   a. Two bonded atoms and no lone pairs
   b. Three bonded atoms and one lone pair

10.14 Predict the shape of a molecule with each of the following:
   a. Four bonded atoms
   b. Two bonded atoms and two lone pairs

10.15 In the molecule PCl_3, the four electron groups around the phosphorus atom are arranged in a tetrahedral geometry. However, the shape of the molecule is called trigonal pyramidal. Why does the shape of the molecule have a different name from the name of the electron group geometry?

10.16 In a molecule of H_2S, the four electron groups around the sulfur atom are arranged in a tetrahedral geometry. However, the shape of the molecule is called bent. Why does the shape of the molecule have a different name from the name of the electron group geometry?

10.17 Compare the electron-dot formulas of BF_3 and NF_3. Why do these molecules have different shapes?

10.18 Compare the electron-dot formulas of CH_4 and H_2O. Why do these molecules have similar bond angles but different names for their shapes?

10.19 Use VSEPR theory to predict the shape of each molecule:
   a. GaH_3 **b.** OF_2
   c. HCN   **d.** CCl_4
   e. SeO_2

10.20 Use VSEPR theory to predict the shape of each molecule:
   a. CF_4 **b.** NCl_3
   c. SCl_2 **d.** CS_2
   e. $BFCl_2$

10.21 Draw the electron-dot formula and predict the shape for each polyatomic ions:
   a. CO_3^{2-} **b.** SO_4^{2-}
   c. BH_4^- **d.** NO_2^+

10.22 Draw the electron-dot formula and predict the shape for each polyatomic ions:
   a. NO_2^- **b.** PO_4^{3-}
   c. ClO_4^- **d.** SF_3^+

10.3 Electronegativity and Polarity

In a compound, the bonding pairs between atoms may or may not be shared equally. In a bond between identical nonmetal atoms, the bonding electrons are shared equally. However, when the bonding electrons occur between atoms of different elements, they are usually shared unequally. Then the shared pairs of electrons are attracted to one atom in the bond more than the other. In ionic compounds, the bonding electrons are lost by a metal atom and gained by a nonmetal atom.

Electronegativity

We can learn more about the chemistry of compounds by looking at how the bonding electrons are shared between atoms. To do this, we use **electronegativity**, which is the relative ability of an atom to attract the shared electrons to itself. Nonmetals have higher electronegativities than do metals, because nonmetals have a greater attraction for electrons than metals. On the electronegativity scale, fluorine was assigned a value of 4.0, and the electronegativities for all other elements were determined relative to the attraction of fluorine for shared electrons. The nonmetals fluorine (4.0) and oxygen (3.5), which are located in the upper right corner of the periodic table, have the highest electronegativities. The metals cesium and francium, which have the lowest electronegativity (0.7), are located in the lower left corner of the periodic table. The electronegativities for the representative elements are shown in Figure 10.1.

LEARNING GOAL

Use electronegativity to determine the polarity of a bond or a molecule.

SELF STUDY ACTIVITY

Electronegativity

FIGURE 10.1 The electronegativities of representative elements, which indicate the ability of atoms to attract shared electrons, increase across a period and decrease going down a group.

Q What element on the periodic table has the strongest attraction for shared electrons?

Electronegativity increases

| H 2.1 |

Electronegativity decreases

1 Group 1A	2 Group 2A		13 Group 3A	14 Group 4A	15 Group 5A	16 Group 6A	17 Group 7A	18 Group 8A
Li 1.0	Be 1.5		B 2.0	C 2.5	N 3.0	O 3.5	F 4.0	
Na 0.9	Mg 1.2		Al 1.5	Si 1.8	P 2.1	S 2.5	Cl 3.0	
K 0.8	Ca 1.0		Ga 1.6	Ge 1.8	As 2.0	Se 2.4	Br 2.8	
Rb 0.8	Sr 1.0		In 1.7	Sn 1.8	Sb 1.9	Te 2.1	I 2.5	
Cs 0.7	Ba 0.9		Tl 1.8	Pb 1.9	Bi 1.9	Po 2.0	At 2.1	

In Chapter 5, Section 5.6, we learned that the sizes of atoms decrease going from left to right across a period. With a greater positive charge on the nuclei, there is an increase in the attraction for the electrons going across each period. This is also seen as an increase in the electronegativities going from left to right across each period of the periodic table. Within each group, the highest electronegativities are at the top and decrease going down the group as the size of the atom increases. The electronegativities for the transition elements are low, but we will not include them in our discussion. There are no electronegativities assigned to the noble gases, because they do not typically form bonds.

CONCEPT CHECK 10.5

Electronegativity

Without referring to the electronegativity values, predict the order of increasing electronegativity for the elements Cl, F, P, and Mg.

ANSWER

We know that electronegativities increase from the lower left corner to the upper right corner of the periodic table. Because Mg is on the left side of Period 3, Mg has the lowest electronegativity of these elements. The element P, which is on the right of Mg but on the left of Cl, would have the next lowest electronegativity. The Cl on the right end of Period 3 would have a higher electronegativity than P. The highest electronegativity is F, which is above Cl, in the upper right corner of the periodic table. The order of increasing electronegativity is Mg, P, Cl, F.

Polarity of Bonds

SELF STUDY ACTIVITY
Bonds and Bond Polarities

In Chapter 6, we discussed bonding as either *ionic,* in which electrons are transferred, or *covalent,* in which electrons are shared. The difference in the electronegativities of two atoms in the bond can be used to predict the type of bond that forms. In H—H, the electronegativity difference is zero (2.1 − 2.1 = 0), which means the bonding electrons are shared equally. A bond between atoms with identical or very similar electronegativities is a **nonpolar covalent bond.** However, most covalent bonds are between atoms that have different electronegativities. When electrons are shared unequally, the bond is a **polar covalent bond.** In H—Cl, an electronegativity difference of 0.9 (3.0 − 2.1) means that the H—Cl bond is polar covalent (see Figure 10.2).

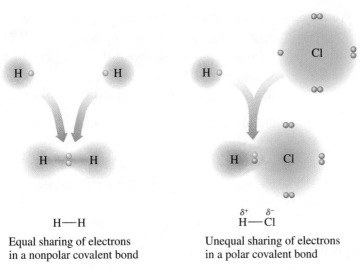

FIGURE 10.2 In the nonpolar covalent bond of H_2, electrons are shared equally. In the polar covalent bond of HCl, electrons are shared unequally.

Q H_2 has a nonpolar covalent bond, but HCl has a polar covalent bond. Explain.

$$H—H$$
Equal sharing of electrons
in a nonpolar covalent bond

$$\overset{\delta^+}{H}—\overset{\delta^-}{Cl}$$
Unequal sharing of electrons
in a polar covalent bond

Dipoles and Bond Polarity

The **polarity** of a bond depends on the electronegativity difference. In a polar covalent bond, the shared electrons are attracted to the more electronegative atom, which gives it a partial negative charge, whereas the atom with the lower electronegativity has a partial positive charge. A bond becomes more *polar* as the electronegativity difference increases. A polar covalent bond that has a separation of charges is called a **dipole**. The positive and negative ends are indicated by the lowercase Greek letter delta with a positive or negative sign, δ^+ and δ^-. Sometimes an arrow pointing from the positive charge to the negative charge $\longmapsto$ is used to indicate the dipole.

Examples of Dipoles in Polar Covalent Bonds

$$\overset{\delta^+}{C}—\overset{\delta^-}{O} \qquad \overset{\delta^+}{N}—\overset{\delta^-}{O} \qquad \overset{\delta^+}{Cl}—\overset{\delta^-}{F}$$

Variations in Bonding

The variations in bonding are continuous; there is no definite point at which one type of bond stops and the next starts. However, we can use some general ranges for predicting the types of bonds between atoms. When electronegativity differences are from 0.0 to 0.4, the electrons are considered to be shared equally in a *nonpolar covalent bond*. For example, H—H ($2.1 - 2.1 = 0$) and C—H ($2.5 - 2.1 = 0.4$) are classified as nonpolar covalent bonds. As electronegativity difference increases, the shared electrons are attracted to the more electronegative atom, which increases the polarity of the bond. When the electronegativity difference is greater than 0.4 but less than 1.8, the bond is a *polar covalent bond*. For example, an O—H bond with an electronegativity difference of 1.4 ($3.5 - 2.1 = 1.4$) is a *polar covalent bond*.

When the difference in electronegativity is 1.8 or greater, electrons are transferred from one atom to another, which gives an *ionic bond*. For example, we would classify K—Cl with an electronegativity difference of $3.0 - 0.8 = 2.2$ as an *ionic bond* (see Table 10.4).

TABLE 10.4 Electronegativity Difference and Types of Bonds

Electronegativity difference	0 0.4 1.8 3.3		
Bond type	Covalent Nonpolar	Covalent Polar	Ionic
Electron bonding	Electrons shared equally	Electrons shared unequally	Electron transfer
		δ^+　δ^-	$+$　$-$

TABLE 10.5 Predicting Bond Type from Electronegativity Differences

Molecule	Bond	Type of Electron Sharing	Electronegativity Difference[a]	Bond Type
H_2	H—H	Shared equally	$2.1 - 2.1 = 0$	Nonpolar covalent
Cl_2	Cl—Cl	Shared equally	$3.0 - 3.0 = 0$	Nonpolar covalent
HBr	$\overset{\delta^+}{H}—\overset{\delta^-}{Br}$	Shared unequally	$2.8 - 2.1 = 0.7$	Polar covalent
HCl	$\overset{\delta^+}{H}—\overset{\delta^-}{Cl}$	Shared unequally	$3.0 - 2.1 = 0.9$	Polar covalent
NaCl	Na^+Cl^-	Electron transfer	$3.0 - 0.9 = 2.1$	Ionic
MgO	$Mg^{2+}O^{2-}$	Electron transfer	$3.5 - 1.2 = 2.3$	Ionic

[a]Values are taken from Figure 10.1.

A summary of predicting bond type is seen in Table 10.5.

CONCEPT CHECK 10.6

■ Bond Polarity

Use electronegativities to classify each of the following bonds as nonpolar covalent, polar covalent, or ionic:

O—H   O—K   Cl—As   N—N

ANSWER

For each of the bonds, we use the electronegativity values to calculate the difference in electronegativities. A difference of 0.0 to 0.4 indicates a nonpolar covalent bond, whereas a difference of 0.5 to 1.7 indicates a polar covalent bond. An ionic bond is predicted when the difference is 1.8 or greater.

Bond	Electronegativity Difference	Type of Bond
O—H	$3.5 - 2.1 = 1.4$	Polar covalent
O—K	$3.5 - 0.8 = 2.7$	Ionic
Cl—As	$3.0 - 2.0 = 1.0$	Polar covalent
N—N	$3.0 - 3.0 = 0.0$	Nonpolar covalent

Polarity of Molecules

We have seen that covalent bonds in molecules can be polar covalent or nonpolar covalent. Now we will look at how those bonds determine the polarity of molecules of covalent compounds, which can be polar or nonpolar.

Nonpolar Molecules

Diatomic molecules of identical atoms, such as H_2 or Cl_2, are nonpolar because they contain one nonpolar covalent bond.

Nonpolar Molecules

H—H   Cl—Cl

Nonpolar covalent bonds

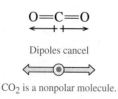

$O{=}C{=}O$

Dipoles cancel

CO_2 is a nonpolar molecule.

Molecules with two or more bonds can be nonpolar if there are no polar bonds or if the polar bonds have a symmetrical arrangement in the molecule. If the polar bonds or dipoles cancel each other, it is a nonpolar molecule. For example, CO_2 contains two polar covalent bonds. Because the CO_2 molecule is linear, the dipoles point in opposite directions. As a result, the dipoles cancel out, which makes a CO_2 molecule nonpolar.

Another example of a nonpolar molecule is CCl_4, which has four chlorine atoms at the corners of a tetrahedron surrounding the central carbon atom. Each of the C—Cl bonds has the same polarity, but because they have a tetrahedral arrangement, their opposing dipoles cancel out. As a result, a molecule of CCl₄ is nonpolar.

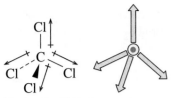

The four C—Cl dipoles cancel out.

Polar Molecules

In a polar molecule, one end of the molecule is more negatively charged than the other end. Polarity in a molecule occurs when the dipoles from the individual polar bonds do not cancel each other. For example, HCl is a polar molecule because it has one covalent bond that is polar.

H — Cl
⊢——→

⊙══⇒

A single dipole does not cancel.

In molecules with two or more electron groups, the shape, such as bent or trigonal pyramidal, determines whether the dipoles cancel or not. For example, we have seen that H_2O has a bent shape. Thus, a water molecule is polar because the individual dipoles do not cancel.

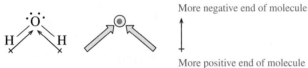

More negative end of molecule

More positive end of molecule

H_2O is a polar molecule because dipoles do not cancel.

The NH₃ molecule has a tetrahedral electron geometry with three bonded atoms, which gives it a trigonal pyramidal shape. Thus, a NH₃ molecule is polar because the individual N—H dipoles do not cancel.

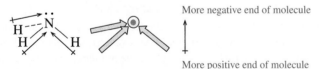

More negative end of molecule

More positive end of molecule

NH₃ is a polar molecule because dipoles do not cancel.

SAMPLE PROBLEM | 10.6

■ Polarity of Molecules

Indicate if each of the following molecules is polar or nonpolar:

a. BF₃   **b.** CH₃F

SOLUTION

a. The molecule BF₃ consists of a central atom, B, with three polar B—F bonds. Because it has a trigonal planar shape, the three B—F dipoles cancel, which makes BF₃ a nonpolar molecule.

b. In a molecule of CH₃F, the atoms bonded to the central atom, C, are not identical. The three C—H bonds are nonpolar, but the C—F bond is polar. Although the bonded atoms have a tetrahedral geometry, the nonpolar C—H bonds do not cancel the polar C—F bond. Thus, the CH₃F molecule is polar.

STUDY CHECK

Is the PCl₃ molecule polar or nonpolar?

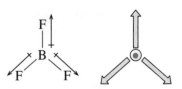

BF₃ is a nonpolar molecule.

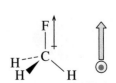

CH₃F is a polar molecule.

QUESTIONS AND PROBLEMS

Electronegativity and Polarity

10.23 Describe the trend in electronegativity going across a period.

10.24 Describe the trend in electronegativity going down a group.

10.25 Approximately what electronegativity difference would you expect for a nonpolar covalent bond?

10.26 Approximately what electronegativity difference would you expect for a polar covalent bond?

10.27 Using the periodic table, arrange the atoms in each of the following sets in order of increasing electronegativity:
    a. Li, Na, K   **b.** Na, P, Cl   **c.** O, Ca, Br

10.28 Using the periodic table, arrange the atoms in each of the following sets in order of increasing electronegativity:
    a. Cl, F, Br   **b.** B, O, N   **c.** Mg, F, S

10.29 For each of the following bonds, indicate the positive δ^+ end and the negative δ^- end. Draw an arrow to show the dipole for each.
    a. N—F   **b.** Si—P   **c.** C—O
    d. P—Br   **e.** B—Cl

10.30 For each of the following bonds, indicate the positive δ^+ end and the negative δ^- end. Draw an arrow to show the dipole for each.
    a. Si—Br   **b.** Se—F   **c.** Br—F
    d. N—H   **e.** N—P

10.31 Predict whether each of the following bonds is ionic, polar covalent, or nonpolar covalent:
    a. Si—Br   **b.** Li—F   **c.** Br—F
    d. Br—Br   **e.** N—P   **f.** C—P

10.32 Predict whether each of the following bonds is ionic, polar covalent, or nonpolar covalent:
    a. Si—O   **b.** K—Cl   **c.** S—F
    d. P—Br   **e.** Li—O   **f.** N—P

10.33 Why is F_2 a nonpolar molecule, but HF is polar?

10.34 Why is CBr_4 a nonpolar molecule, but NBr_3 is polar?

10.35 Identify each of the following molecules as polar or nonpolar:
    a. CS_2 **b.** NF_3
    c. Br_2 **d.** SO_3

10.36 Identify each of the following molecules as polar or nonpolar:
    a. H_2S **b.** PBr_3
    c. $SiCl_4$ **d.** SO_2

10.37 The molecule CO_2 is nonpolar, but CO is polar. Explain.

10.38 The molecules CH_4 and CH_3Cl both have tetrahedral shapes. Why is CH_4 nonpolar whereas CH_3Cl is polar?

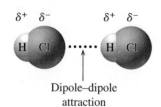

Dipole–dipole attraction

10.4 Attractive Forces in Compounds

In Chapter 6, we learned that the atoms in compounds are held together by ionic and covalent bonds. In this chapter, we will look at the attractive forces that hold molecules and ions together in liquids and solids. If these attractive forces are weak, the substance undergoes a change of state at relatively low melting and boiling points. If the attractive forces are strong, the substance changes state at higher melting and boiling points.

In solids and liquids, there are interactions between the particles to hold them close together. A solid melts and a liquid boils when sufficient heat exceeds the strength of the attractive forces between the particles. In gases, the strength of the attractive forces is minimal, which allows gas molecules to move far apart from each other. Such differences in properties are explained by looking at the various kinds of attractive forces between particles.

Ionic compounds have high melting points. Large amounts of energy are needed to overcome the strong attractive forces between positive and negative ions and to melt the ionic solid. For example, the ionic solid NaCl melts at 801 °C. In solids containing molecules with covalent bonds, there are attractive forces, too, but they are weaker than those of ionic compounds. The attractive forces we will discuss are *dipole–dipole attractions*, *hydrogen bonding*, and *dispersion forces*.

Dipole–Dipole Attractions and Hydrogen Bonds

For polar molecules, attractive forces called **dipole–dipole attractions** occur between the positive end of one molecule and the negative end of another. For a polar molecule with a dipole such as HCl, the partially positive H atom of one HCl molecule attracts the partially negative Cl atom in another molecule.

When a hydrogen atom is attached to highly electronegative atoms of fluorine, oxygen, or nitrogen, there are strong dipole–dipole attractions between the polar covalent molecules. This type of attraction, called a **hydrogen bond**, occurs between the partially positive hydrogen atom of one molecule and a lone pair of electrons on a nitrogen, oxygen, or fluorine atom in another molecule. Hydrogen bonds are the strongest type of attractive forces between polar covalent molecules. They are a major factor in the formation and structure of biological molecules such as proteins and DNA.

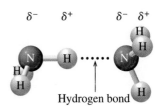

Hydrogen bond

Dispersion Forces

Nonpolar covalent compounds—such as methane, CH_4—can form liquids and solids, but only at very low temperatures. Weak attractive forces called **dispersion forces** occur between nonpolar molecules. Usually, the electrons in a nonpolar covalent molecule are distributed symmetrically. However, the movement of the electrons may place more electrons on one end of the molecule than the other, forming a *temporary dipole*. These momentary dipoles in what is normally a nonpolar substance align the molecules so that the positive end of one molecule is attracted to the negative end of another molecule. Although dispersion forces are very weak, they make it possible for nonpolar molecules to form liquids and solids.

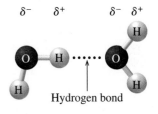

Hydrogen bond

Size, Mass, and Boiling Points

As the size and mass of the nonpolar covalent compound increase, there are more electrons that produce stronger temporary dipoles. In general, larger nonpolar molecules with increased molar masses also have higher boiling points. We see this trend in the boiling points of the first four alkanes shown in Table 10.6.

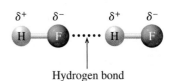

Hydrogen bond

TABLE 10.6 Mass and Boiling Points of the First Four Alkanes

Alkane	Formula		Molar Mass	Boiling Point
Methane	CH_4	CH_4	16.04 g/mol	−162 °C
Ethane	C_2H_6	CH_3—CH_3	30.07 g/mol	−89 °C
Propane	C_3H_8	CH_3—CH_2—CH_3	44.09 g/mol	−42 °C
Butane	C_4H_{10}	CH_3—CH_2—CH_2—CH_3	58.12 g/mol	−1 °C

CONCEPT CHECK 10.7

■ Boiling Points with Increasing Molar Masses

Arrange the following compounds in order of increasing boiling points: CH_4, SiH_4, GeH_4. Assign one of the following boiling points to each compound: −89 °C, −112 °C, or −162 °C.

ANSWER

Because the boiling point of a compound increases with molar mass, we calculate the molar mass of each compound.

   CH_4 16.04 g/mol SiH_4 32.12 g/mol GeH_4 77.67 g/mol

The lowest boiling point of −162 °C is assigned to the compound with the smallest molar mass, CH_4. The highest boiling point of −89 °C is assigned to the compound with the largest molar mass, GeH_4, which leaves −112 °C for SiH_4. In order of increasing boiling points, the three compounds are:

   CH_4 bp = −162 °C SiH_4 bp = −112 °C GeH_4 bp = −89 °C

TABLE 10.7 Comparison of Bonding and Attractive Forces

Type of Force	Particle Arrangement	Energy (kJ/mol)	Example
Between atoms or ions			
Ionic bond		500–5000	$Na^+ \cdots Cl^-$
Covalent bond (X = nonmetal)	X ⋮ X	100–1000	$Cl—Cl$
Between molecules			
Hydrogen bonds (X = F, O, or N)	$\delta^+ \; \delta^-$ $\delta^+ \; \delta^-$ H X ··· H X	10–40	$\delta^+ \;\; \delta^- \;\; \delta^+ \;\; \delta^-$ $H—F \cdots H—F$
Dipole–dipole attractions (X and Y = different nonmetals)	$\delta^+ \; \delta^-$ $\delta^+ \; \delta^-$ Y X ··· Y X	5–20	$\delta^+ \;\; \delta^- \;\; \delta^+ \;\; \delta^-$ $Br—Cl \cdots Br—Cl$
Dispersion forces (Temporary shift of electrons in nonpolar bonds)	$\delta^+ \; \delta^-$ $\delta^+ \; \delta^-$ (temporary dipoles) X ⋮ X ··· X ⋮ X	1–10	$\delta^+ \;\; \delta^- \;\; \delta^+ \;\; \delta^-$ $F—F \cdots F—F$

The various types of attractions within ionic and covalent compounds and between the particles in solids and liquids are summarized in Table 10.7.

Attractive Forces and Melting Points

The melting point of a substance is related to the strength of the attractive forces between its particles. A compound with weak attractive forces such as dispersion forces has a low melting point because only a small amount of energy is needed to separate the molecules and form a liquid. A compound with dipole–dipole attractions requires more energy to break the attractive forces that hold particles together. A compound that can hydrogen bond requires even more energy to overcome the attractive forces that exist between its molecules. However, the highest melting points are seen with ionic compounds that have the very strong attractions between positive and negative ions. More energy is required to break apart the covalent bonds within a molecule or polyatomic ion compared to the energy needed to overcome the attractive forces between molecules. Substances change state by breaking the attractive forces between particles, rather than breaking apart the covalent bonds that hold the atoms together. Table 10.8 compares the melting points of some substances with various kinds of attractive forces.

TABLE 10.8 Melting Points of Selected Substances

Substance	Melting Point (°C)
Ionic bonds	
MgF_2	1248
NaCl	801
Hydrogen bonds	
H_2O	0
NH_3	−78
Dipole–dipole attractions	
HI	−51
HBr	−89
HCl	−115
Dispersion forces	
Br_2	−7
Cl_2	−101
H_2	−259
C_5H_{12}	−130
CH_4	−182

SAMPLE PROBLEM 10.7

■ **Attractive Forces Between Particles**

Indicate which major type of molecular interaction—dipole–dipole attractions, hydrogen bonding, or dispersion forces—is expected of each of the following:

a. H—F   **b.** F—F   **c.** PCl_3

SOLUTION

a. H—F is a polar molecule that interacts with other H—F molecules by hydrogen bonding.
b. Because F—F is nonpolar, only dispersion forces provide attractive forces.
c. The polarity of the PCl_3 molecules provides dipole–dipole attractions.

STUDY CHECK

Why is the boiling point of H_2O higher than that of H_2S?

CHEMISTRY AND HEALTH

Attractive Forces in Biological Compounds

The proteins in our bodies are biological molecules that have many different functions. They are needed for structural components such as cartilage, muscles, hair, and nails and for the formation of enzymes that regulate biological reactions. Other proteins, such as hemoglobin and myoglobin, transport oxygen in the blood and muscle.

Proteins are composed of building blocks called amino acids. Every amino acid has a central carbon atom bonded to an $-NH_3^+$ from an amine, a $-COO^-$ from a carboxylic acid, an H atom, and a side chain called an R group unique for each amino acid.

Ionized form of an amino acid

Several of the amino acids have side chains that contain the functional groups we discussed in Chapter 8: hydroxyl $-OH$, carboxyl $-COOH$, which is ionized as carboxylate, $-COO^-$, carbonyl $-C=O$, and amine, $-NH_2$, which is ionized as ammonium, $-NH_3^+$.

Some important amino acids found in proteins.

Serine (Ser)   Cysteine (Cys)   Asparagine (Asn)   Lysine (Lys)   Aspartic acid (Asp)

In the primary structure of proteins, amino acids are linked together by peptide bonds, which are amide bonds that form between the $-COO^-$ group of one amino acid and the $-NH_3^+$ group of the next amino acid. When there are more than 50 amino acids in a chain, it is called a protein. Every protein in our bodies has a unique sequence of amino acids, the primary structure, that determines its biological function. In addition, proteins have higher levels of structure. In an *alpha helix*, hydrogen bonds form between the hydrogen atom of the N—H groups and the oxygen atom of a C=O group in the next turn. Because many hydrogen bonds form, the backbone of a protein takes a helical shape similar to a corkscrew or a spiral staircase.

Shapes of Biologically Active Proteins

Many proteins have compact, spherical shapes because sections of the chain fold over on top of other sections of the chain. This shape is stabilized by attractive forces between functional groups of side chains (R groups), causing the protein to twist and bend into a specific three-dimensional shape.

Hydrogen bonds form between the side chains within the protein. For example, a hydrogen bond can occur between the $-OH$ groups on two serines or between the $-OH$ of a serine and the $-NH_2$ of asparagine. Hydrogen bonding also occurs between the polar side chains of the amino acids on the outer surface of the protein and $-OH$ or $-H$ of polar water molecules in the external aqueous environment.

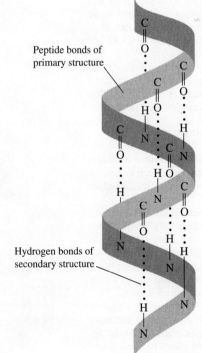

The shape of an alpha helix is stabilized by hydrogen bonds.

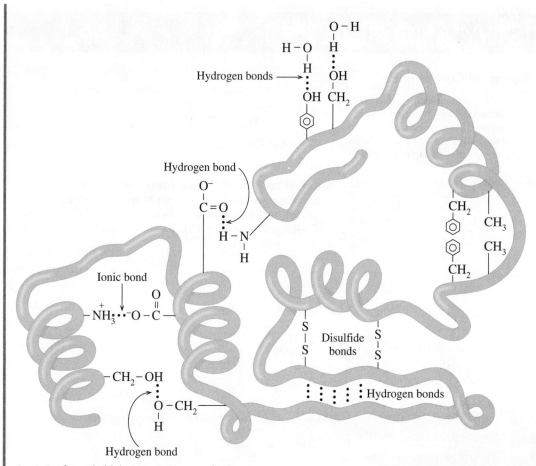

Attractive forces hold the protein in a specific shape.

Ionic bonds can form between the positively and negatively charged R groups of acidic and basic amino acids. For example, an ionic bond can form between the $-NH_3^+$ of lysine and the $-COO^-$ of aspartic acid.

In a disulfide bond ($-S-S-$), a covalent bond forms between the $-SH$ groups of two cysteines in a protein. In some proteins, there are several disulfide bonds, which maintain the three-dimensional shape.

Disruption of Attractive Forces

A protein loses its biological activity when there is a disruption of the attractive forces that stabilize its shape. When there are changes in conditions such as heat or acidity, the shape of a protein is altered; it may unfold and become like a loose piece of cooked spaghetti.

Heat breaks apart hydrogen bonds by increasing the motions of the particles. Generally, temperatures above 50 °C disrupt the shape and biological activity of most proteins. Whenever you cook food, you are using heat to destroy the shape of protein. High temperatures are used to disinfect surgical instruments and gowns by inactivating the proteins of any bacteria present.

The addition of an *acid* or a *base* to a protein breaks down hydrogen bonds and disrupts the ionic bonds. In the preparation of yogurt and cheese, bacteria that produce lactic acid are added to break down milk protein and produce solid casein. Tannic acid is used to coagulate proteins at the site of the burn, forming a protective cover and preventing further loss of fluid from the burn.

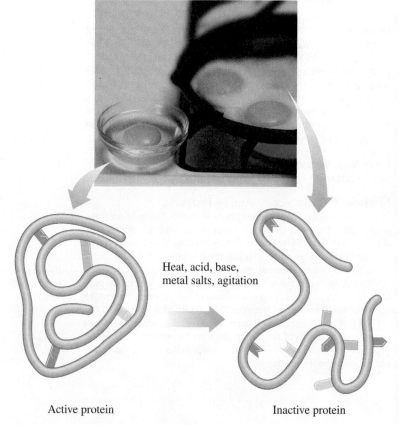

Heat, acid, base, metal salts, agitation

Active protein

Inactive protein

Protein structure is altered by heating.

Organic compounds, such as alcohols like isopropanol (rubbing alcohol), act as disinfectants by forming their own hydrogen bonds with a protein, which disrupt attractive forces. For this reason, an alcohol swab is used before an injection to destroy proteins in bacteria present on the skin. The *metal ions* Ag^+, Pb^{2+}, and Hg^{2+} denature protein by forming bonds with negatively charged side chains or with sulfur in side chains. In hospitals, a dilute (1%) solu-tion of $AgNO_3$ is placed in the eyes of newborn babies to destroy the bacteria that cause gonorrhea.

The whipping of cream and the beating of egg whites are examples of using mechanical *agitation* to disrupt attractive forces in proteins. The whipping action stretches the protein chains until the attractive forces are disrupted.

QUESTIONS AND PROBLEMS

Attractive Forces in Compounds

10.39 Identify the major type of attractive force between particles of each of the following substances:
  a. BrF **b.** KCl
  c. CCl_4 **d.** NF_3
  e. Cl_2

10.40 Identify the major type of attractive force between particles of each of the following substances:
  a. HCl **b.** MgF_2
  c. PBr_3 **d.** Br_2
  e. NH_3

10.41 Identify the strongest attractive forces between molecules of each of the following:
  a. CH_3OH **b.** H_2S
  c. CO **d.** CF_4
  e. $CH_3-CH_2-CH_3$

10.42 Identify the strongest attractive forces between molecules of each of the following:
  a. O_2 **b.** SiH_4
  c. CH_3Cl **d.** H_2O_2
  e. Ne

10.5 Changes of State

In Chapter 2, we described the states and properties of matter: gases, liquids, and solids. We can now discuss how the different types of attractive forces determine when matter under-goes a **change of state** when it is converted from one state to another (see Figure 10.3).

When heat is added to a solid, the particles move faster. At a temperature called the **melting point (mp)**, the particles of a solid gain sufficient energy to overcome the attractive forces that hold them together. The substance is **melting**, changing from a solid to a liquid.

LEARNING GOAL

Describe the changes of state between solids, liquids, and gases; calculate the energy involved.

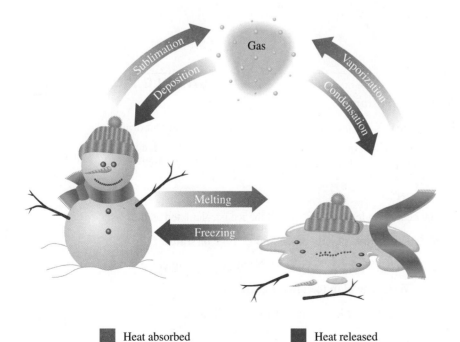

FIGURE 10.3 A summary of the changes of state.

Q Is heat added or released when liquid water freezes?

Heat absorbed   Heat released

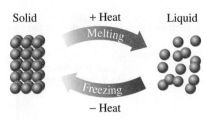

Solid   + Heat   Liquid

Melting

Freezing

– Heat

Melting and freezing are reversible processes.

If the temperature is lowered, the reverse process takes place. Kinetic energy is lost, the particles of a liquid slow down, and attractive forces pull the particles closer together. The substance is **freezing**. A liquid changes to a solid at the **freezing point (fp)**, which is the same temperature as the melting point. Every substance has its own freezing (melting) point: water freezes (melts) at 0 °C; gold freezes (melts) at 1064 °C; and nitrogen freezes (melts) at −210 °C.

During a change of state, the temperature of a substance remains constant. Suppose we have a glass containing ice and water. The ice melts when heat is added at 0 °C, forming more liquid. The liquid freezes when heat is removed at 0 °C, forming more solid. The processes of melting and freezing are reversible at 0 °C.

Heat of Fusion

During melting, energy called the **heat of fusion** is added to separate the particles of a solid. For example, 334 J (80. calories) of heat are needed to melt exactly 1 g of ice at its melting point (0 °C).

(MC)

TUTORIAL

Heat of Vaporization and Heat of Fusion

Heat of Fusion for Water

$$\frac{334 \text{ J}}{1 \text{ g H}_2\text{O}} \qquad \frac{80.\ \text{cal}}{1 \text{ g H}_2\text{O}}$$

The heat of fusion (334 J/g or 80. cal/g) is also the heat that must be removed to freeze 1 g of water at its freezing point (0 °C). Water is sometimes sprayed in fruit orchards during subfreezing weather. If the air temperature drops to 0 °C, the water begins to freeze. Heat is released as the water molecules bond together, which warms the air and protects the fruit.

To determine the heat needed to melt a sample of ice, multiply the mass of the ice by its heat of fusion. There is no temperature change given in the calculation, because the temperature remains constant as long as the ice is melting.

Calculating Heat to Melt (or Freeze) Water

$$\text{Heat} = \text{mass} \times \text{heat of fusion}$$

$$\text{J} = \cancel{g} \times \frac{334 \text{ J}}{\cancel{g}}$$

$$\text{cal} = \cancel{g} \times \frac{80.\ \text{cal}}{\cancel{g}}$$

Guide to Calculations Using Heat of Fusion

STEP 1
List grams of substance and change of state.

STEP 2
Write the plan to convert grams to heat and desired unit.

STEP 3
Write the heat conversion factors and metric factors if needed.

STEP 4
Set up the problem with factors.

SAMPLE PROBLEM | 10.8

■ **Heat of Fusion**

Ice cubes at 0 °C with a mass of 26.0 g are added to your soft drink.

a. How much heat (joules) will be absorbed to melt all the ice at 0 °C?
b. What happens to the temperature of your soft drink? Why?

SOLUTION

a. The heat in joules required to melt the ice is calculated as follows:

STEP 1 List grams of substance and change of state.

    Given 26.0 g of $H_2O(s)$ **Need** joules to melt ice

STEP 2 Write the plan to convert grams to heat.

    grams of ice   Heat of fusion   joules

STEP 3 Write the heat conversion factors and metric factors if needed.

$$1 \text{ g of } H_2O \ (s \rightarrow l) = 334 \text{ J}$$

$$\frac{334 \text{ J}}{1 \text{ g } H_2O} \quad \text{and} \quad \frac{1 \text{ g } H_2O}{334 \text{ J}}$$

STEP 4 Set up the problem with factors.

$$26.0 \text{ g } \cancel{H_2O} \ \times \ \frac{334 \text{ J}}{1 \text{ g } \cancel{H_2O}} \ = \ 8680 \text{ J}$$

b. The soft drink will be colder because heat from the soft drink is providing the energy to melt the ice.

STUDY CHECK

In a freezer, 125 g of water at 0 °C is placed in an ice cube tray. How much heat, in kilojoules, must be removed to form ice cubes at 0 °C?

Boiling and Condensation

Water in a mud puddle disappears, unwrapped food dries out, and clothes hung on a clothesline dry. **Evaporation** is taking place as molecules of liquid water with sufficient energy escape from the liquid surface and enter the gas phase (see Figure 10.4a). The loss of the "hot" water molecules removes heat, which cools the remaining liquid water. As heat is added, more and more water molecules evaporate. At the **boiling point (bp)**, the molecules of a liquid acquire the energy needed to change into a gas. The **boiling** of the liquid occurs as gas bubbles form throughout the liquid, then rise to the surface and escape (see Figure 10.4b).

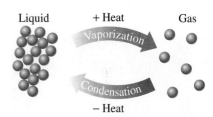

Vaporization and condensation are reversible processes.

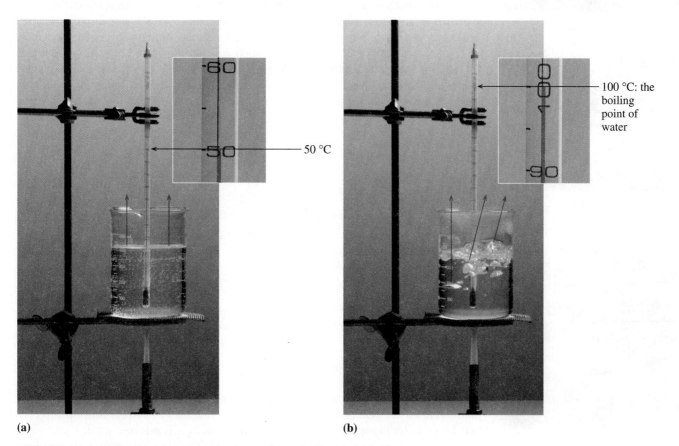

(a)   (b)

FIGURE 10.4 (a) Evaporation occurs at the surface of a liquid. (b) Boiling occurs as bubbles of gas form throughout the liquid.

Q Why does water evaporate faster at 80 °C than at 20 °C?

When heat is removed, a reverse process takes place. In **condensation**, water vapor is converted to liquid as the water molecules lose kinetic energy and slow down. Condensation occurs at the same temperature as boiling but differs because heat is removed. You may have noticed that condensation occurs when you take a hot shower and the water vapor forms water droplets on a mirror. Because a substance loses heat as it condenses, its surroundings become warmer. That is why, when a rainstorm is approaching, we notice a warming of the air as gaseous water molecules condense to rain.

With enough time, water in an open container will all evaporate. However, if a tight-fitting cover is placed on the container, the water level will go down just a little as the water molecules evaporate from the surface. Then, some of the vapor molecules begin to condense and return to liquid. Eventually, the number of evaporating molecules is equal to the number condensing. The reverse processes of evaporation and condensation have equalized. As a result, the water level does not go any lower.

Sublimation

In a process called **sublimation**, the particles on the surface of a solid change directly to a gas with no temperature change and without going through the liquid state. In the reverse process of sublimation, called **deposition**, gas particles change directly to solid.

Heat of Sublimation for Water

$$\frac{2590 \text{ J}}{1 \text{ g H}_2\text{O}} \qquad \frac{620. \text{ cal}}{1 \text{ g H}_2\text{O}}$$

For example, dry ice, which is solid carbon dioxide, undergoes sublimation at −78 °C. It is called "dry" because it does not form a liquid as it warms. In extremely cold areas, snow does not melt but sublimes directly to water vapor.

When frozen foods are left in the freezer for a long time, so much water sublimes that foods, especially meats, become dry and shrunken, a condition called *freezer burn*. Deposition occurs in a freezer when water vapor forms ice crystals on the surface of freezer bags and frozen food.

Freeze-dried foods prepared by sublimation are convenient for long-term storage and for camping and hiking. A food that has been frozen is placed in a vacuum chamber where it dries as the ice sublimes. The dried food retains all of its nutritional value and needs only water to be edible. A food that is freeze-dried does not need refrigeration, because bacteria cannot grow without moisture.

Solid   + Heat   Gas
Sublimation
Deposition
− Heat

Sublimation and deposition are reversible processes.

Dry ice sublimes at −78 °C.

Water vapor will change to solid on contact with a cold surface.

Freeze-dried food is prepared by sublimation.

CONCEPT CHECK 10.8

■ Identifying Changes of State

Give the change of state described in each of the following:

a. particles on the surface of a liquid escaping to form vapor
b. a liquid changing to a solid
c. gas bubbles forming throughout a liquid

ANSWER
a. evaporation   **b.** freezing   **c.** boiling

Heat of Vaporization

The energy that must be added to convert exactly 1 g of liquid to gas at its boiling point is called the **heat of vaporization**. For water, 2260 J or 540 cal is needed to convert 1 g of water to vapor at 100 °C. This amount of heat is released when 1 g of water vapor (gas) changes to liquid at 100 °C. Therefore, 2260 J or 540 cal/g is also the *heat of condensation* of water.

Heat of Vaporization for Water

$$\frac{2260 \text{ J}}{1 \text{ g H}_2\text{O}} \qquad \frac{540 \text{ cal}}{1 \text{ g H}_2\text{O}}$$

To calculate the amount of heat added to (or removed from) a sample of water, the mass of the sample is multiplied by the heat of vaporization. As before, no temperature change occurs during a change of state.

Calculating Heat to Vaporize (or Condense) Water

Heat = mass × heat of vaporization

$$J = g \times \frac{2260 \text{ J}}{g}$$

$$cal = g \times \frac{540 \text{ cal}}{g}$$

Just as substances have different melting and boiling points, they also have different heats of fusion and vaporization (see Table 10.9). Typically, the heats of vaporization are much larger than the heats of fusion (see Figure 10.5).

TABLE 10.9 Heats of Fusion and Heats of Vaporization for Selected Substances

Liquid	Formula	Melting Point (°C)	Heat of Fusion (J/g)	Boiling Point (°C)	Heat of Vaporization (J/g)
Water	H_2O	0	334	100	2260
Ethanol	C_2H_5OH	−114	109	78	841
Ammonia	NH_3	−78	351	−33	1380
Acetone	C_3H_6O	−95	98	56	335
Mercury	Hg	−39	11	357	294
Acetic acid	$C_2H_4O_2$	17	192	118	390

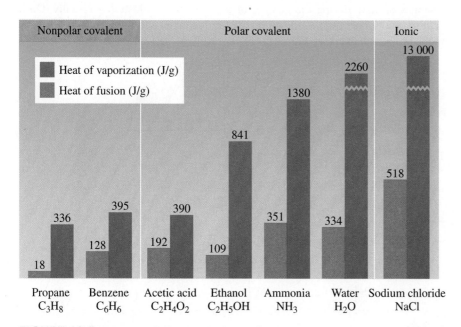

FIGURE 10.5 For any substance, the heat of vaporization is greater than the heat of fusion.

Q Why does the formation of a gas require more energy than the formation of a liquid of the same compound?

SAMPLE PROBLEM | 10.9

■ Using Heat of Vaporization

In a sauna, 122 g of water is converted to steam at 100 °C. How many kilojoules of heat are needed?

SOLUTION

STEP 1 List the grams of substance and change of state.

Given 122 g of $H_2O(l)$ to $H_2O(g)$ **Need** kilojoules to change state

STEP 2 Write the plan to convert grams to heat and desired unit.

g of H_2O | Heat of vaporization ▸ | J | Metric factor | kJ

STEP 3 Write the heat conversion factor and metric factor if needed.

1 g of H_2O $(l \rightarrow g)$ = 2260 J

$$\dfrac{2260 \text{ J}}{1 \text{ g } H_2O} \quad \text{and} \quad \dfrac{1 \text{ g } H_2O}{2260 \text{ J}}$$

1 kJ = 1000 J

$$\dfrac{1000 \text{ J}}{1 \text{ kJ}} \quad \text{and} \quad \dfrac{1 \text{ kJ}}{1000 \text{ J}}$$

STEP 4 Set up the problem with factors.

$$122 \text{ g } H_2O \times \dfrac{2260 \text{ J}}{1 \text{ g } H_2O} \times \dfrac{1 \text{ kJ}}{1000 \text{ J}} = 276 \text{ kJ}$$

STUDY CHECK

When steam from a pan of boiling water reaches a cool window, it condenses. How much heat, in kilojoules (kJ), is released when 25.0 g of steam condenses at 100 °C?

Guide to Using Heat of Vaporization

STEP 1
List the grams of substance and change of state.

STEP 2
Write the plan to convert grams to heat and desired unit.

STEP 3
Write the heat conversion factor and metric factor if needed.

STEP 4
Set up the problem with factors.

Heating and Cooling Curves

All the changes of state during the heating of a solid can be illustrated visually. On a **heating curve**, the temperature is shown on the vertical axis, and the addition of heat is shown on the horizontal axis (see Figure 10.6).

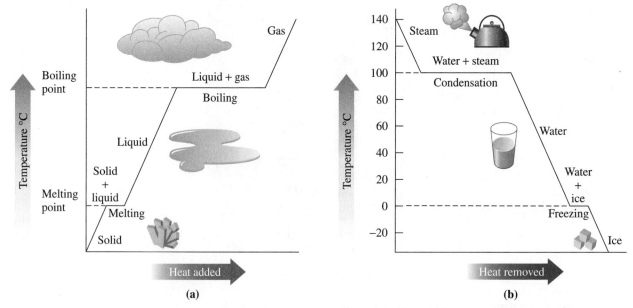

FIGURE 10.6 (a) A heating curve diagrams the temperature increases and changes in state as heat is added. (b) A cooling curve for water.

Q What does the plateau at 100 °C represent on the cooling curve for water?

CHEMISTRY AND HEALTH

Steam Burns

Hot water at 100 °C will cause burns and damage to the skin. However, getting steam on the skin is even more dangerous. If 25 g of hot water at 100 °C falls on a person's skin, the temperature of the water will drop to body temperature, 37 °C. The heat released during cooling burns the skin. The amount of heat can be calculated from the temperature change, 100 °C − 37 °C = 63 °C.

$$25 \text{ g} \times 63 \text{ °C} \times \frac{4.184 \text{ J}}{\text{g °C}} = 6600 \text{ J}$$

For comparison, we can calculate the amount of heat released when 25 g of steam at 100 °C hits the skin. First, the steam condenses to water (liquid) at 100 °C:

$$25 \text{ g} \times \frac{2260 \text{ J}}{1 \text{ g}} = 57\,000 \text{ J}$$

The total amount of heat released from the condensation and cooling of the steam is calculated as follows:

Condensation (100 °C) = 57 000 J

Cooling (100 °C to 37 °C) = 6600 J

Heat released = 57 000 J + 6600 J = 64 000 J (rounded off)

The amount of heat released from steam is 10 times greater than the heat from the same amount of hot water.

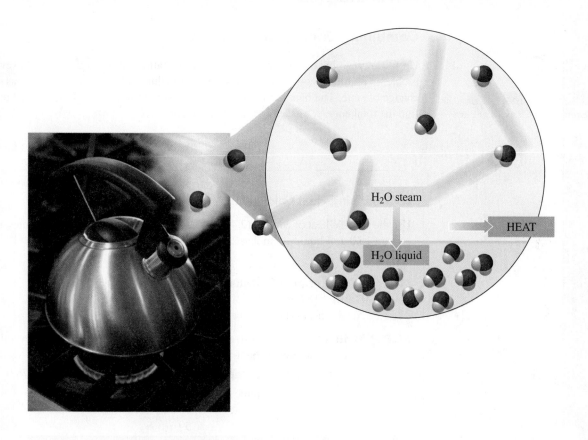

When a gram of steam condenses, 2260 J are released.

Steps on a Heating Curve

The first diagonal line indicates a warming of a solid as heat is added. When the melting temperature is reached, a horizontal line, or plateau, indicates that solid is melting. As melting takes place, the solid is changing to a liquid without any change in temperature (see Figure 10.6a).

Once all of the particles are in the liquid state, heat that is added will increase the temperature of the liquid. This increase is drawn as a diagonal line from the melting point to the boiling point temperature. Once the liquid reaches its boiling point, a horizontal line indicates that the temperature remains constant as liquid changes to gas. Because the heat of vaporization is greater than the heat of fusion, the horizontal line at the boiling point is longer than the line at the melting point. Once all the liquid becomes gas, adding more heat increases the temperature of the gas.

Steps on a Cooling Curve

A **cooling curve** is a diagram of the cooling process. In the diagram, the temperature is plotted on the vertical axis and the removal of heat is plotted on the horizontal axis (see Figure 10.6b). Initially, a diagonal line to the boiling (condensation) point is drawn to show that heat is removed from a substance, cooling the gas until it begins to condense. A horizontal line (plateau) is drawn at the condensation point (same as the boiling point) to indicate the change of state as the gas condenses to form a liquid. After all of the gas has changed into liquid, further cooling lowers the temperature. The decrease in temperature is shown as a diagonal line from the condensation point temperature to the freezing point temperature. At the freezing point, another horizontal line indicates that liquid is changing to solid at the freezing point temperature. Once all of the substance is frozen, a loss of heat decreases the temperature below its freezing point, which is shown as a diagonal line below its freezing point.

Combining Energy Calculations

TUTORIAL

Heat, Energy, and Changes of State

Up to now, we have calculated one step in a heating or cooling curve. However, many problems require a combination of steps that include a temperature change as well as a change of state. The heat is calculated for each step separately and then added together to find the total energy, as seen in Sample Problem 10.10.

SAMPLE PROBLEM 10.10

■ Combining Heat Calculations

Use the specific heat of ethanol (2.46 J/g °C) to calculate the total heat, in joules, needed to convert 15.0 g of liquid ethanol at 25.0 °C to gas at 78.0 °C.

SOLUTION

STEP 1 List the grams of substance and change of state.
  Given 15.0 g of ethanol at 25.0 °C
  Need heat (J) to warm ethanol and change to gas

STEP 2 Write the plan to convert grams to heat in joules. When several changes occur, draw a diagram of heating and changes of state.

  Total heat = joules needed to warm ethanol from 25.0 °C to 78.0 °C
        + joules to change liquid to gas at 78.0 °C

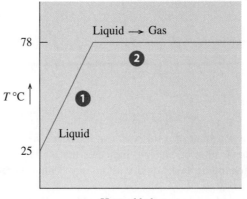

STEP 3   Write the heat conversion factors needed.

$$SH_{Ethanol} = \frac{2.46 \text{ J}}{\text{g }^\circ\text{C}}$$

$$\frac{2.46 \text{ J}}{\text{g }^\circ\text{C}} \quad \text{and} \quad \frac{\text{g }^\circ\text{C}}{2.46 \text{ J}}$$

1 g of ethanol ($l \rightarrow g$) = 841 J

$$\frac{841 \text{ J}}{1 \text{ g ethanol}} \quad \text{and} \quad \frac{1 \text{ g ethanol}}{841 \text{ J}}$$

STEP 4   Set up problem with factors.

$$\Delta T = 78.0 \text{ }^\circ\text{C} - 25.0 \text{ }^\circ\text{C} = 53.0 \text{ }^\circ\text{C}$$

Heat needed to warm ethanol (liquid) at 25.0 °C to ethanol (liquid) at 78.0 °C:

$$15.0 \text{ g} \times 53.0 \text{ }^\circ\text{C} \times \frac{2.46 \text{ J}}{\text{g }^\circ\text{C}} = 1960 \text{ J}$$

Heat needed to change ethanol (liquid) to ethanol (gas) at 78.0 °C:

$$15.0 \text{ g ethanol} \times \frac{841 \text{ J}}{1 \text{ g ethanol}} = 12\,600 \text{ J}$$

Calculate the total heat:

Heating ethanol (25.0 °C to 78.0 °C)	1 960 J
Changing liquid to gas (78.0 °C)	12 600 J
Total heat needed	14 600 J (rounded off)

STUDY CHECK

How many kilojoules (kJ) are released when 75.0 g of steam at 100 °C condenses, cools to 0 °C, and freezes? (*Hint:* The solution will require three energy calculations.)

QUESTIONS AND PROBLEMS

Changes of State

10.43 Calculate the heat needed at 0 °C in each of the following and indicate whether heat was absorbed or released:
a. joules to melt 65.0 g of ice
b. joules to melt 17.0 g of ice
c. kilojoules to freeze 225 g of water
d. kilojoules to freeze 50.0 g of water

10.44 Calculate the heat needed at 0 °C in each of the following and indicate whether heat was absorbed or released:
a. joules to freeze 35.2 g of water
b. joules to freeze 275 g of water
c. kilojoules to melt 145 g of ice
d. kilojoules to melt 5.00 kg of ice

10.45 For each of the following problems, calculate the heat change at 100 °C and indicate whether heat was absorbed or released:
a. joules to vaporize 10.0 g of water
b. kilojoules to vaporize 50.0 g of water
c. joules to condense 8.00 kg of steam
d. kilojoules to condense 175 g of steam

10.46 For each of the following problems, calculate the heat change at 100 °C and indicate whether heat was absorbed or released:
a. joules to condense 10.0 g of steam
b. kilojoules to condense 76.0 g of steam
c. joules to vaporize 44.0 g of water
d. kilojoules to vaporize 5.0 kg of water

10.47 Using the values for the heat of fusion, specific heat of water, and/or heat of vaporization, calculate the amount of heat energy in each of the following:
a. joules needed to warm 20.0 g of water at 15 °C to 72 °C
b. joules needed to melt 50.0 g of ice at 0 °C and to warm the liquid to 65.0 °C
c. kilojoules released when 15.0 g of steam condenses at 100 °C and the liquid cools to 0 °C
d. kilojoules needed to melt 24.0 g of ice at 0 °C, warm the liquid to 100 °C, and change it to steam at 100 °C

10.48 Using the values for the heat of fusion, specific heat of water, and/or heat of vaporization, calculate the amount of heat energy in each of the following:
a. joules to condense 125 g of steam at 100 °C and to cool the liquid to 15.0 °C
b. joules needed to melt a 525-g ice sculpture at 0 °C and to warm the liquid to 15.0 °C
c. kilojoules released when 85.0 g of steam condenses at 100 °C, cool the liquid, and freeze it at 0 °C
d. joules to warm 55.0 mL of water (density = 1.00 g/mL) from 10.0 °C to 100 °C and vaporize it at 100 °C

10.49 An ice bag containing 275 g of ice at 0 °C was used to treat sore muscles. When the bag was removed, the ice had melted and the liquid water had a temperature of 24.0 °C. How many kilojoules of heat were absorbed?

10.50 A 115-g sample of steam at 100 °C is emitted from a volcano. It condenses, cools, and falls as snow at 0 °C. How many kilojoules of heat were released?

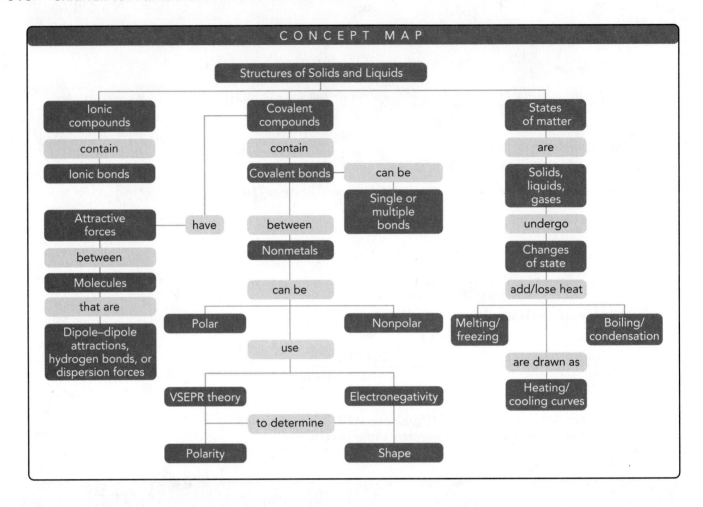

CHAPTER REVIEW

10.1 Electron-Dot Formulas

LEARNING GOAL: *Draw the electron-dot formulas for covalent compounds or polyatomic ions with multiple bonds and show resonance structures.*

In a covalent bond, atoms of nonmetals share valence electrons. In most covalent compounds, atoms achieve a noble gas electron configuration. The total number of valence electrons is determined for all the atoms in the molecule or ion. Any negative charge is added to the total valence electrons, while any positive charge is subtracted. In the electron-dot formulas, a bonding pair is placed between the central atom and attached atoms. The remaining valence electrons are used as lone pairs to complete the octets of the surrounding atoms and then the central atom. When octets are not completed, lone pairs are converted to bonding pairs forming double or triple bonds. Resonance structures are possible when two or more electron-dot formulas can be drawn for a molecule or ion with a multiple bond.

10.2 Shapes of Molecules and Ions (VSEPR Theory)

LEARNING GOAL: *Predict the three-dimensional structure of a molecule or polyatomic ion and classify it as polar or nonpolar.*

The shape of a molecule is determined from the electron-dot formula and the number of bonded atoms and lone pairs. The electron arrangement of two electron groups around a central atom is linear;

three electron groups are trigonal planar; and four are tetrahedral. When all the electron groups are bonded to atoms, the shape has the same name as the electron arrangement. A central atom with two bonded atoms and one or two lone pairs has a bent shape. A central atom with three bonded atoms and one lone pair has a trigonal pyramidal shape.

10.3 Electronegativity and Polarity

LEARNING GOAL: *Use electronegativity to determine the polarity of a bond or a molecule.*

Electronegativity is the ability of an atom to attract the electrons it shares with another atom. In general, the electronegativities of metals are low, while nonmetals have high electronegativities. In a nonpolar covalent bond, atoms share electrons equally. In a polar covalent bond, the electrons are unequally shared because they are attracted to the more electronegative atom. The atom in a polar bond with the lower electronegativity is partially positive (δ^+) and the atom with the higher electronegativity is partially negative (δ^-). Atoms that form ionic bonds have large differences in electronegativities. Nonpolar molecules contain nonpolar covalent bonds or have an arrangement of bonded atoms that causes the dipoles to cancel out. In polar molecules, the dipoles do not cancel because there are nonidentical bonded atoms or lone pairs on the central atom.

10.4 Attractive Forces in Compounds

LEARNING GOAL: *Describe the attractive forces between ions, polar covalent molecules, and nonpolar covalent molecules.*

In ionic solids, oppositely charged ions are held in a rigid structure by ionic bonds. Attractive forces called dipole–dipole attractions and hydrogen bonds hold the solid and liquid states of polar covalent compounds together. Nonpolar compounds form solids and liquids by temporary dipoles called dispersion forces.

10.5 Changes of State

LEARNING GOAL: *Describe the changes of state between solids, liquids, and gases; calculate the energy involved.*

The three states of matter are solid, liquid, and gas. Melting occurs when the particles in a solid absorb enough energy to break apart and form a liquid. The amount of energy required to convert exactly 1 g of solid to liquid is called the heat of fusion. For water, 334 J (80. cal) are needed to melt 1 g of ice or must be removed to freeze 1 g of water. Sublimation is a process whereby a solid changes directly to a gas.

Evaporation occurs when particles in a liquid state absorb enough energy to break apart and form gaseous particles. Boiling is the vaporization of liquid at its boiling point. The heat of vaporization is the amount of heat needed to convert exactly 1 g of liquid to vapor. For water, 2260 J (540 cal) are needed to vaporize 1 g of water or must be removed to condense 1 g of steam.

A heating or cooling curve illustrates the changes in temperature and state as heat is added to or removed from a substance. Plateaus on the graph indicate changes of state. The total heat absorbed or removed from a substance undergoing temperature changes and changes of state is the sum of energy calculations for change(s) of state and change(s) in temperature.

KEY TERMS

bent The shape of a molecule with two bonded atoms and one lone pair or two lone pairs.

boiling The formation of bubbles of gas throughout a liquid.

boiling point (bp) The temperature at which a substance exists as a liquid and gas; liquid changes to gas (boils), and gas changes to liquid (condenses).

change of state The transformation of one state of matter to another, for example, solid to liquid, liquid to solid, liquid to gas.

condensation The change of state of a gas to a liquid.

cooling curve A diagram that illustrates temperature changes and changes of states for a substance as heat is removed.

deposition The change of a gas directly to a solid; the reverse of sublimation.

dipole The separation of positive and negative charge in a polar bond indicated by an arrow that is drawn from the more positive atom to the more negative atom.

dipole–dipole attractions Attractive forces between oppositely charged ends of polar molecules.

dispersion forces Weak dipole bonding that results from a momentary polarization of nonpolar molecules in a substance.

double bond A sharing of two pairs of electrons by two atoms.

electronegativity The relative ability of an element to attract electrons in a bond.

evaporation The formation of a gas (vapor) by the escape of high-energy molecules from the surface of a liquid.

freezing A change of state from liquid to solid.

freezing point (fp) The temperature at which the solid and liquid forms of a substance are in equilibrium; a liquid changes to a solid (freezes), a solid changes to a liquid (melts).

heat of fusion The energy required to melt exactly 1 g of a substance. For water, 334 J are needed to melt 1 g of ice; 334 J are released when 1 g of water freezes.

heat of vaporization The energy required to vaporize 1 g of a substance. For water, 2260 J are needed to vaporize exactly 1 g of liquid; 1 g of steam gives off 2260 J when it condenses.

heating curve A diagram that shows the temperature changes and changes of state of a substance as it is heated.

hydrogen bond The attraction between a partially positive H and a strongly electronegative atom of F, O, or N.

linear The shape of a molecule that has two bonded atoms and no lone pairs.

melting The conversion of a solid to a liquid.

melting point (mp) The temperature at which a solid becomes a liquid (melts). It is the same temperature as the freezing point.

nonpolar covalent bond A covalent bond in which the electrons are shared equally.

polar covalent bond A covalent bond in which the electrons are shared unequally.

polarity A measure of the unequal sharing of electrons, indicated by the difference in electronegativities.

resonance structures Two or more electron-dot formulas that can be written for a molecule or ion by placing a multiple bond between different atoms.

sublimation The change of state in which a solid is transformed directly to a gas without forming a liquid.

tetrahedral The shape of a molecule with four bonded atoms.

trigonal planar The shape of a molecule with three bonded atoms and no lone pairs.

trigonal pyramidal The shape of a molecule that has three bonded atoms and one lone pair.

triple bond A sharing of three pairs of electrons by two atoms.

valence shell electron-pair repulsion (VSEPR) theory A theory that predicts the shape of a molecule by moving the electron groups on a central atom as far apart as possible to minimize the repulsion of the negative regions.

UNDERSTANDING THE CONCEPTS

10.51 Identify the major attractive force between each of the following atoms or molecules:

    a. SiH_4 **b.** NO_2

    c. CH_3-NH_2 **d.** Ar

10.52 Identify the major attractive force between each of the following atoms or molecules:

    a. He   **b.** HBr

    c. SnH_4 **d.** $CH_3-CH_2-CH_2-OH$

10.53 Why would BCl_3 be a nonpolar molecule when PCl_3 is a polar molecule?

10.54 Why would CO_2 be a nonpolar molecule when SO_2 is a polar molecule?

10.55 Use your knowledge of changes of state to explain the following:
  a. How does perspiration during heavy exercise cool the body?
  b. Why do towels dry more quickly on a hot summer day than on a cold winter day?
  c. Why do wet clothes stay wet in a plastic bag?

Perspiration forms on the skin during heavy exercise.

10.56 Use your knowledge of changes of state to explain the following:

A spray is used to numb a sports injury.

  a. Why is a spray that evaporates quickly, such as ethyl chloride, used to numb a sports injury during a game?
  b. Why does water in a wide, flat, shallow dish evaporate more quickly than the same amount of water in a tall, narrow vase?
  c. Why does a sandwich on a plate dry out faster than a sandwich in plastic wrap?

10.57 Draw a heating curve for a sample of ice that is heated from $-20\ °C$ to $150\ °C$. Indicate the segment of the graph that corresponds to each of the following:
  a. solid   **b.** melting
  c. liquid   **d.** boiling
  e. gas

10.58 Draw a cooling curve for a sample of steam that cools from $110\ °C$ to $-10\ °C$. Indicate the segment of the graph that corresponds to each of the following:
  a. solid   **b.** freezing
  c. liquid   **d.** condensing
  e. gas

10.59 The following is a heating curve for chloroform, a solvent for fats, oils, and waxes.

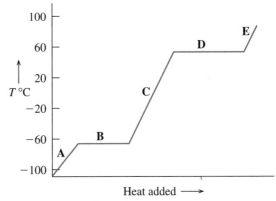

  a. What is the melting point of chloroform?
  b. What is the boiling point of chloroform?
  c. On the heating curve, identify the segments A, B, C, D, and E as solid, liquid, gas, melting, or boiling.
  d. At the following temperatures, is chloroform a solid, liquid, or gas? $-80\ °C$; $-40\ °C$; $25\ °C$; $80\ °C$?

10.60 Associate the contents of the beakers (1–5) with segments (A–E) on the heating curve for water.

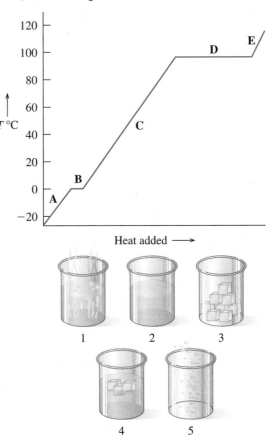

ADDITIONAL QUESTIONS AND PROBLEMS

For instructor-assigned homework, go to
www.masteringchemistry.com.

10.61 Determine the total number of valence electrons in each of the following:
a. CS_2 **b.** CH_3CHO **c.** PH_4^+
d. BCl_3 **e.** SO_3^{2-}

10.62 Determine the total number of valence electrons in each of the following:
a. $COCl_2$ **b.** N_2O **c.** ClO_2^-
d. $SeCl_2$ **e.** PBr_3

10.63 Draw the electron-dot formula for each of the following:
a. BF_4^- **b.** Cl_2O
c. H_2NOH (N is the central atom)
d. NO_2^+ **e.** H_2CCCl_2

10.64 Draw the electron-dot formula for each of the following:
a. H_3COCH_3 (the atoms are in the order C O C)
b. CS_2 (the atoms are in the order S C S)
c. ClO_2^-
d. SF_3^+
e. H_2CCHCN (the atoms are in the order C C C)

10.65 Draw resonance structures for each of the following:
a. N_3^- **b.** NO_2^+ **c.** CNS^-

10.66 Draw resonance structures for each of the following:
a. NO_3^- **b.** CO_3^{2-} **c.** SCN^-

10.67 Use the periodic table to arrange the following atoms in order of increasing electronegativity:
a. I, F, Cl   **b.** Li, K, S, Cl
c. Mg, Sr, Ba, Be

10.68 Use the periodic table to arrange the following atoms in order of increasing electronegativity:
a. Cl, Br, Se   **b.** Na, Cs, O, S
c. O, F, B, Li

10.69 Select the more polar bond in each of the following pairs:
a. C—N or C—O   **b.** N—F or N—Br
c. Br—Cl or S—Cl   **d.** Br—Cl or Br—I
e. N—S or N—O

10.70 Select the more polar bond in each of the following pairs:
a. C—C or C—O   **b.** P—Cl or P—Br
c. Si—S or Si—Cl   **d.** F—Cl or F—Br
e. P—O or P—S

10.71 Show the dipole arrow for each of the following bonds:
a. Si—Cl   **b.** C—N   **c.** F—Cl
d. C—F   **e.** N—O

10.72 Show the dipole arrow for each of the following bonds:
a. P—O   **b.** N—F   **c.** O—Cl
d. S—Cl   **e.** P—F

10.73 Classify each of the following bonds as nonpolar covalent, polar covalent, or ionic:
a. Si—Cl   **b.** C—C   **c.** Na—Cl
d. C—H   **e.** F—F

10.74 Classify each of the following bonds as nonpolar covalent, polar covalent, or ionic:
a. C—N   **b.** Cl—Cl   **c.** K—Br
d. H—H   **e.** N—F

10.75 Draw the electron-dot formula and determine the shape for each of the following:
a. NF_3 **b.** $SiBr_4$
c. $BeCl_2$ **d.** SO_2

10.76 Draw the electron-dot formula and determine the shape for each of the following:
a. NH_4^+
b. O_2^{2-}
c. $COCl_2$ (C is the central atom)
d. BCl_3

10.77 Use the electron-dot formula to determine the shape for each of the following molecules and ions:
a. BrO_2^- **b.** H_2O **c.** CO_3^{2-}
d. CF_4 **e.** CS_2 **f.** PO_3^{3-}

10.78 Use the electron-dot formula to determine the shape for each of the following molecules and ions:
a. PH_3 **b.** NO_3^- **c.** HCN
d. SO_3^{2-} **e.** SF_3^+ **f.** ClO_4^-

10.79 Classify each of the following molecules as polar or nonpolar:
a. HBr   **b.** SiO_2 **c.** NCl_3
d. CH_3Cl **e.** NI_3 **f.** H_2O

10.80 Classify each of the following molecules as polar or nonpolar:
a. GeH_4 **b.** I_2 **c.** CF_3Cl
d. PCl_3 **e.** BCl_3 **f.** SCl_2

10.81 Predict the shape and polarity of each of the following molecules. Assume all bonds are polar.
a. a central atom with three identical bonded atoms and no lone pairs
b. a central atom with two bonded atoms and one lone pair
c. a central atom with two identical atoms and no lone pairs

10.82 Predict the shape and polarity of each of the following molecules. Assume all bonds are polar.
a. a central atom with four identical bonded atoms and no lone pairs
b. a central atom with three identical bonded atoms and one lone pair
c. a central atom with four bonded atoms that are not identical and no lone pairs

10.83 Indicate the major type of attractive force—(1) ionic bonds, (2) dipole–dipole attractions, (3) hydrogen bonding, (4) dispersion forces—that occurs between particles of the following substances:
a. NH_2—OH   **b.** Kr   **c.** CH_4
d. $CHCl_3$ **e.** H_2O **f.** LiCl

10.84 Describe the type of compound that could have each of the following types of attractive forces:
a. dipole–dipole attractions   **b.** hydrogen bonding
c. dispersion forces

10.85 When it rains or snows, the air temperature seems warmer. Explain.

10.86 a. Water is sprayed on the ground of an orchard when temperatures are near freezing to keep the fruit from freezing. Explain.
b. How many kilojoules of energy are released if 5.0 kg of water at 15 °C is sprayed on the ground and cools and freezes at 0 °C?

10.87 An ice cube tray holds 325 g of water. If the water initially has a temperature of 25 °C, how many kilojoules of heat must be removed to cool and freeze the water at 0 °C?

10.88 An ice cube at 0 °C with a mass of 115 g is added to H_2O in a beaker that has a temperature of 64.0 °C. If the final temperature of the mixture is 24.0 °C, what was the initial mass of the warm water?

CHALLENGE QUESTIONS

10.89 Complete the electron-dot formula for each of the following:

a. H   O
   │   ‖
H—N—C—H

b.   H
      │
Cl—C—C—N
      │
      H

c. H—N—N—H

d. O   H
   ‖   │
Cl—C—O—C—H
          │
          H

10.90 Identify the errors in each of the following electron-dot formulas and draw the correct formula:

a. :C̈l═O═C̈l:

b.   :Ö:
      │
   H—C—H

c. H—N̈═Ö—H
      │
      H

10.91 Predict the shape of each of the following molecules or ions:
a. NH_2Cl
b. PH_4^+
c. SCN^-
d. SO_3

10.92 Classify each of the following molecules as polar or nonpolar:
a. BF_3
b. N_2
c. CS_2
d. NH_2Cl

10.93 A 3.0-kg block of lead is taken from a furnace at 300. °C and placed on a large block of ice at 0 °C. The specific heat of lead is 0.13 J/g °C. If all the heat given up by the lead is used to melt ice, how much ice is melted when the temperature of the lead drops to 0 °C?

10.94 The melting point of benzene is 5.5 °C and its boiling point is 80.1 °C. Sketch a heating curve for benzene from 0 °C to 100 °C.
a. What is the state of benzene at 15 °C?
b. What happens on the curve at 5.5 °C?
c. What is the state of benzene at 63 °C?
d. What is the state of benzene at 98 °C?
e. At what temperature will both liquid and gas be present?

10.95 A 45.0-g piece of ice at 0.0 °C is added to a sample of water at 8.0 °C. All of the ice melts and the temperature of the water decreases to 0.0 °C. How many grams of water were in the sample?

10.96 Identify the most important type of attraction between molecules for each of the following and identify the compound with the highest boiling point and the compound with the lowest boiling point:
a. propane, C_3H_8
b. methanol, CH_3OH
c. hydrogen bromide, HBr
d. iodine bromide, IBr

ANSWERS

ANSWERS TO STUDY CHECKS

10.1 :C̈l:Ö:C̈l: or   :C̈l—Ö—C̈l:

10.2 $\left[H:\ddot{N}:H \right]^-$ or $\left[H—\ddot{N}—H \right]^-$

10.3 H:C:::N: or   H—C≡N: In HCN, there is a triple bond between C and N atoms.

10.4 :Ö—S═Ö: ⟷ :Ö—S—Ö: ⟷ :Ö═S—Ö:
         │   ‖   │
        :Ö: :Ö: :Ö:

10.5 With two bonded atoms and two lone pairs, the shape of ClO_2^- is bent.

10.6 The PCl_3 molecule is trigonal pyramidal with polar P—Cl bonds, which makes it a polar molecule.

10.7 H_2O forms hydrogen bonds, but H_2S has only dipole–dipole attractions. Because more energy is needed to break apart hydrogen bonds than dipole–dipole attractions, H_2O has a higher boiling point.

10.8 41.8 kJ removed

10.9 56.5 kJ released

10.10 226 kJ released

ANSWERS TO SELECTED QUESTIONS AND PROBLEMS

10.1 **a.** eight valence electrons   **b.** 14 valence electrons
c. 32 valence electrons   **d.** eight valence electrons

10.3 **a.** HF (8 e^-) H:F̈: or H—F̈:

b. SF_2 (20 e^-) :F̈:S̈:F̈: or :F̈—S̈—F̈:

c. NBr_3 (26 e^-) :Br̈:N̈:Br̈: or :Br̈—N̈—Br̈: (with :Br̈: on top)

d. BH_4^- (8 e^-) [H:B:H with H top and bottom]$^-$ or [H—B—H with H top and bottom]$^-$

e. CH_3OH (14 e^-) H:C:Ö:H (with H top and bottom) or H—C—Ö—H (with H top and bottom)

f. N_2H_4 (14 e^-) H:N̈:N̈:H (with H) or H—N—N—H (with H)

10.5 If complete octets cannot be formed by using all the valence electrons, it is necessary to draw multiple bonds.

10.7 Resonance occurs when we can draw two or more electron-dot formulas for the same molecule or ion.

10.9 **a.** CO (10 e^-) :C:::O: or :C≡O:

b. H_2CCH_2 (12 e^-) H:C::C:H (with H) or H—C=C—H (with H)

c. H_2CO (12 e^-) H:C:H (with :O: top) or H—C—H (with :O: double bond top)

10.11
a. $ClNO_2$:C̈l—N—Ö: (with :O: top) ⟷ :C̈l—N=Ö: (with :O: top)

b. OCN^- [:Ö=C=N̈:]$^-$ ⟷ [:Ö—C≡N:]$^-$ ⟷ [:O≡C—N̈:]$^-$

10.13 **a.** linear   **b.** trigonal pyramidal

10.15 The four electron groups in PCl_3 have a tetrahedral arrangement, but three bonded atoms and one lone pair around a central atom give a trigonal pyramidal shape.

10.17 In BF_3, the central atom B has three bonded atoms and no lone pairs, which gives BF_3 a trigonal planar shape. In NF_3, the central atom N has three bonded atoms and one lone pair, which gives NF_3 a trigonal pyramidal shape.

10.19 **a.** trigonal planar   **b.** bent ($< 109.5°$)
c. linear   **d.** tetrahedral
e. bent ($< 120°$)

10.21
a. CO_3^{2-} (24 e^-) [:Ö—C=Ö: with :O: top]$^{2-}$ trigonal planar

b. SO_4^{2-} (32 e^-) [:Ö—S—Ö: with :O: top and :O: bottom]$^{2-}$ tetrahedral

c. BH_4^- (8 e^-) [H—B—H with H top and bottom]$^-$ tetrahedral

d. NO_2^+ (16 e^-) [:Ö=N=Ö:]$^+$ linear

10.23 The electronegativity increases going across a period.

10.25 A nonpolar covalent bond would have an electronegativity difference of 0.0 to 0.4.

10.27 **a.** K, Na, Li   **b.** Na, P, Cl
c. Ca, Br, O

10.29 **a.** $\overset{\delta^+ \quad \delta^-}{N—F}$ →   **b.** $\overset{\delta^+ \quad \delta^-}{Si—P}$ →

c. $\overset{\delta^+ \quad \delta^-}{C—O}$ →   **d.** $\overset{\delta^+ \quad \delta^-}{P—Br}$ ←

e. $\overset{\delta^+ \quad \delta^-}{B—Cl}$ →

10.31 **a.** polar covalent   **b.** ionic
c. polar covalent   **d.** nonpolar covalent
e. polar covalent   **f.** nonpolar covalent

10.33 Electrons are shared equally between two identical atoms and unequally between nonidentical atoms.

10.35 **a.** nonpolar   **b.** polar
c. nonpolar   **d.** nonpolar

10.37 In the molecule CO_2, the two C—O dipoles cancel; in CO, there is only one dipole.

10.39 **a.** dipole–dipole attractions   **b.** ionic bonds
c. dispersion forces   **d.** dipole–dipole attractions
e. dispersion forces

10.41 **a.** hydrogen bonding   **b.** dipole–dipole attractions
c. dipole–dipole attractions   **d.** dispersion forces
e. dispersion forces

10.43 **a.** 21 700 J; absorbed   **b.** 5680 J; absorbed
c. 75.2 kJ; released   **d.** 16.7 kJ; released

10.45 **a.** 22 600 J; absorbed   **b.** 113 kJ; absorbed
c. 1.81×10^7 J; released   **d.** 396 kJ; released

10.47 **a.** 4800 J   **b.** 30 300 J
c. 40.2 kJ   **d.** 72.3 kJ

10.49 119.5 kJ

10.51 **a.** dispersion forces
b. dipole–dipole attractions
c. hydrogen bonds
d. dispersion forces

10.53 BCl_3 is trigonal planar; all dipoles cancel, and BCl_3 is a nonpolar molecule. PCl_3 is trigonal pyramidal; the dipoles do not cancel, and PCl_3 is a polar molecule.

10.55 a. The heat from the skin is used to evaporate the water (perspiration). Therefore, the skin is cooled.
  b. On a hot day, there are more molecules with sufficient energy to become water vapor.
  c. In a closed bag, some molecules evaporate, but they cannot escape and will condense back to liquid; the clothes will not dry.

10.57

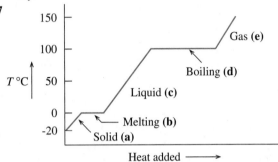

10.59 a. about $-60\ °C$ **b.** about $60\ °C$
  c. A represents the solid state. B represents the change from solid to liquid or melting of the substance. C represents the liquid state as temperature increases. D represents the change from liquid to gas or boiling of the liquid. E represents the gas state.
  d. At $-80\ °C$, solid; at $-40\ °C$, liquid; at $25\ °C$, liquid; $80\ °C$, gas

10.61 a. $4 + 2(6) = 16$ valence electrons
  b. $2(4) + 4(1) + 6 = 18$ valence electrons
  c. $5 + 4(1) - 1 = 8$ valence electrons
  d. $3 + 3(7) = 24$ valence electrons
  e. $6 + 3(6) + 2 = 26$ valence electrons

10.63 a. BF_4^- $(32\ e^-)$ $\left[\begin{array}{c}:\!\ddot{F}\!: \\ :\!\ddot{F}\!:\!B\!:\!\ddot{F}\!: \\ :\!\ddot{F}\!:\end{array}\right]^-$ or $\left[\begin{array}{c}:\!\ddot{F}\!: \\ :\!\ddot{F}\!-\!B\!-\!\ddot{F}\!: \\ :\!\ddot{F}\!:\end{array}\right]^-$

  b. Cl_2O $(20\ e^-)$ $:\!\ddot{C}l\!:\!\ddot{O}\!:\!\ddot{C}l\!:$ or $:\!\ddot{C}l\!-\!\ddot{O}\!-\!\ddot{C}l\!:$

  c. H_2NOH $(14\ e^-)$ $H\!:\!\ddot{N}\!:\!\ddot{O}\!:\!H$ or $H\!-\!\overset{\displaystyle H}{\underset{}{\ddot{N}}}\!-\!\ddot{O}\!-\!H$

  d. NO_2^+ $(16\ e^-)$ $\left[:\!\ddot{O}\!:\!:\!N\!:\!:\!\ddot{O}\!:\right]^+$ or $\left[:\!\ddot{O}\!=\!N\!=\!\ddot{O}\!:\right]^+$

  e. H_2CCCl_2 $(24\ e^-)$ $H\!:\!\overset{\displaystyle H}{\underset{}{C}}\!:\!\overset{\displaystyle :\!\ddot{C}l\!:}{\underset{}{C}}\!:\!\ddot{C}l\!:$ or
  $H\!-\!\overset{\displaystyle H}{\underset{}{C}}\!=\!\overset{\displaystyle :\!\ddot{C}l\!:}{\underset{}{C}}\!-\!\ddot{C}l\!:$

10.65 a. 16 valence electrons $\left[:\!\ddot{N}\!:\!:\!N\!:\!:\!\ddot{N}\!:\right]^-$ or
  $\left[:\!\ddot{N}\!-\!N\!\equiv\!N\!:\right]^- \longleftrightarrow \left[:\!\ddot{N}\!=\!N\!=\!\ddot{N}\!:\right]^- \longleftrightarrow \left[:\!N\!\equiv\!N\!-\!\ddot{N}\!:\right]^-$

  b. 16 valence electrons $\left[:\!\ddot{O}\!:\!:\!N\!:\!:\!\ddot{O}\!:\right]^+$ or
  $\left[:\!\ddot{O}\!-\!N\!\equiv\!O\!:\right]^+ \longleftrightarrow \left[:\!\ddot{O}\!=\!N\!=\!\ddot{O}\!:\right]^+ \longleftrightarrow \left[:\!O\!\equiv\!N\!-\!\ddot{O}\!:\right]^+$

c. 16 valence electrons $\left[:\!C\!:\!:\!:\!N\!:\!\ddot{S}\!:\right]^-$ or
  $\left[:\!\ddot{C}\!-\!N\!\equiv\!S\!:\right]^- \longleftrightarrow \left[:\!C\!\equiv\!N\!-\!\ddot{S}\!:\right]^- \longleftrightarrow \left[:\!\ddot{C}\!=\!N\!=\!\ddot{S}\!:\right]^-$

10.67 a. I, Cl, F   **b.** K, Li, S, Cl   **c.** Ba, Sr, Mg, Be

10.69 a. C—O   **b.** N—F   **c.** S—Cl
  d. Br—I   **e.** N—O has the same degree of polarity as N—S

10.71 a. Si—Cl   **b.** C—N   **c.** F—Cl
  d. C—F   **e.** N—O

10.73 a. polar covalent   **b.** nonpolar covalent
  c. ionic   **d.** nonpolar covalent
  e. nonpolar covalent

10.75 a. NF_3 $:\!\ddot{F}\!-\!\overset{\displaystyle \ddot{N}}{\underset{\displaystyle :\!\ddot{F}\!:}{}}\!-\!\ddot{F}\!:$ trigonal pyramidal

  b. $SiBr_4$ $:\!\ddot{B}r\!-\!\overset{\displaystyle :\!\ddot{B}r\!:}{\underset{\displaystyle :\!\ddot{B}r\!:}{Si}}\!-\!\ddot{B}r\!:$ tetrahedral

  c. $BeCl_2$ $:\!\ddot{C}l\!-\!Be\!-\!\ddot{C}l\!:$ linear

  d. SO_2 $:\!\ddot{O}\!=\!\ddot{S}\!-\!\ddot{O}\!: \longleftrightarrow :\!\ddot{O}\!-\!\ddot{S}\!=\!\ddot{O}\!:$ bent $(<120°)$

10.77 a. bent $(<109.5°)$ **b.** bent $(<109.5°)$
  c. trigonal planar   **d.** tetrahedral
  e. linear   **f.** trigonal pyramidal

10.79 a. polar   **b.** nonpolar   **c.** nonpolar
  d. polar   **e.** polar   **f.** polar

10.81 a. trigonal planar, nonpolar   **b.** bent, polar
  c. linear, nonpolar

10.83 a. 3   **b.** 4   **c.** 4
  d. 2   **e.** 3   **f.** 1

10.85 When water vapor condenses or liquid water freezes, heat is released, which warms the air.

10.87 $34\ kJ + 109\ kJ = 143\ kJ$ removed

10.89
  a. $H\!-\!\overset{\displaystyle H}{\underset{}{\ddot{N}}}\!-\!\overset{\displaystyle :\!\ddot{O}\!:}{\underset{}{C}}\!-\!H$ **b.** $:\!\ddot{C}l\!-\!\overset{\displaystyle H}{\underset{\displaystyle H}{C}}\!-\!C\!\equiv\!N\!:$

  c. $H\!-\!\ddot{N}\!=\!\ddot{N}\!-\!H$ **d.** $:\!\ddot{C}l\!-\!\overset{\displaystyle :\!O\!:}{\underset{}{C}}\!-\!\ddot{O}\!-\!\overset{\displaystyle H}{\underset{\displaystyle H}{C}}\!-\!H$

10.91 a. trigonal pyramidal   **b.** tetrahedral
  c. linear   **d.** trigonal planar

10.93 360 g of ice will melt.

10.95 450 g of water

COMBINING IDEAS FROM CHAPTERS 8 TO 10

CI.13 In an experiment, the mass of a piece of copper is determined to be 8.56 g. Then the copper is reacted with sufficient oxygen gas to produce solid copper(II) oxide.

       —Cu

       8.56 g

a. If copper has a density of 8.94 g/cm^3, what is the volume (cm^3) of the copper?

b. How many copper atoms are in the sample?

c. Write the balanced chemical equation for the reaction.

d. Classify the type of reaction.

e. How many grams of oxygen are required to completely react with the copper?

f. How many grams of copper(II) oxide will result from the reaction of 8.56 g of Cu and 3.72 g of oxygen?

g. How many grams of copper(II) oxide will result in part **f.** if the yield for the reaction is 85.0 percent?

CI.14 One of the alkanes in gasoline is octane, C$_8$H$_{18}$, which has a density of 0.803 g/cm^3 and a ΔH of -5510 kJ/mol. Suppose a hybrid car has a fuel tank with a capacity of 11.9 gal and has a gas mileage of 45 mi/gal.

a. Draw the condensed structural formula for octane.

b. Write a balanced chemical equation for the complete combustion of octane including the heat of reaction.

c. What is the energy, in kilojoules, produced from one tank of fuel assuming it is all octane?

d. How many molecules of C$_8$H$_{18}$ are present in one tank of fuel assuming it is all octane?

e. If the total mileage of this hybrid car for one year is 24 500 miles, how many kilograms of carbon dioxide would be produced from the combustion of the fuel assuming it is all octane?

CI.15 When clothes have stains, bleach may be added to the wash to react with the soil and make the stains colorless. The bleach solution is prepared by bubbling chlorine gas into a solution of sodium hydroxide to produce a solution of sodium hypochlorite, sodium chloride, and water. One brand of bleach contains 5.25% sodium hypochlorite by mass (active ingredient) with a density of 1.08 g/mL.

a. What is the formula and molar mass of sodium hypochlorite?

b. Draw the electron-dot formula for the hypochlorite ion.

c. How many hypochlorite ions are present in 1.00 gallon of bleach solution?

d. Write the balanced chemical equation for the preparation of bleach.

e. How many grams of NaOH are required to produce the sodium hypochlorite for 1.00 gallon of bleach?

f. If 165 g of Cl$_2$ is passed through a solution containing 275 g of NaOH and 162 g of sodium hypochlorite is produced, what is the percent yield for the reaction?

CI.16 Ethanol, CH$_3$—CH$_2$—OH, is obtained from renewable crops such as corn, which use the Sun as their source of energy. In the United States, automobiles can now use a fuel known as E 85 that contains 85.0% ethanol and 15.0% gasoline by volume. Ethanol has a melting point of -115 °C, a boiling point of 78 °C, a heat of fusion of 98.7 J/g, and a heat of vaporization of 841 J/g. Liquid ethanol has a density of 0.796 g/mL and a specific heat of 2.46 J/g °C.

Fuel that is E 85 contains 85% ethanol.

a. Draw a heating curve for ethanol from $-150\ °C$ to $100\ °C$.

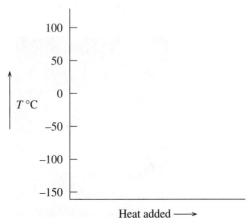

b. When 20.0 g of ethanol at $-62\ °C$ is heated and completely vaporized at $78\ °C$, how much energy (kJ) is required?

c. If a 15.0-gallon gas tank is filled with E 85, how many liters of ethanol are in the gas tank?

d. Write the balanced chemical equation for the complete combustion of ethanol.

e. How many kilograms of CO_2 are produced by the complete combustion of the ethanol in a full 15.0-gal gas tank?

f. What is the strongest attractive force between liquid ethanol molecules?

CI.17 Chloral hydrate, a sedative and hypnotic, was the first drug used to treat insomnia. Chloral hydrate has a melting point of $57\ °C$. At its boiling point of $98\ °C$, it breaks down to chloral and water.

Chloral hydrate   Chloral

a. Draw the electron-dot formulas for chloral hydrate and chloral.

b. What functional groups are in chloral hydrate and chloral?

c. What are the empirical formulas of chloral hydrate and chloral?

d. What is the percent by mass of Cl in chloral hydrate?

CI.18 Ethylene glycol, $C_2H_6O_2$, used as a coolant and antifreeze, has a density of 1.11 g/mL. As a sweet-tasting liquid, it can be appealing to pets and small children, but it can be toxic, with an LD_{50} of 4700 mg/kg. Its accidental ingestion can cause kidney damage and difficulty with breathing. In the body, ethylene glycol is converted to another toxic substance, oxalic acid, $H_2C_2O_4$.

Ethylene glycol is added to a radiator to prevent freezing or boiling.

a. What are the empirical formulas of ethylene glycol and oxalic acid?

b. If ethylene glycol has a C—C single bond with two H atoms attached to each C atom, what is its electron-dot formula?

c. Which bonds in ethylene glycol are polar and which are nonpolar?

d. How many milliliters of ethylene glycol could be toxic for an 11.0-lb cat?

e. What would be the strongest attractive force in ethylene glycol?

f. If oxalic acid has two carboxylic acid groups attached by a C—C single bond, what is its electron-dot formula?

g. Write the balanced chemical equation for the reaction of ethylene glycol and oxygen (O_2) to give oxalic acid and water.

CI.19 Acetone (propanone), a clear liquid solvent with an acrid odor, is used to remove nail polish, paints, and resins. It has a low boiling point and is highly flammable. Acetone has a density of 0.786 g/mL and a heat of combustion of 1790 kJ/mol.

Acetone consists of carbon atoms (black), hydrogen atoms (white), and an oxygen atom (red).

a. Draw the condensed structural formula of acetone.

b. What is the molecular formula and molar mass of acetone?

c. Write the equation for the complete combustion reaction of acetone including the heat of reaction.

d. Is the combustion of acetone an endothermic or exothermic reaction?

e. How much heat, in kilojoules, is released if 2.58 g of acetone reacts completely with oxygen?

f. How many grams of oxygen gas are needed to react with 15.0 mL of acetone?

CI.20 The compound dihydroxyacetone (DHA) is used in "sunless" tanning lotions, which darken the skin by reacting with the amino acids in the outer surface of the skin. A typical drugstore lotion contains 4.0 % (mass/volume) DHA.

A sunless tanning lotion contains DHA to darken the skin.

a. Draw the condensed structural formula of DHA.

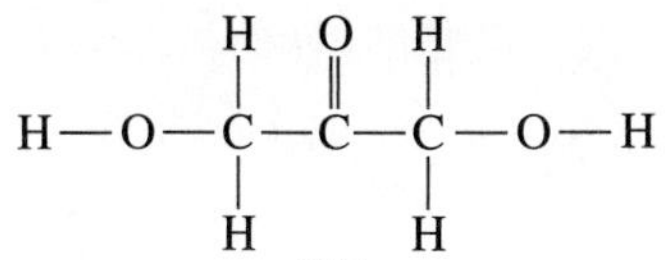

H—O—C—C—C—O—H (structure with H, O, H atoms; DHA)

DHA

b. Identify the functional groups in DHA.
c. What are the molecular formula and molar mass of DHA?
d. A bottle of sunless tanning lotion contains 177 mL of lotion. How many milligrams of DHA are in a bottle?

■ANSWERS

CI.13 **a.** 0.957 cm^3
    b. 8.11×10^{22} atoms of copper
    c. $2Cu(s) + O_2(g) \longrightarrow 2CuO(s)$
    d. combination reaction
    e. $2.16 \text{ g of } O_2$
    f. 10.7 g of CuO
    g. 9.10 g of CuO

CI.15 **a.** NaOCl, 74.44 g/mol
    b.

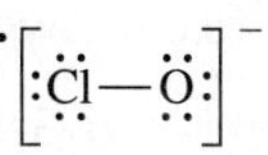

$$\left[:\ddot{C}l - \ddot{O}: \right]^-$$

    c. $1.74 \times 10^{24} \text{ OCl}^-$ ions
    d. $2NaOH(aq) + Cl_2(g) \longrightarrow$
                    $NaOCl(aq) + NaCl(aq) + H_2O(l)$
    e. 231 g of NaOH
    f. 93.6%

CI.17 **a.**

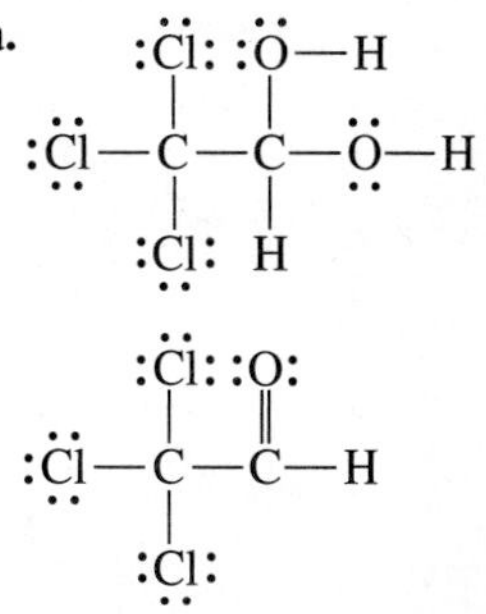

(Lewis structures for chloral hydrate and chloral)

    b. chloral hydrate: alcohol (two); chloral: aldehyde
    c. chloral hydrate: $C_2H_3O_2Cl_3$
        chloral: C_2HOCl_3
    d. 64.33% Cl (by mass)

CI.19 **a.**

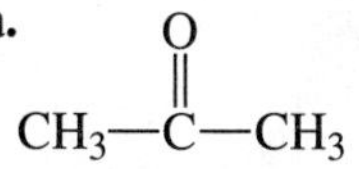

$$CH_3 - \overset{\overset{\textstyle O}{\|}}{C} - CH_3$$

    b. C_3H_6O; 58.08 g/mol
    c. $C_3H_6O(g) + 4O_2(g) \overset{\Delta}{\longrightarrow} 3CO_2(g) + 3H_2O(g) + 1790 \text{ kJ}$
    d. exothermic
    e. 79.5 kJ
    f. $26.0 \text{ g of } O_2$

11 Gases

"When oxygen levels in the blood are low, the cells in the body don't get enough oxygen," says Sunanda Tripathi, registered nurse, Santa Clara Valley Medical Center. "We use a nasal cannula to give supplemental oxygen to a patient. At a flow rate of 2 liters per minute, a patient breathes in a gaseous mixture that is about 28% oxygen compared to 21% in ambient air."

When a patient has a breathing disorder, the flow and volume of oxygen into and out of the lungs is measured. A ventilator may be used if a patient has difficulty breathing. When pressure is increased, the lungs expand. When the pressure of the incoming gas is reduced, the lung volume contracts to expel carbon dioxide. These relationships—known as gas laws—are an important part of ventilation and breathing.

W e all live at the bottom of a sea of gases called the atmosphere. The most important of these gases is oxygen, which constitutes about 21% of the atmosphere. Without oxygen, life on this planet would be impossible: Oxygen is vital to all life processes of plants and animals. Ozone (O_3), formed in the upper atmosphere by the interaction of oxygen with ultraviolet light, absorbs some of the harmful radiation before it can strike Earth's surface. The other gases in the atmosphere include nitrogen (78%), argon, carbon dioxide (CO_2), and water vapor. Carbon dioxide gas, a product of combustion and metabolism, is used by plants in photosynthesis, which produces the oxygen that is essential for humans and animals.

The atmosphere has become a dumping ground for other gases, such as methane, chlorofluoro-carbons (CFCs), and nitrogen oxides, as well as volatile organic compounds (VOCs), which are gases from paints, paint thinners, and cleaning supplies. These gases in the atmosphere are contributing to air pollution, ozone depletion, global warming, and acid rain. These changes can seriously affect our health and the way we live. An understanding of gases and some of the laws that govern gas behavior can help us understand the nature of matter and allow us to make decisions concerning important environmental and health issues.

11.1 Properties of Gases

We are surrounded by gases but are not often aware of their presence. Of the elements on the periodic table, only a handful exist as gases at room temperature: H_2, N_2, O_2, F_2, Cl_2, and the noble gases. Another group of gases includes the oxides of the nonmetals on the upper corner of the periodic table, such as CO, CO_2, NO, NO_2, SO_2, and SO_3. As we learned in Chapter 6, organic compounds with small molar masses are gases at room temperatures. These include the first four alkanes of methane, ethane, propane, and butane, and their related alkenes and alkynes. Generally, molecules that are gases at room temperature have fewer than five atoms from the first or second period.

The behavior of gases is quite different from that of liquids and solids. As we learned in Chapter 3, gas particles are far apart, whereas particles of both liquids and solids are held close together. A gas has no definite shape or volume and will completely fill any container. As we learned in Chapter 10, the attractive forces between gas particles are very minimal. Thus, there are great distances between gas particles, which make a gas less dense than a solid or liquid, and easy to compress. A model for the behavior of a gas, called the **kinetic molecular theory of gases**, helps us understand gas behavior.

LEARNING GOAL

Use the kinetic molecular theory of gases to describe the properties of gases.

TUTORIAL

Properties of Gases

SELF STUDY ACTIVITY

Properties of Gases

Kinetic Molecular Theory of Gases

1. **A gas consists of small particles (atoms or molecules) that move randomly with high velocities.** Gas molecules moving in random directions at high speeds cause a gas to fill the entire volume of a container.

2. **The attractive forces between the particles of a gas are usually very small.** Gas particles are far apart and fill a container of any size and shape.

3. **The actual volume occupied by gas molecules is extremely small compared to the volume that the gas occupies.** The volume of the gas is considered equal to the volume of the container. Most of the volume of a gas is empty space, which allows gases to be easily compressed.

4. **Gas particles are in constant motion, moving rapidly in straight paths.** When gas particles collide, they rebound and travel in new directions. Every time they hit the walls of the container, they exert pressure. An increase in the number or force of collisions against the walls of the container causes an increase in the pressure of the gas.

5. **The average kinetic energy of gas molecules is proportional to the Kelvin temperature.** Gas particles move faster as the temperature increases. At higher temperatures, gas particles hit the walls of the container more often and with more force, producing higher pressures.

The kinetic molecular theory helps explain some of the characteristics of gases. For example, we can quickly smell perfume when a bottle is opened on the other side of a room, because its particles move rapidly in all directions. At room temperatures, the molecules of air are moving at about 1000 miles per hour. They move faster at higher temperatures and more slowly at lower temperatures. Sometimes tires and gas-filled containers explode when temperatures are too high. From the kinetic molecular theory, we know that gas particles move faster when heated, hit the walls of a container with more force, and cause a buildup of pressure inside a container.

When we talk about a gas, we describe it in terms of four properties: pressure, volume, temperature, and the amount of gas.

Pressure (P)   Gas particles are extremely small and move rapidly. When they hit the walls of a container, they exert a force known as pressure (see Figure 11.1). If we heat the container, the molecules move faster and smash into the walls of the container more often and with increased force, thus increasing the pressure. The gas particles in the air, mostly oxygen and nitrogen, exert a pressure on us called **atmospheric pressure** (see Figure 11.2). As you go to higher altitudes, the atmospheric pressure is less because there are fewer particles in the air. The most common units used for gas measurement are the atmosphere (atm) and millimeters of mercury (mmHg). On the TV weather report, you may hear or see the atmospheric pressure given in inches of mercury, or kilopascals in countries other than the United States. In a chemistry lab, the unit torr may be used.

Volume (V)   The volume of gas equals the size of the container in which the gas is placed. When you inflate a tire or a basketball, you are adding more gas particles. The increase in the number of particles hitting the walls of the tire or basketball increases the volume.

FIGURE 11.1 Gas particles move in straight lines within a container. The gas particles exert pressure when they collide with the walls of the container.

Q Why does heating the container increase the pressure of the gas within it?

FIGURE 11.2 A column of air extending from the upper atmosphere to the surface of Earth produces a pressure of about 1 atm. While there is a lot of pressure on the body, it is balanced by the pressure inside the body.

Q Why is there less pressure at higher altitudes?

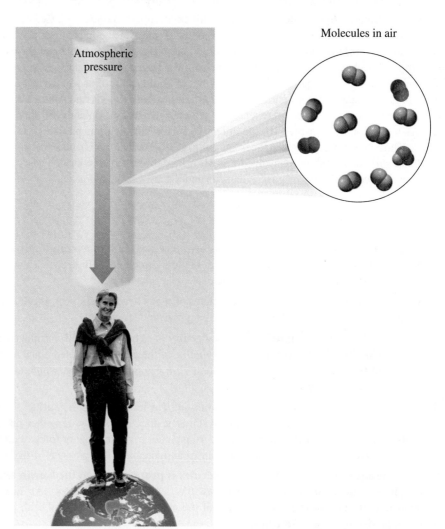

Atmospheric pressure

Molecules in air

Sometimes, on a cold morning, a tire looks flat. The volume of the tire has decreased because a lower temperature decreases the speed of the molecules, which in turn reduces the force of their impacts on the walls of the tire. The most common units for volume measurement are liters (L) and milliliters (mL).

Temperature (T)   The temperature of a gas is related to the kinetic energy of its particles. For example, if we have a gas at 200 K in a rigid container and heat it to a temperature of 400 K, the gas particles will have twice the kinetic energy that they did at 200 K. This also means that the gas at 400 K exerts twice the pressure of the gas at 200 K. Although you measure gas temperature using a Celsius thermometer, all comparisons of gas behavior and all calculations related to temperature must use the Kelvin temperature scale. No one has quite created the conditions for absolute zero (0 K), but we predict that the particles will have zero kinetic energy and exert zero pressure at absolute zero.

Amount of Gas (n)   When you add air to a bicycle tire, you increase the amount of gas, which results in a higher pressure in the tire. Usually, we measure the amount of gas by its mass (grams). In gas law calculations, we need to change the grams of gas to moles.

A summary of the four properties of a gas is given in Table 11.1.

TABLE 11.1 Properties That Describe a Gas

Property	Description	Unit(s) of Measurement
Pressure (P)	The force exerted by a gas against the walls of the container	atmosphere (atm); millimeters of mercury (mmHg); torr; pascal (Pa)
Volume (V)	The space occupied by a gas	liter (L); milliliter (mL); cubic meter (m^3)
Temperature (T)	The determining factor of the kinetic energy and rate of motion of gas particles	degree Celsius (°C); kelvin (K) *is required in calculations*
Amount (n)	The quantity of gas present in a container	grams (g); moles (n) *is required in calculations*

CONCEPT CHECK 11.1

■ **Properties of Gases**

Use the kinetic molecular theory to explain why a gas completely fills a container of any size and shape.

ANSWER

Gas particles move at high speeds in random directions, moving as far apart as possible until they hit the walls of a container. Thus gas particles completely fill a container of any size and shape.

SAMPLE PROBLEM | 11.1 |

■ **Properties of Gases**

Identify the property of a gas that is described by each of the following:

a. increases the kinetic energy of gas particles
b. the force of the gas particles hitting the walls of the container
c. the space that is occupied by a gas

SOLUTION

a. temperature   **b.** pressure   **c.** volume

STUDY CHECK

When helium is added to a balloon, the mass of the gas, in grams, increases. What property of a gas is described?

CHEMISTRY AND HEALTH

Measuring Blood Pressure

The measurement of your blood pressure is one of the important measurements a doctor or nurse makes during a physical examination. Acting like a pump, the heart contracts to create the pressure that pushes blood through the circulatory system. During contraction, the blood pressure is called *systolic* and is at its highest. When the heart muscles relax, the blood pressure is called *diastolic* and falls. The normal range for systolic pressure is 100–120 mmHg, and for diastolic pressure it is 60–80 mmHg, usually expressed as a ratio such as 100/80. These values are somewhat higher in older people. When blood pressures are elevated, such as 140/90, there is a greater risk of stroke, heart attack, or kidney damage. Low blood pressure prevents the brain from receiving adequate oxygen, causing dizziness and fainting.

The blood pressures are measured by a sphygmomanometer, an instrument consisting of a stethoscope and an inflatable cuff connected to a tube of mercury called a manometer. After the cuff is wrapped around the upper arm, it is pumped up with air until it cuts off the flow of blood. With the stethoscope over the artery, the air is slowly released from the cuff. When the pressure equals the systolic pressure, blood starts to flow again, and the noise it makes is heard through the stethoscope. As air continues to be released, the cuff deflates until no sound is heard in the artery. That second pressure reading is noted as the diastolic pressure, the pressure when the heart is not contracting.

The use of digital blood pressure monitors is becoming more common. However, they have not been validated for use in all situations and can sometimes give inaccurate readings.

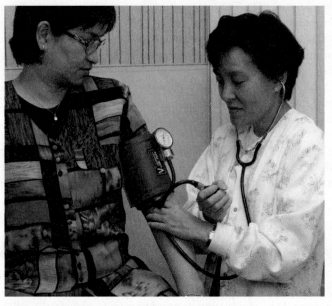

The measurement of blood pressure is part of a routine health checkup.

QUESTIONS AND PROBLEMS

Properties of Gases

11.1 Use the kinetic molecular theory of gases to explain each of the following:
  a. Gases move faster at higher temperatures.
  b. Gases can be compressed much more easily than liquids or solids.
  c. Gases have low densities.

11.2 Use the kinetic molecular theory of gases to explain each of the following:
  a. A container of nonstick cooking spray explodes when thrown into a fire.
  b. The air in a hot-air balloon is heated to make the balloon rise.
  c. You can smell the odor of cooking onions from far away.

11.3 Identify the property of a gas that is measured in each of the following:
  a. 350 K
  b. the space occupied by a gas
  c. 2.00 g of O_2
  d. the force of gas particles striking the walls of the container

11.4 Identify the property of a gas that is measured in each of the following:
  a. 425 K
  b. 1.0 atm
  c. 10.0 L
  d. 0.50 mol of He

11.2 Gas Pressure

LEARNING GOAL

Describe the units of measurement used for pressure, and change from one unit to another.

TUTORIAL
Converting Between Units of Pressure

When billions and billions of gas particles hit against the walls of a container, they exert **pressure**, which is defined as a force acting on a certain area.

$$\text{Pressure } (P) = \frac{\text{force}}{\text{area}}$$

The atmospheric pressure can be measured using a barometer, as shown in Figure 11.3. At a pressure of exactly 1 atmosphere (atm), the mercury column would be exactly 760 mm high. One **atmosphere (atm)** is defined as *exactly* 760 mmHg (millimeters of mercury). One atmosphere is also 760 **torr**, a pressure unit named to honor Evangelista Torricelli, the inventor of the barometer. Because they are equal, units of torr and mmHg are used interchangeably.

1 atm = 760 mmHg = 760 torr (exact)

1 mmHg = 1 torr (exact)

In SI units, pressure is measured in pascals (Pa); 1 atm is equal to 101 325 Pa. Because a pascal is a very small unit, pressures are usually reported in kilopascals.

1 atm = 1.01325×10^5 Pa = 101.325 kPa

The U.S. equivalent of 1 atm is 14.7 lb/in.^{2} (psi). When you use a pressure gauge to check the air pressure in the tires of a car, it may read 30–35 psi. This measurement is actually 30–35 psi above the pressure that the atmosphere exerts on the outside of the tire. Table 11.2 summarizes the various units used in the measurement of pressure.

TABLE 11.2 Units for Measuring Pressure

Unit	Abbreviation	Unit Equivalent to 1 atm
Atmosphere	atm	1 atm (exact)
Millimeters of Hg	mmHg	760 mmHg (exact)
Torr	torr	760 torr
Inches of Hg	in. Hg	29.9 in. Hg
Pounds per square inch	lb/in.^{2} (psi)	14.7 lb/in.^{2}
Pascal	Pa	101 325 Pa
Kilopascal	kPa	101.325 kPa

If you have a barometer in your home, it probably gives pressure in inches of mercury. One atmosphere is equal to the pressure of a column of mercury that is 29.9 in. high. Atmospheric pressure changes with variations in weather and altitude. On a hot, sunny day, a column of air has more particles, which increases the pressure on the mercury surface. The mercury column rises, which indicates a higher atmospheric pressure. On a rainy day, the atmosphere exerts less pressure, which causes the mercury column to fall. In the weather report, this type of weather is called a low-pressure system. Above sea level, the density of the gases in the air decreases, which causes lower atmospheric pressures; the atmospheric pressure is greater than 760 mmHg at the Dead Sea because it is below sea level (see Table 11.3).

Divers must be concerned about increasing pressures on their ears and lungs when they dive below the surface of the ocean. Because water is more dense than air, the pressure on a diver increases rapidly as the diver descends. At a depth of 33 ft below the surface of the ocean, an additional 1 atm of pressure is exerted by the water on a diver, which gives a total pressure of 2 atm. At 100 ft, there is a total pressure of 4 atm on a diver. The regulator that a diver uses continuously adjusts the pressure of the breathing mixture to match the increase in pressure.

TABLE 11.3 Altitude and Atmospheric Pressure

Location	Altitude (km)	Atmospheric Pressure (mmHg)
Dead Sea	−0.40	800
Sea level	0	760
Los Angeles	0.09	752
Las Vegas	0.70	700
Denver	1.60	630
Mount Whitney	4.50	440
Mount Everest	8.90	253

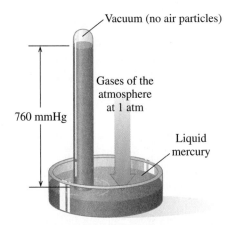

FIGURE 11.3 **A barometer:** the pressure exerted by the gases in the atmosphere is equal to the downward pressure of a mercury column in a closed glass tube. The height of the mercury column measured in mmHg is called atmospheric pressure.

Q Why does the height of the mercury column change from day to day?

CASE STUDY
Scuba Diving and Blood Gases

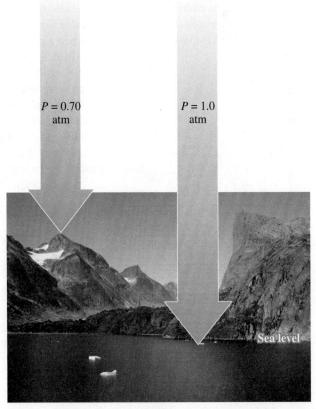

The atmospheric pressure decreases as the altitude increases.

■ **Units of Pressure**

A sample of neon gas has a pressure of 0.50 atm. Calculate the pressure, in mmHg, of the neon.

ANSWER

The equality 1 atm = 760 mmHg can be written as two conversion factors:

$$\frac{760 \text{ mmHg}}{1 \text{ atm}} \quad \text{and} \quad \frac{1 \text{ atm}}{760 \text{ mmHg}}$$

Using the conversion factor that cancels atm, the problem is set up as

$$0.50 \text{ atm} \times \frac{760 \text{ mmHg}}{1 \text{ atm}} = 380 \text{ mmHg}$$

QUESTIONS AND PROBLEMS

Gas Pressure

11.5 What units are used to measure the pressure of a gas?

11.6 Which of the following statement(s) describes the pressure of a gas?
 a. the force of the gas particles on the walls of the container
 b. the number of gas particles in a container
 c. the volume of the container
 d. 3.00 atm
 e. 750 torr

11.7 An oxygen tank contains oxygen (O_2) at a pressure of 2.00 atm. What is the pressure in the tank in terms of the following units?
 a. torr **b.** lb/in.^{2}
 c. mmHg **d.** kPa

11.8 On a climb up Mt. Whitney, the atmospheric pressure drops to 467 mmHg. What is the pressure in terms of the following units?
 a. atm **b.** torr
 c. in. Hg **d.** Pa

11.3 Pressure and Volume (Boyle's Law)

Imagine that you can see air particles hitting the walls inside a bicycle tire pump. What happens to the pressure inside the pump as we push down on the handle? As the volume decreases, there is a decrease in the surface area of the container. The air particles are crowded together, more collisions occur, and the pressure increases within the container.

When a change in one property (in this case, volume) causes a change in another property (in this case, pressure), the properties are related. If the change occurs in opposite directions, the properties have an **inverse relationship**. The inverse relationship between the pressure and volume of a gas is known as **Boyle's law**. The law states that the volume (V) of a sample of gas changes inversely with the pressure (P) of the gas as long as there is no change in the temperature (T) or amount of gas (n), as illustrated in Figure 11.4.

If the volume or pressure of a gas changes without any change occurring in the temperature or in the amount of the gas, the new pressure and volume will give the same PV product as the initial pressure and volume. Then we can set the initial and final PV products equal to each other.

Boyle's Law

$$P_1V_1 = P_2V_2 \qquad \text{No change in number of moles and temperature}$$

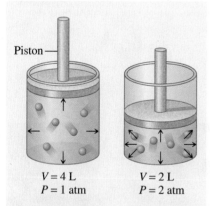

Piston

$V = 4$ L
$P = 1$ atm

$V = 2$ L
$P = 2$ atm

FIGURE 11.4 Boyle's law: As volume decreases, gas molecules become more crowded, which causes the pressure to increase. Pressure and volume are inversely related.

Q If the volume of a gas increases, what will happen to its pressure?

■ **Boyle's Law**

State and explain the reason for the change (*increases, decreases*) in a gas that occurs in each of the following when n and T do not change:

	Pressure	Volume
a.		Decreases
b.		Increases

ANSWER

a. When the volume of a gas decreases at constant n and T, the gas particles are closer together, which increases the number of collisions with the container walls. Therefore, the *pressure increases* when volume decreases with no change in n and T.

b. When the volume of a gas increases with no change in n and T, the gas particles move farther apart, which decreases the number of collisions with the container wall. Therefore the *pressure decreases* when the volume increases with no change in n and T.

Pressure	Volume
a. Increases	Decreases
b. Decreases	Increases

TUTORIAL

Pressure and Volume

SAMPLE PROBLEM | 11.2

■ Calculating Pressure When Volume Changes

A sample of hydrogen gas (H_2) has a volume of 5.0 L and a pressure of 1.0 atm. What is the new pressure if its volume decreases to 2.0 L at constant temperature?

SOLUTION

STEP 1 **Organize the data in a table of initial and final conditions.** In this problem, we want to know the final pressure (P_2) for the change in volume. In calculations with gas laws, it is helpful to organize the data in a table. Because we know that the volume decreases, we can predict that the pressure will increase.

Conditions 1	Conditions 2	Know	Predict
$V_1 = 5.0$ L	$V_2 = 2.0$ L	V decreases	
$P_1 = 1.0$ atm	$P_2 = ?$		P increases

STEP 2 **Rearrange the gas law for the unknown.** For a PV relationship, we use Boyle's law and solve for P_2 by dividing both sides by V_2.

$$P_1 V_1 = P_2 V_2$$

$$\frac{P_1 V_1}{V_2} = \frac{P_2 \cancel{V_2}}{\cancel{V_2}}$$

$$P_2 = P_1 \times \frac{V_1}{V_2}$$

STEP 3 **Substitute values into the gas law to solve for the unknown.** When we substitute in the values, we see that the ratio of the volumes is greater than 1, which increases the pressure. The final pressure (P_2) has increased as we predicted in Step 1, which increases the pressure. Note that the units of volume (L) cancel to give the final pressure in atmospheres.

$$P_2 = 1.0 \text{ atm} \times \frac{5.0 \cancel{L}}{2.0 \cancel{L}} = 2.5 \text{ atm}$$

Volume factor increases pressure

Guide to Using the Gas Laws

STEP 1
Organize the data in a table of initial and final conditions.

STEP 2
Rearrange the gas law for the unknown.

STEP 3
Substitute values into the gas law to solve for the unknown.

STUDY CHECK

A sample of helium gas has a volume of 312 mL at 648 torr. If the volume expands to 825 mL at constant temperature, what is the new pressure in torr?

A gauge indicates the pressure of a gas in a tank.

SAMPLE PROBLEM   11.3

■ Calculating Volume When Pressure Changes

The gauge on a 12-L tank of compressed oxygen reads 3800 mmHg. How many liters would this gas occupy at a pressure of 0.75 atm at constant temperature?

SOLUTION

STEP 1 Organize the data in a table of initial and final conditions. To match the units for pressure, we can convert atm to mmHg, or mmHg to atm.

$$0.75 \ \text{atm} \times \frac{760 \ \text{mmHg}}{1 \ \text{atm}} = 570 \ \text{mmHg}$$

$$3800 \ \text{mmHg} \times \frac{1 \ \text{atm}}{760 \ \text{mmHg}} = 5.0 \ \text{atm}$$

Placing our information using units of mmHg or atm for pressure in a table, we know that pressure decreases. According to Boyle's law, a decrease in the pressure will cause an increase in the volume. (We could have both pressures in atm as well.)

Conditions 1	Conditions 2	Know	Predict
$P_1 = 3800 \ \text{mmHg}$ (5.0 atm)	$P_2 = 570 \ \text{mmHg}$ (0.75 atm)	P decreases	
$V_1 = 12 \ \text{L}$	$V_2 = ?$		V increases

STEP 2 Rearrange the gas law for the unknown. For a PV relationship, we use Boyle's law and solve for V_2 by dividing both sides by P_2.

$$P_1V_1 = P_2 \ V_2$$

$$\frac{P_1V_1}{P_2} = \frac{P_2 V_2}{P_2}$$

$$V_2 = V_1 \times \frac{P_1}{P_2}$$

STEP 3 Substitute values into the gas law to solve for the unknown. When we substitute in the values with pressures in units of mmHg or atm, the ratio of pressures (pressure factor) is greater than 1, which increases the volume as predicted in Step 1.

$$V_2 = 12 \ \text{L} \times \frac{3800 \ \text{mmHg}}{570 \ \text{mmHg}} = 80. \ \text{L}$$

Pressure factor increases volume

Or:

$$V_2 = 12 \ \text{L} \times \frac{5.0 \ \text{atm}}{0.75 \ \text{atm}} = 80. \ \text{L}$$

Pressure factor increases volume

STUDY CHECK

A sample of methane gas, CH_4, has a volume of 125 mL at 0.600 atm pressure and 25 °C. How many milliliters will it occupy at a pressure of 1.50 atm and 25 °C?

CHEMISTRY AND HEALTH

Pressure–Volume Relationship in Breathing

The importance of Boyle's law becomes more apparent when you consider the mechanics of breathing. Our lungs are elastic, balloon-like structures contained within an airtight chamber called the thoracic cavity. The diaphragm, a muscle, forms the flexible floor of the cavity.

Inspiration

The process of taking a breath of air begins when the diaphragm flattens and the rib cage expands, causing an increase in the volume of the thoracic cavity. The elasticity of the lungs allows them to expand when the thoracic cavity expands. According to Boyle's law, the pressure inside the lungs decreases when their volume increases, causing the pressure inside the lungs to fall below the pressure of the atmosphere. This difference in pressures produces a *pressure gradient* between the lungs and the atmosphere. In a pressure gradient, molecules flow from an area of greater pressure to an area of lower pressure. Thus, we inhale as air flows into the lungs (*inspiration*), until the pressure within the lungs becomes equal to the pressure of the atmosphere.

Expiration

Expiration, or the exhalation phase of breathing, occurs when the diaphragm relaxes and moves back up into the thoracic cavity to its resting position. The volume of the thoracic cavity decreases, which

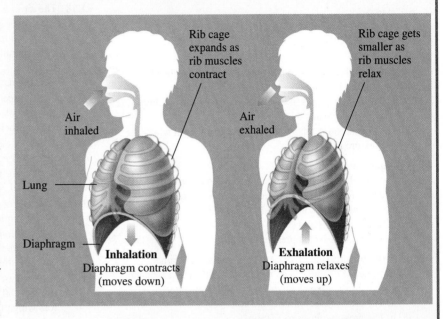

squeezes the lungs and decreases their volume. Now the pressure in the lungs is greater than the pressure of the atmosphere, so air flows out of the lungs. Thus, breathing is a process in which pressure gradients are continuously created between the lungs and the environment because of the changes in the volume and pressure.

QUESTIONS AND PROBLEMS

Pressure and Volume (Boyle's Law)

11.9 Why do scuba divers need to exhale air when they ascend to the surface of the water?

11.10 Why does a sealed bag of chips expand when you take it to a higher altitude?

11.11 Use the words *inspiration* and *expiration* to describe the part of the breathing cycle that occurs because of each of the following:
a. The diaphragm contracts (flattens out).
b. The volume of the lungs decreases.
c. The pressure within the lungs is less than the atmosphere.

11.12 Use the words *inspiration* and *expiration* to describe the part of the breathing cycle that occurs because of each of the following:
a. The diaphragm relaxes, moving up into the thoracic cavity.
b. The volume of the lungs expands.
c. The pressure within the lungs is greater than the atmosphere.

11.13 The air in a cylinder with a piston has a volume of 220 mL and a pressure of 650 mmHg.
a. If a change results in a higher pressure inside the cylinder, does cylinder A or B represent the final volume? Explain your choice.

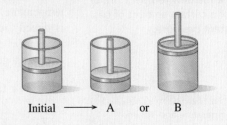

Initial ⟶ A   or   B

b. If the pressure inside the cylinder increases to 1.2 atm, what is the final volume of the cylinder? Complete the following table:

Property	Conditions 1	Conditions 2	Know	Predict
Pressure (P)				
Volume (V)				

11.14 A balloon is filled with helium gas. When the following changes are made at constant temperature, which of these diagrams (A, B, or C) shows the new volume of the balloon?

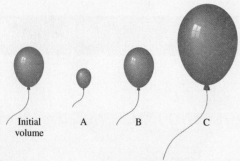

Initial volume   A   B   C

a. The balloon floats to a higher altitude where the outside pressure is lower.

b. The balloon is taken inside the house, but the atmospheric pressure remains the same.

c. The balloon is put in a hyperbaric chamber in which the pressure is increased.

11.15 A gas with a volume of 4.0 L is in a closed container. Indicate the changes in its pressure when the volume undergoes the following changes at constant temperature:

a. The volume is compressed to 2 L.

b. The volume is allowed to expand to 12 L.

c. The volume is compressed to 0.40 L.

11.16 A gas at a pressure of 2.0 atm is in a closed container. Indicate the changes in its volume when the pressure undergoes the following changes at constant temperature:

a. The pressure increases to 6.0 atm.

b. The pressure drops to 1.0 atm.

c. The pressure drops to 0.40 atm.

11.17 A 10.0-L balloon contains helium gas at a pressure of 655 mmHg. What is the new pressure, in mmHg, of the helium gas at each of the following volumes if there is no change in temperature?

a. 20.0 L   **b.** 2.50 L
c. 13 800 mL   **d.** 1250 mL

11.18 The air in a 5.00-L tank has a pressure of 1.20 atm. What is the new pressure, in atm, of the air when the air is placed in tanks that have the following volumes if there is no change in temperature?

a. 1.00 L   **b.** 2500. mL
c. 750. mL   **d.** 8.0 L

11.19 Cyclopropane, C_3H_6, is a general anesthetic. A 5.0-L sample has a pressure of 5.0 atm. What is the volume of the anesthetic given to a patient at a pressure of 1.0 atm?

11.20 A tank of oxygen holds 20.0 L of oxygen (O_2) at a pressure of 15.0 atm. When the gas is released, it provides 300. L of oxygen. What is the pressure of this same gas at a volume of 300. L and constant temperature?

11.21 A sample of nitrogen (N_2) has a volume of 50.0 L at a pressure of 760. mmHg. What is the volume of the gas at each of the following pressures if there is no change in temperature?

a. 1500 mmHg   **b.** 2.0 atm
c. 0.500 atm   **d.** 850 torr

11.22 A sample of methane (CH_4) has a volume of 25 mL at a pressure of 0.80 atm. What is the volume of the gas at each of the following pressures if there is no change in temperature?

a. 0.40 atm   **b.** 2.00 atm
c. 2500 mmHg   **d.** 80.0 torr

11.4 Temperature and Volume (Charles's Law)

LEARNING GOAL

Use the temperature–volume relationship (Charles's law) to determine the new temperature or volume of a certain amount of gas at a constant pressure.

TUTORIAL

Temperature and Volume

Suppose that you are going to take a ride in a hot-air balloon. The captain turns on a propane burner to heat the air inside the balloon. As the air is heated, it expands and becomes less dense than the air outside, causing the balloon and its passengers to lift off. In 1787, Jacques Charles, a balloonist as well as a physicist, proposed that the volume of a gas is related to the temperature. This proposal became **Charles's law**, which states that the volume (V) of a gas is directly related to the temperature (K) when there is no change in the pressure (P) or amount (n) of gas. A **direct relationship** is one in which the related properties increase or decrease together. For two conditions, initial and final, we can write Charles's law as follows:

Charles's Law

$$\frac{V_1}{T_1} = \frac{V_2}{T_2}$$ No change in number of moles and pressure

All temperatures used in gas law calculations must be converted to their corresponding Kelvin (K) temperatures.

To determine the effect of changing temperature on the volume of a gas, the pressure and the amount of gas are kept constant. If we increase the temperature of a gas sample, we know from the kinetic molecular theory that the motion (kinetic energy) of the gas particles will also increase. To keep the pressure constant, the volume of the container must increase (see Figure 11.5). If the temperature of the gas is decreased, the volume of the container must also decrease to maintain the same pressure.

As the gas in a hot-air balloon is heated, it expands.

CONCEPT CHECK 11.4

■ **Charles's Law**

State and explain the reason for the change (*increases, decreases*) in a gas that occurs in each of the following when n and P do not change:

Temperature	Volume
a. Increases	
b. Decreases	

ANSWER

a. When the temperature of a gas increases at constant n and P, the gas particles move faster. To keep the pressure constant, the volume of the container must increase when temperature increases with no change in n and P.

b. When the temperature of a gas decreases at constant n and P, the gas particles move more slowly. To keep the pressure constant, the volume of the container must decrease when the temperature decreases with no change in n and P.

Temperature	Volume
a. Increases	Increases
b. Decreases	Decreases

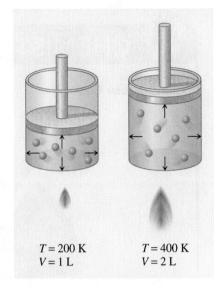

$T = 200 \text{ K}$
$V = 1 \text{ L}$

$T = 400 \text{ K}$
$V = 2 \text{ L}$

FIGURE 11.5 Charles's law: The Kelvin temperature of a gas is directly related to the volume of the gas when there is no change in the pressure. When the temperature increases, making the molecules move faster, the volume must increase to maintain constant pressure.

Q If the temperature of a gas decreases at a constant pressure, how will the volume change?

SAMPLE PROBLEM | **11.4**

■ **Calculating Volume When Temperature Changes**

A sample of argon gas has a volume of 5.40 L and a temperature of 15 °C. Find the new volume of the gas after the temperature has been increased to 42 °C at constant pressure.

SOLUTION

STEP 1 Organize the data in a table of initial and final conditions. When the temperatures are given in degrees Celsius, they must be changed to kelvins.

$$T_1 = 15\,°C + 273 = 288 \text{ K}$$
$$T_2 = 42\,°C + 273 = 315 \text{ K}$$

Conditions 1	Conditions 2	Know	Predict
$T_1 = 288 \text{ K}$	$T_2 = 315 \text{ K}$	T increases	
$V_1 = 5.40 \text{ L}$	$V_2 = ?$		V increases

STEP 2 Rearrange the gas law for the unknown. In this problem, we want to know the final volume (V_2) when the temperature increases. Using Charles's law, we solve for V_2 by multiplying both sides by T_2.

$$\frac{V_1}{T_1} = \frac{V_2}{T_2}$$

$$T_2 \times \frac{V_1}{T_1} = \cancel{T_2} \times \frac{V_2}{\cancel{T_2}}$$

$$V_2 = V_1 \times \frac{T_2}{T_1}$$

STEP 3 Substitute values into the gas law to solve for the unknown. From the table, we see that the temperature has increased. Because temperature is directly related to volume, the volume must increase. When we substitute in the values, we see that the ratio of the temperatures is greater than 1, which increases the volume, as predicted in Step 1:

$$V_2 = 5.40 \text{ L} \times \frac{315 \text{ K}}{288 \text{ K}} = 5.91 \text{ L}$$

Temperature factor increases volume

STUDY CHECK

A mountain climber inhales 486 mL of air at a temperature of −8 °C. What volume, in mL, will the air occupy in the lungs if the climber's body temperature is 37 °C?

CHEMISTRY AND THE ENVIRONMENT

Greenhouse Gases

Percentages of Greenhouse Gases in the Atmosphere

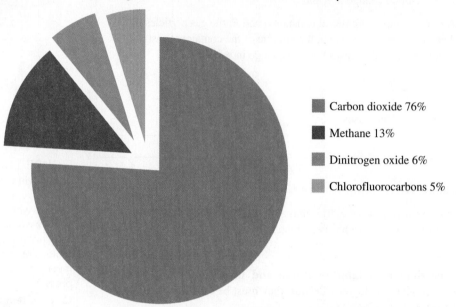

- ■ Carbon dioxide 76%
- ■ Methane 13%
- ■ Dinitrogen oxide 6%
- ■ Chlorofluorocarbons 5%

The term "greenhouse gases" was first used during the early 1800s for the gases in the atmosphere that trap heat. Among the greenhouse gases are carbon dioxide (CO_2), methane (CH_4), dinitrogen oxide (N_2O), and chlorofluorocarbons (CFCs). The molecules of greenhouse gases consist of more than two atoms that vibrate when heat is absorbed. By contrast, oxygen and nitrogen are not greenhouse gases. Because the two atoms in their molecules are tightly bonded, they do not absorb heat.

Greenhouses gases are beneficial in keeping the average surface temperature for Earth at 15 °C. Without greenhouse gases, it is estimated that the average surface temperature of Earth would be −18 °C. Most scientists say that the concentration of greenhouse gases in the atmosphere and the surface temperature of Earth are increasing because of human activities. As we discussed in Chapter 3, the increase in atmospheric carbon dioxide is mostly a result of the burning of fossil fuels and wood.

Methane (CH_4) is a colorless, odorless gas that is released by livestock, rice farming, the decomposition of organic plant material in landfills, and the mining, drilling, and transport of coal and oil. The level of methane in the atmosphere has increased about 150% since industrialization. In one year, as much as 5×10^{11} kg of methane is added to the atmosphere. Livestock produces about 20% of the greenhouse gases. The contribution from livestock comes from the breakdown of organic material in the digestive tracts of cows, sheep, and camels. In one day, one cow emits about 200 g of methane. For a global population of 1.5 billion livestock, a total of 3×10^8 kg of methane is produced every day. In the past few years, methane levels have stabilized due to improvements in the recovery of methane. Methane remains in the atmosphere for about ten years, but its molecular structure causes it to trap 20 times more heat than does carbon dioxide.

Dinitrogen oxide (N_2O), commonly called nitrous oxide, is a colorless greenhouse gas that has a sweet odor. Most people recognize it as an anesthetic used in dentistry called "laughing gas." Although some dinitrogen oxide is released naturally from soil bacteria, its primary increases are from agricultural and industrial processes. Atmospheric dinitrogen oxide has increased by about 15% since industrialization, caused by the extensive use of fertilizers, sewage treatment plants, and car exhaust. Each year, 1×10^{10} kg of dinitrogen oxide is added to the atmosphere. Dinitrogen oxide released today will remain in the atmosphere for about 150–180 years where it has a greenhouse effect that is 300 times greater than that of carbon dioxide.

Chlorofluorinated gases (CFCs) are synthetic compounds containing chlorine, fluorine, and carbon. Chlorofluorocarbons were used as propellants in aerosol cans and refrigerants in refrigerators and air conditioners. During the 1970s, scientists determined that CFCs in the atmosphere were destroying the protective ozone layer. Since then, many countries banned the production and use of CFCs, and their levels in the atmosphere have declined slightly. Hydrofluorocarbons (HFCs), in which hydrogen atoms replace chlorine atoms, are now used as refrigerants. Although HFCs do not destroy the ozone layer, they are greenhouse gases because they trap heat in the atmosphere.

Based on current trends and climate models, scientists estimate that levels of atmospheric carbon dioxide will increase by about 2% each year up to 2025. As long as the greenhouse gases trap more heat than is reflected back into space, average surface temperatures on Earth will continue to rise. Efforts are taking place around the world to slow or decrease the emissions of greenhouse gases into the atmosphere. It is anticipated that temperature will stabilize only when the amount of energy that reaches the surface of Earth is equal to the heat that is reflected back into space.

In 2007, former U.S. Vice President Al Gore and the United Nations Panel on Climate Change were awarded the Nobel Prize for increasing global awareness of the relationship between human activities and global warming.

QUESTIONS AND PROBLEMS

Temperature and Volume (Charles's Law)

11.23 Select the diagram that shows the new volume of a balloon when the following changes are made at constant pressure:

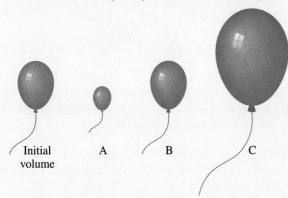

Initial volume   A   B   C

   a. The temperature is changed from 100 K to 300 K.
   b. The balloon is placed in a freezer.
   c. The balloon is first warmed, and then returned to its starting temperature.

11.24 Indicate whether the final volume of gas in each of the following is the same, larger, or smaller than the initial volume:
   a. A volume of 505 mL of air on a cold winter day at −5 °C is breathed into the lungs, where body temperature is 37 °C.

   b. The heater used to heat 1400 L of air in a hot-air balloon is turned off.
   c. A balloon filled with helium at the amusement park is left in a car on a hot day.

11.25 A sample of neon initially has a volume of 2.50 L at 15 °C. What is the new temperature, in °C, when the volume of the sample is changed at constant pressure to each of the following?
   a. 5.00 L   **b.** 1250 mL
   c. 7.50 L   **d.** 3550 mL

11.26 A gas has a volume of 4.00 L at 0 °C. What final temperature, in °C, is needed to cause the volume of the gas to change to the following, if n and P are not changed?
   a. 100. L   **b.** 1200 mL
   c. 250 L   **d.** 50.0 mL

11.27 A balloon contains 2500 mL of helium gas at 75 °C. What is the new volume, in mL, of the gas when the temperature changes to the following, if n and P are not changed?
   a. 55 °C   **b.** 680. K
   c. −25 °C   **d.** 240. K

11.28 An air bubble has a volume of 0.500 L at 18 °C. If the pressure does not change, what is the volume, in liters, at each of the following temperatures?
   a. 0 °C   **b.** 425 K
   c. −12 °C   **d.** 575 K

11.5 Temperature and Pressure (Gay-Lussac's Law)

If we could watch the molecules of a gas as the temperature rises, we would notice that they move faster and hit the sides of the container more often and with greater force. If we keep the volume of the container the same, we would observe an increase in the pressure. A temperature–pressure relationship, also known as **Gay-Lussac's law**, states that the pressure of a gas is directly related to its Kelvin temperature. This means that an increase in temperature increases the pressure of a gas, and a decrease in temperature decreases the pressure of the gas, provided the volume and number of moles of the gas remain the same (see Figure 11.6). The ratio of pressure (P) to temperature (T) is the same under all conditions as long as volume (V) and amount of gas (n) do not change.

LEARNING GOAL

Use the temperature–pressure relationship (Gay-Lussac's law) to determine the new temperature or pressure of a certain amount of gas at a constant volume.

Gay-Lussac's Law

$$\frac{P_1}{T_1} = \frac{P_2}{T_2}$$ No change in number of moles and volume

All temperatures used in gas law calculations must be converted to their corresponding Kelvin (K) temperatures.

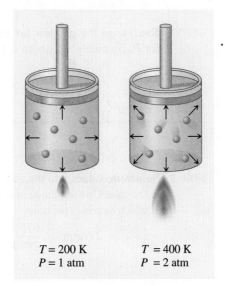

$T = 200$ K $T = 400$ K
$P = 1$ atm $P = 2$ atm

FIGURE 11.6 Gay-Lussac's law: The pressure of a gas is directly related to the temperature of the gas. When the Kelvin temperature of a gas is doubled, the pressure is doubled at constant volume.

Q How does a decrease in the temperature of a gas affect its pressure at constant volume?

■ **Gay-Lussac's Law**

State and explain the reason for the change (*increases, decreases*) in a gas that occurs in each of the following when *n* and *V* do not change:

Temperature	Pressure
a. Increases	
b. Decreases	

ANSWER

a. When the temperature of a gas increases with no change in *n* and *V*, the particles of gas move faster. At constant volume, the gas particles collide more often with the container walls and with more force, increasing the pressure.

b. When the temperature of a gas decreases, the particles of gas move more slowly. At constant volume, the gas particles collide less often with the container walls and with less force, decreasing the pressure.

Temperature	Pressure
a. Increases	Increases
b. Decreases	Decreases

SAMPLE PROBLEM | 11.5

■ **Calculating Pressure When Temperature Changes**

Aerosol containers can be dangerous if they are heated because they can explode. Suppose a container of hair spray with a pressure of 4.0 atm at a room temperature of 25 °C is thrown into a fire. If the temperature of the gas inside the aerosol can reaches 402 °C, what will be its pressure? The aerosol container may explode if the pressure inside exceeds 8.0 atm. Would you expect it to explode?

SOLUTION

STEP 1 **Organize the data in a table of initial and final conditions.** We must first change the temperatures to kelvins.

$$T_1 = 25\,°C + 273 = 298\ K$$
$$T_2 = 402\,°C + 273 = 675\ K$$

Conditions 1	Conditions 2	Know	Predict
$P_1 = 4.0$ atm	$P_2 = ?$		*P* increases
$T_1 = 298$ K	$T_2 = 675$ K	*T* increases	

STEP 2 **Rearrange the gas law for the unknown.** Using Gay-Lussac's law, we solve for P_2 by multiplying both sides by T_2.

$$\frac{P_1}{T_1} = \frac{P_2}{T_2}$$
$$T_2 \times \frac{P_1}{T_1} = T_2 \times \frac{P_2}{T_2}$$
$$P_2 = P_1 \times \frac{T_2}{T_1}$$

STEP 3 **Substitute values into the gas law to solve for the unknown.** When we substitute in the values, we see the ratio of the temperatures (temperature factor) is greater than 1, which increases pressure.

$$P_2 = 4.0\,\text{atm} \times \frac{675\ K}{298\ K} = 9.1\ \text{atm}$$

Temperature factor
increases volume

Because the calculated pressure of 9.1 atm exceeds 8.0 atm, we expect the aerosol can to explode.

In a storage area where the temperature has reached 55 °C, the pressure of oxygen gas in a 15.0-L steel cylinder is 965 torr. To what Celsius temperature would the gas have to be cooled to reduce the pressure to 850. torr?

Vapor Pressure and Boiling Point

In Chapter 10, we learned that liquid molecules with sufficient kinetic energy can break away from the surface of the liquid as they become gas particles or vapor. In an open container, all the liquid will eventually evaporate. In a closed container, the vapor accumulates and creates pressure called **vapor pressure**. Each liquid exerts its own vapor pressure at a given temperature. As temperature increases, more vapor forms, and vapor pressure increases. Table 11.4 lists the vapor pressure of water at various temperatures.

A liquid reaches its boiling point when its vapor pressure becomes equal to the external pressure. As boiling occurs, bubbles of the gas form within the liquid and quickly rise to the surface. For example, at an atmospheric pressure of 760 mmHg, water will boil at 100 °C, the temperature at which its vapor pressure reaches 760 mmHg.

At higher altitudes, atmospheric pressures are lower and the boiling point of water is lower than 100 °C. For example, a typical atmospheric pressure in Denver is 630 mmHg, which means that a vapor pressure of 630 mmHg is required for water to boil in Denver. From Table 11.5, we see that water has a vapor pressure of 630 mmHg at 95 °C. Therefore, water boils at 95 °C in Denver.

TABLE 11.4 Vapor Pressure of Water

Temperature (°C)	Vapor Pressure (mmHg)
0	5
10	9
20	18
30	32
37[a]	47
40	55
50	93
60	149
70	234
80	355
90	528
100	760

[a]Body temperature.

TABLE 11.5 Pressure and the Boiling Point of Water

Pressure (mmHg)	Boiling Point (°C)
270	70
467	87
630	95
752	99
760	100
800	100.4
1075	110
1520 (2 atm)	120
2026	130
7600 (10 atm)	180

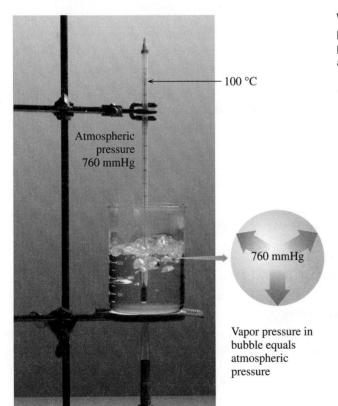

Water boils when its vapor pressure is equal to the pressure of the atmosphere.

100 °C

Atmospheric pressure 760 mmHg

760 mmHg

Vapor pressure in bubble equals atmospheric pressure

TUTORIAL
Vapor Pressure and Boiling Point

An autoclave used to sterilize equipment attains a temperature higher than 100 °C.

People who live at high altitudes often use pressure cookers to obtain higher temperatures when preparing food. When the external pressure is greater than 1 atm, a temperature higher than 100 °C is needed to boil water. Laboratories and hospitals use devices called autoclaves to sterilize laboratory and surgical equipment. An autoclave, like a pressure cooker, is a closed container that increases the total pressure above the liquid so it will boil at higher temperatures. Table 11.5 shows how the boiling point of water increases as pressure increases.

QUESTIONS AND PROBLEMS

Temperature and Pressure (Gay-Lussac's Law)

11.29 Solve for the new pressure, in mmHg, for each of the following with n and V constant:
    a. A gas with an initial pressure of 1200 torr at 155 °C is cooled to 0 °C.
    b. A gas in an aerosol can at an initial pressure of 1.40 atm at 12 °C is heated to 35 °C.

11.30 Solve for the new pressure, in atm, for each of the following with n and V constant:
    a. A gas with an initial pressure of 1.20 atm at 75 °C is cooled to −22 °C.
    b. A sample of N_2 with an initial pressure of 780 mmHg at −75 °C is heated to 28 °C.

11.31 Solve for the new temperature, in °C, for each of the following with n and V constant:
    a. A sample of xenon at 25 °C and 740 mmHg is cooled to give a pressure of 620 mmHg.
    b. A tank of argon gas with a pressure of 0.950 atm at −18 °C is heated to give a pressure of 1250 torr.

11.32 Solve for the new temperature, in °C, for each of the following with n and V constant:
    a. A 10.0-L container of helium gas with a pressure of 250 torr at 0 °C is heated to give a pressure of 1500 torr.
    b. A 500.0-mL sample of air at 40. °C and 740 mmHg is cooled to give a pressure of 680 mmHg.

11.33 Match the terms *vapor pressure, atmospheric pressure,* and *boiling point* to the following descriptions:
    a. the temperature at which bubbles of vapor appear within the liquid
    b. the pressure exerted by a gas above the surface of its liquid
    c. the pressure exerted on Earth by the particles in the air
    d. the temperature at which the vapor pressure of a liquid becomes equal to the external pressure

11.34 In which pair(s) of the following list would boiling occur?

	Atmospheric Pressure	Vapor Pressure
a.	760 mmHg	700 mmHg
b.	480 torr	480 mmHg
c.	1.2 atm	912 mmHg
d.	1020 mmHg	760 mmHg
e.	740 torr	1.0 atm

11.35 Explain each of the following observations:
    a. Water boils at 87 °C on the top of Mount Whitney.
    b. Food cooks more quickly in a pressure cooker than in an open pan.

11.36 Explain each of the following observations:
    a. Boiling water at sea level is hotter than boiling water in the mountains.
    b. Water used to sterilize surgical equipment is heated to 120 °C at 2.0 atm in an autoclave.

11.6 The Combined Gas Law

LEARNING GOAL

Use the combined gas law to find the new pressure, volume, or temperature of a gas when changes in two of these properties are given.

(MC)

TUTORIAL

The Combined Gas Law

All of the pressure–volume–temperature relationships for gases that we have studied may be combined into a single relationship called the **combined gas law**. This expression is useful for studying the effect of changes in two of these variables on the third as long as the amount of gas (number of moles) remains constant.

Combined Gas Law

$$\frac{P_1V_1}{T_1} = \frac{P_2V_2}{T_2}$$
    No change in number of moles of gas

By using the combined gas law, we can derive any of the gas laws by omitting those properties that do not change, as seen in Table 11.6.

TABLE 11.6 Summary of Gas Laws

Combined Gas Law	Properties Held Constant	Relationship	Name of Gas Law
$\dfrac{P_1V_1}{\cancel{T_1}} = \dfrac{P_2V_2}{\cancel{T_2}}$	T, n	$P_1V_1 = P_2V_2$	Boyle's law
$\dfrac{\cancel{P_1}V_1}{T_1} = \dfrac{\cancel{P_2}V_2}{T_2}$	P, n	$\dfrac{V_1}{T_1} = \dfrac{V_2}{T_2}$	Charles's law
$\dfrac{P_1\cancel{V_1}}{T_1} = \dfrac{P_2\cancel{V_2}}{T_2}$	V, n	$\dfrac{P_1}{T_1} = \dfrac{P_2}{T_2}$	Gay-Lussac's law

CONCEPT CHECK 11.6

■ **The Combined Gas Law**

State and explain the reason for the change (*increases, decreases, no change*) in a gas that occurs for the following when n does not change:

	Pressure	Volume	Temperature (K)
a.		Twice as large	Half the Kelvin temperature
b.	Twice as large		Twice as large

ANSWER

a. Pressure decreases by one-half when the volume (at constant n) doubles. If the temperature in Kelvin is halved, the pressure is also halved. The changes in both V and T decrease the pressure to one-fourth its initial value.

b. No change. When the Kelvin temperature of a gas (at constant n) is doubled, the volume is doubled. But when the pressure is twice as much, the volume must decrease to one-half. The changes offset each other, and no change occurs in the volume.

SAMPLE PROBLEM | 11.6

■ **Using the Combined Gas Law**

A 25.0-mL bubble is released from a diver's air tank at a pressure of 4.00 atm and a temperature of 11 °C. What is the volume (mL) of the bubble when it reaches the ocean surface, where the pressure is 1.00 atm and the temperature is 18 °C?

SOLUTION

STEP 1 Organize the data in a table of initial and final conditions. We must first change the temperature to kelvins.

$$T_1 = 11\,°C + 273 = 284\ K$$

$$T_2 = 18\,°C + 273 = 291\ K$$

Conditions 1	Conditions 2
$P_1 = 4.00$ atm	$P_2 = 1.00$ atm
$V_1 = 25.0$ mL	$V_2 = ?$
$T_1 = 284$ K	$T_2 = 291$ K

STEP 2 Rearrange the gas law for the unknown. For changes in two conditions, pressure and temperature, we use the combined gas law to solve for V_2.

$$\frac{P_1 V_1}{T_1} = \frac{P_2 V_2}{T_2}$$

$$\frac{P_1 V_1}{T_1} \times \frac{T_2}{P_2} = \frac{P_2 V_2 \times T_2}{T_2 \times P_2}$$

$$V_2 = V_1 \times \frac{P_1}{P_2} \times \frac{T_2}{T_1}$$

STEP 3 Substitute the values into the gas law to solve for the unknown. From the data table, we determine that both the pressure decrease and the temperature increase will increase the volume. However, when one change decreases the unknown, but the second change increases the unknown, it is difficult to predict the overall change for the unknown.

$$V_2 = 25.0\ \text{mL} \ \times \ \frac{4.00\ \text{atm}}{1.00\ \text{atm}} \times \frac{291\ K}{284\ K} = 102\ \text{mL}$$

Pressure factor increases volume   Temperature factor increases volume

Under water, the pressure on a diver is greater than the atmospheric pressure.

STUDY CHECK

A weather balloon is filled with 15.0 L of helium at a temperature of 25 °C and a pressure of 685 mmHg. What is the pressure, in mmHg, of the helium in the balloon in the upper atmosphere when the temperature is −35 °C and the volume becomes 34.0 L?

QUESTIONS AND PROBLEMS

The Combined Gas Law

11.37 Write the expression for the combined gas law. What gas laws are combined to make the combined gas law?

11.38 Rearrange the variables in the combined gas law to give an expression for the following:
a. T_2 **b.** P_2

11.39 A sample of helium gas has a volume of 6.50 L at a pressure of 845 mmHg and a temperature of 25 °C. What is the pressure of the gas, in atm, when the volume and temperature of the gas sample are changed to each of the following?
a. 1850 mL and 325 K **b.** 2.25 L and 12 °C
c. 12.8 L and 47 °C

11.40 A sample of argon gas has a volume of 735 mL at a pressure of 1.20 atm and a temperature of 112 °C. What is the volume of the gas, in milliliters, when the pressure and temperature of the gas sample are changed to each of the following?
a. 658 mmHg and 281 K **b.** 0.55 atm and 75 °C
c. 15.4 atm and −15 °C

11.41 A 124-mL bubble of hot gas initially at 212 °C and 1.80 atm is emitted from an active volcano. What is the new temperature of the gas outside the volcano if the new volume of the bubble is 138 mL and the pressure is 0.800 atm?

11.42 A scuba diver 40 ft below the ocean surface inhales 50.0 mL of compressed air mixture in a scuba tank at a pressure of 3.00 atm and a temperature of 8 °C. What is the pressure of air in the lungs if the gas expands to 150. mL at a body temperature of 37 °C?

11.7 Volume and Moles (Avogadro's Law)

LEARNING GOAL

Use Avogadro's law to describe the relationship between the amount of a gas and its volume, and use this relationship in calculations.

In our study of the gas laws, we have looked at changes in properties for a specified amount (n) of gas. Now we will consider how the properties of a gas change when there is a change in the number of moles or grams.

When you blow up a balloon, its volume increases because you add more air molecules. If the balloon has a small hole in it, air leaks out, causing its volume to decrease. In 1811, Amedeo Avogadro formulated **Avogadro's law**, which states that the volume of a gas is directly related to the number of moles of a gas when temperature and pressure are not changed. For example, if the number of moles of a gas is doubled, then the volume will also double as long as we do not change the pressure or the temperature (see Figure 11.7). At constant pressure and temperature, we can write Avogadro's law:

$n = 1$ mol
$V = 1$ L
$n = 2$ mol
$V = 2$ L

FIGURE 11.7 Avogadro's law: The volume of a gas is directly related to the number of moles of the gas. If the number of moles is doubled, the volume must double at constant temperature and pressure.

Q If a balloon has a leak, what happens to its volume?

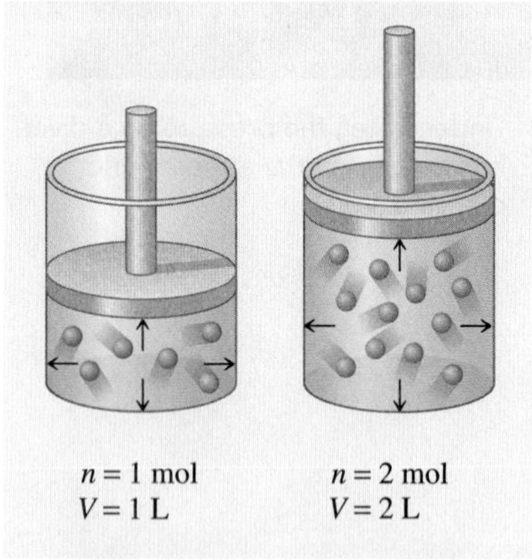

Avogadro's Law

$$\frac{V_1}{n_1} = \frac{V_2}{n_2} \quad \text{No change in pressure or temperature}$$

SAMPLE PROBLEM 11.7

Calculating Volume for a Change in Moles

A weather balloon with a volume of 44 L is filled with 2.0 mol of helium. To what volume (mL) will the balloon expand if 3.0 mol of helium are added, to give a total of 5.0 mol of helium (the pressure and temperature do not change)?

SOLUTION

STEP 1 **Organize the data in a table of initial and final conditions.** A data table for our given information can be set up as follows:

Conditions 1	Conditions 2	Know	Predict
$V_1 = 44$ L	$V_2 = ?$		V increases
$n_1 = 2.0$ mol	$n_2 = 5.0$ mol	n increases	

STEP 2 **Rearrange the gas law for the unknown.** Using Avogadro's law, we can solve for V_2.

$$\frac{V_1}{n_1} = \frac{V_2}{n_2}$$

$$n_2 \times \frac{V_1}{n_1} = \frac{V_2}{\cancel{n_2}} \times \cancel{n_2}$$

$$\boxed{V_2} = V_1 \times \frac{n_2}{n_1}$$

STEP 3 **Substitute the values into the gas law to solve for the unknown.** When we substitute in the values, we see the ratio of the moles (mole factor) is greater than 1, which increases volume as predicted in Step 1.

$$V_2 = 44 \text{ L} \times \frac{5.0 \text{ } \cancel{\text{mol}}}{2.0 \text{ } \cancel{\text{mol}}} = 110 \text{ L}$$

New   Initial   Mole factor
volume   volume   increases volume

STUDY CHECK

A sample containing 8.00 g of oxygen has a volume of 5.00 L. What is the volume (L) after 4.00 g of oxygen is added to the 8.00 g of oxygen in the balloon, if the temperature and pressure do not change?

STP and Molar Volume

Using Avogadro's law, we can say that any two gases will have equal volumes if they contain the same number of moles of gas at the same temperature and pressure. To help us make comparisons between different gases, arbitrary conditions called *standard temperature* (273 K) and *standard pressure* (1 atm) together abbreviated **STP**, were selected by scientists:

STP Conditions

Standard temperature is *exactly* 0 °C (273 K).

Standard temperature is *exactly* 1 atm (760 mmHg).

At STP, one mole of any gas occupies a volume of 22.4 L, which is about the same as the volume of three basketballs. This volume, 22.4 L, of any gas is called the **molar volume** (see Figure 11.8).

The molar volume of a gas is about the same as the volume of three basketballs.

$V = 22.4$ L

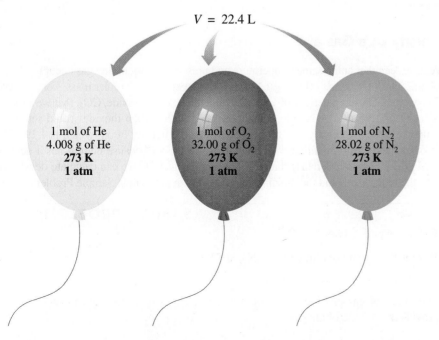

1 mol of He	1 mol of O_2	1 mol of N_2
4.008 g of He	32.00 g of O_2	28.02 g of N_2
273 K	**273 K**	**273 K**
1 atm	**1 atm**	**1 atm**

FIGURE 11.8 Avogadro's law indicates that 1 mole of any gas at STP has a volume of 22.4 L.

Q What volume of gas is occupied by 16.0 g of methane gas, CH_4, at STP?

When a gas is at STP conditions (0 °C and 1 atm), its molar volume can be used as a conversion factor to convert between the number of moles of gas and its volume, in liters.

Molar Volume Conversion Factors

1 mol of gas (STP) = 22.4 L

$$\frac{1 \text{ mol gas (STP)}}{22.4 \text{ L}} \quad \text{and} \quad \frac{22.4 \text{ L}}{1 \text{ mol gas (STP)}}$$

SAMPLE PROBLEM | 11.8

■ **Using Molar Volume to Find Volume at STP**

What is the volume, in liters, of 64.0 g of O_2 gas at STP?

SOLUTION

Once we convert the mass of O_2 to moles of O_2, the molar volume of a gas at STP can be used to calculate the volume (L) of O_2.

STEP 1   Given 64.0 g of $O_2(g)$ at STP   **Need** volume in liters (L)

STEP 2   Write a plan. The grams of O_2 are converted to moles using molar mass. Then a molar volume conversion factor is used to convert the number of moles to volume (L).

grams of O_2 Molar mass   moles of O_2 Molar volume   liters of O_2

STEP 3   Write conversion factors.

1 mol of O_2 = 32.00 g	1 mol of O_2 (STP) = 22.4 L
$\dfrac{32.00 \text{ g } O_2}{1 \text{ mol } O_2}$ and $\dfrac{1 \text{ mol } O_2}{32.00 \text{ g } O_2}$	$\dfrac{22.4 \text{ L } O_2}{1 \text{ mol } O_2}$ and $\dfrac{1 \text{ mol } O_2}{22.4 \text{ L } O_2}$

STEP 4   Set up problem with factors to cancel units.

$$64.0 \text{ g } O_2 \times \frac{1 \text{ mol } O_2}{32.00 \text{ g } O_2} \times \frac{22.4 \text{ L } O_2}{1 \text{ mol } O_2} = 44.8 \text{ L of } O_2 \text{ (STP)}$$

STUDY CHECK

What is the volume (L) of 5.10 g of He at STP?

Guide to Using Molar Volume

STEP 1
Identify given and needed.

STEP 2
Write a plan.

STEP 3
Write conversion factors.

STEP 4
Set up problem with factors to cancel units.

Density of a Gas at STP

We have seen that at the same temperature and pressure, 1 mol of any gas occupies the same volume. Thus, the density (D = g/L) of any gas depends on its molar mass. For example, at STP, oxygen, O_2, has a density of 1.43 g/L while the carbon dioxide, CO_2, that we exhale has a density of 1.96 g/L. A bubble or balloon filled with carbon dioxide would settle to the ground because the density of CO_2 is greater than the density of air, which is 1.29 g/L. On the other hand, balloons filled with helium rise in the air because helium has a density of 0.179 g/L, which is less dense than air. For any gas at STP, we can calculate density (g/L) using the molar mass and the molar volume, as shown in the next Sample Problem.

SAMPLE PROBLEM | 11.9

■ **Density of a Gas at STP**

What is the density of nitrogen gas (N_2) at STP?

SOLUTION

The moles of gas provide the grams for the density expression, and molar volume will provide the volume of the gas.

Balloons rise in the air because helium is less dense than air.

STEP 1   **Given** $N_2(g)$ **Need** density (g/L) of N_2 at STP

STEP 2   **Write a plan.** At STP, the density (g/L) of any gas can be calculated by dividing its molar mass by the molar volume.

$$\text{Density} = \frac{\text{molar mass}}{\text{molar volume}} = \frac{\text{g/mol}}{\text{L/mol}} = \frac{\text{g}}{\text{L}}$$

STEP 3   **Write conversion factors.**

1 mol of N_2 = 28.02 g of N_2

$$\frac{28.02 \text{ g } N_2}{1 \text{ mol } N_2} \quad \text{and} \quad \frac{1 \text{ mol } N_2}{28.02 \text{ g } N_2}$$

1 mol of N_2 (STP) = 22.4 L

$$\frac{22.4 \text{ L } N_2}{1 \text{ mol } N_2} \quad \text{and} \quad \frac{1 \text{ mol } N_2}{22.4 \text{ L } N_2}$$

STEP 4   **Set up the problem with factors to cancel units.**

$$\text{Density (g/L) of } N_2 = \frac{\text{mass}}{\text{volume}} = \frac{\dfrac{28.02 \text{ g } N_2}{1 \text{ mol } N_2}}{\dfrac{22.4 \text{ L } N_2}{1 \text{ mol } N_2}} = 1.25 \text{ g/L}$$

STUDY CHECK

What is the density of hydrogen gas (H_2) at STP?

QUESTIONS AND PROBLEMS

Volume and Moles (Avogadro's Law)

11.43 What happens to the volume of a bicycle tire or a basketball when you use an air pump to add air?

11.44 Sometimes when you blow up a balloon and release it, it flies around the room. What is happening to the air that was in the balloon and its volume?

11.45 A sample containing 1.50 mol of neon gas has a volume of 8.00 L. What is the new volume of the gas, in liters, when the following changes occur in the quantity of the gas at constant pressure and temperature?
a. A leak allows one half of the neon atoms to escape.
b. A sample of 3.50 mol of neon is added to the 1.50 mol of neon gas in the container.
c. A sample of 25.0 g of neon is added to the 1.50 mol of neon gas in the container.

11.46 A sample containing 4.80 g of O_2 gas has a volume of 15.0 L. Pressure and temperature remain constant.
a. What is the new volume if 0.500 mol of O_2 gas is added?
b. Oxygen is released until the volume is 10.0 L. How many moles of O_2 are removed?
c. What is the volume after 4.00 g of He is added to the 4.80 g of O_2 gas in the container?

11.47 Use molar volume to solve each of the following at STP:
a. the number of moles of O_2 in 44.8 L of O_2 gas
b. the number of moles of CO_2 in 4.00 L of CO_2 gas
c. the volume (L) of 6.40 g of O_2
d. the volume (mL) occupied by 50.0 g of neon

11.48 Use molar volume to solve each of the following at STP:
a. the volume (L) occupied by 2.50 mol of N_2
b. the volume (mL) occupied by 0.420 mol of He
c. the number of grams of neon contained in 11.2 L of Ne gas
d. the number of moles of H_2 in 1620 mL of H_2 gas

11.49 Calculate the densities of each of the following gases in g/L at STP:
a. F_2 **b.** CH_4
c. Ne   **d.** SO_2

11.50 Calculate the densities of each of the following gases in g/L at STP:
a. C_3H_8 **b.** NH_3
c. Cl_2 **d.** Ar

11.8 The Ideal Gas Law

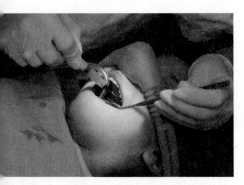

Dinitrogen oxide is used as an anesthetic in dentistry.

The four properties used in the measurement of a gas—pressure (*P*), volume (*V*), temperature (*T*), and amount of a gas (*n*)—can be combined to give a single expression called the **ideal gas law**. The ideal gas law includes all the relationships we have studied for two conditions: Boyle's, Charles's, Gay-Lussac's, and Avogadro's law. The ideal gas law is written as follows:

Ideal Gas Law

$$PV = nRT$$

Rearranging the ideal gas law shows that the four gas properties equal a constant, *R*.

$$\frac{PV}{nT} = R$$

To calculate the value of *R*, we substitute the STP conditions for molar volume into the expression: 1 mol of any gas occupies 22.4 L at STP (273 K and 1 atm).

$$R = \frac{(1.00 \text{ atm})(22.4 \text{ L})}{(1.00 \text{ mol})(273 \text{ K})} = \frac{0.0821 \text{ L} \cdot \text{atm}}{\text{mol} \cdot \text{K}}$$

The value for the **universal gas constant**, *R*, is 0.0821 L · atm per mol · K. If we use 760 mmHg for the pressure, we obtain another useful value for *R* of 62.4 L · mmHg per mol · K.

$$R = \frac{(760 \text{ mmHg})(22.4 \text{ L})}{(1.00 \text{ mol})(273 \text{ K})} = \frac{62.4 \text{ L} \cdot \text{mmHg}}{\text{mol} \cdot \text{K}}$$

The ideal gas law is used when you are given the quantities for any three of the four properties of a gas. Although real gases show some deviations in behavior, the ideal gas law closely approximates the behavior of real gases at typical conditions. In applying the ideal gas law, the units of each gas property must match the units in the universal gas constant, *R*.

Universal Gas Constant (*R*)	$\dfrac{0.0821 \text{ L} \cdot \text{atm}}{\text{mol} \cdot \text{K}}$	$\dfrac{62.4 \text{ L} \cdot \text{mmHg}}{\text{mol} \cdot \text{K}}$
Pressure (*P*)	atm	mmHg
Volume (*V*)	L	L
Amount (*n*)	mol	mol
Temperature (*T*)	K	K

SAMPLE PROBLEM | 11.10

■ Using the Ideal Gas Law

An anesthetic used in dentistry is dinitrogen oxide, N_2O ("laughing gas"). What is the pressure, in atm, of 0.350 mol of N_2O at 22 °C in a 5.00-L container?

SOLUTION

STEP 1 Organize the data given for the gas. When data is given for three of the four quantities (*P*, *V*, *n*, and *T*), we use the ideal gas law to solve for the unknown quantity. The units of *V* and *n* match the units of *R* (0.0821 L · atm/mol · K). However, the temperature in degrees Celsius must be changed to kelvins to match the unit of *T* in the gas constant *R*:

$$P = ? \quad V = 5.00 \text{ L} \quad n = 0.350 \text{ mol}$$
$$R = 0.0821 \frac{\text{L} \cdot \text{atm}}{\text{mol} \cdot \text{K}} \quad T = 22 \text{ °C} + 273 = 295 \text{ K}$$

STEP 2 Solve the ideal gas law for the unknown. By dividing both sides of the ideal gas law by *V*, we can solve for pressure, *P*:

$$P\,V = nRT \quad \text{Ideal gas law}$$

$$P\,\frac{\cancel{V}}{\cancel{V}} = \frac{nRT}{V}$$

$$P = \frac{nRT}{V}$$

STEP 3 Substitute gas data and calculate the unknown quantity.

$$P = \frac{0.350 \ \cancel{mol} \times 0.0821 \ \dfrac{\cancel{L} \cdot atm}{\cancel{mol} \cdot \cancel{K}} \times 295 \ \cancel{K}}{5.00 \ \cancel{L}} = 1.70 \ atm$$

STUDY CHECK

Chlorine gas, Cl_2, is used to purify the water in swimming pools. How many moles of chlorine gas are in a 7.00-L tank if the gas has a pressure of 865 mmHg and a temperature of 24 °C?

Many times we need to know the amount of gas, in grams, involved in a reaction. Then the ideal gas equation can be rearranged to solve for the amount (n) of gas, which is converted to mass in grams using its molar mass as shown in Sample Problem 11.11.

SAMPLE PROBLEM | 11.11 |

■ **Calculating Mass Using the Ideal Gas Law**

Butane, C_4H_{10}, is used as a fuel for barbecues and as an aerosol propellant. If you have 108 mL of butane at 715 mmHg and 25 °C, what is the mass, in grams, of butane?

SOLUTION

STEP 1 Organize the data given for the gas. When three of the quantities (P, V, and T) are known, we use the ideal gas law to solve for moles (n). Because the pressure is given in mmHg, we will use R that has pressure in units of mmHg. Placing the given quantities in a table, we convert volume to liters (L) and temperature to kelvins (K).

	Initial Quantities	Adjusted to Match Units in R
R	$\dfrac{62.4 \ L \cdot mmHg}{mol \cdot K}$	$\dfrac{62.4 \ L \cdot mmHg}{mol \cdot K}$
P	715 mmHg	715 mmHg
V	108 mL	$108 \ \cancel{mL} \times \dfrac{1 \ L}{1000 \ \cancel{mL}} = 0.108 \ L$
n	? mol of C_4H_{10}	? mol of C_4H_{10}
T	25 °C	25 °C + 273 = 298 K

STEP 2 Solve the ideal gas law for the unknown. To solve the ideal gas law for moles, n, divide both sides of the ideal gas law by RT:

$$PV = n \ RT \qquad \text{Ideal gas law}$$

$$\frac{PV}{RT} = n \ \frac{\cancel{RT}}{\cancel{RT}}$$

$$n = \frac{PV}{RT}$$

STEP 3 Substitute gas data and calculate the unknown quantity.

$$n = \frac{715 \ \cancel{mmHg} \times 0.108 \ \cancel{L}}{\dfrac{62.4 \ \cancel{L} \cdot \cancel{mmHg}}{mol \cdot \cancel{K}} \times 298 \ \cancel{K}} = 0.00415 \ mol \ (4.15 \times 10^{-3} \ mol)$$

Now we can convert the moles of butane to grams using its molar mass of 58.12 g/mol:

$$0.00415 \ \cancel{mol \ C_4H_{10}} \times \frac{58.12 \ g \ C_4H_{10}}{1 \ \cancel{mol \ C_4H_{10}}} = 0.241 \ g \ of \ C_4H_{10}$$

STUDY CHECK

What is the volume of 1.20 g of carbon monoxide at 8 °C if it has a pressure of 724 mmHg?

Molar Mass of a Gas

Another typical use of the ideal gas law is to determine the molar mass of a gas. If the mass of the gas is known, the number of moles can be calculated using the ideal gas law. Then the molar mass (g/mol) can be determined.

SAMPLE PROBLEM   11.12

■ **Molar Mass of a Gas Using the Ideal Gas Law**

What is the molar mass of a gas if a 3.16-g sample of gas at 0.750 atm and 45 °C occupies a volume of 2.05 L?

SOLUTION

STEP 1 **Organize the data given for the gas.** When the mass of a gas is given, it is combined with the moles of gas using the ideal gas law, to determine its molar mass.

$$P, V, T \quad \boxed{\text{Ideal Gas Law}} \quad \text{moles } (n) \text{ gas}$$

$$\text{molar mass} = \frac{3.16\text{-g sample}}{\text{moles } (n) \text{ gas}}$$

Initial Quantities	
R	$\dfrac{0.0821 \text{ L} \cdot \text{atm}}{\text{mol} \cdot \text{K}}$
P	0.750 atm
V	2.05 L
n	? mol
T	45 °C + 273 = 318 K
mass	3.16 g

STEP 2 **Solve the ideal gas law for the unknown.** To solve the ideal gas law for moles, n, divide both sides of the ideal gas law by RT:

$$PV = n\,RT \qquad \text{Ideal gas law}$$

$$\frac{PV}{RT} = n\,\frac{RT}{RT}$$

$$n = \frac{PV}{RT}$$

STEP 3 **Substitute gas data and calculate the unknown quantity.**

$$n = \frac{0.750 \text{ atm} \, (2.05 \text{ L})}{\dfrac{0.0821 \text{ L} \cdot \text{atm}}{\text{mol} \cdot \text{K}} \times 318 \text{ K}} = 0.0589 \text{ mol}$$

The molar mass of the gas is obtained by dividing the mass of the gas, in grams, by the moles.

$$\text{Molar mass} = \frac{\text{mass}}{\text{moles}} = \frac{3.16 \text{ g}}{0.0589 \text{ mol}} = 53.7 \text{ g/mol}$$

STUDY CHECK

What is the molar mass of an unknown gas in a 1.50-L container if 0.488 g of the gas has a pressure of 0.0750 atm at 19.0 °C?

QUESTIONS AND PROBLEMS

The Ideal Gas Law

11.51 Calculate the pressure, in atmospheres, of 2.00 mol of helium gas in a 10.0-L container at 27 °C.

11.52 What is the volume, in liters, of 4.00 mol of methane gas, CH_4, at 18 °C and 1.40 atm?

11.53 A tank of oxygen gas has a volume of 20.0 L. How many grams of oxygen are in the container if the gas has a pressure of 845 mmHg at 22 °C?

11.54 A 10.0-g sample of krypton has a temperature of 25 °C at 575 mmHg. What is the volume, in milliliters, of the krypton gas?

11.55 A 25.0-g sample of nitrogen, N_2, has a volume of 50.0 L and a pressure of 630. mmHg. What is the temperature of the gas?

11.56 A 0.226-g sample of carbon dioxide, CO_2, has a volume of 525 mL and a pressure of 455 mmHg. What is the temperature of the gas?

11.57 Determine the molar mass of each of the following gases:
a. 0.84 g of a gas that occupies 450 mL at 0 °C and 1.00 atm (STP)
b. 1.28 g of a gas that occupies 1.00 L at 0 °C and 760 mmHg (STP)
c. 1.48 g of a gas that occupies 1.00 L at 685 mmHg and 22 °C
d. 2.96 g of a gas that occupies 2.30 L at 0.95 atm and 24 °C

11.58 Determine the molar mass of each of the following gases:
a. 2.90 g of a gas that occupies 0.500 L at 0 °C and 1.00 atm (STP)
b. 1.43 g of a gas that occupies 2.00 L at 0 °C and 760 mmHg (STP)
c. 0.726 g of a gas that occupies 855 mL at 1.20 atm and 18 °C
d. 2.32 g of a gas that occupies 1.23 L at 685 mmHg and 25 °C

11.9 Gas Laws and Chemical Reactions

Gases are involved as reactants and products in many chemical reactions. For example, we have seen that the combustion of organic fuels with oxygen gas produces carbon dioxide gas and water vapor. In combination reactions, we have seen that hydrogen gas and nitrogen gas react to form ammonia gas, and hydrogen gas and oxygen gas produce water. Typically, the information given for a gas in a reaction is its pressure (P), volume (V), and temperature (T). Then we can use the ideal gas law to determine the moles of a gas in a reaction. If we are given the number of moles for one of the gases in a reaction, we can use a mole–mole factor to determine the moles of any other substance, as we have done before.

LEARNING GOAL

Determine the mass or volume of a gas that reacts or forms in a chemical reaction.

SAMPLE PROBLEM | 11.13 |

■ **Ideal Gas Law and Chemical Equations**

Limestone ($CaCO_3$) reacts with HCl to produce aqueous calcium chloride, water, and carbon dioxide gas.

$$CaCO_3(s) + 2HCl(aq) \longrightarrow CaCl_2(aq) + CO_2(g) + H_2O(l)$$

How many liters of CO_2 are produced at 752 mmHg and 24 °C from a 25.0-g sample of limestone?

TUTORIAL
The Ideal Gas Law and Stoichiometry

SOLUTION

STEP 1 **Calculate the moles of given using molar mass or ideal gas law.** We use the molar mass of limestone, $CaCO_3$, to calculate the moles of $CaCO_3$.

$$1 \text{ mol of } CaCO_3 = 100.09 \text{ g of } CaCO_3$$
$$\frac{100.09 \text{ g } CaCO_3}{1 \text{ mol } CaCO_3} \quad \text{and} \quad \frac{1 \text{ mol } CaCO_3}{100.09 \text{ g } CaCO_3}$$

$$25.0 \text{ g } CaCO_3 \times \frac{1 \text{ mol } CaCO_3}{100.09 \text{ g } CaCO_3} = 0.250 \text{ mol of } CaCO_3$$

Guide to Reactions Involving Gases

STEP 1
Calculate the moles of given using molar mass or ideal gas law.

STEP 2
Determine the moles of needed using a mole–mole factor.

STEP 3
Convert the moles of needed to mass or volume using molar mass or ideal gas law.

STEP 2 **Determine the moles of needed using a mole–mole factor.**

$$1 \text{ mol of } CaCO_3 = 1 \text{ mol of } CO_2$$
$$\frac{1 \text{ mol } CaCO_3}{1 \text{ mol } CO_2} \quad \text{and} \quad \frac{1 \text{ mol } CO_2}{1 \text{ mol } CaCO_3}$$

$$0.250 \text{ mol } CaCO_3 \times \frac{1 \text{ mol } CO_2}{1 \text{ mol } CaCO_3} = 0.250 \text{ mol of } CO_2$$

Steps 1 and 2 can be combined as

$$25.0 \text{ g } \cancel{CaCO_3} \times \frac{1 \text{ mol } \cancel{CaCO_3}}{100.09 \text{ g } \cancel{CaCO_3}} \times \frac{1 \text{ mol } CO_2}{1 \text{ mol } \cancel{CaCO_3}} = 0.250 \text{ mol of } CO_2$$

STEP 3 **Convert the moles of needed to mass or volume using molar mass or ideal gas law.** Now the ideal gas law is rearranged to solve for the volume (L) of gas.

$$V = \frac{nRT}{P}$$

$$V = \frac{(0.250 \cancel{\text{ mol}}) \times \dfrac{62.4 \text{ L} \cdot \cancel{\text{mmHg}}}{\cancel{K} \cdot \cancel{\text{mol}}} \times 297 \cancel{K}}{752 \cancel{\text{mmHg}}} = 6.16 \text{ L of } CO_2$$

STUDY CHECK

If 12.8 g of aluminum reacts with HCl, how many liters of H_2 would be formed at 715 mmHg and 19 °C?

$$2Al(s) + 6HCl(aq) \longrightarrow 2AlCl_3(aq) + 3H_2(g)$$

QUESTIONS AND PROBLEMS

Gas Laws and Chemical Reactions

11.59 Mg metal reacts with HCl to produce hydrogen gas:

$$Mg(s) + 2HCl(aq) \longrightarrow MgCl_2(aq) + H_2(g)$$

a. What volume of hydrogen at 0 °C and 1.00 atm (STP) is released when 8.25 g of Mg reacts?
b. How many grams of magnesium are needed to prepare 5.00 L of H_2 at 735 mmHg and 18 °C?

11.60 When heated to 350 °C at 0.950 atm, ammonium nitrate decomposes to produce nitrogen, water, and oxygen gases:

$$2NH_4NO_3(s) \xrightarrow{\Delta} 2N_2(g) + 4H_2O(g) + O_2(g)$$

a. How many liters of water vapor are produced when 25.8 g of NH_4NO_3 decomposes?
b. How many grams of NH_4NO_3 are needed to produce 10.0 L of oxygen?

11.61 Butane is used to fill gas tanks for heating. The following equation describes its combustion:

$$2C_4H_{10}(g) + 13O_2(g) \xrightarrow{\Delta} 8CO_2(g) + 10H_2O(g)$$

If a tank contains 55.2 g of butane, what volume, in liters, of oxygen is needed to burn all the butane at 0.850 atm and 25 °C?

11.62 What volume, in liters, of O_2 at 35 °C and 1.19 atm can be produced from the decomposition of 50.0 g of KNO_3?

$$2KNO_3(s) \longrightarrow 2KNO_2(s) + O_2(g)$$

11.63 Aluminum oxide is formed from its elements:

$$4Al(s) + 3O_2(g) \longrightarrow 2Al_2O_3(s)$$

How many liters of oxygen at 0 °C and 760 mmHg (STP) are required to completely react with 5.4 g of aluminum?

11.64 Nitrogen dioxide reacts with water to produce oxygen and ammonia:

$$4NO_2(g) + 6H_2O(g) \longrightarrow 7O_2(g) + 4NH_3(g)$$

At a temperature of 415 °C and a pressure of 725 mmHg, how many grams of NH_3 can be produced when 4.00 L of NO_2 reacts?

11.10 Partial Pressures (Dalton's Law)

LEARNING GOAL

Use partial pressures to calculate the total pressure of a mixture of gases.

Many gas samples are a mixture of gases. For example, the air you breathe is a mixture of mostly oxygen and nitrogen gases. In ideal gas mixtures, scientists observed that all gas particles behave in the same way. Therefore, the total pressure of the gases in a mixture is a result of the collisions of the gas particles regardless of what type of gas they are.

In a gas mixture, each gas exerts its **partial pressure**, which is the pressure it would exert if it were the only gas in the container. **Dalton's law** states that the total pressure of a gas mixture is the sum of the partial pressures of the gases in the mixture.

Dalton's Law

$$P_{\text{total}} = P_1 + P_2 + P_3 + \cdots$$

Total pressure = Sum of the partial pressures
of a gas mixture of the gases in the mixture

Suppose we have two separate tanks, one filled with helium at a pressure of 2.0 atm and the other filled with argon at a pressure of 4.0 atm. When the gases are combined in a single tank with the same volume and temperature, the number of gas molecules, not the type of gas, determines the pressure in a container. There the pressure of the gas mixture would be 6.0 atm, which is the sum of their individual or partial pressures.

$$
\begin{aligned}
P_{total} &= P_{He} + P_{Ar} \\
&= 2.0 \text{ atm} + 4.0 \text{ atm} \\
&= 6.0 \text{ atm}
\end{aligned}
$$

$P_{He} = 2.0 \text{ atm}$ $P_{Ar} = 4.0 \text{ atm}$

The total pressure of two gases is the sum of their partial pressures.

TUTORIAL
Mixtures of Gases

CONCEPT CHECK 11.7

■ Pressure of a Gas Mixture

A scuba tank is filled with Trimix, a breathing gas mixture for deep scuba diving. The tank contains oxygen with a partial pressure of 20. atm, nitrogen with a partial pressure of 40. atm, and helium with a partial pressure of 140. atm. What is the total pressure of the breathing mixture, in atmospheres?

ANSWER

Using Dalton's law of partial pressures, we add together the partial pressures of oxygen, nitrogen, and helium present in the mixture.

$$P_{total} = P_{oxygen} + P_{nitrogen} + P_{helium}$$

P_{total} = 20. atm + 40. atm + 140. atm

= 200. atm

Therefore, when oxygen, nitrogen, and helium are placed in the same container, the sum of their partial pressures is the total pressure of the mixture, which is 200. atm.

Air Is a Gas Mixture

The air you breathe is a mixture of gases. What we call the atmospheric pressure is actually the sum of the partial pressures of the gases in the air. Table 11.7 lists partial pressures for the gases in air on a typical day.

SAMPLE PROBLEM 11.14

■ Partial Pressure of a Gas in a Mixture

A Heliox breathing mixture of oxygen and helium is prepared for a scuba diver who is going to descend 200 ft below the ocean surface. At that depth, the diver breathes a gas mixture that has a total pressure of 7.00 atm. If the partial pressure of the oxygen in the tank at that depth is 1140 mmHg, what is the partial pressure (atm) of the helium in the breathing mixture?

TABLE 11.7 Typical Composition of Air

Gas	Partial Pressure (mmHg)	Percentage (%)
Nitrogen, N_2	594	78.2
Oxygen, O_2	160.	21.0
Carbon dioxide, CO_2 Argon, Ar Water, H_2O	6	0.8
Total air	760.	100

Guide to Solving for Partial Pressure

STEP 1
Write the equation for sum of partial pressures.

STEP 2
Solve for the unknown pressure.

STEP 3
Substitute known pressures and calculate the unknown.

SOLUTION

STEP 1 **Write the equation for the sum of the partial pressures.** From Dalton's law of partial pressures, we know that the total pressure is equal to the sum of the partial pressures.

$$P_{total} = P_{O_2} + P_{He}$$

STEP 2 **Solve for the unknown pressure.**

$$P_{total} = P_{O_2} + P_{He}$$
$$P_{He} = P_{total} - P_{O_2}$$

Convert units to match.

$$P_{O_2} = 1140 \; \cancel{mmHg} \times \frac{1 \; atm}{760 \; \cancel{mmHg}} = 1.50 \; atm$$

STEP 3 **Substitute known pressures and calculate the unknown.**

$$P_{He} = P_{total} - P_{O_2}$$
$$P_{He} = 7.00 \; atm - 1.50 \; atm = 5.50 \; atm$$

STUDY CHECK

An anesthetic consists of a mixture of cyclopropane gas, C_3H_6, and oxygen gas, O_2. If the mixture has a total pressure of 1.09 atm, and the partial pressure of the cyclopropane is 73 mmHg, what is the partial pressure of the oxygen in the anesthetic?

Gases Collected over Water

In the laboratory, gases are often collected by bubbling them through water into a container (see Figure 11.9). Suppose we allow magnesium (Mg) to react with HCl to form $MgCl_2$ and H_2 gas.

$$Mg(s) + 2HCl(aq) \longrightarrow MgCl_2(aq) + H_2(g)$$

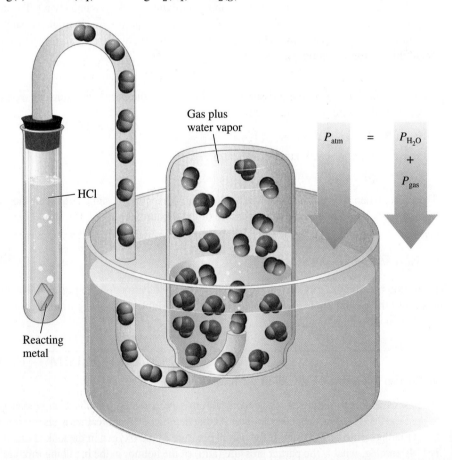

Gas plus water vapor

$$P_{atm} = P_{H_2O} + P_{gas}$$

HCl

Reacting metal

FIGURE 11.9 A gas from a reaction is collected by bubbling through water. Due to evaporation of water, the total pressure is equal to the partial pressure of the gas and the vapor pressure of water.

Q How is the pressure of the dry gas determined?

As hydrogen is produced during the reaction, it displaces some of the water in the container. Because of the vapor pressure of water, the gas that is collected is a mixture of hydrogen and water vapor. For our calculation, we need the pressure of the dry hydrogen gas. We use the vapor pressure of water (Table 11.4) at the experimental temperature, and subtract it from the total gas pressure. Then we can use the ideal gas law to determine the moles or grams of the hydrogen gas that were collected.

SAMPLE PROBLEM   11.15

■ Moles of Gas Collected over Water

When magnesium reacts with HCl, a volume of 355 mL of hydrogen gas is collected over water at 26 °C. The vapor pressure of water at 26 °C is 25 mmHg:

$$Mg(s) + 2HCl(aq) \longrightarrow MgCl_2(aq) + H_2(g)$$

If the total pressure is 752 mmHg, how many moles of $H_2(g)$ were collected?

SOLUTION

STEP 1 **Obtain the vapor pressure of water.** The vapor pressure of water at 26 °C is 25 mmHg.

STEP 2 **Subtract vapor pressure from total P of gas mixture to give partial pressure of needed gas.** Using Dalton's law of partial pressures, determine the partial pressure of H_2.

$$P_{total} = P_{H_2} + P_{H_2O}$$

Solving for the partial pressure of H_2 gives

$$P_{H_2} = P_{total} - P_{H_2O}$$
$$P_{H_2} = 752 \text{ mmHg} - 25 \text{ mmHg}$$
$$= 727 \text{ mmHg}$$

STEP 3 **Use ideal gas law to convert P_{gas} to moles or grams of gas collected.** By dividing both sides of the ideal gas law by RT, we solve for moles, n, of gas:

$$PV = nRT \quad \text{Ideal gas law}$$
$$\frac{PV}{RT} = n\frac{RT}{RT}$$
$$n = \frac{PV}{RT}$$

Solve for the moles of H_2 gas by placing the partial pressure of H_2 (727 mmHg), volume of gas container (0.355 L), temperature (26 °C + 273 = 299 K), and R, using mmHg, into the ideal gas law:

$$n_{H_2} = \frac{727 \text{ mmHg} \times 0.355 \text{ L}}{\dfrac{62.4 \text{ L} \cdot \text{mmHg}}{\text{mol} \cdot \text{K}} \times 299 \text{ K}} = 0.0138 \text{ mol of } H_2 \left(1.38 \times 10^{-2} \text{ mol of } H_2\right)$$

STUDY CHECK

A 456-mL sample of oxygen gas (O_2) was collected over water at a pressure of 744 mmHg and a temperature of 20. °C. How many grams of dry O_2 were collected?

Guide to Gases Collected over Water

STEP 1
Obtain the vapor pressure of water.

STEP 2
Subtract vapor pressure from total P of gas mixture to give partial pressure of needed gas.

STEP 3
Use ideal gas law to convert P_{gas} to moles or grams of gas collected.

CHEMISTRY AND HEALTH

Blood Gases

Our cells continuously use oxygen and produce carbon dioxide. Both gases move in and out of the lungs through the membranes of the alveoli, the tiny air sacs at the ends of the airways in the lungs. An exchange of gases occurs in which oxygen from the air diffuses into the lungs and into the blood, while carbon dioxide produced in the cells is carried to the lungs to be exhaled. In Table 11.8, partial pressures are given for the gases in air that we inhale (inspired air), air in the alveoli, and the air that we exhale (expired air). The partial pressure of water vapor increases within the lungs because the vapor pressure of water is 47 mmHg at body temperature.

TABLE 11.8 Partial Pressures of Gases During Breathing

Gas	Partial Pressure (mmHg)		
	Inspired Air	Expired Air	Alveolar Air
Nitrogen, N_2	594	569	573
Oxygen, O_2	160	116	100
Carbon dioxide, CO_2	0.3	28	40
Water vapor, H_2O	5.7	47	47
Total	760	760	760

At sea level, oxygen normally has a partial pressure of 100 mmHg in the alveoli of the lungs. Because the partial pressure of oxygen in venous blood is 40 mmHg, oxygen diffuses from the alveoli into the bloodstream. The oxygen combines with hemoglobin, which carries it to the tissues of the body where the partial pressure of oxygen can be very low, less than 30 mmHg. Oxygen diffuses from the blood, where the partial pressure of O_2 is high, into the tissues, where O_2 pressure is low.

As oxygen is used in the cells of the body during metabolic processes, carbon dioxide is produced, so the partial pressure of CO_2 may be as high as 50 mmHg or more. Carbon dioxide diffuses from the tissues into the bloodstream and is carried to the lungs. There it diffuses out of the blood, where CO_2 has a partial pressure of 46 mmHg, into the alveoli, where CO_2 is at 40 mmHg and is exhaled. Table 11.9 summarizes the partial pressures of blood gases in the tissues and in oxygenated and deoxygenated blood.

TABLE 11.9 Partial Pressures of Oxygen and Carbon Dioxide in Blood and Tissues

Gas	Partial Pressure (mmHg)		
	Oxygenated Blood	Deoxygenated Blood	Tissues
O_2	100	40	30 or less
CO_2	40	46	50 or greater

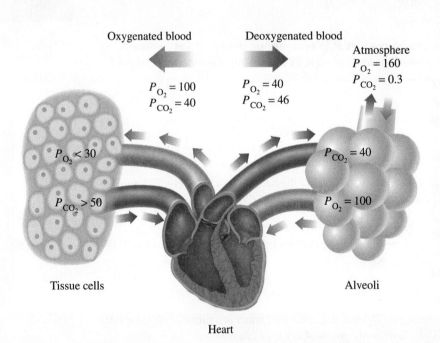

CHEMISTRY AND HEALTH

Hyperbaric Chambers

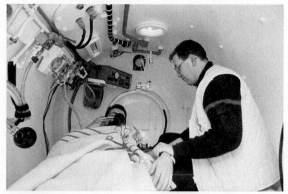

A hyperbaric chamber is used in the treatment of certain diseases.

A burn patient may undergo treatment for burns and infections in a hyperbaric chamber, a device in which pressures can be obtained that are two to three times greater than atmospheric pressure. A greater oxygen pressure increases the level of dissolved oxygen in the blood and tissues, where it fights bacterial infections. High levels of oxygen are toxic to many strains of bacteria. The hyperbaric

chamber may also be used during surgery, to help counteract carbon monoxide (CO) poisoning, and to treat some cancers.

The blood is normally capable of dissolving up to 95 percent of the oxygen. Thus, if the pressure of the oxygen is 2280 mmHg (3 atm), 95 percent of that or 2170 mmHg of oxygen can dissolve in the blood, where it saturates the tissues. In the case of carbon monoxide poisoning, this oxygen can replace the carbon monoxide that has attached to the hemoglobin.

A patient undergoing treatment in a hyperbaric chamber must also undergo decompression (reduction of pressure) at a rate that slowly reduces the concentration of dissolved oxygen in the blood. If decompression is too rapid, the oxygen dissolved in the blood may form gas bubbles in the circulatory system.

If a scuba diver does not decompress slowly, a condition called the "bends" may occur. While below the surface of the ocean, a diver uses breathing mixtures with higher pressures. If there is nitrogen in the mixture, higher quantities of nitrogen gas dissolve in the blood. If the diver ascends to the surface too quickly, the dissolved nitrogen becomes less soluble. Then the nitrogen forms gas bubbles that can produce life-threatening blood clots, or form in the joints and tissues of the body, which can be quite painful. A diver suffering from the bends is placed immediately into a decompression chamber where pressure is first increased and then slowly decreased. The dissolved nitrogen can then diffuse through the lungs until atmospheric pressure is reached. See also Chapter 3, Chemistry and Health *Breathing Mixtures for Scuba.*

QUESTIONS AND PROBLEMS

Partial Pressures (Dalton's Law)

11.65 When solid $KClO_3$ is heated, it decomposes to give solid KCl and O_2 gas. A volume of 256 mL of gas is collected over water at a total pressure of 765 mmHg and 24 °C. The vapor pressure of water at 24 °C is 22 mmHg:

$$2KClO_3(s) \xrightarrow{\Delta} 2KCl(s) + 3O_2(g)$$

a. What was the partial pressure of the O_2 gas?
b. How many moles of O_2 gas were in the gas sample?

11.66 When solid $CaCO_3$ is heated, it decomposes to give solid CaO and CO_2 gas. A volume of 425 mL of gas is collected over water at a total pressure of 758 mmHg and 16 °C. The vapor pressure of water at 16 °C is 14 mmHg.

$$CaCO_3(s) \xrightarrow{\Delta} CaO(s) + CO_2(g)$$

a. What was the partial pressure of the CO_2 gas?
b. How many moles of CO_2 gas were in the CO_2 gas sample?

11.67 A typical air sample in the lungs contains oxygen at 100 mmHg, nitrogen at 573 mmHg, carbon dioxide at 40 mmHg, and water vapor at 47 mmHg. Why are the pressures of oxygen, nitrogen, and carbon dioxide in the air sample called partial pressures?

11.68 Suppose a mixture contains helium and oxygen gases. If the partial pressure of helium is the same as the partial pressure of oxygen, what do you know about the number of helium atoms compared to the number of oxygen molecules? Explain.

11.69 In a gas mixture, the partial pressures are nitrogen 425 torr, oxygen 115 torr, and helium 225 torr. What is the total pressure (torr) exerted by the gas mixture?

11.70 In a gas mixture, the partial pressures are argon 415 mmHg, neon 75 mmHg, and nitrogen 125 mmHg. What is the total pressure (atm) exerted by the gas mixture?

11.71 A gas mixture containing oxygen, nitrogen, and helium exerts a total pressure of 925 torr. If the partial pressures are oxygen 425 torr and helium 75 torr, what is the partial pressure (torr) of the nitrogen in the mixture?

11.72 A gas mixture containing oxygen, nitrogen, and neon exerts a total pressure of 1.20 atm. If helium added to the mixture increases the pressure to 1.50 atm, what is the partial pressure (atm) of the helium?

CONCEPT MAP

```
                              Gases

   are described by      behave according to     volume and moles are related by

   Kinetic                  Gas laws              Avogadro's law
   molecular                                      V and n
   theory
                           that relate
                                                  at STP gives
   states that gas particles
                            P and V                Molar
   Are far apart            Boyle's law            volume

                            V and T          for a gas or    used to find
   Move fast                Charles's law
                                             Gas             Moles or volume
   and                      P and T          mixture         of gases
                            Gay-Lussac's law
                                             where gases exert   in
   Exert                    P, V, and T
   pressure                 Combined gas law  Partial          Reactions
                                              pressures        of gases
                            P, V, n, and T
                            Ideal gas law
```

 # CHAPTER REVIEW

11.1 Properties of Gases
LEARNING GOAL: *Use the kinetic molecular theory of gases to describe the properties of gases.*
In a gas, particles are far apart and moving very fast. A gas is described by the physical properties of pressure (P), volume (V), temperature (T), and amount in moles (n).

11.2 Gas Pressure
LEARNING GOAL: *Describe the units of measurement used for pressure, and change from one unit to another.*
A gas exerts pressure, the force of the gas particles striking the surface of a container. Gas pressure is measured in units such as torr, mmHg, atm, and Pa.

11.3 Pressure and Volume (Boyle's Law)
LEARNING GOAL: *Use the pressure–volume relationship (Boyle's law) to determine the new pressure or volume of a certain amount of gas at a constant temperature.*
The volume (V) of a gas changes inversely with the pressure (P) of the gas if there is no change in the amount and temperature: $P_1V_1 = P_2V_2$. This means that the pressure increases if volume decreases; pressure decreases if volume increases.

11.4 Temperature and Volume (Charles's Law)
LEARNING GOAL: *Use the temperature–volume relationship (Charles's law) to determine the new temperature or volume of a certain amount of gas a constant pressure.*
The volume (V) of a gas is directly related to its Kelvin temperature (T) when there is no change in the amount and pressure of the gas:

$$\frac{V_1}{T_1} = \frac{V_2}{T_2}$$

Therefore, if temperature increases, the volume of the gas increases; if temperature decreases, volume decreases.

11.5 Temperature and Pressure (Gay-Lussac's Law)
LEARNING GOAL: *Use the temperature–pressure relationship (Gay-Lussac's law) to determine the new temperature or pressure of a certain amount of gas at a constant volume.*
The pressure (P) of a gas is directly related to its Kelvin temperature (T):

$$\frac{P_1}{T_1} = \frac{P_2}{T_2}$$

This means that an increase in temperature (T) increases the pressure of a gas, or a decrease in temperature decreases the pressure, as long as the amount and volume stay constant. Vapor pressure is the pressure of the gas that forms when a liquid evaporates. At the boiling point of a liquid, the vapor pressure equals the external pressure.

11.6 The Combined Gas Law
LEARNING GOAL: *Use the combined gas law to find the new pressure, volume, or temperature of a gas when changes in two of these properties are given.*
Gas laws combine into a relationship of pressure (P), volume (V), and temperature (T):

$$\frac{P_1V_1}{T_1} = \frac{P_2V_2}{T_2}$$

This relationship is used to determine the effect of changes in two of the variables on the third.

11.7 Volume and Moles (Avogadro's Law)

LEARNING GOAL: *Use Avogadro's law to describe the relationship between the amount of a gas and its volume, and use this relationship in calculations.*

The volume (V) of a gas is directly related to the number of moles (n) of the gas when the pressure and temperature of the gas do not change:

$$\frac{V_1}{n_1} = \frac{V_2}{n_2}$$

If the moles of gas are increased, the volume must increase, or if the moles of gas are decreased, the volume must decrease. At standard temperature (273 K) and standard pressure (1 atm), abbreviated STP, one mol of any gas has a volume of 22.4 L. The density of a gas at STP is the ratio of the molar mass to the molar volume.

11.8 The Ideal Gas Law

LEARNING GOAL: *Use the ideal gas law to solve for P, V, T, or n of a gas when given three of the four values in the ideal gas law. Calculate density, molar mass, or volume of a gas in a chemical reaction.*

The ideal gas law gives the relationship of all the quantities P, V, n, and T that describe and measure a gas: $PV = nRT$. Any of the four vari-

ables can be calculated if the other three are known. The molar mass of a gas can be calculated using molar volume at STP or the ideal gas law.

11.9 Gas Laws and Chemical Reactions

LEARNING GOAL: *Determine the mass or volume of a gas that reacts or forms in a chemical reaction.*

The ideal gas law is used to convert the quantities (P, V, and T) of gases to moles in a chemical reaction. The moles of gases can be used to determine the number of moles of other substances in the reaction. The pressure and volume of other gases in the reaction can be calculated using the ideal gas law.

11.10 Partial Pressures (Dalton's Law)

LEARNING GOAL: *Use partial pressures to calculate the total pressure of a mixture of gases.*

In a mixture of two or more gases, the total pressure is the sum of the partial pressures of the individual gases.

$$P_{\text{total}} = P_1 + P_2 + P_3 + \cdots$$

The partial pressure of a gas in a mixture is the pressure it would exert if it were the only gas in the container. For gases collected over water, the vapor pressure of water is subtracted from the total pressure of the gas mixture to obtain the partial pressure of the dry gas.

■ KEY TERMS

atmosphere (atm) A unit equal to the pressure exerted by a column of mercury 760 mm high.

atmospheric pressure The pressure exerted by the atmosphere.

Avogadro's law A gas law stating that the volume of a gas changes directly with the number of moles of gas when pressure and temperature do not change.

Boyle's law A gas law stating that the pressure of a gas is inversely related to the volume when temperature and moles of the gas do not change; that is, if volume decreases, pressure increases.

Charles's law A gas law stating that the volume of a gas changes directly with a change in Kelvin temperature when pressure and moles of the gas do not change.

combined gas law A relationship that combines several gas laws relating pressure, volume, and temperature:

$$\frac{P_1V_1}{T_1} = \frac{P_2V_2}{T_2}$$

Dalton's law A gas law stating that the total pressure exerted by a mixture of gases in a container is the sum of the partial pressures that each gas would exert alone.

direct relationship A relationship in which two properties increase or decrease together.

Gay-Lussac's law A gas law stating that the pressure of a gas changes directly with a change in Kelvin temperature when the number of moles of a gas and its volume do not change.

ideal gas law A law that combines the four measured properties of a gas: $PV = nRT$.

inverse relationship A relationship in which two properties change in opposite directions.

kinetic molecular theory of gases A model used to explain the behavior of gases.

molar volume A volume of 22.4 L occupied by 1 mol of a gas at STP conditions of 0 °C (273 K) and 1 atm.

partial pressure The pressure exerted by a single gas in a gas mixture.

pressure The force exerted by gas particles that hit the walls of a container.

STP Standard conditions of exactly 0 °C (273 K) temperature and 1 atm pressure used for the comparison of gases.

torr A unit of pressure equal to 1 mmHg; 760 torr = 1 atm.

universal gas constant (R) A numerical value that relates the quantities P, V, n, and T in the ideal gas law, $PV = nRT$.

vapor pressure The pressure exerted by the particles of vapor above a liquid.

■ UNDERSTANDING THE CONCEPTS

11.73 At 100 °C, which of the following gas samples exerts
a. the lowest pressure?   **b.** the highest pressure?

1.   2.   3.

11.74 Indicate which diagram represents the volume of the gas sample in a flexible container when each of the following changes takes place:

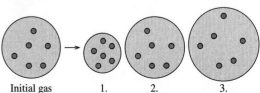

Initial gas   1.   2.   3.

a. Temperature increases at constant pressure.
b. Temperature decreases at constant pressure.
c. Pressure decreases at constant temperature.
d. Doubling the pressure and doubling the Kelvin temperature.

11.75 A balloon is filled with helium gas with a pressure of 1.00 atm and neon gas with a pressure of 0.50 atm. For each of the following changes of the initial balloon, select the diagram (A, B or C) that shows the final (new) volume of the balloon:

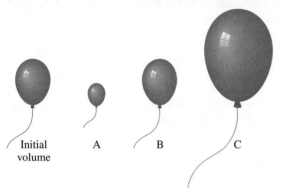

Initial volume   A   B   C

a. The balloon is put in a cold storage unit (*P* and *n* constant).
b. The balloon floats to a higher altitude where the pressure is less (*n*, *T* constant).
c. All of the helium gas is removed (*T* and *P* constant).
d. The Kelvin temperature doubles and 1/2 of the gas atoms leak out (*P* constant).
e. 2.0 mol of O_2 gas is added at constant *T* and *P*.

11.76 Indicate if pressure increases, decreases, or stays the same in each of the following:

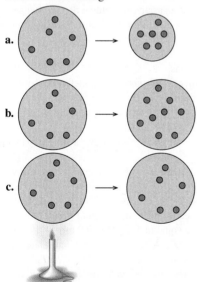

a.
b.
c.

11.77 Your spaceship has docked at a space station above Mars. The temperature inside the space station is a carefully controlled 24 °C at a pressure of 745 mmHg. A balloon with a volume of 425 mL drifts into the airlock where the temperature is −95 °C and the pressure is 0.115 atm. What is the new volume of the balloon? Assume that the balloon is very elastic.

11.78 At a restaurant, a customer chokes on a piece of food. You put your arms around the person's waist and use your fists to push up on the person's abdomen, an action called the Heimlich maneuver.
a. How would this action change the volume of the chest and lungs?
b. Why does it cause the person to expel the food item from the airway?

11.79 In 1783, Jacques Charles launched his first balloon filled with hydrogen gas because it was lighter than air. The balloon had a volume of 31 000 L when it reached an altitude of 1000 m where the pressure was 658 mmHg and the temperature was −8 °C.

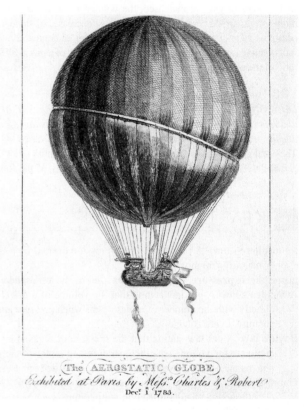

The AEROSTATIC GLOBE
Exhibited at Paris by Meß.ʳ Charles & Robert
Decʳ 1 1783.

How many kilograms of hydrogen would be needed to fill the balloon?

11.80 An airplane is pressurized to 650 mmHg, which is the atmospheric pressure at a ski resort at 13 000 ft altitude.
a. If air is 21% oxygen, what is the partial pressure of oxygen on the plane?
b. If the partial pressure of oxygen drops below 100 mmHg, passengers become drowsy. If this happens, oxygen masks are released. What is the total cabin pressure at which oxygen masks are dropped?

ADDITIONAL QUESTIONS AND PROBLEMS

11.81 A gas sample has a volume of 4250 mL at 15 °C and 745 mmHg. What is the new temperature (°C) after the sample is transferred to a new container with a volume of 2.50 L and a pressure of 1.20 atm?

11.82 A weather balloon has a volume of 750 L when filled with helium at 8 °C at a pressure of 380 torr. What is the new volume of the balloon where the pressure is 0.20 atm and the temperature is −45 °C?

11.83 A 2.00-L container is filled with methane gas, CH_4, at a pressure of 2500 mmHg and a temperature of 18 °C. How many grams of methane are in the container?

11.84 A steel cylinder with a volume of 15.0 L is filled with 50.0 g of nitrogen gas at 25 °C. What is the pressure of the N_2 gas in the cylinder?

11.85 A sample of gas with a mass of 1.62 g occupies a volume of 941 mL at a pressure of 748 torr and a temperature of 20.0 °C. What is the molar mass of the gas?

11.86 What is the molar mass of a gas if 1.15 g of the gas has a volume of 225 mL at 0 °C and 1.00 atm (STP)?

11.87 A gaseous compound has an empirical formula CH_2. When the gas is at 23 °C and 752 mmHg, a volume of 782 mL of the gas has a mass of 2.23 g. What are the molar mass and the molecular formula of the gas?

11.88 A sample of an unknown gas with a mass of 3.24 g occupies a volume of 1.88 L at a pressure of 748 mmHg and a temperature of 20.0 °C.
a. What is the molar mass of the gas?
b. If the unknown gas is composed of 2.78 g of carbon and the rest is hydrogen, what is its molecular formula?

11.89 How many molecules of CO_2 are in 35.0 L of $CO_2(g)$ at 1.2 atm and 5 °C?

11.90 A container is filled with 4.0×10^{22} O_2 molecules at 5 °C and 845 mmHg. What is the volume, in mL, of the container?

11.91 How many liters of H_2 gas can be produced at 0 °C and 1.00 atm (STP) from 25.0 g of Zn?
$$Zn(s) + 2HCl(aq) \longrightarrow ZnCl_2(aq) + H_2(g)$$

11.92 In the formation of smog, nitrogen and oxygen gas react to form nitrogen dioxide. How many grams of NO_2 will be produced when 2.0 L of nitrogen at 840 mmHg and 24 °C are completely reacted?
$$N_2(g) + 2O_2(g) \longrightarrow 2NO_2(g)$$

11.93 Nitrogen dioxide reacts with water to produce oxygen and ammonia:
$$4NO_2(g) + 6H_2O(g) \longrightarrow 7O_2(g) + 4NH_3(g)$$

a. How many liters of O_2 at STP are produced when 2.5×10^{23} molecules of NO_2 react?
b. A 5.00-L sample of $H_2O(g)$ reacts at a temperature of 375 °C and a pressure of 725 mmHg. How many grams of NH_3 can be produced?

11.94 Hydrogen gas can be produced in the laboratory through the reaction of magnesium metal with hydrochloric acid:
$$Mg(s) + 2HCl(aq) \longrightarrow MgCl_2(aq) + H_2(g)$$

What is the volume, in liters, of H_2 gas produced at 24 °C and 835 mmHg, from the reaction of 12.0 g of Mg?

11.95 A gas mixture with a total pressure of 2400 torr is used by a scuba diver. If the mixture contains 2.0 mol of helium and 6.0 mol of oxygen, what is the partial pressure of each gas in the sample?

11.96 What is the total pressure, in mmHg, of a gas mixture containing argon gas at 0.25 atm, helium gas at 350 mmHg, and nitrogen gas at 360 torr?

11.97 A gas mixture contains oxygen and argon at partial pressures of 0.60 atm and 425 mmHg. If nitrogen gas added to the sample increases the total pressure to 1250 torr, what is the partial pressure, in torr, of the nitrogen added?

11.98 A gas mixture contains helium and oxygen at partial pressures of 255 torr and 0.450 atm. What is the total pressure, in mmHg, of the mixture after it is placed in a container one-half the volume of the original container?

11.99 Solid aluminum reacts with aqueous H_2SO_4 to form H_2 gas and aluminum sulfate. When a sample of Al is allowed to react, 415 mL of gas is collected over water at 23 °C, at a pressure of 755 mmHg. At 23 °C, the vapor pressure of water is 21 mmHg:
$$2Al(s) + 3H_2SO_4(aq) \longrightarrow 3H_2(g) + Al_2(SO_4)_3(aq)$$

a. What is the pressure of the dry H_2 gas?
b. How many moles of H_2 were produced?
c. How many grams of Al were reacted?

11.100 When heated, $KClO_3$ solid forms solid KCl and O_2 gas. A sample of $KClO_3$ is heated and 226 mL of gas with a pressure of 744 mmHg is collected over water, at 26 °C. At 26 °C, the vapor pressure of water is 25 mmHg:
$$2KClO_3(s) \xrightarrow{\Delta} 2KCl(s) + 3O_2(g)$$

a. What is the pressure of the dry O_2 gas?
b. How many moles of O_2 were produced?
c. How many grams of $KClO_3$ were reacted?

CHALLENGE QUESTIONS

11.101 Two flasks of equal volume and at the same temperature contain different gases. One flask contains 10.0 g of Ne and the other flask contains 10.0 g of He. Which of the following statements are correct? Explain.
a. Both flasks contain the same number of atoms.
b. The pressures in the flasks are the same.
c. The flask that contains helium has a higher pressure than the flask that contains neon.
d. The densities of the gases are the same.

11.102 A 92.0-g sample of a liquid is placed in a 25.0-L flask. At 140 °C, the liquid evaporates completely to give a pressure of 0.900 atm.
a. What is the molar mass of the gas?
b. If the flask can withstand pressures up to 1.30 atm, calculate the maximum temperature to which the gas can be heated without breaking.

11.103 A sample of a carbon–hydrogen compound is found to contain 9.60 g of carbon and 2.42 g of hydrogen. At STP, 762 mL

of the gas has a mass of 1.02 g. What is the molecular formula for the compound?

11.104 When sensors in a car detect a collision, they cause the reaction of sodium azide, NaN_3, which generates nitrogen gas to fill the air bags within 0.03 s.

$$2NaN_3(s) \longrightarrow 2Na(s) + 3N_2(g)$$

How many liters of N_2 are produced at STP if the air bag contains 132 g of NaN_3?

11.105 Glucose, $C_6H_{12}O_6$, is metabolized in living systems according to the reaction

$$C_6H_{12}O_6(s) + 6O_2(g) \longrightarrow 6CO_2(g) + 6H_2O(l)$$

How many grams of water can be produced from the reaction of 18.0 g of glucose and 7.50 L of O_2 at 1.00 atm and 37 °C?

11.106 2.00 L of N_2, at 25 °C and 1.08 atm, is mixed with 4.00 L of O_2, at 25 °C and 0.118 atm, and the mixture allowed to react. How much NO, in grams, is produced?

$$N_2(g) + O_2(g) \longrightarrow 2NO(g)$$

ANSWERS

ANSWERS TO STUDY CHECKS

11.1 The mass in grams gives the amount of gas.

11.2 245 torr

11.3 50.0 mL

11.4 569 mL

11.5 16 °C

11.6 241 mmHg

11.7 7.50 L

11.8 28.5 L

11.9 0.0900 g/L

11.10 0.327 mol of Cl_2

11.11 1.04 L of CO

11.12 104 g/mol

11.13 18.1 L

11.14 755 mmHg

11.15 0.579 g of O_2

ANSWERS TO SELECTED QUESTIONS AND PROBLEMS

11.1 **a.** At a higher temperature, gas particles have greater kinetic energy, which makes them move faster.
b. Because there are great distances between the particles of a gas, they can be pushed closer together and still remain a gas.
c. Gas particles are very far apart, which means that the mass of a gas in a certain volume is very small, resulting in a low density.

11.3 **a.** temperature **b.** volume
c. amount **d.** pressure

11.5 atmospheres (atm), mmHg, torr, lb/in.^{2}, pascals, kilopascals, in. Hg

11.7 **a.** 1520 torr **b.** 29.4 lb/in.^{2}
c. 1520 mmHg **d.** 203 kPa

11.9 As a diver ascends to the surface, external pressure decreases. If the air in the lungs were not exhaled, its volume would expand and severely damage the lungs. The pressure in the lungs must adjust to changes in the external pressure.

11.11 **a.** inspiration **b.** expiration
c. inspiration

11.13 **a.** The pressure is greater in cylinder A. According to Boyle's law, a decrease in volume pushes the gas particles closer together, which will cause an increase in the pressure.
b.

Property	Conditions 1	Conditions 2	Know	Predict
Pressure (P)	650 mmHg	1.2 atm	P increases	
Volume (V)	220 mL	160 mL		V decreases

11.15 **a.** The pressure doubles.
b. The pressure falls to one-third the initial pressure.
c. The pressure increases to ten times the original pressure.

11.17 **a.** 328 mmHg **b.** 2620 mmHg
c. 475 mmHg **d.** 5240 mmHg

11.19 25 L of cyclopropane

11.21 **a.** 25 L **b.** 25 L
c. 100. L **d.** 45 L

11.23 **a.** C **b.** A
c. B

11.25 **a.** 303 °C **b.** −129 °C
c. 591 °C **d.** 136 °C

11.27 **a.** 2400 mL **b.** 4900 mL
c. 1800 mL **d.** 1700 mL

11.29 **a.** 770 mmHg **b.** 1150 mmHg

11.31 **a.** −23 °C **b.** 168 °C

11.33 **a.** boiling point **b.** vapor pressure
c. atmospheric pressure **d.** boiling point

11.35 **a.** On top of a mountain, water boils below 100 °C because the atmospheric (external) pressure is less than 1 atm.

b. Because the pressure inside a pressure cooker is greater than 1 atm, water boils above 100 °C. At a higher temperature, food cooks faster.

11.37 $\dfrac{P_1 V_1}{T_1} = \dfrac{P_2 V_2}{T_2}$

Boyle's, Charles's, and Gay-Lussac's laws are combined to make this law.

11.39 **a.** 4.26 atm   **b.** 3.07 atm
c. 0.606 atm

11.41 −33 °C

11.43 The volume increases because the number of gas particles is increased.

11.45 **a.** 4.00 L   **b.** 26.7 L
c. 14.6 L

11.47 **a.** 2.00 mol of O_2 **b.** 0.179 mol of O_2
c. 4.48 L   **d.** 55 500 mL

11.49 **a.** 1.70 g/L   **b.** 0.716 g/L
c. 0.901 g/L   **d.** 2.86 g/L

11.51 4.93 atm

11.53 29.4 g of O_2

11.55 566 K (293 °C)

11.57 **a.** 42 g/mol   **b.** 28.7 g/mol
c. 39.8 g/mol   **d.** 33 g/mol

11.59 **a.** 7.60 L of H_2 **b.** 4.92 g of Mg

11.61 178 L of O_2

11.63 3.4 L of O_2

11.65 **a.** 743 mmHg   **b.** 0.0103 mol of O_2

11.67 In a gas mixture, the pressure that each gas exerts as part of the total pressure is called the partial pressure of that gas. Because the air sample is a mixture of gases, the total pressure is the sum of the partial pressures of each gas in the sample.

11.69 765 torr

11.71 425 torr

11.73 **a.** 2 Fewest number of gas particles exerts the lowest pressure.
b. 1 Greatest number of gas particles exerts the highest pressure.

11.75 **a.** A: Volume decreases when temperature decreases.
b. C: Volume increases when pressure decreases.
c. A: Volume decreases when the moles of gas decrease.
d. B: Doubling the temperature, in kelvins, would double the volume, but when half of the gas escapes, the volume would decrease by half. These two opposing effects cancel each other and there is no change in the volume.
e. C: Increasing the moles increases the volume to keep T and P constant.

11.77 2170 mL

11.79 2.5 kg of H_2

11.81 −66 °C

11.83 4.4 g of CH_4

11.85 42.1 g/mol

11.87 70.1 g/mol; C_5H_{10}

11.89 1.1×10^{24} molecules of CO_2

11.91 8.56 L of H_2

11.93 **a.** 16 L of O_2 **b.** 1.02 g of NH_3

11.95 He 600 torr, O_2 1800 torr

11.97 370 torr

11.99 **a.** 734 mmHg   **b.** 0.0165 mol of H_2
c. 0.297 g of Al

11.101 **a.** False. The flask containing helium has more moles of helium and thus more helium atoms.
b. False. There are different numbers of moles in the flasks, which means the pressures are different.
c. True. There are more moles of helium, which makes the pressure of helium greater than that of neon.
d. True. The mass and volume of each are the same, which means the mass/volume ratio or density is the same in both flasks.

11.103 C_2H_6

11.105 5.31 g of water

12 Solutions

"Chemistry is very important when taking care of patients in the hospital," says Dr. Denise Gee, physician, Boston Medical Center. "Blood tests can tell us the amount of various cations and anions in the body. This includes sodium, potassium, chloride, and bicarbonate, among others. An abnormal value can sometimes help diagnose disease or may signal that a patient is getting sicker. In the healthcare setting, chemistry is essential in monitoring overall patient health."

Doctors who are internists, family physicians, or pediatricians are directly involved in caring for people. Research doctors develop new therapies for cancer, genetic disorders, and infectious diseases. Other doctors teach medical students or work for pharmaceutical or health insurance companies.

Visit **www.masteringchemistry.com** for self-study help and instructor-assigned homework.

olutions are everywhere around us. Most of the gases, liquids, and solids we see are mixtures of at least one substance dissolved in another. The air we breathe is a solution that is primarily oxygen and nitrogen gases. Carbon dioxide gas dissolved in water makes our carbonated drinks. When we make solutions of coffee or tea, we use hot water to dissolve substances from coffee beans or tea leaves. The ocean is also a solution, consisting of many salts, such as sodium chloride, dissolved in water. In your medicine cabinet, the antiseptic tincture of iodine is a solution of iodine dissolved in ethanol.

Because the individual components in any mixture are not bonded to each other, the composition of those components can vary. Also, some of the physical properties of the individual components are still noticeable. For example, in ocean water, we detect the dissolved sodium chloride by the salty taste. The flavor we associate with coffee is due to the dissolved components. There are different types of solution. In a homogeneous solution, the components cannot be distinguished one from the other. Syrup is a homogeneous solution of sugar and water: The sugar cannot be distinguished from the water. However, in an aquarium, a heterogeneous mixture, all the components are observable, including the sand on the bottom, the fish, the plants, and the water.

12.1 Solutions

A **solution** is a homogenous mixture in which one substance, called the **solute**, is uniformly dispersed in another substance, called the **solvent**. Because the solute and the solvent do not react with each other, they can be mixed in varying proportions. A little salt dissolved in water tastes slightly salty. When more salt is added, the water tastes very salty. The solute (in this case, salt) is the substance present in the smaller amount, whereas the solvent (in this case, water) is present in the larger amount. In a solution, the particles of the solute are evenly dispersed among the molecules within the solvent (see Figure 12.1).

LEARNING GOAL

Define solute and solvent; describe the formation of a solution.

Solute: The substance present in lesser amount

Salt

Water

Solvent: The substance present in greater amount

A solution consists of at least one solute dispersed in a solvent.

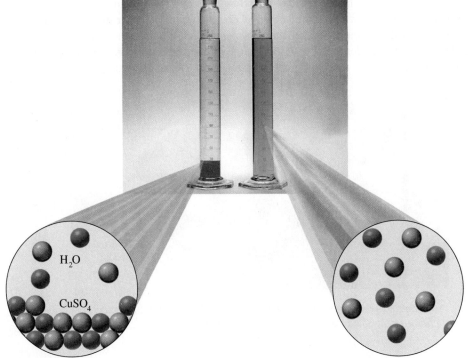

H_2O

$CuSO_4$

FIGURE 12.1 A solution of copper(II) sulfate ($CuSO_4$) forms as particles of solute dissolve, move away from the crystals, and become evenly dispersed among the solvent (water) molecules.

Q What does the uniform blue color indicate about the $CuSO_4$ solution?

Types of Solutes and Solvents

Solutes and solvents may be solids, liquids, or gases. The solution that forms has the same physical state as the solvent. When sugar crystals are dissolved in water, the resulting sugar solution is liquid. Sugar is the solute, and water is the solvent. Soda water and soft drinks are prepared by dissolving carbon dioxide gas in water. The carbon dioxide gas is the solute, and water is the solvent. Table 12.1 lists some solutes and solvents and their solutions.

TABLE 12.1 Some Examples of Solutions

Type	Example	Primary Solute	Solvent
Gas Solutions			
Gas in a gas	Air	Oxygen (gas)	Nitrogen (gas)
Liquid Solutions			
Gas in a liquid	Soda water	Carbon dioxide (gas)	Water (liquid)
	Household ammonia	Ammonia (gas)	Water (liquid)
Liquid in a liquid	Vinegar	Acetic acid (liquid)	Water (liquid)
Solid in a liquid	Seawater	Sodium chloride (solid)	Water (liquid)
	Tincture of iodine	Iodine (solid)	Ethanol (liquid)
Solid Solutions			
Liquid in a solid	Dental amalgam	Mercury (liquid)	Silver (solid)
Solid in a solid	Brass	Zinc (solid)	Copper (solid)
	Steel	Carbon (solid)	Iron (solid)

Water as a Solvent

SELF STUDY ACTIVITY
Hydrogen Bonding

Water is one of the most common substances in nature. In the H_2O molecule, an oxygen atom shares electrons with two hydrogen atoms. Because oxygen is much more electronegative than hydrogen, the O—H bonds are polar. In each polar bond, the oxygen atom has a partial negative (δ^-) charge and the hydrogen atom has a partial positive (δ^+) charge. Because the water molecule has a bent shape, water is a *polar solvent.*

In Chapter 10, we learned that *hydrogen bonds* occur between molecules where partially positive hydrogen is attached to the strongly electronegative atoms O, N, or F. In the diagram, the hydrogen bonds are shown as dots between the water molecules. Although hydrogen bonds are much weaker than covalent or ionic bonds, there are many of them linking water molecules together. As a result, hydrogen bonding plays an important role in the properties of water and biological compounds such as proteins and DNA.

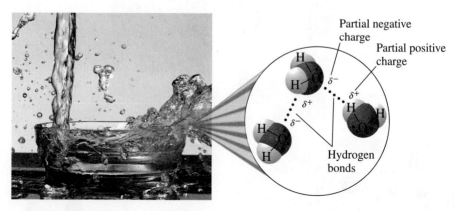

In water, hydrogen bonds form between the lone pairs of an oxygen in one water molecule and the hydrogen in another.

CHEMISTRY AND HEALTH

Water in the Body

The average adult contains about 60% water by mass, and the average infant about 75%. About 60% of the body's water is contained

24 Hours

Water gain	
Liquid	1000 mL
Food	1200 mL
Metabolism	300 mL
Total	2500 mL

Water loss	
Urine	1500 mL
Perspiration	300 mL
Breath	600 mL
Feces	100 mL
Total	2500 mL

The water lost from the body is replaced by the intake of fluids.

within the cells as intracellular fluids; the other 40% makes up extracellular fluids, which include the interstitial fluid in tissue and the plasma in the blood. These external fluids carry nutrients and waste materials between the cells and the circulatory system.

Every day, you lose between 1500 and 3000 mL of water from the kidneys as urine, from the skin as perspiration, from the lungs as you exhale, and from the gastrointestinal tract. Serious dehydration can occur in an adult if there is a 10% net loss in total body fluid, and a 20% loss of fluid can be fatal. An infant suffers severe dehydration with a 5–10% loss in body fluid.

Water loss is continually replaced by the liquids and foods in the diet and from metabolic processes that produce water in the cells of the body. Table 12.2 lists the percent water by mass contained in some foods.

TABLE 12.2 Percentage of Water in Some Foods

Food	Water (% by mass)	Food	Water (% by mass)
Vegetables/Fruits		**Meats/Fish**	
Carrot	88	Chicken, cooked	71
Celery	94	Hamburger, broiled	60
Cucumber	96	Salmon	71
Cantaloupe	91	**Milk Products**	
Orange	86	Cottage cheese	78
Strawberry	90	Milk, whole	87
Watermelon	93	Yogurt	88

Formation of Solutions

The interactions between solute and solvent will determine whether or not a solution will form. Initially, energy is needed to separate the particles in the solute and to separate the solvent particles. Then energy is released as solute particles move between the solvent particles to form a solution. However, there must be attractions between the solute and the solvent particles to provide the energy for the initial separation. These attractions occur when the solute and the solvent have similar polarities. If there are no attractions between a solute and a solvent, there is not sufficient energy to form a solution (see Table 12.3).

TABLE 12.3 Possible Combinations of Solutes and Solvents

Solutions Will Form		Solutions Will Not Form	
Solute	Solvent	Solute	Solvent
Polar	Polar	Polar	Nonpolar
Nonpolar	Nonpolar	Nonpolar	Polar

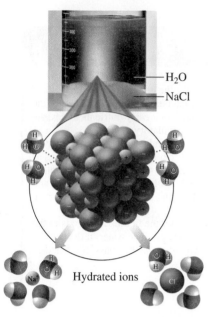

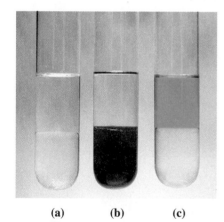

FIGURE 12.2 Ions on the surface of a crystal of NaCl dissolve in water as they are attracted to the polar water molecules that pull the ions into solution and surround them.

Q What helps keep Na^+ and Cl^- ions in solution?

FIGURE 12.4 Like dissolves like. **(a)** The test tubes contain an upper layer of the polar solvent water and a lower layer of the nonpolar solvent CH_2Cl_2. **(b)** The nonpolar solute I_2 (purple) is soluble in the nonpolar solvent CH_2Cl_2. **(c)** The ionic solute $Ni(NO_3)_2$ (green) is soluble in the polar solvent water.

Q In which solvent would polar molecules of sugar be soluble?

Solutions with Ionic and Polar Solutes

In ionic solutes such as sodium chloride, NaCl, there are strong solute–solute attractions between positively charged Na^+ ions and negatively charged Cl^- ions. In water, a polar solvent, the hydrogen bonds provide strong solvent–solvent attractions. When NaCl crystals are placed in water, partially negative oxygen atoms in water molecules attract positive Na^+ ions, and the partially positive hydrogen atoms in other water molecules attract negative Cl^- ions (see Figure 12.2). As soon as the Na^+ ions and the Cl^- ions form a solution, they undergo **hydration** as water molecules surround each ion. Hydration of the ions diminishes their attraction to other ions and keeps them in solution. The strong solute–solvent attractions between Na^+ and Cl^- ions and the polar water molecules provide energy needed to form the solution. In the equation for the formation of the NaCl solution, the solid and aqueous NaCl are shown with the formula H_2O over the arrow, which indicates that water is needed for the dissociation process but is not a reactant.

$$NaCl(s) \xrightarrow{H_2O} Na^+(aq) + Cl^-(aq)$$

In another example, we find that a polar covalent compound such as methanol, CH_3-OH, is soluble in water because methanol has a polar $-OH$ group that forms hydrogen bonds with water (see Figure 12.3). Polar solutes require polar solvents for a solution to form. The expression *"like dissolves like"* is a way of saying that the polarities of a solute and a solvent must be similar in order to form a solution.

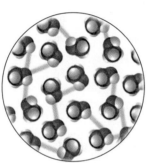
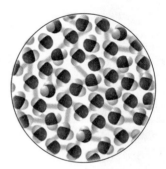
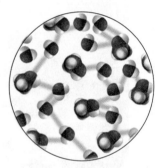
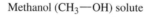

Methanol (CH_3-OH) solute   Water solvent   Methanol–water solution with hydrogen bonding

FIGURE 12.3 Molecules of the polar covalent compound methanol, CH_3-OH, form hydrogen bonds with polar H_2O molecules to form a methanol–water solution.

Q Why is the methanol–water solution an example of "like dissolves like"?

Solutions with Nonpolar Solutes

Compounds containing nonpolar molecules, such as iodine (I_2), oil, or grease, do not dissolve in water because there are little or no attractions between the particles of a nonpolar solute and the polar solvent. Figure 12.4 illustrates the formation of some polar and nonpolar solutions.

CONCEPT CHECK 12.1

■ **Polar and Nonpolar Solutes**

Indicate whether each of the following will form solutions with water. Explain.

a. $KCl(s)$

b. hexane, $CH_3-CH_2-CH_2-CH_2-CH_2-CH_3$

c. ethanol, CH_3-CH_2-OH

ANSWER

a. KCl is an ionic compound. The solute–solvent attractions between K^+ and Cl^- and polar water will release the energy needed to break solute–solute and solvent–solvent bonds. Thus, a KCl solution will form.

b. Hexane is a nonpolar compound, which means it does not form a solution with water. There are no nonpolar solute–polar solvent attractions, and no energy is released to form a solution.

c. The functional group —OH in ethanol makes it a polar compound. Because attractions between a polar solute and the polar solvent water release energy to break solute–solute and solvent–solvent bonds, ethanol will form a solution with water.

QUESTIONS AND PROBLEMS

Solutions

12.1 Identify the solute and the solvent in each solution composed of the following:
  a. 10.0 g of NaCl and 100.0 g of H_2O
  b. 50.0 mL of ethanol, CH_3—CH_2—OH (*l*) and 10.0 mL of H_2O
  c. 0.20 L of O_2 and 0.80 L of N_2

12.2 Identify the solute and the solvent in each solution composed of the following:
  a. 50.0 g of silver and 4.0 g of mercury
  b. 100.0 mL water and 5.0 g of sugar
  c. 1.0 g of I_2 and 50.0 mL of ethanol(*l*)

12.3 Water is a polar solvent; CCl_4 is a nonpolar solvent. In which solvent is each of the following more likely to be soluble?
  a. $NaNO_3$, ionic   **b.** I_2, nonpolar
  c. sucrose (table sugar), polar   **d.** octane, nonpolar

12.4 Water is a polar solvent; hexane is a nonpolar solvent. In which solvent is each of the following more likely to be soluble?
  a. vegetable oil, nonpolar   **b.** benzene, nonpolar
  c. $LiNO_3$, ionic   **d.** Na_2SO_4, ionic

12.5 Describe the formation of an aqueous KI solution.

12.6 Describe the formation of an aqueous LiBr solution.

12.2 Electrolytes and Nonelectrolytes

Solutes can be classified by their ability to conduct an electrical current. When **electrolytes** dissolve in water, they separate into ions forming solutions that are able to conduct electricity. When **nonelectrolytes** dissolve in water, they do not separate into ions and their solutions do not conduct electricity.

To test solutions for the presence of ions, we can use an apparatus that consists of a battery and a pair of electrodes connected by wires to a light bulb. The light bulb glows when electricity can flow, which can only happen when electrolytes provide ions that move to each electrode to complete the circuit.

Electrolytes

Electrolytes can be further classified as *strong electrolytes* and *weak electrolytes*. For all electrolytes, in a process called **dissociation**, some or all of the solute that dissolves produces ions. For a **strong electrolyte**, such as sodium chloride (NaCl), there is 100% dissociation of the solute into ions. When the electrodes from the light bulb apparatus are placed in the NaCl solution, the light bulb is very bright.

In an equation for dissociation, the electrical charges must balance. For example, magnesium nitrate dissociates to give one magnesium ion for every two nitrate ions. However, only the ionic bonds between Mg^{2+} and NO_3^- are broken, not the covalent bonds within the polyatomic ion. The dissociation for $Mg(NO_3)_2$ is written as follows:

$$Mg(NO_3)_2(s) \xrightarrow{H_2O} Mg^{2+}(aq) + 2NO_3^-(aq)$$

LEARNING GOAL

Identify solutes as electrolytes or nonelectrolytes.

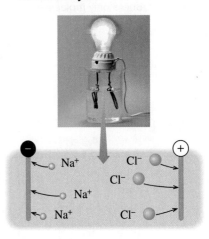

Strong electrolyte

(a)

A strong electrolyte in an aqueous solution completely dissociates into ions.

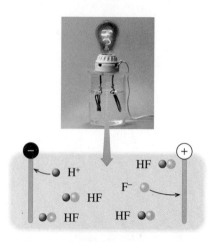

Weak electrolyte

(b)

A weak electrolyte in an aqueous solution forms mostly molecules and a few ions.

For a **weak electrolyte** such as HF, only a small percentage of the dissolved solute dissociates into ions. Most of a weak electrolyte is present in a solution as undissociated molecules. When the electrodes are placed in a solution of a weak electrolyte, the glow of the light bulb is very dim.

Thus, an aqueous solution of the weak electrolyte HF consists of mostly HF molecules and only a few H^+ and F^- ions. As HF molecules dissociate into ions, some of H^+ and F^- ions recombine to form HF molecules. These forward and reverse reactions of molecules to ions and back again are indicated by two arrows that point in opposite directions:

$$HF(aq) \underset{\text{Recombination}}{\overset{\text{Dissociation}}{\rightleftharpoons}} H^+(aq) + F^-(aq)$$

Nonelectrolytes

A nonelectrolyte such as sucrose (sugar) dissolves in water as molecules, which do not dissociate into ions. When electrodes are placed in a solution of a nonelectrolyte, the light bulb does not glow, because the solution does not conduct electricity.

$$C_{12}H_{22}O_{11}(s) \xrightarrow{H_2O} C_{12}H_{22}O_{11}(aq)$$

Table 12.4 summarizes the classification of solutes in aqueous solutions.

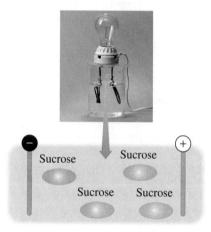

Nonelectrolyte

(c)

A nonelectrolyte in an aqueous solution produces only molecules.

Sucrose (table sugar) in an aqueous solution forms only molecules.

TABLE 12.4 Classification of Solutes in Aqueous Solutions

Types of Solute	Dissociation	Particles in Solution	Conducts Electricity?	Examples
Strong electrolyte	Completely	Ions only	Yes	Ionic compounds such as NaCl, KBr, $MgCl_2$, $NaNO_3$, NaOH, KOH, HCl, HBr, HI, HNO_3, $HClO_4$, H_2SO_4
Weak electrolyte	Partially	Mostly molecules and a few ions	Weakly	HF, H_2O, NH_3, CH_3—COOH (acetic acid)
Nonelectrolyte	None	Molecules only	No	Organic compounds such as CH_3—OH (methanol), CH_3—CH_2—OH (ethanol), $C_{12}H_{22}O_{11}$ (sucrose), CH_4N_2O (urea)

CONCEPT CHECK 12.2

■ **Electrolytes and Nonelectrolytes**

Identify the components in each of the following aqueous solutions, and write the equation for the formation of a solution:

a. ammonium bromide, a strong electrolyte
b. urea, CH_4N_2O, a nonelectrolyte
c. hypobromous acid, HBrO, a weak electrolyte

ANSWER

a. An aqueous solution of the strong electrolyte NH_4Br contains NH_4^+ and Br^- ions and the solvent H_2O molecules:

$$NH_4Br(s) \xrightarrow{H_2O} NH_4^+(aq) + Br^-(aq)$$

b. An aqueous solution of the nonelectrolyte CH_4N_2O contains only polar molecules of urea, CH_4N_2O, and the solvent H_2O molecules:

$$CH_4N_2O(s) \xrightarrow{H_2O} CH_4N_2O(aq)$$

c. An aqueous solution of the weak electrolyte HBrO contains mostly HBrO molecules, a few H^+ ions, a few BrO^- ions, and the solvent H_2O molecules:

$$HBrO(aq) \underset{}{\overset{H_2O}{\rightleftharpoons}} H^+(aq) + BrO^-(aq)$$

SAMPLE PROBLEM | 12.1

■ **Solutions of Electrolytes and Nonelectrolytes**

Indicate whether aqueous solutions of each of the following contain only ions, only molecules, or mostly molecules and a few ions:

a. Na_2SO_4, a strong electrolyte
b. CH_3—CH_2—CH_2—OH, a nonelectrolyte

SOLUTION

a. An aqueous solution of the strong electrolyte Na_2SO_4 contains only the ions Na^+ and SO_4^{2-}.
b. A nonelectrolyte such as CH_3—CH_2—CH_2—OH produces only molecules when it dissolves in water.

STUDY CHECK

Boric acid, H_3BO_3, is a weak electrolyte. Would you expect a boric acid solution to contain only ions, only molecules, or mostly molecules and a few ions?

QUESTIONS AND PROBLEMS

Electrolytes and Nonelectrolytes

12.7   KF is a strong electrolyte, and HF is a weak electrolyte. How does their dissociation in water differ?

12.8   NaOH is a strong electrolyte, and $CH_3—OH$ is a nonelectrolyte. How does their dissociation in water differ?

12.9   Write a balanced equation for the dissociation of each of the following strong electrolytes in water:
  a. KCl   **b.** $CaCl_2$ **c.** K_3PO_4 **d.** $Fe(NO_3)_3$

12.10   Write a balanced equation for the dissociation of each of the following strong electrolytes in water:
  a. LiBr   **b.** $NaNO_3$ **c.** $CuCl_2$ **d.** K_2CO_3

12.11   Indicate whether aqueous solutions of each of the following solutes will contain only ions, only molecules, or mostly molecules and a few ions:
  a. acetic acid, $CH_3—COOH$, a weak electrolyte
  b. NaBr, a strong electrolyte
  c. fructose, $C_6H_{12}O_6$, a nonelectrolyte

12.12   Indicate whether aqueous solutions of each of the following solutes will contain only ions, only molecules, or mostly molecules and a few ions:
  a. NH_4Cl, a strong electrolyte
  b. ethanol, $CH_3—CH_2—OH$, a nonelectrolyte
  c. HCN, hydrocyanic acid, a weak electrolyte

12.13   Classify each solute represented in the following equations as a strong, weak, or nonelectrolyte:

  a. $K_2SO_4(s) \xrightarrow{H_2O} 2K^+(aq) + SO_4^{2-}(aq)$

  b. $NH_3(g) + H_2O(l) \rightleftharpoons NH_4^+(aq) + OH^-(aq)$

  c. $C_6H_{12}O_6(s) \xrightarrow{H_2O} C_6H_{12}O_6(aq)$

12.14   Classify each solute represented in the following equations as a strong, weak, or nonelectrolyte:

  a. $CH_3—OH(l) \xrightarrow{H_2O} CH_3—OH(aq)$

  b. $MgCl_2(s) \xrightarrow{H_2O} Mg^{2+}(aq) + 2Cl^-(aq)$

  c. $HClO(aq) \rightleftharpoons H^+(aq) + ClO^-(aq)$

LEARNING GOAL

Define solubility; distinguish between an unsaturated and a saturated solution. Identify an insoluble salt.

TUTORIAL
Solubility

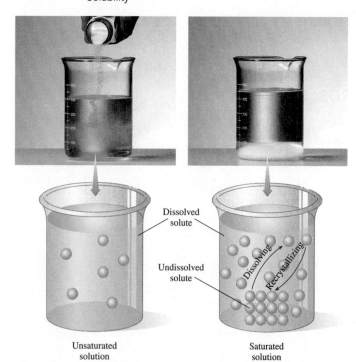

Unsaturated
solution

Saturated
solution

12.3 Solubility

The term **solubility** is used to describe the amount of a solute that can dissolve in a given amount of solvent. Many factors, such as the type of solute, the type of solvent, and temperature, affect the solubility of a solute. Solubility, usually expressed in grams of solute in 100 g of solvent, is the maximum amount of solute that can be dissolved at a certain temperature. If a solute readily dissolves when added to the solvent, the solution does not contain the maximum amount of solute. We call this solution an **unsaturated solution**.

A solution that contains all the solute that can dissolve is a **saturated solution**. When a solution is saturated, the rate at which the solute dissolves becomes equal to the rate of recrystallization. Then there is no further change in the amount of solute dissolved in the solution.

$$\text{Solid solute} \underset{\text{Crystallizes}}{\overset{\text{Dissolves}}{\rightleftharpoons}} \text{saturated solution}$$

We can prepare a saturated solution by adding an amount of solute greater than that needed for solubility. Stirring the solution will dissolve the maximum amount of solute and leave the excess on the bottom of the container. Once we have a saturated solution, the addition of more solute will only increase the amount of undissolved solute.

More solute can dissolve in an unsaturated solution but not in a saturated solution.

CONCEPT CHECK 12.3

■ Saturated Solutions

At 20 °C, the solubility of KCl is 34 g/100 g of water. In the laboratory, a student mixes 75 g of KCl with 200. g of water at a temperature of 20 °C.

a. How much of the KCl can dissolve?
b. Is the solution saturated or unsaturated?
c. What is the mass, in grams, of any solid KCl left undissolved?

ANSWER

a. KCl has a solubility of 34 g of KCl in 100 g of water. Using the solubility as a conversion factor, we can calculate the maximum amount of KCl that can dissolve in 200. g of water as follows:

$$200. \; \text{g H}_2\text{O} \times \frac{34 \; \text{g KCl}}{100 \; \text{g H}_2\text{O}} = 68 \; \text{g of KCl}$$

b. Because 75 g of KCl exceeds the amount that can dissolve in 200. g of water, the KCl solution is saturated.

c. If we add 75 g of KCl to 200. g of water and only 68 g of KCl can dissolve, there is 7 g (75 g − 68 g) of solid (undissolved) KCl on the bottom of the container.

Effect of Temperature on Solubility

The solubility of most solids is greater as temperature increases, which means that solutions usually can contain more dissolved solute at higher temperatures. A few substances show little change in solubility at higher temperatures, and a few are less soluble (see Figure 12.5). For example, when you add sugar to iced tea, some undissolved sugar may form on the bottom of the glass. But if you add sugar to hot tea, many teaspoons of sugar are needed before solid sugar appears. Hot tea dissolves more sugar than does cold tea because the solubility of sugar is much greater at a higher temperature. When a saturated solution is carefully cooled, it might become a *supersaturated solution* because it contains more solute than the solubility allows. Such a solution is unstable, and if the solution is agitated or if a solute crystal is added, the excess solute will crystallize to give a saturated solution again.

The solubility of a gas in water decreases as the temperature increases. At higher temperatures, more gas molecules have the energy to escape from the solution. Perhaps you have observed the bubbles escaping from a cold carbonated soft drink as it warms. At high temperatures, bottles containing carbonated solutions may burst as more gas molecules leave the solution and increase the gas pressure inside the bottle. Biologists have found that increased temperatures in rivers and lakes cause the amount of dissolved oxygen to decrease until the warm water can no longer support a biological community. Electricity-generating plants are required to have their own ponds to use with their cooling towers to lessen the threat of thermal pollution to surrounding waterways.

Henry's Law

Henry's law states that the solubility of gas in a liquid is directly related to the pressure of that gas above the liquid. At higher pressures, there are more gas molecules available to enter and dissolve in the liquid. A can of soda is carbonated by using CO_2 gas at high pressure to increase the solubility of the CO_2 in the beverage. When you open the can at atmospheric pressure, the pressure of the CO_2 drops, decreasing the solubility of CO_2. As a result, bubbles of CO_2 rapidly escape from the solution. The burst of bubbles is even more noticeable when you open a warm can of soda.

TUTORIAL
Solubility of Gases and Solids in Water

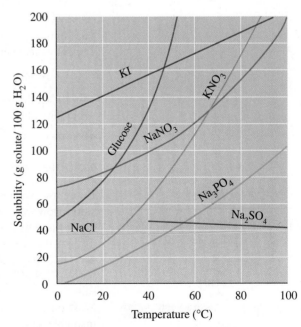

FIGURE 12.5 In water, most common solids are more soluble as the temperature increases.

Q Compare the solubility of $NaNO_3$ at 20 °C and 60 °C.

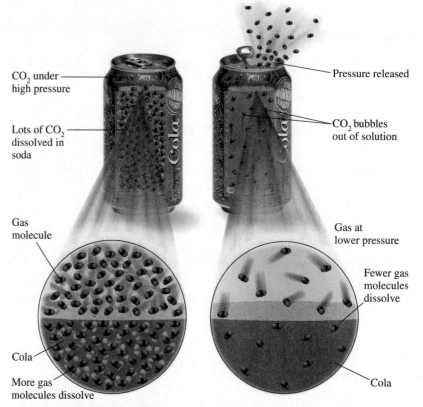

When the pressure of a gas above a solution decreases, the solubility of that gas in the solution also decreases.

SAMPLE PROBLEM | 12.2

■ Factors Affecting Solubility

Indicate whether there is an increase or decrease in each of the following:

a. the solubility of sugar in water at 45 °C compared to its solubility in water at 25 °C
b. the solubility of O_2 in a lake as it warms

SOLUTION

a. An increase in the temperature increases the solubility of the sugar.
b. An increase in the temperature decreases the solubility of O_2 gas.

STUDY CHECK

At 40 °C, the solubility of KNO_3 is 65 g of KNO_3/100 g of H_2O. Would you expect the solubility of KNO_3 to be higher or lower at 80 °C? Explain.

Soluble and Insoluble Salts

Up to now, we have considered ionic compounds that dissolve in water; they are **soluble salts**. However, some ionic compounds do not separate into ions in water. They are **insoluble salts** that remain as solids even in contact with water.

Salts that are soluble in water typically contain at least one of the following ions: Li^+, Na^+, K^+, NH_4^+, NO_3^- or CH_3—COO^- (acetate). Most salts containing Cl^- are soluble, but $AgCl$, $PbCl_2$, or Hg_2Cl_2 are not; they are insoluble chloride salts. Similarly, most salts containing SO_4^{2-} are soluble, but a few are insoluble, as shown in Table 12.5. Most other salts are insoluble and do not dissolve in water (see Figure 12.6). In an insoluble salt, attractions between its positive and negative ions are too strong for the polar water molecules to break. We can use the solubility rules to predict whether a salt (a solid ionic compound) would be expected to dissolve in water. Table 12.6 illustrates the use of these rules.

TUTORIAL
Solubility

SELF STUDY ACTIVITY
Solubility

CHEMISTRY AND HEALTH

Gout and Kidney Stones: A Problem of Saturation in Body Fluids

The conditions of gout and kidney stones involve compounds in the body that exceed their solubility levels and form solids. Gout affects adults, primarily men, over the age of 40. Attacks of gout may occur when the concentration of uric acid in blood plasma exceeds its solubility, which is 7 mg/100 mL of plasma at 37 °C. Insoluble deposits of needlelike crystals of uric acid can form in the cartilage, tendons, and soft tissues, where they cause painful gout attacks. They may also form in the tissues of the kidneys, where they can cause renal damage. High levels of uric acid in the body can be caused by an increase in uric acid production, failure of the kidneys to remove uric acid, or by a diet with an overabundance of foods containing purines, which are metabolized to uric acid in the body. Foods in the diet that contribute to high levels of uric acid include certain meats, sardines, mushrooms, asparagus, and beans. Drinking alcoholic beverages may also significantly increase uric acid levels and bring about gout attacks.

Treatment for gout involves diet changes and drugs. Depending on the levels of uric acid, a medication is used such as probenecid, which helps the kidneys eliminate uric acid, or allopurinol, which blocks the production of uric acid by the body.

Kidney stones are solid materials that form in the urinary tract. Most kidney stones are composed of calcium phosphate and calcium oxalate, although they can be solid uric acid. The excessive ingestion of minerals and insufficient water intake can cause the concentration of mineral salts to exceed their solubility and lead to the formation of kidney stones. When a kidney stone passes through the urinary tract, it causes considerable pain and discomfort, necessitating the use of painkillers and surgery. Sometimes ultrasound is used to break up kidney stones. Persons prone to kidney stones are advised to drink six to eight glasses of water every day to prevent saturation levels of minerals in the urine.

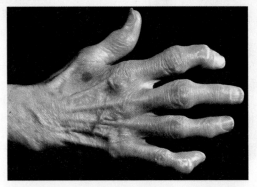

Gout occurs when uric acid exceeds its solubility.

Kidney stones form when calcium phosphate exceeds its solubility.

TABLE 12.5 Solubility Rules for Ionic Solids in Water

Soluble If Salt Contains		Insoluble If Salt Contains
NH_4^+, Li^+, Na^+, K^+ NO_3^-, CH_3-COO^- (acetate)	← but are soluble with	CO_3^{2-}, S^{2-}, PO_4^{3-}, OH^-
Cl^-, Br^-, I^-	but are not soluble with →	Ag^+, Pb^{2+}, Hg_2^{2+}
SO_4^{2-}	but are not soluble with →	Ba^{2+}, Pb^{2+}, Ca^{2+}, Sr^{2+}

MC

CASE STUDY
Kidney Stones and Saturated Solutions

CdS

FeS

$PbCrO_4$

$Ni(OH)_2$

FIGURE 12.6 Mixing certain aqueous solutions produces insoluble salts.

Q What ions make each of these salts insoluble in water?

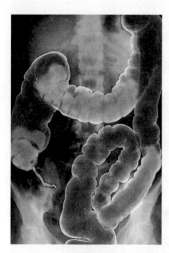

FIGURE 12.7 A barium sulfate-enhanced X-ray of the abdomen shows the large intestine.

Q Is $BaSO_4$ a soluble or an insoluble substance?

TABLE 12.6 Using Solubility Rules

Ionic Compound	Solubility in Water	Reasoning
K_2S	Soluble	Contains K^+
$Ca(NO_3)_2$	Soluble	Contains NO_3^-
$PbCl_2$	Insoluble	Is an insoluble chloride
$NaOH$	Soluble	Contains Na^+
$AlPO_4$	Insoluble	Contains no soluble ions

In medicine, the insoluble salt $BaSO_4$ is used as an opaque substance to enhance X-rays of the gastrointestinal tract. $BaSO_4$ is so insoluble that it does not dissolve in gastric fluids (see Figure 12.7). Other barium salts cannot be used, because they would dissolve in water, releasing Ba^{2+}, which is poisonous.

CONCEPT CHECK 12.4

■ **Soluble and Insoluble Salts**

Predict whether each of the following salts is soluble in water and explain why:

a. Na_3PO_4 **b.** $CaCO_3$

ANSWER

a. The salt Na_3PO_4 is soluble in water because a compound that contains Na^+ is soluble.
b. The salt $CaCO_3$ is not soluble. The compound does not contain a soluble positive ion, which means that a calcium salt containing CO_3^{2-} is not soluble.

Formation of a Solid

We can use solubility rules to predict whether a solid called a *precipitate* forms when two solutions of ionic compounds are mixed. A solid forms when two ions of an insoluble salt come in contact with one another. For example, when a solution of $AgNO_3$ (Ag^+ and NO_3^-) is mixed with a solution of $NaCl$ (Na^+ and Cl^-), the white insoluble salt $AgCl$ is produced. We can write the equation as a double replacement reaction. However, the chemical equation does not show the individual ions to help us decide which, if any, insoluble salt would form. To help us determine any insoluble salt, we can first write the reactants to show all the ions present when the two solutions are mixed:

$$Ag^+(aq) + NO_3^-(aq) + Na^+(aq) + Cl^-(aq) \longrightarrow$$

Then we look at the cations and anions to see if any of the combinations would form an insoluble salt. The combination $AgCl$ is an insoluble salt.

STEP 1 Write the ions of the reactants.

**Reactants
(initial combinations)**

$$Ag^+(aq) + NO_3^-(aq)$$

$$Na^+(aq) + Cl^-(aq)$$

TUTORIAL

Solubility and Precipitation Reactions

STEP 2 **Write the combinations of ions and determine if any are insoluble.** The combination of AgCl is insoluble in water, but the combination of $NaNO_3$ is soluble.

Mixture (combinations)	Product	Soluble?
$Ag^+(aq) + Cl^-(aq)$	$AgCl(s)$	No
$Na^+(aq) + NO_3^-(aq)$	$NaNO_3$	Yes

STEP 3 **Write the ionic equation including any solid.** In the **ionic equation**, we show that a precipitate of AgCl forms, while the ions Na^+ and NO_3^- remain in solution.

$$Ag^+(aq) + NO_3^-(aq) + Na^+(aq) + Cl^-(aq) \longrightarrow AgCl(s) + Na^+(aq) + NO_3^-(aq)$$

STEP 4 **Write the net ionic equation.** Now we remove the Na^+ and NO_3^- ions, known as *spectator ions,* because they are unchanged during the reaction:

$$Ag^+(aq) + \underbrace{NO_3^-(aq) + Na^+(aq)}_{\text{Spectator ions}} + Cl^-(aq) \longrightarrow AgCl(s) + \underbrace{Na^+(aq) + NO_3^-(aq)}_{\text{Spectator ions}}$$

Finally, a **net ionic equation** is written for the chemical reaction that occurred.

$$Ag^+(aq) + Cl^-(aq) \longrightarrow AgCl(s) \quad \text{Net ionic equation}$$

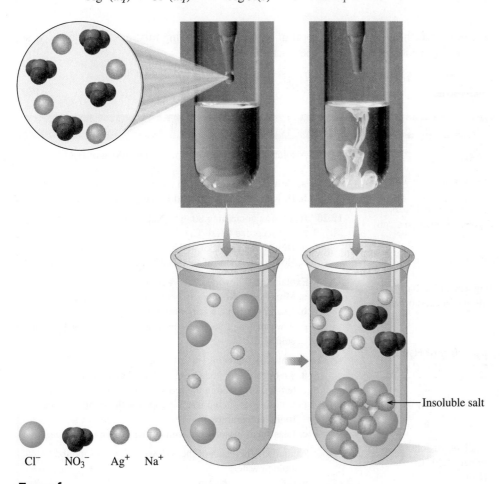

Insoluble salt

Cl^- NO_3^- Ag^+ Na^+

Type of Equation				
Chemical	$AgNO_3(aq)$	$+ NaCl(aq) \longrightarrow$	$AgCl(s) +$	$NaNO_3(aq)$
Ionic	$Ag^+(aq) + NO_3^-$	$+ Na^+(aq) + Cl^-(aq) \longrightarrow$	$AgCl(s) +$	$Na^+(aq) + NO_3^-$
Net ionic	$Ag^+(aq)$	$+ Cl^-(aq) \longrightarrow$	$AgCl(s)$	

Guide to Writing Net Ionic Equations for Formation of an Insoluble Salt

STEP 1
Write the ions of the reactants.

STEP 2
Write the combinations of ions and determine if any are insoluble.

STEP 3
Write the ionic equation including any solid.

STEP 4
Write the net ionic equation.

SAMPLE PROBLEM   12.3

■ Formation of an Insoluble Salt

When solutions of $BaCl_2$ and K_2SO_4 are mixed, a white solid forms. Write the net ionic equation and identify the white solid that forms.

SOLUTION

STEP 1   Write the ions of the reactants.

$$Ba^{2+}(aq) + Cl^-(aq) + K^+(aq) + SO_4^{2-}(aq)$$

STEP 2   Write the combinations of ions and determine if any are insoluble. Of the combinations KCl and $BaSO_4$, $BaSO_4(s)$ is insoluble.

STEP 3   Write the ionic equation including any solid. The balanced ionic equation is written:

$$Ba^{2+}(aq) + 2Cl^-(aq) + 2K^+(aq) + SO_4^{2-}(aq) \longrightarrow BaSO_4(s) + 2Cl^-(aq) + 2K^+(aq)$$

STEP 4   Write the net ionic equation. We remove the spectator ions K^+ and Cl^-, which gives the net ionic equation:

$$Ba^{2+}(aq) + SO_4^{2-}(aq) \longrightarrow BaSO_4(s) \quad \text{Net ionic equation}$$

$BaSO_4$ is the white solid.

STUDY CHECK

Predict whether a solid might form in each of the following mixtures of solutions. If so, write the net ionic equation for the reaction.

a. $NH_4Cl(aq)$ and $Ca(NO_3)_2(aq)$ **b.** $Pb(NO_3)_2(aq)$ and $KCl(aq)$

QUESTIONS AND PROBLEMS

Solubility

12.15 State whether each of the following refers to a saturated or unsaturated solution:
a. A crystal added to a solution does not change in size.
b. A sugar cube completely dissolves when added to a cup of coffee.

12.16 State whether each of the following refers to a saturated or unsaturated solution:
a. A spoonful of salt added to boiling water dissolves.
b. A layer of sugar forms on the bottom of a glass of tea as ice is added.

Use this table for Problems 12.17–12.20.

Substance	Solubility (g/100 g of H_2O)	
	20 °C	**50 °C**
KCl	34	43
$NaNO_3$	88	110
$C_{12}H_{22}O_{11}$ (sucrose)	204	260

12.17 Using the table, determine whether each of the following solutions will be saturated or unsaturated at 20 °C:
a. adding 25 g of KCl to 100. g of H_2O
b. adding 11 g of $NaNO_3$ to 25 g of H_2O
c. adding 400. g of sucrose to 125 g of H_2O

12.18 Using the table, determine whether each of the following solutions will be saturated or unsaturated at 50 °C:
a. adding 25 g of KCl to 50. g of H_2O
b. adding 150. g of $NaNO_3$ to 75 g of H_2O
c. adding 80. g of sucrose to 25 g of H_2O

12.19 A solution containing 80. g of KCl in 200. g of H_2O at 50 °C is cooled to 20 °C.
a. How many grams of KCl remain in solution at 20 °C?
b. How many grams of solid KCl crystallized after cooling?

12.20 A solution containing 80. g of $NaNO_3$ in 75 g of H_2O at 50 °C is cooled to 20 °C.
a. How many grams of $NaNO_3$ remain in solution at 20 °C?
b. How many grams of solid $NaNO_3$ crystallized after cooling?

12.21 Explain the following observations:
a. More sugar dissolves in hot tea than in iced tea.
b. Champagne in a warm room goes flat.
c. A warm can of soda has more spray when opened than a cold one.

12.22 Explain the following observations:
a. An open can of soda loses its "fizz" more quickly at room temperature than in the refrigerator.
b. Chlorine gas in tap water escapes as the sample warms to room temperature.
c. Less sugar dissolves in iced coffee than in hot coffee.

12.23 Predict whether each of the following ionic compounds is soluble in water:
a. LiCl   **b.** AgCl
c. $BaCO_3$ **d.** K_2O
e. $Fe(NO_3)_3$

12.24 Predict whether each of the following ionic compounds is soluble in water:
a. PbS   **b.** NaI
c. Na_2S **d.** Ag_2O
e. $CaSO_4$

12.25 Determine whether a solid forms when solutions containing the following salts are mixed. If so, write the net ionic equation for the reaction.
  a. $KCl(aq)$ and $Na_2S(aq)$
  b. $AgNO_3(aq)$ and $K_2S(aq)$
  c. $CaCl_2(aq)$ and $Na_2SO_4(aq)$
  d. $CuCl_2(aq)$ and $Li_3PO_4(aq)$

12.26 Determine whether a solid forms when solutions containing the following salts are mixed. If so, write the net ionic equation for the reaction.
  a. $Na_3PO_4(aq)$ and $AgNO_3(aq)$
  b. $K_2SO_4(aq)$ and $Na_2CO_3(aq)$
  c. $Pb(NO_3)_2(aq)$ and $Na_2CO_3(aq)$
  d. $BaCl_2(aq)$ and $KOH(aq)$

12.4 Percent Concentration

The amount of solute dissolved in a certain amount of solution is called the **concentration** of the solution. We will look at concentrations that are a ratio of a certain amount of solute in a given amount of solution.

$$\text{Concentration of a solution} = \frac{\text{amount of solute}}{\text{amount of solution}}$$

Mass Percent

Mass percent (m/m) describes the mass of the solute in grams for exactly 100 g of solution. In the calculation of mass percent (m/m), the units of mass of the solute and solution must be the same. If the mass of the solute is given as grams, then the mass of the solution must also be grams. The mass of the solution is the sum of the mass of the solute and the mass of the solvent.

$$\text{Mass percent (m/m)} = \frac{\text{mass of solute (g)}}{\text{mass of solute (g)} + \text{mass of solvent (g)}} \times 100\%$$

$$= \frac{\text{mass of solute (g)}}{\text{mass of solution (g)}} \times 100\%$$

Suppose we prepared a solution by mixing 8.00 g of KCl (solute) with 42.00 g of water (solvent). Together, the mass of the solute and mass of the solvent give the mass of the solution (8.00 g + 42.00 g = 50.00 g). Mass percent is calculated by substituting the values into the mass percent expression:

$$\frac{8.00 \text{ g KCl}}{50.00 \text{ g solution}} \times 100\% = 16.0\% \text{ (m/m)}$$

$$\underbrace{8.00 \text{ g KCl} + 42.00 \text{ g } H_2O}$$
(Solute + Solvent)

Add 8.00 g of KCl

Add water until the solution weighs 50.00 g

CONCEPT CHECK 12.5

■ **Mass Percent**

A NaBr solution is prepared by adding 10.0 g of NaBr to 100.0 g of H_2O. Is the final concentration of the NaBr solution equal to 9.09% (m/m), 10.0% (m/m), or 90.0% (m/m)? Explain your reasoning.

ANSWER

The final concentration of the NaBr is equal to 9.09% (m/m). The mass of the solute is 10.0 g of NaBr, and the mass of the solution is 110.0 g (10.0 g of NaBr + 100.0 g of H_2O).

$$\frac{10.0 \text{ g of NaBr}}{110.0 \text{ g of solution}} \times 100\% = 9.09\% \text{ (m/m) NaBr solution}$$

SAMPLE PROBLEM | **12.4**

■ **Calculating Mass Percent Concentration**

What is the mass percent (m/m) of a solution prepared by dissolving 30.0 g of NaOH in 120.0 g of H_2O?

**Guide to Calculating
Solution Concentration**

STEP 1
Determine quantities of
solute and solution.

STEP 2
Write the % concentration expression.

STEP 3
Substitute solute and solution
quantities into the expression.

SOLUTION

STEP 1 Determine quantities of solute and solution.

$$\begin{array}{r} \text{Mass of solute} = 30.0 \text{ g of NaOH} \\ \underline{\text{Mass of solvent} = 120.0 \text{ g of H}_2\text{O}} \\ \text{Mass of solution} = 150.0 \text{ g of solution} \end{array}$$

STEP 2 Write the % concentration expression.

$$\text{Mass percent (m/m)} = \frac{\text{grams of solute}}{\text{grams of solution}} \times 100\%$$

STEP 3 Substitute solute and solution quantities into the expression.

$$\text{Mass percent (m/m)} = \frac{30.0 \text{ g NaOH}}{150.0 \text{ g solution}} \times 100\%$$

$$= 20.0\% \text{ (m/m) NaOH solution}$$

STUDY CHECK

What is the mass percent (m/m) of NaCl in a solution made by dissolving 2.0 g of NaCl in 56.0 g of H_2O?

Volume Percent

Because the volumes of liquids or gases are easily measured, the concentrations of their solutions are often expressed as **volume percent (v/v)**. The units of volume used in the ratio must be the same—for example, both in milliliters or both in liters.

$$\text{Volume percent (v/v)} = \frac{\text{volume of solute}}{\text{volume of solution}} \times 100\%$$

We interpret a volume/volume percent as the volume of solute in exactly 100 mL of solution. In the wine industry, a label that reads 12% (v/v) means 12 mL of ethanol, CH_3—CH_2—OH, in 100 mL of wine.

SAMPLE PROBLEM | 12.5

■ **Calculating Volume Percent Concentration**

A student prepared a solution by adding water to 5.0 mL of ethanol (CH_3—CH_2—OH) to give a final volume of 250.0 mL. What is the volume percent (v/v) of the ethanol solution?

SOLUTION

STEP 1 Determine quantities of solute and solution.
Given 5.0 mL of CH_3—CH_2—OH in 250.0 mL of solution
Need volume percent (v/v) of CH_3—CH_2—OH

STEP 2 Write the % concentration expression.

$$\text{Volume percent (v/v)} = \frac{\text{volume of solute}}{\text{volume of solution}} \times 100\%$$

STEP 3 Substitute solute and solution quantities into the expression.

$$\text{Volume percent (v/v)} = \frac{5.0 \text{ mL CH}_3\text{—CH}_2\text{—OH}}{250.0 \text{ mL solution}} \times 100\%$$

$$= 2.0\% \text{ (v/v) CH}_3\text{—CH}_2\text{—OH solution}$$

STUDY CHECK

What is the volume percent, v/v, of Br_2 in a solution prepared by dissolving 12 mL of bromine (Br_2) in enough carbon tetrachloride to make 250. mL of solution?

Water added to
make a solution   250.0 mL

5.0 mL of   2.0% (v/v)
CH_3—CH_2—OH CH_3—CH_2—OH
solution

Percent Concentrations as Conversion Factors

TUTORIAL
Percent Concentration as a Conversion Factor

In the preparation of solutions, we often need to calculate the amount of solute or solution. Then the percent concentration is useful as a conversion factor. The value of 100 in the denominator of a percent expression is an *exact* number. Some examples of percent concentrations, their meanings, and possible conversion factors are given in Table 12.7.

TABLE 12.7 Conversion Factors from Percent Concentrations

Percent Concentration	Meaning	Conversion Factors		
15% (m/m) KCl	15 g of KCl in 100 g of solution	$\dfrac{15 \text{ g KCl}}{100 \text{ g solution}}$	and	$\dfrac{100 \text{ g solution}}{15 \text{ g KCl}}$
12% (v/v) ethanol	12 mL of ethanol in 100 mL of solution	$\dfrac{12 \text{ mL ethanol}}{100 \text{ mL solution}}$	and	$\dfrac{100 \text{ mL solution}}{12 \text{ mL ethanol}}$

SAMPLE PROBLEM | 12.6

■ **Using Mass Percent**

An antibiotic ointment is 3.5% (m/m) neomycin. How many grams of neomycin are in a tube containing 64 grams of ointment?

SOLUTION

STEP 1 State the given and needed quantities.
  Given 3.5% (m/m) neomycin   **Need** grams of neomycin

STEP 2 Write a plan to calculate mass or volume.

  grams of ointment | Mass % factor | > grams of neomycin

STEP 3 Write equalities and conversion factors. The mass percent (m/m) indicates the grams of a solute in every 100 g of a solution. The mass percent (3.5% m/m) of neomycin can be written as two conversion factors.

$$100 \text{ g of ointment} = 3.5 \text{ g of neomycin}$$

$$\dfrac{3.5 \text{ g neomycin}}{100 \text{ g ointment}} \quad \text{and} \quad \dfrac{100 \text{ g ointment}}{3.5 \text{ g neomycin}}$$

STEP 4 Set up problem to calculate mass or volume. The grams of the ointment solution are converted to grams of solute using the % conversion factor.

$$64 \text{ g ointment} \times \dfrac{3.5 \text{ g neomycin}}{100 \text{ g ointment}} = 2.2 \text{ g of neomycin}$$

Guide to Using Concentration to Calculate Mass or Volume

STEP 1
State the given and needed quantities.

STEP 2
Write a plan to calculate mass or volume.

STEP 3
Write equalities and conversion factors.

STEP 4
Set up problem to calculate mass or volume.

STUDY CHECK
Calculate the grams of KCl and grams of water in 225 g of an 8.00% (m/m) KCl solution.

QUESTIONS AND PROBLEMS

Percent Concentration

12.27 How would you prepare 250. g of a 5.00% (m/m) glucose solution?

12.28 What is the difference between a 10% (v/v) methanol (CH_3OH) solution and a 10% (m/m) methanol solution?

12.29 Calculate the mass percent (m/m) for the solute in each of the following solutions:
  a. 25 g of KCl and 125 g of H_2O
  b. 8.0 g of $CaCl_2$ in 80.0 g of $CaCl_2$ solution
  c. 12 g of sucrose in 225 g of sucrose solution

12.30 Calculate the mass percent (m/m) for the solute in each of the following solutions:
  a. 75 g of NaOH in 325 g of NaOH solution
  b. 2.0 g of KOH in 20.0 g of H_2O
  c. 48.5 g of Na_2CO_3 in 250.0 g of Na_2CO_3 solution

12.31 Calculate the amount of solute (g or mL) needed to prepare the following solutions:
  a. 50.0 g of a 5.0% (m/m) KCl solution
  b. 1250 g of a 4.0% (m/m) NH_4Cl solution
  c. 250 mL of a 10.0% (v/v) acetic acid solution

12.32 Calculate the amount of solute (g or mL) needed to prepare the following solutions:
a. 150. g of a 40.0% (m/m) LiBr solution
b. 450 g of a 2.0% (m/m) KCl solution
c. 225 mL of a 15% (v/v) isopropyl alcohol solution

12.33 A mouthwash contains 22.5% alcohol by volume. If the bottle of mouthwash contains 355 mL, what is the volume, in milliliters, of the alcohol?

12.34 A bottle of champagne is 11% alcohol by volume. If there are 750 mL of champagne in the bottle, how many milliliters of alcohol are present?

12.35 Calculate the amount of solution (g or mL) that contains each of the following amounts of solute:
a. 5.0 g of $LiNO_3$ from a 25% (m/m) $LiNO_3$ solution
b. 40.0 g of KOH from a 10.0% (m/m) KOH solution
c. 2.0 mL of formic acid from a 10.0% (v/v) formic acid solution

12.36 Calculate the amount of solution (g or mL) that contains each of the following amounts of solute:
a. 7.50 g of NaCl from a 2.0% (m/m) NaCl solution
b. 4.0 g of NaOH from a 25% (m/m) NaOH solution
c. 20.0 mL of ethanol from a 8.0% (v/v) ethanol solution

12.5 Molarity and Dilution

LEARNING GOAL

Calculate the molarity of a solution; use molarity as a conversion factor to calculate the moles of solute or the volume needed to prepare a solution.

When the solutes of solutions take part in reactions, chemists are often interested in the number of reacting particles. For this purpose, chemists use **molarity (M)**, a concentration that states the number of moles of solute in exactly 1 L of solution. The molarity of a solution can be calculated knowing the moles of solute and the volume of solution in liters.

$$\text{Molarity (M)} = \frac{\text{moles of solute}}{\text{liters of solution}} = \frac{\text{mol of solute}}{\text{L of soln}}$$

For example, if 1.0 mol of NaCl were dissolved in enough water to prepare 1.0 L of solution, the resulting NaCl solution has a molarity of 1.0 M. The abbreviation M indicates the units of moles per liter (mol/L).

$$M = \frac{\text{moles of solute}}{\text{liters of solution}} = \frac{1.0 \text{ mol of NaCl}}{1.0 \text{ L of solution}} = 1.0 \text{ M NaCl solution}$$

SAMPLE PROBLEM | **12.7**

■ **Calculating Molarity**

What is the molarity (M) of 60.0 g of NaOH in 0.250 L of solution?

SOLUTION

STEP 1 State the given and needed quantities.

Given 60.0 g of NaOH in 0.250 L of solution
Need molarity (mol/L)

STEP 2 Write a plan to calculate molarity. The calculation of molarity requires the moles of NaOH and the volume of the solution in liters.

$$\text{molarity (M)} = \frac{\text{moles of solute}}{\text{liters of solution}}$$

grams of NaOH ⟶ Molar mass ⟶ $\dfrac{\text{mol NaOH}}{\text{volume (L)}}$ = M of NaOH solution

Guide to Calculating Molarity

STEP 1
State the given and needed quantities.

STEP 2
Write a plan to calculate molarity.

STEP 3
Write equalities and conversion factors needed.

STEP 4
Set up problem to calculate molarity.

STEP 3 Write equalities and conversion factors needed.

$$1 \text{ mol of NaOH} = 40.01 \text{ g of NaOH}$$

$$\frac{1 \text{ mol NaOH}}{40.01 \text{ g NaOH}} \quad \text{and} \quad \frac{40.01 \text{ g NaOH}}{1 \text{ mol NaOH}}$$

STEP 4 Set up problem to calculate molarity.

$$\text{moles of NaOH} = 60.0 \text{ g NaOH} \times \frac{1 \text{ mol NaOH}}{40.01 \text{ g NaOH}} = 1.50 \text{ mol of NaOH}$$

The molarity is calculated by dividing the moles of NaOH by the volume in liters.

$$\frac{1.50 \text{ mol NaOH}}{0.250 \text{ L solution}} = \frac{6.00 \text{ mol NaOH}}{1 \text{ L solution}} = 6.00 \text{ M NaOH solution}$$

STUDY CHECK

What is the molarity of a solution that contains 75.0 g of KNO_3 dissolved in 0.350 L of solution?

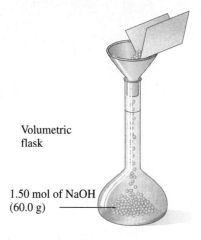

Volumetric flask

1.50 mol of NaOH (60.0 g)

Add water until 250-mL (0.250 L) mark is reached.

Mix

A 6.00 molar (M) NaOH solution

Molarity as a Conversion Factor

When we need to calculate the moles of solute or the volume of solution, the molarity is used as a conversion factor. Examples of conversion factors from molarity are given in Table 12.8.

Using the molarity of the solution with the molar mass of the solute, we can calculate the volume of solution needed, as illustrated in Sample Problem 12.8.

TABLE 12.8 Examples of Molar Solutions and Conversion Factors

Molarity	Meaning	Conversion Factors	
6.0 M HCl	6.0 mol of HCl in 1 L of solution	$\dfrac{6.0 \text{ mol HCl}}{1 \text{ L solution}}$ and	$\dfrac{1 \text{ L solution}}{6.0 \text{ mol HCl}}$
0.20 M NaOH	0.20 mol of NaOH in 1 L of solution	$\dfrac{0.20 \text{ mol NaOH}}{1 \text{ L solution}}$ and	$\dfrac{1 \text{ L solution}}{0.20 \text{ mol NaOH}}$

SAMPLE PROBLEM | **12.8**

■ **Using Molarity as a Conversion Factor**

How many liters of a 2.00 M NaCl solution are needed to provide 67.3 g of NaCl?

SOLUTION

STEP 1 State the given and needed quantities.
Given 67.3 g of NaCl from a 2.00 M NaCl solution
Need liters of NaCl

STEP 2 Write a plan to calculate mass or volume. The volume of NaCl is calculated using the moles of NaCl and molarity of the NaCl solution:

grams of NaCl [Molar mass] moles of NaCl [Molarity] liters of NaCl solution

STEP 3 Write equalities and conversion factors.

$$1 \text{ mol of NaCl} = 58.44 \text{ g of NaCl}$$
$$\frac{1 \text{ mol NaCl}}{58.44 \text{ g NaCl}} \quad \text{and} \quad \frac{58.44 \text{ g NaCl}}{1 \text{ mol NaCl}}$$

The molarity of any solution can be written as two conversion factors:

$$1 \text{ L of solution} = 2.00 \text{ mol of NaCl}$$
$$\frac{1 \text{ L solution}}{2.00 \text{ mol NaCl}} \quad \text{and} \quad \frac{2.00 \text{ mol NaCl}}{1 \text{ L solution}}$$

STEP 4 Set up problem to calculate mass or volume.

$$\text{liters of NaCl} = 67.3 \text{ g NaCl} \times \frac{1 \text{ mol NaCl}}{58.44 \text{ g NaCl}} \times \frac{1 \text{ L solution}}{2.00 \text{ mol NaCl}}$$

$$= 0.576 \text{ L of solution}$$

STUDY CHECK

How many milliliters of a 2.25 M HCl solution will provide 4.12 g of HCl?

To prepare a solution, we must convert the number of moles of solute needed into grams. Using the volume and the molarity of the solution with the molar mass of the solute, we can calculate the number of grams of solute necessary. This type of calculation is illustrated in Sample Problem 12.9.

SAMPLE PROBLEM | 12.9

■ Using Molarity

How many grams of KCl would you need to weigh out to prepare 0.250 L of a 2.00 M KCl solution?

SOLUTION

STEP 1 **State the given and needed quantities.**
Given 0.250 L of 2.00 M KCl solution   Need grams of KCl

STEP 2 **Write a plan to calculate mass or volume.** The grams of KCl are calculated by finding the moles of KCl using the volume and molarity of the KCl solution.

liters of solution   Molarity   moles of KCl   Molar mass   grams of KCl

STEP 3 **Write equalities and conversion factors.**

1 L of solution = 2.00 mol of KCl	1 mol of KCl = 74.55 g of KCl
$\dfrac{1 \text{ L solution}}{2.00 \text{ mol KCl}}$ and $\dfrac{2.00 \text{ mol KCl}}{1 \text{ L solution}}$	$\dfrac{1 \text{ mol KCl}}{74.55 \text{ g KCl}}$ and $\dfrac{74.55 \text{ g KCl}}{1 \text{ mol KCl}}$

STEP 4 **Set up problem to calculate mass or volume.**

$$\text{moles of KCl} = 0.250 \text{ L solution} \times \frac{2.00 \text{ mol KCl}}{1 \text{ L solution}} = 0.500 \text{ mol of KCl}$$

The mass of KCl is calculated by multiplying the moles of KCl by the molar mass.

$$\text{grams of KCl} = 0.500 \text{ mol KCl} \times \frac{74.55 \text{ g KCl}}{1 \text{ mol KCl}} = 37.3 \text{ g of KCl}$$

Combining the steps, we can write the problem setup as follows:

$$0.250 \text{ L solution} \times \frac{2.00 \text{ mol KCl}}{1 \text{ L solution}} \times \frac{74.55 \text{ g KCl}}{1 \text{ mol KCl}} = 37.3 \text{ g of KCl}$$

STUDY CHECK

How many grams of $NaHCO_3$ are in 325 mL of a 4.50 M $NaHCO_3$ solution?

Dilution

TUTORIAL
Dilution

In chemistry and biology, we often prepare diluted solutions from more concentrated solutions. In a process called **dilution**, a solvent, usually water, is added to a solution, which increases the volume. In an everyday example, you are making a dilution when you add three cans of water to a can of concentrated orange juice.

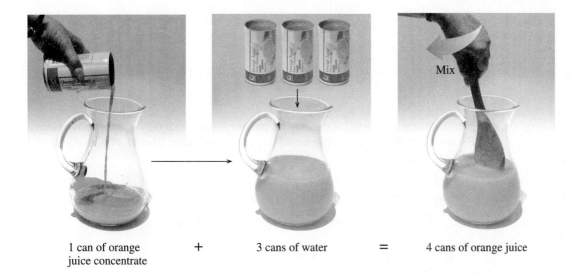

1 can of orange juice concentrate + 3 cans of water = 4 cans of orange juice

Although the addition of solvent increases the volume, the number of moles of solute before dilution is equal to the number of moles of solute in the diluted solution (see Fig. 12.8).

Moles of solute in = moles of solute in
concentrated solution   diluted solution

When the concentration is given as molarity (M), the moles of solute are obtained from the molarity and the volume, in liters.

Moles of solute = molarity × volume (L)

Expressing the number of moles in the concentrated solution as M_1V_1 and the number of moles in the diluted solution as M_2V_2, the equality is written as follows:

M_1V_1 = M_2V_2
Concentrated   Diluted
solution   solution

If we are given any 3 of the 4 variables, we can rearrange the expression to solve for the unknown quantity, as seen in Sample Problem 12.10.

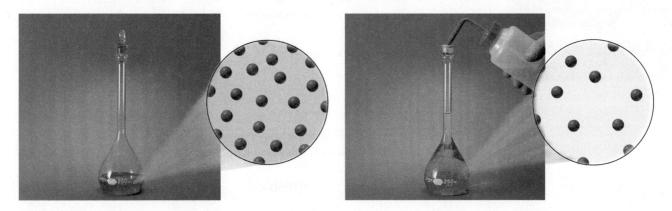

FIGURE 12.8 When water is added to a concentrated solution, there is no change in the number of particles, but the solute particles can spread out as the volume of the diluted solution increases.

Q What is the concentration of the diluted solution after an equal volume of water is added to a sample of a 6 M HCl solution?

■ **Volume of a Diluted Solution**

What volume (mL) of a 0.20 M $SrCl_2$ solution can be prepared by diluting 50.0 mL of a 1.0 M $SrCl_2$ solution?

ANSWER

We can organize the given information by making a table with the molar concentrations and volumes of the initial and diluted solutions. The unknown quantity is the volume of the diluted solution, V_2:

Initial solution: $M_1 = 1.0$ M $SrCl_2$ $V_1 = 50.0$ mL

Diluted solution: $M_2 = 0.20$ M $SrCl_2$ $V_2 = ?$ mL

Now we can rearrange the dilution expression to solve for V_2:

$$M_1 V_1 = M_2 V_2$$

$$\frac{M_1 V_1}{M_2} = \frac{\cancel{M_2} V_2}{\cancel{M_2}}$$

$$V_2 = V_1 \times \frac{M_1}{M_2}$$

Now we can place the known quantities into the dilution expression and calculate V_2:

$$V_2 = 50.0 \text{ mL} \times \frac{1.0 \text{ } \cancel{M SrCl_2}}{0.20 \text{ } \cancel{M SrCl_2}} = 250 \text{ mL of diluted } SrCl_2 \text{ solution}$$

SAMPLE PROBLEM |12.10|

■ **Molarity of a Diluted Solution**

What is the molarity of a solution prepared when 75.0 mL of a 4.00 M KCl solution is diluted to a volume of 0.500 L?

SOLUTION

STEP 1 **Prepare a table of the initial and diluted volumes and concentrations.** Units for volume must be the same, but both could be in milliliters or liters:

Guide to Calculating Dilution Quantities

STEP 1
Prepare a table of the initial and diluted volumes and concentrations.

STEP 2
Solve the dilution expression for the unknown quantity.

STEP 3
Set up problem by placing known quantities in the dilution expression.

Initial solution: $M_1 = 4.00$ M KCl $V_1 = 75.0$ mL $= 0.0750$ L

Diluted solution: $M_2 = ?$ M KCl $V_2 = 0.500$ L

STEP 2 **Solve the dilution expression for the unknown quantity.** In this problem, we need to solve the dilution expression for M_2.

$$M_1 V_1 = M_2 V_2$$

$$\frac{M_1 V_1}{V_2} = \frac{M_2 \cancel{V_2}}{\cancel{V_2}}$$

$$M_2 = M_1 \times \frac{V_1}{V_2}$$

STEP 3 **Set up problem by placing known quantities in the dilution expression.** The diluted concentration is calculated by placing the values from the table into the dilution expression.

$$M_2 = 4.00 \text{ M} \times \frac{0.0750 \text{ } \cancel{L}}{0.500 \text{ } \cancel{L}} = 0.600 \text{ M KCl} \quad \text{(diluted solution)}$$

STUDY CHECK

You need to prepare 600. mL of a 2.00 M NaOH solution from a 10.0 M NaOH solution. What volume of the 10.0 M NaOH solution do you need?

QUESTIONS AND PROBLEMS

Molarity and Dilution

12.37 Calculate the molarity of each of the following solutions:
a. 2.00 mol of glucose in 4.00 L of a glucose solution
b. 5.85 g of NaCl in 40.0 mL of a NaCl solution
c. 4.00 g of KOH in 2.00 L of a KOH solution

12.38 Calculate the molarity of each of the following solutions:
a. 0.500 mol of sucrose in 0.200 L of a sucrose solution
b. 30.4 g of LiBr in 350. mL of a LiBr solution
c. 73.0 g of HCl in 2.00 L of a HCl solution

12.39 Calculate the grams of solute needed to prepare each of the following solutions:
a. 2.00 L of a 1.50 M NaOH solution
b. 125 mL of a 0.200 M KCl solution
c. 25.0 mL of a 3.50 M HCl solution

12.40 Calculate the grams of solute needed to prepare each of the following solutions:
a. 2.00 L of a 5.00 M NaOH solution
b. 325 mL of a 0.100 M $CaCl_2$ solution
c. 15.0 mL of a 0.500 M $LiNO_3$ solution

12.41 Calculate the volume, in milliliters, of each of the following solutions that provides the given amount of solute:
a. 12.5 g of Na_2CO_3 from a 0.120 M solution
b. 0.850 mol of $NaNO_3$ from a 0.500 M solution
c. 30.0 g of LiOH from a 2.70 M solution

12.42 Calculate the volume, in liters, of each of the following solutions that provides the given amount of solute:
a. 5.00 mol of NaOH from a 12.0 M solution
b. 15.0 g of Na_2SO_4 from a 4.00 M solution
c. 28.0 g of $NaHCO_3$ from a 1.50 M solution

12.43 Calculate the final concentration of the solution in each of the following:
a. Water is added to 0.150 L of a 6.00 M HCl solution to give a volume of 0.500 L.
b. A 10.0-mL sample of a 2.50 M KCl solution is diluted with water to 0.250 L.
c. Water is added to 0.250 L of a 12.0 M KBr solution to give a volume of 1.00 L.

12.44 Calculate the final concentration of the solution in each of the following:
a. Water added to 10.0 mL of a 3.50 M KNO_3 solution gives a volume of 0.250 L.
b. A 5.00-mL sample of a 18.0 M sucrose solution is diluted with water to 100. mL.
c. Water is added to 0.250 L of a 12.0 M KBr solution to give a volume of 1.00 L.

12.45 Determine the final volume (mL) for each of the following:
a. diluting 50.0 mL of a 12.0 M NH_4Cl solution to give a 2.00 M NH_4Cl solution
b. diluting 18.0 mL of a 15.0 M $NaNO_3$ solution to give a 1.50 M $NaNO_3$ solution
c. diluting 4.50 mL of an 18.0 M H_2SO_4 solution to give a 2.50 M H_2SO_4 solution

12.46 Determine the final volume (mL) for each of the following:
a. diluting 2.50 mL of an 8.00 M KOH solution to give a 2.00 M KOH solution
b. diluting 50.0 mL of a 12.0 M NH_4Cl solution to give a 2.00 M NH_4Cl solution
c. diluting 75.0 mL of a 6.00 M HCl solution to give a 0.200 M HCl solution

12.47 Determine the volume (mL) required to prepare each of the following:
a. 255 mL of a 0.200 M HNO_3 solution from a 4.00 M HNO_3 solution
b. 715 mL of a 0.100 M $MgCl_2$ solution using a 6.00 M $MgCl_2$ solution
c. 0.100 L of a 0.150 M KCl solution using an 8.00 M KCl solution

12.48 Determine the volume (mL) required to prepare each of the following:
a. 20.0 mL of a 0.250 M KNO_3 solution from a 6.00 M KNO_3 solution
b. 25.0 mL of 2.50 M H_2SO_4 solution using a 12.0 M H_2SO_4 solution
c. 0.500 L of a 1.50 M NH_4Cl solution using a 10.0 M NH_4Cl solution

12.49 You need to dilute 25.0 mL of a 3.00 M HCl solution to make a 0.150 M HCl solution. What is the volume of diluted solution after you add water?

12.50 You need to dilute 30.0 mL of a 2.50 M NaCl solution to make a 0.500 M NaCl solution. What is the volume of diluted solution after you add water?

12.6 Solutions in Chemical Reactions

When chemical reactions involve aqueous solutions, we use the balanced chemical equation, as well as the molarity and volume, to determine the moles of the reactants or products. In other problems, we use the molarity and moles to determine the volume of a reactant or product solution.

LEARNING GOAL

Given the volume and molarity of a solution, calculate the amount of another reactant or product in the reaction.

SAMPLE PROBLEM 12.11

■ **Volume of a Solution in a Reaction**

Zinc reacts with HCl to produce $ZnCl_2$ and hydrogen gas, H_2.

$$Zn(s) + 2HCl(aq) \longrightarrow ZnCl_2(aq) + H_2(g)$$

TUTORIAL

Solution Stoichiometry

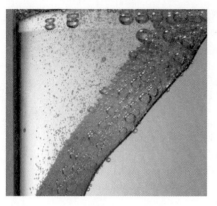

Zinc reacts when placed in a HCl solution.

Guide to Calculations Involving Solutions in Chemical Reactions

STEP 1
State the given and needed quantities.

STEP 2
Write a plan to calculate needed quantity or concentration.

STEP 3
Write equalities and conversion factors including mole–mole and concentration factors.

STEP 4
Set up problem to calculate needed quantity or concentration.

How many liters of a 1.50 M HCl solution completely react with 5.32 g of zinc?

SOLUTION

STEP 1 State the given and needed quantities.

Given 5.32 g of Zn and a 1.50 M HCl solution
Need liters of HCl solution

STEP 2 Write a plan to calculate needed quantity or concentration. We start the problem with the grams of Zn given and use its molar mass to calculate moles. Then we use the mole–mole factor from the equation and the molarity of the HCl solution as conversion factors to obtain the liters of HCl solution.

grams of Zn ⟩ Molar mass ⟩ moles of Zn ⟩ Mole–mole factor ⟩ moles of HCl

⟩ Molarity ⟩ liters of HCl solution

STEP 3 Write equalities and conversion factors including mole–mole and concentration factors.

$$1 \text{ mol of Zn} = 65.41 \text{ g of Zn}$$
$$\frac{1 \text{ mol Zn}}{65.41 \text{ g Zn}} \text{ and } \frac{65.41 \text{ g Zn}}{1 \text{ mol Zn}}$$

$$1 \text{ mol of Zn} = 2 \text{ mol of HCl}$$
$$\frac{1 \text{ mol Zn}}{2 \text{ mol HCl}} \text{ and } \frac{2 \text{ mol HCl}}{1 \text{ mol Zn}}$$

$$1 \text{ L of solution} = 1.50 \text{ mol of HCl}$$
$$\frac{1 \text{ L solution}}{1.50 \text{ mol HCl}} \text{ and } \frac{1.50 \text{ mol HCl}}{1 \text{ L solution}}$$

STEP 4 Set up problem to calculate needed quantity or concentration. We can write the problem setup as seen in our plan:

$$5.32 \text{ g Zn} \times \frac{1 \text{ mol Zn}}{65.41 \text{ g Zn}} \times \frac{2 \text{ mol HCl}}{1 \text{ mol Zn}} \times \frac{1 \text{ L solution}}{1.50 \text{ mol HCl}} = 0.108 \text{ L of HCl solution}$$

STUDY CHECK

Using the reaction in Sample Problem 12.11, how many grams of zinc can react with 225 mL of a 0.200 M HCl solution?

SAMPLE PROBLEM │12.12│

■ **Volume of a Reactant**

How many mL of a 0.250 M $BaCl_2$ solution is needed to react with 32.5 mL of a 0.160 M Na_2SO_4 solution?

$$Na_2SO_4(aq) + BaCl_2(aq) \longrightarrow BaSO_4(s) + 2NaCl(aq)$$

SOLUTION

STEP 1 State the given and needed quantities.
Given 32.5 mL (0.0325 L) of a 0.160 M Na_2SO_4 solution and a 0.250 M $BaCl_2$ solution
Need milliliters of $BaCl_2$ solution

STEP 2 Write a plan to calculate needed quantity or concentration. We start the problem with the volume and molarity of the Na_2SO_4 solution to calculate moles. Then we use the mole–mole factor from the equation and the molarity of $BaCl_2$ solution to calculate the liters and milliliters.

liters of Na_2SO_4 solution ⟩ Molarity ⟩ moles of Na_2SO_4 ⟩ Mole–mole factor ⟩ moles of $BaCl_2$

⟩ Molarity ⟩ liters of $BaCl_2$ solution ⟩ Metric factor ⟩ mL of $BaCl_2$ solution

When a $BaCl_2$ solution is added to a Na_2SO_4 solution, $BaSO_4$, a white solid, forms.

STEP 3 **Write equalities and conversion factors including mole–mole and concentration factors.**

$$1 \text{ L of solution} = 0.160 \text{ mol of Na}_2\text{SO}_4$$

$$\frac{1 \text{ L solution}}{0.160 \text{ mol Na}_2\text{SO}_4} \quad \text{and} \quad \frac{0.160 \text{ mol Na}_2\text{SO}_4}{1 \text{ L solution}}$$

$$1 \text{ mol of Na}_2\text{SO}_4 = 1 \text{ mol of BaCl}_2$$

$$\frac{1 \text{ mol Na}_2\text{SO}_4}{1 \text{ mol BaCl}_2} \quad \text{and} \quad \frac{1 \text{ mol BaCl}_2}{1 \text{ mol Na}_2\text{SO}_4}$$

$$1 \text{ L of solution} = 0.250 \text{ mol of BaCl}_2$$

$$\frac{1 \text{ L solution}}{0.250 \text{ mol BaCl}_2} \quad \text{and} \quad \frac{0.250 \text{ mol BaCl}_2}{1 \text{ L solution}}$$

$$1 \text{ L} = 1000 \text{ mL}$$

$$\frac{1 \text{ L}}{1000 \text{ mL}} \quad \text{and} \quad \frac{1000 \text{ mL}}{1 \text{ L}}$$

STEP 4 **Set up problem to calculate needed quantity or concentration.** The problem is set up using the plan and appropriate conversion factors:

$$0.0325 \text{ L solution} \times \frac{0.160 \text{ mol Na}_2\text{SO}_4}{1 \text{ L solution}} \times \frac{1 \text{ mol BaCl}_2}{1 \text{ mol Na}_2\text{SO}_4} \times \frac{1 \text{ L solution}}{0.250 \text{ mol BaCl}_2} \times \frac{1000 \text{ mL BaCl}_2 \text{ solution}}{1 \text{ L solution}}$$

$$= 20.8 \text{ mL of BaCl}_2 \text{ solution}$$

STUDY CHECK

For the reaction in Sample Problem 12.12, how many milliliters of a 0.330 M Na_2SO_4 solution are needed to react with 26.8 mL of a 0.216 M $BaCl_2$ solution?

SAMPLE PROBLEM 12.13

■ **Volume of a Gas from a Solution**

Acid rain results from the reaction of nitrogen dioxide with water in the air:

$$3NO_2(g) + H_2O(l) \longrightarrow 2HNO_3(aq) + NO(g)$$

At STP, how many liters of NO_2 gas are required to produce 0.275 L of a 0.400 M HNO_3 solution?

SOLUTION

STEP 1 **State the given and needed quantities.**
Given 0.275 L of a 0.400 M HNO_3 solution
Need liters of NO_2 gas at STP

STEP 2 **Write a plan to calculate needed quantity or concentration.** We start the problem with the volume and molarity of the HNO_3 solution to calculate moles. Then we can use the mole–mole factor and the molar volume to calculate the liters of NO_2 gas.

liters of solution **Molarity** ▸ moles of HNO_3 **Mole–mole factor** ▸ moles of NO_2 **Molar volume** ▸ liters of NO_2 gas

STEP 3 **Write equalities and conversion factors including mole–mole and concentration factors.**

$$1 \text{ L of solution} = 0.400 \text{ mol of HNO}_3$$

$$\frac{1 \text{ L solution}}{0.400 \text{ mol HNO}_3} \quad \text{and} \quad \frac{0.400 \text{ mol HNO}_3}{1 \text{ L solution}}$$

$$3 \text{ mol of NO}_2 = 2 \text{ mol of HNO}_3$$

$$\frac{2 \text{ mol HNO}_3}{3 \text{ mol NO}_2} \quad \text{and} \quad \frac{3 \text{ mol NO}_2}{2 \text{ mol HNO}_3}$$

$$1 \text{ mol of NO}_2 = 22.4 \text{ L of NO}_2 \text{ at STP}$$

$$\frac{22.4 \text{ L NO}_2}{1 \text{ mol NO}_2} \quad \text{and} \quad \frac{1 \text{ mol NO}_2}{22.4 \text{ L NO}_2}$$

STEP 4 **Set up problem to calculate needed quantity or concentration.** We can write the problem setup as seen in our plan:

$$0.275 \text{ L solution} \times \frac{0.400 \text{ mol HNO}_3}{1 \text{ L solution}} \times \frac{3 \text{ mol NO}_2}{2 \text{ mol HNO}_3} \times \frac{22.4 \text{ L NO}_2}{1 \text{ mol NO}_2} = 3.70 \text{ L of NO}_2 \text{ gas}$$

STUDY CHECK

Using the equation in Sample Problem 12.13, determine the volume of NO produced at 100 °C and 1.20 atm, when 2.20 L of a 1.50 M HNO_3 solution is also produced.

Figure 12.9 gives a summary of the pathways and conversion factors needed for substances including solutions involved in chemical reactions.

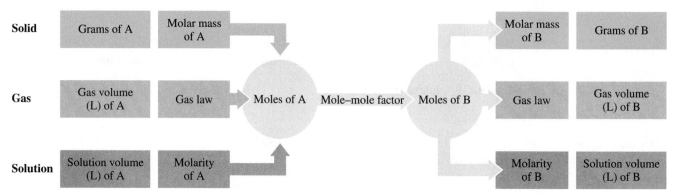

FIGURE 12.9 In calculations involving chemical reactions, substance A is converted to moles of A using molar mass (if solid), gas laws (if gas), or molarity (if solution). Then moles of A are converted to moles of substance B, which are converted to grams of solid, liters of gas, or liters of solution, as needed.

Q What sequence of conversion factors would you use to calculate the number of grams of $CaCO_3$ needed to react with 1.50 L of a 2.00 M HCl solution in the reaction $2HCl(aq) + CaCO_3(s) \longrightarrow CaCl_2(aq) + CO_2(g) + H_2O(l)$?

QUESTIONS AND PROBLEMS

Solutions in Chemical Reactions

12.51 Given the reaction

$Pb(NO_3)_2(aq) + 2KCl(aq) \longrightarrow PbCl_2(s) + 2KNO_3(aq)$

a. How many grams of $PbCl_2$ will be formed from 50.0 mL of a 1.50 M KCl solution?

b. How many milliliters of a 2.00 M $Pb(NO_3)_2$ solution can react with 50.0 mL of a 1.50 M KCl solution?

c. What is the molarity of 20.0 mL of a KCl solution that reacts completely with 30.0 mL of a 0.400 M $Pb(NO_3)_2$ solution?

12.52 In the reaction

$NiCl_2(aq) + 2NaOH(aq) \longrightarrow Ni(OH)_2(s) + 2NaCl(aq)$

a. How many milliliters of a 0.200 M NaOH solution are needed to react with 18.0 mL of a 0.500 M $NiCl_2$ solution?

b. How many grams of $Ni(OH)_2$ are produced from the reaction of 35.0 mL of a 1.75 M NaOH solution and excess $NiCl_2$?

c. What is the molarity of 30.0 mL of a $NiCl_2$ solution if this volume of solution reacts completely with 10.0 mL of a 0.250 M NaOH solution?

12.53 In the reaction

$Mg(s) + 2HCl(aq) \longrightarrow MgCl_2(aq) + H_2(g)$

a. How many milliliters of a 6.00 M HCl solution are required to react with 15.0 g of magnesium?

b. How many liters of hydrogen gas at STP can form when 0.500 L of a 2.00 M HCl solution reacts with excess magnesium?

c. What is the molarity of a HCl solution if the reaction of 45.2 mL of the HCl solution reacts with excess magnesium to produce 5.20 L of H_2 gas at 735 mmHg and 25 °C?

12.54 The calcium carbonate in limestone reacts with a HCl solution to produce a calcium chloride solution, carbon dioxide, and water.

$CaCO_3(s) + 2HCl(aq) \longrightarrow CaCl_2(aq) + H_2O(l) + CO_2(g)$

a. How many milliliters of a 0.200 M HCl solution can react with 8.25 g of $CaCO_3$?

b. How many liters of CO_2 gas can form at STP when 15.5 mL of a 3.00 M HCl solution reacts with excess $CaCO_3$?

c. What is the molarity of a HCl solution if the reaction of 200. mL of the HCl solution with excess $CaCO_3$ produces 12.0 L of CO_2 gas at 725 mmHg and 18 °C?

12.7 Properties of Solutions

The solute particles in a solution play an important role in determining the properties of that solution. In the solutions discussed up to now, the solute was dissolved as small particles that are uniformly dispersed throughout the solvent to give a homogeneous solution. The particles in a solution are so small that they go through filters and *semipermeable membranes* such as cell walls in the body.

LEARNING GOAL

Identify a mixture as a solution, a colloid, or a suspension. Describe how particles of a solution affect the freezing point, boiling point, and osmotic pressure of a solution.

Colloids

In *colloids*, the solute particles are large molecules, such as proteins, or groups of molecules. Colloids, similar to solutions, are homogeneous mixtures that do not separate or settle out. Colloidal particles are small enough to pass through filters but too large to pass through semipermeable membranes. Table 12.9 lists several examples of colloids.

TABLE 12.9 Examples of Colloids

	Substance Dispersed	Dispersing Medium
Fog, clouds, sprays	Liquid	Gas
Dust, smoke	Solid	Gas
Shaving cream, whipped cream, soapsuds	Gas	Liquid
Mayonnaise, butter, homogenized milk	Liquid	Liquid
Cheese, butter	Liquid	Solid

In the body, colloids from food are too large to pass through the intestinal membrane and remain in the intestinal tract. Digestion breaks down these colloids, such as starch and protein, into smaller solution particles, such as glucose and amino acids, that can pass through the intestinal membrane and enter the circulatory system. However fiber, such as bran, cannot be broken down by our digestive processes and moves through the intestine intact.

Suspensions

Suspensions are heterogeneous, nonuniform mixtures containing very large particles that are trapped by filters and do not pass through semipermeable membranes. If you stir muddy water, it mixes but then quickly separates as the suspension particles settle to the bottom. You may use suspensions such as Kaopectate, calamine lotion, antacid mixtures, and liquid penicillin. It is important to "shake well before using" to suspend all the particles before giving a medication that is a suspension.

Water-treatment plants make use of the properties of suspensions to purify water. When chemicals such as aluminum sulfate or iron(III) sulfate are added to untreated water, they react with small particles to form large suspension particles called *floc*. In the water-treatment plant, a system of filters traps the suspension particles but clean water passes through.

Table 12.10 compares the different types of mixtures, and Figure 12.10 illustrates some properties of solutions, colloids, and suspensions.

TABLE 12.10 Comparison of Solutions, Colloids, and Suspensions

Type of Mixture	Type of Particle	Settling Out	Separation
Solution	Small particles such as atoms, ions, or small molecules	Particles do not settle out	Particles cannot be separated by filters or semipermeable membranes
Colloid	Larger molecules or groups of molecules or ions	Particles do not settle out	Particles can be separated by semipermeable membranes but not by filters
Suspension	Very large particles that may be visible	Particles settle out rapidly	Particles can be separated by filters

FIGURE 12.10 Properties of different types of mixtures: (a) suspensions settle out; (b) suspensions are separated by a filter; (c) solution particles pass through a semipermeable membrane, but colloids and suspensions do not.

Q Why is a filter paper used to separate suspension particles from a solution, but a semipermeable membrane is needed to separate colloids from a solution?

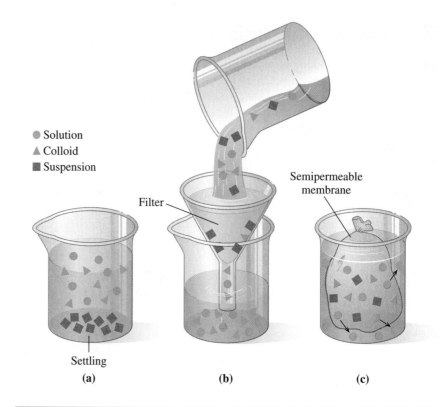

● Solution
▲ Colloid
■ Suspension

Filter

Semipermeable membrane

Settling

(a)   (b)   (c)

CONCEPT CHECK 12.7

■ Classifying Types of Mixtures

Classify each of the following as a solution, colloid, or suspension:

a. a mixture that has particles that settle out upon standing

b. a mixture whose solute particles pass through both filters and membranes

c. an enzyme, which is a large protein molecule that cannot pass through cellular membranes but does pass through a filter

ANSWER

a. A suspension has very large particles that settle out upon standing.

b. A solution contains particles small enough to pass through both filters and membranes.

c. A colloid contains particles that are small enough to pass through a filter but too large to pass through a membrane.

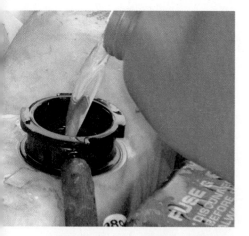

Ethylene glycol is added to a radiator to form an aqueous solution that has a lower freezing point than water.

Freezing Point Lowering and Boiling Point Elevation

When a solute is added to water, the physical properties such as freezing point and boiling point change. Therefore, an aqueous solution will have a lower freezing point and a higher boiling point than pure water. These types of changes in physical properties known as **colligative properties** depend on the number of solute particles in the solution.

Probably one familiar example is the process of spreading salt on icy sidewalks and roads when temperatures drop below freezing. The particles from the salt combine with water to lower the freezing point, which causes the ice to melt. Another example is the addition of antifreeze, such as ethylene glycol $HO-CH_2-CH_2-OH$, to the water in a car radiator. Ethylene glycol is an organic compound with two alcohol functional groups, which form hydrogen bonds to make it very soluble in water. If the ethylene glycol and water mixture is about 50%–50% by mass, it does not freeze until the temperature drops to about −34 °F, and does not boil unless the temperature reaches about 255 °F. The solution in the radiator prevents the water in the radiator from forming ice in cold weather and boiling over on a hot desert highway.

Molality

The calculation for freezing point lowering or boiling point elevation uses a concentration unit of *molality*. The **molality**, abbreviation *m*, of a solution is the number of moles of solute particles per kilogram of solvent. This may seem similar to molarity, but the denominator for molality refers to the mass of the solvent, not the volume of the solution.

$$\text{Molality } (m) = \frac{\text{moles of solute}}{\text{kilograms of solvent}}$$

CONCEPT CHECK 12.8

■ Calculating Molality

Calculate the molality of a solution containing 35.5 g of glucose ($C_6H_{12}O_6$) in 0.400 kg of water.

ANSWER

STEP 1 Given 35.5 g of glucose ($C_6H_{12}O_6$) in 0.400 kg of water
Need molality (moles of glucose/kg of water)

STEP 2 Plan We start the problem using the molar mass of glucose to calculate moles. Then we place moles of glucose and kilograms of water into the molality relationship to calculate the molality (*m*) of the glucose solution.

grams of glucose $\boxed{\text{Molar mass}}$ moles of glucose

$$\text{molality } (m) \text{ of glucose solution} = \frac{\text{moles of glucose}}{\text{kilogram of water}}$$

STEP 3 Equalities/Conversion Factors

1 mol of glucose = 180.2 g of glucose

$$\frac{1 \text{ mol glucose}}{180.2 \text{ g glucose}} \quad \text{and} \quad \frac{180.2 \text{ g glucose}}{1 \text{ mol glucose}}$$

STEP 4 Set Up Problem We set up the problem using our plan to calculate the molality of the glucose solution.

$$\text{moles of glucose} = 35.5 \text{ g glucose} \times \frac{1 \text{ mol glucose}}{180.2 \text{ g glucose}} = 0.197 \text{ mol of glucose}$$

$$\text{molality } (m) = \frac{0.197 \text{ mol glucose}}{0.400 \text{ kg water}} = 0.493 \ m$$

Freezing Point and Boiling Point Changes

The lowering of the freezing point (ΔT_f) and the raising of the boiling point (ΔT_b) of water occur because the solute particles disrupt the formation of the solid ice structure as well as the formation of water vapor. Thus, a lower temperature is required to freeze the water in the solution. The greater the solute concentration, the lower the freezing point will be. When there is 1 mol of particles in 1 kg of water, the freezing point depression (ΔT_f) is the molality of the solution multiplied by the freezing point constant, K_f. For water, the K_f is 1.86 °C for a 1 *m* solution.

$$\Delta T_f = mK_f = \frac{m \, (1.86 \, °C)}{m} = 1.86 \, °C$$

Thus in a 1 *m* solution, the freezing point of water drops from 0 °C to −1.86 °C. If there are 2 mol of particles dissolved in 1 kg of water, the freezing point drops twice that much to −3.72 °C. A similar change occurs with the boiling point of water. The boiling point

TUTORIAL
Freezing Point Depression and Boiling Point Elevation

constant (K_b) for water is 0.52 °C, which means that 1 mol of particles in 1 kg of water raises the boiling point (ΔT_b) by 0.52 °C. Thus, the temperature rises from 100. °C to 100.52 °C.

$$\Delta T_b = mK_b = \frac{\cancel{m}(0.52\ °C)}{\cancel{m}} = 0.52\ °C$$

The typical solute in antifreeze, ethylene glycol, $HO—CH_2—CH_2—OH$, is a nonelectrolyte that dissolves in water only as molecules.

Nonelectrolyte: 1 mol of $C_2H_6O_2(l)$ = 1 mol of $C_2H_6O_2(aq)$

However, a strong electrolyte, such as NaCl or $CaCl_2$, dissolves in water producing two or three ions, which behave as two or three mol of particles in solution. For example, the dissociation of 1 mol of $CaCl_2$ gives 3 mol of particles, which lowers the freezing point of water three times as much as 1 mol of ethylene glycol.

Strong electrolytes: 1 mol of NaCl(s) = $\underbrace{\text{1 mol of Na}^+(aq) + \text{1 mol of Cl}^-(aq)}_{\text{2 mol of particles }(aq)}$

1 mol of $CaCl_2$(s) = $\underbrace{\text{1 mol of Ca}^{2+}(aq) + \text{2 mol of Cl}^-(aq)}_{\text{3 mol of particles }(aq)}$

The effect of some solutes on freezing and boiling point is summarized in Table 12.11.

TABLE 12.11 Effect of Solute Concentration on Freezing and Boiling Points of 1 kg of Water

			Freezing Point	**Boiling Point**
Pure water			0 °C	100 °C
Temperature change constant (water)			$K_f = 1.86\ °C/m$	$K_b = 0.52\ °C/m$
Examples of Solutions	**Type of Solute**	**Moles of Solute Particles**	**Lowered Freezing Point**	**Raised Boiling Point**
1 m ethylene glycol	Nonelectrolyte	1 mol	−1.86 °C	100.52 °C
1 m NaCl	Strong electrolyte	2 mol	−3.72 °C	101.04 °C
1 m CaCl$_2$	Strong electrolyte	3 mol	−5.58 °C	101.56 °C

A truck spreads calcium chloride on the road to melt ice and snow.

SAMPLE PROBLEM 12.14

■ Calculating Freezing-Point Lowering

In the northeastern United States during freezing temperatures, $CaCl_2$ is spread on icy highways to melt the ice. Calculate the freezing-point lowering and freezing point of a solution containing 225 g of $CaCl_2$ in 500. g of water.

SOLUTION

STEP 1   Given 225 g of $CaCl_2$ 500. g of water (solvent) = 0.500 kg of water
Need freezing-point lowering ΔT_f and freezing point of $CaCl_2$ solution

STEP 2   Plan We use molar mass to calculate the moles of $CaCl_2$. Then we multiply by three to obtain number of moles of ions (particles) produced by 1 mol of $CaCl_2$ in solution. The molality (m) of the particles in solution is obtained by dividing the moles of particles by the number of kilograms of water in the solution. The freezing-point lowering is calculated using the molality and the freezing-point constant. Finally, the freezing-point lowering is subtracted from 0 °C to obtain the new freezing point of the $CaCl_2$ solution.

grams of $CaCl_2$ Molar mass   moles of $CaCl_2$ $\times \dfrac{\text{3 mol particles}}{\text{1 mol CaCl}_2}$ = moles of particles

$$m = \frac{\text{moles of particles}}{\text{kilograms of water}} \quad \text{and } m \text{ Freezing point factor} \quad \Delta T_f \text{ of CaCl}_2 \text{ solution}$$

final temperature $= 0\,°C$ (water) $- \Delta T_f$

STEP 3 Equalities/Conversion Factors

1 mol of CaCl$_2$ = 110.98 g of CaCl$_2$	1 m = 1.86 °C	1 mol of CaCl$_2$ = 3 mol of particles
$\dfrac{1 \text{ mol CaCl}_2}{110.98 \text{ g CaCl}_2}$ and $\dfrac{110.98 \text{ g CaCl}_2}{1 \text{ mol CaCl}_2}$	$\dfrac{1.86\,°C}{1\,m}$ and $\dfrac{1\,m}{1.86\,°C}$	$\dfrac{1 \text{ mol CaCl}_2}{3 \text{ mol particles}}$ and $\dfrac{3 \text{ mol particles}}{1 \text{ mol CaCl}_2}$

STEP 4 Set Up Problem We use our plan and conversion factors to calculate the molality of the CaCl$_2$ solution and the new freezing point.

$$\text{moles of particles} = 225 \text{ g CaCl}_2 \times \frac{1 \text{ mol CaCl}_2}{110.98 \text{ g CaCl}_2} \times \frac{3 \text{ mol particles}}{1 \text{ mol CaCl}_2} = 6.08 \text{ mol of particles}$$

$$\text{molality } (m) = \frac{6.08 \text{ mol particles}}{0.500 \text{ kg water}} = 12.2\, m$$

$$\Delta T_f = 12.2\, m \times \frac{1.86\,°C}{1\, m} = 22.7\,°C$$

freezing point $= 0.0\,°C - 22.7\,°C = -22.7\,°C$

STUDY CHECK

Ethylene glycol, $HO-CH_2-CH_2-OH$, a nonelectrolyte, is added to the water in a radiator to give a solution containing 515 g of ethylene glycol in 565 g of water (solvent). What is the boiling point elevation and the boiling point of the solution?

Osmotic Pressure

The movement of water into and out of the cells of plants as well as our own bodies is an important biological process that also depends on the solute concentration. In a process called osmosis, the water molecules move through a semipermeable membrane from the solution with the lower concentration of solute into a solution with the higher solute concentration. In an osmosis apparatus, water is placed on one side of a semipermeable membrane and a sucrose (sugar) solution on the other side. The semipermeable membrane allows water molecules to flow back and forth but blocks the sucrose molecules because they are too large to pass through the membrane. Because the sucrose solution has a higher solute concentration, more water molecules flow into the sucrose solution than out of the sucrose solution. The volume level of the sucrose solution rises as the volume level on the water side falls. The increase of water dilutes the sucrose solution to equalize (or attempt to equalize) the concentrations on both sides of the membrane.

Eventually the height of the sucrose solution creates sufficient pressure to equalize the flow of water between the two compartments. This pressure, called **osmotic pressure**, prevents the flow of additional water into the more concentrated solution. Then there is no further change in the volumes of the two solutions. The osmotic pressure depends on the concentration of solute particles in the solution. The greater the number of particles dissolved, the higher its osmotic pressure. In this example, the sucrose solution has a higher osmotic pressure than pure water, which has an osmotic pressure of zero.

In a process called *reverse osmosis*, a pressure greater than the osmotic pressure is applied to a solution. The flow of water is reversed so that water flows out of the solution with the higher solute concentration. This process of reverse osmosis is used in desalination plants to obtain pure water from sea (salt) water.

SELF STUDY ACTIVITY
Diffusion

TUTORIAL
Osmosis

Water flows into the solution with a higher solute concentration until the flow of water becomes equal in both directions.

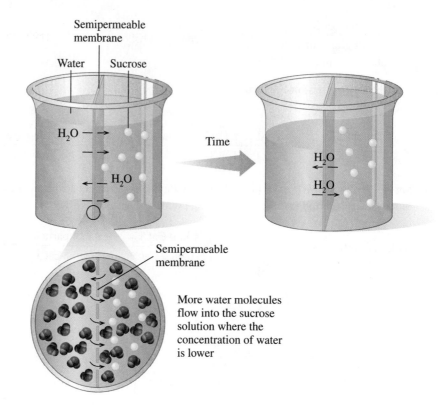

Semipermeable membrane

Water   Sucrose

H_2O →

← H_2O

Time

H_2O ←

H_2O →

Semipermeable membrane

More water molecules flow into the sucrose solution where the concentration of water is lower

---

CONCEPT CHECK 12.9

■ **Osmotic Pressure**

A 2% (m/m) sucrose solution and an 8% (m/m) sucrose solution are separated by a semipermeable membrane.

a. Which sucrose solution exerts the greater osmotic pressure?
b. In what direction does water flow initially?
c. Which solution will have the higher level of liquid at equilibrium?

ANSWER

a. The 8% (m/m) sucrose solution has the higher solute concentration, more solute particles, and the greater osmotic pressure.
b. Initially, water will flow out of the 2% (m/m) solution into the more concentrated 8% (m/m) solution.
c. The level of the 8% (m/m) solution will be higher.

Isotonic Solutions

Because the cell membranes in biological systems are semipermeable, osmosis is an ongoing process. The solutes in body solutions such as blood, tissue fluids, lymph, and plasma all exert osmotic pressure. Most intravenous (IV) solutions used in a hospital are *isotonic* solutions, which exert the same osmotic pressure as body fluids such as blood. The percent concentration typically used in IV solutions is similar to the types of percent concentrations we have already discussed, except that the concentration of IV solutions is *mass/volume percent (m/v)*. The most typical isotonic solutions are 0.9% (m/v) NaCl solution, or 0.9 g NaCl/100 mL of solution, and 5% (m/v) glucose, or 5 g glucose/100 mL of solution. Although they do not contain the same kinds of particles, a 0.9% (m/v) NaCl solution as well as a 5% (m/v) glucose solution are both 0.3 M (Na^+ and Cl^- ions or glucose molecules). A red blood cell placed in an isotonic solution retains its volume because there is an equal flow of water into and out of the cell (see Fig. 12.11a).

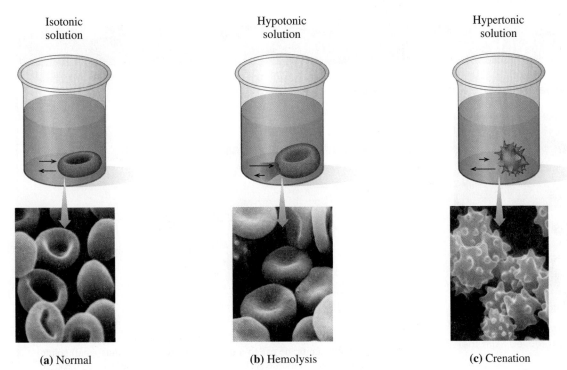

| Isotonic solution | Hypotonic solution | Hypertonic solution |
| (a) Normal | (b) Hemolysis | (c) Crenation |

Hypotonic and Hypertonic Solutions

If a red blood cell is placed in a solution that is not isotonic, the differences in osmotic pressure inside and outside the cell can drastically alter the volume of the cell. When a red blood cell is placed in a *hypotonic* solution, which has a lower solute concentration (*hypo* means "lower than"), water flows into the cell by osmosis. The increase in fluid causes the cell to swell and possibly burst—a process called *hemolysis* (see Fig. 12.11b). A similar process occurs when you place dehydrated food, such as raisins or dried fruit, in water. The water enters the cells, and the food becomes plump and smooth.

If a red blood cell is placed in a *hypertonic* solution, which has a higher solute concentration (*hyper* means "greater than"), water flows out of the cell into the hypertonic solution by osmosis. Suppose a red blood cell is placed in a 10% (m/v) NaCl solution. Because the osmotic pressure in the cell is equal to that of a 0.9% (m/v) NaCl solution, the cell shrinks, a process called *crenation* (see Fig. 12.11c). A similar process occurs when making pickles, which uses a hypertonic salt solution that causes the cucumbers to shrivel as they lose water.

FIGURE 12.11 **(a)** In an isotonic solution, a red blood cell retains its normal volume. **(b)** Hemolysis: In a hypotonic solution, water flows into a red blood cell, causing it to swell and burst. **(c)** Crenation: In a hypertonic solution, water leaves the red blood cell, causing it to shrink.

Q What happens to a red blood cell placed in a 4% NaCl solution?

SAMPLE PROBLEM | 12.15

■ **Isotonic, Hypotonic, and Hypertonic Solutions**

Describe each of the following solutions as isotonic, hypotonic, or hypertonic. Indicate whether a red blood cell placed in each solution will undergo hemolysis, crenation, or no change.

a. a 5.0% (m/v) glucose solution   **b.** a 0.2% (m/v) NaCl solution

SOLUTION

a. A 5.0% (m/v) glucose solution is isotonic. A red blood cell will not undergo any change.
b. A 0.2% (m/v) NaCl solution is hypotonic. A red blood cell will undergo hemolysis.

STUDY CHECK

What is the effect of a 10% (m/v) glucose solution on a red blood cell?

CHEMISTRY AND HEALTH

Dialysis by the Kidneys and the Artificial Kidney

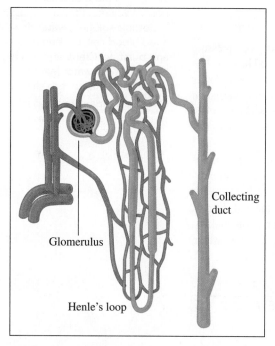

During dialysis, waste products and excess water are removed from the blood.

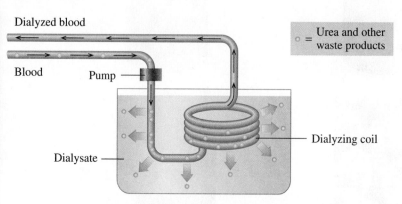

Dialyzed blood

Blood

Pump

Dialysate

Dialyzing coil

○ = Urea and other waste products

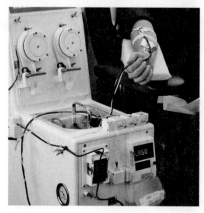

Collecting duct

Glomerulus

Henle's loop

In the kidneys, the nephrons each contain a glomerulus where urea and waste products are removed to form urine.

Dialysis is a process that is similar to osmosis. In dialysis, a semipermeable membrane called a dialyzing membrane permits small solute molecules and ions as well as solvent water molecules to pass through, but it retains large particles, such as colloids. Dialysis is a way to separate solution particles from colloids.

The fluids of the body undergo dialysis by the membranes of the kidneys, which remove waste materials, excess salts, and water. In an adult, each kidney contains about 2 million nephrons. At the top of each nephron, there is a network of arterial capillaries called the glomerulus.

As blood flows into the glomerulus, small particles, such as amino acids, glucose, urea, water, and certain ions, will move through the capillary membranes into the nephron. As this solution moves through the nephron, substances still of value to the body (such as amino acids, glucose, certain ions, and 99% of the water) are reabsorbed. The major waste product, urea, is excreted in the urine.

If the kidneys fail to dialyze waste products, increased levels of urea can become life-threatening in a relatively short time. A person with kidney failure must use an artificial kidney, which cleanses the blood by *hemodialysis*.

A typical artificial-kidney machine contains a large tank filled with about 100 L of water containing selected electrolytes. In the center of this dialyzing bath (dialysate), there is a dialyzing coil or membrane made of cellulose tubing. As the patient's blood flows through the dialyzing coil, the highly concentrated waste products dialyze out of the blood. No blood is lost, because the membrane is not permeable to large particles such as red blood cells.

Dialysis patients do not produce much urine. As a result, they retain large amounts of water between dialysis treatments, which produces a strain on the heart. The intake of fluids for a dialysis patient may be restricted to as little as a few teaspoons of water a day. In the dialysis procedure, the pressure of the blood is increased as it circulates through the dialyzing coil so water can be squeezed out of the blood. For some dialysis patients, 2–10 L of water may be removed during one treatment. Dialysis patients typically have from two to three treatments a week, each treatment lasting about 5 to 7 hours. Some of the newer treatments require less time. For many patients, dialysis is done at home with a home dialysis unit.

TUTORIAL

Dialysis

QUESTIONS AND PROBLEMS

Properties of Solutions

12.55 Identify the following as characteristic of a solution, colloid, or suspension:
  a. a mixture that cannot be separated by a semipermeable membrane
  b. a mixture that settles out upon standing

12.56 Identify the following as characteristic of a solution, colloid, or suspension:
  a. Particles of this mixture remain inside a semipermeable membrane, but pass through filters.
  b. The particles of solute in this mixture are very large and visible.

12.57 How many moles of each of the following strong electrolytes are needed to give the same freezing point lowering as 1.2 mol of the nonelectrolyte ethylene glycol in 1 kg of water?
  a. NaCl   **b.** K_3PO_4

12.58 How many moles of each of the following strong electrolytes are needed to give the same freezing point lowering as 3.0 mol of the nonelectrolyte ethylene glycol in 1 kg of water?
  a. CH_3—OH, a nonelectrolyte
  b. KNO_3

12.59 Calculate the molality (*m*) of the following solutions:
  a. 325 g of CH_3—OH, a nonelectrolyte, added to 455 g of water

b. 640. g of the antifreeze propylene glycol, $C_3H_8O_2$, a nonelectrolyte, dissolved in 1.22 kg of water

12.60 Calculate the molality (*m*) of the following solutions:
  a. 65.0 g of glucose, $C_6H_{12}O_6$, a nonelectrolyte, added to 0.112 kg of water
  b. 0.320 mol of sucrose, $C_{12}H_{22}O_{11}$, a nonelectrolyte, dissolved in 1.50 kg of water

12.61 Calculate the freezing point and boiling point of each of the solutions in Problem 12.59.

12.62 Calculate the freezing point and boiling point of each of the solutions in Problem 12.60.

12.63 Indicate the compartment (A or B) that will increase in volume for each of the following pairs of solutions separated by a semipermeable membrane:

A	B
a. 5.0% (m/v) starch	10% (m/v) starch
b. 4% (m/v) albumin	8% (m/v) albumin
c. 10% (m/v) sucrose	0.1% (m/v) sucrose

12.64 Indicate the compartment (A or B) that will increase in volume for each of the following pairs of solutions separated by a semipermeable membranes

A	B
a. 20% (m/v) starch	10% (m/v) starch
b. 10% (m/v) albumin	2% (m/v) albumin
c. 0.5% (m/v) sucrose	5% (m/v) sucrose

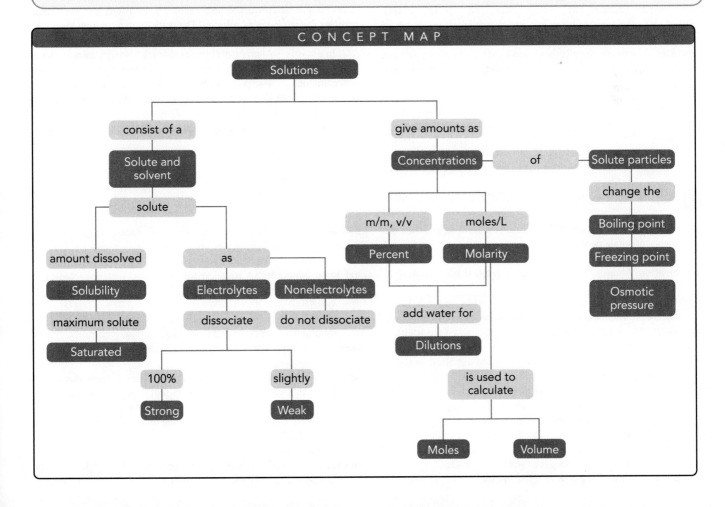

CONCEPT MAP

CHAPTER REVIEW

12.1 Solutions

LEARNING GOAL: *Define solute and solvent; describe the formation of a solution.*

A solution forms when a solute dissolves in a solvent. In a solution, the particles of solute are evenly dispersed in the solvent. The solute and the solvent may be solid, liquid, or gas. The polar O—H bond leads to hydrogen bonding between water molecules. An ionic solute dissolves in water, a polar solvent, because the polar water molecules attract and pull the ions into solution, where they become hydrated. The expression "like dissolves like" means that a polar or ionic solute dissolves in a polar solvent and a nonpolar solute requires a nonpolar solvent.

12.2 Electrolytes and Nonelectrolytes

LEARNING GOAL: *Identify solutes as electrolytes or nonelectrolytes.*

Substances that release ions are called electrolytes because the solution will conduct an electrical current. Strong electrolytes are completely ionized, whereas weak electrolytes are only partially ionized. Nonelectrolytes are substances that dissolve in water to produce molecules and solutions that cannot conduct electrical currents.

12.3 Solubility

LEARNING GOAL: *Define solubility; distinguish between an unsaturated and a saturated solution. Identify an insoluble salt.*

A solution that contains the maximum amount of dissolved solute is a saturated solution. The solubility of a solute is the maximum amount of a solute that can dissolve in 100 g of solvent. A solution containing less than the maximum amount of dissolved solute is unsaturated. An increase in temperature increases the solubility of most solids in water but decreases the solubility of gases in water. Salts that are soluble in water usually contain Li^+, Na^+, K^+, NH_4^+, NO_3^-, or acetate, CH_3—COO^-. An ionic equation consists of writing all the dissolved substances in an equation for the formation of an insoluble salt as individual ions. A net ionic equation is written by removing all the ions not involved in the chemical change (spectator ions) from the ionic equation.

12.4 Percent Concentration

LEARNING GOAL: *Calculate the percent concentration of a solute in a solution; use percent concentration to calculate the amount of solute or solution.*

The concentration of a solution is the amount of solute dissolved in a certain amount of solution. Mass percent (m/m) expresses the ratio of the mass of solute to the mass of solution multiplied by 100. Percent concentration can also be expressed as a volume/volume (v/v) ratio. In calculations of grams or milliliters of solute or solution, the percent concentration is used as a conversion factor.

12.5 Molarity and Dilution

LEARNING GOAL: *Calculate the molarity of a solution; use molarity as a conversion factor to calculate the moles of solute or the volume needed to prepare a solution.*

Molarity is the moles of solute per liter of solution. Units of molarity, moles/liter, are used in conversion factors to solve for moles of solute or volume of solution. In dilution, a solvent such as water is added to a solution, which increases the volume and decreases the concentration.

12.6 Solutions in Chemical Reactions

LEARNING GOAL: *Given the volume and molarity of a solution, calculate the amount of another reactant or product in the reaction.*

When solutions are involved in chemical reactions, the moles of a substance in solution can be determined from the volume and molarity of the solution. When mass, volume, and molarities of substances in a reaction are given, the balanced equation is used to determine the quantities or concentrations of other substances in the reaction.

12.7 Properties of Solutions

LEARNING GOAL: *Identify a mixture as a solution, a colloid, or a suspension. Describe how particles of a solution affect the freezing point, boiling point, and osmotic pressure of a solution.*

Colloids contain particles that do not settle out and pass through most filters but not semipermeable membranes. Suspensions have very large particles that settle out of solution. The particles in a solution lower the freezing point, raise the boiling point, and increase the osmotic pressure.

In osmosis, solvent (water) passes through a semipermeable membrane from a solution with a lower osmotic pressure (lower solute concentration) to a solution with a higher osmotic pressure (higher solute concentration). Isotonic solutions have osmotic pressures equal to that of body fluids. A red blood cell maintains its volume in an isotonic solution but swells in a hypotonic solution and shrinks in a hypertonic solution.

KEY TERMS

colligative property A property of solution which depends on the number of solute particles in solution but not the specific type of particle.

concentration A measure of the amount of solute that is dissolved in a specified amount of solution.

dilution A process by which water (solvent) is added to a solution to increase the volume and decrease (dilute) the solute concentration.

dissociation The separation of a solute into ions when the solute is dissolved in water.

electrolyte A substance that produces ions when dissolved in water; its solution conducts electricity.

Henry's law The solubility of a gas in a liquid is directly related to the pressure of that gas above the liquid.

hydration The process of surrounding dissolved ions by water molecules.

insoluble salt An ionic compound that does not dissolve in water.

ionic equation An equation for a reaction in solution that gives all the individual ions, both reacting ions and spectator ions.

mass percent (m/m) The grams of solute in exactly 100 g of solution.

molality (m) The number of moles of solute in exactly 1 kg of solvent.

molarity (M) The number of moles of solute in exactly 1 L of solution.

net ionic equation An equation for a reaction that gives only the reactants undergoing chemical change and leaves out spectator ions.

nonelectrolyte A substance that dissolves in water as molecules; its solution will not conduct an electrical current.

osmotic pressure The pressure that prevents the flow of water into the more concentrated solution.

saturated solution A solution containing the maximum amount of solute that can dissolve at a given temperature. Any additional solute will remain undissolved in the container.

solubility The maximum amount of solute that can dissolve in exactly 100 g of solvent, usually water, at a given temperature.

soluble salt An ionic compound that dissolves in water.

solute A substance that is the smaller amount uniformly dispersed in another substance called the solvent.

solution A homogeneous mixture in which the solute is made up of small particles (ions or molecules).

solvent The substance in which the solute dissolves; usually the component present in greater amount.

strong electrolyte A compound that ionizes completely when it dissolves in water; its solution is a good conductor of electricity.

unsaturated solution A solution that contains less solute than can be dissolved.

volume percent (v/v) A percent concentration that relates the volume of the solute to the volume of the solution.

weak electrolyte A substance that produces only a few ions along with many molecules when it dissolves in water; its solution is a weak conductor of electricity.

■ UNDERSTANDING THE CONCEPTS

12.65 Select the diagram that represents the solution formed by a solute 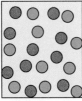 that is a

    a. nonelectrolyte   **b.** weak electrolyte
    c. strong electrolyte

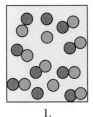

      1.   2.   3.

12.66 Match the diagrams with the following:
    a. a polar solute and a polar solvent
    b. a nonpolar solute and a polar solvent
    c. a nonpolar solute and a nonpolar solvent

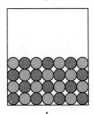

      1.   2.

12.67 Select the container that represents the dilution of a 4% (m/m) KCl solution to each of the following:
    a. 2% (m/m) KCl solution
    b. 1% (m/m) KCl solution

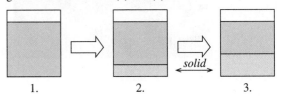

4% (m/m) KCl   1.   2.   3.

12.68 Do you think solution (1) has undergone heating or cooling to give the solid shown in (2) and (3)?

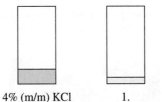

      1.   2.   3.

Use the following beakers and solutions for Problems 12.69 and 12.70:

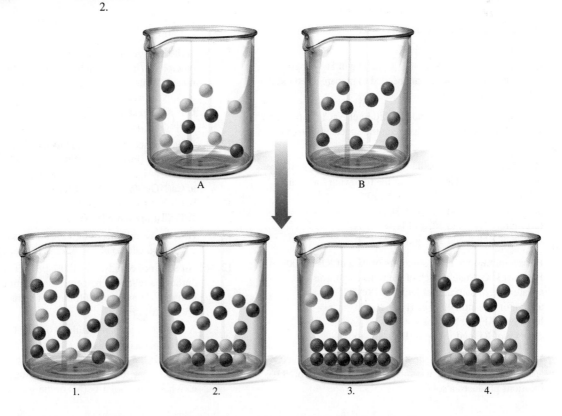

A   B

1.   2.   3.   4.

12.69 Use the following:

Na⁺   Cl⁻   Ag⁺ ● NO₃⁻

a. Select the beaker (1, 2, 3, or 4) that contains the products after the solutions in beakers A and B are mixed.
b. If an insoluble salt forms, write the ionic equation.
c. If a reaction occurs, write the net ionic equation.

12.70 Use the following:

K⁺ ● NO₃⁻ ● NH₄⁺ ● Br⁻ ●

a. Select the beaker (1, 2, 3, or 4) that contains the products after the solutions in beakers A and B are mixed.
b. If an insoluble salt forms, write the ionic equation.
c. If a reaction occurs, write the net ionic equation.

12.71 A pickle is made by soaking a cucumber in brine, a saltwater solution. What makes the smooth cucumber become wrinkled like a prune?

12.72 Why do lettuce leaves in a salad wilt after a vinaigrette dressing containing salt is added?

ADDITIONAL QUESTIONS AND PROBLEMS

For instructor-assigned homework, go to
www.masteringchemistry.com.

12.73 Why does iodine dissolve in hexane, but not in water?

12.74 How do temperature and pressure affect the solubility of solids and gases in water?

12.75 If NaCl has a solubility of 36.0 g in 100 g of H₂O at 20 °C, how many grams of water are needed to prepare a saturated solution containing 80.0 g of NaCl?

12.76 If the solid NaCl in a saturated solution of NaCl continues to dissolve, why is there no change in the concentration of the NaCl solution?

12.77 Potassium nitrate has a solubility of 32 g of KNO₃ in 100 g of H₂O at 20 °C. State if each of the following forms an unsaturated or saturated solution at 20 °C:
a. 32 g of KNO₃ and 200. g of H₂O
b. 19 g of KNO₃ and 50. g of H₂O
c. 68 g of KNO₃ and 150. g of H₂O

12.78 Potassium fluoride has a solubility of 92 g of KF in 100 g of H₂O at 18 °C. State if each of the following forms an unsaturated or saturated solution at 18 °C:
a. 46 g of KF and 100. g of H₂O
b. 46 g of KF and 50. g of H₂O
c. 184 g of KF and 150. g of H₂O

12.79 Why would a solution made by mixing solutions of NaNO₃ and KCl be clear, while a combination of KCl and Pb(NO₃)₂ solutions produces a solid?

12.80 Indicate whether each of the following ionic compounds is soluble in water:
a. KCl   **b.** MgSO₄
c. CuS   **d.** AgNO₃
e. Ca(OH)₂

12.81 Write the net ionic equation to show the formation of a precipitate (insoluble salt) when the following solutions are mixed. Write *none* if there is not a precipitate.
a. AgNO₃(*aq*) and LiCl(*aq*)
b. NaCl(*aq*) and KNO₃(*aq*)
c. Na₂SO₄(*aq*) and BaCl₂(*aq*)

12.82 Write the net ionic equation to show the formation of a precipitate (insoluble salt) when the following solutions are mixed. Write *none* if there is not a precipitate.
a. Ca(NO₃)₂(*aq*) and Na₂S(*aq*)
b. Na₃PO₄(*aq*) and Pb(NO₃)₂(*aq*)
c. FeCl₃(*aq*) and NH₄NO₃(*aq*)

12.83 How many milliliters of a 12% (v/v) propyl alcohol solution would you take to obtain 4.5 mL of propyl alcohol?

12.84 An 80-proof brandy is a 40.0% (v/v) ethanol solution. The "proof" is twice the percent concentration of alcohol in the beverage. How many milliliters of alcohol are present in 750. mL of brandy?

12.85 A solution is prepared with 70.0 g of HNO₃ and 130.0 g of H₂O. It has a density of 1.21 g/mL.
a. What is the mass percent (m/m) of the HNO₃ solution?
b. What is the total volume (mL) of the solution?
c. What is the molarity (M) of the solution?

12.86 A solution is prepared by dissolving 22.0 g of NaOH in 118.0 g of water. The NaOH solution has a density of 1.15 g/mL.
a. What is the mass percent (m/m) of the NaOH solution?
b. What is the total volume (mL) of the solution?
c. What is the molarity (M) of the solution?

12.87 How many liters of a 2.50 M KNO_3 solution can be prepared from 60.0 g of KNO_3?

12.88 How many liters of a 4.00 M NaCl solution will provide 25.0 g of NaCl?

12.89 If you were in the laboratory, how would you prepare 250. mL of a 2.00 M KCl solution?

12.90 What is the molarity (M) of a solution containing 15.6 g of KCl in 274 mL of solution?

12.91 How many grams of solute are in each of the following solutions?
a. 2.52 L of a 3.00 M KNO_3 solution
b. 75.0 mL of a 0.506 M Na_2SO_4 solution
c. 45.2 mL of a 1.80 M HCl solution

12.92 How many grams of solute are in each of the following solutions?
a. 428 mL of a 0.450 M Na_2SO_4 solution
b. 10.5 mL of a 2.50 M $AgNO_3$ solution
c. 28.4 mL of a 6.00 M H_3PO_4 solution

12.93 The antacid Amphogel contains aluminum hydroxide $Al(OH)_3$. How many milliliters of a 6.00 M HCl solution are required to react with 60.0 mL of a 2.00 M $Al(OH)_3$ solution?

$$Al(OH)_3(s) + 3HCl(aq) \longrightarrow AlCl_3(aq) + 3H_2O(l)$$

12.94 Calcium carbonate, $CaCO_3$, reacts with stomach acid (HCl, hydrochloric acid) according to the following equation:

$$CaCO_3(s) + 2HCl(aq) \longrightarrow CaCl_2(aq) + H_2O(l) + CO_2(g)$$

One tablet of Tums, an antacid, contains 500.0 mg of $CaCO_3$. If one tablet of Tums is added to 20.0 mL of a 0.100 M HCl solution how many liters of CO_2 gas are produced at STP?

12.95 A 355-mL sample of a HCl solution reacts with excess Mg to produce 4.20 L of H_2 gas measured at 745 mmHg and 35 °C. What is the molarity (M) of the HCl solution?

$$Mg(s) + 2HCl(aq) \longrightarrow MgCl_2(aq) + H_2(g)$$

12.96 A 255-mL sample of a HCl solution reacts with excess Ca to produce 14.0 L of H_2 gas at STP. What is the molarity (M) of the HCl solution?

$$Ca(s) + 2HCl(aq) \longrightarrow CaCl_2(aq) + H_2(g)$$

12.97 Calculate the molarity (M) of the solution when water is added to prepare each of the following:
a. 25.0 mL of a 0.200 M NaBr solution is diluted to 50.0 mL
b. 15.0 mL of a 1.20 M K_2SO_4 solution is diluted to 40.0 mL
c. 75.0 mL of a 6.00 M NaOH solution is diluted to 255 mL

12.98 Calculate the molarity (M) of the solution when water is added to prepare each of the following:
a. 25.0 mL of a 18.0 M HCl solution is diluted to 500. mL
b. 50.0 mL of a 1.50 M NaCl solution is diluted to 125 mL
c. 4.50 mL of a 8.50 M KOH solution is diluted to 75.0 mL

12.99 What is the final volume, in mL, when 25.0 mL of a 5.00 M HCl solution is diluted to each of the following concentrations?
a. 2.50 M HCl   **b.** 1.00 M HCl
c. 0.500 M HCl

12.100 What is the final volume, in mL, when 5.00 mL of a 12.0 M NaOH solution is diluted to each of the following concentrations?
a. 0.600 M NaOH   **b.** 1.00 M NaOH
c. 2.50 M NaOH

12.101 Why would solutions with a high salt content be used to prepare dried flowers?

12.102 Why can't you drink seawater if you were stranded on a desert island?

CHALLENGE QUESTIONS

12.103 Indicate whether each of the following ionic compounds is soluble or insoluble in water:
a. Na_3PO_4 **b.** $PbBr_2$
c. $ZnCl_2$ **d.** $(NH_4)_2S$
e. $MgCO_3$ **f.** $FePO_4$

12.104 Write the net ionic equation to show the formation of a precipitate (insoluble salt) when the following solutions are mixed. Write *none* if no insoluble salt forms.
a. $AgNO_3(aq) + Na_2SO_4(aq)$
b. $KCl(aq) + Pb(NO_3)_2(aq)$
c. $CaCl_2(aq) + (NH_4)_3PO_4(aq)$
d. $K_2SO_4(aq) + BaCl_2(aq)$

12.105 In a laboratory experiment, a 10.0-mL sample of NaCl solution is poured into an evaporating dish with a mass of 24.10 g. The combined mass of the evaporating dish and the NaCl solution is 36.15 g. After heating, the evaporating dish and dry NaCl have a combined mass of 25.50 g.
a. What is the mass percent (m/m) of the NaCl solution?
b. What is the molarity (M) of the NaCl solution?
c. If water is added to 10.0 mL of the initial NaCl solution to give a final volume of 60.0 mL, what is the molarity of the diluted NaCl solution?

12.106 A solution contains 4.56 g of KCl in 175 mL of solution. If the density of the KCl solution is 1.12 g/mL, what are the mass percent (m/m) and molarity (M) for the KCl solution?

12.107 How many milliliters of a 1.75 M LiCl solution contain 15.2 g of LiCl?

12.108 How many grams of NaBr are contained in 75.0 mL of a 1.50 M NaBr solution?

12.109 Cadmium reacts with HCl to produce cadmium chloride and hydrogen gas.

$$Cd(s) + 2HCl(aq) \longrightarrow CdCl_2(aq) + H_2(g)$$

What is the molarity (M) of the HCl solution if 250. mL of the HCl solution reacts with excess cadmium to produce 4.20 L of H_2 gas measured at STP?

12.110 How many liters of NO gas can be produced at STP from 80.0 mL of a 4.00 M HNO_3 solution and 10.0 g of Cu?

$$3Cu(s) + 8HNO_3(aq) \longrightarrow 3Cu(NO_3)_2(aq) + 4H_2O(l) + 2NO(g)$$

12.111 The boiling point of a NaCl solution is 101.04 °C.
   a. What is the molality (*m*) of the NaCl solution?
   b. What is the freezing point of the NaCl solution?

12.112 The freezing point of a $CaCl_2$ solution is −25 °C.
   a. What is the molality (*m*) of the $CaCl_2$ solution?
   b. What is the boiling point of the $CaCl_2$ solution?

ANSWERS

ANSWERS TO STUDY CHECKS

12.1 A solution of a weak electrolyte will contain mostly molecules and a few ions.

12.2 The solubility of KNO_3 is expected to be higher because the solubility of most solids increases when the temperature increases.

12.3 **a.** No solid forms.
   b. $Pb^{2+}(aq) + 2Cl^-(aq) \longrightarrow PbCl_2(s)$

12.4 3.4% (m/m) NaCl solution

12.5 4.8% (v/v) Br_2 in CCl_4

12.6 18 g of KCl and 207 g of H_2O

12.7 2.12 M KNO_3 solution

12.8 50.2 mL

12.9 123 g of $NaHCO_3$

12.10 120. mL of a 10.0 M NaOH solution

12.11 1.47 g of Zn

12.12 17.5 mL of a Na_2SO_4 solution

12.13 42.1 L of NO

12.14 $\Delta T_b = 7.6$ °C; boiling point is 107.6 °C

12.15 The red blood cell will undergo crenation.

ANSWERS TO SELECTED QUESTIONS AND PROBLEMS

12.1 **a.** NaCl, solute; water, solvent
   b. water, solute; ethanol, solvent
   c. oxygen, solute; nitrogen, solvent

12.3 **a.** water   **b.** CCl_4
   c. water   **d.** CCl_4

12.5 The polar water molecules pull the K^+ and I^- ions away from the solid and into solution, where they are hydrated.

12.7 In a solution of KF, only the ions of K^+ and F^- are present in the solvent. In an HF solution, there are a few ions of H^+ and F^- present but mostly dissolved HF molecules.

12.9 **a.** $KCl(s) \xrightarrow{H_2O} K^+(aq) + Cl^-(aq)$

   b. $CaCl_2(s) \xrightarrow{H_2O} Ca^{2+}(aq) + 2Cl^-(aq)$

   c. $K_3PO_4(s) \xrightarrow{H_2O} 3K^+(aq) + PO_4^{3-}(aq)$

   d. $Fe(NO_3)_3(s) \xrightarrow{H_2O} Fe^{3+}(aq) + 3NO_3^-(aq)$

12.11 **a.** mostly molecules and a few ions   **b.** ions only
   c. molecules only

12.13 **a.** strong electrolyte   **b.** weak electrolyte
   c. nonelectrolyte

12.15 **a.** saturated   **b.** unsaturated

12.17 **a.** unsaturated   **b.** unsaturated
   c. saturated

12.19 **a.** 68 g of KCl   **b.** 12 g of KCl

12.21 **a.** The solubility of solid solutes typically increases as temperature increases.
   b. The solubility of a gas is less at a higher temperature.
   c. Gas solubility is less at a higher temperature and the CO_2 pressure in the can is increased.

12.23 **a.** soluble   **b.** insoluble
   c. insoluble   **d.** soluble
   e. soluble

12.25 **a.** no solid forms
   b. $2Ag^+(aq) + S^{2-}(aq) \longrightarrow Ag_2S(s)$
   c. $Ca^{2+}(aq) + SO_4^{2-}(aq) \longrightarrow CaSO_4(s)$
   d. $3Cu^{2+}(aq) + 2PO_4^{3-}(aq) \longrightarrow Cu_3(PO_4)_2(s)$

12.27 12.5 g of glucose is added to 237.5 g of water to make 250. g of a 5.00% (m/m) glucose solution.

12.29 **a.** 17% (m/m) KCl   **b.** 10.% (m/m) $CaCl_2$
   c. 5.3% (m/m) sucrose

12.31 **a.** 2.5 g of KCl   **b.** 50. g of NH_4Cl
   c. 25 mL of acetic acid

12.33 79.9 mL of alcohol

12.35 **a.** 20. g   **b.** 400. g
   c. 20. mL

12.37 **a.** 0.500 M glucose solution
   b. 2.50 M NaCl solution
   c. 0.0356 M KOH solution

12.39 **a.** 120. g of NaOH   **b.** 1.86 g of KCl
   c. 3.19 g of HCl

12.41 **a.** 983 mL   **b.** 1700 mL (1.70×10^3 mL)
   c. 464 mL

12.43 **a.** 1.80 M HCl solution   **b.** 0.100 M KCl solution
   c. 3.00 M KBr solution

12.45 **a.** 300. mL
   b. 180. mL
   c. 32.4 mL

12.47 **a.** 12.8 mL of the HNO_3 solution
   b. 11.9 mL of the $MgCl_2$ solution
   c. 1.88 mL of the KCl solution

12.49 500. mL of HCl solution

12.51 **a.** 10.4 g of $PbCl_2$ **b.** 18.8 mL of $Pb(NO_3)_2$ solution
   c. 1.20 M KCl solution

12.53 **a.** 206 mL of HCl solution   **b.** 11.2 L of H_2 gas
   c. 9.09 M HCl solution

12.55 **a.** solution   **b.** suspension

12.57 **a.** 0.60 mol of NaCl   **b.** 0.30 mol of K_3PO_4

12.59 **a.** 22.3 m **b.** 6.89 m

12.61 **a.** freezing point: -41.5 °C; boiling point: 111.6 °C
   b. freezing point: -12.8 °C; boiling point: 103.6 °C

12.63 **a.** B 10% (m/v) starch solution
   b. B 8% (m/v) albumin solution
   c. A 10% (m/v) sucrose solution

12.65 **a.** 3 (no dissociation)
   b. 1 (some dissociation, a few ions)
   c. 2 (all ionized)

12.67 **a.** 2; to halve the % concentration, the volume would double.
   b. 3; to go to one-fourth the % concentration, the volume would be four times the initial volume.

12.69 **a.** beaker 3
   b. $Na^+(aq) + Cl^-(aq) + Ag^+(aq) + NO_3^-(aq) \longrightarrow$
   $AgCl(s) + Na^+(aq) + NO_3^-(aq)$
   c. $Ag^+(aq) + Cl^-(aq) \longrightarrow AgCl(s)$

12.71 The skin of the cucumber acts like a semipermeable membrane and water from the more dilute solution inside flows into the more concentrated brine solution.

12.73 Because iodine is a nonpolar molecule, it will dissolve in hexane, a nonpolar solvent. Iodine does not dissolve in water because water is a polar solvent.

12.75 222 g of water

12.77 **a.** unsaturated solution   **b.** saturated solution
   c. saturated solution

12.79 When solutions of $NaNO_3$ and KCl are mixed, no insoluble products are formed; all the combinations of salts are soluble. When KCl and $Pb(NO_3)_2$ solutions are mixed, the insoluble salt $PbCl_2$ forms.

12.81 **a.** $Ag^+(aq) + Cl^-(aq) \longrightarrow AgCl(s)$
   b. none
   c. $Ba^{2+}(aq) + SO_4^{2-}(aq) \longrightarrow BaSO_4(s)$

12.83 38 mL of propyl alcohol solution

12.85 **a.** 35.0% (m/m) HNO_3 solution
   b. 165 mL
   c. 6.73 M HNO_3 solution

12.87 **a.** 0.237 L of KNO_3 solution

12.89 To make a 2.00 M KCl solution, weigh out 37.3 g of KCl (0.500 mol) and place in a volumetric flask. Add water to dissolve the KCl and give a final volume of 0.250 L.

12.91 **a.** 764 g of KNO_3 **b.** 5.39 g of Na_2SO_4
   c. 2.97 g of HCl

12.93 60.0 mL of HCl solution

12.95 0.917 M HCl solution

12.97 **a.** 0.100 M NaBr solution   **b.** 0.450 M K_2SO_4 solution
   c. 1.76 M NaOH solution

12.99 **a.** 50.0 mL   **b.** 125 mL
   c. 250. mL

12.101 The solution will dehydrate the flowers because water will flow out of the cells of the flowers into the more concentrated salt solution.

12.103 **a.** Na^+ salts are soluble.
   b. A halide salt containing Pb^{2+} is insoluble.
   c. Most halide salts are soluble.
   d. A salt containing NH_4^+ ions is soluble.
   e. A salt containing Mg^{2+} and CO_3^{2-} is insoluble.
   f. A salt containing Fe^{3+} and PO_4^{3-} is insoluble.

12.105 **a.** 11.6% (m/m) NaCl solution
   b. 2.40 M NaCl solution
   c. 0.400 M NaCl solution

12.107 205 mL of LiCl solution

12.109 1.50 M HCl solution

12.111 **a.** 2.0 m **b.** -3.7 °C

13 Chemical Equilibrium

"I use radioactive isotopes to understand the cycling of elements like carbon and phosphorus in the ocean," explains Claudia Benitez-Nelson, a chemical oceanographer and assistant professor of geological sciences at the University of South Carolina. *"For example, I use thorium-234 to trace how and when particles are formed and transported to the bottom of the ocean. I also examine the biological consumption of the nutrient phosphorus by measuring fluctuations over time in the levels of the naturally occurring radioactive isotopes of phosphorus. My knowledge of chemistry is essential for understanding nutrient biogeochemistry and carbon sequestration in the oceans."*

Oceanographers study the oceans and the plants and animals that live in the ocean. They study marine life, the chemical compounds in the ocean, the shape and composition of the ocean floor, and the effects of waves and tides.

Visit **www.masteringchemistry.com** for self-study materials and instructor-assigned homework.

arlier we looked at chemical reactions and determined the amounts of substances that react and the products that form. Now we are interested in how fast a reaction goes. If we know how fast a medication acts on the body, we can adjust the time over which the medication is taken. In construction, substances are added to cement to make it dry faster so work can continue. Some reactions such as explosions or the formation of precipitates in a solution are very fast. When we roast a turkey or bake a cake, the reaction is slow. Some reactions such as the tarnishing of silver and the aging of the body are much slower (see Fig. 13.1). We will see that some reactions need energy while other reactions produce energy. We burn gasoline in our automobile engines to produce energy to make our cars move. In this chapter, we will also look at the effect of changing the concentrations of reactants or products on the rate of reaction.

Reaction rate increases

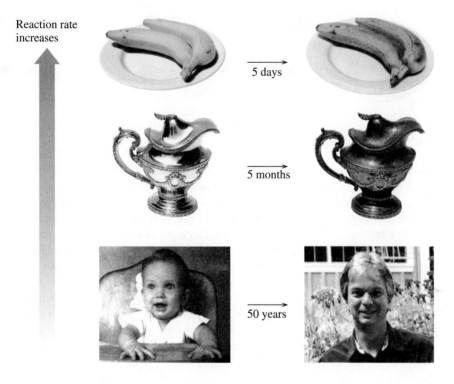

5 days

5 months

50 years

FIGURE 13.1 Reaction rates vary greatly for everyday processes. A banana ripens in a few days, silver tarnishes in a few months, while the aging process of humans takes many years.

Q How would you compare the rates of the reaction that forms sugars in plants by photosynthesis with the reactions that digest sugars in the body?

Up to now, we have considered a reaction as proceeding in a forward direction from reactants to products. However, in many reactions a reverse reaction also takes place as products collide to re-form reactants. When the forward and reverse reactions take place at the same rate, the amounts of reactants and products stay the same. When this balance in the rates of the forward and reverse reactions is reached, we say that the reaction has reached *equilibrium*. At equilibrium both reactants and products are present. Some reaction mixtures contain mostly reactants and form only a few products, while others contain mostly products and a few reactants.

13.1 Rates of Reactions

For a chemical reaction to take place, the molecules of the reactants must come in contact with each other. The **collision theory** indicates that a reaction takes place only when molecules collide with the proper orientation and sufficient energy. Many collisions can occur, but only a few actually lead to the formation of product. For example, consider the reaction of nitrogen and oxygen molecules (see Fig. 13.2). To form nitrogen oxide product, the collisions between N_2 and O_2 molecules must place the atoms in the proper alignment. If the molecules are not aligned properly, no reaction takes place.

LEARNING GOAL

Describe how temperature, concentration, and catalysts affect the rate of a reaction.

FIGURE 13.2 Reacting molecules must collide, have a minimum amount of energy, and have the proper orientation to form products.

Q What happens when reacting molecules collide with the minimum energy but don't have the proper orientation?

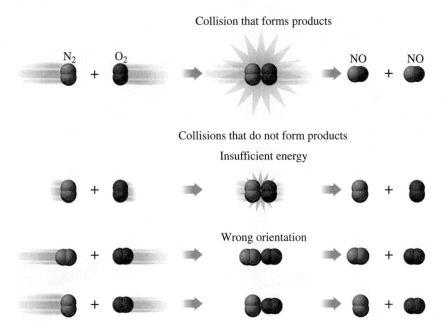

Collision that forms products

Collisions that do not form products

Insufficient energy

Wrong orientation

Activation Energy

Even when a collision has the proper orientation, there still must be sufficient energy to break the bonds between the atoms of the reactants. The **activation energy** is the amount of energy required to break the bonds between atoms of the reactants. In Figure 13.3, activation energy appears as an energy hill. The concept of activation energy is analogous to climbing a hill. To reach a destination on the other side, we must have the energy needed to climb to the top of the hill. Once we are at the top, we can run down the other side. The energy needed to get us from our starting point to the top of the hill would be the activation energy.

In the same way, a collision must provide enough energy to push the reactants to the top of the energy hill. Then the reactants may be converted to products. If the energy provided by the collision is less than the activation energy, the molecules simply bounce apart and no reaction occurs. The features that lead to a successful reaction are summarized next.

FIGURE 13.3 The activation energy is the energy needed to convert the colliding molecules into product.

Q What happens in a collision of reacting molecules that have the proper orientation, but not the energy of activation?

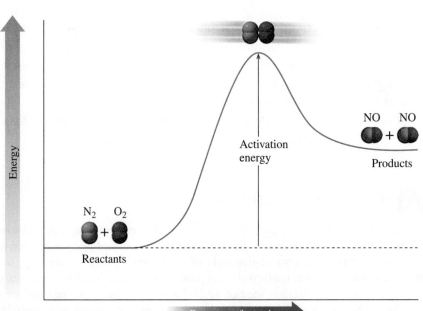

Three Conditions Required for a Reaction to Occur

1. **Collision**   The reactants must collide.
2. **Orientation**   The reactants must align properly to break and form bonds.
3. **Energy**   The collision must provide the energy of activation.

Reaction Rates

The **rate** (or speed) **of reaction** is determined by measuring the amount of a reactant used up, or the amount of a product formed, in a certain period of time.

$$\text{Rate of reaction} = \frac{\text{Change in concentration}}{\text{Change in time}}$$

We can describe the rate of reaction with the analogy of eating a pizza. When we start to eat, we have a whole pizza. As time goes by, there are fewer slices of pizza left. If we know how long it took to eat the pizza, we could determine the rate at which the pizza was consumed. Let's assume 4 slices are eaten every 8 minutes. That gives a rate of $\frac{1}{2}$ slice per minute. After 16 minutes, all 8 slices are gone.

Rate at Which Pizza Slices Are Eaten

Slices Eaten	0	4 slices	6 slices	8 slices
Time (min)	0	8 min	12 min	16 min

$$\text{Rate} = \frac{4 \text{ slices}}{8 \text{ min}} = \frac{1 \text{ slice}}{2 \text{ min}} = \frac{\frac{1}{2} \text{ slice}}{1 \text{ min}}$$

Factors That Affect the Rate of a Reaction

Some reactions go very fast, while others are very slow. For any reaction, the rate is affected by changes in temperature, changes in the concentration of the reactants, and the addition of catalysts.

Temperature

At higher temperatures, the increase in kinetic energy makes the reacting molecules move faster. As a result, more collisions occur and more colliding molecules have sufficient energy to react and form products. If we want food to cook faster, we use more heat to raise the temperature. When body temperature rises, there is an increase in the pulse rate, rate of breathing, and metabolic rate. On the other hand, we slow down reactions by lowering the temperature. We refrigerate perishable foods to retard spoilage and make them last longer. For some injuries, we apply ice to lessen the bruising process.

Concentrations of Reactants

The rate of a reaction increases when the concentrations of the reactants increases. When there are more reacting molecules, more collisions can occur, and the reaction goes faster (see Fig. 13.4). For example, a person having difficulty breathing may be given oxygen. The increase in the number of oxygen molecules in the lungs increases the rate at which oxygen combines with hemoglobin and helps the person breathe more easily.

TUTORIAL
Factors That Affect Rate

Reactants in reaction flask	Possible collisions
● ⟶ ●	1
● ⟋ ●	2
● ⤬ ●	4

FIGURE 13.4 Increasing the concentration of a reactant increases the number of collisions that are possible.

Q How many collisions are possible if one more red reactant is added?

Catalysts

Another way to speed up a reaction is to lower the energy of activation. We saw that the energy of activation is the energy needed to break apart the bonds of the reacting molecules. If a collision provides less than the activation energy, the bonds do not break and the molecules bounce apart. A **catalyst** speeds up a reaction by providing an alternative pathway that has a lower energy of activation. When activation energy is lowered, more collisions provide sufficient energy for reactants to form product. During a reaction, a catalyst is not changed or used up.

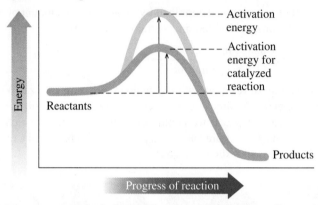

Catalysts have many uses in industry. In the manufacturing of margarine, hydrogen (H_2) is added to vegetable oils. Normally, the reaction is very slow because it has a high activation energy. However, when platinum (Pt) is used as a catalyst, the reaction occurs rapidly. In the body, biocatalysts called enzymes make most metabolic reactions proceed at rates necessary for proper cellular activity. A summary of the factors affecting reaction rates is given in Table 13.1.

TABLE 13.1 Factors That Increase Reaction Rate

Factor	Reason
Increasing reactant concentration	More collisions
Increasing temperature	More collisions, more collisions with energy of activation
Adding a catalyst	Lowers energy of activation

CONCEPT CHECK 13.1

■ **Rate of Reactions**

Will a decrease in the concentration of a reactant cause an increase or decrease in the rate of a reaction? Explain.

ANSWER

When the concentration of a reactant is decreased, there are fewer collisions between the reactant molecules, which slow or decrease the rate of reaction.

SAMPLE PROBLEM | 13.1

■ **Factors That Affect the Rate of Reaction**

Indicate whether the following changes will increase, decrease, or have no effect upon the rate of reaction:

a. increasing the temperature   **b.** adding a catalyst

SOLUTION

a. A higher temperature increases the kinetic energy of the particles, which increases the number of collisions and makes more collisions effective, causing an increase in the rate of reaction.

b. Adding a catalyst increases the rate of reaction by lowering the activation energy, which increases the number of collisions that form product.

STUDY CHECK

How does the lowering of temperature affect the rate of reaction?

CHEMISTRY AND THE ENVIRONMENT

Catalytic Converters

For over 20 years, manufacturers have been required to include catalytic converters on gasoline automobile engines. When gasoline burns, the products found in the exhaust of a car contain high levels of pollutants. These include carbon monoxide (CO) from incomplete combustion, hydrocarbons such as C_8H_{18} (octane) from unburned fuel, and nitrogen oxide (NO) from the reaction of N_2 and O_2 at the high temperatures reached within the engine. Carbon monoxide is toxic, and nitrogen oxide is involved in the formation of smog and acid rain.

The purpose of a catalytic converter is to lower the activation energy for reactions that convert each of these pollutants into substances such as CO_2, N_2, O_2, and H_2O, which are already present in the atmosphere.

$$2CO(g) + O_2(g) \longrightarrow 2CO_2(g)$$
$$2C_8H_{18}(g) + 25O_2(g) \longrightarrow 16CO_2(g) + 18H_2O(g)$$
$$2NO(g) \longrightarrow N_2(g) + O_2(g)$$

A catalytic converter consists of solid-particle catalysts, such as platinum (Pt) and palladium (Pd), on a ceramic honeycomb that provides a large surface area and facilitates contact with pollutants. As the pollutants pass through the converter, they react with the catalysts. Today, we all use unleaded gasoline because lead interferes with the ability of the Pt and Pd catalysts in the converter to react with the pollutants.

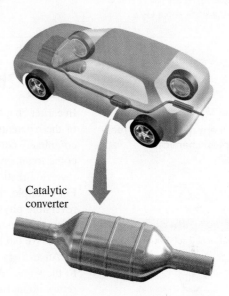

Catalytic converter

$$2NO(g) \longrightarrow N_2(g) + O_2(g)$$

NO absorbed on catalyst

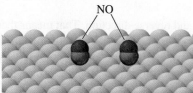

Surface of metal (Pt, Pd) catalyst

NO dissociates

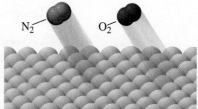

$$2CO(g) + O_2(g) \longrightarrow 2CO_2(g)$$

CO and O_2 absorbed on catalyst

Surface of metal (Pt, Pd) catalyst

O_2 dissociates

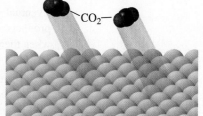

QUESTIONS AND PROBLEMS

Rates of Reactions

13.1 **a.** What is meant by the rate of a reaction?
b. Why does bread grow mold more quickly at room temperature than in the refrigerator?

13.2 **a.** How does a catalyst affect the activation energy?
b. Why is pure oxygen used in respiratory distress?

13.3 In the following reaction, what happens to the number of collisions when more $Br_2(g)$ molecules are added?

$$H_2(g) + Br_2(g) \longrightarrow 2HBr(g)$$

13.4 In the following reaction, what happens to the number of collisions when the temperature of the reaction is decreased?

$$2H_2(g) + CO(g) \longrightarrow CH_3OH(g)$$

13.5 How would each of the following change the rate of the reaction shown here?

$$2SO_2(g) + O_2(g) \longrightarrow 2SO_3(g)$$

a. adding $SO_2(g)$
b. raising the temperature
c. adding a catalyst
d. removing some $O_2(g)$

13.6 How would each of the following change the rate of the reaction shown here?

$$2NO(g) + 2H_2(g) \longrightarrow N_2(g) + 2H_2O(g)$$

a. adding $NO(g)$
b. lowering the temperature
c. removing some $H_2(g)$
d. adding a catalyst

13.2 Chemical Equilibrium

LEARNING GOAL

Use the concept of reversible reactions to explain chemical equilibrium.

TUTORIAL
Chemical Equilibrium

SELF STUDY ACTIVITY
Equilibrium

In earlier chapters, we considered the *forward reaction* in an equation and assumed that all of the reactants were converted to products. However, most of the time reactants are not completely converted to products because a *reverse reaction* takes place in which products come together and form the reactants. When a reaction proceeds in both a forward and reverse direction, it is said to be reversible. We have looked at other reversible processes. For example, the melting of solids to form liquids and the freezing of liquids to solids is a reversible physical change. Even in our daily life we have reversible events. We go from home to school and we return from school to home. We go up an escalator and come back down. We put money in our bank account and take money out.

An analogy for a forward and reverse reaction can be found in the phrase "We are going to the store." Although we mention our trip in one direction, we know that we will also return home from the store. Because our trip has both a forward and reverse direction, we can say the trip is reversible. It is not very likely that we would stay at the store forever.

Reversible Chemical Reactions

A **reversible reaction** proceeds in both the forward and reverse direction. That means there are two reaction rates: one is the rate of the forward reaction and the other is the rate of the reverse reaction. When molecules begin to react, the rate of the forward reaction is faster than the rate of the reverse reaction. As reactants are consumed and products accumulate, the rate of the forward reaction decreases and the rate of the reverse reaction increases.

Equilibrium

Eventually, the rates of the forward and reverse reactions become equal; the reactants form products as often as the products form reactants. A reaction reaches **chemical equilibrium** when there is no further change in the concentrations of the reactants and products.

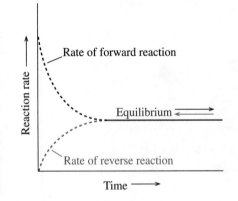

At Equilibrium:
The rate of the forward reaction is equal to the rate of the reverse reaction.
No further changes occur in the concentrations of reactants and products, even though the two reactions continue at equal but opposite rates.

Let us look at the process as the reaction of H_2 and I_2 proceeds to equilibrium. Initially, only H_2 and I_2 are present. Soon, a few molecules of HI are produced by the forward reaction. With more time, additional HI molecules are produced. As the concentration of HI increases, more HI molecules collide and react in the reverse direction (see Fig. 13.5).

Forward reaction: $H_2(g) + I_2(g) \longrightarrow 2HI(g)$

Reverse reaction: $2HI(g) \longrightarrow H_2(g) + I_2(g)$

As more HI forms, the rate of the reverse reaction increases while the rate of the forward reaction decreases. Eventually, the rates become equal, which means the reaction has reached equilibrium. Even though the concentrations remain constant at equilibrium, the forward and reverse reactions continue to occur. The forward and reverse reactions are

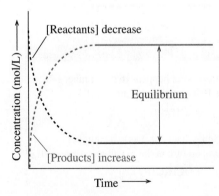

$$H_2(g) + I_2(g) \rightleftharpoons 2HI(g)$$

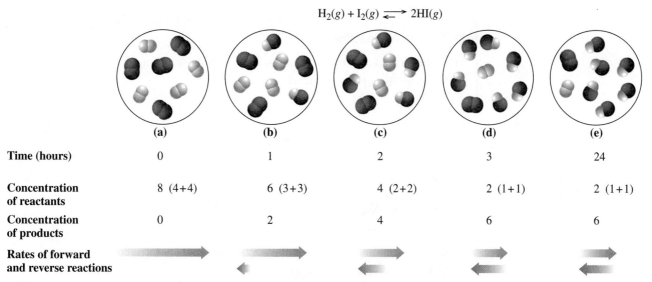

Time (hours)	0	1	2	3	24
	(a)	(b)	(c)	(d)	(e)
Concentration of reactants	8 (4+4)	6 (3+3)	4 (2+2)	2 (1+1)	2 (1+1)
Concentration of products	0	2	4	6	6
Rates of forward and reverse reactions					

FIGURE 13.5 **(a)** Initially, the reaction flask contains only the reactants H_2 (white) and I_2 (purple). **(b)** The forward reaction between H_2 and I_2 begins to produce HI. **(c)** As the reaction proceeds, there are fewer molecules of H_2 and I_2 and more molecules of HI, which increases the rate of the reverse reaction. **(d)** At equilibrium, the concentrations of reactants H_2 and I_2 and product HI are constant. **(e)** The reaction continues with the rate of the forward reaction equal to the rate of the reverse reaction.

Q How do the rates of the forward and reverse reactions compare once a chemical reaction reaches equilibrium?

usually shown together in a single equation by using a double arrow. A reversible reaction is two opposing reactions that occur at the same time.

Forward reaction

$$H_2(g) + I_2(g) \rightleftharpoons 2HI(g)$$

Reverse reaction

CONCEPT CHECK 13.2

■ **Reaction Rates and Equilibrium**

Complete each of the following with equal or not equal, faster or slower, change or do not change:

a. Before equilibrium is reached, the concentrations of the reactants and products

_____ .

b. Initially, reactants placed in a container have a _____ rate of reaction than the rate of reaction of the products.

c. At equilibrium, the rate of the forward reaction is _____ to the rate of the reverse reaction.

d. At equilibrium, the concentrations of the reactants and products _____ .

ANSWER

a. Until the rates of the forward and reverse reactions become equal, the concentrations of the reactants and products will *change.*

b. When there are only reactants in the container, the rate of the forward reaction will be *faster* than the rate of the reverse reaction.

c. Equilibrium is reached when the rates of the forward and reverse reactions become *equal.*

d. At equilibrium, when the rates of the forward and reverse reactions become equal, the concentrations of the reactants and products *do not change.*

■ **Reversible Reactions**

Write the forward and reverse reactions for each of the following:

a. $N_2(g) + 3H_2(g) \rightleftharpoons 2NH_3(g)$
b. $2CO(g) + O_2(g) \rightleftharpoons 2CO_2(g)$

SOLUTION

The equations are separated into forward and reverse reactions.

a. Forward reaction: $N_2(g) + 3H_2(g) \longrightarrow 2NH_3(g)$
Reverse reaction: $2NH_3(g) \longrightarrow N_2(g) + 3H_2(g)$
b. Forward reaction: $2CO(g) + O_2(g) \longrightarrow 2CO_2(g)$
Reverse reaction: $2CO_2(g) \longrightarrow O_2(g) + 2CO(g)$

STUDY CHECK

Write the equilibrium equation for the reaction that contains the following reverse reaction:

$$2HBr(g) \longrightarrow H_2(g) + Br_2(g)$$

QUESTIONS AND PROBLEMS

Chemical Equilibrium

13.7 What is meant by the term "reversible reaction"?

13.8 When does a reversible reaction reach equilibrium?

13.9 Which of the following processes are reversible?
a. breaking a glass   **b.** melting ice
c. heating a pan

13.10 Which of the following processes are at equilibrium?
a. Opposing rates of reaction are equal.
b. The rate of the forward reaction is faster than the rate of the reverse reaction.
c. Concentrations of reactants and products do not change.

13.3 Equilibrium Constants

LEARNING GOAL

Calculate the equilibrium constant for a reversible reaction given the concentrations of reactants and products at equilibrium.

TUTORIAL

Equilibrium Constant

When the number of people skiing becomes constant, we say the system is at equilibrium.

At equilibrium, reactions occur in opposite directions at the same rate, which means the concentrations of the reactants and products remain constant. We can use a ski lift as an analogy. Early in the morning, skiers at the bottom of the mountain begin to ride the ski lift up to the slopes. After the skiers reach the top of the mountain, they ski down. Eventually, the number of people riding up the ski lift becomes equal to the number of people skiing down the mountain. There is no further change in the number of skiers on the slopes; the system is at equilibrium.

Equilibrium Constant Expression

Because the concentrations in a reaction at equilibrium no longer change, they can be used to set up a relationship between the products and the reactants. Suppose we write a general equation for reactants A and B that form products C and D. The small italic letters are the coefficients in the balanced equation.

$$aA + bB \rightleftharpoons cC + dD$$

An **equilibrium constant expression** for the reaction multiplies the concentrations of the products together and divides by the concentrations of the reactants. Each concentration is raised to a power that is equal to its coefficient in the balanced chemical reaction. A square bracket is drawn around each substance to indicate the concentration in moles per liter (M). The **equilibrium constant**, K_c, is the numerical value obtained by substituting molar concentrations at equilibrium into the equilibrium constant expression. For our general reaction, this is written as:

Equilibrium constant   Equilibrium constant expression

$$K_c = \frac{\text{Products}}{\text{Reactants}} = \frac{[C]^c [D]^d}{[A]^a [B]^b} \quad \text{Coefficients}$$

We can write the equilibrium constant expression for the reaction of H_2 and I_2 using the balanced equation

$$H_2(g) + I_2(g) \rightleftharpoons 2HI(g)$$

STEP 1 Write the balanced equilibrium equation.

$$H_2(g) + I_2(g) \rightleftharpoons 2HI(g)$$

STEP 2 Write the products in brackets as the numerator and reactants in brackets as the denominator.

$$\frac{\text{Products} \longrightarrow}{\text{Reactants} \longrightarrow} \quad \frac{[HI]}{[H_2][I_2]}$$

STEP 3 Write the coefficient of each substance as an exponent.

$$K_c = \frac{[HI]^2}{[H_2][I_2]}$$

CONCEPT CHECK 13.3

■ **Writing an Equilibrium Constant Expression**

Select the correctly written equilibrium constant expression for the following reaction, and explain your choice:

$$CH_4(g) + H_2O(g) \rightleftharpoons CO(g) + 3H_2(g)$$

a. $K_c = \dfrac{[CO][3H_2]}{[CH_4][H_2O]}$ 　　　　　**b.** $K_c = \dfrac{[CO][H_2]^3}{[CH_4][H_2O]}$

c. $K_c = \dfrac{[CH_4][H_2O]}{[CO][H_2]^3}$ 　　　　　**d.** $K_c = \dfrac{[CO][H_2]}{[CH_4][H_2O]}$

ANSWER

The correctly written equilibrium constant expression is **b**. The products are written in the numerator and the reactants are written in the denominator. Because H_2 has a coefficient of 3 in the balanced equation, an exponent of 3 is used with the concentration of H_2.

SAMPLE PROBLEM 13.3

■ **Writing Equilibrium Constant Expressions**

Write the equilibrium constant expression for the following reaction:

$$2SO_2(g) + O_2(g) \rightleftharpoons 2SO_3(g)$$

SOLUTION

STEP 1 Write the balanced equilibrium equation.

$$2SO_2(g) + O_2(g) \rightleftharpoons 2SO_3(g)$$

STEP 2 Write the products in brackets as the numerator and reactants in brackets as the denominator.

$$\frac{\text{Products} \longrightarrow}{\text{Reactants} \longrightarrow} \quad \frac{[SO_3]}{[SO_2][O_2]}$$

STEP 3 Write the coefficient of each substance as an exponent.

$$K_c = \frac{[SO_3]^2}{[SO_2]^2[O_2]}$$

STUDY CHECK

Write the balanced chemical equation that would give the following equilibrium constant expression:

$$K_c = \frac{[NO_2]^2}{[NO]^2[O_2]}$$

Guide to Writing the K_c Expression

STEP 1
Write the balanced equilibrium equation.

STEP 2
Write the products in brackets as the numerator and reactants in brackets as the denominator. Do not include pure solids or liquids.

STEP 3
Write the coefficient of each substance as an exponent.

$$CaCO_3(s) \rightleftharpoons CaO(s) + CO_2(g)$$

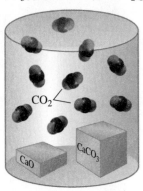

$T = 800\ °C$

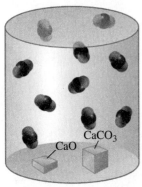

$T = 800\ °C$

FIGURE 13.6 At equilibrium at constant temperature, the concentration of CO_2 is the same regardless of the amounts of $CaCO_3(s)$ and $CaO(s)$ in the container.

Q Why are the concentrations of $CaO(s)$ and $CaCO_3(s)$ not included in K_c for the decomposition of $CaCO_3$?

TUTORIAL

Equilibrium Constant

Heterogeneous Equilibrium

Up to now, our examples have been reactions that involve only gases. A reaction in which all the reactants and products are in the same state reaches **homogeneous equilibrium**. When the reactants and products are in two or more states, the equilibrium is termed a **heterogeneous equilibrium**. For example, in the following reaction, the decomposition of calcium carbonate reaches heterogeneous equilibrium with calcium oxide and carbon dioxide (see Fig. 13.6).

$$CaCO_3(s) \rightleftharpoons CaO(s) + CO_2(g)$$

In contrast to gases, the concentrations of pure solids and pure liquids in a heterogeneous equilibrium are constant; they do not change. Therefore, pure solids and liquids are not included in the equilibrium constant expression. For this heterogeneous equilibrium, the K_c expression does not include the concentration of $CaCO_3(s)$ or $CaO(s)$. It is written as $K_c = [CO_2]$.

CONCEPT CHECK 13.4

■ Equilibrium Constant for a Heterogeneous Equilibrium

Select the correctly written equilibrium constant expression for the following reaction:

$$CO(g) + 2H_2(g) \rightleftharpoons CH_3OH(l)$$

a. $\dfrac{[CH_3OH]}{[CO][H_2]^2}$ 　　　　　　　**b.** $[CO][H_2]^2$

c. $\dfrac{1}{[CO][H_2]^2}$

ANSWER

In a heterogeneous equilibrium, only the gases are included in the equilibrium constant expression. The concentration of the liquid CH_3OH remains constant; it is not included in the equilibrium constant expression. Therefore, the correct answer **c** gives the concentrations of the gases, CO and H_2, in the denominator.

SAMPLE PROBLEM　13.4

■ Heterogeneous Equilibrium Constant Expression

Write the equilibrium constant expression for each of the following reactions at equilibrium:

a. $Si(s) + 2Cl_2(g) \rightleftharpoons SiCl_4(g)$　　　　**b.** $2Mg(s) + O_2(g) \rightleftharpoons 2MgO(s)$

SOLUTION

In the equilibrium constant expressions for heterogeneous reactions, the concentrations of the pure solids are not included.

a. $K_c = \dfrac{[SiCl_4]}{[Cl_2]^2}$　　　　　　**b.** $K_c = \dfrac{1}{[O_2]}$

STUDY CHECK

Solid iron(II) oxide and carbon monoxide gas are in equilibrium with solid iron and carbon dioxide gas. Write the equilibrium equation and the equilibrium constant expression for the reaction.

Calculating Equilibrium Constants

The numerical value of the equilibrium constant is calculated from the equilibrium constant expression by substituting experimentally measured concentrations of the reactants and products at equilibrium into the equilibrium constant expression. For example, the equilibrium constant expression for the reaction of H_2 and I_2 is written

$$H_2(g) + I_2(g) \rightleftharpoons 2HI(g) \qquad K_c = \dfrac{[HI]^2}{[H_2][I_2]}$$

In the first experiment, the molar concentrations for the reactants and products at equilibrium are found to be $[H_2] = 0.10$ M, $[I_2] = 0.20$ M, and $[HI] = 1.04$ M. When we substitute these values into the equilibrium constant expression, we obtain the numerical value of the equilibrium constant.

Reactants

$[H_2] = 0.10$ M

$[I_2] = 0.20$ M

Products

$[HI] = 1.04$ M

$$K_c = \frac{[HI]^2}{[H_2][I_2]} = \frac{[1.04]^2}{[0.10][0.20]} = 54$$

In experiments 2 and 3, we can see different equilibrium concentrations of reactants and products for the H_2, I_2, and HI system at equilibrium at the same temperature. When the concentrations of reactants and products are measured in each equilibrium sample and used to calculate K_c for the reaction, the same value of K_c is obtained (see Table 13.2). Thus, a reaction at a specific temperature can have only one value for the equilibrium constant. Although all of the concentrations in experiments 1, 2, and 3 are in moles/liter (M), we do not attach units to the value of K_c.

TABLE 13.2 Equilibrium Constant for $H_2(g) + I_2(g) \rightleftharpoons 2HI(g)$ at 427 °C

Experiment	$[H_2]$	$[I_2]$	$[HI]$	$K_c = \dfrac{[HI]^2}{[H_2][I_2]}$
1	0.10 M	0.20 M	1.04 M	54
2	0.20 M	0.20 M	1.47 M	54
3	0.30 M	0.17 M	1.66 M	54

SAMPLE PROBLEM | 13.5

■ Calculating an Equilibrium Constant

The decomposition of dinitrogen tetroxide forms nitrogen dioxide.

$$N_2O_4(g) \rightleftharpoons 2NO_2(g)$$

What is the value of K_c at 100 °C if a reaction mixture at equilibrium contains $[N_2O_4] = 0.45$ M and $[NO_2] = 0.31$ M?

SOLUTION

Given reactant: $[N_2O_4] = 0.45$ M product: $[NO_2] = 0.31$ M
Need K_c value

STEP 1 Write the K_c expression for the equilibrium. Use the coefficients in the balanced equation as exponents of the molar concentrations:

$$K_c = \frac{[NO_2]^2}{[N_2O_4]}$$

STEP 2 Substitute equilibrium (molar) concentrations and calculate K_c.

$$K_c = \frac{[0.31]^2}{[0.45]} = 0.21$$

Guide to Calculating the K_c Value

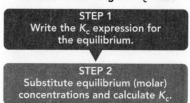

STEP 1
Write the K_c expression for the equilibrium.

STEP 2
Substitute equilibrium (molar) concentrations and calculate K_c.

STUDY CHECK

Ammonia decomposes when heated to give hydrogen and nitrogen:

$$2NH_3(g) \rightleftharpoons 3H_2(g) + N_2(g)$$

Calculate the equilibrium constant if an equilibrium mixture contains $[NH_3] = 0.040$ M, $[N_2] = 0.20$ M, and $[H_2] = 0.60$ M.

QUESTIONS AND PROBLEMS

Equilibrium Constants

13.11 Write the equilibrium constant expression, K_c, for each of the following reactions:
   a. $CH_4(g) + 2H_2S(g) \rightleftharpoons CS_2(g) + 4H_2(g)$
   b. $2NO(g) \rightleftharpoons N_2(g) + O_2(g)$
   c. $2SO_3(g) + CO_2(g) \rightleftharpoons CS_2(g) + 4O_2(g)$

13.12 Write the equilibrium constant expression, K_c, for each of the following reactions:
   a. $2HBr(g) \rightleftharpoons H_2(g) + Br_2(g)$
   b. $2BrNO(g) \rightleftharpoons Br_2(g) + 2NO(g)$
   c. $CH_4(g) + H_2O(g) \rightleftharpoons CO(g) + 3H_2(g)$

13.13 Identify each of the following as a homogeneous or heterogeneous equilibrium:
   a. $2O_3(g) \rightleftharpoons 3O_2(g)$
   b. $2NaHCO_3(s) \rightleftharpoons Na_2CO_3(s) + CO_2(g) + H_2O(g)$
   c. $CH_4(g) + H_2O(g) \rightleftharpoons 3H_2(g) + CO(g)$
   d. $4HCl(g) + O_2(g) \rightleftharpoons 2H_2O(l) + 2Cl_2(g)$

13.14 Identify each of the following as a homogeneous or heterogeneous equilibrium:
   a. $CO(g) + H_2(g) \rightleftharpoons C(s) + H_2O(g)$
   b. $NH_4Cl(s) \rightleftharpoons NH_3(g) + HCl(g)$
   c. $CS_2(g) + 4H_2(g) \rightleftharpoons CH_4(g) + 2H_2S(g)$
   d. $Br_2(g) + Cl_2(g) \rightleftharpoons 2BrCl(g)$

13.15 Write the equilibrium constant expression for each of the reactions in Problem 13.13.

13.16 Write the equilibrium constant expression for each of the reactions in Problem 13.14.

13.17 What is the value of K_c for the following reaction at equilibrium

$$N_2O_4(g) \rightleftharpoons 2NO_2(g)$$

if $[NO_2] = 0.21$ M and $[N_2O_4] = 0.030$ M?

13.18 What is the value of K_c for the following reaction at equilibrium

$$CO_2(g) + H_2(g) \rightleftharpoons CO(g) + H_2O(g)$$

if $[CO] = 0.20$ M, $[H_2O] = 0.30$ M, $[CO_2] = 0.30$ M, and $[H_2] = 0.033$ M?

13.19 What is the value of K_c for the following reaction at equilibrium

$$CO(g) + 3H_2(g) \rightleftharpoons CH_4(g) + H_2O(g)$$

if $[CH_4] = 1.8$ M, $[H_2O] = 2.0$ M, $[CO] = 0.51$ M, and $[H_2] = 0.30$ M?

13.20 What is the value of K_c for the following reaction at equilibrium

$$N_2(g) + 3H_2(g) \rightleftharpoons 2NH_3(g)$$

if $[NH_3] = 2.2$ M, $[N_2] = 0.44$ M, and $[H_2] = 0.40$ M?

13.4 Using Equilibrium Constants

LEARNING GOAL

Use an equilibrium constant to predict the extent of reaction and to calculate equilibrium concentrations.

TUTORIAL

Calculations Using the Equilibrium Constant

We have seen that the values of K_c can be large or small. We can now look at the K_c values to predict how far the reaction proceeds to products at equilibrium. When K_c is large, the numerator (products) is greater than the concentrations of the reactants in the denominator.

$$\frac{[Products]}{[Reactants]} = \text{Large } K_c$$

When K_c is small, the numerator is smaller than the denominator, which means that the reaction favors the reactants.

$$\frac{[Products]}{[Reactants]} = \text{Small } K_c$$

Using a general reaction and its equilibrium constant expression, we can look at the relative concentrations of reactant A and product B.

$$A(g) \rightleftharpoons B(g) \qquad K_c = \frac{[B]}{[A]}$$

When the K_c value is large, [B] is greater than [A]. For example, if the K_c is 1×10^3, or 1000, [B] would be 1000 times greater than [A] at equilibrium.

$$K_c = \frac{[B]}{[A]} = 1000 \quad \text{or rearranged} \quad [B] = 1000\,[A]$$

When the K_c value is small, [A] is greater than [B]. For example, if the K_c is 1×10^{-2}, [A] is 100 times greater than [B] at equilibrium (see Fig. 13.7).

$$K_c = \frac{[B]}{[A]} = \frac{1}{100} \quad \text{or rearranged} \quad [A] = 100\,[B]$$

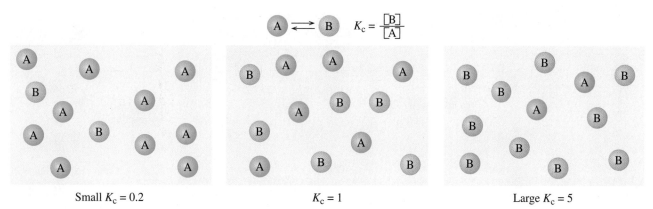

Small $K_c = 0.2$ $\qquad\qquad\qquad$ $K_c = 1$ $\qquad\qquad\qquad$ Large $K_c = 5$

FIGURE 13.7 A reaction with $K_c < 1$ contains a higher concentration of reactant A than product B. A reaction with K_c of about 1 has about the same concentration of product B as reactant A. A reaction with $K_c > 1$ has a higher concentration of product B than reactant A.

Q Does a reaction in which [A] = 10 [B] at equilibrium have a K_c greater than, about equal to, or less than 1?

Equilibrium with a Large K_c

A reaction with a large K_c forms a substantial amount of product by the time equilibrium is established. The greater the value of K_c, the more the equilibrium favors the products. A reaction with a very large K_c essentially goes to completion to give mostly products. Consider the reaction of SO_2 and O_2, which has a large K_c. At equilibrium, the reaction mixture contains mostly product and very little reactants:

$$2SO_2(g) + O_2(g) \rightleftharpoons 2SO_3(g)$$

$$K_c = \frac{[SO_3]^2}{[SO_2]^2[O_2]} \quad \frac{\text{Mostly product}}{\text{Few reactants}} = 3.4 \times 10^2 \quad \text{Reaction favors product}$$

We can start the reaction with only the reactants SO_2 and O_2 or we can start the reaction with just the product SO_3 (see Fig. 13.8). In one reaction, SO_2 and O_2 form SO_3 and in the other, SO_3 reacts to form SO_2 and O_2. However, in both equilibrium mixtures, the concentration of SO_3 is much higher than the concentrations of SO_2 and O_2 (see Fig. 13.9). Because there is more product than reactants at equilibrium, the energy of activation for the forward reaction must be lower than the energy of activation for the reverse reaction.

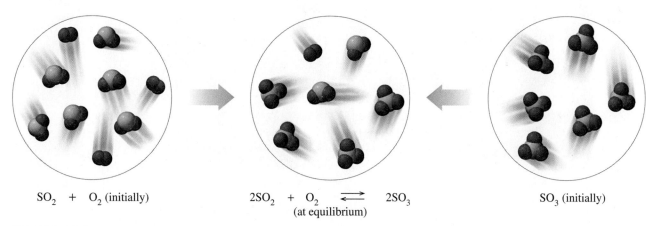

SO$_2$ + O$_2$ (initially) $\qquad\qquad$ 2SO$_2$ + O$_2$ $\rightleftharpoons$ 2SO$_3$ $\qquad\qquad$ SO$_3$ (initially)
$\qquad\qquad\qquad\qquad\qquad\qquad$ (at equilibrium)

FIGURE 13.8 One sample initially contains $SO_2(g)$ and $O_2(g)$, while another sample contains only $SO_3(g)$. At equilibrium, mostly $SO_3(g)$ and only small amounts of $SO_2(g)$ and $O_2(g)$ are present in the equilibrium mixture.

Q Why is the same equilibrium mixture obtained from $SO_2(g)$ and $O_2(g)$ as from $SO_3(g)$?

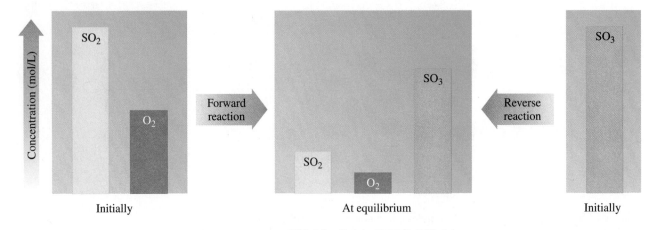

$$2SO_2(g) + O_2(g) \rightleftharpoons 2SO_3(g)$$

FIGURE 13.9 In the reaction of $SO_2(g)$ and $O_2(g)$, the equilibrium favors the formation of product $SO_3(g)$, which results in a large K_c.

Q Why is an equilibrium mixture obtained after starting with $SO_3(g)$?

Equilibrium with a Small K_c

For a reaction with a small K_c, the equilibrium mixture contains very small concentrations of products. Consider the reaction for the formation of $NO(g)$ from $N_2(g)$ and $O_2(g)$, which has a small K_c (see Fig. 13.10).

$$N_2(g) + O_2(g) \rightleftharpoons 2NO(g)$$

$$K_c = \frac{[NO]^2}{[N_2][O_2]} \quad \frac{\text{Few products}}{\text{Mostly reactants}} = 2 \times 10^{-9} \quad \text{Reaction favors reactants}$$

Thus a reaction starting with N_2 and O_2 or with only NO forms an equilibrium mixture that contains mostly N_2 and O_2, and very little NO. Because there is more reactant than product at equilibrium, the energy of activation for the forward reaction must be greater than the energy of activation for the reverse reaction. Reactions with very small K_c values produce very small quantities of products.

Reactions with equilibrium constants close to 1 have about the same concentrations of reactants and products (see Fig. 13.11). Table 13.3 lists some equilibrium constants and the extent of their reaction.

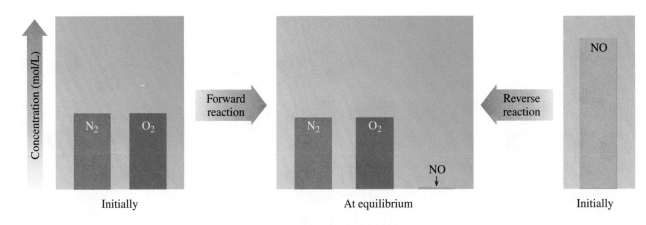

$$N_2(g) + O_2(g) \rightleftharpoons 2NO(g)$$

FIGURE 13.10 At equilibrium, the reaction $N_2(g) + O_2(g) \rightleftharpoons 2NO(g)$ favors the reactants and the reaction mixture at equilibrium contains mostly $N_2(g)$ and $O_2(g)$, which results in a small K_c.

Q Starting with only N_2 and O_2 in a closed container, how do the forward and reverse reactions change as equilibrium is reached?

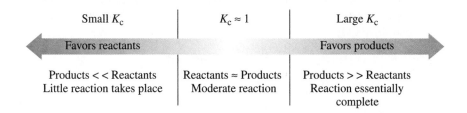

FIGURE 13.11 The equilibrium constant K_c indicates how far a reaction goes to products. A reaction with a large K_c contains mostly products; a reaction with a small K_c contains mostly reactants.

Q Does a reaction with $K_c = 1.2 \times 10^{15}$ contain mostly reactants or products at equilibrium?

TABLE 13.3 Examples of Reactions with Large and Small K_c Values

Reactants	Products	K_c	Equilibrium Favors
$2CO(g) + O_2(g) \rightleftharpoons 2CO_2(g)$		2×10^{11}	Products
$2H_2(g) + S_2(g) \rightleftharpoons 2H_2S(g)$		1.1×10^7	Products
$N_2(g) + 3H_2(g) \rightleftharpoons 2NH_3(g)$		1.6×10^2	Products
$PCl_5(g) \rightleftharpoons PCl_3(g) + Cl_2(g)$		1.2×10^{-2}	Reactants
$N_2(g) + O_2(g) \rightleftharpoons 2NO(g)$		2×10^{-9}	Reactants

---

CONCEPT CHECK 13.5

■ **Extent of Reaction**

Predict whether the equilibrium favors the reactants or products for each of the following reactions:

a. $2H_2(g) + O_2(g) \rightleftharpoons 2H_2O(g)$ $K_c = 2.9 \times 10^{82}$
b. $N_2O_4(g) \rightleftharpoons 2NO_2(g)$ $K_c = 5.9 \times 10^{-3}$

ANSWER

a. When K_c has a large value, it indicates that there are high concentrations of products in the numerator and low concentrations of reactants in the denominator. Thus, this equilibrium with a large K_c favors the products.
b. When K_c has a small value, it indicates that there are low concentrations of products in the numerator and high concentrations of reactants in the denominator. Thus, this equilibrium with a small K_c favors the reactants.

Calculating Concentrations at Equilibrium

When a reaction goes essentially all to products, we can use the mole–mole factors we studied earlier to calculate the quantity of a product. However, many reactions reach equilibrium without using up all the reactants. If this is the case, then we need to use the equilibrium constant to calculate the amount of a product that is formed in the reaction. For example, if we know the equilibrium constant for a reaction and all the concentrations except one, we can calculate the unknown concentration using the equilibrium constant expression.

SAMPLE PROBLEM 13.6

■ **Calculating Concentration Using an Equilibrium Constant**

For the reaction of carbon dioxide and hydrogen, the equilibrium concentrations are $[CO_2] = 0.25$ M, $[H_2] = 0.80$ M, and $[H_2O] = 0.50$ M. What is the equilibrium concentration of $CO(g)$?

$$CO_2(g) + H_2(g) \rightleftharpoons CO(g) + H_2O(g) \qquad K_c = 0.11$$

SOLUTION

STEP 1 Write the K_c expression for the equilibrium equation. From the equation, the equilibrium constant expression is written as:

$$K_c = \frac{[CO][H_2O]}{[CO_2][H_2]}$$

Guide to Using K_c Value

STEP 1
Write the K_c expression for the equilibrium equation.

STEP 2
Solve the K_c expression for the unknown concentration.

STEP 3
Substitute the known values into the rearranged K_c expression.

STEP 4
Check answer by using the calculated concentration in the K_c expression.

STEP 2 Solve the K_c expression for the unknown concentration. Rearrange K_c to solve for [CO] by multiplying both sides by $[CO_2][H_2]$ and then dividing both sides by $[H_2O]$:

$$K_c \times [CO_2][H_2] = \frac{[CO][H_2O]}{[CO_2][H_2]} \times [CO_2][H_2]$$

$$K_c [CO_2][H_2] = [CO][H_2O]$$

$$K_c \frac{[CO_2][H_2]}{[H_2O]} = \frac{[CO][H_2O]}{[H_2O]}$$

$$[CO] = K_c \frac{[CO_2][H_2]}{[H_2O]}$$

STEP 3 Substitute the known values into the rearranged K_c expression. To calculate the [CO], substitute K_c and the concentrations given at equilibrium:

$$[CO] = K_c \frac{[CO_2][H_2]}{[H_2O]} = 0.11 \frac{[0.25][0.80]}{[0.50]}$$

$$[CO] = 0.044 \text{ M}$$

STEP 4 Check answer by using the calculated concentration in the K_c expression.

$$K_c = \frac{[CO][H_2O]}{[CO_2][H_2]} = \frac{[0.044][0.50]}{[0.25][0.80]}$$

$$K_c = 0.11$$

STUDY CHECK

When the alkene ethene (C_2H_4) reacts with water vapor, ethanol, an alcohol, is produced. At 327 °C, the K_c is 9.0×10^3.

$$C_2H_4(g) + H_2O(g) \rightleftharpoons C_2H_5OH(g)$$

If an equilibrium mixture has concentrations of $[C_2H_4] = 0.020$ M and $[H_2O] = 0.015$ M, what is the equilibrium concentration of C_2H_5OH?

QUESTIONS AND PROBLEMS

Using Equilibrium Constants

13.21 Indicate whether each of the following equilibrium mixtures contains mostly products or mostly reactants:
   a. $Cl_2(g) + NO(g) \rightleftharpoons 2NOCl(g)$ $K_c = 3.7 \times 10^8$
   b. $2H_2(g) + S_2(g) \rightleftharpoons 2H_2S(g)$ $K_c = 1.1 \times 10^7$
   c. $3O_2(g) \rightleftharpoons 2O_3(g)$ $K_c = 1.7 \times 10^{-56}$

13.22 Indicate whether each of the following equilibrium mixtures contains mostly products or mostly reactants:
   a. $CO(g) + Cl_2(g) \rightleftharpoons COCl_2(g)$ $K_c = 5.0 \times 10^{-9}$
   b. $2HF(g) \rightleftharpoons H_2(g) + F_2(g)$ $K_c = 1.0 \times 10^{-95}$
   c. $2NO(g) + O_2(g) \rightleftharpoons 2NO_2(g)$ $K_c = 6.0 \times 10^{13}$

13.23 The equilibrium constant, K_c, for this reaction is 54.

$$H_2(g) + I_2(g) \rightleftharpoons 2HI(g)$$

If the equilibrium mixture contains 0.030 M HI and 0.015 M I_2, what is $[H_2]$?

13.24 The equilibrium constant, K_c, for this reaction is 4.6×10^{-3}.

$$N_2O_4(g) \rightleftharpoons 2NO_2(g)$$

If the equilibrium mixture contains 0.050 M NO_2, what is $[N_2O_4]$?

13.25 The K_c at 100 °C is 2.0 for the reaction

$$2NOBr(g) \rightleftharpoons 2NO(g) + Br_2(g)$$

If the system at equilibrium contains $[NO] = 2.0$ M and $[Br_2] = 1.0$ M, what is $[NOBr]$?

13.26 An equilibrium mixture at 225 °C contains 0.14 M NH_3 and 0.18 M H_2 for the reaction

$$3H_2(g) + N_2(g) \rightleftharpoons 2NH_3(g)$$

If the K_c at this temperature is 1.7×10^2, what is $[N_2]$?

13.5 Changing Equilibrium Conditions: Le Châtelier's Principle

LEARNING GOAL

Use Le Châtelier's principle to describe the changes made in equilibrium concentrations when reaction conditions change.

We have seen that when a reaction reaches equilibrium, the rates of the forward and reverse reactions are equal and the concentrations remain constant. Now we will look at what happens to a system at equilibrium when changes occur in reaction conditions, such as changes in temperature, concentration, and pressure.

Le Châtelier's Principle

TUTORIAL
Le Châtelier's Principle

In the previous section, we saw that a system at equilibrium consists of forward and reverse reactions occurring at equal rates. Thus, at equilibrium, the concentrations of the substances do not change. However, any changes that occur in the reaction conditions will disturb the equilibrium. Concentrations can be changed by adding or removing one of the substances, the volume (pressure) can be changed, or the temperature can be changed. When we alter any of the conditions of a system at equilibrium, the rates of the forward and reverse reactions will no longer be equal. We say that a *stress* is placed on the equilibrium. We use Le Châtelier's principle to determine the direction that the equilibrium must shift to relieve that stress and reestablish equilibrium.

> **Le Châtelier's Principle**
> When a stress (change in conditions) is placed on a reaction at equilibrium, the equilibrium will shift in the direction that relieves the stress.

Effect of Concentration Changes

We will use the equilibrium for the reaction of PCl_5 to illustrate the stress caused by a change in concentration and how the system reacts to the stress. Consider the following reaction, which has a K_c of 0.042 at 250 °C:

$$PCl_5(g) \rightleftharpoons PCl_3(g) + Cl_2(g)$$

Every reaction at a given temperature has only one equilibrium constant. Even if there are changes in the concentrations of the components, the K_c value does not change. What will change are the concentrations of the other components in the reaction in order to relieve the stress. For example, we can see that an equilibrium mixture that contains 1.20 M PCl_5, 0.20 M PCl_3, and 0.25 M Cl_2 has K_c of 0.042.

$$K_c = \frac{[PCl_3][Cl_2]}{[PCl_5]} = \frac{[0.20][0.25]}{[1.20]} = 0.042$$

Suppose that now we add PCl_5 to the equilibrium mixture to increase $[PCl_5]$ to 2.00 M. If we substitute the concentrations into the equilibrium expression at this point, the ratio of products to reactants is 0.025, which is smaller than K_c of 0.042.

$$\frac{\text{Products}}{\text{Reactants (added)}} \quad \frac{[PCl_3][Cl_2]}{[PCl_5]} = \frac{[0.20][0.25]}{[2.00]} = 0.025 < K_c$$

Add PCl_5

$$PCl_5(g) \rightleftharpoons PCl_3(g) + Cl_2(g)$$

Because K_c does not change for a reaction at a given temperature, adding more PCl_5 places a stress on the system (see Fig. 13.12). Forming more of the products can relieve this stress. According to Le Châtelier's principle, adding reactants causes the equilibrium to *shift* toward the products.

In our experiment, equilibrium is reestablished with new concentrations of $[PCl_5] = 1.94$ M, $[PCl_3] = 0.26$ M, and $[Cl_2] = 0.31$ M. The resulting equilibrium mixture now contains more reactants and products than initially, but the new concentrations in the equilibrium expression are once again equal to K_c:

$$K_c = \frac{[PCl_3][Cl_2]}{[PCl_5]} = \frac{[0.26][0.31]}{[1.94]} = 0.042 = K_c \quad \text{New higher concentrations}$$

Suppose that in another experiment some PCl_5 is removed from the original equilibrium mixture, which lowers $[PCl_5]$ to 0.76 M. Now the ratio of the products to the reactants is greater than the K_c value of 0.042. The removal of some of the reactant has placed a stress on the equilibrium.

$$\frac{\text{Products}}{\text{Reactant (removed)}} \quad \frac{[PCl_3][Cl_2]}{[PCl_5]} = \frac{[0.20][0.25]}{[0.76]} = 0.066 > K_c$$

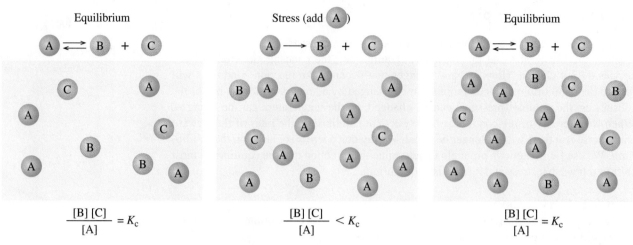

Equilibrium

Stress (add A)

Equilibrium

$$\frac{[B][C]}{[A]} = K_c \qquad \frac{[B][C]}{[A]} < K_c \qquad \frac{[B][C]}{[A]} = K_c$$

FIGURE 13.12 The addition of A places stress on the equilibrium A ⇌ B + C. To relieve the stress, the forward reaction converts some A to B and C to reestablish the equilibrium.

Q When C is added, does the equilibrium shift toward products or reactants? Why?

In this case, the stress is relieved as the reverse reaction converts some products to reactants. Using Le Châtelier's principle, we see that removing some reactant *shifts* the equilibrium toward the reactants.

> **Remove PCl₅**

$$PCl_5(g) \rightleftharpoons PCl_3(g) + Cl_2(g)$$

In this experiment, equilibrium is reestablished with new concentrations of $[PCl_5] = 0.80\,M$, $[PCl_3] = 0.16\,M$, and $[Cl_2] = 0.21\,M$. The resulting equilibrium mixture now contains lower concentrations of reactants and products, but their new concentrations in the equilibrium expression once again are equal to K_c:

$$K_c = \frac{[PCl_3][Cl_2]}{[PCl_5]} = \frac{[0.16][0.21]}{[0.80]} = 0.042 = K_c \qquad \text{New lower concentrations}$$

There can also be changes in the concentrations of other components in this reaction. We could add or remove one of the products in this reaction. Suppose the $[Cl_2]$ is doubled, which makes the product/reactant ratio greater than K_c.

$$\frac{[PCl_3][Cl_2]}{[PCl_5]} = \frac{[0.20][0.50]}{[1.20]} = 0.083 > K_c$$

With an increase in the concentration of Cl_2, the rate of the reverse reaction increases and converts some of the products to reactants. Using Le Châtelier's principle, we see that the addition of a product causes a *shift* toward the reactants.

> **Add Cl₂**

$$PCl_5(g) \rightleftharpoons PCl_3(g) + Cl_2(g)$$

On the other hand, we could remove some Cl_2, which would decrease $[Cl_2]$ and *shift* the equilibrium toward the products.

> **Remove Cl₂**

$$PCl_5(g) \rightleftharpoons PCl_3(g) + Cl_2(g)$$

In summary, Le Châtelier's principle indicates that a stress caused by adding a substance at equilibrium is relieved by shifting the reaction away from that substance. When a substance is removed, the equilibrium shifts toward that substance. These features of Le Châtelier's principle are summarized in Table 13.4.

TABLE 13.4 Effect of Concentration Changes on Equilibrium
$PCl_5(g) \rightleftharpoons PCl_3(g) + Cl_2(g)$

Stress	Shift	Equilibrium Changes		
		$PCl_5(g)$	$PCl_3(g)$	$Cl_2(g)$
Increase PCl_5	Toward products	Added	More	More
Decrease PCl_5	Toward reactants	Removed	Less	Less
Increase PCl_3	Toward reactants	More	Added	Less
Decrease PCl_3	Toward products	Less	Removed	More
Increase Cl_2	Toward reactants	More	Less	Added
Decrease Cl_2	Toward products	Less	More	Removed

Catalysts

Sometimes a catalyst is added to a reaction. Earlier we showed that a catalyst speeds up a reaction by lowering the activation energy. As a result, the rates of the forward and reverse reactions both increase. The time required to reach equilibrium is shorter, but the same ratios of products and reactants are attained. Therefore, a catalyst speeds up the forward and reverse reactions, but it has no effect on the equilibrium constant.

CONCEPT CHECK 13.6

■ Effect of Changes in Concentrations

Describe the effect of each change on the equilibrium system.

$CO(g) + H_2O(g) \rightleftharpoons CO_2(g) + H_2(g)$

a. increasing [CO]   **b.** increasing [H_2]
c. decreasing [H_2O]   **d.** decreasing [CO_2]
e. adding a catalyst

ANSWER
According to Le Châtelier's principle, when a stress is applied to a reaction at equilibrium, the equilibrium will shift in the direction that relieves that stress.
a. When the reactant [CO] increases, the rate of the forward reaction increases to shift the equilibrium toward the products.
b. When the product [H_2] increases, the rate of the reverse reaction increases to shift the equilibrium toward the reactants.
c. When the reactant [H_2O] decreases, the rate of the forward reaction decreases to shift the equilibrium toward the reactants.
d. When the product [CO_2] decreases, the rate of the reverse reaction decreases to shift the equilibrium toward the products.
e. When a catalyst is added, it changes the rates of the forward and reverse reactions equally, but does not cause a shift in the equilibrium.

Effect of Volume (Pressure) Changes on Equilibrium

Reactions that involve gases exert pressure. Although the volume and therefore pressure can change, the value of the equilibrium constant does not change at a given temperature. Using the gas laws we discussed in Chapter 11, we know that increasing the volume of the container decreases the pressure, while decreasing the volume increases the pressure.

According to Le Châtelier's principle, decreasing the number of moles of gas relieves the stress of increased pressure. This means that the reaction shifts toward the fewer number of

CHEMISTRY AND HEALTH

Oxygen–Hemoglobin Equilibrium and Hypoxia

The transport of oxygen involves an equilibrium between hemoglobin (Hb), oxygen, and oxyhemoglobin (HbO$_2$):

$$Hb(aq) + O_2(g) \rightleftharpoons HbO_2(aq)$$

When the O$_2$ level is high in the alveoli of the lung, the reaction favors the product HbO$_2$. In the tissues where O$_2$ concentration is low, the reverse reaction releases the oxygen from the hemoglobin. The equilibrium expression is written

$$K_c = \frac{[HbO_2]}{[Hb][O_2]}$$

Since the concentrations of Hb and HbO$_2$ can change, they are included in K_c. At normal atmospheric pressure, oxygen diffuses into the blood because the partial pressure of oxygen in the alveoli is higher than that in the blood. At altitudes above 8000 ft (2400 m), the decrease in the amount of oxygen in the air results in a significant reduction of oxygen to the blood and body tissues. At an altitude of 18 000 ft (5500 m), a person will obtain 29 percent less oxygen. When oxygen levels are lowered, a person may experience hypoxia, which has symptoms that include increased respiratory rate, headache, decreased mental acuteness, fatigue, decreased physical coordination, nausea, vomiting, and cyanosis. A similar problem occurs in persons with a history of lung disease that impairs gas diffusion in the alveoli or in persons who have a reduced number of red blood cells, which occurs in smokers.

From the equilibrium expression, we see that a decrease in oxygen will shift the equilibrium toward the reactants. Such a shift depletes the concentration of HbO$_2$ and causes the hypoxia condition:

$$Hb(aq) + O_2(g) \longleftarrow HbO_2(aq)$$

Immediate treatment of altitude sickness includes hydration, rest, and if necessary, descending to a lower altitude. The adaptation to

At high altitudes, a reduction in oxygen may cause hypoxia.

lowered oxygen levels requires about 10 days. During this time the bone marrow increases red blood cell production, providing more red blood cells and more hemoglobin. A person living at a high altitude can have 50 percent more red blood cells than someone at sea level. This increase in hemoglobin causes a shift in the equilibrium back toward HbO$_2$ product. Eventually, the higher concentration of HbO$_2$ will provide more oxygen to the tissues and the symptoms of hypoxia will lessen:

$$Hb(aq) + O_2(g) \longrightarrow HbO_2(aq)$$

For travellers at high altitudes, it is important to take time to acclimatize. At very high altitudes, it may be necessary to use an oxygen tank.

moles of gas. Let's look at the effect of decreasing the volume of the equilibrium mixture that originally contained 1.20 M PCl$_5$, 0.20 M PCl$_3$, and 0.25 M Cl$_2$ with K_c of 0.042:

$$PCl_5(g) \rightleftharpoons PCl_3(g) + Cl_2(g)$$

$$K_c = \frac{[PCl_3][Cl_2]}{[PCl_5]} = \frac{[0.20][0.25]}{[1.20]} = 0.042$$

If we decrease the volume by half, all the molar concentrations are doubled. In the equation there are more moles of products than reactants, so there is an increase in the product/reactant ratio.

$$\frac{[PCl_3][Cl_2]}{[PCl_5]} = \frac{[0.40][0.50]}{[2.40]} = 0.083 > K_c$$

To relieve the stress, the equilibrium shifts toward the reactant PCl$_5$, which increases the [PCl$_5$] and decreases the [PCl$_3$] and [Cl$_2$] of the products (see Fig. 13.13).

Decrease V

$$PCl_5(g) \rightleftharpoons PCl_3(g) + Cl_2(g)$$
1 mol   2 mol

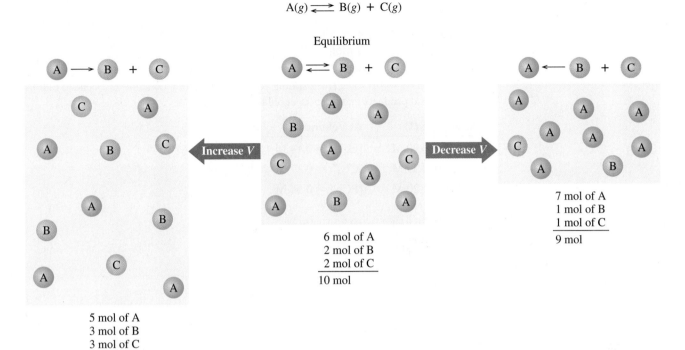

FIGURE 13.13 The decrease in the volume of the container places stress on the equilibrium A(g) ⇌ B(g) + C(g). To relieve the stress, the reverse reaction converts some products to reactant, which gives a smaller number of moles of gas, reduces pressure, and reestablishes the equilibrium. When the volume increases, the forward reaction converts reactant to products, which increases the moles of gas and relieves the stress.

Q If you want to increase the products, would you increase or decrease the volume of the reaction container?

When equilibrium is reestablished, the new concentrations are $[PCl_5] = 2.52$ M, $[PCl_3] = 0.28$ M, and $[Cl_2] = 0.38$ M. The resulting equilibrium mixture contains new concentrations of reactant and products that are now equal to the K_c value, as the following shows:

At Volume (1)	At Volume (2)
$[PCl_3] = 0.20$ M	$[PCl_3] = 0.28$ M
$[Cl_2]\ \ = 0.25$ M	$[Cl_2]\ \ = 0.38$ M
$[PCl_5] = 1.20$ M	$[PCl_5] = 2.52$ M

$$K_c = \frac{[PCl_3][Cl_2]}{[PCl_5]} = \frac{[0.20][0.25]}{[1.20]} = 0.042$$

At volume 1

$$K_c = \frac{[PCl_3][Cl_2]}{[PCl_5]} = \frac{[0.28][0.38]}{[2.52]} = 0.042$$

At volume 2

On the other hand, when volume increases and pressure decreases, the reaction shifts toward the greater number of moles of gas. Suppose that the volume is doubled. Then the molar concentrations of all the gases decrease by half. Because there are more moles of products than reactants, there is a decrease in the product/reactant ratio.

$$\frac{[PCl_3][Cl_2]}{[PCl_5]} = \frac{[0.100][0.125]}{[0.600]} = 0.021 < K_c$$

Now the equilibrium has to shift toward the products to relieve the stress; this will increase the concentrations of the products.

Increase V

$$PCl_5(g) \rightleftharpoons PCl_3(g) + Cl_2(g)$$
1 mol   2 mol

When equilibrium is reestablished, the new concentrations are $[PCl_5] = 0.56$ M, $[PCl_3] = 0.14$ M, and $[Cl_2] = 0.17$ M. The resulting equilibrium mixture contains new concentrations of reactants and products that are now equal to K_c:

At Volume (1)	At Volume (2)
$[PCl_3] = 0.20$ M	$[PCl_3] = 0.14$ M
$[Cl_2]\ \ = 0.25$ M	$[Cl_2]\ \ = 0.17$ M
$[PCl_5] = 1.20$ M	$[PCl_5] = 0.56$ M

$$K_c = \frac{[PCl_3][Cl_2]}{[PCl_5]} = \frac{[0.20][0.25]}{[1.20]} = 0.042$$

At volume 1

$$K_c = \frac{[PCl_3][Cl_2]}{[PCl_5]} = \frac{[0.14][0.17]}{[0.56]} = 0.042$$

At volume 2

When a reaction has the same number of moles of gases in the reactants as products, a volume change does not affect the equilibrium. There is no effect on equilibrium because the molar concentrations of the reactants and products change in the same way. Consider the reaction of H_2 and I_2 to form HI, which has K_c of 54:

$$H_2(g) + I_2(g) \rightleftharpoons 2HI(g)$$

2 mol of gas   2 mol of gas

Suppose we start with $[H_2] = 0.060$ M, $[I_2] = 0.015$ M, and $[HI] = 0.22$ M.

$$K_c = \frac{[HI]^2}{[H_2][I_2]} = \frac{[0.22]^2}{[0.060][0.015]} = 54$$

If the volume is decreased by half, the pressure will double, and all the molar concentrations will double. However, the equation has the same number of moles of products as reactants, so there is no effect on the equilibrium. The product/reactant ratio stays the same. We can see this by substituting the increased concentrations into the equilibrium constant expression, as follows:

At Volume (1)	At Volume (2)
$[HI] = 0.22$ M	$[HI] = 0.44$ M
$[H_2] = 0.060$ M	$[H_2] = 0.12$ M
$[I_2]\ \ = 0.015$ M	$[I_2]\ \ = 0.030$ M

$$K_c = \frac{[HI]^2}{[H_2][I_2]} = \frac{[0.22]^2}{[0.060][0.015]} = 54$$

At volume 1

$$K_c = \frac{[HI]^2}{[H_2][I_2]} = \frac{[0.44]^2}{[0.12][0.030]} = 54$$

At volume 2

SAMPLE PROBLEM 13.7

■ Effect of Changes in Volume

Indicate whether the effect of decreasing the volume for each of the following causes the number of moles of product to increase, decrease, or not change:

a. $C_2H_2(g) + 2H_2(g) \rightleftharpoons C_2H_6(g)$

b. $2NO_2(g) \rightleftharpoons 2NO(g) + O_2(g)$

c. $CO(g) + H_2O(g) \rightleftharpoons CO_2(g) + H_2(g)$

SOLUTION

To relieve the stress of decreasing the volume, the equilibrium shifts toward the side with the fewer moles of gaseous components.

a. The equilibrium shifts toward C_2H_6 product to reduce the number of moles of gas. The number of moles of product is increased.

$$C_2H_2(g) + 2H_2(g) \longrightarrow C_2H_6(g)$$

3 mol of gas   1 mol of gas

b. The equilibrium shifts toward NO_2 reactant to reduce the number of moles of gas. The number of moles of product is decreased.

$$2NO_2(g) \longleftarrow 2NO(g) + O_2(g)$$

2 mol of gas   3 mol of gas

c. There is no shift in equilibrium because there is no change in the number of moles; the moles of reactant are equal to the moles of product. The number of moles of product does not change.

$$CO(g) + H_2O(g) \rightleftharpoons CO_2(g) + H_2(g)$$

2 mol of gas   2 mol of gas

STUDY CHECK

Suppose you want to increase the yield of product in the following reaction. Would you increase or decrease the volume of the reaction container?

$$CO(g) + 2H_2(g) \rightleftharpoons CH_3OH(g)$$

Effect of a Change in Temperature on Equilibrium

When changes occur to a system at equilibrium, the system shifts to reestablish the same value of the equilibrium constant. However, if we change the temperature of a system at equilibrium, there will be a change in the value of K_c. When the temperature of an equilibrium system increases, the favored reaction is the one that removes heat. Adding heat to an endothermic reaction causes the equilibrium to shift toward products, which uses up heat. Then the value of K_c increases because there is also an increase in the concentrations of products and a decrease in the concentrations of reactants.

Increase T; increase K_c

$$N_2(g) + O_2(g) + heat \rightleftharpoons 2NO(g)$$

If the temperature is lowered, the equilibrium shifts to increase the concentrations of the reactants, and the value of K_c decreases (see Table 13.5).

TABLE 13.5 Equilibrium Shifts for Temperature Changes in an Endothermic Reaction

K_c	Temperature Change	Equilibrium Shift	Change in K_c Value
$\dfrac{[NO]^2}{[N_2][O_2]}$	⬆ Increases	More product $\dfrac{[NO]^2}{[N_2][O_2]}$ Less reactant	⬆ Increases
$\dfrac{[NO]^2}{[N_2][O_2]}$	⬇ Decreases	Less product $\dfrac{[NO]^2}{[N_2][O_2]}$ More reactant	⬇ Decreases

TABLE 13.6 Equilibrium Shifts for Temperature Changes in an Exothermic Reaction

K_c	Temperature Change	Equilibrium Shift	Change in K_c Value
$\dfrac{[SO_3]^2}{[SO_2]^2\,[O_2]}$	⬆ Increases	Less product $\dfrac{[SO_3]^2}{[SO_2]^2\,[O_2]}$ More reactant	⬇ Decreases
$\dfrac{[SO_3]^2}{[SO_2]^2\,[O_2]}$	⬇ Decreases	More product $\dfrac{[SO_3]^2}{[SO_2]^2\,[O_2]}$ Less reactant	⬆ Increases

Decrease T; decrease K_c

$$N_2(g) + O_2(g) + \text{heat} \rightleftharpoons 2NO(g)$$

For an exothermic reaction, the addition of heat favors the reverse reaction, which uses up heat. The value of K_c for an exothermic reaction decreases when the temperature increases.

Increase T; decrease K_c

$$2SO_2(g) + O_2(g) \rightleftharpoons 2SO_3(g) + \text{heat}$$

If heat is removed, the equilibrium of an exothermic reaction favors the products, which provides heat (see Table 13.6).

Decrease T; increase K_c

$$2SO_2(g) + O_2(g) \rightleftharpoons 2SO_3(g) + \text{heat}$$

SAMPLE PROBLEM | 13.8

■ Effect of Temperature Change on Equilibrium

Indicate the change in the concentration of products and the change in K_c when the temperature of each of the following reactions at equilibrium is increased:

a. $N_2(g) + 3H_2(g) \rightleftharpoons 2NH_3(g) + 92\ \text{kJ}$
b. $N_2(g) + O_2(g) + 180\ \text{kJ} \rightleftharpoons 2NO(g)$

SOLUTION

a. The addition of heat shifts an exothermic reaction toward reactants, which decreases the concentration of the products. The K_c will decrease.
b. The addition of heat shifts an endothermic reaction toward products, which increases the concentration of the products. The K_c will increase.

STUDY CHECK

Indicate the change in the concentration of reactants and the change in K_c when there is a decrease in the temperature of each of the reactions at equilibrium in Sample Problem 13.8.

Table 13.7 summarizes the ways we can use Le Châtelier's principle to determine the shift in equilibrium that relieves a stress caused by change in a condition.

CHEMISTRY AND HEALTH

Homeostasis: Regulation of Body Temperature

In a physiological system of equilibrium called *homeostasis*, changes in our environment are balanced by changes in our bodies. It is crucial to our survival that we balance heat gain with heat loss. If we do not lose enough heat, our body temperature rises. At high temperatures, the body can no longer regulate our metabolic reactions. If we lose too much heat, body temperature drops. At low temperatures, essential functions proceed too slowly.

The skin plays an important role in the maintenance of body temperature. When the outside temperature rises, receptors in the skin send signals to the brain. The temperature-regulating part of the brain stimulates the sweat glands to produce perspiration. As perspiration evaporates from the skin, heat is removed and the body temperature is lowered.

In cold temperatures, epinephrine is released, causing an increase in metabolic rate, which increases the production of heat. Receptors on the skin signal the brain to constrict the blood vessels. Less blood flows through the skin, and heat is conserved. The production of perspiration stops to lessen the heat lost by evaporation.

Blood vessels dilate
• sweat production increases
• sweat evaporates
• skin cools

Blood vessels constrict and epinephrine is released
• metabolic activity increases
• muscular activity increases
• shivering occurs
• sweat production stops

TABLE 13.7 Effects of Condition Changes on Equilibrium

Condition	Change (Stress)	Reaction to Remove Stress
Concentration	Add reactant	Forward
	Remove reactant	Reverse
	Add product	Reverse
	Remove product	Forward
Volume (container)	Decrease	Toward fewer moles in the gas phase
	Increase	Toward more moles in the gas phase
Temperature	**Endothermic reaction**	
	Raise T	Forward, larger value for K_c
	Lower T	Reverse, smaller value for K_c
	Exothermic reaction	
	Raise T	Reverse, smaller value for K_c
	Lower T	Forward, larger value for K_c
Catalyst	Increases rates equally	No effect

QUESTIONS AND PROBLEMS

Changing Equilibrium Conditions: Le Châtelier's Principle

13.27 **a.** Does the addition of reactant to an equilibrium mixture cause the product/reactant ratio to be higher or lower than K_c?
  b. According to Le Châtelier's principle, how is equilibrium in part **a** reestablished?

13.28 **a.** What is the effect on K_c when the temperature of an exothermic reaction is lowered?
  b. According to Le Châtelier's principle, how is equilibrium in part **a** reestablished?

13.29 Oxygen is converted to ozone (O_3) by the energy provided from an electric spark.

$$3O_2(g) + \text{heat} \rightleftharpoons 2O_3(g)$$

For each of the following changes at equilibrium, indicate whether the equilibrium shifts toward product or reactant or does not shift:
  a. adding $O_2(g)$
  b. adding $O_3(g)$
  c. raising the temperature
  d. decreasing the volume of the container
  e. adding a catalyst

13.30 Ammonia is produced by reacting nitrogen gas and hydrogen gas.

$$N_2(g) + 3H_2(g) \rightleftharpoons 2NH_3(g) + 92 \text{ kJ}$$

For each of the following changes at equilibrium, indicate whether the equilibrium shifts toward product or reactants or does not shift:

a. removing $N_2(g)$ **b.** lowering the temperature
c. adding $NH_3(g)$ **d.** adding $H_2(g)$
e. increasing the volume of the container

13.31 Hydrogen chloride can be made by reacting hydrogen gas and chlorine gas.

$$H_2(g) + Cl_2(g) + \text{heat} \rightleftharpoons 2HCl(g)$$

For each of the following changes at equilibrium, indicate whether the equilibrium shifts toward product or reactants or does not shift:

a. adding $H_2(g)$ **b.** increasing the temperature
c. removing $HCl(g)$ **d.** adding a catalyst
e. removing $Cl_2(g)$

13.32 When heated, carbon reacts with water to produce carbon monoxide and hydrogen.

$$C(s) + H_2O(g) + \text{heat} \rightleftharpoons CO(g) + H_2(g)$$

For each of the following changes at equilibrium, indicate whether the equilibrium shifts toward products or reactants or does not shift:

a. increasing the temperature
b. adding $C(s)$
c. removing $CO(g)$ as its forms
d. adding $H_2O(g)$
e. decreasing the volume of the container

13.6 Equilibrium in Saturated Solutions

LEARNING GOAL

Write the solubility product expression for a slightly soluble salt and calculate the K_{sp}; use the K_{sp} to determine the solubility.

Until now, we have looked primarily at equilibrium in the context of gases. However, there are also equilibrium systems that involve aqueous solutions, some of which are saturated solutions or contain insoluble salts. Everyday examples of solubility equilibrium in solution are found in tooth decay and kidney stones. When bacteria in the mouth react with sugars in food, acids are produced that dissolve the enamel of a tooth, which is made of a mineral called hydroxyapatite, $Ca_5(PO_4)_3OH$. Kidney stones are composed of calcium salts such as calcium oxalate, CaC_2O_4, and calcium phosphate, $Ca_3(PO_4)_2$, which are rather insoluble. When the concentrations of the Ca^{2+} ions and oxalate $C_2O_4^{2-}$ ions exceed their solubility in the kidneys, they form solid CaC_2O_4. A similar reaction takes place when the concentrations of the Ca^{2+} and PO_4^{3-} ions exceed their solubilities.

$$Ca^{2+}(aq) + C_2O_4^{2-}(aq) \rightleftharpoons CaC_2O_4(s)$$
$$3Ca^{2+}(aq) + 2PO_4^{3-}(aq) \rightleftharpoons Ca_3(PO_4)_2(s)$$

To understand the role of solubility in biology and the environment, we can look at the equilibrium that occurs in saturated solutions.

Solubility Product Constant

In Chapter 12, we learned that a saturated solution contains some undissolved solute in contact with the maximum amount of dissolved solute. A saturated solution is a dynamic system in which the rate of dissolution for a solute has become equal to the rate of solute recrystallization out of solution. As long as the temperature remains constant, the concentration of the ions in the saturated solution is constant. Let us look at the equilibrium equation for CaC_2O_4, which is written with the solid solute on the left and the ions in solution on the right:

$$CaC_2O_4(s) \rightleftharpoons Ca^{2+}(aq) + C_2O_4^{2-}(aq)$$

At equilibrium, the concentrations of Ca^{2+} and $C_2O_4^{2-}$ are constant. We represent the solubility of CaC_2O_4 by a solubility product expression, which is the product of the ion concentrations. The numerical value of the solubility product expression is the **solubility product constant**, K_{sp}. As in other heterogeneous equilibria (Section 13.3), the concentration of the solid is constant and not included in the K_{sp} expression. As before, the brackets mean the molar concentration of the ions:

$$K_{sp} = [Ca^{2+}][C_2O_4^{2-}]$$

In another example, we consider the equilibrium of solid calcium fluoride and its ions Ca^{2+} and F^-:

$$CaF_2(s) \rightleftharpoons Ca^{2+}(aq) + 2F^-(aq)$$

At equilibrium, the rate of dissolving for CaF_2 is equal to the rate of its recrystallization, which means the concentrations of the ions remain constant. The solubility product expression for this solubility is written as the product of the ion concentrations. As with other equilibrium expressions, $[F^-]$ is raised to the power of 2 because there is a coefficient of 2 in the equilibrium equation:

$$K_{sp} = [Ca^{2+}][F^-]^2$$

CONCEPT CHECK 13.7

■ **Writing the Solubility Product Expression**

For each of the following slightly soluble salts, write the equilibrium equation and the solubility product expression:

a. AgBr **b.** Li_2CO_3

ANSWER

a. In the equation, the solid salt is written on the left in equilibrium with the ions Ag^+ and Br^- in aqueous solution; the solubility product expression, K_{sp}, gives the product of the molar concentrations of the ions:

$$AgBr(s) \rightleftharpoons Ag^+(aq) + Br^-(aq) \quad K_{sp} = [Ag^+][Br^-]$$

b. In the equation, the solid salt is written on the left in equilibrium with the ions $2Li^+$ and $CO_3{}^{2-}$ in aqueous solution; the solubility product expression, K_{sp}, gives the molar concentrations of the ions with $[Li^+]$ having an exponent of 2, which is its coefficient in the balanced equation:

$$Li_2CO_3(s) \rightleftharpoons 2Li^+(aq) + CO_3{}^{2-}(aq) \quad K_{sp} = [Li^+]^2[CO_3{}^{2-}]$$

TUTORIAL
Solubility Product Constant Expression

Calculating Solubility Product Constant

Experiments in the lab can measure the concentrations of ions in a saturated solution. For example, we can make a saturated solution of $CaCO_3$ by adding solid $CaCO_3$ to water and stirring until equilibrium is reached. Then we would measure the concentrations of Ca^{2+} and $CO_3{}^{2-}$ in solution. Suppose that in a saturated solution of $CaCO_3$, $[Ca^{2+}] = 7.1 \times 10^{-5}$ M and $[CO_3{}^{2-}] = 7.1 \times 10^{-5}$ M.

STEP 1 Write the equilibrium equation for the dissociation of the ionic compound.
$$CaCO_3(s) \rightleftharpoons Ca^{2+}(aq) + CO_3{}^{2-}(aq)$$

STEP 2 Write the solubility product expression (K_{sp}).
$$K_{sp} = [Ca^{2+}][CO_3{}^{2-}]$$

STEP 3 Substitute the molarity of each ion into the K_{sp} expression and calculate.
$$K_{sp} = [7.1 \times 10^{-5}][7.1 \times 10^{-5}] = 5.0 \times 10^{-9}$$

Table 13.8 gives values of K_{sp} for a selected group of ionic compounds at 25 °C.

TABLE 13.8 Solubility Product Constants (K_{sp}) for Selected Ionic Compounds (25 °C)

Formula	K_{sp}
AgCl	1.8×10^{-10}
Ag_2SO_4	1.2×10^{-5}
$BaCO_3$	2.0×10^{-9}
$BaSO_4$	1.1×10^{-10}
CaF_2	3.2×10^{-11}
$Ca(OH)_2$	6.5×10^{-6}
$CaSO_4$	2.4×10^{-5}
$PbCl_2$	1.5×10^{-6}
$PbCO_3$	7.4×10^{-14}

SAMPLE PROBLEM | 13.9

■ **Calculating the Solubility Product Constant**

In a saturated solution of strontium fluoride, SrF_2, $[Sr^{2+}] = 8.7 \times 10^{-4}$ M and $[F^-] = 1.7 \times 10^{-3}$ M. What is the value of the K_{sp} for SrF_2?

SOLUTION

STEP 1 Write the equilibrium equation for the dissociation of the ionic compound.
$$SrF_2(s) \rightleftharpoons Sr^{2+}(aq) + 2F^-(aq)$$

STEP 2 Write the solubility product expression (K_{sp}).
$$K_{sp} = [Sr^{2+}][F^-]^2$$

STEP 3 Substitute the molarity of each ion into the K_{sp} expression and calculate.
$$K_{sp} = [8.7 \times 10^{-4}][1.7 \times 10^{-3}]^2 = 2.5 \times 10^{-9}$$

STUDY CHECK

What is the K_{sp} of silver bromide, AgBr, if a saturated solution has $[Ag^+] = 7.1 \times 10^{-7}$ M and $[Br^-] = 7.1 \times 10^{-7}$ M?

Guide to Calculating K_{sp}

STEP 1
Write the equilibrium equation for the dissociation of the ionic compound.

STEP 2
Write the solubility product expression (K_{sp}).

STEP 3
Substitute the molarity of each ion into the K_{sp} expression and calculate.

Molar Solubility, S

The molar solubility, S, of a slightly soluble salt is the number of moles of solute that dissolves in 1 liter of solution. For example, the solubility of CdS is found experimentally to be 1×10^{-12} M. This means that 1×10^{-12} mol of CdS in 1 L dissociates into Cd^{2+} and S^{2-} ions: $[Cd^{2+}] = 1 \times 10^{-12}$ M and $[S^{2-}] = 1 \times 10^{-12}$ M.

$$CdS(s) \rightleftharpoons Cd^{2+}(aq) + S^{2-}(aq)$$

$$S = 1 \times 10^{-12} \text{ mol/L}$$

$$[Cd^{2+}] = [S^{2-}] = 1 \times 10^{-12} \text{ M}$$

If we know the K_{sp} of a slightly soluble salt, we can calculate the molarity of each ion and determine its molar solubility as shown in Sample Problem 13.10.

SAMPLE PROBLEM 13.10

■ Calculating the Molar Solubility

Calculate the molar solubility, S, of $PbSO_4$ if it has a $K_{sp} = 1.6 \times 10^{-8}$.

SOLUTION

STEP 1 Write the equilibrium equation for the dissociation of the ionic compound.

$$PbSO_4(s) \rightleftharpoons Pb^{2-}(aq) + SO_4^{2-}(aq)$$

STEP 2 Write the solubility product expression (K_{sp}).

$$K_{sp} = [Pb^{2-}][SO_4^{2-}]$$

STEP 3 Substitute S for the molarity of each ion into the K_{sp} expression.

$$K_{sp} = S \times S = S^2 = 1.6 \times 10^{-8}$$

STEP 4 Calculate the molar solubility (S).

$$S^2 = 1.6 \times 10^{-8}$$

$$S = \sqrt{1.6 \times 10^{-8}} = 1.3 \times 10^{-4} \text{ M}$$

STUDY CHECK

Calculate the molar solubility, S, of NiS if it has a $K_{sp} = 4 \times 10^{-20}$.

Guide to Calculating Molar Solubility from K_{sp}

STEP 1
Write the equilibrium equation for the dissociation of the ionic compound.

STEP 2
Write the solubility product expression K_{sp}.

STEP 3
Substitute S for the molarity of each ion into the K_{sp} expression.

STEP 4
Calculate the molar solubility (S).

Effect of Adding a Common Ion

We have seen when a slightly soluble salt such as $MgCO_3$ dissolves in water that small amounts of Mg^{2+} and CO_3^{2-} ions are produced in equal quantities.

$$MgCO_3(s) \rightleftharpoons Mg^{2+}(aq) + CO_3^{2-}(aq) \qquad K_{sp} = 3.5 \times 10^{-8}$$

Then the solubility of $MgCO_3$ is 1.9×10^{-4} M and the concentrations of both Mg^{2+} and CO_3^{2-} are the same (1.9×10^{-4} M).

However, we can change the solubility of $MgCO_3$ by adding a soluble salt containing a common ion, which is one of the ions already present, Mg^{2+} or CO_3^{2-}. Suppose $MgCl_2$, a soluble salt, is added to the above solution. The soluble salt dissociates into Mg^{2+} and $2Cl^-$, which increases the concentration of the common ion, Mg^{2+}. Because the K_{sp} for $MgCO_3$ must stay the same, some Mg^{2+} combines with CO_3^{2-} forming solid $MgCO_3$, which decreases $[CO_3^{2-}]$. Thus the solubility of a slightly soluble salt decreases when a common ion is added to the solution.

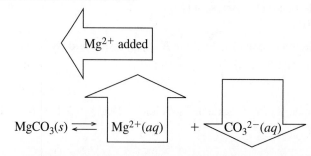

The same thing will happen if Na_2CO_3 is added to the solution of Mg^{2+} and CO_3^{2-}. Then some CO_3^{2-} combines with Mg^{2+} forming solid $MgCO_3$. Therefore, the concentration of Mg^{2+} decreases due to the addition of the common ion CO_3^{2-}.

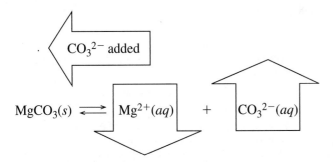

Earlier we described kidney stones as crystals composed of calcium salts such as calcium oxalate, CaC_2O_4. If the product of the concentrations of Ca^{2+} and $C_2O_4^{2-}$ exceeds the K_{sp}, solid CaC_2O_4 will form. If we measure the $C_2O_4^{2-}$ concentration in body fluid, we can calculate the maximum concentration of Ca^{2+} required to form a saturated CaC_2O_4 solution as seen in Sample Problem 13.11.

SAMPLE PROBLEM | 13.11 |

■ **Calculating the Concentration of an Ion**

CaC_2O_4 has a K_{sp} of 2.7×10^{-9}. If $Na_2C_2O_4$ is added to the solution to give the common ion $[C_2O_4^{2-}] = 3.5 \times 10^{-2}$ M, what is the $[Ca^{2+}]$?

SOLUTION

STEP 1 Write the equilibrium equation for the dissociation of the ionic compound.

$$CaC_2O_4(s) \rightleftharpoons Ca^{2+}(aq) + C_2O_4^{2-}(aq)$$

STEP 2 Write the solubility product expression (K_{sp}).

$$K_{sp} = [Ca^{2+}][C_2O_4^{2-}]$$

STEP 3 Substitute the known concentration into the solubility product expression.

$$K_{sp} = [Ca^{2+}][3.5 \times 10^{-2}] = 2.7 \times 10^{-9}$$

STEP 4 Calculate the unknown concentration.

$$[Ca^{2+}] = \frac{2.7 \times 10^{-9}}{[3.5 \times 10^{-2}]} = 7.7 \times 10^{-8} \text{ M}$$

The concentration of Ca^{2+} is 7.7×10^{-8} M.

STUDY CHECK

Nickel(II) carbonate, $NiCO_3$, has a K_{sp} of 1.3×10^{-7}. What is the concentration of Ni^{2+} if Na_2CO_3 is added to give $[CO_3^{2-}] = 4.2 \times 10^{-2}$ M?

Kidney stones form when $[Ca^{2+}][C_2O_4^{2-}]$ is equal to or greater than the solubility product. Normally, the urine contains substances such as magnesium and citrate ions that prevent the formation of kidney stones. Some of the factors that contribute to the formation of kidney stones are drinking too little water, limited physical activity, consuming foods with high levels of oxalate, and some metabolic diseases. Prevention includes drinking large quantities of water and decreasing consumption of foods high in oxalate, such as spinach, rhubarb, and soybean products.

QUESTIONS AND PROBLEMS

Equilibrium in Saturated Solutions

13.33 For each of the following slightly soluble salts, write the equilibrium equation for dissociation and the solubility product expression:
   a. $MgCO_3$ **b.** CaF_2 **c.** Ag_3PO_4

13.34 For each of the following slightly soluble salts, write the equilibrium equation for dissociation and the solubility product expression:
   a. Ag_2S **b.** $Al(OH)_3$ **c.** BaF_2

13.35 A saturated solution of barium sulfate, $BaSO_4$, has $[Ba^{2+}] = 1 \times 10^{-5}$ M and $[SO_4^{2-}] = 1 \times 10^{-5}$ M. What is the value of K_{sp} for $BaSO_4$?

13.36 A saturated solution of copper(II) sulfide, CuS, has $[Cu^{2+}] = 1.1 \times 10^{-18}$ M and $[S^{2-}] = 1.1 \times 10^{-18}$ M. What is the value of K_{sp} for CuS?

13.37 A saturated solution of silver carbonate, Ag_2CO_3, has $[Ag^+] = 2.6 \times 10^{-4}$ M and $[CO_3^{2-}] = 1.3 \times 10^{-4}$ M. What is the value of K_{sp} for Ag_2CO_3?

13.38 A saturated solution of barium fluoride, BaF_2, has $[Ba^{2+}] = 3.6 \times 10^{-3}$ M and $[F^-] = 7.2 \times 10^{-3}$ M. What is the value of K_{sp} for BaF_2?

13.39 What are $[Cu^+]$ and $[I^-]$ in a saturated CuI solution if K_{sp} of CuI is 1×10^{-12}?

13.40 What are $[Sn^{2+}]$ and $[S^{2-}]$ in a saturated SnS solution if K_{sp} of SnS is 1×10^{-26}?

13.41 What is $[Cl^-]$ in a saturated solution of AgCl if $AgNO_3$, with a common ion, is added to give $[Ag^+] = 2.0 \times 10^{-3}$ M (see Table 13.8 for K_{sp})?

13.42 What is $[Pb^{2+}]$ in a saturated solution of $PbCO_3$ if Na_2CO_3, with a common ion, is added to give $[CO_3^{2-}] = 3.0 \times 10^{-4}$ M (see Table 13.8 for K_{sp})?

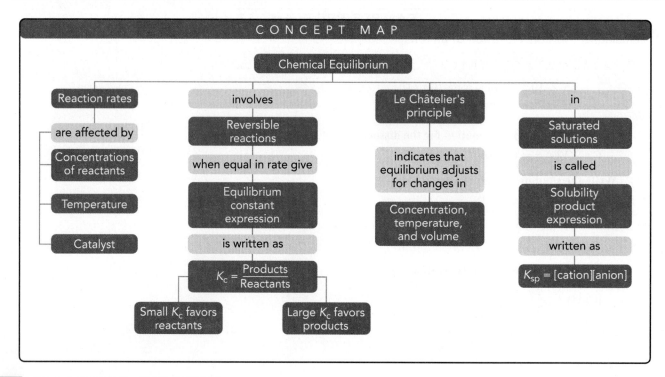

CONCEPT MAP

CHAPTER REVIEW

13.1 Rates of Reactions

LEARNING GOAL: *Describe how temperature, concentration, and catalysts affect the rate of a reaction.*

The rate of a reaction is the speed at which the reactants are converted to products. Increasing the concentrations of reactants, raising the temperature, or adding a catalyst can increase the rate of a reaction.

13.2 Chemical Equilibrium

LEARNING GOAL: *Use the concept of reversible reactions to explain chemical equilibrium.*

Chemical equilibrium is a reversible reaction in which the rate of the forward reaction is equal to the rate of the reverse reaction. At equilib-

rium, no further change occurs in the concentrations of the reactants and products as the forward and reverse reactions continue.

13.3 Equilibrium Constants

LEARNING GOAL: *Calculate the equilibrium constant for a reversible reaction given the concentrations of reactants and products at equilibrium.*

An equilibrium constant, K_c, is the ratio of the concentrations of the products to the concentrations of the reactants with each concentration raised to a power equal to its coefficient in the chemical equation. For heterogeneous reactions, only the molar concentrations of gases are placed in the equilibrium expression.

13.4 Using Equilibrium Constants

LEARNING GOAL: *Use an equilibrium constant to predict the extent of reaction and to calculate equilibrium concentrations.*

A large value of K_c indicates that an equilibrium favors the products and could go nearly to completion, whereas a small value of K_c indicates that the equilibrium favors the reactants. Equilibrium constants can be used to calculate the concentration of a component in the equilibrium mixture.

13.5 Changing Equilibrium Conditions: Le Châtelier's Principle

LEARNING GOAL: *Use Le Châtelier's principle to describe the changes made in equilibrium concentrations when reaction conditions change.*

The addition of reactants or removal of products favors the forward reaction. The removal of reactants or addition of products favors the reverse reaction. A decrease in the volume of a reaction container changes the pressure of gases at equilibrium causing a shift toward the side with the fewer number of moles. Raising or lowering the temperature for exothermic and endothermic reactions changes the value of K_c and shifts the equilibrium for a reaction.

13.6 Equilibrium in Saturated Solutions

LEARNING GOAL: *Write the solubility product expression for a slightly soluble salt and calculate the K_{sp}; use the K_{sp} to determine the solubility.*

In a saturated solution of a slightly soluble salt, the rate of dissolving the solute is equal to the rate of recrystallization. In a saturated solution, the concentrations of the ions from the solute are constant and can be used to calculate the solubility product constant, K_{sp}, for the salt. If the K_{sp} for a slightly soluble salt is known, its solubility can be calculated.

■ KEY TERMS

activation energy The energy that must be provided by a collision to break apart the bonds of the reacting molecules.

catalyst A substance that increases the rate of reaction by lowering the activation energy.

chemical equilibrium The point at which the forward and reverse reactions take place at the same rate so that there is no further change in concentrations of reactants and products.

collision theory A model for a chemical reaction that states that molecules must collide with sufficient energy and proper orientation in order to form products.

equilibrium constant, K_c The numerical value obtained by substituting the equilibrium concentrations of the components into the equilibrium constant expression.

equilibrium constant expression The ratio of the concentrations of products to the concentrations of reactants with each component raised to an exponent equal to the coefficient of that compound in the chemical equation.

heterogeneous equilibrium An equilibrium system in which the components are in different states.

homogeneous equilibrium An equilibrium system in which all components are in the same state.

Le Châtelier's principle When a stress is placed on a system at equilibrium, the equilibrium shifts to relieve that stress.

rate of reaction The speed at which reactants are used to form product(s).

reversible reaction A reaction in which a forward reaction occurs from reactants to products, and a reverse reaction occurs from products back to reactants.

solubility product constant, K_{sp} The product of the concentrations of the ions in a saturated solution of a slightly soluble salt with each concentration raised to a power equal to its coefficient in the equilibrium equation.

■ UNDERSTANDING THE CONCEPTS

13.43 Write the equilibrium constant expression for each of the following reactions:
  a. $CH_4(g) + 2O_2(g) \rightleftharpoons CO_2(g) + 2H_2O(g)$
  b. $4NH_3(g) + 3O_2(g) \rightleftharpoons 2N_2(g) + 6H_2O(g)$
  c. $C(s) + 2H_2(g) \rightleftharpoons CH_4(g)$

13.44 Write the equilibrium constant expression for each of the following reactions:
  a. $2C_2H_6(g) + 7O_2(g) \rightleftharpoons 4CO_2(g) + 6H_2O(g)$
  b. $NH_4HS(s) \rightleftharpoons NH_3(g) + H_2S(g)$
  c. $4NH_3(g) + 5O_2(g) \rightleftharpoons 4NO(g) + 6H_2O(g)$

13.45 Would the reaction shown in the diagrams have a large or small equilibrium constant?

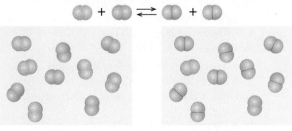

Initial   Equilibrium

13.46 Would the reaction shown in the diagrams have a large or small equilibrium constant?

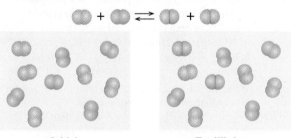

Initial   Equilibrium

13.47 **a.** Would T_2 be higher or lower than T_1 for the reaction shown in the diagrams?
b. Would the K_c at T_2 be larger or smaller than the K_c at T_1?

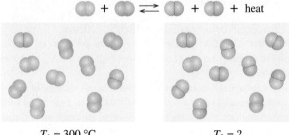

$T_1 = 300 \, °C$ $T_2 = ?$

13.48 **a.** Would the reaction shown in the diagrams be exothermic or endothermic?
b. To increase K_c for this reaction, would you raise or lower the temperature?

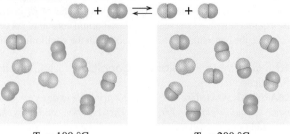

$T_1 = 100 \, °C$ $T_2 = 200 \, °C$

ADDITIONAL QUESTIONS AND PROBLEMS

For instructor-assigned homework, go to
www.masteringchemistry.com.

13.49 Consider the following reaction at equilibrium:

$$C_2H_4(g) + Cl_2(g) \rightleftharpoons C_2H_4Cl_2(g) + heat$$

Indicate how each of the following will shift the equilibrium:
a. raising the temperature of the reaction
b. decreasing the volume of the reaction container
c. adding a catalyst
d. adding $Cl_2(g)$

13.50 Consider the following reaction at equilibrium:

$$N_2(g) + O_2(g) + heat \rightleftharpoons 2NO(g)$$

Indicate how each of the following will shift the equilibrium:
a. raising the temperature of the reaction
b. decreasing the volume of the reaction container
c. adding a catalyst
d. adding $N_2(g)$

13.51 For each of the following reactions at equilibrium indicate if the equilibrium mixture contains mostly products, mostly reactants, or both products and reactants:
a. $H_2(g) + Cl_2(g) \rightleftharpoons 2HCl(g)$ $K_c = 1.3 \times 10^{34}$
b. $2NOBr(g) \rightleftharpoons 2NO(g) + Br_2(g)$ $K_c = 2.0$
c. $2NOCl(g) \rightleftharpoons Cl_2(g) + 2NO(g)$ $K_c = 2.7 \times 10^{-9}$

13.52 For each of the following reactions at equilibrium indicate if the equilibrium mixture contains mostly products, mostly reactants, or both products and reactants:
a. $2H_2O(g) \rightleftharpoons 2H_2(g) + O_2(g)$ $K_c = 4 \times 10^{-48}$
b. $N_2(g) + 3H_2(g) \rightleftharpoons 2NH_3(g)$ $K_c = 0.30$
c. $2SO_2(g) + O_2(g) \rightleftharpoons 2SO_3(g)$ $K_c = 1.2 \times 10^9$

13.53 Write the balanced chemical equation for each of the following equilibrium constant expressions:
a. $K_c = \dfrac{[SO_2][Cl_2]}{[SO_2Cl_2]}$ **b.** $K_c = \dfrac{[BrCl]^2}{[Br_2][Cl_2]}$

c. $K_c = \dfrac{[CH_4][H_2O]}{[CO][H_2]^3}$ **d.** $K_c = \dfrac{[N_2O][H_2O]^3}{[O_2]^2[NH_3]^2}$

13.54 Write the balanced chemical equation for each of the following equilibrium constant expressions:
a. $K_c = \dfrac{[CO_2][H_2]}{[CO][H_2O]}$ **b.** $K_c = \dfrac{[H_2][F_2]}{[HF]^2}$

c. $K_c = \dfrac{[O_2][HCl]^4}{[Cl_2]^2[H_2O]^2}$ **d.** $K_c = \dfrac{[CS_2][H_2]^4}{[CH_4][H_2S]^2}$

13.55 Consider the reaction

$$2NH_3(g) \rightleftharpoons N_2(g) + 3H_2(g)$$

a. Write the equilibrium constant expression for K_c.
b. What is the K_c for the reaction if at equilibrium the concentrations are $[N_2] = 3.0$ M, $[H_2] = 0.50$ M, and $[NH_3] = 0.20$ M?

13.56 Consider the reaction

$$2SO_2(g) + O_2(g) \rightleftharpoons 2SO_3(g)$$

a. Write the equilibrium constant expression for K_c.
b. What is the K_c for the reaction if at equilibrium the concentrations are $[SO_2] = 0.10$ M, $[O_2] = 0.12$ M, and $[SO_3] = 0.60$ M?

13.57 The equilibrium constant for the following reaction is 5.0 at 100 °C. If an equilibrium mixture contains $[NO_2] = 0.50$ M, what is the $[N_2O_4]$?

$$2NO_2(g) \rightleftharpoons N_2O_4(g)$$

13.58 The equilibrium constant for the following reaction is 0.20 at 1000 °C. If an equilibrium mixture contains solid carbon, $[H_2O] = 0.40$ M, and $[CO] = 0.40$ M, what is the $[H_2]$?

$$C(s) + H_2O(g) \rightleftharpoons CO(g) + H_2(g)$$

13.59 According to Le Châtelier's principle, does the equilibrium shift toward products or reactants when O_2 is added to the equilibrium mixture of each of the following reactions?
a. $3O_2(g) \rightleftharpoons 2O_3(g)$
b. $2CO_2(g) \rightleftharpoons 2CO(g) + O_2(g)$
c. $P_4(g) + 5O_2(g) \rightleftharpoons P_4O_{10}(s)$
d. $2NO_2(g) \rightleftharpoons N_2(g) + 2O_2(g)$

13.60 According to Le Châtelier's principle, what is the effect on the products when N_2 is added to the equilibrium mixture of each of the following reactions?
a. $2NH_3(g) \rightleftharpoons 3H_2(g) + N_2(g)$
b. $N_2(g) + O_2(g) \rightleftharpoons 2NO(g)$
c. $2NO_2(g) \rightleftharpoons N_2(g) + 2O_2(g)$
d. $4NH_3(g) + 3O_2(g) \rightleftharpoons 2N_2(g) + 6H_2O(g)$

13.61 Would decreasing the volume of the equilibrium mixture of each of the following reactions cause the equilibrium to shift, and if so, will the shift be toward products or reactants?
a. $3O_2(g) \rightleftharpoons 2O_3(g)$
b. $2CO_2(g) \rightleftharpoons 2CO(g) + O_2(g)$
c. $P_4(g) + 5O_2(g) \rightleftharpoons P_4O_{10}(s)$
d. $2SO_2(g) + 2H_2O(g) \rightleftharpoons 2H_2S(g) + 3O_2(g)$

13.62 Would increasing the volume of the equilibrium mixture of each of the following reactions cause the equilibrium to shift, and if so, will the shift be toward products or reactants?
a. $2NH_3(g) \rightleftharpoons 3H_2(g) + N_2(g)$
b. $N_2(g) + O_2(g) \rightleftharpoons 2NO(g)$
c. $2NO_2(g) \rightleftharpoons N_2(g) + 2O_2(g)$
d. $4NH_3(g) + 3O_2(g) \rightleftharpoons 2N_2(g) + 6H_2O(g)$

13.63 For each of the following slightly soluble salts, write the equilibrium equation for dissociation and the solubility product expression:
a. $CuCO_3$ **b.** PbF_2 **c.** $Fe(OH)_3$

13.64 For each of the following slightly soluble salts, write the equilibrium equation for dissociation and the solubility product expression:
a. CuS **b.** Ag_2SO_4 **c.** $Zn(OH)_2$

13.65 A saturated solution of iron(II) sulfide, FeS, has $[Fe^{2+}] = 7.7 \times 10^{-10}$ M and $[S^{2-}] = 7.7 \times 10^{-10}$ M. What is the value of K_{sp} for FeS?

13.66 A saturated solution of copper(I) chloride, CuCl, has $[Cu^+] = 1.1 \times 10^{-3}$ M and $[Cl^-] = 1.1 \times 10^{-3}$ M. What is the value of K_{sp} for CuCl?

13.67 A saturated solution of manganese(II) hydroxide, $Mn(OH)_2$, has $[Mn^{2+}] = 3.7 \times 10^{-5}$ M and $[OH^-] = 7.4 \times 10^{-5}$ M. What is the value of K_{sp} for $Mn(OH)_2$?

13.68 A saturated solution of silver chromate, Ag_2CrO_4, has $[Ag^+] = 1.3 \times 10^{-4}$ M and $[CrO_4^{2-}] = 6.5 \times 10^{-5}$ M. What is the value of K_{sp} for Ag_2CrO_4?

13.69 What are $[Cd^{2+}]$ and $[S^{2-}]$ in a saturated CdS solution if the K_{sp} of CdS is 1.0×10^{-24}?

13.70 What are $[Cu^{2+}]$ and $[CO_3^{2-}]$ in a saturated $CuCO_3$ solution if the K_{sp} of $CuCO_3$ is 1×10^{-26}?

13.71 A soluble salt, $BaCl_2$, with a common ion is added to a saturated solution of $BaSO_4$ to give $[Ba^{2+}] = 1.0 \times 10^{-3}$ M. What is the $[SO_4^{2-}]$ (see Table 13.8 for the K_{sp})?

13.72 A soluble salt, $AgNO_3$, with a common ion is added to a saturated solution of AgCl to give $[Ag^+] = 2.0 \times 10^{-2}$ M. What is the $[Cl^-]$ (see Table 13.8 for the K_{sp})?

CHALLENGE QUESTIONS

13.73 For each of the following K_c values indicate whether the equilibrium mixture contains mostly reactants, mostly products, or similar amounts of reactants and products:
a. $N_2(g) + O_2(g) \rightleftharpoons 2NO(g)$ $K_c = 1 \times 10^{-30}$
b. $H_2(g) + Br_2(g) \rightleftharpoons 2HBr(g)$ $K_c = 2.0 \times 10^{19}$

13.74 The K_c at 250 °C is 4.2×10^{-2} for the reaction
$$PCl_5(g) \rightleftharpoons PCl_3(g) + Cl_2(g)$$
a. Write the equilibrium constant expression.
b. Initially, 0.60 mol of PCl_5 is placed in a 1.00-L flask. At equilibrium, there is 0.16 mol of PCl_3 in the flask. What are the equilibrium concentrations of the PCl_5 and Cl_2?
c. What is the equilibrium constant for the reaction?
d. If 0.20 mol of Cl_2 is added to the equilibrium mixture, will $[PCl_5]$ increase or decrease?

13.75 The K_c at 100 °C is 2.0 for the reaction
$$2NOBr(g) \rightleftharpoons 2NO(g) + Br_2(g)$$
In an experiment, 1.0 mol of each substance is placed in a 1.0-L container.
a. What is the equilibrium constant expression for the reaction?

b. Is the system at equilibrium?
c. If not, will the rate of the forward or reverse reaction initially speed up?
d. Which concentrations will increase and which will decrease when the system has come to equilibrium?

13.76 For the reaction
$$C(s) + CO_2(g) \rightleftharpoons 2CO(g)$$
the equilibrium mixture contains solid carbon, $[CO] = 0.030$ M, and $[CO_2] = 0.060$ M.
a. What is the value of K_c for the reaction?
b. What is the effect of adding more CO_2 to the equilibrium mixture?
c. What is the effect of decreasing the volume of the container?

13.77 The antacid milk of magnesia, which contains $Mg(OH)_2$, is used to neutralize excess stomach acid. If the solubility of $Mg(OH)_2$ in water is 9.7×10^{-3} g/L, what is the K_{sp}?

13.78 A soluble salt, NaF, with a common ion is added to a saturated solution of CaF_2 to give $[F^-] = 2.2 \times 10^{-3}$ M. What is $[Ca^{2+}]$? (See Table 13.8 for the K_{sp}.)

ANSWERS

ANSWERS TO STUDY CHECKS

13.1 Lowering the temperature will decrease the rate of reaction.

13.2 $H_2(g) + Br_2(g) \rightleftharpoons 2HBr(g)$

13.3 $2NO(g) + O_2(g) \rightleftharpoons 2NO_2(g)$

13.4 $FeO(s) + CO(g) \rightleftharpoons Fe(s) + CO_2(g)$
$$K_c = \frac{[CO_2]}{[CO]}$$

13.5 $K_c = 27$

13.6 $[C_2H_5OH] = 2.7$ M

13.7 Decreasing the volume of the reaction container will shift the equilibrium toward the product side, which has fewer moles of gas.

13.8 **a.** A decrease in temperature will decrease the concentration of reactants and increase the K_c value.
b. A decrease in temperature will increase the concentration of reactants and decrease the K_c value.

13.9 $K_{sp} = 5.0 \times 10^{-13}$

13.10 Molar solubility, $S = 2 \times 10^{-10}$ M

13.11 $[Ni^{2+}] = 3.1 \times 10^{-6}$ M

ANSWERS TO SELECTED QUESTIONS AND PROBLEMS

13.1 **a.** The rate of the reaction indicates how fast the products form or how fast the reactants are used up.
b. At room temperature, the reactions involved in the growth of bread mold will proceed at a faster rate than at the lower temperature of the refrigerator.

13.3 The number of collisions will increase when the number of Br_2 molecules is increased.

13.5 **a.** increase **b.** increase
c. increase **d.** decrease

13.7 A reversible reaction is one in which a forward reaction converts reactants to products, whereas a reverse reaction converts products to reactants.

13.9 **a.** not reversible **b.** reversible
c. reversible

13.11 **a.** $K_c = \dfrac{[CS_2][H_2]^4}{[CH_4][H_2S]^2}$ **b.** $K_c = \dfrac{[N_2][O_2]}{[NO]^2}$

c. $K_c = \dfrac{[CS_2][O_2]^4}{[SO_3]^2[CO_2]}$

13.13 **a.** homogeneous equilibrium
b. heterogeneous equilibrium
c. homogeneous equilibrium
d. heterogeneous equilibrium

13.15 **a.** $K_c = \dfrac{[O_2]^3}{[O_3]^2}$ **b.** $K_c = [CO_2][H_2O]$

c. $K_c = \dfrac{[H_2]^3[CO]}{[CH_4][H_2O]}$ **d.** $K_c = \dfrac{[Cl_2]^2}{[HCl]^4[O_2]}$

13.17 $K_c = 1.5$

13.19 $K_c = 260$

13.21 **a.** mostly products **b.** mostly products
c. mostly reactants

13.23 $[H_2] = 1.1 \times 10^{-3}$ M

13.25 $[NOBr] = 1.4$ M

13.27 **a.** When more reactant is added to an equilibrium mixture, the product/reactant ratio is initially less than K_c.
b. According to Le Châtelier's principle, equilibrium is reestablished when the forward reaction forms more products to make the product/reactant ratio equal the K_c again.

13.29 **a.** Equilibrium shifts toward product.
b. Equilibrium shifts toward reactant.
c. Equilibrium shifts toward product.
d. Equilibrium shifts toward product.
e. No shift in equilibrium occurs.

13.31 **a.** Equilibrium shifts toward product.
b. Equilibrium shifts toward product.
c. Equilibrium shifts toward product.
d. No shift in equilibrium occurs.
e. Equilibrium shifts toward reactant.

13.33 **a.**
$MgCO_3(s) \rightleftharpoons Mg^{2+}(aq) + CO_3{}^{2-}(aq); K_{sp} = [Mg^{2+}][CO_3{}^{2-}]$

b. $CaF_2(s) \rightleftharpoons Ca^{2+}(aq) + 2F^-(aq); K_{sp} = [Ca^{2+}][F^-]^2$

c. $Ag_3PO_4(s) \rightleftharpoons 3Ag^+(aq) + PO_4{}^{3-}(aq); K_{sp} = [Ag^+]^3[PO_4{}^{3-}]$

13.35 $K_{sp} = 1 \times 10^{-10}$

13.37 $K_{sp} = 8.8 \times 10^{-12}$

13.39 $[Cu^+] = 1 \times 10^{-6}$ M; $[I^-] = 1 \times 10^{-6}$ M

13.41 $[Cl^-] = 9.0 \times 10^{-8}$ M

13.43 **a.** $K_c = \dfrac{[CO_2][H_2O]^2}{[CH_4][O_2]^2}$ **b.** $K_c = \dfrac{[N_2]^2[H_2O]^6}{[NH_3]^4[O_2]^3}$

c. $K_c = \dfrac{[CH_4]}{[H_2]^2}$

13.45 The equilibrium constant for the reaction would have a large value.

13.47 **a.** T_2 is lower than T_1.
b. K_c at T_2 is larger than K_c at T_1.

13.49 **a.** shift toward reactants **b.** shift toward products
c. no change **d.** shift toward products

13.51 **a.** mostly products **b.** both products and reactants
c. mostly reactants

13.53 **a.** $SO_2Cl_2(g) \rightleftharpoons SO_2(g) + Cl_2(g)$
b. $Br_2(g) + Cl_2(g) \rightleftharpoons 2BrCl(g)$
c. $CO(g) + 3H_2(g) \rightleftharpoons CH_4(g) + H_2O(g)$
d. $2O_2(g) + 2NH_3(g) \rightleftharpoons N_2O(g) + 3H_2O(g)$

13.55 **a.** $K_c = \dfrac{[N_2][H_2]^3}{[NH_3]^2}$ **b.** $K_c = 9.4$

13.57 $[N_2O_4] = 1.3$ M

13.59 **a.** Equilibrium shifts toward products.
b. Equilibrium shifts toward reactants.
c. Equilibrium shifts toward products.
d. Equilibrium shifts toward reactants.

13.61 **a.** Equilibrium shifts toward products.
b. Equilibrium shifts toward reactants.
c. Equilibrium shifts toward products.
d. Equilibrium shifts toward reactants.

13.63 **a.** $CuCO_3(s) \rightleftharpoons Cu^{2+}(aq) + CO_3{}^{2-}(aq); K_{sp} = [Cu^{2+}][CO_3{}^{2-}]$

b. $PbF_2(s) \rightleftharpoons Pb^{2+}(aq) + 2F^-(aq); K_{sp} = [Pb^{2+}][F^-]^2$

c. $Fe(OH)_3(s) \rightleftharpoons Fe^{3+}(aq) + 3OH^-(aq); K_{sp} = [Fe^{3+}][OH^-]^3$

13.65 $K_{sp} = 5.9 \times 10^{-19}$

13.67 $K_{sp} = 2.0 \times 10^{-13}$

13.69 $[Cd^{2+}] = 1.0 \times 10^{-12}$ M; $[S^{2-}] = 1.0 \times 10^{-12}$ M

13.71 $[SO_4{}^{2-}] = 1.1 \times 10^{-7}$ M

13.73 **a.** A small K_c indicates that the equilibrium mixture contains mostly reactants.
b. A large K_c indicates that the equilibrium mixture contains mostly products.

13.75 **a.** $K_c = \dfrac{[Br_2][NO]^2}{[NOBr]^2}$

b. When the concentrations are placed in the expression, the result is 1.0, which is not equal to K_c. The system is not at equilibrium.
c. The rate of the forward reaction will increase.
d. The $[Br_2]$ and $[NO]$ will increase and $[NOBr]$ will decrease.

13.77 $K_{sp} = 2.0 \times 10^{-11}$

Acids and Bases

14

Visit **www.masteringchemistry.com**
for self-study materials and instructor-assigned homework.

"In our toxicology lab, we measure the drugs in samples of urine or blood," says Penny Peng, assistant supervisor of chemistry in the toxicology lab of the Santa Clara Valley Medical Center. "But first we extract the drugs from the fluid and concentrate them so they can be detected in the machine we use. We extract the drugs by using different organic solvents, such as methanol, ethyl acetate, or methylene chloride, and by changing the pH. We evaporate most of the organic solvent to concentrate any drugs it may contain. A small sample of the concentrate is placed into a machine called a gas chromatograph. As the gas moves over a column, the drugs in it are separated. From the results, we can identify as many as 10 to 15 different drugs from one urine sample."

Acids and bases are important substances in health, industry, and the environment. One of the most common characteristics of acids is their sour taste. Lemons and grapefruits are sour because they contain organic acids such as citric and ascorbic acid (vitamin C). Vinegar tastes sour because it contains acetic acid. We produce lactic acid in our muscles when we exercise. Acid from bacteria turns milk sour in the production of yogurt and cottage cheese. We have hydrochloric acid in our stomachs that helps us digest food. Sometimes we take antacids, which are bases such as sodium bicarbonate or milk of magnesia, to neutralize the effects of too much stomach acid.

Acids and bases have many uses in the chemical industry. Sulfuric acid, H_2SO_4, is the world's most widely produced chemical. It is used to produce fertilizers and plastics, to manufacture detergents, and to conduct electricity in lead-acid storage batteries for automobiles. Sodium hydroxide, NaOH, is used in the production of pulp and paper, the manufacture of soaps, in the textile industries, and in the manufacture of glass.

In the environment, the acidity, or pH, of rain, water, and soil can have significant effects. When rain becomes too acidic, it can dissolve marble statues and accelerate the corrosion of metals. In lakes and ponds, the acidity of water can affect the ability of plants and fish to survive. The acidity of soil around plants affects their growth. If the soil pH is too acidic or too basic, the roots of the plant cannot take up some nutrients. Most plants thrive in soil with a nearly neutral pH, although certain plants, such as orchids, camellias, and blueberries, require a more acidic soil.

14.1 Acids and Bases

LEARNING GOAL

Describe and name Arrhenius, Brønsted–Lowry, and organic acids and bases.

The term *acid* comes from the Latin word *acidus*, which means "sour." We are familiar with the sour tastes of vinegar and lemons and other common acids in foods.

In 1887 the Swedish chemist Svante Arrhenius was the first to describe **acids** as substances that produce hydrogen ions (H^+) when they dissolve in water. For example, hydrogen chloride ionizes in water to give hydrogen ions, H^+, and chloride ions, Cl^-. It is the hydrogen ions that give acids a sour taste, change the blue litmus indicator to red, and corrode some metals.

$$HCl(g) \xrightarrow{\text{H}_2\text{O}} H^+(aq) + Cl^-(aq)$$

Polar covalent   Ionization   Hydrogen
compound   in water   ion

TUTORIAL

Definitions of Acids and Bases

Naming Acids

Acids, which are strong or weak electrolytes, dissolve in water to produce hydrogen ions along with a negative ion that may be a simple nonmetal anion or a polyatomic ion.

When an acid dissolves in water to produce a hydrogen ion and a simple nonmetal anion, the prefix *hydro* is used before the name of the nonmetal, and its *ide* ending is changed to *ic acid*. For example, hydrogen chloride (HCl) dissolves in water to form HCl(*aq*), which is named hydrochloric acid. An exception is hydrogen cyanide (HCN), which as an acid is named hydrocyanic acid, HCN(*aq*). When the anion is an oxygen-containing polyatomic ion, the *ate* in the name of the polyatomic anion is replaced by *ic acid*. If the acid contains a polyatomic ion with an *ite* ending, its name ends in *ous acid*.

$$HNO_3(aq) \xrightarrow{\text{H}_2\text{O}} H^+(aq) + NO_3^-(aq)$$

Nitric acid   Nitrate ion

$$HNO_2(aq) \xrightarrow{\text{H}_2\text{O}} H^+(aq) + NO_2^-(aq)$$

Nitrous acid   Nitrite ion

The halogens in Group 7A (17) can form more than two oxygen-containing acids. For chlorine, the common form is chloric acid, $HClO_3$, which contains the chlorate polyatomic ion (ClO_3^-). For the acid that contains one more oxygen atom than the common form, the prefix *per* is used; $HClO_4$ is named *per*chloric *acid*. When the polyatomic ion in the acid has one oxygen atom less than the common form, the suffix *ous* is used. Thus, $HClO_2$ is named

TABLE 14.1 Naming Common Acids

Acid	Name of Acid	Anion	Name of Anion
HCl	**Hydro**chloric acid	Cl⁻	Chloride
HBr	**Hydro**bromic acid	Br⁻	Bromide
HCN	**Hydro**cyanic acid	CN⁻	Cyanide
HNO₃	Nitric acid	NO₃⁻	Nitrate
HNO₂	Nitrous acid	NO₂⁻	Nitrite
H₂SO₄	Sulfuric acid	SO₄²⁻	Sulfate
H₂SO₃	Sulfurous acid	SO₃²⁻	Sulfite
H₂CO₃	Carbonic acid	CO₃²⁻	Carbonate
H₃PO₄	Phosphoric acid	PO₄³⁻	Phosphate
HClO₄	**Per**chloric acid	ClO₄⁻	**Per**chlorate
HClO₃	Chloric acid	ClO₃⁻	Chlorate
HClO₂	Chlorous acid	ClO₂⁻	Chlorite
HClO	**Hypo**chlorous acid	ClO⁻	**Hypo**chlorite
CH₃—COOH	Acetic acid	CH₃—COO⁻	Acetate

chlor*ous acid*; it contains the chlorite ion, ClO₂⁻. The prefix *hypo* is used for the acid that has two oxygen atoms less than the common form; HClO is named *hypo*chlor*ous acid*. The names of some common acids and their anions are listed in Table 14.1.

CONCEPT CHECK 14.1

■ Naming Acids

a. If H₃PO₄ is named phosphoric acid, what is the name of H₃PO₃? Why?
b. If HBrO₃ is named bromic acid, what is the name of HBrO? Why?

ANSWER

a. H₃PO₄ is named phosphoric acid because it is the common form that contains the phosphate ion. If the polyatomic ion has one less oxygen than the common form, its name ends in *ite*. Then the acid is named as an *ous acid*. Thus, H₃PO₃ is named phosphorous acid.
b. HBrO is named hypobromous acid. The prefix *hypo* is used for the acid that has two oxygen atoms less than the common form HBrO₃.

Naming Carboxylic Acids

As seen in Chapter 8, Section 8.4, a *carboxylic acid* contains a *carboxyl group*, which is a hydroxyl group attached to a carbonyl group. Many carboxylic acids have common names, which are derived from their natural sources. Formic acid is injected under the skin from bee or red ant stings and other insect bites. Acetic acid is produced when ethanol in wines and apple cider reacts with the oxygen in the air. Propionic acid is obtained from the fats in dairy products. Butyric acid gives the foul odor to rancid butter (see Table 14.2).

The IUPAC names of carboxylic acids are based on the alkane names of the corresponding carbon chains. We identify the longest carbon chain containing the carboxyl group and replace the final *e* with *oic acid* (see Figure 14.1).

Red ants inject formic acid under the skin, which causes burning and irritation.

Methanoic acid
(formic acid)

Ethanoic acid
(acetic acid)

Propanoic acid
(propionic acid)

FIGURE 14.1 The IUPAC names of carboxylic acids use the alkane names but replace *e* with *oic acid*.

Q What is the IUPAC and common name of a carboxylic acid with a chain of four carbons?

TABLE 14.2 Names and Natural Sources of Carboxylic Acids

Condensed Structural Formula	IUPAC Name	Common Name	Occurs In
$\overset{\displaystyle O}{\underset{\displaystyle \parallel}{H-C-OH}}$	Methanoic acid	Formic acid	Ant and bee stings (Latin *formica*, "ant")
$\overset{\displaystyle O}{\underset{\displaystyle \parallel}{CH_3-C-OH}}$	Ethanoic acid	Acetic acid	Vinegar (Latin *acetum*, "vinegar")
$\overset{\displaystyle O}{\underset{\displaystyle \parallel}{CH_3-CH_2-C-OH}}$	Propanoic acid	Propionic acid	Dairy products (Greek *pro*, "first," *pion*, "fat")
$\overset{\displaystyle O}{\underset{\displaystyle \parallel}{CH_3-CH_2-CH_2-C-OH}}$	Butanoic acid	Butyric acid	Rancid butter (Latin *butyrum*, "butter")

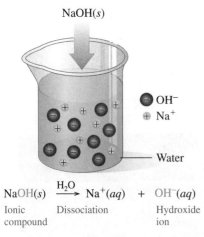

NaOH(s)

OH⁻
Na⁺

Water

$$NaOH(s) \xrightarrow{H_2O} Na^+(aq) + OH^-(aq)$$

Ionic compound · Dissociation · Hydroxide ion

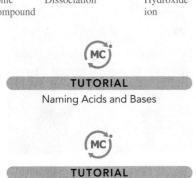

MC

TUTORIAL
Naming Acids and Bases

MC

TUTORIAL
Acid and Base Formulas

MC

TUTORIAL
Properties of Acids and Bases

Bases

You may be familiar with some bases such as antacids, drain openers, and oven cleaners. According to the Arrhenius theory, **bases** are ionic compounds that dissociate into cations and hydroxide ions (OH⁻) when they dissolve in water. For example, sodium hydroxide is an Arrhenius base that dissociates in water to give sodium ions, Na⁺, and hydroxide ions, OH⁻.

Most Arrhenius bases are from Groups 1A (1) and 2A (2), such as NaOH, KOH, LiOH, and Ba(OH)₂. Bases such as Ca(OH)₂, Al(OH)₃, and Fe(OH)₃ are strong, but they are not very soluble in water. The hydroxide ions (OH⁻) produced by Arrhenius bases give these bases common characteristics, such as a bitter taste and a slippery feel. A base turns the litmus indicator blue and phenolphthalein indicator pink. Table 14.3 compares some characteristics of acids and bases.

Naming Bases

Typical Arrhenius bases are named as hydroxides.

Base	Name
NaOH	Sodium **hydroxide**
KOH	Potassium **hydroxide**
Ca(OH)₂	Calcium **hydroxide**
Al(OH)₃	Aluminum **hydroxide**

TABLE 14.3 Some Characteristics of Acids and Bases

Characteristic	Acids	Bases
Arrhenius	Produce H⁺	Produce OH⁻
Electrolytes	Yes	Yes
Taste	Sour	Bitter, chalky
Feel	May sting	Soapy, slippery
Litmus	Red	Blue
Phenolphthalein	Colorless	Pink
Neutralization	Neutralize bases	Neutralize acids

CHEMISTRY AND HEALTH

Salicylic Acid and Aspirin

Chewing on a piece of willow bark was used as a way of relieving pain for many centuries. By the 1800s, chemists discovered that salicylic acid was the agent in the bark responsible for pain relief. However, salicylic acid, which has both a carboxylic group and a hydroxyl group, irritates the stomach lining. A less irritating ester made from salicylic acid and acetic acid, called acetylsalicylic acid or "aspirin", was prepared in 1899 by the Bayer chemical company in Germany. In some aspirin preparations, a buffer is added to neutralize the carboxylic acid group and lessen its irritation of the stomach. Aspirin is used as an analgesic (pain reliever), antipyretic (fever reducer), and anti-inflammatory agent. Many people take a daily low-dose aspirin, which is believed to lower the risk of heart attack and stroke.

Salicylic acid   Acetic acid

Acetylsalicylic acid, aspirin

Oil of wintergreen, or methyl salicylate, has a spearmint odor and flavor. Because it can pass through the skin, methyl salicylate is used in skin ointments where it acts as a counterirritant, producing heat to soothe sore muscles.

Salicylic acid   Methyl alcohol

Methyl salicylate, oil of wintergreen

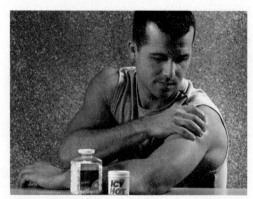

Aspirin, made from salicylic acid, is a widely used analgesic.

CONCEPT CHECK 14.2

■ Dissociation of an Arrhenius Base

Write an equation for the dissociation of barium hydroxide in water.

ANSWER

When barium hydroxide, which has the formula $Ba(OH)_2$, dissolves in water, the solution contains barium ions (Ba^{2+}) and twice as many hydroxide ions (OH^-). The equation is written

$$Ba(OH)_2(s) \xrightarrow{H_2O} Ba^{2+}(aq) + 2OH^-(aq)$$

SAMPLE PROBLEM | 14.1

■ Names and Formulas of Acids and Bases

1. Name each of the following as an acid or base:
   a. H_3PO_4 **b.** NaOH **c.** CH_3-CH_2-COOH
2. Write the formula of each of the following acids:
   a. nitrous acid **b.** bromic acid **c.** pentanoic acid

SOLUTION

1. **a.** phosphoric acid **b.** sodium hydroxide
   c. propanoic acid (IUPAC); propionic acid (common)

2. a. HNO_2 **b.** $HBrO_3$

c. $CH_3—CH_2—CH_2—CH_2—COOH$

STUDY CHECK

Give the name for H_2SO_4. Write the formula for potassium hydroxide.

QUESTIONS AND PROBLEMS

Acids and Bases

14.1   Indicate whether each of the following statements indicates an acid or a base:
a. causes vinegar to have a sour taste
b. neutralizes bases
c. produces H^+ ions in water
d. is named barium hydroxide

14.2   Indicate whether each of the following statements indicates an acid or a base:
a. makes a red ant sting burn
b. produces OH^- in water
c. has a soapy feel
d. turns litmus red

14.3   Name each of the following as an acid or base:
a. HCl   **b.** $Ca(OH)_2$ **c.** H_2CO_3 **d.** HNO_3
e. H_2SO_3 **f.** $HBrO_2$ **g.**
$$CH_3—\overset{\displaystyle O}{\overset{\|}{C}}—OH$$

14.4   Name each of the following as an acid or base:
a. $Al(OH)_3$ **b.** HBr   **c.** H_2SO_4 **d.** KOH
e. HNO_2 **f.** $HClO_2$ **g.**
$$CH_3—CH_2—\overset{\displaystyle O}{\overset{\|}{C}}—OH$$

14.5   Write formulas for the following acids and bases:
a. magnesium hydroxide   **b.** hydrofluoric acid
c. formic acid   **d.** lithium hydroxide
e. ammonium hydroxide   **f.** periodic acid

14.6   Write formulas for the following acids and bases:
a. barium hydroxide   **b.** hydroiodic acid
c. nitric acid   **d.** strontium hydroxide
e. acetic acid   **f.** hypochlorous acid

14.2 Brønsted–Lowry Acids and Bases

LEARNING GOAL

Identify conjugate acid–base pairs for Brønsted–Lowry acids and bases.

In 1923 J. N. Brønsted in Denmark and T. M. Lowry in Great Britain expanded the definition of acids and bases. A **Brønsted–Lowry acid** donates a proton (hydrogen ion, H^+) to another substance, and a **Brønsted–Lowry base** accepts a proton.

> A Brønsted–Lowry acid is a proton (H^+) donor.
> A Brønsted–Lowry base is a proton (H^+) acceptor.

A free, dissociated proton (H^+) does not actually exist in water. Its attraction to polar water molecules is so strong that the proton bonds to a water molecule and forms a **hydronium ion, H_3O^+**.

$$H—\overset{..}{\underset{\displaystyle H}{O}}: + H^+ \longrightarrow \left[H—\overset{..}{\underset{\displaystyle H}{O}}—H \right]^+$$

Water   Proton   Hydronium ion

We write the formation of a hydrochloric acid solution as a transfer of a proton from hydrogen chloride to water. By accepting a proton in the reaction, water is acting as a base according to the Brønsted–Lowry concept.

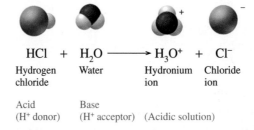

HCl	+	H_2O	⟶	H_3O^+	+	Cl^-
Hydrogen chloride		Water		Hydronium ion		Chloride ion
Acid (H^+ donor)		Base (H^+ acceptor)		(Acidic solution)		

Ammonia, (NH_3), acts as a base by accepting a proton when it reacts with water. Because the nitrogen of NH_3 has a stronger attraction for a proton than the oxygen of water, water acts as an acid and donates a proton.

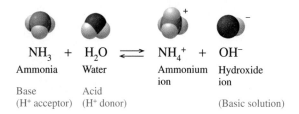

NH_3	+	H_2O	$\rightleftharpoons$	NH_4^+	+	OH^-
Ammonia		Water		Ammonium ion		Hydroxide ion
Base		Acid				
(H^+ acceptor)		(H^+ donor)				(Basic solution)

SAMPLE PROBLEM | **14.2**

■ Acids and Bases

In each of the following equations, identify the reactant that is an acid (H^+ donor) and the reactant that is a base (H^+ acceptor):

a. $HBr(aq) + H_2O(l) \longrightarrow H_3O^+(aq) + Br^-(aq)$
b. $H_2O(l) + CN^-(aq) \rightleftharpoons HCN(aq) + OH^-(aq)$

SOLUTION

a. HBr, acid; H_2O, base   **b.** H_2O, acid; CN^-, base

STUDY CHECK

When HNO_3 reacts with water, water acts as a base (H^+ acceptor). Write the equation for the reaction.

Conjugate Acid–Base Pairs

According to the Brønsted–Lowry theory, a **conjugate acid–base pair** consists of molecules or ions related by the loss or gain of one H^+. Every acid–base reaction contains two conjugate acid–base pairs because protons are transferred in both the forward and reverse directions. When the acid HA donates H^+, the conjugate base A^- forms. When the base B accepts the H^+, it forms the conjugate acid, BH^+. We can write a general equation for a Brønsted–Lowry acid–base reaction as follows:

TUTORIAL
Identifying Conjugate Acid–Base Pairs

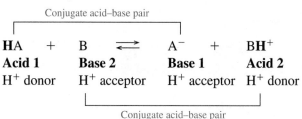

Conjugate acid–base pair

HA	+	**B**	$\rightleftharpoons$	**A$^-$**	+	**BH$^+$**
Acid 1		**Base 2**		**Base 1**		**Acid 2**
H^+ donor		H^+ acceptor		H^+ acceptor		H^+ donor

Conjugate acid–base pair

Now we can identify the conjugate acid–base pairs in a reaction such as hydrofluoric acid and water. Because the reaction is reversible, the conjugate acid H_3O^+ can transfer a proton to the conjugate base F^- and re-form the acid HF. Using the relationship of loss and gain of H^+, we identify the conjugate acid–base pairs as HF/F^- and H_3O^+/H_2O.

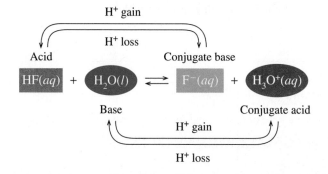

TABLE 14.4 Some Conjugate Acid–Base Pairs

Acid		Conjugate Base	
Strong Acids			
Perchloric acid	$HClO_4$	Perchlorate ion	ClO_4^-
Sulfuric acid	H_2SO_4	Hydrogen sulfate ion	HSO_4^-
Hydroiodic acid	HI	Iodide ion	I^-
Hydrobromic acid	HBr	Bromide ion	Br^-
Hydrochloric acid	HCl	Chloride ion	Cl^-
Nitric acid	HNO_3	Nitrate ion	NO_3^-
Weak Acids			
Hydronium ion	H_3O^+	Water	H_2O
Hydrogen sulfate ion	HSO_4^-	Sulfate ion	SO_4^{2-}
Phosphoric acid	H_3PO_4	Dihydrogen phosphate ion	$H_2PO_4^-$
Hydrofluoric acid	HF	Fluoride ion	F^-
Nitrous acid	HNO_2	Nitrite ion	NO_2^-
Formic acid	$HCOOH$	Formate ion	$HCOO^-$
Acetic acid	$CH_3\!-\!COOH$	Acetate ion	$CH_3\!-\!COO^-$
Carbonic acid	H_2CO_3	Bicarbonate ion	HCO_3^-
Hydrosulfuric acid	H_2S	Hydrogen sulfide ion	HS^-
Dihydrogen phosphate	$H_2PO_4^-$	Hydrogen phosphate ion	HPO_4^{2-}
Ammonium ion	NH_4^+	Ammonia	NH_3
Hydrocyanic acid	HCN	Cyanide ion	CN^-
Bicarbonate ion	HCO_3^-	Carbonate ion	CO_3^{2-}
Methylammonium ion	$CH_3\!-\!NH_3^+$	Methylamine	$CH_3\!-\!NH_2$
Hydrogen sulfide ion	HS^-	Sulfide ion	S^{2-}
Water	H_2O	Hydroxide ion	OH^-

Increasing acid strength

Increasing base strength

In another proton-transfer reaction, ammonia, NH_3, accepts H^+ from H_2O to form the conjugate acid NH_4^+ and conjugate base OH^-. Each of these conjugate acid–base pairs, NH_4^+/NH_3 and H_2O/OH^-, is related by the loss and gain of one H^+.

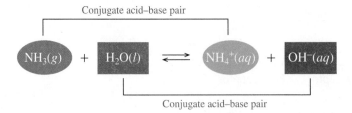

Conjugate acid–base pair

$$NH_3(g) \;+\; H_2O(l) \;\rightleftharpoons\; NH_4^+(aq) \;+\; OH^-(aq)$$

Conjugate acid–base pair

Table 14.4 gives some more examples of conjugate acid–base pairs. In these two examples, we see that water can act as an acid when it donates H^+ or as a base when it accepts H^+. Substances that can act as both acids and bases are **amphoteric**. For water, the most common amphoteric substance, the acidic or basic behavior depends on whether the other reacting substance is a stronger acid or base. Water donates H^+ when it reacts with a stronger base, and it accepts H^+ when it reacts with a stronger acid.

CONCEPT CHECK 14.3

■ **Conjugate Acid–Base Pairs**

Write the formula of the conjugate base of each of the following Brønsted–Lowry acids:

a. $HClO_3$ **b.** H_2CO_3 **c.** $HCOOH$

ANSWER

When a substance, acting as an acid, donates H^+, the result is its conjugate base.

a. When the acid $HClO_3$ loses H^+, it forms its conjugate base, which has the formula ClO_3^-.

b. When the acid H_2CO_3 loses H^+, it forms its conjugate base, which has the formula HCO_3^-.

c. When $HCOOH$ loses H^+, it forms its conjugate base, which has the formula $HCOO^-$.

SAMPLE PROBLEM | 14.3 |

■ Identifying Conjugate Acid–Base Pairs

Write the Brønsted–Lowry equation for the reaction between HBr and CH_3—NH_2. Identify the conjugate acid–base pairs.

SOLUTION

In the reaction, the acid HBr donates H^+ to the base CH_3—NH_2. The product Br^- is the conjugate base, and CH_3—NH_3^+ is the conjugate acid.

$$HBr(aq) \;+\; CH_3\text{—}NH_2(aq) \;\rightleftharpoons\; Br^-(aq) \;+\; CH_3\text{—}NH_3^+(aq)$$

The conjugate acid–base pairs are HBr/Br^-, along with CH_3—NH_3^+/CH_3—NH_2.

STUDY CHECK

In the following reaction, identify the conjugate acid–base pairs:

$$HCN(aq) \;+\; SO_4^{2-}(aq) \;\rightleftharpoons\; CN^-(aq) \;+\; HSO_4^-(aq)$$

QUESTIONS AND PROBLEMS

Brønsted–Lowry Acids and Bases

14.7 Identify the acid (proton donor) and base (proton acceptor) for the reactants in each of the following:
a. $HI(aq) + H_2O(l) \longrightarrow H_3O^+(aq) + I^-(aq)$
b. $F^-(aq) + H_2O(l) \rightleftharpoons HF(aq) + OH^-(aq)$
c. $H_2S(aq) + CH_3\text{—}CH_2\text{—}NH_2(aq) \rightleftharpoons$
    $HS^-(aq) + CH_3\text{—}CH_2\text{—}NH_3^+(aq)$

14.8 Identify the acid (proton donor) and base (proton acceptor) for the reactants in each of the following:
a. $CO_3^{2-}(aq) + H_2O(l) \rightleftharpoons HCO_3^-(aq) + OH^-(aq)$
b. $H_2SO_4(aq) + H_2O(l) \longrightarrow H_3O^+(aq) + HSO_4^-(aq)$
c. $CH_3\text{—}COO^-(aq) + H_3O^+(aq) \rightleftharpoons$
    $H_2O(aq) + CH_3\text{—}COOH(aq)$

14.9 Write the formula of the conjugate base for each of the following acids:
a. HF **b.** H_2O
c. $H_2PO_3^-$ **d.** HSO_4^-
e. $HClO_2$

14.10 Write the formula of the conjugate base for each of the following acids:
a. HCO_3^- **b.** $CH_3\text{—}NH_3^+$
c. HPO_4^{2-} **d.** HNO_2
e. $HBrO$

14.11 Write the formula of the conjugate acid for each of the following bases:
a. CO_3^{2-} **b.** H_2O
c. $H_2PO_4^-$ **d.** Br^-
e. ClO_4^-

14.12 Write the formula of the conjugate acid for each of the following bases:
a. SO_4^{2-} **b.** CN^-
c. $CH_3\text{—}CH_2\text{—}COO^-$ **d.** ClO_2^-
e. HS^-

14.13 Identify the acid and base on the left side of the following equations and identify their conjugate species on the right side:
a. $H_2CO_3(aq) + H_2O(l) \rightleftharpoons H_3O^+(aq) + HCO_3^-(aq)$
b. $NH_4^+(aq) + H_2O(l) \rightleftharpoons H_3O^+(aq) + NH_3(aq)$
c. $HCN(aq) + NO_2^-(aq) \rightleftharpoons CN^-(aq) + HNO_2(aq)$
d. $CH_3\text{—}COO^-(aq) + HF(aq) \rightleftharpoons$
    $F^-(aq) + CH_3\text{—}COOH(aq)$

14.14 Identify the acid and base on the left side of the following equations and identify their conjugate species on the right side:
a. $H_3PO_4(aq) + H_2O(l) \rightleftharpoons H_3O^+(aq) + H_2PO_4^-(aq)$
b. $CO_3^{2-}(aq) + H_2O(l) \rightleftharpoons OH^-(aq) + HCO_3^-(aq)$
c. $H_3PO_4(aq) + NH_3(aq) \rightleftharpoons NH_4^+(aq) + H_2PO_4^-(aq)$
d. $HNO_2(aq) + CH_3\text{—}CH_2\text{—}NH_2(aq) \rightleftharpoons$
    $NO_2^-(aq) + CH_3\text{—}CH_2\text{—}NH_3^+(aq)$

14.15 When ammonium chloride dissolves in water, the ammonium ion NH_4^+ donates a proton to water. Write a balanced equation for the reaction of the ammonium ion with water.

14.16 When sodium carbonate dissolves in water, the carbonate ion CO_3^{2-} acts as a base. Write a balanced equation for the reaction of the carbonate ion with water.

14.3 Strengths of Acids and Bases

The *strength* of acids is determined by the moles of H_3O^+ that are produced for each mole of acid that dissolves. The *strength* of bases is determined by the moles of OH^- that are produced for each mole of base that dissolves. In the process called **dissociation**, an acid or base separates into or produces ions in water. Acids and bases vary greatly in their ability to produce H_3O^+ or OH^-. Strong acids and strong bases dissociate completely. In water, weak acids and weak bases dissociate only slightly, leaving most of the initial acid or base undissociated.

Strong and Weak Acids

Strong acids are examples of strong electrolytes because they donate protons so easily that their dissociation in water is virtually complete. For example, when HCl, a strong acid, dissociates in water, H^+ is transferred to H_2O; the resulting solution contains essentially only the ions H_3O^+ and Cl^-. We consider the reaction of HCl in H_2O as going nearly 100% to products. Therefore, the equation for a strong acid such as HCl is written with a single arrow to the products:

$$HCl(g) + H_2O(l) \longrightarrow H_3O^+(aq) + Cl^-(aq)$$

There are only six common strong acids. All other acids are weak. Table 14.4 lists the strong acids along with some common some weak acids.

Weak acids are weak electrolytes because they dissociate slightly in water, which means that only a small percentage of the weak acid donates H^+ to H_2O. Thus, a weak acid reacts with water to form only a small amount of H_3O^+ ions. Even at high concentrations, weak acids produce low concentrations of H_3O^+ ions (see Figure 14.2). In carbonated soft drinks, CO_2 dissolves in water to form carbonic acid, H_2CO_3. A weak acid such as H_2CO_3 reaches equilibrium between the mostly undissociated H_2CO_3 molecules and the ion H_3O^+ and HCO_3^-. Such a reaction is written with a double arrow. A longer reverse arrow may be used to indicate that the equilibrium favors the undissociated reactants:

$$H_2CO_3(aq) + H_2O(l) \rightleftharpoons H_3O^+(aq) + HCO_3^-(aq)$$

Acids produce hydrogen ions in aqueous solution.

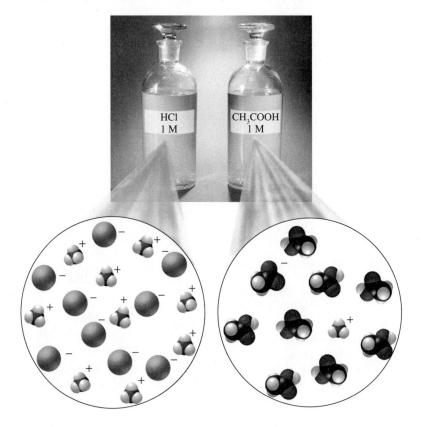

FIGURE 14.2 A strong acid such as HCl is completely dissociated ($\approx$100%), whereas a weak acid such as CH_3COOH contains mostly molecules and a few ions.

Q What is the difference between a strong acid and a weak acid?

Organic acids such as citric acid and acetic acid are weak acids. Citric acid is a weak acid found in fruits and fruit juices such as lemons, oranges, and grapefruit. In the vinegar used in salad dressings, acetic acid is present typically as a 5% acetic acid solution.

$$CH_3{-}COOH(aq) + H_2O(l) \rightleftharpoons CH_3{-}COO^-(aq) + H_3O^+(aq)$$

In summary, if HA is a strong acid in water, the solution consists of the ions H_3O^+ and A^-. However, if HA is a weak acid, the aqueous solution consists of mostly undissociated HA and only a few H_3O^+ and A^- ions (see Figure 14.2).

Strong acid: $HA(aq) + H_2O(l) \longrightarrow H_3O^+(aq) + A^-(aq)$ (100% dissociated)

Weak acid: $HA(aq) + H_2O(l) \rightleftharpoons H_3O^+(aq) + A^-(aq)$ (small % dissociated)

Strong and Weak Bases

Strong bases are strong electrolytes that dissociate completely in water to give an aqueous solution of a metal ion and hydroxide ion. The Group 1A (1) hydroxides are very soluble in water, which can give high concentrations of OH^- ions. The other strong bases are less soluble in water, but they also dissolve completely as ions. For example, when KOH forms a KOH solution, it contains only the ions K^+ and OH^-.

$$KOH(s) \xrightarrow{\text{H}_2\text{O}} K^+(aq) + OH^-(aq)$$

The Arrhenius bases of Groups 1A (1) and 2A (2), such as LiOH, KOH, NaOH, and $Ba(OH)_2$, are strong bases. Sodium hydroxide, NaOH (also known as lye), is used in household products to remove grease from ovens and to clean drains. Because high concentrations of hydroxide ions cause severe damage to the skin and eyes, directions must be followed carefully when such products are used in the home and in the chemistry laboratory. If you spill an acid or a base on your skin or get some in your eyes, be sure to flood the area immediately with water for at least 10 minutes and seek medical attention.

Weak bases are weak electrolytes that are poor acceptors of protons and produce very few ions in solution. A typical weak base, ammonia, NH_3, is found in window cleaners. In an aqueous solution, only a few ammonia molecules accept protons to form NH_4^+ and OH^-:

$$NH_3(g) + H_2O(l) \rightleftharpoons NH_4^+(aq) + OH^-(aq)$$

Bases in household products are used to remove grease and to open drains.

Direction of Reaction

There is a relationship between the components in each conjugate acid–base pair. Strong acids have weak conjugate bases that do not readily accept protons. As the strength of the acid decreases, the strength of its conjugate base increases.

In any acid–base reaction, there are two acids and two bases. However, one acid is stronger than the other acid, and one base is stronger than the other base. By comparing their relative strengths, we can determine the direction of the reaction. For example, the strong acid H_2SO_4 readily gives up protons to water. The hydronium ion H_3O^+ produced is a weaker acid than H_2SO_4, and the conjugate base HSO_4^- is a weaker base than water.

$$H_2SO_4(aq) + H_2O(l) \longrightarrow H_3O^+(aq) + HSO_4^-(aq) \quad \text{Favors products}$$

Stronger   Stronger   Weaker   Weaker
acid   base   acid   base

Let's look at another reaction in which water donates a proton to carbonate, CO_3^{2-}, to form HCO_3^- and OH^-. From Table 14.4, we see that HCO_3^- is a stronger acid than H_2O. We also see that OH^- is a stronger base than CO_3^{2-}. The equilibrium favors the reaction of the strong acid and base reactants to form the weaker acid and weaker base, which is indicated by the long arrow for the reverse reaction:

$$CO_3^{2-}(aq) + H_2O(l) \rightleftharpoons HCO_3^-(aq) + OH^-(aq) \quad \text{Favors reactants}$$

Weaker   Weaker   Stronger   Stronger
base   acid   acid   base

CONCEPT CHECK 14.4

■ **Strengths of Acids and Bases**

For each of the following questions, select from HCO_3^-, HSO_4^-, or HNO_2:

a. Which is the strongest acid?

b. Which acid has the strongest conjugate base?

ANSWER

a. The strongest acid in this group is the acid listed closest to the top of Table 14.4, which is HSO_4^-.

b. The weakest acid, the acid listed nearest the bottom of Table 14.4, HCO_3^-, has the strongest conjugate base CO_3^{2-}.

SAMPLE PROBLEM   14.4

■ **Direction of Reaction**

Does equilibrium favor the reactants or products in the following reaction?

$$HF(aq) + H_2O(l) \rightleftharpoons H_3O^+(aq) + F^-(aq)$$

SOLUTION

From Table 14.4, we see that HF is a weaker acid than H_3O^+, and H_2O is a weaker base than F^-. Equilibrium favors the reverse direction and therefore the reactants.

$$HF(aq) + H_2O(l) \rightleftharpoons H_3O^+(aq) + F^-(aq)$$

| Weaker acid | Weaker base | Stronger acid | Stronger base |

STUDY CHECK

Does the reaction of nitric acid and water favor the reactants or the products?

QUESTIONS AND PROBLEMS

Strengths of Acids and Bases

14.17 What is meant by the phrase "A strong acid has a weak conjugate base"?

14.18 What is meant by the phrase "A weak acid has a strong conjugate base"?

14.19 Identify the stronger acid in each of the following pairs:
   a. HBr or HNO_2
   b. H_3PO_4 or HSO_4^-
   c. HCN or H_2CO_3

14.20 Identify the stronger acid in each of the following pairs:
   a. NH_4^+ or H_3O^+
   b. H_2SO_4 or HCN
   c. H_2O or H_2CO_3

14.21 Identify the weaker acid in each of the following pairs:
   a. HCl or HSO_4^-
   b. HNO_2 or HF
   c. HCO_3^- or NH_4^+

14.22 Identify the weaker acid in each of the following pairs:
   a. HNO_3 or HCO_3^-
   b. HSO_4^- or H_2O
   c. H_2SO_4 or H_2CO_3

14.23 Predict whether the equilibrium favors the reactants or the products for each of the following reactions:
   a. $H_2CO_3(aq) + H_2O(l) \rightleftharpoons H_3O^+(aq) + HCO_3^-(aq)$
   b. $NH_4^+(aq) + H_2O(l) \rightleftharpoons H_3O^+(aq) + NH_3(aq)$
   c. $HCl(aq) + NH_3(aq) \rightleftharpoons Cl^-(aq) + NH_4^+(aq)$
   d. $CH_3-COO^-(aq) + CH_3-NH_3^+(aq) \rightleftharpoons$
   $CH_3-COOH(aq) + CH_3-NH_2(aq)$

14.24 Predict whether the equilibrium favors the reactants or the products for each of the following reactions:
   a. $H_3PO_4(aq) + H_2O(l) \rightleftharpoons H_3O^+(aq) + H_2PO_4^-(aq)$
   b. $CO_3^{2-}(aq) + H_2O(l) \rightleftharpoons OH^-(aq) + HCO_3^-(aq)$
   c. $HS^-(aq) + H_2O(l) \rightleftharpoons H_3O^+(aq) + S^{2-}(aq)$
   d. $HCN(aq) + CH_3-NH_2(aq) \rightleftharpoons$
   $CN^-(aq) + CH_3-NH_3^+(aq)$

14.25 Write an equation for the acid–base reaction between ammonium ion and sulfate ion. Why does the equilibrium favor the reactants?

14.26 Write an equation for the acid–base reaction between nitrous acid and hydroxide ion. Why does the equilibrium favor the products?

14.4 Dissociation Constants

We have seen that reactions of weak acids in water reach equilibrium. If HA is a weak acid, the concentrations of H_3O^+ and A^- ions are small, which means that the equilibrium favors the reactants (see Figure 14.3).

$$HA(aq) + H_2O(l) \rightleftharpoons H_3O^+(aq) + A^-(aq)$$

Dissociation Constants for Weak Acids and Weak Bases

As we have seen, acids and bases have different strengths depending on how much they dissociate in water. Because the dissociation of strong acids in water is essentially complete, the reaction is not considered to be an equilibrium situation. However, because weak acids in water dissociate only slightly, the ion products reach equilibrium with the undissociated weak acid molecules. Thus, an equilibrium expression can be written for weak acids that gives the ratio of the concentrations of products to the weak acid reactants. As with other equilibrium expressions, the molar concentration of the products is divided by the molar concentration of the reactants:

$$\frac{[H_3O^+][A^-]}{[HA][H_2O]}$$

Because water is a pure liquid, its concentration, which is constant, is omitted from the equilibrium expression, which gives the **acid dissociation constant, K_a** (or acid ionization constant). Thus, for a weak acid, the K_a is written

$$K_a = \frac{[H_3O^+][A^-]}{[HA]} \quad \text{Acid dissociation constant}$$

Let's consider the equilibrium of carbonic acid, which dissociates in water to form bicarbonate ion and hydronium ion:

$$H_2CO_3(aq) + H_2O(l) \rightleftharpoons HCO_3^-(aq) + H_3O^+(aq)$$

LEARNING GOAL

Write the expression for the dissociation constant of a weak acid or weak base.

TUTORIAL

Using Dissociation Constants

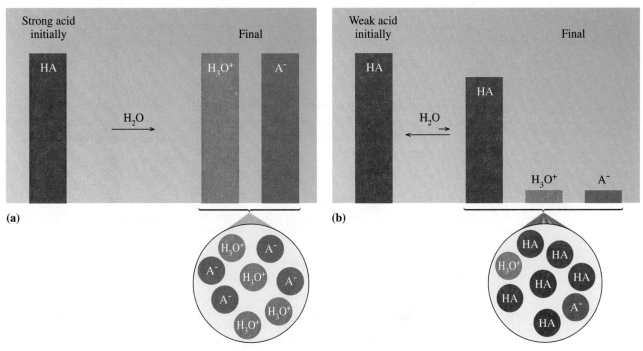

FIGURE 14.3 After dissociation in water, **(a)** a strong acid (HA) has a high concentration of H_3O^+ and A^-, and **(b)** a weak acid (HA) has a high concentration of HA and low concentrations of H_3O^+ and A^-.

Q How does the height of H_3O^+ and A^- in the bar diagram change for a weak acid?

TABLE 14.5 K_a and K_b Values for Selected Weak Acids and Bases

Acids (K_a)		
Phosphoric acid	H_3PO_4	7.5×10^{-3}
Hydrofluoric acid	HF	7.2×10^{-4}
Nitrous acid	HNO_2	4.5×10^{-4}
Formic acid	HCOOH	1.8×10^{-4}
Acetic acid	CH_3-COOH	1.8×10^{-5}
Carbonic acid	H_2CO_3	4.3×10^{-7}
Hydrosulfuric acid	H_2S	9.1×10^{-8}
Dihydrogen phosphate	$H_2PO_4^-$	6.2×10^{-8}
Hydrocyanic acid	HCN	4.9×10^{-10}
Hydrogen phosphate	HPO_4^{2-}	2.2×10^{-13}
Bases (K_b)		
Methylamine	CH_3-NH_2	4.4×10^{-4}
Carbonate	CO_3^{2-}	2.2×10^{-4}
Ammonia	NH_3	1.8×10^{-5}

The K_a expression for carbonic acid is

$$K_a = \frac{[H_3O^+][HCO_3^-]}{[H_2CO_3]} = 4.3 \times 10^{-7}$$

We can conclude that weak acids have small K_a values because their equilibria favor the reactants. On the other hand, strong acids, which are essentially 100% dissociated, have large K_a values, although these values are not usually measured. Table 14.5 gives K_a values for selected weak acids. Recall that the concentration units are omitted for equilibrium constants.

Let us now consider the equilibrium expression for the weak base methylamine:

$$CH_3-NH_2(aq) + H_2O(l) \rightleftharpoons CH_3-NH_3^+(aq) + OH^-(aq)$$

As we did with the acid dissociation expression, the concentration of water is omitted from the equilibrium expression, which gives the **base dissociation constant, K_b** (or base ionization constant). Thus, for a weak base such as methylamine, the K_b is written

$$K_b = \frac{[CH_3-NH_3^+][OH^-]}{[CH_3-NH_2]} = 4.4 \times 10^{-4}$$

The K_b for methylamine is small because the equilibrium favors the reactants. The smaller the K_a or K_b value, the weaker the acid or base (see Table 14.5). We have described strong and weak acids in several ways. Table 14.6 summarizes the characteristics of acids and bases in terms of strength and equilibrium position.

TABLE 14.6 Characteristics of Acids and Bases

Characteristic	Strong Acid	Weak Acid
Equilibrium position	Toward ionized products	Toward unionized reactants
K_a	Large	Small
$[H_3O^+]$ and $[A^-]$	100% of [HA] reacts	Small percent of [HA] reacts
Conjugate base	Weak	Strong
Characteristic	**Strong Base**	**Weak Base**
Equilibrium position	Toward ionized products	Toward unionized reactants
K_b	Large	Small
$[BH^+]$ and $[OH^-]$	100% of [B] reacts	Small percent of [B] reacts
Conjugate acid	Weak	Strong

CONCEPT CHECK 14.5

■ **Acid Dissociation Constants**

Acid HX has a K_a of 4.0×10^{-4}, and acid HY has K_a of 8.0×10^{-6}. If each acid has a 0.10 M concentration, which solution has the higher concentration of H_3O^+?

ANSWER

Acid HX has a larger K_a value than acid HY. When acid HX dissolves in water, there is more dissociation of HX, which gives a higher concentration of H_3O^+ and X^- ions in solution.

SAMPLE PROBLEM | 14.5

■ **Writing an Acid Dissociation Constant Expression**

Write the acid dissociation constant expression for nitrous acid.

SOLUTION

The equation for the dissociation of nitrous acid is written

$$HNO_2(aq) + H_2O(l) \rightleftharpoons H_3O^+(aq) + NO_2^-(aq)$$

The acid dissociation constant expression is written as the concentrations of the products divided by the concentration of the undissociated weak acid.

$$K_a = \frac{[H_3O^+][NO_2^-]}{[HNO_2]}$$

STUDY CHECK

What is the value of the acid dissociation constant for nitrous acid?

QUESTIONS AND PROBLEMS

Dissociation Constants

14.27 Consider the following acids and their dissociation constants:

$$H_2SO_3(aq) + H_2O(l) \rightleftharpoons H_3O^+(aq) + HSO_3^-$$
$$K_a = 1.2 \times 10^{-2}$$

$$HS^-(aq) + H_2O(l) \rightleftharpoons H_3O^+(aq) + S^{2-}(aq)$$
$$K_a = 1.3 \times 10^{-19}$$

a. Which is the stronger acid, H_2SO_3 or HS^-?
b. What is the conjugate base of H_2SO_3?
c. Which acid has the weaker conjugate base?
d. Which acid has the stronger conjugate base?
e. Which acid produces more ions?

14.28 Consider the following acids and their dissociation constants:

$$HPO_4^{2-}(aq) + H_2O(l) \rightleftharpoons H_3O^+(aq) + PO_4^{3-}(aq)$$
$$K_a = 2.2 \times 10^{-13}$$

$$HCOOH(aq) + H_2O(l) \rightleftharpoons H_3O^+(aq) + HCOO^-(aq)$$
$$K_a = 1.8 \times 10^{-4}$$

a. Which is the weaker acid, HPO_4^{2-} or HCOOH?
b. What is the conjugate base of HPO_4^{2-}?
c. Which acid has the weaker conjugate base?
d. Which acid has the stronger conjugate base?
e. Which acid produces more ions?

14.29 Phosphoric acid dissociates to form dihydrogen phosphate and hydronium ion. Phosphoric acid has K_a of 7.5×10^{-3}. Write the equation and the acid dissociation constant expression for phosphoric acid.

14.30 Aniline, an organic amine, $C_6H_5NH_2$, a weak base with K_b of 4.0×10^{-10}, reacts with water to form its conjugate acid, $C_6H_5NH_3^+$. Write the equation and the base dissociation constant expression for aniline.

14.5 Ionization of Water

We have seen that in some acid–base reactions, water is amphoteric, which means that water can be both an acid and a base. This is exactly what happens with water molecules in pure water. Let's see how this happens. One water molecule acts as an acid by

LEARNING GOAL

Use the ion product of water to calculate the $[H_3O^+]$ and $[OH^-]$ in an aqueous solution.

TUTORIAL

Ionization of Water

donating H^+ to another water molecule, which is acting as a base. The products are H_3O^+ and OH^-. Let's take a look at the conjugate acid–base pairs of water:

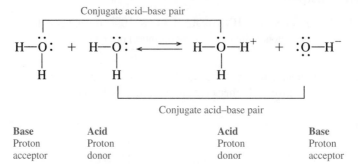

Base	**Acid**	**Acid**	**Base**
Proton	Proton	Proton	Proton
acceptor	donor	donor	acceptor

In the ionization of water, there is both a forward and reverse reaction:

$$H_2O(l) + H_2O(l) \rightleftharpoons H_3O^+(aq) + OH^-(aq)$$

Every time H^+ is transferred between two water molecules, the products are one H_3O^+ and one OH^-. Experiments have determined that in pure water, the concentrations of H_3O^+ and OH^- at 25 °C are each 1.0×10^{-7} M. Square brackets around the symbols indicate their concentrations in moles per liter (M).

$$\text{Pure water } [H_3O^+] = [OH^-] = 1.0 \times 10^{-7} \text{ M}$$

When these concentrations are multiplied, the expression and value is called the **ion product constant of water, K_w**, which is 1.0×10^{-14}. The concentration units are omitted in the K_w value.

$$K_w = [H_3O^+][OH^-]$$
$$= (1.0 \times 10^{-7} \text{ M})(1.0 \times 10^{-7} \text{ M}) = 1.0 \times 10^{-14}$$

The K_w value (1.0×10^{-14}) applies to any aqueous solution at 25 °C because all aqueous solutions contain both H_3O^+ and OH^-.

When the $[H_3O^+]$ and $[OH^-]$ in a solution are equal, the solution is **neutral**. However, most solutions are not neutral and have different concentrations of $[H_3O^+]$ and $[OH^-]$. If an acid is added to water, there is an increase in $[H_3O^+]$ and a decrease in $[OH^-]$, which makes an acidic solution. If base is added, $[OH^-]$ increases and $[H_3O^+]$ decreases, which makes a basic solution (see Figure 14.4). However, for any aqueous solution, whether it is neutral, acidic, or basic, the product $[H_3O^+][OH^-]$ is equal to K_w (1.0×10^{-14}).

FIGURE 14.4 In a neutral solution, $[H_3O^+]$ and $[OH^-]$ are equal. In acidic solutions, the $[H_3O^+]$ is greater than the $[OH^-]$. In basic solutions, the $[OH^-]$ is greater than the $[H_3O^+]$.

Q If the $[H_3O^+] = 1.0 \times 10^{-3}$ M, is the solution acidic, basic, or neutral?

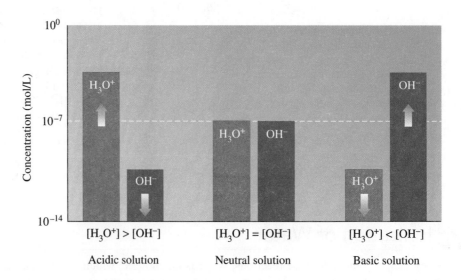

When the $[H_3O^+]$ is given, the K_w is rearranged to calculate $[OH^-]$. If the $[OH^-]$ is given, the K_w is rearranged to calculate $[H_3O^+]$.

$$K_w = [H_3O^+][OH^-]$$

$$[OH^-] = \frac{K_w}{[H_3O^+]} \qquad [H_3O^+] = \frac{K_w}{[OH^-]}$$

To illustrate these calculations, we can calculate $[H_3O^+]$ for a solution that has an $[OH^-] = 1.0 \times 10^{-6}$ M.

STEP 1 Write the K_w for water.

$$K_w = [H_3O^+][OH^-] = 1.0 \times 10^{-14}$$

STEP 2 Solve the K_w for the unknown $[H_3O^+]$. Rearrange the K_w by dividing through by the $[OH^-]$.

$$\frac{K_w}{[OH^-]} = \frac{[H_3O^+][\cancel{OH^-}]}{[\cancel{OH^-}]} = \frac{1.0 \times 10^{-14}}{[OH^-]}$$

$$[H_3O^+] = \frac{1.0 \times 10^{-14}}{[OH^-]}$$

STEP 3 Substitute the known $[OH^-]$ and calculate.

$$[H_3O^+] = \frac{1.0 \times 10^{-14}}{[1.0 \times 10^{-6}]} = 1.0 \times 10^{-8} \text{ M}$$

Because the $[OH^-]$ of 1.0×10^{-6} M is larger than the $[H_3O^+]$ of 1.0×10^{-8} M, the solution is basic (see Table 14.7).

TABLE 14.7 Examples of $[H_3O^+]$ and $[OH^-]$ in Neutral, Acidic, and Basic Solutions

Type of Solution	$[H_3O^+]$	$[OH^-]$	K_w
Neutral	1.0×10^{-7} M	1.0×10^{-7} M	1.0×10^{-14}
Acidic	1.0×10^{-2} M	1.0×10^{-12} M	1.0×10^{-14}
Acidic	2.5×10^{-5} M	4.0×10^{-10} M	1.0×10^{-14}
Basic	1.0×10^{-8} M	1.0×10^{-6} M	1.0×10^{-14}
Basic	5.0×10^{-11} M	2.0×10^{-4} M	1.0×10^{-14}

SAMPLE PROBLEM | 14.6

■ **Calculating $[H_3O^+]$ and $[OH^-]$ in Solution**

A vinegar solution has a $[H_3O^+] = 2.0 \times 10^{-3}$ M at 25 °C. What is the $[OH^-]$ of the vinegar solution? Is the solution acidic, basic, or neutral?

SOLUTION

STEP 1 Write the K_w for water.

$$K_w = [H_3O^+][OH^-] = 1.0 \times 10^{-14}$$

STEP 2 Solve the K_w for the unknown $[OH^-]$. Rearranging the K_w by dividing through by $[H_3O^+]$ gives

$$\frac{K_w}{[H_3O^+]} = \frac{[\cancel{H_3O^+}]\,[OH^-]}{[\cancel{H_3O^+}]} = \frac{1.0 \times 10^{-14}}{[H_3O^+]}$$

$$[OH^-] = \frac{1.0 \times 10^{-14}}{[H_3O^+]}$$

Guide to Calculating $[H_3O^+]$ and $[OH^-]$ in Aqueous Solutions

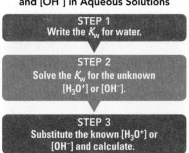

STEP 1
Write the K_w for water.

STEP 2
Solve the K_w for the unknown $[H_3O^+]$ or $[OH^-]$.

STEP 3
Substitute the known $[H_3O^+]$ or $[OH^-]$ and calculate.

STEP 3 Substitute the known $[H_3O^+]$ and calculate.

$$[OH^-] = \frac{1.0 \times 10^{-14}}{[2.0 \times 10^{-3}]} = 5.0 \times 10^{-12} \, M$$

Because the $[H_3O^+]$ of $2.0 \times 10^{-3} \, M$ is much larger than the $[OH^-]$ of $5.0 \times 10^{-12} \, M$, the solution is acidic.

STUDY CHECK

What is the $[H_3O^+]$ of an ammonia cleaning solution with an $[OH^-] = 4.0 \times 10^{-4} \, M$? Is the solution acidic, basic, or neutral?

QUESTIONS AND PROBLEMS

Ionization of Water

14.31 Why are the concentrations of H_3O^+ and OH^- equal in pure water?

14.32 What is the meaning and value of K_w at 25 °C?

14.33 In an acidic solution, how does the concentration of H_3O^+ compare to the concentration of OH^-?

14.34 If a base is added to pure water, why does the $[H_3O^+]$ decrease?

14.35 Indicate whether each of the following solutions are acidic, basic, or neutral at 25 °C:
a. $[H_3O^+] = 2.0 \times 10^{-5} \, M$ **b.** $[H_3O^+] = 1.4 \times 10^{-9} \, M$
c. $[OH^-] = 8.0 \times 10^{-3} \, M$ **d.** $[OH^-] = 3.5 \times 10^{-10} \, M$

14.36 Indicate whether each of the following solutions are acidic, basic, or neutral at 25 °C:
a. $[H_3O^+] = 6.0 \times 10^{-12} \, M$
b. $[H_3O^+] = 1.4 \times 10^{-4} \, M$
c. $[OH^-] = 5.0 \times 10^{-12} \, M$
d. $[OH^-] = 4.5 \times 10^{-2} \, M$

14.37 Calculate the $[H_3O^+]$ of each aqueous solution with the following $[OH^-]$ at 25 °C:
a. coffee, $1.0 \times 10^{-9} \, M$ **b.** soap, $1.0 \times 10^{-6} \, M$
c. cleanser, $2.0 \times 10^{-5} \, M$ **d.** lemon juice, $4.0 \times 10^{-13} \, M$

14.38 Calculate the $[H_3O^+]$ of each aqueous solution with the following $[OH^-]$ at 25 °C:
a. NaOH, $1.0 \times 10^{-2} \, M$
b. aspirin, $1.8 \times 10^{-11} \, M$
c. milk of magnesia, $1.0 \times 10^{-5} \, M$
d. seawater, $2.0 \times 10^{-6} \, M$

14.39 Calculate the $[OH^-]$ of each aqueous solution with the following $[H_3O^+]$ at 25 °C:
a. vinegar, $1.0 \times 10^{-3} \, M$
b. urine, $5.0 \times 10^{-6} \, M$
c. ammonia, $1.8 \times 10^{-12} \, M$
d. NaOH, $4.0 \times 10^{-13} \, M$

14.40 Calculate the $[OH^-]$ of each aqueous solution with the following $[H_3O^+]$ at 25 °C:
a. baking soda, $1.0 \times 10^{-8} \, M$
b. orange juice, $2.0 \times 10^{-4} \, M$
c. milk, $5.0 \times 10^{-7} \, M$
d. bleach, $4.8 \times 10^{-12} \, M$

LEARNING GOAL

Calculate pH from $[H_3O^+]$; given the pH, calculate $[H_3O^+]$ and $[OH^-]$ of a solution.

CASE STUDY

Hyperventilation and Blood pH

SELF STUDY ACTIVITY

The pH Scale

14.6 The pH Scale

Many kinds of careers, such as respiratory therapy, food processing, medicine, agriculture, and wine making, require the measurement of $[H_3O^+]$ and $[OH^-]$. The proper level of acidity is necessary to evaluate the functioning of the lungs and kidneys, to control bacterial growth in foods, and to prevent the growth of pests in food crops.

On the pH scale, a number between 0 and 14 represents the H_3O^+ concentration for most solutions. A neutral solution has a pH of 7.0 at 25 °C. An acidic solution has a pH value less than 7.0 A basic solution has a pH value greater than 7.0 (see Figure 14.5).

In the laboratory, a pH meter is commonly used to determine the pH of a solution. There are also indicators and pH papers that turn specific colors when placed in solutions of different pH values. The pH is found by comparing the colors to a color chart (see Figure 14.6).

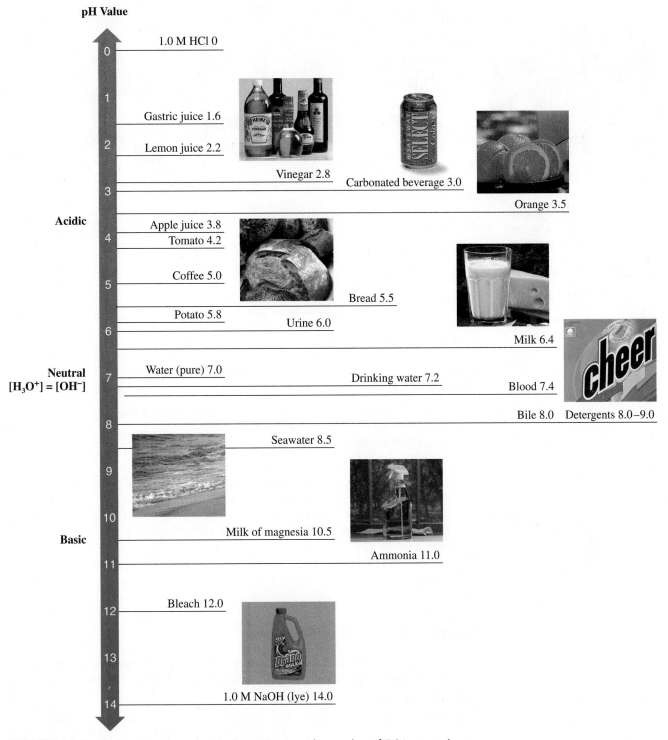

pH Value

0	1.0 M HCl 0
1	
	Gastric juice 1.6
2	Lemon juice 2.2
	Vinegar 2.8 Carbonated beverage 3.0
3	Orange 3.5

Acidic

4	Apple juice 3.8 Tomato 4.2
5	Coffee 5.0 Bread 5.5
	Potato 5.8 Urine 6.0
6	Milk 6.4

Neutral
[H$_3$O$^+$] = [OH$^-$]

7	Water (pure) 7.0 Drinking water 7.2 Blood 7.4
8	Bile 8.0 Detergents 8.0–9.0
	Seawater 8.5
9	
10	

Basic

	Milk of magnesia 10.5
11	Ammonia 11.0
12	Bleach 12.0
13	
14	1.0 M NaOH (lye) 14.0

FIGURE 14.5 On the pH scale, values below 7.0 are acidic, a value of 7.0 is neutral, and values above 7.0 are basic.

Q Is apple juice an acidic, basic, or a neutral solution?

Acidic solution	pH < 7.0	[H$_3$O$^+$] > 1.0 × 10^{-7} M
Neutral solution	pH = 7.0	[H$_3$O$^+$] = 1.0 × 10^{-7} M
Basic solution	pH > 7.0	[H$_3$O$^+$] < 1.0 × 10^{-7} M

(a)

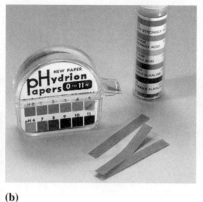

(b)

(c)

FIGURE 14.6 The pH of a solution can be determined using **(a)** a pH meter, **(b)** pH paper, and **(c)** indicators that turn different colors corresponding to different pH values.

Q If a pH meter reads 4.00, is the solution acidic, basic, or neutral?

CONCEPT CHECK 14.6

■ **pH of Solutions**

Consider the pH of the following items:

Item	pH
Root beer	5.8
Kitchen cleaner	10.9
Pickles	3.5
Glass cleaner	7.6
Cranberry juice	2.9

a. Place the pH values of the items on the list in order of most acidic to most basic.
b. Which item has the highest $[H_3O^+]$?
c. Which item has the highest $[OH^-]$?

ANSWER
a. The most acidic item is the one with the lowest pH, and the most basic is the item with the highest pH: cranberry juice (2.9), pickles (3.5), root beer (5.8), glass cleaner (7.6), kitchen cleaner (10.9).
b. The item with the highest $[H_3O^+]$ would have the lowest pH value, which is cranberry juice.
c. The item with the highest $[OH^-]$ would have the highest pH value, which is kitchen cleaner.

Calculating the pH of Solutions

The pH scale is a logarithmic scale that corresponds to the hydrogen-ion concentrations of aqueous solutions. Mathematically, **pH** is the negative logarithm (base 10) of the $[H_3O^+]$.

$$pH = -\log[H_3O^+]$$

Essentially, the negative powers of 10 in the molar concentrations are converted to positive numbers. For example, a lemon juice solution with $[H_3O^+] = 1.0 \times 10^{-2}$ M has a pH of 2.00. This can be calculated using the pH equation:

$$pH = -\log[1.0 \times 10^{-2}]$$
$$pH = -(-2.00)$$
$$= 2.00$$

The number of *decimal places* in the pH value is the same as the number of significant figures in the $[H_3O^+]$.

TUTORIAL
The pH Scale

$$[H_3O^+] = \mathbf{1.0} \times 10^{-2} \qquad pH = \mathbf{2.00}$$

Two significant figures   Two decimal places

CONCEPT CHECK 14.7

■ Calculating pH

Indicate if the pH values given for each of the following are correct or incorrect and why:

a. $[H_3O^+] = 1 \times 10^{-6}$ pH = -6.0
b. $[OH^-] = 1.0 \times 10^{-10}$ pH = 10.00
c. $[H_3O^+] = 1.0 \times 10^{-6}$ pH = 6.00
d. $[OH^-] = 1 \times 10^{-2}$ pH = 12.0

ANSWER

a. Incorrect. The pH of this solution is 6.0, which has a positive value, not negative.
b. Incorrect. The pH is calculated from the $[H_3O^+]$, not $[OH^-]$. This solution has a $[H_3O^+]$ of 1.0×10^{-4} M, which has a pH of 4.00. There are two zeros after the decimal point to match the two significant figures in the coefficient of the molarity.
c. The pH is correctly calculated from the $[H_3O^+]$ because there are two zeros after the decimal point to match the two significant figures in the coefficient of the molarity.
d. The pH is correctly calculated since $[H_3O^+] = 1 \times 10^{-12}$ M with one zero after the decimal point to match the one significant figure in the coefficient of the molarity.

Steps for a pH Calculation

The pH of a solution is determined using the *log* key and the *change sign* key. For example, to calculate the pH of a vinegar solution with $[H_3O^+] = 2.4 \times 10^{-3}$ M you can use the following steps:

STEP 1 Enter the $[H_3O^+]$ value.
Enter 2.4 and press [EE or EXP].

Display
$2.4^{\,00}$ or $2.4\ 00$ or $2.4\ E00$

Enter 3 and press [+/–] to change the sign. (For calculators without a change sign key, consult the instructions for the calculator.)

Display
$2.4^{\,-03}$ or $2.4{-}03$ or $2.4\ E{-}03$

TUTORIAL
Logarithms

STEP 2 Press the [log] key and change the sign.

-2.619789

2.619789

The steps can be combined to give the calculator sequence as follows:

pH = $-\log[2.4 \times 10^{-3}]$ = 2.4 [EE or EXP] 3 [+/–] [log] [+/–]

= 2.619789

Be sure to check the instructions for your calculator. On some calculators, the log key is used first, followed by the concentration.

STEP 3 Adjust the number of significant figures on the right of the decimal point to equal the SFs in the coefficient. In a pH value, the number to the *left* of the decimal point is an *exact* number derived from the power of 10. The number of digits to the *right* of the decimal point is equal to the number of significant figures in the coefficient.

Coefficient	**Power of 10**	

$[H_3O^+] = \underset{\substack{\text{Two significant} \\ \text{figures (2 SFs)}}}{2.4} \times \underset{\text{Exact}}{10^{-3}} \text{ M}$ $\text{pH} = -\log[2.4 \times 10^{-3}] = \underset{\substack{\text{Exact} \quad \text{Two decimal} \\ \text{places}}}{2.62}$

Because pH is a log scale, a change of one pH unit corresponds to a tenfold change in $[H_3O^+]$. It is important to note that the pH decreases as the $[H_3O^+]$ increases. For example, a solution with a pH of 2.00 has a $[H_3O^+]$ 10 times higher than a solution with a pH of 3.00, and 100 times higher than a solution with a pH of 4.00.

SAMPLE PROBLEM   14.7

■ **Calculating pH**

Determine the pH for a solution with $[H_3O^+] = 5 \times 10^{-8}$ M.

SOLUTION

Guide to Calculating pH of an Aqueous Solution

STEP 1
Enter the $[H_3O^+]$ value.

↓

STEP 2
Press the *log* key and change the sign.

↓

STEP 3
Adjust the number of significant figures on the *right* of the decimal point to equal the SFs in the coefficient.

STEP 1   Enter the $[H_3O^+]$ value.

5 $\boxed{\text{EE or EXP}}$ 8 $\boxed{+/-}$

Display

5^{-08} or $5-08$ or $5E-08$

STEP 2   Press the *log* key and change the sign.

$\boxed{\text{log}}$ $\boxed{+/-}$

7.301029

STEP 3   Adjust the number of significant figures on the *right* of the decimal point to equal the SFs in the coefficient.

5×10^{-8} M   pH = 7.3

1 SF ⟶ 1 SF on the *right* of the decimal point

STUDY CHECK

What is the pH of bleach with $[H_3O^+] = 4.2 \times 10^{-12}$ M?

SAMPLE PROBLEM   14.8

■ **Calculating pH from $[OH^-]$**

What is the pH of an ammonia solution with $[OH^-] = 3.7 \times 10^{-3}$ M?

SOLUTION

STEP 1   Enter the $[H_3O^+]$ value. Because $[OH^-]$ is given for the ammonia solution, we have to calculate $[H_3O^+]$ using the ion product constant of water, K_w. Divide through by $[OH^-]$ to give $[H_3O^+]$.

$$\frac{K_w}{[OH^-]} = \frac{[H_3O^+]\,[\cancel{OH^-}]}{[\cancel{OH^-}]} = \frac{1.0 \times 10^{-14}}{[OH^-]}$$

$$[H_3O^+] = \frac{1.0 \times 10^{-14}}{[3.7 \times 10^{-3}]} = 2.7 \times 10^{-12} \text{ M}$$

Display

$\text{pH} = -\log[2.7 \times 10^{-12}] = 2.7\ \boxed{\text{EE or EXP}}\ 12\ \boxed{+/-} = 2.7^{-12}$ or $2.7-12$ or $2.7E-12$

STEP 2   Press the *log* key and change the sign.

$\boxed{\text{log}}$ $\boxed{+/-}$

11.56863

STEP 3   Adjust the number of significant figures on the *right* of the decimal point to equal the SFs in the coefficient.

2.7×10^{-12} M   pH = 11.57

2 SFs ⟶ 2 SFs on the *right* of the decimal point

STUDY CHECK

Calculate the pH of a sample of acid rain that has $[OH^-] = 2 \times 10^{-10}$ M.

---

pOH

The **pOH** scale is similar to the pH scale except that pOH is associated with the $[OH^-]$ of an aqueous solution.

$$\text{pOH} = -\log[OH^-]$$

Solutions with high [OH⁻] have low pOH values; solutions with low [OH⁻] have high pOH values. In any aqueous solution, the sum of the pH and pOH is equal to 14.00, which is the negative logarithm of the K_w:

$$pH + pOH = 14.00$$

For example, if the pH of a solution is 3.50, the pOH can be calculated as follows:

$$pH + pOH = 14.00$$
$$pOH = 14.00 - pH = 14.00 - 3.50 = 10.50$$

Calculating [H₃O⁺] from pH

In another calculation, we are given the pH of the solution and asked to determine the [H₃O⁺]. This is a reverse of the pH calculation:

$$[H_3O^+] = 10^{-pH}$$

For pH values that are not whole numbers, the calculation requires the use of the *10ˣ* key, which is usually a *2ⁿᵈ function* key. On some calculators, this operation is done using the *inverse* key and the *log* key.

SAMPLE PROBLEM | 14.9

■ Calculating [H₃O⁺] from pH

Calculate [H₃O⁺] for a solution of baking soda with a pH of 8.25.

SOLUTION

STEP 1 Enter the pH value and change the sign.   **Display**

    8.25 ⌊+/−⌋   −8.25

STEP 2 Convert −pH to [H₃O⁺]. Press the *2ⁿᵈ function* key and then the *10ˣ* key or press the *inverse* key and then the *log* key.

    ⌊2ⁿᵈ⌋ ⌊10ˣ⌋   5.62341⁻⁰⁹ or 5.62341−09 or 5.62341 E−09

    ⌊inv⌋ ⌊log⌋   5.62341⁻⁰⁹ or 5.62341−09 or 5.62341 E−09

Write the display in scientific notation with units of concentration: 5.62341×10^{-9} M

STEP 3 Adjust the significant figures in the coefficient. Because the pH value of 8.25 has two digits on the *right* of the decimal point, the [H₃O⁺] is written with two significant figures:

$$[H_3O^+] = 5.6 \times 10^{-9} \text{ M}$$

STUDY CHECK

What are the [H₃O⁺] and [OH⁻] of beer that has a pH of 4.50?

CHEMISTRY AND HEALTH

Stomach Acid, HCl

When a person sees, smells, thinks about, or tastes food, the gastric glands in the stomach begin to secrete a strongly acidic solution of HCl. In a single day, a person may secrete as much as 2000 mL of gastric juice.

The HCl in the gastric juice activates a digestive enzyme called *pepsin*, which breaks down proteins in food entering the stomach.

The secretion of HCl continues until the stomach has a pH of about 2, which is the optimum pH for activating the digestive enzymes without ulcerating the stomach lining. Normally, large quantities of viscous mucus are secreted within the stomach to protect its lining from acid and enzyme damage.

A comparison of $[H_3O^+]$, $[OH^-]$, and their corresponding pH and pOH values is given in Table 14.8.

TABLE 14.8 A Comparison of $[H_3O^+]$, $[OH^-]$, and Corresponding pH Values at 25 °C

$[H_3O^+]$	pH	$[OH^-]$	pOH	
10^0	0	10^{-14}	14	
10^{-1}	1	10^{-13}	13	
10^{-2}	2	10^{-12}	12	
10^{-3}	3	10^{-11}	11	Acidic
10^{-4}	4	10^{-10}	10	
10^{-5}	5	10^{-9}	9	
10^{-6}	6	10^{-8}	8	
10^{-7}	7	10^{-7}	7	Neutral
10^{-8}	8	10^{-6}	6	
10^{-9}	9	10^{-5}	5	
10^{-10}	10	10^{-4}	4	
10^{-11}	11	10^{-3}	3	Basic
10^{-12}	12	10^{-2}	2	
10^{-13}	13	10^{-1}	1	
10^{-14}	14	10^0	0	

QUESTIONS AND PROBLEMS

The pH Scale

14.41 Why does a neutral solution have a pH of 7.00?

14.42 If you know the $[OH^-]$, how can you determine the pH of a solution?

14.43 State whether each of the following solutions is acidic, basic, or neutral:
a. blood, pH 7.38   b. vinegar, pH 2.8
c. drain cleaner, pOH 2.8 d. coffee, pH 5.52
e. tomatoes, pH 4.2   f. chocolate cake, pH 7.6

14.44 State whether each of the following solutions is acidic, basic, or neutral:
a. soda, pH 3.22   b. shampoo, pOH 8.3
c. laundry detergent, pOH 4.56   d. rain, pH 5.8
e. honey, pH 3.9   f. cheese, pH 7.4

14.45 A solution with a pH of 3 is 10 times more acidic than a solution with pH 4. Explain.

14.46 A solution with a pH of 10 is 100 times more basic than a solution with pH 8. Explain.

14.47 Calculate the pH of each solution given the following $[H_3O^+]$ or $[OH^-]$ values:
a. $[H_3O^+] = 1.0 \times 10^{-4}$ M
b. $[H_3O^+] = 3.0 \times 10^{-9}$ M
c. $[OH^-] = 1.0 \times 10^{-5}$ M
d. $[OH^-] = 2.5 \times 10^{-11}$ M
e. $[H_3O^+] = 6.7 \times 10^{-8}$ M
f. $[OH^-] = 8.2 \times 10^{-4}$ M

14.48 Calculate the pH of each solution given the following $[H_3O^+]$ or $[OH^-]$ values:
a. $[H_3O^+] = 1.0 \times 10^{-8}$ M
b. $[H_3O^+] = 5.0 \times 10^{-6}$ M
c. $[OH^-] = 4.0 \times 10^{-2}$ M
d. $[OH^-] = 8.0 \times 10^{-3}$ M
e. $[H_3O^+] = 4.7 \times 10^{-2}$ M
f. $[OH^-] = 3.9 \times 10^{-6}$ M

14.49 Complete the following table:

$[H_3O^+]$	$[OH^-]$	pH	pOH	Acidic, Basic, or Neutral?
	1.0×10^{-6} M			
		3.49		
2.8×10^{-5} M				
			2.00	

14.50 Complete the following table:

$[H_3O^+]$	$[OH^-]$	pH	pOH	Acidic, Basic, or Neutral?
		10.00		
				Neutral
			5.66	
6.4×10^{-12} M				

14.7 Reactions of Acids and Bases

Typical reactions of acids and bases include the reactions of acids with metals, bases, and carbonate or bicarbonate ions. For example, when you drop an antacid tablet in water, the bicarbonate ion and citric acid in the tablet react to produce carbon dioxide bubbles, a *salt*, and water. A **salt** is an ionic compound that does not have H^+ as the cation or OH^- as the anion. $NaCl$, CaF_2, and NH_4NO_3 are examples of some salts.

LEARNING GOAL

Write balanced equations for reactions of acids and bases.

Acids and Metals

Acids react with certain metals known as *active metals* to produce hydrogen gas (H_2) and a salt. Active metals include potassium, sodium, calcium, magnesium, aluminum, zinc, iron, and tin. In these single replacement reactions, the metal ion replaces the hydrogen in the acid.

$$Mg(s) + 2HCl(aq) \longrightarrow MgCl_2(aq) + H_2(g)$$

Metal   Acid   Salt   Hydrogen

$$Zn(s) + 2HNO_3(aq) \longrightarrow Zn(NO_3)_2(aq) + H_2(g)$$

Metal   Acid   Salt   Hydrogen

(MC)

SELF STUDY ACTIVITY

Nature of Acids and Bases

Acids, Carbonates, and Bicarbonates

When an acid is added to a carbonate or bicarbonate, the products are carbon dioxide gas, water, and a salt. The acid reacts with CO_3^{2-} or HCO_3^- to produce carbonic acid, H_2CO_3, which breaks down rapidly to CO_2 and H_2O.

$$HBr(aq) + NaHCO_3(aq) \longrightarrow CO_2(g) + H_2O(l) + NaBr(aq)$$

Acid   Bicarbonate   Carbon   Water   Salt
                               dioxide

$$2HCl(aq) + Na_2CO_3(aq) \longrightarrow CO_2(g) + H_2O(l) + 2NaCl(aq)$$

Acid   Carbonate   Carbon   Water   Salt
                               dioxide

Acids and Bases: Neutralization

Neutralization is a reaction between an acid and a base to produce a salt and water. The H^+ of an acid and the OH^- of a strong base combine to form water as one product. The salt is the cation from the base and the anion from the acid. We can write the following equation for the neutralization reaction between HCl and NaOH:

$$HCl(aq) + NaOH(aq) \longrightarrow NaCl(aq) + H_2O(l)$$

Acid   Base   Salt   Water

If we write the strong acid HCl and the strong base NaOH as ions, we see that H^+ combines with OH^- to form water, leaving the ions Na^+ and Cl^- in solution.

$$\mathbf{H^+}(aq) + Cl^-(aq) + Na^+(aq) + \mathbf{OH^-}(aq) \longrightarrow Na^+(aq) + Cl^-(aq) + \mathbf{H_2O}(l)$$

When we omit the ions that do not change during the reaction (spectator ions), we obtain the *net ionic equation*.

$$\mathbf{H^+}(aq) + \cancel{Cl^-(aq)} + \cancel{Na^+(aq)} + \mathbf{OH^-}(aq) \longrightarrow \cancel{Na^+(aq)} + \cancel{Cl^-(aq)} + \mathbf{H_2O}(l)$$

The net ionic equation for the neutralization is the reaction of H^+ and OH^- to form H_2O.

$$H^+(aq) + OH^-(aq) \longrightarrow H_2O(l) \qquad \text{Net ionic equation}$$

CHEMISTRY AND THE ENVIRONMENT

Acid Rain

Natural rain is slightly acidic, with a pH of 5.6. In the atmosphere, carbon dioxide combines with water to form carbonic acid, a weak acid, which dissociates to give hydronium ions and bicarbonate:

$$CO_2(g) + H_2O(l) \rightleftharpoons H_2CO_3(aq)$$
$$H_2CO_3(aq) + H_2O(l) \rightleftharpoons H_3O^+(aq) + HCO_3^-(aq)$$

However, in many parts of the world, rain has become considerably more acidic. *Acid rain* is a term given to precipitation such as rain, snow, hail, or fog, in which the water has a pH that is less than 5.6. In the United States, pH values of rain have decreased to about 4–4.5. In some parts of the world, pH values have been reported as low as 2.6, which is about as acidic as lemon juice or vinegar. Because the calculation of pH involves powers of 10, a pH value of 2.6 would be 1000 times more acidic than natural rain.

Although natural sources such as volcanoes and forest fires release SO_2, the primary sources of acid rain today are from the burning of fossil fuels in automobiles and coal in industrial plants. When coal and oil are burned, the sulfur impurities combine with oxygen in the air to produce SO_2 and then SO_3. The reaction of SO_3 with water forms sulfuric acid, H_2SO_4, a strong acid.

$$S(s) + O_2(g) \longrightarrow O_2(g)$$
$$2SO_2(g) + O_2(g) \longrightarrow 2SO_3(g)$$
$$SO_3(g) + H_2O(l) \longrightarrow H_2SO_4(aq)$$

In an effort to decrease the formation of acid rain, legislation has required a reduction in SO_2 emissions. Coal-burning plants have installed equipment called "scrubbers" that absorb SO_2 before it is emitted. In a smokestack, "scrubbing" removes 95% of the SO_2 as the flue gases containing SO_2 pass through limestone ($CaCO_3$) and water. The end product, $CaSO_4$, also called "gypsum," is used in agriculture and to prepare cement products.

Nitrogen oxide forms at high temperatures in the engines of automobiles as air containing nitrogen and oxygen gases is burned. As nitrogen oxide is emitted into the air, it combines with more oxygen to form nitrogen dioxide, which is responsible for the brown color of smog. When nitrogen dioxide dissolves in water in the atmosphere, nitric acid forms.

A marble statue in Washington Square Park that has been eroded by acid rain.

$$N_2(g) + O_2(g) \longrightarrow 2NO(g)$$
$$2NO(g) + O_2(g) \longrightarrow 2NO_2(g)$$
$$3NO_2(g) + H_2O(g) \longrightarrow 2HNO_3(aq) + NO(g)$$

Air currents in the atmosphere carry the sulfuric acid and nitric acid many thousands of kilometers before they precipitate in areas far away from the site of the contamination. The acids in acid rain have detrimental effects on marble and limestone structures, lakes, and forests. Throughout the world, monuments made of marble (a form of $CaCO_3$) are deteriorating as acid rain dissolves the marble:

$$CaCO_3(s) + H_2SO_4(aq) \longrightarrow CaSO_4(aq) + H_2O(l) + CO_2(g)$$

Acid rain is changing the pH of many lakes and streams in parts of the United States and Europe. When the pH of a lake falls below 4.5–5, most fish and plant life cannot survive. As the soil near a lake becomes more acidic, aluminum becomes more soluble. Increased levels of aluminum ion in lakes are toxic to fish and other water animals.

Trees and forests are susceptible to acid rain, too. Acid rain breaks down the protective waxy coating on leaves and interferes with photosynthesis. Tree growth is impaired as nutrients and minerals in the soil dissolve and wash away. In Eastern Europe, acid rain is causing an environmental disaster. Nearly 70% of the forests in the Czech Republic have been severely damaged, and some parts of the land are so acidic that crops will not grow.

Acid rain has severely damaged forests in Eastern Europe.

Balancing Neutralization Equations

In a neutralization reaction, one H^+ always reacts with one OH^-. Therefore, the coefficients in the neutralization equation must be chosen so that the H^+ from the acid is balanced by the OH^- in the base. We balance the neutralization of HCl and $Ba(OH)_2$ as follows:

STEP 1 **Write the reactants and products.**

$$HCl(aq) + Ba(OH)_2(s) \longrightarrow H_2O(l) + salt$$

STEP 2 **Balance the H^+ in the acid with the OH^- in the base.** Placing a 2 in front of the HCl provides two H^+ for the two OH^- in $Ba(OH)_2$.

$$2HCl(aq) + Ba(OH)_2(s) \longrightarrow H_2O(l) + salt$$

STEP 3 **Balance the H_2O with the H^+ and the OH^-.** Use a coefficient of 2 in front of H_2O to balance $2H^+$ and $2OH^-$.

$$2HCl(aq) + Ba(OH)_2(s) \longrightarrow 2H_2O(l) + salt$$

STEP 4 **Write the salt from the remaining ions.** The ions Ba^{2+} and $2Cl^-$ are used to write the formula of the salt as $BaCl_2$.

$$2HCl(aq) + \mathbf{Ba(OH)_2}(s) \longrightarrow 2H_2O(l) + \mathbf{BaCl_2}(aq)$$

Guide to Balancing an Equation for Neutralization

STEP 1
Write the reactants and products.

STEP 2
Balance the H^+ in the acid with the OH^- in the base.

STEP 3
Balance the H_2O with the H^+ and the OH^-.

STEP 4
Write the salt from the remaining ions.

CONCEPT CHECK 14.8

■ Reactions of Acids

Write a balanced equation for the reaction of HCl(aq) with each of the following:

a. Al(s)   **b.** K_2CO_3(s)

ANSWER

a. Al

When an active metal reacts with an acid, the products are $H_2(g)$ and a salt:

$$Al(s) + HCl(aq) \longrightarrow H_2(g) + salt$$

The metal ion Al^{3+} and the anion Cl^- combine to form a salt $AlCl_3(aq)$:

$$\mathbf{Al}(s) + HCl(aq) \longrightarrow H_2(g) + \mathbf{AlCl_3}(aq)$$

Now coefficients are added to balance the equation:

$$2Al(s) + 6HCl(aq) \longrightarrow 3H_2(g) + 2AlCl_3(aq)$$

b. K_2CO_3

When a carbonate reacts with an acid, the products are $CO_2(g)$, $H_2O(l)$, and a salt:

$$K_2CO_3(s) + HCl(aq) \longrightarrow CO_2(g) + H_2O(l) + salt$$

The metal ion K^+ and the anion Cl^- combine to form a salt KCl:

$$\mathbf{K_2CO_3}(s) + HCl(aq) \longrightarrow CO_2(g) + H_2O(l) + \mathbf{KCl}(aq)$$

Now coefficients are added to balance the equation:

$$K_2CO_3(s) + 2HCl(aq) \longrightarrow CO_2(g) + H_2O(l) + 2KCl(aq)$$

QUESTIONS AND PROBLEMS

Reactions of Acids and Bases

14.51 Complete and balance the equations for each of the following reactions:
  a. $ZnCO_3(s) + HBr(aq) \longrightarrow$
  b. $Zn(s) + HCl(aq) \longrightarrow$
  c. $HCl(aq) + NaHCO_3(s) \longrightarrow$
  d. $H_2SO_4(aq) + Mg(OH)_2(s) \longrightarrow$

14.52 Complete and balance the equations for each of the following reactions:
  a. $KHCO_3(s) + HBr(aq) \longrightarrow$
  b. $Ca(s) + H_2SO_4(aq) \longrightarrow$
  c. $H_2SO_4(aq) + Ca(OH)_2(s) \longrightarrow$
  d. $Na_2CO_3(s) + H_2SO_4(aq) \longrightarrow$

14.53 Balance each of the following neutralization reactions:

a. $HCl(aq) + Mg(OH)_2(s) \longrightarrow MgCl_2(aq) + H_2O(l)$

b. $H_3PO_4(aq) + LiOH(aq) \longrightarrow Li_3PO_4(aq) + H_2O(l)$

14.54 Balance each of the following neutralization reactions:

a. $HNO_3(aq) + Ba(OH)_2(s) \longrightarrow$
$$Ba(NO_3)_2(aq) + H_2O(l)$$

b. $H_2SO_4(aq) + Al(OH)_3(s) \longrightarrow$
$$Al_2(SO_4)_3(aq) + H_2O(l)$$

14.55 Write a balanced equation for the neutralization of each of the following:

a. $H_2SO_4(aq)$ and $NaOH(aq)$

b. $HCl(aq)$ and $Fe(OH)_3(s)$

c. $H_2CO_3(aq)$ and $Mg(OH)_2(s)$

14.56 Write a balanced equation for the neutralization of each of the following:

a. $H_3PO_4(aq)$ and $NaOH(aq)$

b. $HI(aq)$ and $LiOH(aq)$

c. $HNO_3(aq)$ and $Ca(OH)_2(s)$

CHEMISTRY AND HEALTH

Antacids

Antacids are substances used to neutralize excess stomach acid (HCl). Some antacids are mixtures of aluminum hydroxide and magnesium hydroxide. These hydroxides are not very soluble in water, so the levels of available OH^- are not damaging to the intestinal tract. However, aluminum hydroxide has the side effects of producing constipation and binding phosphate in the intestinal tract, which may cause weakness and loss of appetite. Magnesium hydroxide has a laxative effect. These side effects are less likely when a combination of the antacids is used.

$$Al(OH)_3(s) + 3HCl(aq) \longrightarrow AlCl_3(aq) + 3H_2O(l)$$
$$Mg(OH)_2(s) + 2HCl(aq) \longrightarrow MgCl_2(aq) + 2H_2O(l)$$

Antacids neutralize excess stomach acid.

Some antacids use calcium carbonate to neutralize excess stomach acid. About 10 percent of the calcium is absorbed into the bloodstream, where it elevates the levels of serum calcium. Calcium carbonate is not recommended for patients who have peptic ulcers or a tendency to form kidney stones.

$$CaCO_3(s) + 2HCl(aq) \longrightarrow H_2O(l) + CO_2(g) + CaCl_2(aq)$$

Still other antacids contain sodium bicarbonate. This type of antacid has a tendency to increase blood pH and elevate sodium levels in the body fluids. It also is not recommended in the treatment of peptic ulcers.

$$NaHCO_3(s) + HCl(aq) \longrightarrow NaCl(aq) + CO_2(g) + H_2O(l)$$

The neutralizing substances in some antacid preparations are given in Table 14.9.

TABLE 14.9 Basic Compounds in Some Antacids

Antacid	Base(s)
Amphojel	$Al(OH)_3$
Milk of magnesia	$Mg(OH)_2$
Mylanta, Maalox, Di-Gel, Gelusil, Riopan	$Mg(OH)_2$, $Al(OH)_3$
Bisodol, Rolaids	$CaCO_3$, $Mg(OH)_2$
Titralac, Tums, Pepto-Bismol	$CaCO_3$
Alka-Seltzer	$NaHCO_3$, $KHCO_3$

14.8 Acid–Base Titration

LEARNING GOAL

Calculate the molarity or volume of an acid or base from titration information.

Suppose we need to find the molarity of a solution of HCl, which has an unknown concentration. We can do this by a laboratory procedure called **titration** in which we neutralize an acid sample with a known amount of base. In our titration, we first place a measured volume of the acid in a flask and add a few drops of an **indicator**, such as phenolphthalein. An indicator is a compound that dramatically changes color when pH of the solution changes. In an acidic solution, phenolphthalein is colorless. Then we fill a buret with a solution of NaOH with a known molarity and carefully add NaOH to the acid in the flask, as shown in Figure 14.7.

FIGURE 14.7 The titration of an acid. A known volume of an acid is placed in a flask with an indicator and titrated with a measured volume of a base, such as NaOH, to the neutralization endpoint.

Q What data are needed to determine the molarity of the acid in the flask?

In the titration, we neutralize the acid by adding a volume of base that contains a matching number of moles of OH^-. We know that neutralization has taken place when the phenolphthalein in the solution changes from colorless to pink. This is called the neutralization **endpoint**. From the volume added and molarity of the NaOH, we can calculate the number of moles of NaOH and then the concentration of the acid.

SAMPLE PROBLEM | 14.10

Guide to Calculations for an Acid–Base Titration

STEP 1
State the given and needed quantities and concentrations.

STEP 2
Write a plan to calculate molarity or volume.

STEP 3
State equalities and conversion factors including concentrations.

STEP 4
Set up problem to calculate needed quantity.

▓ Titration of an Acid

A 25.0-mL sample of an HCl solution is placed in a flask with a few drops of phenolphthalein (indicator). If 32.6 mL of a 0.185 M NaOH solution is needed to reach the endpoint, what is the concentration (M) of the HCl solution?

$$NaOH(aq) + HCl(aq) \longrightarrow NaCl(aq) + H_2O(l)$$

SOLUTION

STEP 1 **State the given and needed quantities and concentrations.**
Given 32.6 mL of a 0.185 M NaOH solution; 25.0 mL of HCl = 0.0250 L of HCl
Need molarity of HCl solution

STEP 2 **Write a plan to calculate molarity.**

| mL of solution | Metric factor | L | Molarity | moles of NaOH | Mole–mole factor | moles of HCl | Divide by liters | molarity of HCl solution |

STEP 3 **State equalities and conversion factors including concentrations.**

1 L of NaOH = 1000 mL of NaOH   1 L of NaOH = 0.185 mol of NaOH

$$\frac{1 \text{ L NaOH}}{1000 \text{ mL NaOH}} \text{ and } \frac{1000 \text{ mL NaOH}}{1 \text{ L NaOH}} \qquad \frac{1 \text{ L NaOH}}{0.185 \text{ mol NaOH}} \text{ and } \frac{0.185 \text{ mol NaOH}}{1 \text{ L NaOH}}$$

1 mol of HCl = 1 mol of NaOH

$$\frac{1 \text{ mol HCl}}{1 \text{ mol NaOH}} \text{ and } \frac{1 \text{ mol NaOH}}{1 \text{ mol HCl}}$$

STEP 4 **Set up problem to calculate needed quantity.**

$$32.6 \text{ mL NaOH} \times \frac{1 \text{ L NaOH}}{1000 \text{ mL NaOH}} \times \frac{0.185 \text{ mol NaOH}}{1 \text{ L NaOH}} \times \frac{1 \text{ mol HCl}}{1 \text{ mol NaOH}} = 0.006\ 03 \text{ mol of HCl}$$

$$\text{molarity of HCl} = \frac{0.00603 \text{ mol HCl}}{0.0250 \text{ L HCl}} = 0.241 \text{ M HCl solution}$$

STUDY CHECK

What is the molarity of an HCl solution if 28.6 mL of a 0.175 M NaOH solution is needed to neutralize a 25.0-mL sample of the HCl solution?

TUTORIAL
Acid–Base Titrations

SAMPLE PROBLEM 14.11

■ Volume of Base for Titration

What volume, in milliliters, of a 0.115 M NaOH solution would neutralize 25.0 mL of a 0.106 M H_2SO_4 solution?

$$2NaOH(aq) + H_2SO_4(aq) \longrightarrow Na_2SO_4(aq) + 2H_2O(l)$$

SOLUTION

STEP 1 State the given and needed quantities and concentrations.

Given 25.0 mL of 0.106 M H_2SO_4 solution and a 0.115 M NaOH solution
Need volume (mL) of NaOH solution

STEP 2 Write a plan to calculate molarity.

mL of H_2SO_4 solution	Metric factor	L of solution	Molarity	moles of H_2SO_4	Mole–mole factor	Moles of NaOH	Molarity	mL of NaOH solution

STEP 3 State equalities and conversion factors including concentrations.

$$1\ L\ H_2SO_4 = 1000\ mL\ H_2SO_4$$
$$\frac{1\ L\ H_2SO_4}{1000\ mL\ H_2SO_4} \text{ and } \frac{1000\ mL\ H_2SO_4}{1\ L\ H_2SO_4}$$

$$1\ L\ H_2SO_4 = 0.106\ mol\ H_2SO_4$$
$$\frac{1\ L\ H_2SO_4}{0.106\ mol\ H_2SO_4} \text{ and } \frac{0.106\ mol\ H_2SO_4}{1\ L\ H_2SO_4}$$

$$2\ mol\ NaOH = 1\ mol\ H_2SO_4$$
$$\frac{1\ mol\ H_2SO_4}{2\ mol\ NaOH} \text{ and } \frac{2\ mol\ NaOH}{1\ mol\ H_2SO_4}$$

$$1000\ mL\ NaOH = 0.115\ mol\ NaOH$$
$$\frac{1000\ mL\ NaOH}{0.115\ mol\ NaOH} \text{ and } \frac{0.115\ mol\ NaOH}{1000\ mL\ NaOH}$$

STEP 4 Set up problem to calculate needed quantity.

$$25.0\ mL\ H_2SO_4 \times \frac{1\ L\ H_2SO_4}{1000\ mL\ H_2SO_4} \times \frac{0.106\ mol\ H_2SO_4}{1\ L\ H_2SO_4} \times \frac{2\ mol\ NaOH}{1\ mol\ H_2SO_4} \times \frac{1000\ mL\ NaOH}{0.115\ mol\ NaOH}$$

$$= 46.1\ mL\ of\ NaOH\ solution$$

STUDY CHECK

What volume, in milliliters, of a 0.158 M KOH solution is needed to neutralize 50.0 mL of a 0.212 M HCl solution?

QUESTIONS AND PROBLEMS

Acid–Base Titration

14.57 If you need to determine the molarity of a formic acid solution, HCOOH, how would you proceed?

$$HCOOH(aq) + H_2O(l) \rightleftharpoons H_3O^+(aq) + HCOO^-(aq)$$

14.58 If you need to determine the molarity of an acetic acid solution, CH_3COOH, how would you proceed?

$$CH_3COOH(aq) + H_2O(l) \rightleftharpoons H_3O^+(aq) + CH_3COO^-(aq)$$

14.59 What is the molarity of a solution of HCl if 5.00 mL of the HCl solution is titrated with 28.6 mL of 0.145 M NaOH solution?

$$HCl(aq) + NaOH(aq) \longrightarrow NaCl(aq) + H_2O(l)$$

14.60 If 29.7 mL of a 0.205 M KOH solution is required to completely neutralize 25.0 mL of a solution of CH_3COOH, what is the molarity of the acetic acid solution?

$$CH_3COOH(aq) + KOH(aq) \longrightarrow KCH_3COO(aq) + H_2O(l)$$

14.61 If 38.2 mL of a 0.163 M KOH solution is required to neutralize completely 25.0 mL of a H_2SO_4 solution, what is the molarity of the H_2SO_4 solution?

$$H_2SO_4(aq) + 2KOH(aq) \longrightarrow K_2SO_4(aq) + 2H_2O(l)$$

14.62 A solution of 0.162 M NaOH is used to neutralize 25.0 mL of a H_2SO_4 solution. If 32.8 mL of the NaOH solution is required to reach the endpoint, what is the molarity of the H_2SO_4 solution?

$$H_2SO_4(aq) + 2NaOH(aq) \longrightarrow Na_2SO_4(aq) + 2H_2O(l)$$

14.63 A solution of 0.204 M NaOH is used to neutralize 50.0 mL of a 0.0224 M H_3PO_4 solution. What volume, in milliliters, of the NaOH solution is required to reach the endpoint?

$$H_3PO_4(aq) + 3NaOH(aq) \longrightarrow Na_3PO_4(aq) + 3H_2O(l)$$

14.64 A solution of 0.312 M KOH is used to neutralize 15.0 mL of a 0.186 M H_3PO_4 solution. What volume, in milliliters, of the KOH solution is required to reach the endpoint?

$$H_3PO_4(aq) + 3KOH(aq) \longrightarrow K_3PO_4(aq) + 3H_2O(l)$$

14.9 Acid–Base Properties of Salt Solutions

When a salt dissolves in water, it dissociates into cations and anions. Solutions of salts can be acidic, basic, or neutral. Anions and cations from strong acids and bases do not affect pH; however, anions from weak acids and cations from weak bases change the pH of an aqueous solution.

LEARNING GOAL

Predict whether a salt will form an acidic, basic, or neutral solution.

Salts That Form Neutral Solutions

A solution of a salt containing a cation from a strong base and an anion from a strong acid will be neutral. For example, a salt such as $NaNO_3$ will form a neutral solution.

$$NaNO_3(s) \xrightarrow{H_2O} Na^+(aq) + NO_3^-(aq)$$

Does not change H^+ Does not attract H^+ from water   Neutral solution (pH = 7.0)

The cation, Na^+, from a strong base does not change H^+, and the anion, NO_3^-, from a strong acid does not attract H^+ from water. Thus there is no effect on the pH of water; the solution is neutral with a pH of 7.0. Salts such as NaCl, KCl, $NaNO_3$, KNO_3, and KBr contain cations from strong bases and anions from strong acids and form neutral solutions.

TUTORIAL
Salts of Weak Acids and Bases

Some Components of Neutral Salt Solutions

Cations from strong bases: Group 1A (1): Li^+, Na^+, K^+

Group 2A (2): Mg^{2+}, Ca^{2+}, Sr^{2+}, Ba^{2+}

Anions from strong acids: Cl^-, Br^-, I^-, NO_3^-, ClO_4^-

Salts That Form Basic Solutions

A salt solution containing the cation from a strong base and the anion from a weak acid produces a basic solution. Suppose we have a solution of the salt NaF, which contains Na^+ and F^- ions:

$$NaF(s) \xrightarrow{H_2O} Na^+(aq) + F^-(aq)$$

Does not change H^+ Attracts H^+ from water

The metal ion Na^+ has no effect on the pH of the solution. However, F^- is the conjugate base of the weak acid HF. Thus, F^- will attract a proton from water and form OH^- in solution, which makes it basic:

$$F^-(aq) + H_2O(l) \rightleftharpoons HF(aq) + OH^-(aq)$$ Basic solution (pH > 7.0)

Salts with anions from weak acids such as NaCN, KNO_2, and Na_2SO_4 produce basic solutions.

Some Components of Basic Salt Solutions

Cations from strong bases: Group 1A (1): Li^+, Na^+, K^+

Group 2A (2): Mg^{2+}, Ca^{2+}, Sr^{2+}, Ba^{2+}

Anions from weak acids: F^-, NO_2^-, CN^-, CO_3^{2-}, SO_4^{2-}, CH_3COO^-, S^{2-}, PO_4^{3-}

Salts That Form Acidic Solutions

A salt solution containing the cation from a weak base and the anion from a strong acid produces an acidic solution. Suppose we have a solution of the salt NH_4Cl, which contains NH_4^+ and Cl^- ions:

$$NH_4Cl(s) \xrightarrow{H_2O} NH_4^+(aq) + Cl^-(aq)$$

Donates H^+ to water   Does not attract H^+ from water

The anion Cl^- has no effect on the pH of the solution. However, as a weak acid, the cation NH_4^+ donates a proton to water, which produces H_3O^+:

$$NH_4^+(aq) + H_2O(l) \rightleftharpoons NH_3(aq) + H_3O^+(aq)$$ Acidic solution (pH < 7.0)

TABLE 14.10 Cations and Anions of Salts in Neutral, Basic, and Acidic Salt Solutions

Type of Solution	Cations	Anions	pH
Neutral	From strong bases: Group 1A (1): Li^+, Na^+, K^+ Group 2A (2): Mg^{2+}, Ca^{2+}, Sr^{2+}, Ba^{2+} (but not Be^{2+})	From strong acids: Cl^-, Br^-, I^-, NO_3^-, ClO_4^-	7.0
Basic	From strong bases: Group 1A (1): Li^+, Na^+, K^+ Group 2A (2): Mg^{2+}, Ca^{2+}, Sr^{2+}, Ba^{2+} (but not Be^{2+})	From weak acids: F^-, NO_2^-, CN^-, CO_3^{2-}, SO_4^{2-}, CH_3COO^-, S^{2-}, PO_4^{3-}	> 7.0
Acidic	From weak bases: NH_4^+, Be^{2+}, Al^{3+}, Zn^{2+}, Cr^{3+}, Fe^{3+} (small, highly charged metal ions)	From strong acids: Cl^-, Br^-, I^-, NO_3^-, ClO_4^-	< 7.0

Some Components of Acidic Salt Solutions
Cations of weak bases: NH_4^+ and Be^{2+}, Al^{3+}, Zn^{2+}, Cr^{3+}, Fe^{3+}
(small, highly charged metal ions)
Anions of strong acids: Cl^-, Br^-, I^-, NO_3^-, ClO_4^-

Table 14.10 summarizes the cations and anions of salts that form neutral, basic, and acidic solutions.

Sometimes a salt contains the cation of a weak base and the anion of a weak acid. For example, when NH_4F dissociates in water, it produces NH_4^+ and F^- ions. We have seen that NH_4^+ forms an acidic solution and F^- forms a basic solution. The ion that reacts to a greater extent with water determines whether the solution is acidic or basic. The salt solution will be neutral only if the ions react with water to the same extent. The determination of these reactions is complex and will not be considered in this text.

Table 14.11 summarizes the acid–base properties of some typical salts in water.

TABLE 14.11 Acid–Base Properties of Some Salt Solutions

Typical Salts	Types of Ions	pH	Solution
$NaCl$, $MgBr_2$, KNO_3	Cation from a strong base Anion from a strong acid	7.0	Neutral
NaF, $MgCO_3$, KNO_2	Cation from a strong base Anion from a weak acid	> 7.0	Basic
NH_4Cl, $FeBr_3$, $Al(NO_3)_3$	Cation from a weak base Anion from a strong acid	< 7.0	Acidic

CONCEPT CHECK 14.9

■ **Acid–Base Properties of a Salt Solution**

Predict whether a solution of KCN would be an acidic, basic, or neutral solution.

ANSWER

There are six strong acids; all other acids are weak. Bases with cations from Groups 1A (1) and 2A (2) are strong; all other bases are weak. For the salt, KCN, the cation is from a strong base (KOH), but the anion is from a weak acid (HCN).

$$KCN(s) \xrightarrow{H_2O} K^+(aq) + CN^-(aq)$$

The cation K^+ has no effect on pH. However, the anion CN^- will attract protons from water to produce a basic solution.

$$CN^-(aq) + H_2O(l) \rightleftharpoons HCN(aq) + OH^-(aq)$$

SAMPLE PROBLEM 14.12

■ **Predicting the Acid–Base Properties of Salt Solutions**

Predict whether solutions of each of the following salts would be acidic, basic, or neutral:

a. NH_4Br **b.** $NaNO_3$

SOLUTION

a. NH_4Br

In the salt NH_4Br, the cation is from a weak base, NH_3, but the anion is from a strong acid, HBr.

$$NH_4Br(s) \xrightarrow{H_2O} NH_4^+(aq) + Br^-(aq)$$

The anion Br^- has no effect on pH because it is from HBr, a strong acid. However, the cation NH_4^+ donates protons to water to produce an acidic solution.

$$NH_4^+(aq) + H_2O(l) \rightleftharpoons NH_3(aq) + H_3O^+(aq)$$

b. $NaNO_3$

The salt $NaNO_3$ contains a cation from a strong base, NaOH, and an anion from a strong acid, HNO_3. Thus, there is no change of pH; the salt solution is neutral.

$$NaNO_3(s) \xrightarrow{H_2O} Na^+(aq) + NO_3^-(aq)$$

STUDY CHECK

Would a solution of Na_3PO_4 be acidic, basic, or neutral?

QUESTIONS AND PROBLEMS

Acid–Base Properties of Salt Solutions

14.65 Why does a salt containing a cation from a strong base and an anion from a weak acid form a basic solution?

14.66 Why does a salt containing a cation from a weak base and an anion from a strong acid form an acidic solution?

14.67 Predict whether each of the following salts will form an acidic, basic, or neutral solution. For acidic and basic solutions, write an equation for the reaction that takes place.
a. $MgCl_2$ **b.** NH_4NO_3
c. Na_2CO_3 **d.** K_2S

14.68 Predict whether each of the following salts will form an acidic, basic, or neutral solution. For acidic and basic solutions, write an equation for the reaction that takes place.
a. Na_2SO_4 **b.** KBr
c. $BaCl_2$ **d.** NH_4I

14.10 Buffers

The pH of water and most solutions changes drastically when a small amount of acid or base is added. However, if a solution is buffered, there is little change in pH. A **buffer solution** is a solution that maintains pH by neutralizing small amounts of added acid or base. For example, blood contains buffers that maintain a consistent pH of about 7.4. If the pH of the blood goes slightly above or below 7.4, changes in our oxygen levels and our metabolic processes can be drastic enough to cause death. Even though we obtain acids and bases from foods and cellular reactions, the buffers in the body absorb those compounds so effectively that the pH of our blood remains essentially unchanged (see Figure 14.8).

In a buffer, an acid must be present to react with any OH^- that is added, and a base must be available to react with any added H_3O^+. However, that acid and base must not be able to neutralize each other. Therefore, a combination of an acid–base conjugate pair is used in

LEARNING GOAL

Describe the role of buffers in maintaining the pH of a solution.

SELF STUDY ACTIVITY

pH and Buffers

FIGURE 14.8 Adding an acid or a base to water changes the pH drastically, but a buffer resists pH change when small amounts of acid or base are added.

Q Why does the pH change several pH units when acid is added to water, but not when acid is added to a buffer?

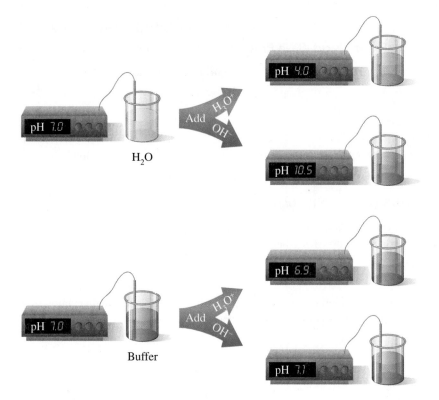

H₂O

Buffer

TUTORIAL

Buffer Solutions

buffers. Most buffer solutions consist of nearly equal concentrations of a weak acid and a salt containing its conjugate base (see Figure 14.9). Buffers may also contain a weak base and the salt of the weak base, which contains its conjugate acid.

For example, a typical buffer contains acetic acid (CH_3COOH) and a salt such as sodium acetate ($NaCH_3COO$). As a weak acid, acetic acid dissociates slightly in water to form H_3O^+ and a very small amount of CH_3COO^-. The presence of the salt provides a much larger concentration of acetate ion (CH_3COO^-), which is necessary for its buffering capability.

$$CH_3COOH(aq) + H_2O(l) \rightleftharpoons H_3O^+(aq) + CH_3COO^-(aq)$$

Large amount   Large amount

Let's see how this buffer solution maintains the $[H_3O^+]$. When a small amount of acid is added, it will combine with the acetate ion (anion) as the equilibrium shifts toward the reactant acetic acid. There will be a small decrease in the $[CH_3COO^-]$ and a small increase in $[CH_3COOH]$, but the $[H_3O^+]$ will not change very much.

$$CH_3COOH(aq) + H_2O(l) \longleftarrow H_3O^+(aq) + CH_3COO^-(aq)$$

FIGURE 14.9 The buffer described here consists of about equal concentrations of acetic acid (CH_3COOH) and its conjugate base acetate ion (CH_3COO^-). Adding H_3O^+ to the buffer uses up some CH_3COO^-, whereas adding OH^- neutralizes some CH_3COOH. The pH of the solution is maintained as long as the added amounts of acid or base are small compared to the concentrations of the buffer components.

Q How does this acetic acid/acetate ion buffer maintain pH?

If a small amount of base is added to this buffer solution, it is neutralized by the acetic acid. The products are acetate ion and water. The [CH_3COOH] decreases slightly and the [CH_3COO^-] increases slightly, but again the [H_3O^+] does not change very much. Thus, pH of the solution is maintained.

$$CH_3COOH(aq) + OH^-(aq) \longrightarrow H_2O(l) + CH_3COO^-(aq)$$

CONCEPT CHECK 14.10

■ Identifying Buffer Solutions

Indicate whether each of the following would make a buffer solution:

a. HCl (a strong acid) and NaCl
b. H_3PO_4 (a weak acid)
c. HF (a weak acid) and NaF

ANSWER
a. No. A strong acid will change the pH of a solution. A buffer requires a weak acid and a salt containing its conjugate base.
b. No. A weak acid is part of a buffer, but the salt containing the conjugate base of the weak acid is also needed.
c. Yes. This mixture would be a buffer because it contains a weak acid and a salt containing its conjugate base.

Calculating the pH of a Buffer

By rearranging the K_a expression to solve for [H_3O^+], we obtain the ratio of the acetic acid/acetate buffer.

$$K_a = \frac{[H_3O^+][CH_3COO^-]}{[CH_3COOH]}$$

Solving for [H_3O^+] gives:

$$[H_3O^+] = K_a \times \frac{[CH_3COOH]}{[CH_3COO^-]}$$

Because K_a is a constant at a given temperature, the [H_3O^+] is determined by the [CH_3COOH]/[CH_3COO^-] ratio. As long as the addition of small amounts of either acid or base changes the ratio of [CH_3COOH]/[CH_3COO^-] only slightly, the changes in [H_3O^+] will be small and the pH will be maintained. It is important to note that the amount of acid or base that is added must be small compared to the supply of the buffer components CH_3COOH and CH_3COO^-. If a large amount of acid or base is added, the buffering capacity of the system may be exceeded.

Other buffers can be prepared from conjugate acid–base pairs such as $H_2PO_4^-$/HPO_4^{2-}, HPO_4^{2-}/PO_4^{3-}, HCO_3^-/CO_3^{2-}, or NH_4^+/NH_3. The pH of the buffer solution will depend on the acid–base pair chosen.

TUTORIAL
Calculating the pH of a Buffer

TUTORIAL
Preparing Buffer Solutions

SAMPLE PROBLEM 14.13

■ pH of a Buffer

The K_a for acetic acid, CH_3COOH, is 1.8×10^{-5}. What is the pH of a buffer prepared with 1.0 M CH_3COOH and 1.0 M CH_3COO^-?

$$CH_3COOH(aq) + H_2O(l) \rightleftharpoons H_3O^+(aq) + CH_3COO^-(aq)$$

SOLUTION
STEP 1 Write the K_a expression.

$$K_a = \frac{[H_3O^+][CH_3COO^-]}{[CH_3COOH]}$$

Guide to Calculating pH of a Buffer

STEP 1
Write the K_a or K_b expression.

STEP 2
Rearrange the K_a or K_b for [H_3O^+].

STEP 3
Substitute in the [HA] and [A⁻].

STEP 4
Use [H_3O^+] to calculate pH.

STEP 2 **Rearrange the K_a for [H_3O^+].**

$$[H_3O^+] = K_a \times \frac{[CH_3COOH]}{[CH_3COO^-]}$$

STEP 3 **Substitute in [HA] and [A^-].** Substituting these values in the expression for [H_3O^+] gives

$$[H_3O^+] = 1.8 \times 10^{-5} \times \frac{[1.0]}{[1.0]}$$

$$[H_3O^+] = 1.8 \times 10^{-5} \, M$$

STEP 4 **Use [H_3O^+] to calculate pH.** Using the [H_3O^+] in the pH expression gives the pH of the buffer.

$$pH = -\log [1.8 \times 10^{-5}] = 4.74$$

STUDY CHECK

One of the conjugate acid–base pairs that buffers the blood is $H_2PO_4^-/HPO_4^{2-}$ with a K_a of 6.2×10^{-8}. What is the pH of a buffer that is 0.10 M $H_2PO_4^-$ and 0.50 M HPO_4^{2-}?

CHEMISTRY AND HEALTH

Buffers in the Blood

The arterial blood has a normal pH of 7.35–7.45. If changes in H_3O^+ lower the pH below 6.8 or raise it above 8.0, cells cannot function properly and death may result. In the cells of the body, CO_2 is continually produced as an end product of cellular metabolism. Some CO_2 is carried to the lungs for elimination, and the rest dissolves in body fluids such as plasma and saliva, forming carbonic acid, H_2CO_3. As a weak acid, carbonic acid dissociates to give bicarbonate, HCO_3^-, and H_3O^+. More of the anion HCO_3^- is supplied by the kidneys to give an important buffer system in the body fluid—the H_2CO_3/HCO_3^- buffer:

$$CO_2(g) + H_2O(l) \rightleftarrows H_2CO_3(aq) \rightleftarrows H_3O^+(aq) + HCO_3^-(aq)$$

Excess H_3O^+ entering the body fluids reacts with the HCO_3^- and excess OH^- reacts with the carbonic acid:

$$H_2CO_3(aq) + H_2O(l) \longleftarrow H_3O^+(aq) + HCO_3^-(aq)$$
Equilibrium shifts left

$$H_2CO_3(aq) + OH^-(aq) \longrightarrow H_2O(l) + HCO_3^-(aq)$$
Equilibrium shifts right

For carbonic acid, we can write the equilibrium expression as

$$K_a = \frac{[H_3O^+][HCO_3^-]}{[H_2CO_3]}$$

To maintain the normal blood pH (7.35–7.45), the ratio of H_2CO_3/HCO_3^- needs to be about 1 to 10, which is obtained by the concentrations in the blood of 0.0024 M H_2CO_3 and 0.024 M HCO_3^-.

$$[H_3O^+] = K_a \times \frac{[H_2CO_3]}{[HCO_3^-]}$$

$$= 4.3 \times 10^{-7} \times \frac{[0.0024]}{[0.024]} = 4.3 \times 10^{-7} \times 0.10$$

$$= 4.3 \times 10^{-8} \, M$$

$$pH = -\log[4.3 \times 10^{-8}] = 7.37$$

In the body, the concentration of carbonic acid is closely associated with the partial pressure of CO_2, P_{CO_2}. Table 14.12 lists the normal values for arterial blood. If the CO_2 level rises, increasing [H_2CO_3], the equilibrium shifts to produce more H_3O^+, which lowers the pH. This condition is called *acidosis*. Difficulty with ventilation or gas diffusion can lead to respiratory acidosis, which can happen in emphysema or when an accident or depressive drugs affect the medulla of the brain.

A lowering of the CO_2 level leads to a high blood pH, a condition called *alkalosis*. Excitement, trauma, or a high temperature may cause a person to hyperventilate, which expels large amounts of CO_2. As the partial pressure of CO_2 in the blood falls below normal, the equilibrium shifts from H_2CO_3 to CO_2 and H_2O. This shift decreases the [H_3O^+] and raises the pH. The kidneys also regulate H_3O^+ and HCO_3^-, but they do so more slowly than the adjustment made by the lungs during ventilation.

TABLE 14.12 Normal Values for Blood Buffer in Arterial Blood

P_{CO_2}	40 mmHg
H_2CO_3	2.4 mmol/L of plasma
HCO_3^-	24 mmol/L of plasma
pH	7.35–7.45

QUESTIONS AND PROBLEMS

Buffers

14.69 Which of the following represent a buffer system? Explain.
  a. NaOH and NaCl   **b.** H_2CO_3 and $NaHCO_3$
  c. HF and KF   **d.** KCl and NaCl

14.70 Which of the following represent a buffer system? Explain.
  a. H_3PO_3
  b. $NaNO_3$
  c. CH_3COOH and $NaCH_3COO$
  d. HCl and NaOH

14.71 Consider the buffer system of hydrofluoric acid, HF, and its salt, NaF:

$$HF(aq) + H_2O(l) \rightleftharpoons H_3O^+(aq) + F^-(aq)$$

  a. What is the purpose of the buffer system?
  b. Why does a buffer require a salt that contains the same conjugate base as the acid?
  c. How does the buffer react when some H_3O^+ is added?
  d. How does the buffer react when some OH^- is added?

14.72 Consider the buffer system of nitrous acid, HNO_2, and its salt, $NaNO_2$:

$$HNO_2(aq) + H_2O(l) \rightleftharpoons H_3O^+(aq) + NO_2^-(aq)$$

  a. What is the purpose of a buffer system?
  b. What is the purpose of $NaNO_2$ in the buffer?
  c. How does the buffer react when some H_3O^+ is added?
  d. How does the buffer react when some OH^- is added?

14.73 Nitrous acid has a K_a of 4.5×10^{-4}. What is the pH of a buffer solution containing 0.10 M HNO_2 and 0.10 M NO_2^-?

14.74 Acetic acid has a K_a of 1.8×10^{-5}. What is the pH of a buffer solution containing 0.15 M CH_3COOH (acetic acid) and 0.15 M CH_3COO^-?

14.75 Compare the pH of a HF buffer that contains 0.10 M HF and 0.10 M NaF with another HF buffer that contains 0.060 M HF and 0.120 M NaF (see Table 14.5).

14.76 Compare the pH of a H_2CO_3 buffer that contains 0.10 M H_2CO_3 and 0.10 M $NaHCO_3$ with another H_2CO_3 buffer that contains 0.15 M H_2CO_3 and 0.050 M $NaHCO_3$ (see Table 14.5).

CONCEPT MAP

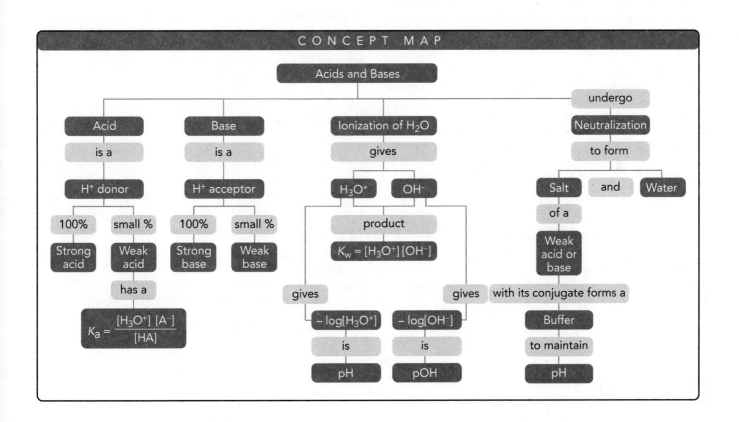

CHAPTER REVIEW

14.1 Acids and Bases

LEARNING GOAL: *Describe and name Arrhenius, Brønsted–Lowry, and organic acids and bases.*

An Arrhenius acid produces H^+ and an Arrhenius base produces OH^- in aqueous solutions. Acids taste sour, may sting, and neutralize bases. Bases taste bitter, feel slippery, and neutralize acids. Acids containing a simple anion use a *hydro* prefix, whereas acids with oxygen-containing polyatomic anions are named as *ic* or *ous acids*. Organic acids use common names such as formic acid and acetic acid. The IUPAC names are derived from their alkane names by replacing the *e* with *oic acid*.

14.2 Brønsted–Lowry Acids and Bases

LEARNING GOAL: *Identify conjugate acid–base pairs for Brønsted–Lowry acids and bases.*

According to the Brønsted–Lowry theory, acids are proton (H^+) donors and bases are proton acceptors. Two conjugate acid–base pairs are present in an acid–base reaction. Each acid–base pair is related by the loss or gain of a H^+. For example, when the acid HF donates H^+, the F^- is its conjugate base. The other acid–base pair would be H_3O^+/H_2O.

$$HF(aq) + H_2O(l) \rightleftharpoons H_3O^+(aq) + F^-(aq)$$

14.3 Strengths of Acids and Bases

LEARNING GOAL: *Write equations for the dissociation of strong and weak acids; identify the direction of reaction.*

In strong acids, all the H^+ in the acid is donated to H_2O; in a weak acid, only a small percentage of acid molecules produces H_3O^+. Strong bases are hydroxides of Groups 1A (1) and 2A (2) that dissociate completely in water. An important weak base is ammonia, NH_3.

14.4 Dissociation Constants

LEARNING GOAL: *Write the expression for the dissociation constant of a weak acid or weak base.*

In water, weak acids and weak bases produce only a few ions when equilibrium is reached. The reaction for a weak acid can be written as $HA + H_2O \rightleftharpoons H_3O^+ + A^-$. The acid dissociation constant expression constant is written as $K_a = \dfrac{[H_3O^+][A^-]}{[HA]}$.

For a weak base, $B + H_2O \rightleftharpoons BH^+ + OH^-$, the base dissociation constant expression K_b is written as $K_b = \dfrac{[BH^+][OH^-]}{[B]}$.

14.5 Ionization of Water

LEARNING GOAL: *Use the ion product of water to calculate the $[H_3O^+]$ and $[OH^-]$ in an aqueous solution.*

In pure water, which is neutral, a few molecules of water transfer protons to other water molecules, producing small, but equal, amounts of $[H_3O^+]$ and $[OH^-]$, which is 1.0×10^{-7} mol/L. The ion product constant, K_w, $[H_3O^+][OH^-] = 1.0 \times 10^{-14}$, applies to all aqueous solutions at 25 °C. In acidic solutions, the $[H_3O^+]$ is greater than the $[OH^-]$. In basic solutions, the $[OH^-]$ is greater than the $[H_3O^+]$.

14.6 The pH Scale

LEARNING GOAL: *Calculate pH from $[H_3O^+]$; given the pH, calculate $[H_3O^+]$ and $[OH^-]$ of a solution.*

The pH scale is a range of numbers typically from 0 to 14, which relates to the $[H_3O^+]$ of the solution. A neutral solution has a pH of 7.0. In acidic solutions, the pH is below 7.0, and in basic solutions the pH is above 7.0. Mathematically, pH is the negative logarithm of the hydronium ion concentration (pH $= -\log[H_3O^+]$). The pOH is the negative log of the hydroxide ion concentration (pOH $= -\log[OH^-]$). The sum of the pH + pOH is 14.00.

14.7 Reactions of Acids and Bases

LEARNING GOAL: *Write balanced equations for reactions of acids and bases.*

When an acid reacts with a metal, hydrogen gas and a salt are produced. The reaction of an acid with a carbonate or bicarbonate produces carbon dioxide, a salt, and water. In a neutralization reaction, an acid reacts with a base to produce a salt and water.

14.8 Acid–Base Titration

LEARNING GOAL: *Calculate the molarity or volume of an acid or base from titration information.*

In a laboratory procedure called titration, an acid sample is neutralized with a known amount of a base. From the volume and molarity of the base, the concentration of the acid is calculated.

14.9 Acid–Base Properties of Salt Solutions

LEARNING GOAL: *Predict whether a salt will form an acidic, basic, or neutral solution.*

A salt of a weak acid contains an anion that removes protons from water and makes the solution basic. A salt of a weak base contains an ion that donates a proton to water, producing an acidic solution. Salts of strong acids and strong bases produce neutral solutions because they contain ions that do not affect the pH.

14.10 Buffers

LEARNING GOAL: *Describe the role of buffers in maintaining the pH of a solution.*

A buffer solution resists changes in pH when small amounts of an acid or a base are added. A buffer contains either a weak acid and its salt or a weak base and its salt. The weak acid reacts with added OH^-, and the anion of the salt picks up added H^+.

KEY TERMS

acid A substance that dissolves in water and produces hydrogen ions (H^+), according to the Arrhenius theory. All acids are proton donors, according to the Brønsted–Lowry theory.

acid dissociation constant, K_a The product of the ions from the dissociation of a weak acid divided by the concentration of the weak acid.

amphoteric Substances that can act as either an acid or a base in water.

base A substance that dissolves in water and produces hydroxide ions (OH^-) according to the Arrhenius theory. All bases are proton acceptors, according to the Brønsted–Lowry theory.

base dissociation constant, K_b The product of the ions from the dissociation of a weak base divided by the concentration of the weak base.

Brønsted–Lowry acids and bases An acid is a proton donor; a base is a proton acceptor.

buffer solution A solution of a weak acid and its conjugate base or a weak base and its conjugate acid that maintains the pH by neutralizing added acid or base.

conjugate acid–base pair An acid and base that differ by one H^+. When an acid donates a proton, the product is its conjugate base, which is capable of accepting a proton in the reverse reaction.

dissociation The separation of an acid or a base into ions in water.

endpoint The point at which an indicator changes color. For the indicator phenolphthalein, the color change occurs when the number of moles of OH^- is equal to the number of moles of H_3O^+ in the sample.

hydronium ion, H_3O^+ The ion formed by the attraction of a proton (H^+) to a water molecule.

indicator A substance added to a titration sample that changes color when the pH of the solution changes.

ion product constant of water, K_w The product of $[H_3O^+]$ and $[OH^-]$ in solution; $K_w = [H_3O^+][OH^-]$.

neutral The term that describes a solution with equal concentrations of $[H_3O^+]$ and $[OH^-]$.

neutralization A reaction between an acid and a base to form a salt and water.

pH A measure of the $[H_3O^+]$ in a solution; $pH = -\log[H_3O^+]$.

pOH A measure of the $[OH^-]$ in a solution; $pOH = -\log[OH^-]$.

salt An ionic compound that contains a metal ion or NH_4^+ and a nonmetal or polyatomic ion other than OH^-.

strong acid An acid that completely ionizes in water.

strong base A base that completely ionizes in water.

titration The addition of base to an acid sample to determine the concentration of the acid.

weak acid An acid that is a poor donor of H^+ and dissociates only slightly in water.

weak base A base that is a poor acceptor of H^+ and produces only a small number of ions in water.

UNDERSTANDING THE CONCEPTS

14.77 In each of the following diagrams of acid solutions, determine if each diagram represents a strong acid or a weak acid. The acid has the formula HX.

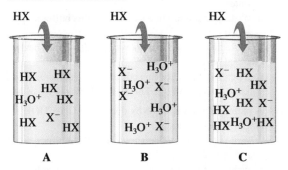

14.78 Adding a few drops of a strong acid to water will lower the pH appreciably. However, adding the same number of drops to a buffer does not appreciably alter the pH. Why?

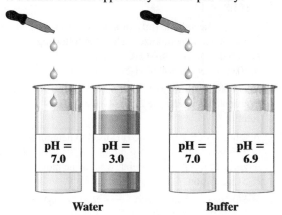

14.79 Sometimes, during stress or trauma, a person can start to hyperventilate. To avoid fainting, a person may breathe into a paper bag.

a. What changes occur in the blood pH during hyperventilation?

b. How does breathing into a paper bag help return blood pH to normal?

Breathing into a paper bag provides carbon dioxide.

14.80 In the blood plasma, pH is maintained by the carbonic acid–bicarbonate buffer system.

a. How is pH maintained when acid is added to the buffer system?

b. How is pH maintained when base is added to the buffer system?

ADDITIONAL QUESTIONS AND PROBLEMS

14.81 Identify each of the following as an acid, base, or salt, and give its name:
a. $HBrO_2$ **b.** RbOH
c. $Mg(NO_3)_2$ **d.** $CH_3—CH_2—CH_2—COOH$
e. $HClO_4$

14.82 Identify each of the following as an acid, base, or salt, and give its name:
a. HI   **b.** $MgBr_2$
c. NH_3 **d.** Li_2SO_3
e. $CH_3—CH_2—COOH$

14.83 Are each of the following solutions acidic, basic, or neutral?
a. rain, pH 5.2   **b.** tears, pH 7.5
c. tea, pH 3.8   **d.** cola, pH 2.5
e. photo developer, pH 12.0

14.84 Are the following solutions acidic, basic, or neutral?
a. saliva, pH 6.8   **b.** urine, pH 5.9
c. pancreatic juice, pH 8.0   **d.** bile, pH 8.4
e. blood, pH 7.45

14.85 One ingredient in some antacids is $Mg(OH)_2$.
a. If the base is not very soluble in water, why is it considered a strong base?
b. Write the neutralization reaction of $Mg(OH)_2$ with stomach acid, HCl.

14.86 Acetic acid, used to prepare vinegar, is a weak acid. Why?

14.87 Using Table 14.4, identify the stronger acid in each of the following pairs:
a. HF or HCN   **b.** H_3O^+ or H_2S
c. HNO_2 or CH_3COOH **d.** H_2O or HCO_3^-

14.88 Using Table 14.4, identify the stronger base in each of the following pairs:
a. H_2O or Cl^- **b.** OH^- or NH_3
c. SO_4^{2-} or NO_2^- **d.** CO_3^{2-} or H_2O

14.89 Determine the pH and pOH for each of the following solutions:
a. $[H_3O^+] = 2.0 \times 10^{-8}\,M$ **b.** $[H_3O^+] = 5.0 \times 10^{-2}\,M$
c. $[OH^-] = 3.5 \times 10^{-4}\,M$ **d.** $[OH^-] = 0.0054\,M$

14.90 Determine the pH and pOH for each of the following solutions:
a. $[OH^-] = 1.0 \times 10^{-7}\,M$ **b.** $[H_3O^+] = 4.2 \times 10^{-3}\,M$
c. $[H_3O^+] = 0.0001\,M$ **d.** $[OH^-] = 8.5 \times 10^{-9}\,M$

14.91 Are the solutions in Problem 14.89 acidic, basic, or neutral?

14.92 Are the solutions in Problem 14.90 acidic, basic, or neutral?

14.93 What are the $[H_3O^+]$ and $[OH^-]$ for a solution with each of the following pH values?
a. 3.00   **b.** 6.48
c. 8.85   **d.** 11.00

14.94 What are the $[H_3O^+]$ and $[OH^-]$ for a solution with each of the following pH values?
a. 10.0   **b.** 5.0
c. 6.5   **d.** 1.82

14.95 Solution A has a pH of 4.5 and solution B has a pH of 6.7.
a. Which solution is more acidic?
b. What is the $[H_3O^+]$ in each?
c. What is the $[OH^-]$ in each?

14.96 Solution X has a pH of 9.5, and solution Y has a pH of 7.5.
a. Which solution is more acidic?
b. What is the $[H_3O^+]$ in each?
c. What is the $[OH^-]$ in each?

14.97 What is the $[OH^-]$ in a solution that contains 0.225 g of NaOH in 0.250 L of solution?

14.98 What is the $[H_3O^+]$ in a solution that contains 1.54 g of HNO_3 in 0.500 L of solution?

14.99 What is the pH and pOH of a solution prepared by dissolving 2.5 g of HCl in water to make 425 mL of solution?

14.100 What is the pH and pOH of a solution prepared by dissolving 1.00 g of $Ca(OH)_2$ in water to make 875 mL of solution?

14.101 Will solutions of the following salts be acidic, basic, or neutral?
a. KF   **b.** NaCN
c. CH_3NH_3Cl **d.** NaBr

14.102 Will solutions of the following salts be acidic, basic, or neutral?
a. K_2SO_4 **b.** KNO_2
c. CaF_2 **d.** NH_4Cl

14.103 A buffer solution is made by dissolving H_3PO_4 and NaH_2PO_4 in water.
a. Write an equation that shows how this buffer neutralizes added acid.
b. Write an equation that shows how this buffer neutralizes added base.
c. Calculate the pH of this buffer if it contains 0.50 M H_3PO_4 and 0.20 M $H_2PO_4^-$. The K_a for H_3PO_4 is 7.5×10^{-3}.

14.104 A buffer solution is made by dissolving CH_3COOH and $NaCH_3COO$ in water.
a. Write an equation that shows how this buffer neutralizes added acid.
b. Write an equation that shows how this buffer neutralizes added base.
c. Calculate the pH of this buffer if it contains 0.20 M CH_3COOH and 0.40 M CH_3COO^-. The K_a for CH_3COOH is 1.8×10^{-5}.

14.105 Calculate the volume (mL) of a 0.150 M NaOH solution that will completely neutralize each of the following:
a. 25.0 mL of a 0.288 M HCl solution
b. 10.0 mL of a 0.560 M H_2SO_4 solution

14.106 How many milliliters of a 0.215 M NaOH solution are needed to completely neutralize 2.50 mL of a 0.825 M H_2SO_4 solution?

14.107 A solution of 0.205 M NaOH is used to neutralize 20.0 mL of a H_2SO_4 solution. If 45.6 mL of the NaOH solution is required to reach the endpoint, what is the molarity of the H_2SO_4 solution?

$$H_2SO_4(aq) + 2NaOH(aq) \longrightarrow Na_2SO_4(aq) + 2H_2O(l)$$

14.108 A 10.0-mL sample of vinegar, which is an aqueous solution of acetic acid, CH_3COOH, requires 16.5 mL of a 0.500 M NaOH solution to reach the endpoint in a titration. What is the molarity of the acetic acid solution?

$$CH_3COOH(aq) + NaOH(aq) \longrightarrow NaCH_3COO(aq) + H_2O(l)$$

CHALLENGE QUESTIONS

14.109 Consider the following:
    1. H_2S **2.** H_3PO_4 **3.** HCO_3^-
    a. For each, write the formula of the conjugate base.
    b. For each, write the K_a expression.
    c. Write the formula of the weakest acid.
    d. Write the formula of the strongest acid.

14.110 Identify the conjugate acid–base pairs in each of the following equations and whether the equilibrium mixture contains mostly products or mostly reactants:
    a. $NH_3(aq) + HNO_3(aq) \rightleftharpoons NH_4^+(aq) + NO_3^-(aq)$
    b. $H_2O(l) + HBr(aq) \rightleftharpoons H_3O^+(aq) + Br^-(aq)$
    c. $HNO_2(aq) + HS^-(aq) \rightleftharpoons H_2S(g) + NO_2^-(aq)$
    d. $Cl^-(aq) + H_2O(l) \rightleftharpoons OH^-(aq) + HCl(aq)$

14.111 Complete and balance each of the following:
    a. $ZnCO_3(s) + H_2SO_4(aq) \longrightarrow$
    b. $Al(s) + HNO_3(aq) \longrightarrow$
    c. $H_3PO_4(aq) + Ca(OH)_2(s) \longrightarrow$
    d. $KHCO_3(s) + HNO_3(aq) \longrightarrow$

14.112 Predict whether a solution of each of the following salts is acidic, basic, or neutral. For salts that form acidic or basic solutions, write a balanced equation for the reaction.
    a. $CH_3NH_3NO_3$
    b. KNO_2
    c. $Mg(NO_3)_2$
    d. BaF_2
    e. $NaHS$

14.113 Determine each of the following for a 0.050 M KOH solution:
    a. $[H_3O^+]$
    b. pH
    c. pOH
    d. the balanced equation for the reaction with H_3PO_4
    e. milliliters of solution required to neutralize 40.0 mL of a 0.035 M H_2SO_4 solution

14.114 Consider the reaction of KOH and HNO_2.
    a. Write the balanced chemical equation.
    b. Calculate the milliliters of a 0.122 M KOH solution required to neutralize 36.0 mL of a 0.250 M HNO_2 solution.
    c. Determine whether the final solution would be acidic, basic, or neutral.

14.115 Globally, more lakes are becoming acidic due to acid rain. Suppose a lake has a pH of 4.2, well below the recommended pH of 6.5.

A helicopter drops calcium carbonate on an acidic lake to increase its pH.

    a. What are the $[H_3O^+]$ and $[OH^-]$ of the lake?
    b. What are the $[H_3O^+]$ and $[OH^-]$ of a lake that has a pH of 6.5?
    c. One way to raise the pH and restore aquatic life of an acidic lake is to add limestone ($CaCO_3$). How many grams of $CaCO_3$ are needed to neutralize 1.0 kL of the acidic water from the lake if the acid is written as HA?

$$2HA + CaCO_3(s) \longrightarrow CaA_2 + CO_2(g) + H_2O(l)$$

14.116 The daily output of stomach acid (gastric juices) is 1000 mL to 2000 mL. Prior to a meal, stomach acid (HCl) typically has a pH of 1.42.
    a. What is the $[H_3O^+]$ of stomach acid?
    b. The antacid Maalox contains 200. mg of $Al(OH)_3$ per tablet. Write the balanced equation for the neutralization and calculate the milliliters of stomach acid neutralized by two tablets of Maalox.
    c. The antacid milk of magnesia contains 400. mg of $Mg(OH)_2$ per teaspoon. Write the balanced equation for the neutralization and calculate the number of milliliters of stomach acid that are neutralized by 1 tablespoon of milk of magnesia (1 tablespoon = 3 teaspoons).

ANSWERS

ANSWERS TO STUDY CHECKS

14.1   Sulfuric acid; KOH

14.2 $HNO_3(aq) + H_2O(l) \longrightarrow H_3O^+(aq) + NO_3^-(aq)$

14.3   The conjugate acid–base pairs are HCN/CN^- and HSO_4^-/SO_4^{2-}.

14.4 $HNO_3(aq) + H_2O(l) \longrightarrow H_3O^+(aq) + NO_3^-(aq)$

    The products are favored because HNO_3 is a stronger acid than H_3O^+.

14.5   Nitrous acid has a $K_a = 4.5 \times 10^{-4}$.

14.6 $[H_3O^+] = 2.5 \times 10^{-11}$ M; basic

14.7   11.38

14.8   pH = 4.3

14.9 $[H_3O^+] = 3.2 \times 10^{-5}$ M; $[OH^-] = 3.1 \times 10^{-10}$ M

14.10   0.200 M HCl solution

14.11   67.1 mL of KOH solution

14.12   The anion PO_4^{3-} attracts protons from H_2O, forming the weak acid HPO_4^{2-} and OH^-, which makes the solution basic.

14.13   pH = 7.91

ANSWERS TO SELECTED QUESTIONS AND PROBLEMS

14.1 **a.** acid
c. acid
b. acid
d. base

14.3 **a.** hydrochloric acid
c. carbonic acid
e. sulfurous acid
g. ethanoic acid (IUPAC); acetic acid (common)
b. calcium hydroxide
d. nitric acid
f. bromous acid

14.5 **a.** $Mg(OH)_2$
c. HCOOH
e. NH_4OH
b. HF
d. LiOH
f. HIO_4

14.7 **a.** HI is the acid (proton donor), and H_2O is the base (proton acceptor).
b. H_2O is the acid (proton donor), and F^- is the base (proton acceptor).
c. H_2S is the acid (proton donor), and $CH_3-CH_2-NH_2$ is the base (proton acceptor).

14.9 **a.** F^-
c. HPO_3^{2-}
e. ClO_2^-
b. OH^-
d. SO_4^{2-}

14.11 **a.** HCO_3^-
c. H_3PO_4
e. $HClO_4$
b. H_3O^+
d. HBr

14.13 **a.** acid H_2CO_3, conjugate base HCO_3^-; base H_2O, conjugate acid H_3O^+
b. acid NH_4^+, conjugate base NH_3; base H_2O, conjugate acid H_3O^+
c. acid HCN, conjugate base CN^-; base NO_2^-, conjugate acid HNO_2
d. acid HF, conjugate base F^-; base CH_3-COO^-, conjugate acid CH_3-COOH

14.15 $NH_4^+(aq) + H_2O(l) \rightleftharpoons NH_3(aq) + H_3O^+(aq)$

14.17 A strong acid is a good proton donor, whereas its conjugate base is a poor proton acceptor.

14.19 **a.** HBr
c. H_2CO_3
b. HSO_4^-

14.21 **a.** HSO_4^-
c. HCO_3^-
b. HNO_2

14.23 **a.** reactants
c. products
b. reactants
d. reactants

14.25 The reactants are favored because NH_4^+ is a weaker acid than HSO_4^- and SO_4^{2-} is a weaker base than NH_3.

$NH_4^+(aq) + SO_4^{2-}(aq) \rightleftharpoons NH_3(aq) + HSO_4^-(aq)$

14.27 **a.** H_2SO_3
c. H_2SO_3
e. H_2SO_3
b. HSO_3^-
d. HS^-

14.29 $H_3PO_4(aq) + H_2O(l) \rightleftharpoons$

$H_3O^+(aq) + H_2PO_4^-(aq)$ $K_a = \dfrac{[H_3O^+][H_2PO_4^-]}{[H_3PO_4]}$

14.31 In pure water, $[H_3O^+] = [OH^-]$ because one of each is produced every time a proton is transferred from one water molecule to another.

14.33 In an acidic solution, the $[H_3O^+]$ is greater than the $[OH^-]$.

14.35 **a.** acidic
c. basic
b. basic
d. acidic

14.37 **a.** 1.0×10^{-5} M
c. 5.0×10^{-10} M
b. 1.0×10^{-8} M
d. 2.5×10^{-2} M

14.39 **a.** 1.0×10^{-11} M
c. 5.6×10^{-3} M
b. 2.0×10^{-9} M
d. 2.5×10^{-2} M

14.41 In a neutral solution, the $[H_3O^+]$ is 1.0×10^{-7} M and the pH is 7.00, which is the negative value of the power of 10.

14.43 **a.** basic
c. basic
e. acidic
b. acidic
d. acidic
f. basic

14.45 An increase or decrease of one pH unit changes the $[H_3O^+]$ by a factor of 10. Thus a pH of 3 is 10 times more acidic than a pH of 4.

14.47 **a.** 4.00
c. 9.00
e. 7.17
b. 8.52
d. 3.40
f. 10.92

14.49

$[H_3O^+]$	$[OH^-]$	pH	pOH	Acidic, Basic, or Neutral?
1.0×10^{-8} M	1.0×10^{-6} M	8.00	6.00	Basic
3.2×10^{-4} M	3.1×10^{-11} M	3.49	10.51	Acidic
2.8×10^{-5} M	3.6×10^{-10} M	4.55	9.45	Acidic
1.0×10^{-12} M	1.0×10^{-2} M	12.00	2.00	Basic

14.51

a. $ZnCO_3(s) + 2HBr(aq) \longrightarrow ZnBr_2(aq) + CO_2(g) + H_2O(l)$
b. $Zn(s) + 2HCl(aq) \longrightarrow ZnCl_2(aq) + H_2(g)$
c. $HCl(aq) + NaHCO_3(s) \longrightarrow NaCl(aq) + H_2O(l) + CO_2(g)$
d. $H_2SO_4(aq) + Mg(OH)_2(s) \longrightarrow MgSO_4(aq) + 2H_2O(l)$

14.53 **a.** $2HCl(aq) + Mg(OH)_2(s) \longrightarrow MgCl_2(aq) + 2H_2O(l)$
b. $H_3PO_4(aq) + 3LiOH(aq) \longrightarrow Li_3PO_4(aq) + 3H_2O(l)$

14.55 **a.** $H_2SO_4(aq) + 2NaOH(aq) \longrightarrow Na_2SO_4(aq) + 2H_2O(l)$
b. $3HCl(aq) + Fe(OH)_3(s) \longrightarrow FeCl_3(aq) + 3H_2O(l)$
c. $H_2CO_3(aq) + Mg(OH)_2(s) \longrightarrow MgCO_3(s) + 2H_2O(l)$

14.57 To a known volume of formic acid, add a few drops of indicator. Place a solution of NaOH of known molarity in a buret. Add base to acid until one drop changes the color of the solution. Use the volume and molarity of NaOH and the volume of formic acid to calculate the concentration of the formic acid in the sample.

14.59 0.830 M HCl solution

14.61 0.124 M H_2SO_4 solution

14.63 16.5 mL

14.65 The anion from the weak acid removes a proton from H_2O to make a basic solution.

14.67 **a.** neutral
b. acidic: $NH_4^+(aq) + H_2O(l) \rightleftharpoons NH_3(aq) + H_3O^+(aq)$
c. basic: $CO_3^{2-}(aq) + H_2O(l) \rightleftharpoons HCO_3^-(aq) + OH^-(aq)$
d. basic: $S^{2-}(aq) + H_2O(l) \rightleftharpoons HS^-(aq) + OH^-(aq)$

14.69 **b** and **c** are buffer systems. **b** contains the weak acid H_2CO_3 and its salt $NaHCO_3$. **c** contains HF, a weak acid, and its salt KF.

14.71 **a.** A buffer system keeps the pH constant.
b. The same conjugate base is used to produce the same acid when the base neutralizes added H_3O^+.
c. The added H_3O^+ reacts with F^- from NaF.
d. The added OH^- is neutralized by the HF.

14.73 pH = 3.35

14.75 The pH of the 0.10 M HF/0.10 M NaF buffer is 3.14.

The pH of the 0.060 M HF/0.120 M NaF buffer is 3.44.

14.77 a. This diagram represents a weak acid; only a few HX molecules separate into H_3O^+ and X^- ions.

b. This diagram represents a strong acid; all the HX molecules separate into H_3O^+ and X^- ions.

c. This diagram represents a weak acid; only a few HX molecules separate into H_3O^+ and X^- ions.

14.79 a. During hyperventilation, a person will lose CO_2 and the blood pH will rise.

b. Breathing into a paper bag will increase the CO_2 concentration and lower the blood pH.

14.81 a. acid, bromous acid

b. base, rubidium hydroxide

c. salt, magnesium nitrate

d. acid, butanoic acid (butyric acid)

e. acid, perchloric acid

14.83 a. acidic   **b.** basic

c. acidic   **d.** acidic

e. basic

14.85 a. The $Mg(OH)_2$ that dissolves is completely dissociated, making it a strong base.

b. $Mg(OH)_2(s) + 2HCl(aq) \longrightarrow MgCl_2(aq) + 2H_2O(l)$

14.87 a. HF   **b.** H_3O^+

c. HNO_2 **d.** HCO_3^-

14.89 a. pH = 7.70; pOH = 6.30

b. pH = 1.30; pOH = 12.70

c. pH = 10.54; pOH = 3.46

d. pH = 11.73; pOH = 2.27

14.91 a. basic   **b.** acidic

c. basic   **d.** basic

14.93 a. $[H_3O^+] = 1.0 \times 10^{-3}$ M; $[OH^-] = 1.0 \times 10^{-11}$ M

b. $[H_3O^+] = 3.3 \times 10^{-7}$ M; $[OH^-] = 3.0 \times 10^{-8}$ M

c. $[H_3O^+] = 1.4 \times 10^{-9}$ M; $[OH^-] = 7.1 \times 10^{-6}$ M

d. $[H_3O^+] = 1.0 \times 10^{-11}$ M; $[OH^-] = 1.0 \times 10^{-3}$ M

14.95 a. Solution A

b. Solution A $[H_3O^+] = 3 \times 10^{-5}$ M; Solution B $[H_3O^+] = 2 \times 10^{-7}$ M

c. Solution A $[OH^-] = 3 \times 10^{-10}$ M; Solution B $[OH^-] = 5 \times 10^{-8}$ M

14.97 $[OH^-] = 0.0225$ M

14.99 pH = 0.80; pOH = 13.20

14.101 a. basic   **b.** basic

c. acidic   **d.** neutral

14.103 a. acid:

$H_2PO_4^-(aq) + H_3O^+(aq) \longrightarrow H_3PO_4(aq) + H_2O(l)$

b. base:

$H_3PO_4(aq) + OH^-(aq) \longrightarrow H_2PO_4^-(aq) + H_2O(l)$

c. pH = 1.72

14.105 a. 48.0 mL of NaOH solution

b. 74.7 mL of NaOH solution

14.107 0.234 M H_2SO_4 solution

14.109 a. 1. HS^- **2.** $H_2PO_4^-$ **3.** CO_3^{2-}

b. 1. $\dfrac{[H_3O^+][HS^-]}{[H_2S]}$ **2.** $\dfrac{[H_3O^+][H_2PO_4^-]}{[H_3PO_4]}$

3. $\dfrac{[H_3O^+][CO_3^{2-}]}{[HCO_3^-]}$

c. HCO_3^- **d.** H_3PO_4

14.111 a. $ZnCO_3(s) + H_2SO_4(aq) \longrightarrow$
$ZnSO_4(aq) + CO_2(g) + H_2O(l)$

b. $2Al(s) + 6HNO_3(aq) \longrightarrow 2Al(NO_3)_3(aq) + 3H_2(g)$

c. $2H_3PO_4(aq) + 3Ca(OH)_2(s) \longrightarrow$
$Ca_3(PO_4)_2(s) + 6H_2O(l)$

d. $KHCO_3(s) + HNO_3(aq) \longrightarrow$
$KNO_3(aq) + CO_2(g) + H_2O(l)$

14.113 a. $[H_3O^+] = 2.0 \times 10^{-13}$ M

b. pH = 12.70

c. pOH = 1.30

d. $3KOH(aq) + H_3PO_4(aq) \longrightarrow K_3PO_4(aq) + 3H_2O(l)$

e. 56 mL of the KOH solution

14.115 a. $[H_3O^+] = 6 \times 10^{-5}$ M; $[OH^-] = 2 \times 10^{-10}$ M

b. $[H_3O^+] = 3 \times 10^{-7}$ M; $[OH^-] = 3 \times 10^{-8}$ M

c. 3 g of $CaCO_3$

COMBINING IDEAS FROM CHAPTERS 11 TO 14

CI.21 Methane, an alkane with one carbon atom covalently bonded to hydrogen, is a major component of purified natural gas used for heating and cooking. When one mole of methane gas burns with oxygen to produce carbon dioxide and water vapor, 883 kJ of heat is produced. Methane gas has a density of 0.715 g/L at STP. For transport, the volume of natural gas is cooled to $-163\ °C$, which gives liquefied natural gas (LNG) with a density of 0.45 g/mL. A tank on a ship can hold 7.0 million gallons of LNG.

a. What is the condensed structural formula of methane?
b. What is the mass, in kilograms, of LNG (assume that LNG is all methane) transported in one tank on a ship?
c. What is the volume, in liters, of LNG (methane) from one tank when the LNG in one tank is converted to gas at STP?
d. Write the balanced combustion equation for the combustion of methane and oxygen in a gas burner.
e. How many kilograms of oxygen are needed to react with all of the methane in one tank of LNG?
f. How much heat, in kilojoules, is released from burning all of the methane in one tank of LNG?

CI.22 A mixture of 25.0 g of CS_2 gas and 30.0 g of O_2 gas is placed in 10.0-L closed container and heated to 125 °C. In the reaction, the products are carbon dioxide gas and sulfur dioxide gas.
a. Write a balanced equation for the reaction.
b. How many grams of CO_2 are produced?
c. What is the partial pressure of the excess reactant?
d. What is the final pressure in the container?

CI.23 Consider the following reaction at equilibrium:
$$2H_2(g) + S_2(g) \rightleftharpoons 2H_2S(g) + \text{heat}$$
In a 10.0-L container, an equilibrium mixture contains 68.2 g of H_2S, 2.02 g of H_2, and 10.3 g of S_2.
a. What is the K_c value for this equilibrium mixture?
b. If H_2 is added to the mixture, how will the equilibrium shift?
c. How will the equilibrium shift if the mixture is placed in a 5.00-L container with no change in temperature?
d. If a 5.00-L container has an equilibrium mixture of 0.300 mol of H_2 and 2.50 mol of H_2S, what is the $[S_2]$ if temperature is the same?
e. Will an increase in temperature increase or decrease the K_c value?

CI.24 In wine-making, sugar ($C_6H_{12}O_6$) from grapes undergoes fermentation in the absence of oxygen to produce ethanol and carbon dioxide. A bottle of vintage port wine has a volume of 750 mL and contains 135 mL of ethanol (CH_3-CH_2-OH). Ethanol has a density of 0.789 g/mL. In 1.5 lb of grapes, there are 26 g of sugar.

a. Calculate the volume percent (v/v) of ethanol.
b. What is the molarity (M) of ethanol in the port wine?
c. Write the balanced equation for the fermentation reaction of sugar in grapes.
d. How many grams of sugar from grapes are required to produce one bottle of port wine?
e. How many bottles of port wine can be produced from 1.0 ton of grapes if 1 ton is equal to 2000 lb?

CI.25 A metal M with a mass of 0.420 g completely reacts with 34.8 mL of a 0.520 M HCl solution to form aqueous MCl_3 and H_2 gas.

a. Write a balanced equation for the reaction of the metal M(s) and HCl(aq).

b. What volume, in mL, of H_2 at 720. mmHg and 24 °C is produced?

c. How many moles of metal M reacted?

d. Use your results from part **c** to determine the molar mass and name of metal M.

e. Write the balanced equation for the reaction.

f. What is the electron configuration of the metal and its cation?

CI.26 A saturated solution of cobalt(II) hydroxide has a pH of 9.36.

a. Write the solubility product expression for cobalt(II) hydroxide.

b. Calculate the value of the K_{sp} for cobalt(II) hydroxide.

c. How many grams of cobalt(II) hydroxide will dissolve in 2.0 L of water?

d. How many grams of cobalt(II) hydroxide will dissolve in 50.0 mL of a 0.0100 M NaOH solution?

■ ANSWERS

CI.21 a. CH_4

b. 1.2×10^7 kg of LNG (methane)

c. 1.7×10^{10} L of LNG (methane)

d. $CH_4(g) + 2O_2(g) \longrightarrow CO_2(g) + 2H_2O(g)$

e. 4.8×10^7 kg of O_2

f. 6.6×10^{11} kJ

CI.23 a. $K_c = 248$

b. If H_2 is added, the equilibrium will shift toward the products.

c. If the volume decreases, the equilibrium shifts toward the products.

d. $[S_2] = 0.280$ M

e. An increase in temperature will decrease the value of K_c.

CI.25 a. $2M(s) + 6HCl(aq) \longrightarrow 2MCl_3(aq) + 3H_2(g)$

b. 233 mL of H_2

c. 6.03×10^{-3} mol of M

d. 69.7 g/mol; gallium

e. $2Ga(s) + 6HCl(aq) \longrightarrow 2GaCl_3(aq) + 3H_2(g)$

f. Ga $1s^2 2s^2 2p^6 3s^2 3p^6 4s^2 3d^{10} 4p^1$;
$Ga^{3+} 1s^2 2s^2 2p^6 3s^2 3p^6 3d^{10}$

15 Oxidation and Reduction

"As a conservator of photographic materials, it is essential to have an understanding of the chemical reactions of many different photographic processes," says Theresa Andrews, Conservator of Photographs at the San Francisco Museum of Modern Art. "For example, the creation of the latent image in many photographs is based upon the light sensitivity of silver halides. Photolytic silver 'prints out' when exposed to a light source such as the Sun, and filamentary silver 'develops out' when an exposed photographic paper is placed in a bath with reducing agents. Photolytic silver particles are much smaller than filamentary silver particles, making them more vulnerable to abrasion and image loss. This knowledge is critical when making recommendations for light levels and for the protection of photographs when they are on exhibition. Conservation treatments require informed decisions based on the reactivity of the materials within the photograph and also the compatibility of materials that might be required for repair or preservation of the photograph."

Mastering**CHEMISTRY**

Visit **www.masteringchemistry.com** for self-study materials and instructor-assigned homework.

Perhaps you have never heard of an oxidation and reduction reaction. However, this type of reaction has many important applications in your everyday life. When you see a rusty nail, tarnish on a silver spoon, or corrosion on metal, you are observing oxidation. Historically, the term *oxidation* was used for reactions of the elements with oxygen to form oxides.

$$4Fe(s) + 3O_2(g) \longrightarrow 2Fe_2O_3(s) \quad \text{Fe is oxidized}$$

Rust

When we turn the lights on in our automobiles, an oxidation–reduction reaction within the car battery provides the electricity. On a cold, wintry day, we might build a fire. As the wood burns, oxygen combines with carbon and hydrogen to produce carbon dioxide, water, and heat. In Chapter 8, Section 8.5, we discussed reactions with oxygen as combustion, which we can now describe as oxidation–reduction reactions. When we eat foods with starches in them, the starches break down to give glucose, which is oxidized in our cells to give energy along with carbon dioxide and water. Every breath we take provides oxygen to carry out oxidation in our cells.

$$C_6H_{12}O_6(aq) + 6O_2(g) \longrightarrow 6CO_2(g) + 6H_2O(l) + \text{energy}$$

Glucose

The term *reduction* was originally used for reactions that removed oxygen from compounds. Metal oxides in ores are reduced to obtain the pure metal. For example, iron metal is obtained by reducing the iron in iron ore with carbon.

$$2Fe_2O_3(l) + 3C(s) \longrightarrow 3CO_2(g) + 4Fe(l) \quad \text{Fe in Fe}_2\text{O}_3 \text{ is reduced}$$

As more was learned about oxidation and reduction reactions, scientists found that they do not always involve oxygen. Today an oxidation–reduction reaction is any reaction that involves the transfer of electrons from one substance to another.

Rust forms on a set of tools when oxygen reacts with iron.

Pure iron is obtained from reducing iron ores.

15.1 Oxidation–Reduction Reactions

In every **oxidation–reduction reaction** (abbreviated *redox*), electrons are transferred from one substance to another. If one substance loses electrons, another substance must gain electrons. **Oxidation** is defined as the *loss* of electrons; **reduction** is the *gain* of electrons. One way to remember these definitions is to use the following:

Oil Rig

Oxidation **I**s **L**oss of electrons

Reduction **I**s **G**ain of electrons

We can now look at the oxidation and reduction reactions that take place when calcium metal reacts with sulfur to produce the ionic compound calcium sulfide.

$$Ca(s) + S(s) \longrightarrow CaS(s)$$

Because calcium has a 2+ charge in CaS, each calcium atom has lost two electrons to form the calcium ion (Ca^{2+}). Therefore, calcium metal has been oxidized in the reaction.

$$Ca \longrightarrow Ca^{2+} + 2\,e^- \quad \text{Oxidation; loss of electrons by Ca}$$

Because sulfur has a 2− charge in CaS, each sulfur atom has gained two electrons to form sulfide ion (S^{2-}). Therefore, sulfur has been reduced in the reaction.

$$S + 2\,e^- \longrightarrow S^{2-} \quad \text{Reduction; gain of electrons by S}$$

LEARNING GOAL

Identify what is oxidized and what is reduced in an oxidation–reduction reaction.

TUTORIAL

Identifying Oxidation–Reduction Reactions

Reduced		Oxidized
Na	Oxidation: lose e^- →	$Na^+ + e^-$
Ca		$Ca^{2+} + 2\,e^-$
$2Br^-$		$Br_2 + 2\,e^-$
Fe^{2+}	← Reduction: gain e^-	$Fe^{3+} + e^-$

Oxidation is a loss of electrons; reduction is a gain of electrons.

FIGURE 15.1 In this single replacement reaction, Zn(s) is oxidized to Zn^{2+} when it provides two electrons to reduce Cu^{2+} to Cu(s):
$$Zn(s) + Cu^{2+}(aq) \longrightarrow Zn^{2+}(aq) + Cu(s)$$
Q In this reaction, is Zn(s) oxidized or reduced?

Therefore, the reaction that forms CaS involves both an oxidation and a reduction.

$$Ca(s) + S(s) \longrightarrow Ca^{2+} + S^{2-} = CaS(s)$$

As we see in the next reaction between zinc and copper(II) sulfate, there is always an oxidation with every reduction (see Figure 15.1).

$$Zn(s) + CuSO_4(aq) \longrightarrow ZnSO_4(aq) + Cu(s)$$

We can rewrite the equation to show the atoms and ions:

$$\mathbf{Zn(s) + Cu^{2+}(aq) + SO_4{}^{2-}(aq) \longrightarrow Zn^{2+}(aq) + SO_4{}^{2-}(aq) + Cu(s)}$$

In this reaction, Zn atoms lose two electrons to form Zn^{2+} ions; Zn(s) is oxidized. At the same time, Cu^{2+} ions gain two electrons to form Cu(s); Cu^{2+} is reduced. The $SO_4{}^{2-}$ ions are spectator ions and do not change.

$$Zn(s) \longrightarrow Zn^{2+}(aq) + 2\ e^- \qquad \text{Oxidation of Zn}$$

$$Cu^{2+}(aq) + 2\ e^- \longrightarrow Cu(s) \qquad \text{Reduction of } Cu^{2+}$$

CONCEPT CHECK 15.1

■ **Loss and Gain of Electrons**

Identify each of the following as an oxidation or a reduction:

a. $Be(s) \longrightarrow Be^{2+}(aq)$
b. $Mg^{2+}(aq) \longrightarrow Mg(s)$
c. $2Cl^-(aq) \longrightarrow Cl_2(g)$

ANSWER

a. When an atom of Be loses two electrons to form a 2+ ion, it is an oxidation.

$$Be(s) \longrightarrow Be^{2+}(aq) + 2\ e^-$$

b. When a magnesium ion with a 2+ charge gains two electrons to form a neutral magnesium atom, it is a reduction.

$$Mg^{2+}(aq) + 2\ e^- \longrightarrow Mg(s)$$

c. When two chloride ions each with a 1− charge lose two electrons to form neutral Cl atoms in a Cl_2 molecule, it is an oxidation.

$$2Cl^-(aq) \longrightarrow Cl_2(g) + 2\ e^-$$

SAMPLE PROBLEM 15.1

■ **Oxidation–Reduction Reactions**

In photographic film, the following decomposition reaction occurs in the presence of light. What is oxidized and what is reduced?

$$2AgBr(s) \xrightarrow{\text{Light}} 2Ag(s) + Br_2(g)$$

SOLUTION

To determine oxidation and reduction, we determine the ions and charges in the reactants and products. In AgBr, there is a silver ion (Ag^+) with a 1+ charge and a bromide ion (Br^-) with a charge of 1−. We can write a balanced reaction as follows:

$$2Ag^+(aq) + 2Br^-(aq) \longrightarrow 2Ag(s) + 2Br_2(g)$$

Now we can compare Ag^+ with the product Ag atom. In this case, each Ag^+ gained an electron; Ag^+ is reduced.

$$2Ag^+(aq) + 2\,e^- \longrightarrow 2Ag(s) \qquad \text{Reduction}$$

Then we compare Br^- with the Br in the product Br_2. In this case, each Br^- lost an electron; Br^- is oxidized.

$$2Br^-(aq) \longrightarrow Br_2(g) + 2\,e^- \qquad \text{Oxidation}$$

STUDY CHECK

In the following equation, which reactant is oxidized and which is reduced?

$$2Al(s) + 3Sn^{2+}(aq) \longrightarrow 2Al^{3+}(aq) + 3Sn(s)$$

A vintage photograph with sepia tone caused by the reaction of light with silver in the film.

QUESTIONS AND PROBLEMS

Oxidation–Reduction Reactions

15.1 Identify each of the following as an oxidation or a reduction:
  a. $Na^+(aq) + e^- \longrightarrow Na(s)$
  b. $Ni(s) \longrightarrow Ni^{2+}(aq) + 2\,e^-$
  c. $Cr^{3+}(aq) + 3\,e^- \longrightarrow Cr(s)$
  d. $2H^+(aq) + 2\,e^- \longrightarrow H_2(g)$

15.2 Identify each of the following as an oxidation or a reduction:
  a. $O_2(g) + 4\,e^- \longrightarrow 2O^{2-}(aq)$
  b. $Al(s) \longrightarrow Al^{3+}(aq) + 3\,e^-$
  c. $Fe^{3+}(aq) + e^- \longrightarrow Fe^{2+}(aq)$
  d. $2Br^-(aq) \longrightarrow Br_2(l) + 2\,e^-$

15.3 In the following reactions, identify the reactant that is oxidized and the reactant that is reduced:
  a. $Zn(s) + Cl_2(g) \longrightarrow ZnCl_2(s)$
  b. $Cl_2(g) + 2NaBr(aq) \longrightarrow 2NaCl(aq) + Br_2(l)$
  c. $2Pb(s) + O_2(g) \longrightarrow 2PbO(s)$
  d. $2Fe^{3+}(aq) + Sn^{2+}(aq) \longrightarrow 2Fe^{2+}(aq) + Sn^{4+}(aq)$

15.4 In the following reactions, identify the reactant that is oxidized and the reactant that is reduced:
  a. $2Li(s) + F_2(g) \longrightarrow 2LiF(s)$
  b. $Cl_2(g) + 2KI(aq) \longrightarrow 2KCl(aq) + I_2(s)$
  c. $Zn(s) + Cu^{2+}(aq) \longrightarrow Zn^{2+}(aq) + Cu(s)$
  d. $Fe(s) + CuSO_4(aq) \longrightarrow FeSO_4(aq) + Cu(s)$

15.2 Oxidation Numbers

In more complex oxidation–reduction reactions, the identification of the substances oxidized and reduced is not as obvious as in the previous section. To help identify the atoms or ions that are oxidized or reduced, we can assign values called **oxidation numbers** (or sometimes oxidation states) to the elements of the reactants and products. It is important to recognize that oxidation numbers do not always represent actual charges, but they help us identify loss or gain of electrons.

Rules for Assigning Oxidation Numbers

The rules for assigning oxidation numbers to the atoms or ions in the reactants and products are given in Table 15.1.

LEARNING GOAL

Assign and use oxidation numbers to identify elements that are oxidized or reduced, and to balance an oxidation–reduction equation.

TUTORIAL

Assigning Oxidation States

TABLE 15.1 Rules for Assigning Oxidation Numbers

1. The sum of the oxidation numbers in a molecule is equal to zero (0), or for a polyatomic ion the sum of the oxidation numbers is equal to its charge.

2. The oxidation number of an element (monatomic or diatomic) is zero (0).

3. The oxidation number of a monatomic ion is equal to its charge.

4. In compounds, the oxidation numbers of Group 1A (1) elements is +1 and of Group 2A (2) elements is +2.

5. In compounds, the oxidation number of fluorine is always −1. Other elements in Group 7A (7) also are −1 except when combined with oxygen or fluorine.

6. In compounds, the oxidation number of oxygen is usually −2 except in OF_2.

7. In compounds with nonmetals, the oxidation number of hydrogen is +1; in compounds with metals, the oxidation number of hydrogen is −1.

We can now look at how these rules are used to assign oxidation numbers. For each formula, the oxidation numbers are written *below* the symbols of the elements (see Table 15.2).

TABLE 15.2 Examples of Using Rules to Assign Oxidation Numbers

Formula	Oxidation Numbers	Explanation
Br_2	Br_2 0	Each Br in the diatomic element bromine has an oxidation number of 0 (Rule 2).
Ba^{2+}	Ba^{2+} +2	Ba^{2+}, a monatomic ion, is assigned an oxidation number of +2 (Rule 3).
CO_2	CO_2 +4−2	In compounds, O has an oxidation number of −2 (Rule 6). Because CO_2 is neutral, the oxidation number of C is calculated as +4 (Rule 1). $1C + 2O = 0$ $C + 2(-2) = 0$ $C = +4$
Al_2O_3	Al_2O_3 +3−2	In compounds, the oxidation number of O is −2 (Rule 6). For Al_2O_3 (neutral), the oxidation number of Al is calculated as +3 (Rule 1). $2Al + 3O = 0$ $2Al + 3(-2) = 0$ $2Al = +6$ $Al = +3$
$HClO_3$	$HClO_3$ +1+5−2	The oxidation number of H is +1 (Rule 7), and O is −2 (Rule 6). For $HClO_3$ (neutral), the oxidation number of Cl is calculated as +5 (Rule 1). $H + Cl + 3O = 0$ $(+1) + Cl + 3(-2) = 0$ $Cl - 5 = 0$ $Cl = +5$
SO_4^{2-}	SO_4^{2-} +6−2	The oxidation number of O is −2 (Rule 6). For the SO_4^{2-} (−2 charge), the oxidation number of S is calculated as +6 (Rule 1). $S + 4O = -2$ $S + 4(-2) = -2$ $S = +6$
CH_2O	CH_2O 0+1−2	The oxidation number of H is +1 (Rule 7), and O is −2 (Rule 6). For CH_2O (neutral), the oxidation number of C is calculated as 0 (Rule 1). $C + 2H + O = 0$ $C + 2(+1) + (-2) = 0$ $C = 0$

SAMPLE PROBLEM   15.2

■ **Assigning Oxidation Numbers**

Assign oxidation numbers to the elements in each of the following:

a. NCl_3 **b.** CO_3^{2-} **c.** SF_6

SOLUTION

a. NCl_3: The oxidation number of Cl is -1 (Rule 5). For NCl_3 (neutral), the sum of the oxidation numbers of N and 3Cl must be equal to zero (Rule 1). Thus, the oxidation number of N is calculated as $+3$.

$$N + 3Cl\ \ = 0$$
$$N + 3(-1) = 0$$
$$N\ \ \ \ \ \ \ \ \ \ = +3$$

The oxidation numbers are:

NCl_3

$+3-1$

b. CO_3^{2-}: The oxidation number of O is -2 (Rule 6). For CO_3^{2-}, the sum of oxidation numbers is equal to -2 (Rule 1). The oxidation number of C is calculated as $+4$.

$$C\ + 3O\ \ \ = -2$$
$$C + 3(-2) = -2$$
$$C\ \ \ \ \ \ \ \ \ = +4$$

The oxidation numbers are:

CO_3^{2-}

$+4-2$

c. SF_6: The oxidation number of F is -1 (Rule 5). For SF_6 (neutral), the oxidation number of S is calculated as $+6$ (Rule 1).

$$S\ \ + 6F\ \ = 0$$
$$S + 6(-1) = 0$$
$$S\ \ \ \ \ \ \ \ \ = +6$$

The oxidation numbers are:

SF_6

$+6-1$

STUDY CHECK

Assign oxidation numbers to the elements in each of the following:

a. H_3PO_4 **b.** MnO_4^-

Using Oxidation Numbers to Identify Oxidation–Reduction

Oxidation numbers can be used to identify the elements that are oxidized and the elements that are reduced in a reaction. In oxidation, the loss of electrons increases the oxidation number so that it is higher (more positive) in the product than in the reactant. In reduction, the gain of electrons decreases the oxidation number so that it is lower (more negative) in the product than in the reactant.

⟵ Reduction: oxidation number decreases

$$-7\ -6\ -5\ -4\ -3\ -2\ -1\ \ \ 0\ +1\ +2\ +3\ +4\ +5\ +6\ +7$$

Oxidation: oxidation number increases ⟶

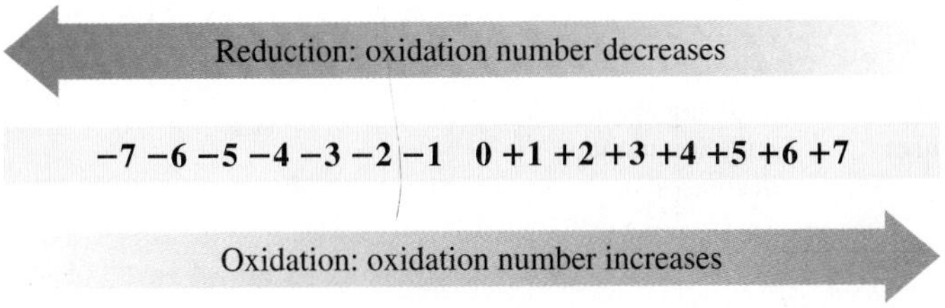

CONCEPT CHECK 15.2

■ **Oxidation Numbers**

For the following equation,

$$CO_2(g) + H_2(g) \longrightarrow CO(g) + H_2O(g)$$

a. determine the oxidation numbers of all the atoms
b. identify the element that is oxidized, and the element that is reduced

ANSWER

a. In H_2, the oxidation number of H is 0. In H_2O, the oxidation number of H is $+1$. In CO_2, CO, and H_2O, the oxidation number of O is -2.

$$CO_2(g) + H_2(g) \longrightarrow CO(g) + H_2O(g)$$

| ? −2 | 0 | ? −2 | +1 −2 | Oxidation numbers |

The oxidation number of C is calculated as $+4$ in CO_2 and $+2$ in CO.

$$CO_2(g) + H_2(g) \longrightarrow CO(g) + H_2O(g)$$

| +4 −2 | 0 | +2 −2 | +1 −2 | Oxidation numbers |

b. In H_2, H is oxidized because its oxidation number increases from 0 in the reactant to $+1$ in the product. In CO_2, C is reduced because its oxidation number decreases from $+4$ to $+2$.

H is oxidized

$$CO_2(g) + H_2(g) \longrightarrow CO(g) + H_2O(g)$$

| +4−2 | 0 | +2−2 | +1−2 | Oxidation numbers |

C is reduced

TUTORIAL
Oxidation, Reduction, Oxidizing Agents,
and Reducing Agents

Oxidizing and Reducing Agents

We have seen that an oxidation reaction must always be accompanied by a reduction reaction. The substance that is oxidized loses electrons, and the substance that is reduced gains those electrons. For example, Zn is oxidized to Zn^{2+} by losing 2 electrons that are transferred to Cl_2 to reduce it to $2Cl^-$.

$$Zn(s) + Cl_2(g) \longrightarrow ZnCl_2(s)$$

In an oxidation–reduction reaction, a **reducing agent** provides electrons and an **oxidizing agent** accepts electrons. In this reaction, Zn is a *reducing agent* because it provides electrons used to reduce Cl_2. At the same time, Cl_2 is an *oxidizing agent* because it accepts electrons provided by the oxidation of Zn. In any oxidation–reduction reaction, the reducing agent is oxidized, and the oxidizing agent is reduced.

$Zn \longrightarrow Zn^{2+} + 2\,e^-$ Zn is oxidized; Zn is the *reducing agent*.

$Cl_2 + 2\,e^- \longrightarrow 2Cl^-$ Cl in Cl_2 is reduced; Cl_2 is the *oxidizing agent*.

We can now summarize the terms we have used to describe oxidation and reduction. Their relationships follow:

Oxidation–Reduction Terminology

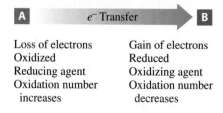

| A | e^- Transfer | B |

Loss of electrons	Gain of electrons
Oxidized	Reduced
Reducing agent	Oxidizing agent
Oxidation number increases	Oxidation number decreases

SAMPLE PROBLEM | 15.3

■ Identifying Oxidized and Reduced Substances

In the following, identify the substance that is oxidized, the substance that is reduced, the oxidizing agent, and the reducing agent:

$$PbO(s) + CO(g) \longrightarrow Pb(s) + CO_2(g)$$

SOLUTION

The O atoms in PbO, CO, and CO_2 have oxidation numbers of -2. The element Pb has an oxidation number of 0.

$$PbO(s) + CO(g) \longrightarrow Pb(s) + CO_2(g)$$
  ? -2 ? -2 0   ? -2 Oxidation numbers

For PbO, the oxidation number of Pb is calculated as $+2$. The oxidation number of C in CO is calculated as $+2$ and in CO_2 as $+4$. Therefore C is oxidized because its oxidation number increases from $+2$ to $+4$; the Pb in PbO is reduced because its oxidation number decreases from $+2$ to 0.

C is oxidized

$$PbO(s) + CO(g) \longrightarrow Pb(s) + CO_2(g)$$
 $+2-2$ $+2-2$ 0 $+4-2$ Oxidation numbers

Pb is reduced

Because the Pb^{2+} in PbO accepts electrons, PbO is the oxidizing agent. The reducing agent is CO because it provides electrons.

STUDY CHECK

Use oxidation numbers to identify the substance that is oxidized, the substance that is reduced, the oxidizing agent, and the reducing agent.

$$Zn(s) + CuCl_2(aq) \longrightarrow ZnCl_2(aq) + Cu(s)$$

Lead(II) oxide, once used in paint, is now banned due to the toxicity of lead.

Using Oxidation Numbers to Balance Oxidation–Reduction Equations

We have seen that the processes of oxidation and reduction always occur simultaneously, which means that the number of electrons lost during oxidation is equal to the number of electrons gained during reduction. This requirement of an equal loss and gain of electrons can now be used to balance oxidation–reduction equations.

In the oxidation number method of balancing, oxidation numbers are assigned to all the elements. We equalize the loss and gain of electrons by multiplying the atoms that are oxidized and reduced by small integers. This oxidation-number method is typically used to balance equations written in the molecular form.

SAMPLE PROBLEM | 15.4

■ Using Oxidation Numbers to Balance Equations

Use oxidation numbers to balance the following equation:

$$FeO(s) + C(s) \longrightarrow Fe(s) + CO_2(g)$$

Iron(II) oxide is a jet-black compound used in the manufacture of steel.

Guide to Balancing Equations Using Oxidation Numbers

STEP 1
Assign oxidation numbers to all the elements.

STEP 2
Identify the oxidized and reduced elements from the changes in oxidation numbers.

STEP 3
Multiply the changes in oxidation numbers by small integers to equalize the increase and decrease.

STEP 4
Balance the remaining elements by inspection.

SOLUTION

Using the guide for balancing with oxidation numbers, we can proceed as follows:

STEP 1 **Assign oxidation numbers to all the elements.**

$$FeO(s) + C(s) \longrightarrow Fe(s) + CO_2(g)$$
$$+2-2 0 0 +4-2$$

STEP 2 **Identify the oxidized and reduced elements from the changes in oxidation numbers.** The oxidation number of C increases from 0 to +4; C is oxidized. The oxidation number of Fe decreases from +2 to 0; Fe is reduced.

Increases by 4

$$FeO(s) + C(s) \longrightarrow Fe(s) + CO_2(g)$$
$$+2-2 0 0 +4-2$$

Decreases by 2

STEP 3 **Multiply the changes in oxidation numbers by small integers to equalize the increase and decrease.** Iron needs a multiplying factor of 2 to equalize the increase and decrease of oxidation numbers.

Equalizing Changes in Oxidation Numbers

	Oxidation Numbers	Change	Multiplying Factor	Total Electron Loss and Gain
Oxidation	C(0) ⟶ C(+4)	4 (increase)	× 1	= 4
Reduction	Fe(+2) ⟶ Fe(0)	2 (decrease)	× 2	= 4

The multiplying factor of 2 is used as a coefficient for FeO and Fe. The coefficient of 1 from the multiplying factor for C and CO_2 is understood.

1 × (Increases by 4)

$$\mathbf{2}FeO(s) + C(s) \longrightarrow \mathbf{2}Fe(s) + CO_2(g)$$
$$+2-2 0 0 +4-2$$

2 × (Decreases by 2)

STEP 4 **Balance the remaining elements by inspection.** All the atoms including the O atoms are balanced. Thus we can write the completely balanced equation as

$$2FeO(s) + C(s) \longrightarrow 2Fe(s) + CO_2(g)$$

STUDY CHECK

Use oxidation numbers to balance the following equation:

$$Li(s) + AlCl_3(aq) \longrightarrow LiCl(aq) + Al(s)$$

SAMPLE PROBLEM | **15.5**

■ **Balancing Equations with Oxidation Numbers**

Use oxidation numbers to balance the equation for the oxidation–reduction reaction of tin and nitric acid.

$$Sn(s) + HNO_3(aq) \longrightarrow SnO_2(s) + NO_2(g) + H_2O(g)$$

SOLUTION

Using the guide for balancing with oxidation numbers, we can proceed as follows:

STEP 1 **Assign oxidation numbers to all the elements.**

$$Sn(s) + HNO_3(aq) \longrightarrow SnO_2(s) + NO_2(g) + H_2O(g)$$
$$0 +1+5-2 +4-2 +4-2 +1-2$$

STEP 2 **Identify the oxidized and reduced elements from the changes in oxidation numbers.** The oxidation number of Sn increases from 0 to +4; Sn is oxidized. The oxidation number of N decreases from +5 to +4; N is reduced.

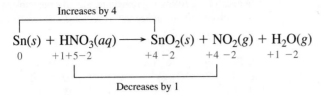

STEP 3 **Multiply the changes in oxidation numbers by small integers to equalize the increase and decrease.** We now multiply the change in the oxidation number of nitrogen by 4 to equalize the changes in oxidation numbers.

Equalizing Changes in Oxidation Number

	Oxidation Numbers	Change	Multiplying Factor	Total Electron Loss and Gain
Oxidation	$Sn(0) \longrightarrow Sn(+4)$	4 (increase)	$\times\,1$	$= 4$
Reduction	$N(+5) \longrightarrow N(+4)$	1 (decrease)	$\times\,4$	$= 4$

The multiplying factor of 4 is used as a coefficient for HNO_3 and NO_2. The coefficient of 1 from the multiplying factor for Sn and SnO_2 is understood.

$$\overset{1\,\times\,\text{(Increases by 4)}}{Sn(s) + 4HNO_3(aq) \longrightarrow SnO_2(s) + 4NO_2(g) + H_2O(g)}$$
$$0 \qquad +1+5-2 \qquad\quad +4-2 \qquad +4-2 \qquad +1-2$$
$$4 \times \text{(Decreases by 1)}$$

STEP 4 **Balance the remaining elements by inspection.** In the equation, the four H atoms in the reactants are balanced by placing a coefficient of 2 in front of H_2O. This also balances the total number of O atoms on the reactant side. We can now write the balanced equation as

$$Sn(s) + 4HNO_3(aq) \longrightarrow SnO_2(s) + 4NO_2(g) + 2H_2O(g)$$

STUDY CHECK

Use oxidation numbers to balance the equation for the oxidation–reduction reaction of iron(III) oxide and carbon to form iron and carbon dioxide.

$$Fe_2O_3(s) + C(s) \longrightarrow Fe(s) + CO_2(g)$$

QUESTIONS AND PROBLEMS

Oxidation Numbers

15.5 Assign oxidation numbers to each of the following:
  a. Cu   **b.** F_2
  c. Fe^{2+} **d.** Cl^-

15.6 Assign oxidation numbers to each of the following:
  a. Al   **b.** Al^{3+}
  c. F^- **d.** N_2

15.7 Assign oxidation numbers to all the elements in each of the following:
  a. KCl   **b.** MnO_2
  c. CO   **d.** Mn_2O_3

15.8 Assign oxidation numbers to all the elements in each of the following:
  a. H_2S **b.** NO_2
  c. CCl_4 **d.** PCl_3

15.9 Assign oxidation numbers to all the elements in each of the following compounds or polyatomic ions:
  a. $AlPO_4$ **b.** SO_3^{2-}
  c. Cr_2O_3 **d.** NO_3^-

15.10 Assign oxidation numbers to all the elements in each of the following compounds or polyatomic ions:
  a. $C_2H_3O_2^-$ **b.** $AlCl_3$
  c. NH_4^+ **d.** $HBrO_4$

15.11 Assign oxidation numbers to all the elements in each of the following compounds or polyatomic ions:
   a. HSO_4^- **b.** H_3PO_3
   c. $Cr_2O_7^{2-}$ **d.** Na_2CO_3

15.12 Assign oxidation numbers to all the elements in each of the following compounds or polyatomic ions:
   a. N_2O **b.** $LiOH$
   c. SbO_2^- **d.** IO_4^-

15.13 What is the oxidation number of the specified element in each compound or polyatomic ion?
   a. N in HNO_3 **b.** C in C_3H_6
   c. P in K_3PO_4 **d.** Cr in CrO_4^{2-}

15.14 What is the oxidation number of the specified element in each compound or polyatomic ion?
   a. C in $ZnCO_3$ **b.** Fe in $Fe(NO_3)_2$
   c. Cl in ClF_4^- **d.** S in $S_2O_3^{2-}$

15.15 Indicate whether each of the following describes the oxidizing agent or the reducing agent in an oxidation–reduction reaction:
   a. the substance that is oxidized
   b. the substance that gains electrons

15.16 Indicate whether each of the following describes the oxidizing agent or the reducing agent in an oxidation–reduction reaction:
   a. the substance that is reduced
   b. the substance that loses electrons

15.17 In Problem 15.3, identify the oxidizing agent and the reducing agent in each reaction.

15.18 In Problem 15.4, identify the oxidizing agent and the reducing agent in each reaction.

15.19 In each of the following reactions, identify the substance that is oxidized, the substance that is reduced, the oxidizing agent, and the reducing agent:
   a. $2NiS(s) + 3O_2(g) \longrightarrow 2NiO(s) + 2SO_2(g)$
   b. $Sn^{2+}(aq) + 2Fe^{3+}(aq) \longrightarrow Sn^{4+}(aq) + 2Fe^{2+}(aq)$
   c. $CH_4(g) + 2O_2(g) \longrightarrow CO_2(g) + 2H_2O(g)$
   d. $2Cr_2O_3(s) + 3Si(s) \longrightarrow 4Cr(s) + 3SiO_2(s)$

15.20 In each of the following reactions, identify the substance that is oxidized, the substance that is reduced, the oxidizing agent, and the reducing agent:
   a. $2HgO(s) \longrightarrow 2Hg(l) + O_2(g)$
   b. $Zn(s) + 2HCl(aq) \longrightarrow ZnCl_2(aq) + H_2(g)$
   c. $2Na(s) + 2H_2O(l) \longrightarrow$
                     $2Na^+(aq) + 2OH^-(aq) + H_2(g)$
   d. $6Fe^{2+}(aq) + Cr_2O_7^{2-}(aq) + 14H^+(aq) \longrightarrow$
                   $6Fe^{3+}(aq) + 2Cr^{3+}(aq) + 7H_2O(l)$

15.21 Use oxidation numbers to balance the following equations:
   a. $Cu_2S(s) + H_2(g) \longrightarrow Cu(s) + H_2S(g)$
   b. $Fe(s) + Cl_2(g) \longrightarrow FeCl_3(s)$
   c. $Al(s) + H_2SO_4(aq) \longrightarrow Al_2(SO_4)_3(aq) + H_2(g)$

15.22 Use oxidation numbers to balance the following equations:
   a. $KClO_3(aq) + HBr(aq) \longrightarrow$
                   $Br_2(l) + KCl(aq) + H_2O(l)$
   b. $Cu(s) + HNO_3(aq) \longrightarrow$
                $Cu(NO_3)_2(aq) + NO_2(g) + H_2O(l)$
   c. $C_2H_6(g) + O_2(g) \longrightarrow CO_2(g) + H_2O(g)$

15.3 Balancing Oxidation–Reduction Equations Using Half-Reactions

LEARNING GOAL

Balance oxidation–reduction equations using the half-reaction method.

TUTORIAL

Balancing Redox Reactions

In the **half-reaction method** for balancing equations, an oxidation–reduction reaction is written as two *half-reactions*. As each half-reaction is balanced for atoms and charge, it becomes apparent which one is oxidation and which one is reduction. Once the loss and gain of electrons are equalized for the half-reactions, they are combined to obtain the overall balanced equation. The half-reaction method is typically used to balance equations that are written as ionic equations. Let us consider the reaction between aluminum metal and a solution of Cu^{2+} as shown in Sample Problem 15.6.

SAMPLE PROBLEM | **15.6**

■ **Using Half-Reactions to Balance Equations**

Use half-reactions to balance the following equation:

$$Al(s) + Cu^{2+}(aq) \longrightarrow Cu(s) + Al^{3+}(aq)$$

SOLUTION

STEP 1 **Write two half-reactions for the equation.** We can write one half-reaction for Al and Al^{3+}, and another for Cu^{2+} and Cu.

$$Al(s) \longrightarrow Al^{3+}(aq)$$
$$Cu^{2+}(aq) \longrightarrow Cu(s)$$

STEP 2 Balance the elements other than H and O in each half-reaction. In each half-reaction, the Al and Cu are already balanced.

$$Al(s) \longrightarrow Al^{3+}(aq)$$
$$Cu^{2+}(aq) \longrightarrow Cu(s)$$

STEP 3 Balance each half-reaction for charge by adding electrons to the side with more positive charge. For the aluminum half-reaction, we need to add three electrons on the product side to balance charge. With a loss of electrons, this is an oxidation.

$$Al(s) \longrightarrow \underline{Al^{3+}(aq) + 3\ e^-} \qquad \text{Oxidation}$$
$$\text{0 charge} \quad = \qquad \text{0 charge}$$

For the Cu^{2+} half-reaction, we need to add two electrons on the reactant side to balance charge. With a gain of electrons, this is a reduction.

$$\underline{Cu^{2+}(aq) + 2\ e^-} \longrightarrow Cu(s) \qquad \text{Reduction}$$
$$\text{0 charge} \qquad = \qquad \text{0 charge}$$

STEP 4 Multiply each half-reaction by factors that equalize the loss and gain of electrons. To obtain the same number of electrons in each half-reaction, we need to multiply the oxidation half-reaction by 2 and the reduction half-reaction by 3.

$$\mathbf{2} \times [Al(s) \longrightarrow Al^{3+}(aq) + 3e^-]$$
$$2Al(s) \longrightarrow 2Al^{3+}(aq) + 6\ e^- \qquad 6\ e^-\ \text{lost}$$
$$\mathbf{3} \times [Cu^{2+}(aq) + 2\ e^- \longrightarrow Cu(s)]$$
$$3Cu^{2+}(aq) + 6\ e^- \longrightarrow 3Cu(s) \qquad 6\ e^-\ \text{gained}$$

STEP 5 Add balanced half-reactions and cancel electrons. Check balance of atoms and charge.

$$2Al(s) \longrightarrow 2Al^{3+}(aq) + 6\ e^-$$
$$3Cu^{2+}(aq) + 6\ e^- \longrightarrow 3Cu(s)$$
$$\overline{2Al(s) + 3Cu^{2+}(aq) + \cancel{6e^-} \longrightarrow 2Al^{3+}(aq) + \cancel{6e^-} + 3Cu(s)}$$

Final balanced equation:

$$2Al(s) + 3Cu^{2+}(aq) \longrightarrow 2Al^{3+}(aq) + 3Cu(s)$$

Check balance of atoms and charge.

Atoms:	2Al = 2Al
	3Cu = 3Cu
Charge:	**6+ = 6+**

STUDY CHECK

Use the half-reaction method to balance the equation

$$Zn(s) + Fe^{3+}(aq) \longrightarrow Zn^{2+}(aq) + Fe^{2+}(aq)$$

Guide to Balancing Redox Equations Using Half-Reactions

> **STEP 1**
> Write two half-reactions for the equation.

> **STEP 2**
> Balance elements other than H and O in each half-reaction. Add H₂O to the side that needs O and add H⁺ to the side that needs H.

> **STEP 3**
> Balance each half-reaction for charge by adding electrons to the side with more positive charge.

> **STEP 4**
> Multiply each half-reaction by factors that equalize the loss and gain of electrons.

> **STEP 5**
> Add balanced half-reactions, cancel electrons, and combine H₂O and H⁺. Check balance of atoms and charge.

Balancing Oxidation–Reduction Equations in Acidic Solution

When we use the half-reaction method for balancing equations for reactions in acidic solution, we balance O by adding H_2O, and balance H by adding H^+ as shown in Sample Problem 15.7.

A dichromate solution (yellow) and an iodide solution (colorless) form a brown solution of Cr^{3+} and iodine.

SAMPLE PROBLEM 15.7

■ **Using Half-Reactions to Balance Equations in Acidic Solutions**

Use half-reactions to balance the following equation for a reaction that takes place in acidic solution:

$$I^-(aq) + Cr_2O_7{}^{2-}(aq) \longrightarrow I_2(s) + Cr^{3+}(aq)$$

SOLUTION

STEP 1 Write two half-reactions for the equation. One of the half-reactions is written for I and the other for Cr.

$$I^-(aq) \longrightarrow I_2(s)$$
$$Cr_2O_7{}^{2-}(aq) \longrightarrow Cr^{3+}(aq)$$

STEP 2 Balance elements other than H and O in each half-reaction. Add H_2O to the side that needs O and add H^+ to the side that needs H. The two I atoms in I_2 are balanced with a coefficient of 2 for I^-.

$$2I^-(aq) \longrightarrow I_2(s)$$

The two Cr atoms are balanced with a coefficient of 2 for Cr^{3+}.

$$Cr_2O_7{}^{2-}(aq) \longrightarrow 2Cr^{3+}(aq)$$

Now add $7H_2O$ to the product side to balance O, and $14H^+$ to the reactant side to balance H:

$$\mathbf{14H^+}(aq) + Cr_2O_7{}^{2-}(aq) \longrightarrow 2Cr^{3+}(aq) + \mathbf{7H_2O}(l)$$

STEP 3 Balance each half-reaction for charge by adding electrons to the side with more positive charge. A charge of −2 is balanced with two electrons on the product side.

$$2I^-(aq) \longrightarrow I_2(s) + \mathbf{2\,e^-} \qquad \text{Oxidation}$$
$$\quad -2 \quad = \quad -2$$

A charge of +6 is obtained on the reactant side by adding six electrons.

$$\underbrace{\mathbf{6\,e^-} + 14H^+(aq) + Cr_2O_7{}^{2-}(aq)}_{+6} \longrightarrow \underset{=\quad +6}{2Cr^{3+}(aq) + 7H_2O(l)} \quad \text{Reduction}$$

STEP 4 Multiply each half-reaction by factors that equalize the loss and gain of electrons. The half-reaction with I is multiplied by 3 to equal the gain of 6 e^- by Cr.

$$\mathbf{3} \times [2I^-(aq) \longrightarrow I_2(s) + \mathbf{2\,e^-}]$$
$$6I^-(aq) \longrightarrow 3I_2(s) + \mathbf{6\,e^-} \qquad \text{6 } e^- \text{ lost}$$
$$\mathbf{6\,e^-} + 14H^+(aq) + Cr_2O_7{}^{2-}(aq) \longrightarrow 2Cr^{3+}(aq) + 7H_2O(l) \qquad \text{6 } e^- \text{ gained}$$

STEP 5 Add balanced half-reactions, cancel electrons, and combine H_2O and H^+. Check balance of atoms and charge.

$$6I^-(aq) \longrightarrow 3I_2(s) + \mathbf{6\,e^-}$$
$$\mathbf{6\,e^-} + 14H^+(aq) + Cr_2O_7{}^{2-}(aq) \longrightarrow 2Cr^{3+}(aq) + 7H_2O(l)$$

$$\cancel{6e^-} + 14H^+(aq) + Cr_2O_7{}^{2-}(aq) + 6I^-(aq) \longrightarrow 2Cr^{3+}(aq) + 3I_2(s) + 7H_2O(l) + \cancel{6e^-}$$

Final balanced equation:

$$14H^+(aq) + Cr_2O_7{}^{2-}(aq) + 6I^-(aq) \longrightarrow 2Cr^{3+}(aq) + 3I_2(s) + 7H_2O(l)$$

Check balance of atoms and charge.

Atoms:
$$6I = 6I$$
$$2Cr = 2Cr$$
$$14H = 14H$$

Charge:
$$6+ = 6+$$

STUDY CHECK

Use half-reactions to balance the following equation in acidic solution:

$$Cu^{2+}(aq) + SO_2(g) \longrightarrow Cu(s) + SO_4^{2-}(aq)$$

Balancing Oxidation–Reduction Equations in Basic Solution

An oxidation–reduction reaction can also take place in a basic solution. In that case, we use the same half-reaction method, but once we have the balanced equation, we will neutralize the H^+ with OH^- to form water. The H^+ is neutralized by adding OH^- to both sides of the equation to form H_2O as shown in Sample Problem 15.8.

SAMPLE PROBLEM | 15.8

■ **Using Half-Reactions to Balance Equations in Basic Solutions**

Use half-reactions to balance the following equation that takes place in basic solution:

$$Fe^{2+}(aq) + MnO_4^-(aq) \longrightarrow MnO_2(s) + Fe^{3+}(aq)$$

SOLUTION

STEP 1 Write two half-reactions for the equation.   We separate the equation into two half-reactions by writing one half-reaction for Fe and one for Mn.

$$Fe^{2+}(aq) \longrightarrow Fe^{3+}(aq)$$
$$MnO_4^-(aq) \longrightarrow MnO_2(s)$$

STEP 2 Balance the elements other than H and O in each half-reaction.   Add H_2O to the side that needs O and H^+ to the side that needs H.

$$Fe^{2+}(aq) \longrightarrow Fe^{3+}(aq)$$
$$MnO_4^-(aq) \longrightarrow MnO_2(s) + \mathbf{2H_2O}(l) \qquad \text{H}_2\text{O balances O}$$
$$\mathbf{4H^+}(aq) + MnO_4^-(aq) \longrightarrow MnO_2(s) + \mathbf{2H_2O}(l) \qquad \text{H}^+ \text{ balances H}$$

STEP 3 Balance each half-reaction for charge by adding electrons to the side with more positive charge.

$$Fe^{2+}(aq) \longrightarrow \underbrace{Fe^{3+}(aq) + \mathbf{1\,e^-}} \qquad \text{Oxidation}$$
$$\quad +2 \qquad\qquad = \qquad\qquad +2$$
$$\underbrace{\mathbf{4H^+}(aq) + MnO_4^-(aq) + \mathbf{3\,e^-}} \longrightarrow MnO_2(s) + \mathbf{2H_2O}(l) \qquad \text{Reduction}$$
$$\qquad\qquad 0 \text{ charge} \qquad = \qquad 0 \text{ charge}$$

STEP 4 Multiply each half-reaction by factors that equalize the loss and gain of electrons.   The half-reaction with Fe is multiplied by 3 to equal the gain of $3\,e^-$ by Mn.

$$\mathbf{3} \times [Fe^{2+}(aq) \longrightarrow Fe^{3+}(aq) + \mathbf{1\,e^-}]$$
$$3Fe^{2+}(aq) \longrightarrow 3Fe^{3+}(aq) + \mathbf{3\,e^-} \qquad 3\,e^- \text{ lost}$$
$$4H^+(aq) + MnO_4^-(aq) + \mathbf{3\,e^-} \longrightarrow MnO_2(s) + 2H_2O(l) \qquad 3\,e^- \text{ gained}$$

STEP 5 Add half-reactions, cancel electrons, and combine any H_2O and H^+. Check balance of atoms and charge.

$$3Fe^{2+}(aq) \longrightarrow 3Fe^{3+}(aq) + 3e^-$$
$$4H^+(aq) + MnO_4^-(aq) + 3e^- \longrightarrow MnO_2(s) + 2H_2O(l)$$

$$\overline{3Fe^{2+}(aq) + 4H^+(aq) + MnO_4^-(aq) + \cancel{3e^-} \longrightarrow 3Fe^{3+}(aq) + \cancel{3e^-} + MnO_2(s) + 2H_2O(l)}$$

Balanced equation (in acidic solution):

$$3Fe^{2+}(aq) + 4H^+(aq) + MnO_4^-(aq) \longrightarrow 3Fe^{3+}(aq) + MnO_2(s) + 2H_2O(l)$$

To write this oxidation–reduction reaction in a basic solution, we will neutralize the H^+ with OH^- to form H_2O. For this equation, we add $4OH^-(aq)$ to both the reactant and product sides:

$$3Fe^{2+}(aq) + 4H^+(aq) + 4OH^-(aq) + MnO_4^-(aq) \longrightarrow 3Fe^{3+}(aq) + MnO_2(s) + 2H_2O(l) + 4OH^-(aq)$$

Combining $4H^+$ and $4OH^-$ gives $4H_2O$:

$$4H^+(aq) + 4OH^-(aq) \longrightarrow 4H_2O(l)$$

$$3Fe^{2+}(aq) + 4H_2O(l) + MnO_4^-(aq) \longrightarrow 3Fe^{3+}(aq) + MnO_2(s) + 2H_2O(l) + 4OH^-(aq)$$

Canceling $2H_2O$ on both the reactant and the product side gives the balanced equation in a basic solution:

Final balanced equation (in basic solution):

$$3Fe^{2+}(aq) + 2H_2O(l) + MnO_4^-(aq) \longrightarrow 3Fe^{3+}(aq) + MnO_2(s) + 4OH^-(aq)$$

Check balance of atoms and charge.

Atoms:	3Fe	=	3Fe
	1Mn	=	1Mn
	4H	=	4H
	6O	=	6O
Charge:	5+	=	5+

STUDY CHECK

Use half-reactions to balance the following equation in basic solution:

$$N_2O(g) + ClO^-(aq) \longrightarrow NO_2^-(aq) + Cl^-(aq)$$

QUESTIONS AND PROBLEMS

Balancing Oxidation–Reduction Equations Using Half-Reactions

15.23 Balance each of the following half-reactions in acidic solution:
a. $Sn^{2+}(aq) \longrightarrow Sn^{4+}(aq)$
b. $Mn^{2+}(aq) \longrightarrow MnO_4^-(aq)$
c. $NO_2^-(aq) \longrightarrow NO_3^-(aq)$
d. $ClO_3^-(aq) \longrightarrow ClO_2(aq)$

15.24 Balance each of the following half-reactions in acidic solution:
a. $Cu(s) \longrightarrow Cu^{2+}(aq)$
b. $SO_4^{2-}(aq) \longrightarrow SO_3^{2-}(aq)$
c. $BrO_3^-(aq) \longrightarrow Br^-(aq)$
d. $IO_3^-(aq) \longrightarrow I_2(s)$

15.25 Use the half-reaction method to balance each of the following in acidic solution except as indicated:
a. $Ag(s) + NO_3^-(aq) \longrightarrow Ag^+(aq) + NO_2(g)$
b. $Fe(s) + CrO_4^{2-}(aq) \longrightarrow Fe_2O_3(s) + Cr_2O_3(s)$
   (basic solution)
c. $NO_3^-(aq) + S(s) \longrightarrow NO(g) + SO_2(g)$
d. $S_2O_3^{2-}(aq) + Cu^{2+}(aq) \longrightarrow S_4O_6^{2-}(aq) + Cu(s)$
e. $PbO_2(s) + Mn^{2+}(aq) \longrightarrow Pb^{2+}(aq) + MnO_4^-(aq)$

15.26 Use the half-reaction method to balance each of the following in acidic solution except as indicated:
a. $Sn^{2+}(aq) + IO_4^-(aq) \longrightarrow Sn^{4+}(aq) + I^-(aq)$
b. $Al(s) + ClO^-(aq) \longrightarrow AlO_2^-(aq) + Cl^-(aq)$
   (basic solution)
c. $Mn(s) + NO_3^-(aq) \longrightarrow Mn^{2+}(aq) + NO_2(g)$
d. $C_2O_4^{2-}(aq) + MnO_4^-(aq) \longrightarrow CO_2(g) + Mn^{2+}(aq)$
e. $ClO_3^-(aq) + SO_3^{2-}(aq) \longrightarrow Cl^-(aq) + SO_4^{2-}(aq)$

15.4 Oxidation of Alcohols

We have seen that oxidation involves the loss of electrons and reduction involves the gain of electrons. For organic and biochemical compounds, oxidation typically involves the addition of oxygen or the loss of hydrogen, and reduction involves the loss of oxygen or the gain of hydrogen. In organic chemistry, oxidation occurs when there is an increase in the number of carbon–oxygen bonds. In a reduction reaction, the product has fewer bonds between carbon and oxygen.

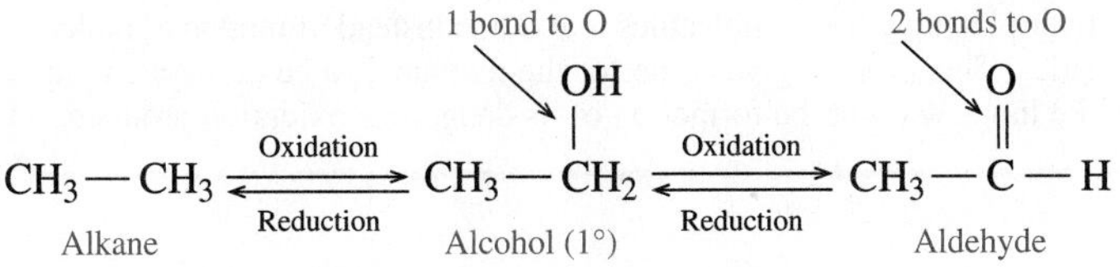

LEARNING GOAL

Classify alcohols as primary, secondary, or tertiary; write equations for the oxidation of alcohols.

$$CH_3{-}CH_3 \underset{\text{Reduction}}{\overset{\text{Oxidation}}{\rightleftharpoons}} CH_3{-}CH_2{-}\overset{\text{1 bond to O}}{\overset{\displaystyle |}{OH}} \underset{\text{Reduction}}{\overset{\text{Oxidation}}{\rightleftharpoons}} CH_3{-}\overset{\text{2 bonds to O}}{\overset{\displaystyle O}{\underset{\displaystyle \|}{C}}}{-}H$$

Alkane · Alcohol (1°) · Aldehyde

Classification of Alcohols

Alcohols are classified by the number of carbon groups attached to the carbon atom bonded to the hydroxyl (—OH) group. A **primary (1°) alcohol** has one carbon group attached to the carbon atom bonded to the —OH, a **secondary (2°) alcohol** has two carbon groups, and a **tertiary (3°) alcohol** has three carbon groups. The simplest alcohol, methanol, which has a carbon atom attached to three H atoms is considered a primary alcohol.

Primary (1°) alcohol   **Secondary (2°) alcohol**   **Tertiary (3°) alcohol**

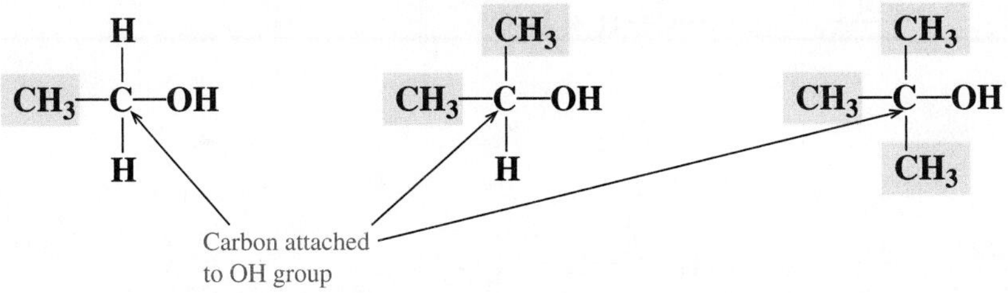

$$CH_3{-}\overset{\displaystyle H}{\underset{\displaystyle H}{C}}{-}OH \qquad CH_3{-}\overset{\displaystyle CH_3}{\underset{\displaystyle H}{C}}{-}OH \qquad CH_3{-}\overset{\displaystyle CH_3}{\underset{\displaystyle CH_3}{C}}{-}OH$$

Carbon attached to OH group

CONCEPT CHECK 15.3

■ Classifying Alcohols

Classify each of the following alcohols as primary, secondary, or tertiary:

a. $CH_3{-}CH_2{-}CH_2{-}OH$

b.
$$CH_3{-}CH_2{-}\overset{\displaystyle OH}{\underset{\displaystyle CH_3}{C}}{-}CH_3$$

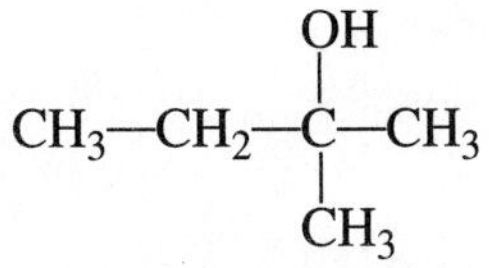

c.
$$CH_3{-}\overset{\displaystyle OH}{CH}{-}CH_3$$

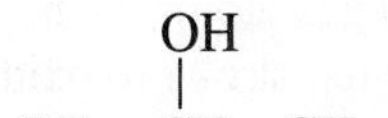

ANSWER

a. One carbon group attached to the carbon atom bonded to the —OH makes this a primary alcohol.

b. Three carbon groups attached to the carbon atom bonded to the —OH makes this a tertiary alcohol.

c. Two carbon groups attached to the carbon atom bonded to the —OH makes this a secondary alcohol.

CHEMISTRY AND HEALTH

Methanol Poisoning

Methanol (methyl alcohol), CH_3—OH, is a highly toxic alcohol present in products such as windshield washer fluid, Sterno, and paint strippers. Methanol is rapidly absorbed in the gastrointestinal tract. In the liver, it is oxidized to formaldehyde and then formic acid, a substance that causes nausea, severe abdominal pain, and blurred vision. Blindness can occur because the intermediate products destroy the retina of the eye. As little as 4 mL of methanol can produce blindness. The formic acid, which is not readily eliminated from the body, lowers blood pH so severely that just 30 mL of methanol can lead to coma and death.

The treatment for methanol poisoning involves giving sodium bicarbonate to neutralize the formic acid in the blood. In some cases, ethanol is given intravenously to the patient. The enzymes in the liver pick up ethanol molecules to oxidize instead of methanol molecules. This process gives time for the methanol to be eliminated via the lungs without the formation of its dangerous oxidation products.

Oxidation of Primary and Secondary Alcohols

In Chapter 8, Section 8.4, we discussed the functional groups of alcohols, aldehydes, and ketones. When a primary alcohol is oxidized, it produces an aldehyde. The oxidation occurs by removing two hydrogen atoms, one from the —OH group and another from the carbon that is bonded to the —OH. To indicate the presence of an oxidizing agent such as $KMnO_4$ and $K_2Cr_2O_7$, reactions are written with the symbol [O] over the reaction arrow:

$$H-\underset{\underset{H}{|}}{\overset{\overset{OH}{|}}{C}}-H \quad \xrightarrow{[O]} \quad H-\overset{\overset{O}{||}}{C}-H + H_2O$$

Methanol   Methanal

$$CH_3-\underset{}{\overset{\overset{OH}{|}}{CH_2}} \quad \xrightarrow{[O]} \quad CH_3-\overset{\overset{O}{||}}{C}-H + H_2O$$

Ethanol   Ethanal

The other product of the oxidation is H_2O, formed when two H atoms from the alcohol combine with an O atom from the oxidizing agent. In the oxidation of secondary alcohols, the products are ketones. One H is removed from the —OH, and another H is from the carbon bonded to the —OH group.

$$CH_3-\underset{\underset{H}{|}}{\overset{\overset{OH}{|}}{C}}-CH_3 \quad \xrightarrow{[O]} \quad CH_3-\overset{\overset{O}{||}}{C}-CH_3 + H_2O$$

2-Propanol   Propanone, acetone

Tertiary alcohols do not oxidize readily because there are no hydrogen atoms on the carbon bonded to the —OH group. Because C—C bonds are usually too strong to oxidize, tertiary alcohols resist oxidation.

No double bond forms   No hydrogen on this carbon

$$CH_3-\underset{\underset{CH_3}{|}}{\overset{\overset{OH}{|}}{C}}-CH_3 \quad \xrightarrow{[O]} \quad \text{No oxidation product readily formed}$$

3° Alcohol

SAMPLE PROBLEM 15.9

■ **Oxidation of Alcohols**

Draw the condensed structural formula of the aldehyde or ketone formed by the oxidation of each of the following:

$$OH$$
a. 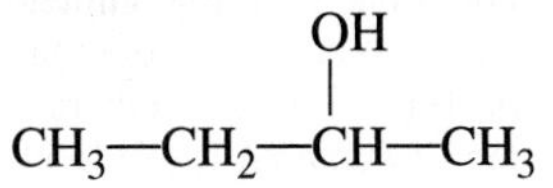 $CH_3-CH_2-CH-CH_3$

b. $CH_3-CH_2-CH_2-OH$

SOLUTION

a. The oxidation of a secondary alcohol produces a ketone.

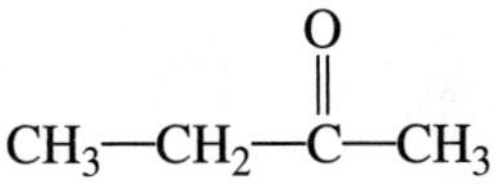

$$O$$
$$\parallel$$
$$CH_3-CH_2-C-CH_3$$

b. The oxidation of a primary alcohol produces an aldehyde.

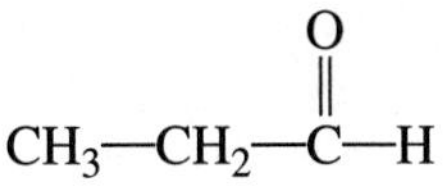

$$O$$
$$\parallel$$
$$CH_3-CH_2-C-H$$

STUDY CHECK

Draw the condensed structural formula of the aldehyde or ketone formed by the oxidation of the following alcohol:

$$CH_3 \quad CH_3$$
$$CH_3-CH-CH-CH_2-CH_2-OH$$

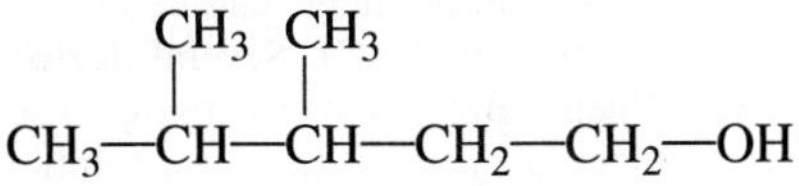

QUESTIONS AND PROBLEMS

Oxidation of Alcohols

15.27 Classify each of the following as a primary, secondary, or tertiary alcohol:

$$CH_3$$
a. 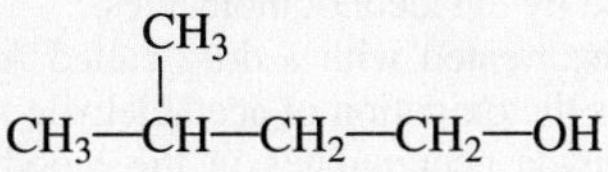 $CH_3-CH-CH_2-CH_2-OH$

b. $CH_3-CH_2-CH_2-CH_2-OH$

$$OH$$
c. 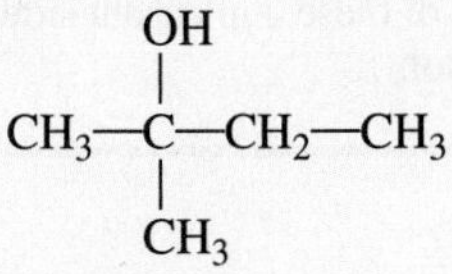 $CH_3-C-CH_2-CH_3$
$$CH_3$$

15.28 Classify each of the following as a primary, secondary, or tertiary alcohol:

$$CH_3$$
a. 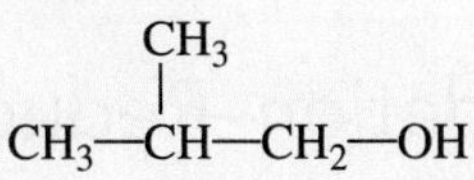 $CH_3-CH-CH_2-OH$

$$OH$$
b. 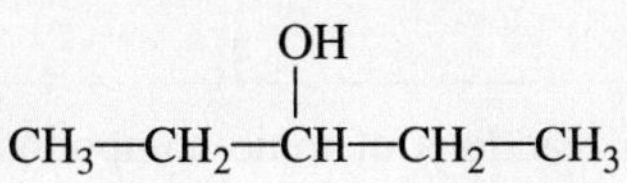 $CH_3-CH_2-CH-CH_2-CH_3$

$$CH_3$$
c. 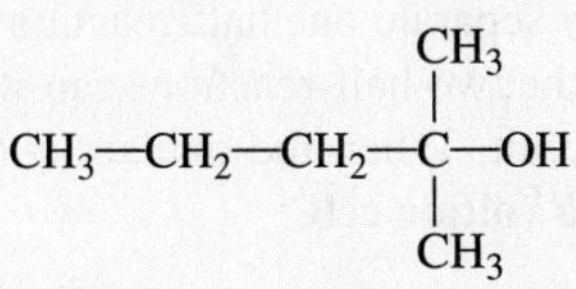 $CH_3-CH_2-CH_2-C-OH$
$$CH_3$$

15.29 Draw the condensed structural formula of the aldehyde or ketone produced when each of the following alcohols is oxidized [O] (if no reaction, write *none*):

a. $CH_3-CH_2-CH_2-CH_2-CH_2-OH$

$$OH$$
$$CH_3-C-CH_2-CH_3$$
b. CH_3

$$OH \qquad CH_3$$
c. $CH_3-CH-CH_2-CH-CH_3$

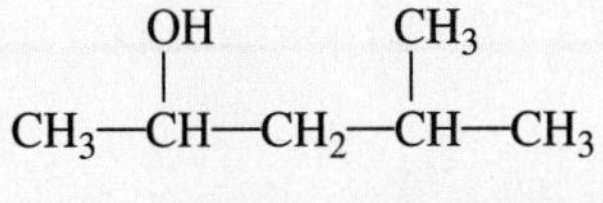

15.30 Draw the condensed structural formula of the aldehyde or ketone produced when each of the following alcohols is oxidized [O] (if no reaction, write *none*):

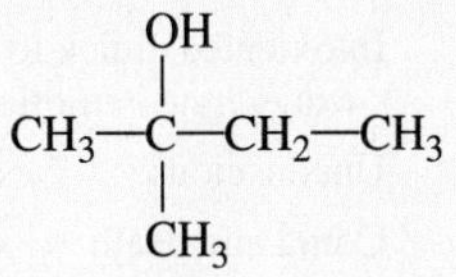

$$CH_3$$
a. 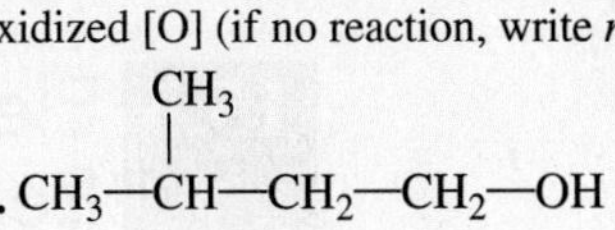 $CH_3-CH-CH_2-CH_2-OH$

$$OH$$
b. 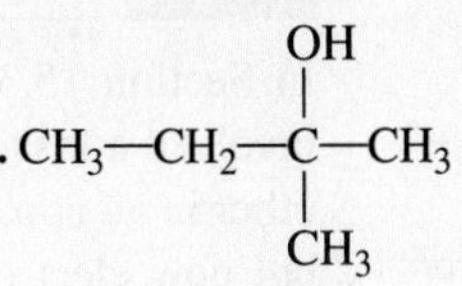 $CH_3-CH_2-C-CH_3$
$$CH_3$$

$$OH$$
c. $CH_3-CH_2-CH-CH_2-CH_3$

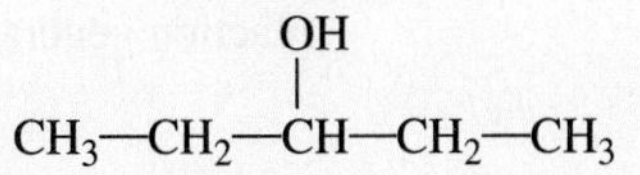

CHEMISTRY AND HEALTH

Oxidation of Alcohol in the Body

Ethanol is the most commonly abused drug in the United States. When ingested in small amounts, ethanol may produce a feeling of euphoria in the body despite the fact that it is a depressant. In the liver, enzymes such as alcohol dehydrogenase oxidize ethanol to acetaldehyde, a substance that impairs mental and physical coordination. If the blood alcohol concentration exceeds 0.4%, coma or death may occur. Table 15.3 gives some of the typical behaviors exhibited at various levels of blood alcohol.

$$CH_3-CH_2-OH \xrightarrow{[O]} CH_3-\overset{\overset{\displaystyle O}{\|}}{C}-H \xrightarrow{[O]} 2CO_2 + H_2O$$

Ethanol (ethyl alcohol) Ethanal (acetaldehyde)

The acetaldehyde produced from ethanol in the liver is further oxidized to acetic acid, which is eventually converted to carbon dioxide and water in the citric acid cycle. However, the intermediate products can cause considerable damage while they are present within the cells of the liver.

TABLE 15.3 Typical Behaviors Exhibited by a 150-lb Person Consuming Alcohol

Number of Beers (12 oz) or Glasses of Wine (5 oz)	Percent Blood Alcohol Concentration (BAC)	Typical Behavior
1	0.025	Slightly dizzy, talkative
2	0.05	Euphoria, loud talking, and laughing
4	0.10	Loss of inhibition, loss of coordination, drowsiness, legally drunk in most states
8	0.20	Intoxicated, quick to anger, exaggerated emotions
12	0.30	Unconscious
16–20	0.40–0.50	Coma and death

A person weighing 150 lb requires about one hour to completely metabolize 12 oz of beer. However, the rate of metabolism of ethanol varies between nondrinkers and drinkers. Typically, nondrinkers and social drinkers can metabolize 12–15 mg of ethanol/dL of blood in one hour, but an alcoholic can metabolize as much as 30 mg of ethanol/dL in one hour. Some effects of alcohol metabolism include an increase in liver lipids ("fatty liver"), gastritis, pancreatitis, ketoacidosis, alcoholic hepatitis, as well as psychological disturbances.

A breathalyzer test is used to determine blood level of ethanol.

When alcohol is present in the blood, it evaporates through the lungs. Thus, the percentage of alcohol in the lungs can be used to calculate the blood alcohol concentration (BAC). Several devices are used to measure the BAC. When a Breathalyzer is used, a suspected drunk driver exhales through a mouthpiece into a solution containing the orange Cr^{6+} ion. Any alcohol present in the exhaled air is oxidized, which reduces the orange Cr^{6+} to green Cr^{3+}:

$$CH_3-CH_2-OH + Cr^{6+} \xrightarrow{[O]} CH_3-\overset{\overset{\displaystyle O}{\|}}{C}-OH + Cr^{3+}$$

Ethanol Orange Acetic acid Green

The Alcosensor uses an oxidation of alcohol in a fuel cell to generate an electric current that is measured. The Intoxilyzer measures the amount of light absorbed by the alcohol molecules.

Sometimes alcoholics are treated with a drug called Antabuse (disulfiram), which prevents the oxidation of acetaldehyde to acetic acid. As a result, acetaldehyde accumulates in the blood, which causes nausea, profuse sweating, headache, dizziness, vomiting, and respiratory difficulties. Because of these unpleasant side effects, the patient is less likely to use alcohol.

CASE STUDY
Toxicity of Alcohol

15.5 Electrical Energy from Oxidation–Reduction Reactions

LEARNING GOAL

Write the half-reactions that occur at the anode and cathode of a voltaic cell; write the shorthand cell notation.

In Section 15.3, we balanced oxidation–reduction reactions of ionic compounds, which involved a loss and gain of electrons. If we physically separate one half-reaction from the other in an apparatus called an **electrochemical cell**, the two half-reactions can still occur, but now electrons must flow through an external circuit. When the oxidation–reduction reaction generates electrical energy, the cell is called a **voltaic cell**.

Voltaic Cells

When a piece of zinc metal is placed in a Cu^{2+} solution, the silvery zinc becomes coated with a rusty-brown coating of Cu while the blue color (Cu^{2+}) of the solution fades. The oxidation of the zinc metal provides electrons for the reduction of the Cu^{2+} ions. We can write the two half-reactions as

$Zn(s) \longrightarrow Zn^{2+}(aq) + 2\,e^-$ Oxidation

$Cu^{2+}(aq) + 2\,e^- \longrightarrow Cu(s)$ Reduction

The overall cell reaction is

$Zn(s) + Cu^{2+}(aq) \longrightarrow Cu(s) + Zn^{2+}(aq)$

As long as the Zn metal and Cu^{2+} ions are in the same container, the electrons are transferred directly from Zn to Cu^{2+}. However, the components of the two half-reactions can be placed in separate containers, called *half-cells*, connected by an external circuit. When the electrons flow from one half-cell to the other, an electrical current is produced. In each half-cell, there is a strip of metal, called an *electrode*, in contact with the ionic solution. The electrode where oxidation takes place is called the **anode**; the **cathode** is where reduction takes place. In this example, the anode is a zinc metal strip placed in a Zn^{2+} ($ZnSO_4$) solution. The cathode is a copper metal strip placed in a Cu^{2+} ($CuSO_4$) solution. In this voltaic cell, the Zn anode and Cu cathode are connected by a wire that allows electrons to move from the oxidation half-cell to the reduction half-cell (see Figure 15.2).

The circuit is completed by a *salt bridge* containing positive and negative ions that is placed in the half-cell solutions. The purpose of the salt bridge is to provide ions, such as Na^+ and SO_4^{2-} ions, to maintain an electrical balance in each half-cell solution. As oxidation occurs at the Zn anode, there is an increase in Zn^{2+} ions, which is balanced by SO_4^{2-} anions from the salt bridge. At the cathode, there is a loss of positive charge as Cu^{2+} is reduced to Cu, which is balanced by SO_4^{2-} in the solution moving into the salt bridge and Na^+ moving out into solution. The complete circuit involves the flow of electrons from the anode to the cathode and the flow of anions from the cathode solution to the anode solution.

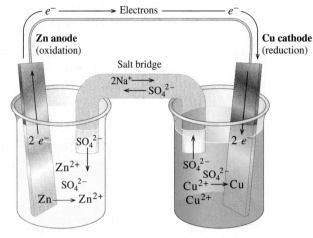

$Zn(s) \rightarrow Zn^{2+}(aq) + 2\,e^-$ $Cu^{2+}(aq) + 2\,e^- \rightarrow Cu(s)$

$Zn(s) + Cu^{2+}(aq) \longrightarrow Cu(s) + Zn^{2+}(aq)$

FIGURE 15.2 In this voltaic cell, the Zn anode is in a Zn^{2+} solution, and the Cu cathode is in a Cu^{2+} solution. Electrons produced by the oxidation of Zn move out of the Zn anode through the wire and into the Cu cathode where they reduce Cu^{2+} to Cu. As electrons flow through the wire, the circuit is completed by the flow of SO_4^{2-} through the salt bridge.

Q Which electrode will be heavier when the reaction ends?

Anode is where

$\left.\begin{array}{l}\text{oxidation takes place} \\ \text{electrons are produced}\end{array}\right\}$ $Zn(s) \longrightarrow Zn^{2+}(aq) + 2\,e^-$

Cathode is where

$\left.\begin{array}{l}\text{reduction takes place} \\ \text{electrons are used up}\end{array}\right\}$ $Cu^{2+}(aq) + 2\,e^- \longrightarrow Cu(s)$

An electrical current is produced as electrons flow from the anode through the wire to the cathode. Eventually, the loss of Zn decreases the mass of the Zn anode, while the formation of Cu increases the mass of the Cu cathode. We can diagram the cell using a shorthand notation as follows:

$Zn(s)\,|\,Zn^{2+}(aq)\,\|\,Cu^{2+}(aq)\,|\,Cu(s)$

The components of the oxidation half-cell (anode) are written on the left side in this shorthand notation, and the components of the reduction half-cell (cathode) are written on the right. A single vertical line separates the solid Zn anode from the ionic Zn^{2+} solution and the Cu^{2+} solution from the Cu cathode. A double vertical line separates the two half-cells.

Reading the notation from left to right indicates that Zn is oxidized to Zn^{2+}, and Cu^{2+} is reduced to Cu as electrons move through the wire from left to right.

Electrons →

Oxidation half-cell   Reduction half-cell

Anode   Cathode

$$Zn(s) \mid Zn^{2+}(aq) \parallel Cu^{2+}(aq) \mid Cu(s)$$

(Salt bridge)

In some voltaic cells, there is no component in the half-reactions that can be used as an electrode. When this is the case, electrodes made of graphite or platinum are used for the transfer of electrons. If there are two ionic components in a cell, their symbols are separated by a comma. For example, suppose a voltaic cell consists of a platinum anode placed in a Sn^{2+} solution such as $Sn(NO_3)_2$ and a silver cathode placed in a Ag^+ solution such as $AgNO_3$. The notation for the cell would be written as

$$Pt(s) \mid Sn^{2+}(aq), Sn^{4+}(aq) \parallel Ag^+(aq) \mid Ag(s)$$

The oxidation reaction at the anode is

$$Sn^{2+}(aq) \longrightarrow Sn^{4+}(aq) + 2\,e^-$$

The reduction reaction at the cathode is

$$Ag^+(aq) + e^- \longrightarrow Ag(s)$$

To balance the overall cell reaction, we multiply the cathode reduction by 2 and combine the two half-reactions.

$$\begin{aligned}
2Ag^+(aq) + \mathbf{2\,e^-} &\longrightarrow 2Ag(s) \\
Sn^{2+}(aq) &\longrightarrow Sn^{4+}(aq) + \mathbf{2\,e^-} \\
\hline
Sn^{2+}(aq) + 2Ag^+(aq) &\longrightarrow Sn^{4+}(aq) + 2Ag(s)
\end{aligned}$$

To operate the cell, a wire connects the Pt anode and the Ag cathode and a salt bridge is placed in the Sn^{2+} and Ag^+ solutions.

SAMPLE PROBLEM   15.10

■ **Diagramming a Voltaic Cell**

A voltaic cell consists of an iron (Fe) anode in a Fe^{2+} solution ($Fe(NO_3)_2$) and a tin (Sn) cathode placed in a Sn^{2+} solution ($Sn(NO_3)_2$). Write the cell notation, the oxidation and reduction half-reactions, and the overall cell reaction.

SOLUTION

The notation for the cell would be written as

$$Fe(s) \mid Fe^{2+}(aq) \parallel Sn^{2+}(aq) \mid Sn(s)$$

The oxidation reaction at the anode is

$$Fe(s) \longrightarrow Fe^{2+}(aq) + 2\,e^-$$

The reduction reaction at the cathode is

$$Sn^{2+}(aq) + 2\,e^- \longrightarrow Sn(s)$$

To write the overall cell reaction, we combine the two half-reactions.

$$\begin{aligned}
Fe(s) &\longrightarrow Fe^{2+}(aq) + 2\,e^- \\
Sn^{2+}(aq) + 2\,e^- &\longrightarrow Sn(s) \\
\hline
Fe(s) + Sn^{2+}(aq) &\longrightarrow Fe^{2+}(aq) + Sn(s)
\end{aligned}$$

To operate the cell, a wire connects the Fe anode and the Sn cathode and a salt bridge is placed in the Fe^{2+} and Sn^{2+} solutions.

STUDY CHECK

Write the half-reactions and the overall cell reaction for the following notation of a voltaic cell:

$$Co(s)\,|\,Co^{2+}(aq)\,\|\,Cu^{2+}(aq)\,|\,Cu(s)$$

Prevention of Corrosion

Corrosion of metals has been a major problem for centuries. Because so many building materials involve iron or iron and carbon (steel), the corrosion of iron is detrimental to the strength of girders, cars, and ships. Buildings can collapse, the hulls of ships get holes, and pipes laid underground crumble. Many billions of dollars are spent each year to prevent corrosion and repair building materials made of iron. One way to prevent corrosion is to paint the bridges, cars, and ships with paints containing materials that seal the iron surface from H_2O and O_2. But it is necessary to repaint often and a scratch in the paint exposes the iron, which then begins to rust.

A more effective way to prevent corrosion is to place the iron in contact with a metal that substitutes for the anode region of iron. Metals such as Zn, Mg, or Al lose electrons more easily than iron. When one of these metals is in contact with iron, that metal acts as the anode instead of iron. For example, in a process called *galvanization*, an object made of iron is coated with zinc. The zinc becomes the anode because zinc loses electrons more easily than Fe. As long as Fe does not act as an anode, rust cannot form.

In a method called *cathodic protection*, structures such as iron pipes and underground storage containers are placed in contact with a piece of metal such as Mg, Al, or Zn. Again, because these metals lose electrons more easily than Fe, they become the anode, thereby preventing the rusting of the iron. As long as iron acts only as a cathode for the reduction of $O_2(g)$, the iron will not corrode. For example, a magnesium plate welded or bolted to a ship's hull loses electrons more easily than iron or steel and protects the hull from rusting. Occasionally, a new magnesium plate is added to replace the magnesium as it is used up. Magnesium stakes placed in the ground are connected to underground pipelines and storage containers to prevent corrosion damage.

$$Mg(s) \longrightarrow Mg^{2+}(aq) + 2\,e^-$$

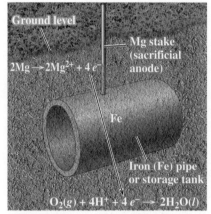

Rusting of an iron pipe is prevented by attaching a piece of magnesium, which oxidizes more easily than iron.

Batteries

Batteries are needed to power your cell phone, watch, and calculator. Batteries also are needed to make cars start and flashlights produce light. Within each of these batteries are voltaic cells that produce electrical energy. Let's look at some examples of commonly used batteries.

Lead Storage Battery

A lead storage battery is used to operate the electrical system in a car. We need a car battery to start the engine, turn on the lights, or operate the radio. If the battery runs down, the car won't start and the lights won't turn on. A car battery or a lead storage battery is a type of voltaic cell. In a typical 12-V battery, there are six voltaic cells linked together. Each of the cells consists of a lead (Pb) plate that acts as the anode and a lead(IV) oxide (PbO_2) plate that acts as the cathode. Both half-cells contain a sulfuric acid (H_2SO_4) solution. When the car battery is producing electrical energy (discharging), the following half-reactions take place:

Anode (oxidation):

$$Pb(s) + SO_4{}^{2-}(aq) \longrightarrow PbSO_4(s) + 2\,e^-$$

Cathode (reduction):

$$PbO_2(s) + 4H^+(aq) + SO_4{}^{2-}(aq) + 2\,e^- \longrightarrow PbSO_4(s) + 2H_2O(l)$$

Overall cell reaction:

$$Pb(s) + PbO_2(s) + 4H^+(aq) + 2SO_4{}^{2-}(aq) \longrightarrow 2PbSO_4(s) + 2H_2O(l)$$

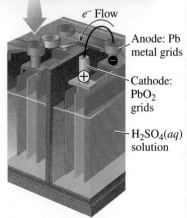

A 12-volt car battery is also known as a lead storage battery.

In both half-reactions, Pb^{2+} is produced, which combines with SO_4^{2-} to form an insoluble salt $PbSO_4(s)$. As a car battery is used, there is a buildup of $PbSO_4$ on the electrodes. At the same time, there is a decrease in the concentrations of the sulfuric acid components, H^+ and SO_4^{2-}. As a car runs, the battery is continuously recharged by an alternator, which is powered by the engine. The recharging reactions restore the Pb and PbO_2 electrodes as well as H_2SO_4. Without recharging, the car battery cannot continue to produce electrical energy.

Dry-Cell Batteries

Dry-cell batteries are used in calculators, watches, flashlights, and battery-operated toys. The term *dry cell* describes a battery that uses a paste rather than an aqueous solution. Dry cells can be *acidic* or *alkaline*. In an acidic dry cell, the anode is a zinc metal case that contains a paste of MnO_2, NH_4Cl, $ZnCl_2$, H_2O, and starch. Within this MnO_2 electrolyte mixture is a graphite cathode.

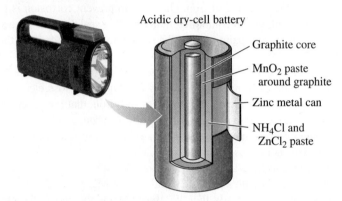

Acidic dry-cell battery

- Graphite core
- MnO_2 paste around graphite
- Zinc metal can
- NH_4Cl and $ZnCl_2$ paste

Anode (oxidation):	$Zn(s) \longrightarrow Zn^{2+}(aq) + 2\,e^-$
Cathode (reduction):	$2MnO_2(s) + 2NH_4^+(aq) + 2\,e^- \longrightarrow Mn_2O_3(s) + 2NH_3(aq) + H_2O(l)$
Overall cell reaction:	$Zn(s) + 2MnO_2(s) + 2NH_4^+(aq) \longrightarrow Zn^{2+}(aq) + Mn_2O_3(s) + 2NH_3(aq) + H_2O(l)$

An alkaline battery has similar components except that NaOH or KOH replaces the NH_4Cl electrolyte. Under basic conditions, the product of oxidation is zinc oxide (ZnO). Alkaline batteries tend to be more expensive but they last longer and produce more power than acidic dry-cell batteries.

Anode (oxidation):	$Zn(s) + 2OH^-(aq) \longrightarrow ZnO(s) + H_2O(l) + 2\,e^-$
Cathode (reduction):	$2MnO_2(s) + H_2O(l) + 2\,e^- \longrightarrow Mn_2O_3(s) + 2OH^-(aq)$
Overall cell reaction:	$Zn(s) + 2MnO_2(s) \longrightarrow ZnO(s) + Mn_2O_3(s)$

Mercury and Lithium Batteries

Mercury and lithium batteries are similar to alkaline dry-cell batteries. For example, a mercury battery has a zinc anode, but the cathode is steel in a mixture of HgO, KOH, and $Zn(OH)_2$. The reduced product Hg is toxic and an environmental hazard. Mercury batteries come with warnings on the label and should be disposed of properly.

Anode (oxidation):	$Zn(s) + 2OH^-(aq) \longrightarrow ZnO(s) + H_2O(l) + 2\,e^-$
Cathode (reduction):	$HgO(s) + H_2O(l) + 2\,e^- \longrightarrow Hg(l) + 2OH^-(aq)$
Overall cell reaction:	$Zn(s) + HgO(s) \longrightarrow ZnO(s) + Hg(l)$

In a lithium battery, the anode is lithium, not zinc. Lithium is much less dense than zinc and a lithium battery can be made very small.

CHEMISTRY AND THE ENVIRONMENT

Corrosion: Oxidation of Metals

Metals used in building materials, such as iron, eventually oxidize, which causes deterioration of the metal. This oxidation process, known as *corrosion*, produces rust on cars, bridges, ships, and underground pipes.

$$4Fe(s) + 3O_2(g) \longrightarrow 2Fe_2O_3(s)$$
$$\text{Rust}$$

The formation of rust requires both oxygen and water. The process of rusting requires an anode and cathode in different places on the surface of a piece of iron. In one area of the iron surface, called the anode region, the oxidation half-reaction takes place (see Figure 15.3).

Anode (oxidation):

$$Fe(s) \longrightarrow Fe^{2+}(aq) + 2\ e^-$$

or

$$2Fe(s) \longrightarrow 2Fe^{2+}(aq) + 4\ e^-$$

The electrons move through the iron metal from the anode to an area called the cathode region where oxygen dissolved in water is reduced to water:

Cathode (reduction):

$$O_2(g) + 4H^+(aq) + 4\ e^- \longrightarrow 2H_2O(l)$$

By combining the half-reactions that occur in the anode and cathode regions, we can write the overall oxidation–reduction equation:

$$2Fe(s) + O_2(g) + 4H^+(aq) \longrightarrow 2Fe^{2+}(aq) + 2H_2O(l)$$

The formation of rust occurs as Fe^{2+} ions move out of the anode region and come in contact with dissolved oxygen (O_2). The Fe^{2+} oxidizes to Fe^{3+}, which reacts with oxygen to form rust.

$$4Fe^{2+}(aq) + O_2(g) + 4H_2O(l) \longrightarrow 2Fe_2O_3(s) + 8H^+(aq)$$
$$\text{Rust}$$

We can write the formation of rust starting with solid Fe reacting with O_2 as follows. There is no H^+ in the overall equation because H^+ is used and produced in equal quantities:

Corrosion of iron

$$4Fe(s) + 3O_2(g) \longrightarrow 2Fe_2O_3(s)$$
$$\text{Rust}$$

Other metals such as aluminum, copper, and silver also undergo corrosion, but at a slower rate than iron. The oxidation of Al on the surface of an aluminum object produces Al^{3+}, which reacts with oxygen in the air to form a protective coating of Al_2O_3. This Al_2O_3 coating can prevent any further oxidation of the aluminum underneath it.

$$Al(s) \longrightarrow Al^{3+}(aq) + 3\ e^-$$

When copper is used on a roof, dome, or a steeple, it oxidizes to Cu^{2+}, which is converted to a green patina of $Cu_2(OH)_2CO_3$.

$$Cu(s) \longrightarrow Cu^{2+}(aq) + 2\ e^-$$

When we use silver dishes and utensils, the Ag^+ ion from oxidation reacts with sulfides in food to form Ag_2S, which we call "tarnish."

$$Ag(s) \longrightarrow Ag^+(aq) + e^-$$

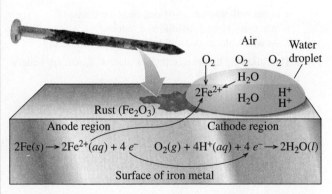

FIGURE 15.3 Rust forms when regions on the surface of iron metal establish an electrochemical cell. Electrons from the oxidation of Fe flow from the anode region to the cathode region where oxygen is reduced. As Fe^{2+} ions come in contact with O_2 and H_2O, rust forms.

Q Why must both O_2 and H_2O be present for the corrosion of iron?

The green patina on copper is due to oxidation.

A NiCad battery in a cell phone can be recharged many times.

Nickel–Cadmium (NiCad) Batteries

Nickel–cadmium (NiCad) batteries can be recharged. They use a cadmium anode and a cathode of solid nickel oxide $NiO(OH)(s)$.

Anode (oxidation): $Cd(s) + 2OH^-(aq) \longrightarrow Cd(OH)_2(s) + 2\ e^-$

Cathode (reduction): $2NiO(OH)(s) + 2H_2O(l) + 2\ e^- \longrightarrow$
$$2Ni(OH)_2(s) + 2OH^-(aq)$$

Overall cell reaction: $Cd(s) + 2NiO(OH)(s) + 2H_2O(l) \longrightarrow$
$$Cd(OH)_2(s) + 2Ni(OH)_2(s)$$

NiCad batteries are expensive, but they can be recharged many times. A charger provides an electrical current that converts the solid $Cd(OH)_2$ and $Ni(OH)_2$ products in the NiCad battery back to the reactants needed for oxidation and reduction.

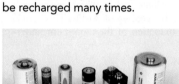

Batteries come in many shapes and sizes.

CONCEPT CHECK 15.4

■ Batteries

The following half-reaction takes place in a dry-cell battery used in portable radios and flashlights:

$$Zn(s) \longrightarrow Zn^{2+}(aq) + 2\ e^-$$

a. Why is this half-reaction an oxidation?
b. At which electrode does this half-reaction occur?

ANSWER
a. This half-reaction is an oxidation because $Zn(s)$ loses electrons.
b. The oxidation of Zn would take place at the anode.

QUESTIONS AND PROBLEMS

Electrical Energy from Oxidation–Reduction Reactions

15.31 Write the half-reactions and the overall cell reaction for each of the following voltaic cells:
   a. $Pb(s) \mid Pb^{2+}(aq) \parallel Cu^{2+}(aq) \mid Cu(s)$
   b. $Cr(s) \mid Cr^{2+}(aq) \parallel Ag^+(aq) \mid Ag(s)$

15.32 Write the half-reactions and the overall cell reaction for each of the following voltaic cells:
   a. $Al(s) \mid Al^{3+}(aq) \parallel Cd^{2+}(aq) \mid Cd(s)$
   b. $Sn(s) \mid Sn^{2+}(aq) \parallel Fe^{3+}(aq), Fe^{2+}(aq) \mid C\ (graphite)$

15.33 Describe the voltaic cell and half-cell components and write the shorthand notation for the following oxidation–reduction reactions:
   a. $Cd(s) + Sn^{2+}(aq) \longrightarrow Cd^{2+}(aq) + Sn(s)$
   b. $Zn(s) + Cl_2(g) \longrightarrow Zn^{2+}(aq) + 2Cl^-(aq)$ (C graphite cathode)

15.34 Describe the voltaic cell and half-cell components and write the shorthand notation for the following oxidation–reduction reactions:
   a. $Mn(s) + Sn^{2+}(aq) \longrightarrow Mn^{2+}(aq) + Sn(s)$
   b. $Ni(s) + 2Ag^+(aq) \longrightarrow Ni^{2+}(aq) + 2Ag(s)$

15.35 The following half-reaction takes place in a nickel–cadmium battery used in a cordless drill:

$$Cd(s) + 2OH^-(aq) \longrightarrow Cd(OH)_2(s) + 2\ e^-$$

a. Is the half-reaction an oxidation or a reduction?
b. What substance is oxidized or reduced?
c. At which electrode would this half-reaction occur?

15.36 The following half-reaction takes place in a mercury battery used in hearing aids:

$$HgO(s) + H_2O(l) + 2\ e^- \longrightarrow Hg(l) + 2OH^-(aq)$$

a. Is the half-reaction an oxidation or a reduction?
b. What substance is oxidized or reduced?
c. At which electrode would this half-reaction occur?

15.37 The following half-reaction takes place in a mercury battery used in pacemakers and watches:

$$Zn(s) + 2OH^-(aq) \longrightarrow ZnO(s) + H_2O(l) + 2\ e^-$$

a. Is the half-reaction an oxidation or a reduction?
b. What substance is oxidized or reduced?
c. At which electrode would this half-reaction occur?

15.38 The following half-reaction takes place in a lead storage battery used in automobiles:

$$Pb(s) + SO_4^{2-}(aq) \longrightarrow PbSO_4(s) + 2\ e^-$$

a. Is the half-reaction an oxidation or a reduction?
b. What substance is oxidized or reduced?
c. At which electrode would this half-reaction occur?

CHEMISTRY AND THE ENVIRONMENT

Fuel Cells: Clean Energy for the Future

Fuel cells are of interest to scientists because they provide an alternative source of electrical energy that is more efficient, does not use up oil reserves, and generates products that do not pollute the atmosphere. Fuels cells are considered to be a clean way to produce energy.

Like other cells, a fuel cell consists of an anode and a cathode connected by a wire. But unlike other cells, the reactants must continuously enter the fuel cell to produce energy; electrical current is generated only as long as the fuels are supplied. One type of hydrogen–oxygen fuel cell has been used in automobile prototypes. In this hydrogen cell, gas enters the fuel cell and comes in contact with a platinum catalyst embedded in a plastic membrane. The catalyst assists in the oxidation of hydrogen atoms to hydrogen ions and electrons (see Figure 15.4).

The electrons produce an electric current as they travel through the wire from the anode to the cathode. The hydrogen ions flow through the plastic membrane to the cathode. At the cathode, oxygen molecules are reduced to oxide ions that combine with the hydrogen ions to form water. The overall hydrogen–oxygen fuel cell reaction can be written as

$$2H_2(g) + O_2(g) \longrightarrow 2H_2O(l)$$

Fuel cells have already been used for power on the space shuttle and may soon be available to produce energy for cars and buses. A major drawback to the practical use of fuel cells is the economic impact of converting cars to fuel cell operation. The storage and cost of producing hydrogen are also problems. Some manufacturers are experimenting with systems that convert gasoline or methanol to hydrogen for immediate use in fuel cells.

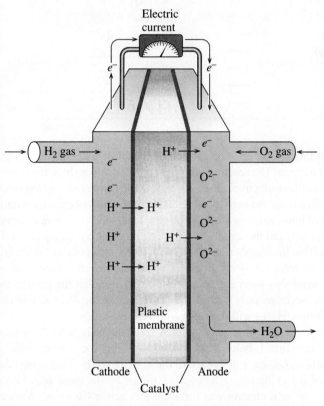

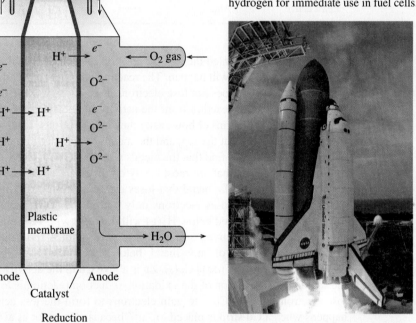

Oxidation
$H_2(g) \longrightarrow 4H^+(aq) + 4\,e^-$

Reduction
$O_2(g) + 4H^+(aq) + 4\,e^- \longrightarrow 2H_2O(l)$

FIGURE 15.4 With a supply of hydrogen and oxygen, a fuel cell can generate electricity continuously.

Q In most electrochemical cells, the electrodes are eventually used up. Is this true for a fuel cell? Why or why not?

Fuel cells are used to supply power on the space shuttle orbiter.

In homes, fuel cells may one day replace the batteries currently used to provide electrical power for cell phones, DVD players, and laptop computers. Fuel cell design is still in the prototype phase, although there is much interest in their development. We already know they can work, but modifications must still be made before they become reasonably priced and part of our everyday lives.

15.6 Oxidation–Reduction Reactions That Require Electrical Energy

LEARNING GOAL

Describe the half-cell reactions and the overall reactions that occur in electrolytic cells.

TUTORIAL
Electrolysis

In the previous section we looked at oxidation–reduction reactions that were spontaneous. In each example, the reactant that lost electrons more easily was oxidized. For example, in the cell $Zn(s)\,|\,Zn^{2+}(aq)\,\|\,Cu^{2+}(aq)\,|\,Cu(s)$, the reaction was spontaneous because Zn is oxidized more easily than Cu. When we placed a zinc metal strip in a solution of Cu^{2+}, reddish-brown Cu metal accumulated on the Zn strip according to the following spontaneous reaction:

$$Zn(s) + Cu^{2+}(aq) \longrightarrow Zn^{2+}(aq) + Cu(s) \qquad \text{Spontaneous}$$

TABLE 15.4 Activity Series for Some Metals

	Metal		Ion
Most active	Li(s)	$\longrightarrow$	$Li^+(aq) + e^-$
	K(s)	$\longrightarrow$	$K^+(aq) + e^-$
	Ca(s)	$\longrightarrow$	$Ca^{2+}(aq) + 2\,e^-$
	Na(s)	$\longrightarrow$	$Na^+(aq) + e^-$
	Mg(s)	$\longrightarrow$	$Mg^{2+}(aq) + 2\,e^-$
	Al(s)	$\longrightarrow$	$Al^{3+}(aq) + 3\,e^-$
	Zn(s)	$\longrightarrow$	$Zn^{2+}(aq) + 2\,e^-$
	Cr(s)	$\longrightarrow$	$Cr^{3+}(aq) + 3\,e^-$
	Fe(s)	$\longrightarrow$	$Fe^{2+}(aq) + 2\,e^-$
	Ni(s)	$\longrightarrow$	$Ni^{2+}(aq) + 2\,e^-$
	Sn(s)	$\longrightarrow$	$Sn^{2+}(aq) + 2\,e^-$
	Pb(s)	$\longrightarrow$	$Pb^{2+}(aq) + 2\,e^-$
	$H_2(g)$	$\longrightarrow$	$2H^+(aq) + 2\,e^-$
	Cu(s)	$\longrightarrow$	$Cu^{2+}(aq) + 2\,e^-$
	Ag(s)	$\longrightarrow$	$Ag^+(aq) + e^-$
Least active	Au(s)	$\longrightarrow$	$Au^{3+}(aq) + 3\,e^-$

TUTORIAL

Predict Spontaneous Reactions Using the Activity Series

— Copper strip

— Zn^{2+} solution

Because Cu is below Zn on the activity series, no oxidation–reduction reaction occurs.

Suppose that we wanted the reverse reaction to occur. If we place a Cu metal strip in a Zn^{2+} solution, nothing will happen. The reaction does not run spontaneously in the reverse direction because Cu does not lose electrons as easily as Zn. We can determine the direction of a spontaneous reaction from the activity series for metals and $H_2(g)$, which ranks the metals and H_2 in terms of how easily they lose electrons. The metals that lose electrons most easily are placed at the top, and the metals that do not lose electrons easily are at the bottom. We would also find that the metals whose ions gain electrons easily are at the bottom. Thus the metals that are most easily oxidized are above the metals whose ions are most easily reduced. The metal that loses electrons most easily is called the *most active* metal; the metal that loses electrons only with difficulty is called the *least active* (see Table 15.4). Metals listed below $H_2(g)$ will not react with H^+ from acids.

According to the activity series, a metal will oxidize spontaneously when it is combined with the ions of any metal below it on the list. For the voltaic cell of $Zn(s)\,|\,Zn^{2+}(aq)\,\|\,Cu^{2+}(aq)\,|\,Cu(s)$, Zn is above Cu in the activity series. This means that the spontaneous direction of the oxidation–reduction reaction is for the more active Zn to lose electrons and for Cu^{2+} to gain electrons to form the less active Cu metal. Nothing happens when a Cu strip is placed in Zn^{2+} because Cu is not as active as Zn.

$$Zn(s) + Cu^{2+}(aq) \longrightarrow Zn^{2+}(aq) + Cu(s) \quad \text{Spontaneous}$$

More active $\qquad \Longrightarrow \qquad$ Less active

We can use the activity series to help us predict the direction of the spontaneous reaction. Suppose we have two beakers. In one we place a Zn strip in a solution with Al^{3+} ions. In the other, we place an Al strip in a solution with Zn^{2+} ions. How can we predict whether or not a reaction will occur? Looking at the activity series we see that Al is listed above Zn, which means that Al is the more active metal and loses electrons more easily than Zn. Therefore, we predict the following half-reactions and overall reaction will occur:

More active metal:	$Al(s)$	$\longrightarrow Al^{3+}(aq) + 3\,e^-$	Spontaneous
Less active metal:	$Zn^{2+}(aq) + 2\,e^-$	$\longrightarrow Zn(s)$	Spontaneous
	$2Al(s) + 3Zn^{2+}(aq)$	$\longrightarrow 2Al^{3+}(aq) + 3Zn(s)$	Spontaneous

More active $\qquad \Longrightarrow \qquad$ Less active

Thus, there will be a coating of Zn on the Al strip as the oxidation–reduction reaction takes place. No reaction will occur spontaneously in the beaker containing the Zn strip and Al^{3+} ions.

We can see the activity of metals when various kinds of metals are placed in hydrochloric acid (HCl). Suppose we placed a Zn strip, a Mg strip, and a Cu strip in three beakers, each containing HCl. In the activity series, Zn and Mg are above H_2, and Cu is below H_2. The strips of Mg and Zn disappear as they are oxidized, while the reduction of H^+ produces lots of H_2 bubbles. The Cu strip does not react with HCl, which means the Cu metal remains intact in the HCl solution and no H_2 bubbles form.

CONCEPT CHECK 15.5

■ Predicting Spontaneous Reactions

A Cr strip is placed in a beaker containing a solution of Ag^+ ions. In another beaker, an Ag strip is placed in a solution with Cr^{3+} ions. Write the half-reactions and the overall equation for the spontaneous reaction that takes place.

ANSWER

Looking at the activity series in Table 15.4, we see that Cr is above Ag, which means that Cr is the more active metal and loses electrons more easily than Ag. Therefore, we predict the following half-reactions and overall reaction will occur:

More active metal:	$Cr(s) \longrightarrow Cr^{3+}(aq) + 3\ e^-$	Spontaneous
Less active metal:	$Ag^+(aq) + e^- \longrightarrow Ag(s)$	Spontaneous
	$Cr(s) + 3Ag^+(aq) \longrightarrow Cr^{3+}(aq) + 3Ag(s)$	Spontaneous

More active $\Longrightarrow$ Less active

Electrolytic Cells

Suppose that we try to reduce Zn^{2+} to Zn in the presence of Cu and Cu^{2+}. When we look at the activity series, we see that Cu is below Zn. This means that the oxidation–reduction reaction in this direction is not spontaneous.

$$Cu(s) + Zn^{2+}(aq) \longrightarrow Zn(s) + Cu^{2+}(aq) \quad \text{Not spontaneous}$$
Less active   More active

To make a nonspontaneous reaction take place, we need to use an electrical current, which is a process known as **electrolysis**. An **electrolytic cell** is an electrochemical cell in which electrical energy is used to drive a nonspontaneous oxidation–reduction reaction. We can think of the reactants in spontaneous oxidation–reduction reactions (voltaic cells) as rolling down a hill from higher to lower energy, which produces electrical energy. In electrolytic cells, energy must be provided by an outside energy source to push the reactants up a hill from lower energy to higher energy (see Figure 15.5).

$\Longrightarrow$ Electrical energy

$$Cu(s) + Zn^{2+}(aq) \longrightarrow Zn(s) + Cu^{2+}(aq) \quad \text{Not spontaneous as written}$$

Electrolysis of Sodium Chloride

When molten sodium chloride is electrolyzed, the products are sodium metal and chlorine gas. In this electrolytic cell, electrodes are placed in the mixture of Na^+ and Cl^- and connected to a battery. In this cell, the products are separated to prevent them from reacting spontaneously with each other. As electrons flow to the cathode, Na^+ is reduced to sodium

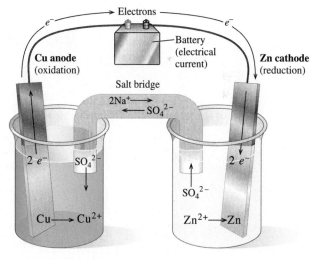

$$Cu(s) \longrightarrow Cu^{2+}(aq) + 2\ e^- \qquad Zn^{2+}(aq) + 2\ e^- \longrightarrow Zn(s)$$
$$Cu(s) + Zn^{2+}(aq) \longrightarrow Zn(s) + Cu^{2+}(aq)$$

FIGURE 15.5 In an electrolytic cell, the Cu anode is in a Cu^{2+} solution, and the Zn cathode is in a Zn^{2+} solution. Electrons provided by a battery reduce Zn^{2+} to Zn and drive the oxidation of Cu to Cu^{2+} at the Cu anode.

Q Why is an electrical current needed to make the reaction of Cu(s) and $Zn^{2+}(aq)$ happen?

metal. At the same time, electrons leave the anode as Cl^- is oxidized to Cl_2. The half-reactions and the overall reactions are

Anode (oxidation): $2Cl^-(l) \longrightarrow Cl_2(g) + 2\,e^-$

Cathode (reduction): $2Na^+(l) + 2\,e^- \longrightarrow 2Na(l)$

$$2Na^+(l) + 2Cl^-(l) \longrightarrow Cl_2(g) + 2Na(l)$$

Electrical energy

Electroplating

In industry, the process of electroplating uses electrolytic cells to coat a metal with a thin layer of a metal such as silver, platinum, or gold. Car bumpers are electroplated with chromium to prevent rusting. Silver-plated utensils, bowls, and platters are made by electroplating objects with a layer of silver.

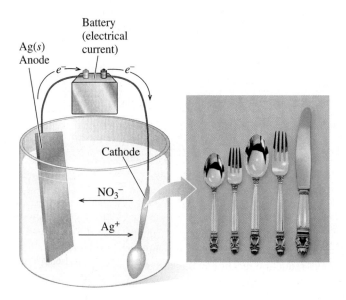

SAMPLE PROBLEM 15.11

■ **Electrolytic Cells**

Electrolysis is used to chrome plate an iron hubcap by placing the hubcap in a Cr^{3+} solution.

a. What half-reaction takes place to plate the hubcap with metallic chromium?
b. Is the iron hubcap the anode or the cathode?

SOLUTION

a. The Cr^{3+} ion in solution would gain electrons (reduction).

$$Cr^{3+}(aq) + 3\,e^- \longrightarrow Cr(s)$$

b. The iron hubcap is the cathode where reduction takes place.

STUDY CHECK

Why is energy needed to chrome plate the iron in Sample Problem 15.11?

QUESTIONS AND PROBLEMS

Oxidation–Reduction Reactions That Require Electrical Energy

15.39 What we call "tin cans" are really iron cans coated with a thin layer of tin. The anode is a bar of tin and the cathode is the iron can. An electrical current is used to oxidize the Sn to Sn^{2+} in solution, which is reduced to produce a thin coating of Sn on the steel can.

a. What half-reaction takes place to tin plate an iron can?
b. Why is the iron can the cathode?
c. Why is the tin bar the anode?

15.40 Electrolysis is used to gold plate jewelry made of stainless steel.
a. What half-reaction takes place to plate Au^{3+} on a stainless steel earring?
b. Is the earring the anode or the cathode?
c. Why is energy needed to gold plate the earring?

15.41 When the tin coating on an iron can is scratched, rust will form. Use the activity series to explain why this happens.

15.42 When the zinc coating on an iron can is scratched, rust does not form. Use the activity series to explain why this happens.

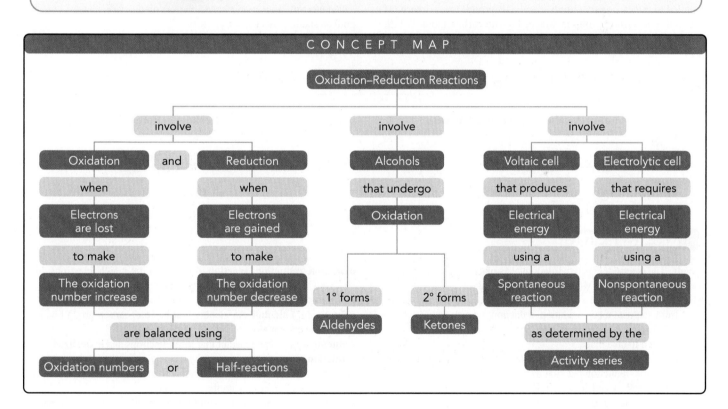

CONCEPT MAP

Oxidation–Reduction Reactions

involve — Oxidation and Reduction
when — Electrons are lost / Electrons are gained
to make — The oxidation number increase / The oxidation number decrease
are balanced using — Oxidation numbers or Half-reactions

involve — Alcohols
that undergo — Oxidation
1° forms Aldehydes / 2° forms Ketones

involve — Voltaic cell / Electrolytic cell
that produces Electrical energy using a Spontaneous reaction
that requires Electrical energy using a Nonspontaneous reaction
as determined by the Activity series

CHAPTER REVIEW

15.1 Oxidation–Reduction Reactions

LEARNING GOAL: *Identify what is oxidized and what is reduced in an oxidation–reduction reaction.*

In oxidation–reduction reactions, electrons are transferred from one reactant to another. The reactant that loses electrons is oxidized, and the reactant that gains electrons is reduced. Oxidation must always occur with reduction. The reducing agent is the substance that provides electrons for reduction. The oxidizing agent is the substance that accepts the electrons from oxidation.

15.2 Oxidation Numbers

LEARNING GOAL: *Assign and use oxidation numbers to identify elements that are oxidized or reduced, and to balance an oxidation–reduction equation.*

Oxidation numbers assigned to elements keep track of the changes in the loss and gain of electrons. Oxidation is an increase in oxidation number; reduction is a decrease in oxidation number. In covalent compounds and polyatomic ions, oxidation numbers are assigned using a set of rules. The oxidation number of an element is zero, and the oxidation number of a monatomic ion is the same as the ionic charge of the ion. The sum of the oxidation numbers for a compound is equal to zero and for a polyatomic ion is equal to the overall charge. Balancing oxidation–reduction equations using oxidation numbers involves the following: (1) assigning oxidation numbers; (2) determining the loss and gain of electrons; (3) equalizing the loss and gain of electrons; and (4) balancing the remaining substances by inspection.

15.3 Balancing Oxidation–Reduction Equations Using Half-Reactions

LEARNING GOAL: *Balance oxidation–reduction equations using the half-reaction method.*

Balancing oxidation–reduction equations using half-reactions involves the following: (1) separating the equation into half-reactions; (2) balancing elements other than H and O, then balancing O with H_2O and H with H^+; (3) balancing charge with electrons; (4) multiplying half-reactions by factors that equalize the loss and gain of electrons; (5) combining half-reactions, canceling electrons, and combining H_2O and H^+.

15.4 Oxidation of Alcohols

LEARNING GOAL: *Classify alcohols as primary, secondary, or tertiary; write equations for the oxidation of alcohols.*

Alcohols are classified according to the number of carbon groups bonded to the carbon that holds the —OH group. In a primary (1°) alcohol, one carbon group is attached to the carbon atom with the —OH group. In a secondary (2°) alcohol, two carbon groups are attached, and in a tertiary (3°) alcohol, there are three carbon groups. Primary alcohols are oxidized to aldehydes. Secondary alcohols are oxidized to ketones. Tertiary alcohols do not oxidize.

15.5 Electrical Energy from Oxidation–Reduction Reactions

LEARNING GOAL: *Write the half-reactions that occur at the anode and cathode of a voltaic cell; write the shorthand cell notation.*

In a voltaic cell, the components of the two half-reactions of a spontaneous oxidation–reduction reaction are placed in separate containers called half-cells. With a wire connecting the half-cells, an electrical current is generated as electrons move from the anode where oxidation takes place to the cathode where reduction takes place.

15.6 Oxidation–Reduction Reactions That Require Electrical Energy

LEARNING GOAL: *Describe the half-cell reactions and the overall reactions that occur in electrolytic cells.*

The activity series, which lists metals with the most easily oxidized metal at the top, is used to predict the direction of a spontaneous reaction. In an electrolytic cell, electrical energy from an external source is used to make reactions take place that are not spontaneous. A method called electrolysis is used to plate chrome on hubcaps, zinc on iron, or gold on stainless steel jewelry.

■ SUMMARY OF REACTIONS

Oxidation of Primary Alcohols to Form Aldehydes

Ethanol   Ethanal

Oxidation of Secondary Alcohols to Form Ketones

2-Propanol   Propanone

■ KEY TERMS

anode The electrode where oxidation takes place.

cathode The electrode where reduction takes place.

electrochemical cell An apparatus that produces electrical energy from a spontaneous oxidation–reduction reaction or uses electrical energy to cause a nonspontaneous oxidation–reduction reaction to take place.

electrolysis The use of electrical energy to run a nonspontaneous oxidation–reduction reaction in an electrolytic cell.

electrolytic cell A cell in which electrical energy is used to make a nonspontaneous oxidation–reduction reaction happen.

half-reaction method A method of balancing oxidation–reduction reactions in which the half-reactions are balanced separately and then combined to give the complete reaction.

oxidation The loss of electrons by a substance.

oxidation number A number equal to zero in an element or the charge of a monatomic ion; in covalent compounds and polyatomic ions, oxidation numbers are assigned using a set of rules.

oxidation–reduction reaction A reaction in which electrons are transferred from one reactant to another.

oxidizing agent The reactant that gains electrons and is reduced.

primary (1°) alcohol An alcohol that has one carbon group bonded to the carbon atom with the —OH group.

reducing agent The reactant that loses electrons and is oxidized.

reduction The gain of electrons by a substance.

secondary (2°) alcohol An alcohol that has two carbon groups bonded to the carbon atom with the —OH group.

tertiary (3°) alcohol An alcohol that has three carbon groups bonded to the carbon atom with the —OH group.

voltaic cell A type of electrochemical cell that uses spontaneous oxidation–reduction reactions to produce electrical energy.

UNDERSTANDING THE CONCEPTS

15.43 Classify each of the following as oxidation or reduction:
a. Electrons are lost.
b. Requires an oxidizing agent.
c. $O_2(g) \longrightarrow OH^-(aq)$
d. $Br_2(l) \longrightarrow 2Br^-(aq)$
e. $Sn^{2+}(aq) \longrightarrow Sn^{4+}(aq)$

15.44 Classify each of the following as oxidation or reduction:
a. Electrons are gained.
b. Requires a reducing agent.
c. $Ni(s) \longrightarrow Ni^{2+}(aq)$
d. $MnO_4^-(aq) \longrightarrow MnO_2(s)$
e. $Sn^{4+}(aq) \longrightarrow Sn^{2+}(aq)$

15.45 Assign oxidation numbers to the elements in each of the following:
a. VO_2 **b.** Ag_2CrO_4
c. $S_2O_8^{2-}$ **d.** $FeSO_4$

15.46 Assign oxidation numbers to the elements in each of the following:
a. $NbCl_3$ **b.** NbO
c. NbO_2 **d.** Nb_2O_5

15.47 Consider the following reaction:
$$Cr_2O_3(s) + Si(s) \longrightarrow Cr(s) + SiO_2(s)$$
a. Identify the substance reduced.
b. Identify the substance oxidized.
c. Identify the oxidizing agent.
d. Identify the reducing agent.
e. Write the balanced equation for the overall reaction.

Chromium(III) oxide and silicon undergo an oxidation–reduction reaction.

15.48 Consider the following reaction in an acidic solution:
$$MnO_4^-(aq) + Cl^-(aq) \longrightarrow Mn^{2+}(aq) + Cl_2(g)$$
a. Identify the substance reduced.
b. Identify the substance oxidized.
c. Identify the oxidizing agent.
d. Identify the reducing agent.
e. Write the balanced equation for the overall reaction in acid.

15.49 Classify each of the following as primary, secondary, or tertiary alcohols:

a. 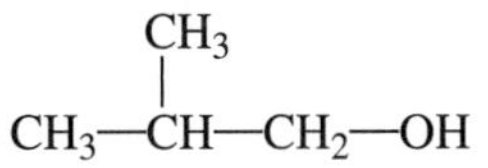
$$CH_3-\overset{\overset{\displaystyle CH_3}{|}}{CH}-CH_2-OH$$

b. 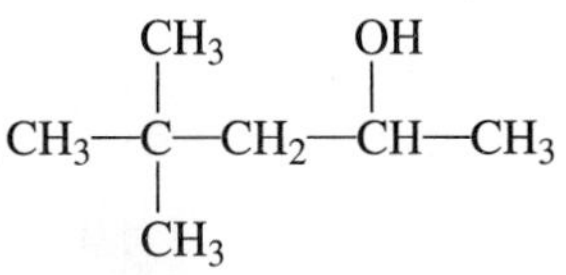
$$CH_3-\overset{\overset{\displaystyle CH_3}{|}}{\underset{\underset{\displaystyle CH_3}{|}}{C}}-CH_2-\overset{\overset{\displaystyle OH}{|}}{CH}-CH_3$$

c. CH_3-OH

15.50 Classify each of the following as primary, secondary, or tertiary alcohols:

a. 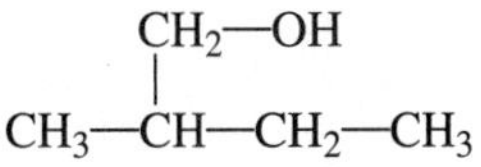
$$CH_3-\overset{\overset{\displaystyle CH_2-OH}{|}}{CH}-CH_2-CH_3$$

b. 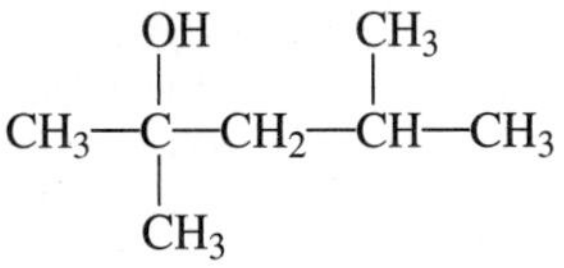
$$CH_3-\overset{\overset{\displaystyle OH}{|}}{\underset{\underset{\displaystyle CH_3}{|}}{C}}-CH_2-\overset{\overset{\displaystyle CH_3}{|}}{CH}-CH_3$$

c. CH_3-CH_2-OH

15.51 Draw the condensed structural formula for the aldehyde or ketone product of each of the following reactions:

a. 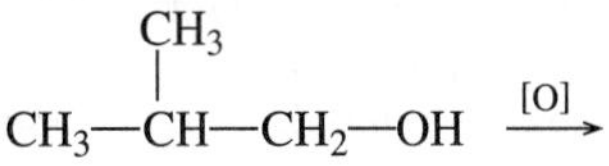
$$CH_3-\overset{\overset{\displaystyle CH_3}{|}}{CH}-CH_2-OH \xrightarrow{[O]}$$

b. 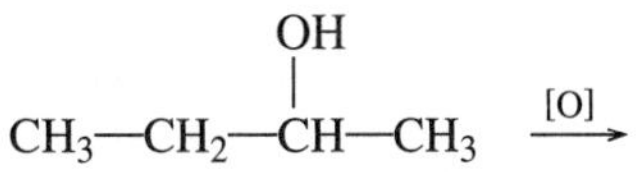
$$CH_3-CH_2-\overset{\overset{\displaystyle OH}{|}}{CH}-CH_3 \xrightarrow{[O]}$$

15.52 Draw the condensed structural formula for the aldehyde or ketone product of each of the following reactions:

a. 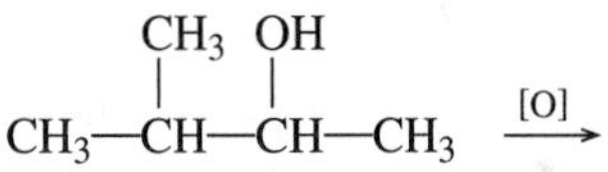
$$CH_3-\overset{\overset{\displaystyle CH_3}{|}}{CH}-\overset{\overset{\displaystyle OH}{|}}{CH}-CH_3 \xrightarrow{[O]}$$

b. 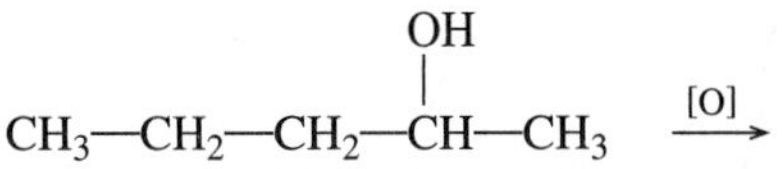
$$CH_3-CH_2-CH_2-\overset{\overset{\displaystyle OH}{|}}{CH}-CH_3 \xrightarrow{[O]}$$

15.53 Consider the following voltaic cell:

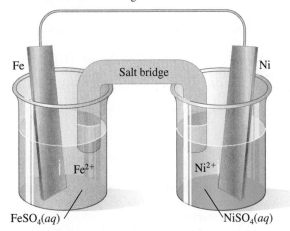

a. What is the oxidation half-reaction?
b. What is the reduction half-reaction?
c. What metal is the anode?
d. What metal is the cathode?
e. What is the direction of electron flow?
f. What is the overall reaction that takes place?
g. Write the shorthand cell notation.

15.54 Consider the following voltaic cell:

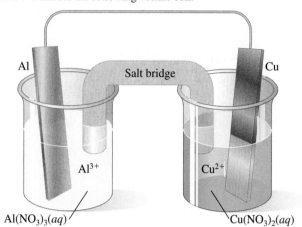

a. What is the oxidation half-reaction?
b. What is the reduction half-reaction?
c. What metal is the anode?
d. What metal is the cathode?
e. What is the direction of electron flow?
f. What is the overall reaction that takes place?
g. Write the shorthand cell notation.

ADDITIONAL QUESTIONS AND PROBLEMS

15.55 Which of the following are oxidation–reduction reactions?
a. $AgNO_3(aq) + NaCl(aq) \longrightarrow AgCl(s) + NaNO_3(aq)$
b. $6Li(s) + N_2(g) \longrightarrow 2Li_3N(s)$
c. $Ni(s) + Pb(NO_3)_2(aq) \longrightarrow Ni(NO_3)_2(aq) + Pb(s)$
d. $2K(s) + 2H_2O(l) \longrightarrow 2KOH(aq) + H_2(g)$

15.56 Which of the following are oxidation–reduction reactions?
a. $Ca(s) + F_2(g) \longrightarrow CaF_2(s)$
b. $Fe(s) + 2HCl(aq) \longrightarrow FeCl_2(aq) + H_2(g)$
c. $2NaCl(aq) + Pb(NO_3)_2(aq) \longrightarrow$
$$PbCl_2(s) + 2NaNO_3(aq)$$
d. $2CuCl(aq) \longrightarrow Cu(s) + CuCl_2(aq)$

15.57 In the mitochondria of human cells, energy is provided by the oxidation and reduction of the iron ions in the cytochromes. Identify each of the following reactions as an oxidation or reduction:
a. $Fe^{3+} + e^- \longrightarrow Fe^{2+}$ **b.** $Fe^{2+} \longrightarrow Fe^{3+} + e^-$

15.58 Chlorine (Cl_2) is used as a germicide to kill microbes in swimming pools. If the product is Cl^-, was the elemental chlorine oxidized or reduced?

15.59 Assign oxidation numbers to all the elements in each of the following:
a. Co_2O_3 **b.** $KMnO_4$
c. $SbCl_5$ **d.** ClO_3^-
e. PO_4^{3-}

15.60 Assign oxidation numbers to all the elements in each of the following:
a. PO_3^{2-} **b.** NH_4^+
c. $Fe(OH)_2$ **d.** HNO_3
e. $Cr_2O_7^{2-}$

15.61 Assign oxidation numbers to all the elements in the following reactions, and identify the reactant that is oxidized and the reactant that is reduced. Balance each equation.
a. $FeCl_2(aq) + Cl_2(g) \longrightarrow FeCl_3(aq)$
b. $H_2S(g) + O_2(g) \longrightarrow H_2O(l) + SO_2(g)$
c. $P_2O_5(s) + C(s) \longrightarrow P(s) + CO(g)$

15.62 Assign oxidation numbers to all the elements in the following reactions, and identify the reactant that is oxidized and the reactant that is reduced. Balance each equation.
a. $Al(s) + O_2(g) \longrightarrow Al_2O_3(s)$
b. $I_2O_5(s) + CO(g) \longrightarrow I_2(g) + CO_2(g)$
c. $Cr_2O_3(s) + C(s) \longrightarrow Cr(s) + CO_2(g)$

15.63 Balance each of the following half-reactions in an acidic solution:
a. $Zn(s) \longrightarrow Zn^{2+}(aq)$
b. $SnO_2^{2-}(aq) \longrightarrow SnO_3^{2-}(aq)$
c. $SO_3^{2-}(aq) \longrightarrow SO_4^{2-}(aq)$
d. $NO_3^-(aq) \longrightarrow NO(g)$

15.64 Balance each of the following half-reactions in an acidic solution:
a. $I_2(s) \longrightarrow I^-(aq)$
b. $MnO_4^-(aq) \longrightarrow Mn^{2+}(aq)$
c. $Br_2(l) \longrightarrow BrO_3^-(aq)$
d. $ClO_3^-(aq) \longrightarrow ClO_4^-(aq)$

15.65 Write the balanced half-reactions and a balanced ionic equation for each of the following reactions in an acidic solution:
a. $Zn(s) + NO_3^-(aq) \longrightarrow Zn^{2+}(aq) + NO_2(g)$
b. $MnO_4^-(aq) + SO_3^{2-}(aq) \longrightarrow Mn^{2+}(aq) + SO_4^{2-}(aq)$
c. $ClO_3^-(aq) + I^-(aq) \longrightarrow I_2(s) + Cl^-(aq)$
d. $Cr_2O_7^{2-}(aq) + C_2O_4^{2-}(aq) \longrightarrow Cr^{3+}(aq) + CO_2(g)$

15.66 Write the balanced half-reactions and a balanced ionic equation for each of the following reactions in an acidic solution:
a. $Sn^{2+}(aq) + IO_4^-(aq) \longrightarrow Sn^{4+}(aq) + I^-(aq)$
b. $S_2O_3^{2-}(aq) + I_2(s) \longrightarrow I^-(aq) + S_4O_6^{2-}(aq)$
c. $Mg(s) + VO_4^{3-}(aq) \longrightarrow Mg^{2+}(aq) + V^{2+}(aq)$
d. $Al(s) + Cr_2O_7^{2-}(aq) \longrightarrow Al^{3+}(aq) + Cr^{3+}(aq)$

15.67 Draw the condensed structural formula of the aldehyde or ketone formed when each of the following is oxidized:
a. $CH_3-CH_2-CH_2-OH$

b. $CH_3-\overset{\displaystyle OH}{\underset{\displaystyle |}{CH}}-CH_2-CH_2-CH_3$

c. $CH_3-CH_2-CH_2-CH_2-OH$

15.68 Draw the condensed structural formula of the aldehyde or ketone formed when each of the following is oxidized:

a. $CH_3-CH_2-\overset{\displaystyle OH}{\underset{\displaystyle |}{CH}}-CH_2-OH$

b. $CH_3-\overset{\displaystyle OH}{\underset{\displaystyle |}{CH}}-CH_3$

c. $CH_3-\overset{\displaystyle CH_3}{\underset{\displaystyle |}{CH}}-CH_2-CH_2-OH$

15.69 Use the activity series to predict whether each of the following reactions will occur spontaneously:

a. $Cu(s) + 2H^+(aq) \longrightarrow Cu^{2+}(aq) + H_2(g)$
b. $Ni^{2+}(aq) + Fe(s) \longrightarrow Ni(s) + Fe^{2+}(aq)$
c. $2Ag(s) + Cu^{2+}(aq) \longrightarrow 2Ag^+(aq) + Cu(s)$
d. $3Ni^{2+}(aq) + 2Cr(s) \longrightarrow 3Ni(s) + 2Cr^{3+}(aq)$
e. $Zn(s) + Cu^{2+}(aq) \longrightarrow Zn^{2+}(aq) + Cu(s)$
f. $Pb^{2+}(aq) + Zn(s) \longrightarrow Pb(s) + Zn^{2+}(aq)$

15.70 Use the activity series to predict whether each of the following reactions will occur spontaneously:

a. $2Ag(s) + 2H^+(aq) \longrightarrow 2Ag^+(aq) + H_2(g)$
b. $Mg(s) + Cu^{2+}(aq) \longrightarrow Mg^{2+}(aq) + Cu(s)$
c. $2Al(s) + 3Cu^{2+}(aq) \longrightarrow 2Al^{3+}(aq) + 3Cu(s)$
d. $Mg^{2+}(aq) + Zn(s) \longrightarrow Mg(s) + Zn^{2+}(aq)$
e. $Al^{3+}(aq) + 3Na(s) \longrightarrow Al(s) + 3Na^+(aq)$
f. $Ni^{2+}(aq) + Mg(s) \longrightarrow Ni(s) + Mg^{2+}(aq)$

15.71 In a voltaic cell, one half-cell consists of nickel metal in a Ni^{2+} solution, and the other half-cell consists of magnesium metal in a Mg^{2+} solution. Identify each of the following:

a. the anode
b. the cathode
c. the half-reaction at the anode
d. the half-reaction at the cathode
e. the overall reaction
f. the shorthand cell notation

15.72 In a voltaic cell, one half-cell consists of a zinc metal in a Zn^{2+} solution, and the other half-cell consists of a copper metal in a Cu^{2+} solution. Identify each of the following:

a. the anode
b. the cathode
c. the half-reaction at the anode
d. the half-reaction at the cathode
e. the overall reaction
f. the shorthand cell notation

15.73 Use the activity series to determine which of the following ions will be reduced when an iron strip is placed in an aqueous solution of that ion:

a. $Ca^{2+}(aq)$ **b.** $Ag^+(aq)$
c. $Ni^{2+}(aq)$ **d.** $Al^{3+}(aq)$
e. $Pb^{2+}(aq)$

15.74 Use the activity series to determine which of the following ions will be reduced when an aluminum strip is placed in an aqueous solution of that ion:

a. $Fe^{2+}(aq)$ **b.** $Au^{3+}(aq)$
c. $Zn^{2+}(aq)$ **d.** $H^+(aq)$
e. $Pb^{2+}(aq)$

15.75 Steel bolts made for sailboats are coated with zinc. Add the necessary components (electrodes, wires, batteries) to this diagram of an electrolytic cell with a zinc nitrate solution to show how it could be used to zinc plate a steel bolt.

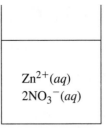

$Zn^{2+}(aq)$
$2NO_3^-(aq)$

a. What is the anode?
b. What is the cathode?
c. What is the half-reaction that takes place at the anode?
d. What is the half-reaction that takes place at the cathode?
e. If steel is mostly iron, what is the purpose of the zinc coating?

15.76 Copper cooking pans are stainless steel pans plated with a layer of copper. Add the necessary components (electrodes, wires, batteries) to this diagram of an electrolytic cell with a copper(II) nitrate solution to show how it could be used to copper plate a stainless steel (iron) pan.

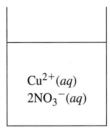

$Cu^{2+}(aq)$
$2NO_3^-(aq)$

a. What is the anode?
b. What is the cathode?
c. What is the half-reaction that takes place at the anode?
d. What is the half-reaction that takes place at the cathode?

15.77 In a lead storage battery, the following unbalanced half-reaction takes place:

$$Pb(s) + SO_4^{2-}(aq) \longrightarrow PbSO_4(s)$$

a. Balance the half-reaction.
b. Is $Pb(s)$ oxidized or reduced?
c. Indicate whether the half-reaction takes place at the anode or cathode.

15.78 In an acidic dry-cell battery, the following unbalanced half-reaction takes place in acidic solution:

$$MnO_2(s) \longrightarrow Mn_2O_3(s)$$

a. Balance the half-reaction.
b. Is $MnO_2(s)$ oxidized or reduced?
c. Indicate whether the half-reaction takes place at the anode or cathode.

CHALLENGE QUESTIONS

15.79 The following unbalanced reaction takes place in acidic solution:

$$Ag(s) + NO_3^-(aq) \longrightarrow Ag^+(aq) + NO(g)$$

a. Write the balanced equation.
b. How many liters of $NO(g)$ are produced at STP when 15.0 g of silver reacts with excess nitric acid?

15.80 The following unbalanced reaction takes place in acidic solution:

$$MnO_4^-(aq) + Fe^{2+}(aq) \longrightarrow Mn^{2+}(aq) + Fe^{3+}(aq)$$

a. Write the balanced equation.
b. How many milliliters of a 0.150 M $KMnO_4$ solution are needed to react with 25.0 mL of a 0.400 M $FeSO_4$ solution?

15.81 A concentrated nitric acid solution is used to dissolve copper(II) sulfide.

$$CuS(s) + HNO_3(aq) \longrightarrow CuSO_4(aq) + NO(g) + H_2O(l)$$

a. Write the balanced equation.
b. How many milliliters of a 16.0 M HNO_3 solution are needed to dissolve 24.8 g of CuS?

15.82 The following unbalanced reaction takes place in acidic solution:

$$Cr_2O_7^{2-}(aq) + Fe^{2+}(aq) \longrightarrow Cr^{3+}(aq) + Fe^{3+}(aq)$$

a. Write the balanced equation.
b. How many milliliters of a 0.211 M $K_2Cr_2O_7$ solution are needed to react with 5.00 g of $FeSO_4$?

15.83 Which of the following are oxidation–reduction reactions?
a. $Ca(s) + 2H_2O(l) \longrightarrow Ca(OH)_2(aq) + H_2(g)$
b. $CaCO_3(s) \longrightarrow CaO(s) + CO_2(g)$
c. $Cl_2(g) + 2NaBr(aq) \longrightarrow Br_2(l) + 2NaCl(aq)$
d. $BaCl_2(aq) + Na_2SO_4(aq) \longrightarrow BaSO_4(s) + 2NaCl(aq)$

15.84 Assign oxidation numbers to all the elements in the following equation and balance:

$$Fe_2O_3(s) + CO(g) \longrightarrow Fe(s) + CO_2(g)$$

a. Which substance is oxidized?
b. Which substance is reduced?
c. Which substance is the oxidizing agent?
d. Which substance is the reducing agent?

15.85 Determine the oxidation number of Br in each of the following:
a. Br_2
b. $HBrO_2$
c. BrO_3^-
d. $NaBrO_4$

15.86 Use half-reactions to balance the following equation in acidic solution:

$$Cr_2O_7^{2-}(aq) + NO_2^-(aq) \longrightarrow Cr^{3+}(aq) + NO_3^-(aq)$$

15.87 Draw a diagram of a voltaic cell for $Ni(s) \,|\, Ni^{2+}(aq) \,\|\, Ag^+(aq) \,|\, Ag(s)$.
a. What is the anode?
b. What is the cathode?
c. What is the half-reaction that takes place at the anode?
d. What is the half-reaction that takes place at the cathode?
e. What is the overall reaction for the cell?

15.88 Using the activity series for metals, indicate whether each of the following can be used to generate an electrical current or will require a battery:
a. $Ca^{2+}(aq) + Zn(s) \longrightarrow Ca(s) + Zn^{2+}(aq)$
b. $2Al(s) + 3Sn^{2+}(aq) \longrightarrow 2Al^{3+}(aq) + 3Sn(s)$
c. $Mg(s) + 2H^+(aq) \longrightarrow Mg^{2+}(aq) + H_2(g)$
d. $Cu(s) + Ni^{2+}(aq) \longrightarrow Cu^{2+}(aq) + Ni(s)$
e. $2Cr(s) + 3Fe^{2+}(aq) \longrightarrow 2Cr^{3+}(aq) + 3Fe(s)$

15.89 In a process called anodizing, tumblers of silvery-colored aluminum are coated in bright colors. The aluminum anode is oxidized to form an oxide coating. When dyes are added, they attach to the coating to produce tumblers with bright colors. At the cathode, hydrogen gas is produced.

a. Balance the half-reaction in acidic solution at the anode:

$$Al(s) \longrightarrow Al_2O_3(s)$$

b. Balance the half-reaction in acidic solution at the cathode:

$$H^+(aq) \longrightarrow H_2(g)$$

c. Write the overall reaction for the formation of the aluminum-oxide coating.

ANSWERS

ANSWERS TO STUDY CHECKS

15.1 Al loses 3 electrons and is oxidized. Sn^{2+} gains 2 electrons and is reduced.

15.2 a. Because H has an oxidation number of +1, and O is -2, P will have an oxidation number of +5 to maintain a neutral charge.

H_3PO_4

$+1 +5 -2$

b. Because O has an oxidation number of -2, Mn must be +7 to give an overall charge of -1.

MnO_4^-

$+7 -2$

15.3 $Zn(s) + CuCl_2(aq) \longrightarrow ZnCl_2(aq) + Cu(s)$

$\quad 0 \qquad +2 -1 \qquad\qquad +2 -1 \qquad 0 \quad$ Oxidation numbers

Because Zn is oxidized, Zn is the reducing agent. Because the Cu^{2+} (in $CuCl_2$) is reduced, $CuCl_2$ is the oxidizing agent.

15.4 $3Li(s) + AlCl_3(aq) \longrightarrow 3LiCl(aq) + Al(s)$

$\quad\quad$ 0 $\quad\quad$ +3$-$1 $\quad\quad\quad$ +1$-$1 $\quad\quad$ 0 Oxidation numbers

15.5 $2Fe_2O_3(s) + 3C(s) \longrightarrow 4Fe(s) + 3CO_2(g)$

$\quad\quad$ +3$-$2 $\quad\quad$ 0 $\quad\quad\quad$ 0 $\quad\quad$ +4$-$2 Oxidation numbers

15.6 $Zn(s) + 2Fe^{3+}(aq) \longrightarrow Zn^{2+}(aq) + 2Fe^{2+}(aq)$

15.7 $Cu^{2+}(aq) + SO_2(g) + 2H_2O(l) \longrightarrow$
$\quad\quad\quad\quad\quad\quad Cu(s) + SO_4{}^{2-}(aq) + 4H^+(aq)$

15.8 $N_2O(g) + 2ClO^-(aq) + 2OH^-(aq) \longrightarrow$
$\quad\quad\quad\quad 2Cl^-(aq) + 2NO_2{}^-(aq) + H_2O(l)$

15.9
$$CH_3-\overset{\overset{\displaystyle CH_3}{|}}{CH}-\overset{\overset{\displaystyle CH_3}{|}}{CH}-CH_2-\overset{\overset{\displaystyle O}{\|}}{C}-H$$

15.10 Anode reaction: $Co(s) \longrightarrow Co^{2+}(aq) + 2e^-$

Cathode reaction: $Cu^{2+}(aq) + 2e^- \longrightarrow Cu(s)$

Overall cell reaction:
$Co(s) + Cu^{2+}(aq) \longrightarrow Co^{2+}(aq) + Cu(s)$

15.11 Since Fe is below Cr in the activity series, the plating of Cr^{3+} onto Fe is not spontaneous. Energy is needed to make the reaction proceed.

ANSWERS TO SELECTED QUESTIONS AND PROBLEMS

15.1 **a.** Na^+ gains electrons; this is a reduction.
b. Ni loses electrons; this is an oxidation.
c. Cr^{3+} gains electrons; this is a reduction.
d. H^+ gains electrons; this is a reduction.

15.3 **a.** Zn loses electrons and is oxidized. Cl_2 gains electrons and is reduced.
b. Br^- (in NaBr) loses electrons and is oxidized. Cl_2 gains electrons and is reduced.
c. Pb loses electrons and is oxidized. O_2 gains electrons and is reduced.
d. Sn^{2+} loses electrons and is oxidized. Fe^{3+} gains electrons and is reduced.

15.5 **a.** 0 $\quad\quad\quad\quad\quad\quad$ **b.** 0
c. +2 $\quad\quad\quad\quad\quad$ **d.** $-$1

15.7 **a.** K is +1, Cl is $-$1. $\quad$ **b.** Mn is +4, O is $-$2.
c. C is +2, O is $-$2. $\quad$ **d.** Mn is +3, O is $-$2.

15.9 **a.** Al is +3, P is +5, and O is $-$2.
b. S is +4, O is $-$2.
c. Cr is +3, O is $-$2.
d. N is +5, O is $-$2.

15.11 **a.** H is +1, S is +6, and O is $-$2.
b. H is +1, P is +3, and O is $-$2.
c. Cr is +6, O is $-$2.
d. Na is +1, C is +4, and O is $-$2.

15.13 **a.** +5 $\quad\quad\quad\quad\quad$ **b.** $-$2
c. +5 $\quad\quad\quad\quad\quad$ **d.** +6

15.15 **a.** The substance that is oxidized is the reducing agent.
b. The substance that gains electrons is reduced and is the oxidizing agent.

15.17 **a.** Zn is the reducing agent. Cl_2 is the oxidizing agent.
b. Br^- (in NaBr) is the reducing agent. Cl_2 is the oxidizing agent.
c. Pb is the reducing agent. O_2 is the oxidizing agent.
d. Sn^{2+} is the reducing agent. Fe^{3+} is the oxidizing agent.

15.19 **a.** S^{2-} (in NiS) is oxidized; O_2 is reduced. NiS is the reducing agent, and O_2 is the oxidizing agent.
b. Sn^{2+} is oxidized; Fe^{3+} is reduced. Sn^{2+} is the reducing agent, and Fe^{3+} is the oxidizing agent.
c. C (in CH_4) is oxidized; O_2 is reduced. CH_4 is the reducing agent, and O_2 is the oxidizing agent.
d. Si is oxidized; Cr^{3+} (in Cr_2O_3) is reduced. Si is the reducing agent, and Cr_2O_3 is the oxidizing agent.

15.21 **a.** $Cu_2S(s) + H_2(g) \longrightarrow 2Cu(s) + H_2S(g)$
b. $2Fe(s) + 3Cl_2(g) \longrightarrow 2FeCl_3(s)$
c. $2Al(s) + 3H_2SO_4(aq) \longrightarrow Al_2(SO_4)_3(aq) + 3H_2(g)$

15.23 **a.** $Sn^{2+}(aq) \longrightarrow Sn^{4+}(aq) + 2e^-$
b. $Mn^{2+}(aq) + 4H_2O(l) \longrightarrow$
$\quad\quad\quad\quad MnO_4{}^-(aq) + 8H^+(aq) + 5e^-$
c. $NO_2{}^-(aq) + H_2O(l) \longrightarrow$
$\quad\quad\quad\quad NO_3{}^-(aq) + 2H^+(aq) + 2e^-$
d. $ClO_3{}^-(aq) + 2H^+(aq) + e^- \longrightarrow ClO_2(aq) + H_2O(l)$

15.25 **a.** $2H^+(aq) + Ag(s) + NO_3{}^-(aq) \longrightarrow$
$\quad\quad\quad\quad Ag^+(aq) + NO_2(g) + H_2O(l)$
b. $2Fe(s) + 2CrO_4{}^{2-}(aq) + 2H_2O(l) \longrightarrow$
$\quad\quad\quad\quad Fe_2O_3(s) + Cr_2O_3(s) + 4OH^-(aq)$
c. $4H^+(aq) + 4NO_3{}^-(aq) + 3S(s) \longrightarrow$
$\quad\quad\quad\quad 4NO(g) + 3SO_2(g) + 2H_2O(l)$
d. $2S_2O_3{}^{2-}(aq) + Cu^{2+}(aq) \longrightarrow S_4O_6{}^{2-}(aq) + Cu(s)$
e. $4H^+(aq) + 5PbO_2(s) + 2Mn^{2+}(aq) \longrightarrow$
$\quad\quad\quad\quad 5Pb^{2+}(aq) + 2MnO_4{}^-(aq) + 2H_2O(l)$

15.27 **a.** primary $\quad\quad\quad\quad$ **b.** primary
c. tertiary

15.29 **a.**
$$CH_3-CH_2-CH_2-CH_2-\overset{\overset{\displaystyle O}{\|}}{C}-H$$

b. none

c.
$$CH_3-\overset{\overset{\displaystyle O}{\|}}{C}-CH_2-\overset{\overset{\displaystyle CH_3}{|}}{CH}-CH_3$$

15.31 **a.** Anode reaction: $Pb(s) \longrightarrow Pb^{2+}(aq) + 2e^-$
Cathode reaction: $Cu^{2+}(aq) + 2e^- \longrightarrow Cu(s)$
Overall cell reaction:
$\quad Cu^{2+}(aq) + Pb(s) \longrightarrow Cu(s) + Pb^{2+}(aq)$
b. Anode reaction: $Cr(s) \longrightarrow Cr^{2+}(aq) + 2e^-$
Cathode reaction: $Ag^+(aq) + e^- \longrightarrow Ag(s)$
Overall cell reaction:
$\quad 2Ag^+(aq) + Cr(s) \longrightarrow 2Ag(s) + Cr^{2+}(aq)$

15.33 **a.** The anode is a Cd metal electrode in a Cd^{2+} solution. The anode reaction is
$\quad Cd(s) \longrightarrow Cd^{2+}(aq) + 2e^-$
The cathode is a Sn metal electrode in a Sn^{2+} solution. The cathode reaction is
$\quad Sn^{2+}(aq) + 2e^- \longrightarrow Sn(s)$

The shorthand notation for this cell is
$\quad Cd(s)\,|\,Cd^{2+}(aq)\,\|\,Sn^{2+}(aq)\,|\,Sn(s)$

b. The anode is a Zn metal electrode in a Zn^{2+} solution. The anode reaction is
$\quad Zn(s) \longrightarrow Zn^{2+}(aq) + 2e^-$

The cathode is a C (graphite) electrode, where Cl_2 gas is reduced to Cl^-. The cathode reaction is
$\quad Cl_2(g) + 2e^- \longrightarrow 2Cl^-(aq)$

The shorthand cell notation is

$$Zn(s) \mid Zn^{2+}(aq) \parallel Cl_2(g), Cl^-(aq) \mid C \text{ (graphite)}$$

15.35 **a.** The half-reaction is an oxidation.
b. Cd metal is oxidized.
c. Oxidation takes place at the anode.

15.37 **a.** The half-reaction is an oxidation.
b. Zn metal is oxidized.
c. Oxidation takes place at the anode.

15.39 **a.** $Sn^{2+}(aq) + 2\,e^- \longrightarrow Sn(s)$
b. The reduction of Sn^{2+} to Sn occurs at the cathode, which is the iron can.
c. The oxidation of Sn to Sn^{2+} occurs at the anode, which is the tin bar.

15.41 Since Fe is above Sn in the activity series, if the Fe is exposed to air and water, Fe will be oxidized and rust will form. To protect iron, Sn would have to be *more* active than Fe and it is not.

15.43 **a.** oxidation
b. reduction
c. reduction
d. reduction
e. oxidation

15.45 **a.** $V = +4, O = -2$
b. $Ag = +1, Cr = +6, O = -2$
c. $S = +7, O = -2$
d. $Fe = +2, S = +6, O = -2$

15.47 **a.** Cr in Cr_2O_3 is reduced.
b. Si is oxidized.
c. Cr_2O_3 is the oxidizing agent.
d. Si is the reducing agent.
e. $2Cr_2O_3(s) + 3Si(s) \longrightarrow 4Cr(s) + 3SiO_2(s)$

15.49 **a.** primary   **b.** secondary   **c.** primary

15.51 **a.**

$$CH_3\!-\!\overset{\displaystyle CH_3}{\underset{}{CH}}\!-\!\overset{\displaystyle O}{\underset{}{C}}\!-\!H$$

b.

$$CH_3\!-\!CH_2\!-\!\overset{\displaystyle O}{\underset{}{C}}\!-\!CH_3$$

15.53 **a.** $Fe(s) \longrightarrow Fe^{2+}(aq) + 2\,e^-$
b. $Ni^{2+}(aq) + 2\,e^- \longrightarrow Ni(s)$
c. Fe is the anode.
d. Ni is the cathode.
e. The electrons flow from Fe to Ni.
f. $Fe(s) + Ni^{2+}(aq) \longrightarrow Fe^{2+}(aq) + Ni(s)$
g. $Fe(s) \mid Fe^{2+}(aq) \parallel Ni^{2+}(aq) \mid Ni(s)$

15.55 Reactions **b**, **c**, and **d** all involve loss and gain of electrons; **b**, **c**, and **d** are oxidation–reduction reactions.

15.57 **a.** Fe^{3+} is gaining electrons; this is a reduction.
b. Fe^{2+} is losing electrons; this is an oxidation.

15.59 **a.** $Co = +3, O = -2$
b. $K = +1, Mn = +7, O = -2$
c. $Sb = +5, Cl = -1$
d. $Cl = +5, O = -2$
e. $P = +5, O = -2$

15.61 **a.** $FeCl_2(aq) + Cl_2(g) \longrightarrow FeCl_3(aq)$

$$\underset{+2\,-1}{FeCl_2} \quad \underset{0}{Cl_2} \quad \underset{+3\,-1}{FeCl_3}$$

Fe in $FeCl_2$ is oxidized and Cl in Cl_2 is reduced.
$$2FeCl_2(aq) + Cl_2(g) \longrightarrow 2FeCl_3(aq)$$

b. $H_2S(g) + O_2(g) \longrightarrow H_2O(l) + SO_2(g)$

$$\underset{+1\,-2}{H_2S} \quad \underset{0}{O_2} \quad \underset{+1\,-2}{H_2O} \quad \underset{+4\,-2}{SO_2}$$

S in H_2S is oxidized and O in O_2 is reduced.
$$2H_2S(g) + 3O_2(g) \longrightarrow 2H_2O(l) + 2SO_2(g)$$

c. $P_2O_5(s) + C(s) \longrightarrow P(s) + CO(g)$

$$\underset{+5\,-2}{P_2O_5} \quad \underset{0}{C} \quad \underset{0}{P} \quad \underset{+2\,-2}{CO}$$

C is oxidized and P in P_2O_5 is reduced.
$$P_2O_5(s) + 5C(s) \longrightarrow 2P(s) + 5CO(g)$$

15.63 **a.** $Zn(s) \longrightarrow Zn^{2+}(aq) + 2\,e^-$
b. $SnO_2{}^{2-}(aq) + H_2O(l) \longrightarrow$
$$SnO_3{}^{2-}(aq) + 2H^+(aq) + 2\,e^-$$
c. $SO_3{}^{2-}(aq) + H_2O(l) \longrightarrow$
$$SO_4{}^{2-}(aq) + 2H^+(aq) + 2\,e^-$$
d. $NO_3{}^-(aq) + 4H^+(aq) + 3\,e^- \longrightarrow NO(g) + 2H_2O(l)$

15.65 **a.** $Zn(s) \longrightarrow Zn^{2+}(aq) + 2\,e^-$;
$NO_3{}^-(aq) + 2H^+(aq) + e^- \longrightarrow NO_2(g) + H_2O(l)$
Overall:
$$Zn(s) + 2NO_3{}^-(aq) + 4H^+(aq) \longrightarrow$$
$$Zn^{2+}(aq) + 2NO_2(g) + 2H_2O(l)$$
b. $MnO_4{}^-(aq) + 8H^+(aq) + 5\,e^- \longrightarrow$
$$Mn^{2+}(aq) + 4H_2O(l);$$
$SO_3{}^{2-}(aq) + H_2O(l) \longrightarrow SO_4{}^{2-}(aq) + 2H^+(aq) + 2\,e^-$
Overall:
$$2MnO_4{}^-(aq) + 5SO_3{}^{2-}(aq) + 6H^+(aq) \longrightarrow$$
$$2Mn^{2+}(aq) + 5SO_4{}^{2-}(aq) + 3H_2O(l)$$
c. $2I^-(aq) \longrightarrow I_2(s) + 2\,e^-$;
$ClO_3{}^-(aq) + 6H^+(aq) + 6\,e^- \longrightarrow Cl^-(aq) + 3H_2O(l)$
Overall:
$$ClO_3{}^-(aq) + 6I^-(aq) + 6H^+(aq) \longrightarrow$$
$$Cl^-(aq) + 3I_2(s) + 3H_2O(l)$$
d. $C_2O_4{}^{2-}(aq) \longrightarrow 2CO_2(g) + 2\,e^-$;
$Cr_2O_7{}^{2-}(aq) + 14H^+(aq) + 6\,e^- \longrightarrow$
$$2Cr^{3+}(aq) + 7H_2O(l)$$
Overall:
$$Cr_2O_7{}^{2-}(aq) + 3C_2O_4{}^{2-}(aq) + 14H^+(aq) \longrightarrow$$
$$2Cr^{3+}(aq) + 6CO_2(g) + 7H_2O(l)$$

15.67

a.
$$CH_3\!-\!CH_2\!-\!\overset{\displaystyle O}{\underset{}{C}}\!-\!H$$

b.
$$CH_3\!-\!\overset{\displaystyle O}{\underset{}{C}}\!-\!CH_2\!-\!CH_2\!-\!CH_3$$

c.
$$CH_3\!-\!CH_2\!-\!CH_2\!-\!\overset{\displaystyle O}{\underset{}{C}}\!-\!H$$

15.69 **a.** Since Cu is below H_2 in the activity series, the reaction will not be spontaneous.
b. Since Fe is above Ni in the activity series, the reaction will be spontaneous.
c. Since Ag is below Cu in the activity series, the reaction will not be spontaneous.
d. Since Cr is above Ni in the activity series, the reaction will be spontaneous.
e. Since Zn is above Cu in the activity series, the reaction will be spontaneous.
f. Since Zn is above Pb in the activity series, the reaction will be spontaneous.

15.71 **a.** The anode is Mg.
b. The cathode is Ni.
c. The half-reaction at the anode is
$$Mg(s) \longrightarrow Mg^{2+}(aq) + 2\,e^-$$

d. The half-reaction at the cathode is
$$Ni^{2+}(aq) + 2\ e^- \longrightarrow Ni(s)$$
e. The overall reaction is
$$Mg(s) + Ni^{2+}(aq) \longrightarrow Mg^{2+}(aq) + Ni(s)$$
f. The shorthand cell notation is
$$Mg(s)\,|\,Mg^{2+}(aq)\,\|\,Ni^{2+}(aq)\,|\,Ni(s)$$

15.73 **a.** $Ca^{2+}(aq)$ will not be reduced by an iron strip.
b. $Ag^+(aq)$ will be reduced by an iron strip.
c. $Ni^{2+}(aq)$ will be reduced by an iron strip.
d. $Al^{3+}(aq)$ will not be reduced by an iron strip.
e. $Pb^{2+}(aq)$ will be reduced by an iron strip.

15.75

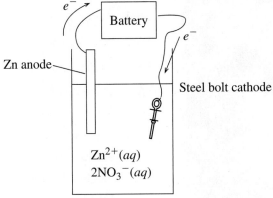

a. The anode is a bar of zinc.
b. The cathode is the steel bolt.
c. $Zn(s) \longrightarrow Zn^{2+}(aq) + 2\ e^-$
d. $Zn^{2+}(aq) + 2\ e^- \longrightarrow Zn(s)$
e. The purpose of the zinc coating is to prevent rusting of the bolt by H_2O and O_2.

15.77 **a.** $Pb(s) + SO_4{}^{2-}(aq) \longrightarrow PbSO_4(s) + 2\ e^-$
b. $Pb(s)$ is oxidized.
c. The half-reaction takes place at the anode.

15.79 **a.** $3Ag(s) + NO_3{}^-(aq) + 4H^+(aq) \longrightarrow$
$$3Ag^+(aq) + NO(g) + 2H_2O(l)$$
b. 1.04 L of $NO(g)$ are produced at STP.

15.81 **a.** $3CuS(s) + 8HNO_3(aq) \longrightarrow$
$$3CuSO_4(aq) + 8NO(g) + 4H_2O(l)$$
b. 43.2 mL of HNO_3 solution

15.83 Reactions **a** and **c** are oxidation–reduction reactions.

15.85 **a.** 0   **b.** +3
          c. +5   **d.** +7

15.87

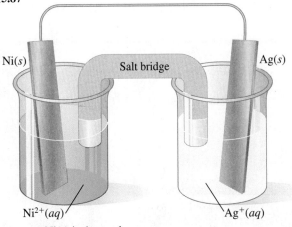

a. $Ni(s)$ is the anode.
b. $Ag(s)$ is the cathode.
c. $Ni(s) \longrightarrow Ni^{2+}(aq) + 2\ e^-$
d. $Ag^+(aq) + e^- \longrightarrow Ag(s)$
e. $Ni(s) + 2Ag^+(aq) \longrightarrow Ni^{2+}(aq) + 2Ag(s)$

15.89 **a.** $2Al(s) + 3H_2O(l) \longrightarrow Al_2O_3(s) + 6H^+(aq) + 6\ e^-$
b. $2H^+(aq) + 2\ e^- \longrightarrow H_2(g)$
c. $2Al(s) + 3H_2O(l) \longrightarrow Al_2O_3(s) + 3H_2(g)$

16 Nuclear Radiation

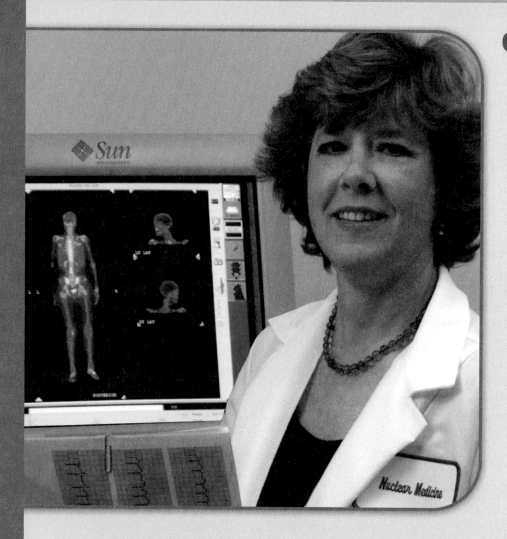

"Everything we do in this department involves radioactive materials," says Julie Goudak, nuclear medicine technologist at Kaiser Hospital. "The radioisotopes are given in several ways. The patient may ingest an isotope, breathe it in, or receive it by an IV injection. We do many diagnostic tests, particularly of the heart function, to determine if a patient needs a cardiac CT scan."

A nuclear medicine technologist administers isotopes that emit radiation to determine the level of function of an organ, such as the thyroid or heart, to detect the presence and size of a tumor, or to treat disease. A radioisotope locates in a specific organ, and its radiation is used by a computer to create an image of that organ. From these data, a physician can make a diagnosis and design a treatment program.

Mastering**CHEMISTRY**

Visit **www.masteringchemistry.com**
for self-study material and
instructor-assigned homework.

With the production of the first artificial radioactive substances in 1934, the field of nuclear medicine was established. In 1937, the first radioactive isotope was used to treat a person with leukemia at the University of California at Berkeley. Major strides in the use of radioactivity in medicine occurred in 1946, when a radioactive iodine isotope was successfully used to diagnose thyroid function and to treat hyperthyroidism and thyroid cancer. In the 1970s and 1980s, a variety of radioactive substances were used to produce images of organs, such as liver, spleen, thyroid gland, kidneys, and the brain, and to detect heart disease. Today procedures in nuclear medicine provide information about the function and structure of every organ in the body, which allows the nuclear physician to diagnose and treat diseases early.

16.1 Natural Radioactivity

LEARNING GOAL

Describe alpha, beta, positron, and gamma radiation.

Most naturally occurring isotopes of elements up to atomic number 19 have stable nuclei. Elements with atomic numbers 20 and higher usually have one or more isotopes that have unstable nuclei in which the nuclear forces cannot offset the repulsions between the protons. An unstable nucleus is *radioactive*, which means that it spontaneously emits small particles of energy called **radiation** to become more stable. Radiation may take the form of alpha (α) and beta (β) particles, positrons (β^+), or pure energy such as gamma (γ) rays. An isotope of an element that emits radiation is called a **radioisotope**. For most types of radiation, there is a change in the number of protons in the nucleus, which means that an atom is converted into an atom of a different element. This kind of nuclear change was not evident to Dalton when he made his predictions about atoms. Elements with atomic numbers of 93 and higher are produced artificially in nuclear laboratories and consist only of radioactive isotopes.

In Chapter 4, Section 4.5, we wrote symbols for the different isotopes of an element. These symbols had the mass number written in the upper left corner and the atomic number in the lower left corner. Recall that the mass number is equal to the number of protons and neutrons in the nucleus and that the atomic number is equal to the number of protons. For example, a radioactive isotope of iodine used in the diagnosis and treatment of thyroid conditions has a mass number of 131 and an atomic number of 53.

Mass number (protons and neutrons)
Element symbol
Atomic number (protons)

$^{131}_{53}\text{I}$

Radioactive isotopes are identified by writing the mass number after the element's name or symbol. Thus, in this example, the isotope is called iodine-131 or I-131. Table 16.1 compares some stable, nonradioactive isotopes with some radioactive isotopes.

TABLE 16.1 Stable and Radioactive Isotopes of Some Elements

Magnesium	Iodine	Uranium
Stable Isotopes		
$^{24}_{12}\text{Mg}$	$^{127}_{53}\text{I}$	None
Magnesium-24	Iodine-127	
Radioactive Isotopes		
$^{23}_{12}\text{Mg}$	$^{125}_{53}\text{I}$	$^{235}_{92}\text{U}$
Magnesium-23	Iodine-125	Uranium-235
$^{27}_{12}\text{Mg}$	$^{131}_{53}\text{I}$	$^{238}_{92}\text{U}$
Magnesium-27	Iodine-131	Uranium-238

TUTORIAL

Types of Radiation

Types of Radiation

By emitting radiation, an unstable nucleus forms a more stable, lower energy nucleus. One type of radiation consists of *alpha particles*. An **alpha particle** is identical to a helium (He) nucleus, which has 2 protons and 2 neutrons. An alpha particle has a mass number of 4, an atomic number of 2, and a charge of 2+. The symbol for an alpha particle is the Greek letter alpha (α) or the symbol of a helium nucleus except that the 2+ charge is omitted.

^4_2He

Alpha (α) particle

Another type of radiation occurs when a radioisotope emits a *beta particle*. A **beta particle**, which is a high-energy electron, has a charge of $1-$, and because its mass is so much less than the mass of a proton, it is given a mass number of 0. It is represented by the Greek letter beta (β) or by the symbol for the electron (e), including the mass number and the charge. A beta particle is formed when a neutron in an unstable nucleus changes into a proton.

$^{0}_{-1}e$

Beta (β) particle

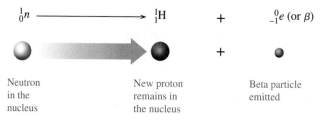

$$^{1}_{0}n \longrightarrow ^{1}_{1}H \quad + \quad ^{0}_{-1}e \text{ (or } \beta)$$

| Neutron in the nucleus | New proton remains in the nucleus | Beta particle emitted |

A **positron**, represented as β^{+}, has a positive charge ($1+$) with a mass number of 0, which makes it similar to a beta (β) particle, but with the opposite charge. When the symbol β is used with no charge, it represents a beta particle rather than a positron. We can write the symbols of a beta particle and a positron as follows:

Beta particle   Positron

Mass number
Charge $^{0}_{-1}e$ $^{0}_{+1}e$

A positron is produced by an unstable nucleus when a proton is transformed into a neutron and a positron.

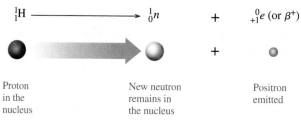

$$^{1}_{1}H \longrightarrow ^{1}_{0}n \quad + \quad ^{0}_{+1}e \text{ (or } \beta^{+})$$

| Proton in the nucleus | New neutron remains in the nucleus | Positron emitted |

A positron is an example of *antimatter*, a term physicists use to describe a particle that is the exact opposite of another particle, in this case, an electron. When an electron and a positron collide, their minute masses are completely converted to energy in the form of *gamma rays*.

$$^{0}_{-1}e + ^{0}_{+1}e \longrightarrow 2^{0}_{0}\gamma$$

Gamma rays are high-energy radiation, released when an unstable nucleus undergoes a rearrangement of its particles to give a more stable, lower energy nucleus. Gamma rays are often emitted along with other types of radiation. A gamma ray is written as the Greek letter gamma (γ). Because gamma rays are energy only, zeros are used to show that a gamma ray has no mass or charge.

$^{0}_{0}\gamma$

Gamma (γ) ray

Table 16.2 summarizes the types of radiation we will use in nuclear equations.

TABLE 16.2 Some Common Forms of Radiation

Type of Radiation	Symbol		Mass Number	Charge
Alpha particle	α	$^{4}_{2}He$	4	2+
Beta particle	β	$^{0}_{-1}e$	0	1−
Positron	β^{+}	$^{0}_{+1}e$	0	1+
Gamma ray	γ	$^{0}_{0}\gamma$	0	0
Proton	p	$^{1}_{1}H$	1	1+
Neutron	n	$^{1}_{0}n$	1	0

CONCEPT CHECK 16.1

■ **Radiation Particles**

Identify and write the symbol for each of the following types of radiation:

a. contains two protons and two neutrons
b. has a mass number of 0 and a $1-$ charge

ANSWER

a. An alpha (α) particle, ^4_2He, has two protons and two neutrons.
b. A beta (β) particle, $^{\ 0}_{-1}e$, is like an electron with a mass number of 0 and a $1-$ charge.

Radiation Protection

SELF STUDY ACTIVITY
Radiation and Its Biological Effects

Radiologists, doctors, and nurses working with radioactive isotopes must use proper radiation protection. Proper *shielding* is necessary to prevent exposure. Alpha particles, the heaviest of the radiation particles, travel only a few centimeters in the air before they collide with air molecules, acquire electrons, and become helium atoms. A piece of paper, clothing, and our skin are protection against alpha particles. Lab coats and gloves will also provide sufficient shielding. However, if ingested or inhaled, alpha emitters can bring about serious internal damage because of the large mass and high charge of the alpha particle.

Beta particles have a very small mass and move much faster and farther than alpha particles, traveling as much as several meters through air. They can pass through paper and penetrate as far as 4–5 mm into body tissue. External exposure to beta particles can burn the surface of the skin, but they are stopped before they can reach the internal organs. Heavy clothing such as lab coats and gloves are needed to protect the skin from beta particles.

Gamma rays travel great distances through the air and pass through many materials, including body tissues. Only very dense shielding, such as lead or concrete, will stop them. Because gamma rays penetrate so deeply, exposure to gamma rays can be extremely hazardous. When preparing radioactive materials, the radiologist wears special gloves and works behind leaded glass windows (see Figure 16.1). Long tongs are used within the work area to pick up vials of radioactive material, keeping them away from the hands and body. Even the syringe used to give an injection of a gamma-emitting radioactive isotope is placed inside a special leaded glass cover. Table 16.3 summarizes the shielding materials required for the various types of radiation.

If you work in an environment such as a nuclear medicine facility, keep the time you spend in a radioactive area to a minimum. Remaining in a radioactive area twice as long exposes you to twice as much radiation.

Keep your distance! The greater the distance from the radioactive source, the lower the intensity of radiation received. Just by doubling your distance from the radiation source, the intensity of radiation drops to $\left(\frac{1}{2}\right)^2$, or one-fourth of its previous value.

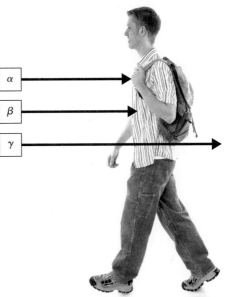

Different types of shielding are needed for different radiation particles.

FIGURE 16.1 A person working with radioisotopes wears protective clothing and gloves and stands behind a shield.

Q What types of radiation are stopped by the lead shield?

CHEMISTRY AND HEALTH

Biological Effects of Radiation

When radiation strikes molecules in its path, electrons may be knocked away, forming unstable ions. For example, when radiation passes through the human body, it may interact with water molecules, removing electrons and producing H_2O^+, which can cause undesirable chemical reactions.

The cells most sensitive to radiation are the ones undergoing rapid division—those of the bone marrow, skin, reproductive organs, and intestinal lining, as well as all cells of growing children. Damaged cells may lose their ability to produce necessary materials. For example, if radiation damages cells of the bone marrow, red blood cells may no longer be produced. If sperm cells, ova, or the cells of a fetus are damaged, birth defects may result. In contrast, cells of the nerves, muscles, liver, and adult bones are much less sensitive to radiation because they undergo little or no cellular division.

Cancer cells are another example of rapidly dividing cells. Because cancer cells are highly sensitive to radiation, large doses of radiation are used to destroy them. The normal tissue that surrounds cancer cells divides at a slower rate and suffers less damage from radiation. However, radiation itself may cause malignant tumors, leukemia, anemia, and genetic mutations.

TABLE 16.3 Properties of Radiation and Shielding Required

Property	Alpha (α) particle	Beta (β) particle	Gamma (γ) ray
Travel distance in air	2–4 cm	200–300 cm	500 m
Tissue depth	0.05 mm	4–5 mm	50 cm or more
Shielding	Paper, clothing	Heavy clothing, lab coats, gloves	Lead, thick concrete
Typical source	Radium-226	Carbon-14	Technetium-99m

SAMPLE PROBLEM | 16.1

■ Radiation Protection

How does the type of shielding for alpha radiation differ from that used for gamma radiation?

SOLUTION

Alpha radiation is stopped by paper and clothing. However, lead or concrete is needed for protection from gamma radiation.

STUDY CHECK

Besides shielding, what other methods help reduce exposure to radiation?

QUESTIONS AND PROBLEMS

Natural Radioactivity

16.1 **a.** How are an alpha particle and a helium nucleus similar?
 b. What symbols are used for alpha particles?
 c. What is the source of an alpha particle?

16.2 **a.** How are a beta particle and an electron similar?
 b. What symbols are used for beta particles?
 c. What is the source of a beta particle?

16.3 Naturally occurring potassium consists of three isotopes: potassium-39, potassium-40, and potassium-41.
 a. Write the atomic symbol for each isotope.
 b. In what ways are the isotopes similar, and in what ways do they differ?

16.4 Naturally occurring iodine is iodine-127. Medically, radioactive isotopes of iodine-125 and iodine-130 are used.
 a. Write the atomic symbol for each isotope.
 b. In what ways are the isotopes similar, and in what ways do they differ?

16.5 Supply the missing information in the following table:

Medical Use	Atomic Symbol	Mass Number	Number of Protons	Number of Neutrons
Heart imaging	$^{201}_{81}Tl$			
Radiation therapy		60	27	
Abdominal scan			31	36
Hyper-thyroidism	$^{131}_{53}I$			
Leukemia treatment		32		17

16.6 Supply the missing information in the following table:

Medical Use	Atomic Symbol	Mass Number	Number of Protons	Number of Neutrons
Cancer treatment	$^{131}_{55}\text{Cs}$			
Brain scan		99	43	
Blood flow		141	58	
Bone scan		85		47
Lung function	$^{133}_{54}\text{Xe}$			

16.7 Write a symbol for each of the following:
   a. alpha particle   **b.** neutron
   c. beta particle   **d.** nitrogen-15
   e. iodine-125

16.8 Write a symbol for each of the following:
   a. proton   **b.** gamma ray
   c. positron   **d.** barium-131
   e. palladium-103

16.9 Identify the symbol for X in each of the following:
   a. $^{0}_{-1}\text{X}$ **b.** $^{4}_{2}\text{X}$
   c. $^{1}_{0}\text{X}$ **d.** $^{24}_{11}\text{X}$
   e. $^{14}_{6}\text{X}$

16.10 Identify the symbol for X in each of the following:
   a. $^{1}_{1}\text{X}$ **b.** $^{32}_{15}\text{X}$
   c. $^{0}_{0}\text{X}$ **d.** $^{59}_{26}\text{X}$
   e. $^{0}_{+1}\text{X}$

16.11 a. Why does beta radiation penetrate farther into body tissue than alpha radiation?
   b. How does radiation cause damage to cells of the body?
   c. Why does the radiation technician leave the room when you receive an X-ray?
   d. What is the purpose of wearing gloves when handling radioisotopes?

16.12 a. As a scientist you sometimes work with radioisotopes. What are three ways you can minimize your exposure to radiation?
   b. Why are cancer cells more sensitive to radiation than nerve cells?
   c. What is the purpose of placing a lead apron on a person who is receiving routine dental X-rays?
   d. Why are the walls in a radiation lab built of thick concrete blocks?

16.2 Nuclear Equations

LEARNING GOAL

Write an equation showing mass numbers and atomic numbers for radioactive decay.

In a process called **radioactive decay**, a nucleus spontaneously breaks down by emitting radiation. The *nuclear equation* is written using the atomic symbols of the original radioactive nucleus, the new nucleus, and the type of radiation emitted. An arrow between the atomic symbols indicates that this is a nuclear equation.

Radioactive nucleus $\longrightarrow$ new nucleus + radiation (α, β, γ, β^{+})

In a nuclear equation, the sum of the mass numbers and the atomic numbers must be equal on both sides. In most nuclear equations, there is a change in the number of protons, which gives a different element.

The changes in mass number and atomic number of an unstable nucleus that undergoes radioactive decay are shown in Table 16.4.

TABLE 16.4 Mass Number and Atomic Number Changes Due to Radiation

Decay Process	Radiation Symbol	Change in Mass Number	Change in Atomic Number	Change in Neutron Number
Alpha decay	$^{4}_{2}\text{He}$	−4	−2	−2
Beta decay	$^{0}_{-1}e$	0	+1	−1
Positron emission	$^{0}_{+1}e$	0	−1	+1
Gamma emission	$^{0}_{0}\gamma$	0	0	0

Alpha Decay

An unstable nucleus undergoes alpha decay by emitting an alpha particle. Because an alpha particle consists of two protons and two neutrons, the mass number decreases by four, and the atomic number decreases by two. For example, uranium-238 emits an alpha particle to form a different nucleus with a mass number of 234. Compared to uranium with 92 protons, the new nucleus has 90 protons, which makes it thorium.

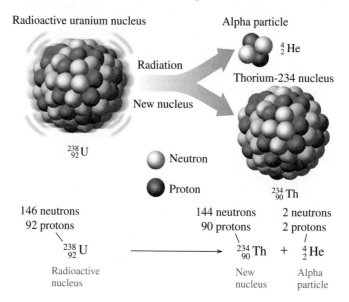

Radioactive uranium nucleus

Alpha particle

$_2^4$He

Radiation

Thorium-234 nucleus

New nucleus

$_{92}^{238}$U

Neutron

Proton

$_{90}^{234}$Th

146 neutrons   144 neutrons   2 neutrons
92 protons   90 protons   2 protons

$_{92}^{238}$U $\longrightarrow$ $_{90}^{234}$Th + $_2^4$He

Radioactive   New   Alpha
nucleus   nucleus   particle

TUTORIAL
Writing Nuclear Equations

Guide to Completing a Nuclear Equation

In another example of radioactive decay, radium-226 emits an alpha particle to form a nucleus that has a new identity with a different mass number and atomic number.

STEP 1 **Write the incomplete nuclear equation.**

$$_{88}^{226}\text{Ra} \longrightarrow ? + _2^4\text{He}$$

STEP 2 **Determine the missing mass number.** In the nuclear equation, the mass number of the radium, 226, is equal to the sum of the mass numbers of the alpha particle and the new nucleus.

226　　 = ? + 4
226 − 4 = ?
222　　 = ? (mass number of new nucleus)

STEP 3 **Determine the missing atomic number.** The atomic number of radium, 88, must equal the sum of the atomic numbers of the alpha particle and the new nucleus.

88　　 = ? + 2
88 − 2 = ?
86　　 = ? (atomic number of new nucleus)

STEP 4 **Determine the symbol of the new nucleus.** On the periodic table, the element that has atomic number 86 is radon, Rn. The nucleus of this isotope of Rn is written as $_{86}^{222}$Rn.

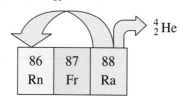

$_2^4$He

86	87	88
Rn	Fr	Ra

STEP 5   Complete the nuclear equation.

$$^{226}_{88}Ra \longrightarrow\ ^{222}_{86}Rn\ +\ ^{4}_{2}He$$

In this nuclear reaction, a radium-226 nucleus decays by releasing an alpha particle and produces a radon-222 nucleus.

CONCEPT CHECK 16.2

■ Alpha Decay

Francium-221 emits an alpha particle when it decays.
a. Does the new nucleus have a larger or smaller mass number? By how much?
b. Does the new nucleus have a larger or smaller atomic number? By how much?

ANSWER

a. The loss of an alpha particle will give a smaller mass number to the new nucleus. Because an alpha particle is a helium nucleus, $^{4}_{2}He$, the mass number of the new nucleus will decrease by four from 221 to 217.
b. The loss of an alpha particle will give a smaller atomic number to the new nucleus. Because an alpha particle is a helium nucleus, $^{4}_{2}He$, the atomic number of the new nucleus will decrease by two from 87 to 85.

SAMPLE PROBLEM 16.2

■ Writing an Equation for Alpha Decay

Smoke detectors that are used in homes and apartments contain americium-241, which undergoes alpha decay. When alpha particles collide with air molecules, charged particles are produced that generate an electrical current. As smoke particles enter the detector, they interfere with the formation of charged particles in the air, and the electric current is interrupted. This causes the alarm to sound and warns the occupants of the danger of fire. Complete the following nuclear equation for the decay of americium-241:

$$^{241}_{95}Am \longrightarrow\ ?\ +\ ^{4}_{2}He$$

SOLUTION

STEP 1   Write the incomplete nuclear equation.

$$^{241}_{95}Am \longrightarrow\ ?\ +\ ^{4}_{2}He$$

STEP 2   Determine the missing mass number.   In the equation, the mass number of the americium, 241, is equal to the sum of the mass numbers of the alpha particle and the new nucleus.

$$241\ \ \ \ \ \ = ?\ +\ 4$$
$$241\ -\ 4 = ?$$
$$237\ \ \ \ \ \ = ?\ \text{(mass number of new nucleus)}$$

STEP 3   Determine the missing atomic number.   The atomic number of americium, 95, must equal the sum of the atomic numbers of the alpha particle and the new nucleus.

$$95\ \ \ \ \ \ = ?\ +\ 2$$
$$95\ -\ 2 = ?$$
$$93\ \ \ \ \ \ = ?\ \text{(atomic number of new nucleus)}$$

STEP 4   Determine the symbol of the new nucleus.   On the periodic table, the element that has atomic number 93 is neptunium, Np. The nucleus of this isotope of Np is written as $^{237}_{93}Np$.

STEP 5   Complete the nuclear equation.

$$^{241}_{95}Am \longrightarrow\ ^{237}_{93}Np\ +\ ^{4}_{2}He$$

STUDY CHECK

Write a balanced nuclear equation for the alpha decay of Po-214.

Guide to Completing a Nuclear Equation

STEP 1
Write the incomplete nuclear equation.

STEP 2
Determine the missing mass number.

STEP 3
Determine the missing atomic number.

STEP 4
Determine the symbol of the new nucleus.

STEP 5
Complete the nuclear equation.

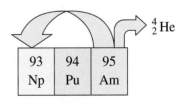

| 93 | 94 | 95 |
| Np | Pu | Am |

$^{4}_{2}He$

CHEMISTRY AND THE ENVIRONMENT

Radon in Our Homes

The presence of radon gas has become a much publicized environmental and health issue because of the radiation danger it poses. Radioactive isotopes, such as radium-226 and uranium-238, are naturally present in many types of rocks and soils. Radium-226 emits an alpha particle and is converted into radon gas, which diffuses out of the rocks and soil:

$$^{226}_{88}Ra \longrightarrow {}^{222}_{86}Rn + {}^{4}_{2}He$$

Outdoors, radon gas poses little danger because it dissipates in the air. However, if the radioactive source is under a house or building, the radon gas can enter the house through cracks in the foundation or other openings. Those who live or work there may inhale the radon. Inside the lungs, radon-222 emits alpha particles to form polonium-218, which is known to cause lung cancer:

$$^{222}_{86}Rn \longrightarrow {}^{218}_{84}Po + {}^{4}_{2}He$$

Some researchers have estimated that 10 percent of all lung cancer deaths in the United States result from radon gas exposure. The Environmental Protection Agency (EPA) recommends that the maximum level of radon not exceed 4 picocuries (pCi) per liter of air in a home. One picocurie (pCi) is equal to 10^{-12} curies (Ci); curies are described in Section 16.3. In California, 1% of all the houses surveyed exceeded the EPA's recommended maximum radon level.

A radon gas detector is used to determine radon levels in buildings with inadequate ventilation.

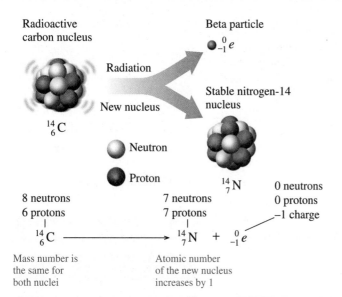

Radioactive carbon nucleus

Radiation

New nucleus

Beta particle

$^{0}_{-1}e$

Stable nitrogen-14 nucleus

$^{14}_{6}C$

Neutron

Proton

$^{14}_{7}N$

8 neutrons
6 protons

$^{14}_{6}C$ $\longrightarrow$ $^{14}_{7}N$ + $^{0}_{-1}e$

7 neutrons
7 protons

0 neutrons
0 protons
−1 charge

Mass number is the same for both nuclei

Atomic number of the new nucleus increases by 1

Beta Decay

In the nuclear equation for beta decay, the mass number of the radioactive nucleus and the mass number of the new nucleus are the same. However, the atomic number of the new nucleus increases by one, indicating a change of one element into another. For example, the beta decay of a carbon-14 nucleus produces a nitrogen-14 nucleus.

SAMPLE PROBLEM | 16.3

■ Writing an Equation for Beta Decay

Write the nuclear equation for the beta decay of cobalt-60.

SOLUTION

STEP 1 Write the incomplete nuclear equation.

$$^{60}_{27}Co \longrightarrow ? + {}^{0}_{-1}e$$

STEP 2 Determine the missing mass number. In the equation, the mass number of the cobalt, 60, is equal to the sum of the mass numbers of the beta particle and the new nucleus.

$$60 = ? + 0$$
$$60 - 0 = ?$$
$$60 = ? \text{ (mass number of new nucleus)}$$

STEP 3 **Determine the missing atomic number.**   The atomic number of cobalt, 27, must equal the sum of the atomic numbers of the beta particle and the new nucleus.

$$27 = ? - 1$$
$$27 + 1 = ?$$
$$28 = ? \text{ (atomic number of new nucleus)}$$

STEP 4 **Determine the symbol of the new nucleus.**   On the periodic table, the element that has atomic number 28 is nickel (Ni). The nucleus of this isotope of Ni is written as $^{60}_{28}\text{Ni}$.

STEP 5 **Complete the nuclear equation.**

$$^{60}_{27}\text{Co} \longrightarrow {}^{60}_{28}\text{Ni} + {}^{0}_{-1}e$$

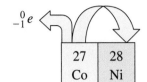

STUDY CHECK

Write the nuclear equation for the beta decay of iodine-131.

Positron Emission

In positron emission, a proton in an unstable nucleus is converted to a neutron and a positron. The neutron remains in the nucleus, but the positron is emitted from the nucleus. In a nuclear equation for positron emission, the mass number of the radioactive nucleus and the mass number of the new nucleus are the same. However, the atomic number of the new nucleus decreases by one indicating a change of one element into another. For example, an aluminum-24 nucleus undergoes positron emission to produce a magnesium-24 nucleus. The atomic number of magnesium (12) and the charge of the positron (1+) give the atomic number of aluminum (13).

$$^{24}_{13}\text{Al} \longrightarrow {}^{24}_{12}\text{Mg} + {}^{0}_{+1}e$$

SAMPLE PROBLEM | 16.4

■ **Writing an Equation for Positron Emission**

Write the nuclear equation for manganese-49, which decays by emitting a positron.

SOLUTION

STEP 1 **Write the incomplete nuclear equation.**

$$^{49}_{25}\text{Mn} \longrightarrow ? + {}^{0}_{+1}e$$

STEP 2 **Determine the missing mass number.**   In the equation, the mass number of the manganese, 49, is equal to the combined mass numbers of the positron and the new nucleus.

$$49 = ? + 0$$
$$49 - 0 = ?$$
$$49 = ? \text{ (mass number of new nucleus)}$$

STEP 3 **Determine the missing atomic number.**   The atomic number of manganese, 25, must equal the sum of the atomic numbers of the positron and the new nucleus.

$$25 = ? + 1$$
$$25 - 1 = ?$$
$$24 = ? \text{ (atomic number of new nucleus)}$$

STEP 4 **Determine the symbol of the new nucleus.**   On the periodic table, the element that has atomic number 24 is chromium, Cr. The nucleus of this isotope of Cr is written as $^{49}_{24}\text{Cr}$.

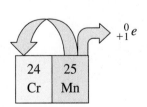

STEP 5 Complete the nuclear equation.

$$^{49}_{25}\text{Mn} \longrightarrow ^{49}_{24}\text{Cr} + ^{0}_{+1}e$$

STUDY CHECK

Write the nuclear equation for xenon-118, which decays by emitting a positron.

Gamma Emission

Pure gamma emitters are rare, although gamma radiation accompanies most alpha and beta radiation. In radiology, one of the most commonly used gamma emitters is technetium (Tc). Because the unstable isotope of technetium decays quickly, it is written as the *metastable* (symbol m) isotope: technetium-99m, Tc-99m, or $^{99m}_{43}$Tc. By emitting energy in the form of gamma rays, the nucleus becomes more stable.

$$^{99m}_{43}\text{Tc} \longrightarrow ^{99}_{43}\text{Tc} + ^{0}_{0}\gamma$$

Figure 16.2 summarizes the changes in the nucleus for alpha, beta, positron, and gamma radiation.

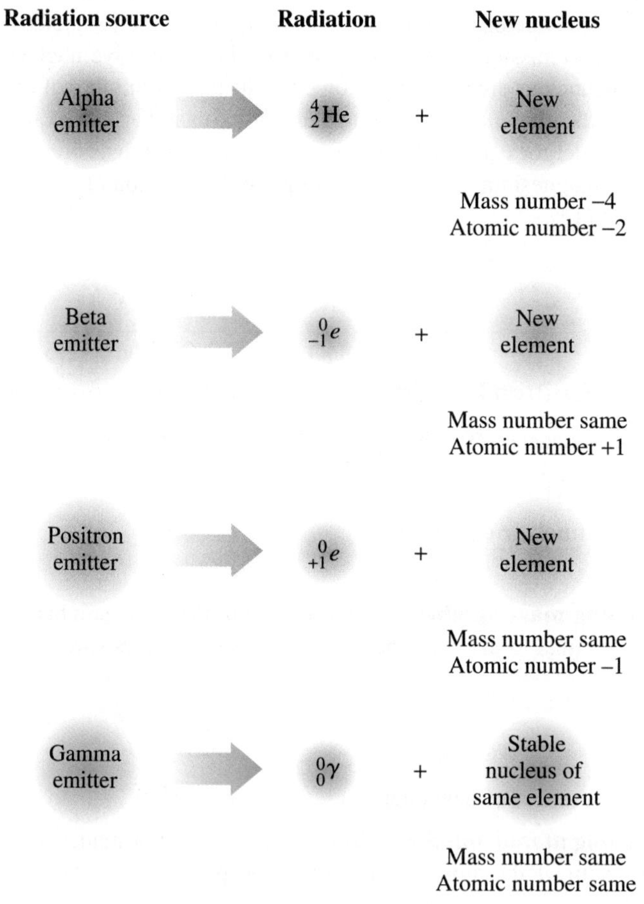

Radiation source	Radiation		New nucleus
Alpha emitter	$^{4}_{2}$He	+	New element
			Mass number −4 Atomic number −2
Beta emitter	$^{0}_{-1}e$	+	New element
			Mass number same Atomic number +1
Positron emitter	$^{0}_{+1}e$	+	New element
			Mass number same Atomic number −1
Gamma emitter	$^{0}_{0}\gamma$	+	Stable nucleus of same element
			Mass number same Atomic number same

FIGURE 16.2 When the nuclei of alpha, beta, positron, and gamma emitters emit radiation, new, more stable nuclei are produced.

Q What changes occur in the number of protons and neutrons when an alpha emitter gives off radiation?

Producing Radioactive Isotopes

Today many radioisotopes are produced in small amounts by converting stable, nonradioactive isotopes into radioactive ones. In a process called *transmutation*, a stable nucleus is bombarded by high-speed particles such as alpha particles, protons, neutrons, and small nuclei. When one of these particles is absorbed, the nucleus becomes a radioactive isotope.

When boron-10, a nonradioactive isotope, is bombarded by an alpha particle, it is converted to nitrogen-13, a radioisotope. In this bombardment reaction, a neutron is emitted.

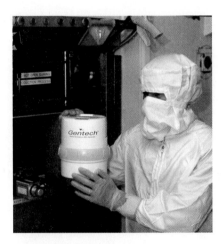

$$\underset{\text{Bombarding}}{\overset{4}{2}\text{He}} + \underset{\text{Stable}}{\overset{10}{5}\text{B}} \longrightarrow \underset{\text{New radioactive}}{\overset{13}{7}\text{N}} + \underset{\text{Neutron}}{\overset{1}{0}n}$$

^4_2He — Bombarding particle
$^{10}_5\text{B}$ — Stable nucleus
$^{13}_7\text{N}$ — New radioactive nucleus
^1_0n — Neutron

All elements that have atomic numbers greater than 92 have been produced by bombardment; none of these elements occurs naturally. Most have been produced in only small amounts and exist for such a short time that it is difficult to study their properties. An example is element 105, dubnium, which is produced when californium-249 is bombarded with nitrogen-15.

$$^{15}_7\text{N} + ^{249}_{98}\text{Cf} \longrightarrow ^{260}_{105}\text{Db} + 4^1_0n$$

Technetium-99m is a radioisotope used in nuclear medicine for several diagnostic procedures, including the detection of brain tumors and the examination of the liver and spleen. The source of technetium-99m is molybdenum-99, which is produced in a nuclear reactor by neutron bombardment of molybdenum-98.

$$^1_0n + ^{98}_{42}\text{Mo} \longrightarrow ^{99}_{42}\text{Mo}$$

Many radiology laboratories have small generators containing the radioactive molybdenum-99, which decays to give the technetium-99m radioisotope.

$$^{99}_{42}\text{Mo} \longrightarrow ^{99m}_{43}\text{Tc} + ^{0}_{-1}e$$

The technetium-99m radioisotope decays by emitting gamma rays. Gamma emission is desirable for diagnostic work because the gamma rays pass through the body to the detection equipment.

$$^{99m}_{43}\text{Tc} \longrightarrow ^{99}_{43}\text{Tc} + ^0_0\gamma$$

A generator is used to prepare technetium-99m.

CONCEPT CHECK 16.3

■ Writing an Isotope Produced by Bombardment

Sulfur-32 is bombarded with a neutron to produce a new isotope and an alpha particle.

$$^1_0n + ^{32}_{16}\text{S} \longrightarrow \text{?} + ^4_2\text{He}$$

What is the name of the new isotope?

ANSWER

To determine the name of the new isotope, we need to calculate its mass number and atomic number. On the left side, the sum of the mass numbers of sulfur, 32, and one neutron, 1, gives a total of 33. On the right side, we subtract the mass number of the alpha particle, 4, which gives a mass number of 29 to the new nucleus.

$$^1_0n + ^{32}_{16}\text{S} \longrightarrow ^{29}_{?}\text{?} + ^4_2\text{He}$$

On the left side, the sum of the atomic numbers of a neutron, 0, and sulfur, 16, gives a total of 16. On the right side, we subtract the atomic number of the alpha particle, 2, to give an atomic number of 14 to the new nucleus. The element with the atomic number of 14 is silicon.

$$^1_0n + ^{32}_{16}\text{S} \longrightarrow ^{29}_{14}\text{Si} + ^4_2\text{He}$$

Thus, the new isotope is named silicon-29.

SAMPLE PROBLEM | 16.5 |

■ **Completing a Nuclear Equation for a Bombardment Reaction**

Write the balanced nuclear equation for the bombardment of nickel-58 by a proton, $_1^1H$, which produces a radioactive isotope and an alpha particle.

SOLUTION

STEP 1 **Write the incomplete nuclear equation.**

$$_1^1H + _{28}^{58}Ni \longrightarrow ? + _2^4He$$

STEP 2 **Determine the missing mass number.** In the equation, the sum of the mass numbers of the proton, 1, and the nickel, 58, must equal the sum of the mass numbers of the alpha particle, 4, and the new nucleus.

$$1 + 58 = ? + 4$$
$$59 - 4 = ?$$
$$55 \quad = ? \text{ (mass number of new nucleus)}$$

STEP 3 **Determine the missing atomic number.** The sum of the atomic numbers of the proton, 1, and of the nickel, 28, must equal the sum of the atomic numbers of the alpha particle, 2, and the new nucleus.

$$1 + 28 = ? + 2$$
$$29 - 2 = ?$$
$$27 \quad = ? \text{ (atomic number of new nucleus)}$$

STEP 4 **Determine the symbol of the new nucleus.** On the periodic table, the element that has atomic number 27 is cobalt, Co. The nucleus of this isotope of Co is written as $_{27}^{55}Co$.

STEP 5 **Complete the nuclear equation.**

$$_1^1H + _{28}^{58}Ni \longrightarrow _{27}^{55}Co + _2^4He$$

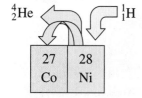

STUDY CHECK

Complete the following bombardment equation:

$$_2^4He + ? \longrightarrow _8^{17}O + _1^1H$$

QUESTIONS AND PROBLEMS

Nuclear Equations

16.13 Write a balanced nuclear equation for the alpha decay of each of the following radioactive isotopes:
a. $_{84}^{208}Po$ **b.** $_{90}^{232}Th$
c. $_{102}^{251}No$ **d.** radon-220

16.14 Write a balanced nuclear equation for the alpha decay of each of the following radioactive isotopes:
a. curium-243 **b.** $_{99}^{252}Es$
c. $_{98}^{251}Cf$ **d.** $_{107}^{261}Bh$

16.15 Write a balanced nuclear equation for the beta decay of each of the following radioactive isotopes:
a. $_{11}^{25}Na$ **b.** $_8^{20}O$
c. strontium-92 **d.** potassium-42

16.16 Write a balanced nuclear equation for the beta decay of each of the following radioactive isotopes:
a. $_{19}^{44}K$ **b.** iron-59
c. iron-60 **d.** $_{56}^{141}Ba$

16.17 Write a balanced nuclear equation for the positron emission of each of the following radioactive isotopes:
a. silicon-26 **b.** cobalt-54
c. $_{37}^{77}Rb$ **d.** $_{45}^{93}Rh$

16.18 Write a balanced nuclear equation for the positron emission of each of the following radioactive isotopes:
a. boron-8 **b.** $_7^{13}N$
c. $_{19}^{40}K$ **d.** xenon-118

16.19 Complete each of the following nuclear equations:
a. $_{13}^{28}Al \longrightarrow ? + _{-1}^0e$ **b.** $? \longrightarrow _{36}^{86}Kr + _0^1n$
c. $_{29}^{66}Cu \longrightarrow _{30}^{66}Zn + ?$ **d.** $? \longrightarrow _2^4He + _{90}^{234}Th$
e. $_{80}^{188}Hg \longrightarrow ? + _{+1}^0e$

16.20 Complete each of the following nuclear equations:
a. $_6^{11}C \longrightarrow _4^7Be + ?$ **b.** $_{16}^{35}S \longrightarrow ? + _{-1}^0e$
c. $_{39}^{90}Y \longrightarrow ? + _{-1}^0e$ **d.** $_{83}^{210}Bi \longrightarrow ? + _2^4He$
e. $? \longrightarrow _{59}^{135}Pr + _{-1}^0e$

16.21 Complete each of the following bombardment reactions:

a. $_0^1n + _4^9\text{Be} \longrightarrow$?

b. ? $+ _{16}^{32}\text{S} \longrightarrow _{15}^{32}\text{P}$

c. $_0^1n + $? $\longrightarrow _{11}^{24}\text{Na} + _2^4\text{He}$

d. $_2^4\text{He} + _{13}^{27}\text{Al} \longrightarrow$? $+ _0^1n$

16.22 Complete each of the following bombardment reactions:

a. ? $+ _{18}^{40}\text{Ar} \longrightarrow _{19}^{43}\text{K} + _1^1\text{H}$

b. $_0^1n + _{92}^{238}\text{U} \longrightarrow$?

c. $_0^1n + $? $\longrightarrow _6^{14}\text{C} + _1^1\text{H}$

d. ? $+ _{28}^{64}\text{Ni} \longrightarrow _{111}^{272}\text{Rg} + _0^1n$

16.3 Radiation Measurement

One of the most common instruments for detecting beta and gamma radiation is the Geiger counter. It consists of a metal tube filled with a gas such as argon. When radiation enters a window on the end of the tube, it produces ions in the gas; the ions produce an electrical current. Each burst of current is amplified to give a click and a reading on a meter.

$$\text{Ar} + \text{radiation} \longrightarrow \text{Ar}^+ + e^-$$

Radiation is measured in several different ways. We can measure the activity of a radioactive sample or determine the impact of radiation on biological tissue.

Measuring Radiation

When a radiology laboratory obtains a radioisotope, the activity of the sample is measured in terms of the number of nuclear disintegrations per second. The **curie (Ci)**, the original unit of activity, was defined as the number of disintegrations that occur in one second for one gram of radium, which is equal to 3.7×10^{10} disintegrations per second. The curie was named for the Polish scientist Marie Curie, who, along with her husband, Pierre, discovered the radioactive elements radium and polonium. The SI unit of radiation activity is the **becquerel (Bq)**, which is one disintegration per second.

The **rad (radiation absorbed dose)** is a unit that measures the amount of radiation absorbed by a gram of material such as body tissue. The SI unit for absorbed dose is the **gray (Gy)**, which is defined as the joules of energy absorbed by one kilogram of body tissue. The gray is equal to 100 rad.

The **rem (radiation equivalent in humans)** measures the biological effects of different kinds of radiation. Although alpha particles do not penetrate the skin, if they should enter the body by some other route, they can cause extensive damage within a short distance. High-energy radiation such as beta particles and high-energy protons and neutrons that penetrate the skin and travel into tissue cause more damage. Gamma rays are damaging because they travel a long way through body tissue.

To determine the **equivalent dose** or rem dose, the absorbed dose (rads) is multiplied by a factor that adjusts for biological damage caused by a particular form of radiation. For beta and gamma radiation the factor is 1, so the biological damage in rems is the same as the absorbed radiation (rads). For high-energy protons and neutrons, the factor is about 10, and for alpha particles it is 20.

Biological damage (rem) = Absorbed dose (rad) × Factor

Often, the measurement for an equivalent dose will be in units of millirems (mrem). One rem is equal to 1000 mrem. The SI unit is the **sievert (Sv)**. One sievert is equal to 100 rem.

People who work in radiology laboratories wear film badges to determine their exposure to radiation. A film badge consists of a piece of photographic film in a container that is attached to clothing. Periodically, the film badges are collected and developed to determine the level of exposure to radiation.

Table 16.5 summarizes the units used to measure radiation.

LEARNING GOAL

Describe the detection and measurement of radiation.

A radiation technician uses a Geiger counter to check radiation levels.

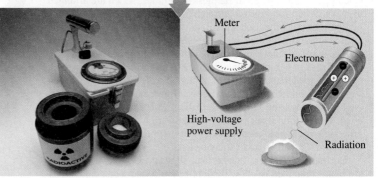

A film badge measures radiation exposure.

A film badge measures radiation exposure.

TABLE 16.5 Units of Radiation Measurement

Measurement	Common Unit	SI Unit	Relationship
Activity	curie (Ci) $1 \text{ Ci} = 3.7 \times 10^{10}$ disintegrations/s	becquerel (Bq) $1 \text{ Bq} = 1$ disintegration/s	$1 \text{ Ci} = 3.7 \times 10^{10} \text{ Bq}$
Absorbed dose	rad	gray (Gy) $1 \text{ Gy} = 1 \text{ J/kg of tissue}$	$1 \text{ Gy} = 100 \text{ rad}$
Biological damage	rem = rad × factor	sievert (Sv)	$1 \text{ Sv} = 100 \text{ rem}$

CHEMISTRY AND THE ENVIRONMENT

Radiation and Food

Food-borne illnesses caused by pathogenic bacteria such as *Salmonella*, *Listeria*, and *Escherichia coli* have become a major health concern in the United States. The Centers for Disease Control and Prevention estimates that each year *E. coli* in contaminated foods infects 20 000 people in the United States and that 500 people die. *E. coli* has been responsible for outbreaks of illness from contaminated ground beef, fruit juices, lettuce, and alfalfa sprouts.

The Food and Drug Administration (FDA) has approved the use of 0.3 kilogray (0.3 kGy) to 1 kGy of ionizing radiation produced by cobalt-60 or cesium-137 for the treatment of foods. The irradiation technology is much like that used to sterilize medical supplies. Cobalt pellets are placed in stainless steel tubes, which are arranged in racks. When food moves through the series of racks, the gamma rays pass through the food and kill the bacteria.

It is important for consumers to understand that when food is irradiated, it never comes in contact with the radioactive source. The gamma rays pass through the food to kill bacteria, but that does not make the food radioactive. The radiation kills bacteria because it stops their ability to divide and grow. We cook or heat food thoroughly for the same purpose. Radiation, as well as heat, has little effect on the food itself because its cells are no longer dividing or growing. Thus irradiated food is not harmed, although a small amount of vitamin B_1 and C may be lost.

Currently, tomatoes, blueberries, strawberries, and mushrooms are being irradiated to allow them to be harvested when completely ripe and extend their shelf life (see Figure 16.3). The FDA has also approved the irradiation of pork, poultry, and beef in order to decrease potential infections and to extend shelf life. Currently, irradiated vegetable and meat products are available in retail markets in

South Africa. *Apollo 17* astronauts ate irradiated foods on the moon, and some U.S. hospitals and nursing homes now use irradiated poultry to reduce the possibility of infections among residents. The extended shelf life of irradiated food also makes it useful for campers and military personnel. Soon consumers concerned about food safety will have a choice of irradiated meats, fruits, and vegetables at the market.

(a)

(b)

FIGURE 16.3 (a) The FDA requires this symbol to appear on irradiated retail foods. **(b)** After 2 weeks, the irradiated strawberries on the right show no spoilage. Mold is on the nonirradiated ones on the left.

Q Why are irradiated foods used on spaceships and in nursing homes?

CASE STUDY
Food Irradiation

CONCEPT CHECK 16.4

■ Radiation Measurement

One treatment of bone pain involves intravenous administration of the radioisotope phosphorus-32, which is incorporated into bone. A typical dose of 7 mCi can produce up to 450 rad in the bone. What is the difference between the units of mCi and rad?

ANSWER

The millicuries (mCi) indicate the activity of the P-32 in terms of nuclei that break down in 1 second. The radiation absorbed dose (rad) is a measure of amount of radiation absorbed by the bone.

SAMPLE PROBLEM 16.6

■ **Measuring Activity**

A sample of phosphorus-32, a beta emitter, has an activity of 2 millicuries (mCi). How many beta particles are emitted by this sample of P-32 in 1 s?

SOLUTION

STEP 1 Given activity = 2 mCi; time = 1 s
Need number of beta particles

STEP 2 Plan

mCi Metric factor Ci Curie factor β particles/s Time β particles

STEP 3 Equalities/Conversion Factors

$$1 \text{ Ci} = 1000 \text{ mCi}$$
$$\frac{1000 \text{ mCi}}{1 \text{ Ci}} \quad \text{and} \quad \frac{1 \text{ Ci}}{1000 \text{ mCi}}$$

$$1 \text{ Ci} = 3.7 \times 10^{10} \beta \text{ particles/s}$$
$$\frac{3.7 \times 10^{10} \beta \text{ particles/s}}{1 \text{ Ci}} \quad \text{and} \quad \frac{1 \text{ Ci}}{3.7 \times 10^{10} \beta \text{ particles/s}}$$

STEP 4 Set Up Problem We calculate the number of beta particles emitted in 1 s from the activity of the radioisotope.

$$2 \text{ mCi} \times \frac{1 \text{ Ci}}{1000 \text{ mCi}} \times \frac{3.7 \times 10^{10} \beta \text{ particles/s}}{1 \text{ Ci}} \times 1 \text{ s} = 7 \times 10^{7} \beta \text{ particles}$$

STUDY CHECK

An iodine-131 source has an activity of 0.25 Ci. How many radioactive atoms will disintegrate in 1.0 min?

Exposure to Radiation

Every day, we are exposed to low levels of radiation from naturally occurring radioactive isotopes in the buildings where we live and work, in our food and water, and in the air we breathe. For example, potassium-40, a naturally occurring radioactive isotope, is present in all potassium-containing food. Other naturally occurring radioisotopes in air and food are carbon-14, radon-222, strontium-90, and iodine-131. The average person in the United States is exposed to about 360 mrem of radiation annually. Table 16.6 lists some common sources of radiation.

Another source of background radiation is cosmic radiation produced in space by the Sun. People who live at high altitudes or travel by airplane receive a greater amount of cosmic radiation because there are fewer molecules in the atmosphere to absorb the radiation. For example, a person living in Denver receives about twice the cosmic radiation as a person living in Los Angeles. A person living close to a nuclear power plant normally does not receive much additional radiation, perhaps 0.1 mrem in one year. However, in the accident at the Chernobyl nuclear power plant in 1986 in Ukraine, it is estimated that people in a nearby town received as much as 1 rem/h.

Medical sources of radiation including dental, hip, spine and chest X-rays, and mammograms add to our radiation exposure.

Radiation Sickness

The larger the dose of radiation received at one time, the greater the effect on the body. Exposure to radiation less than 25 rem usually cannot be detected. Whole-body exposure of about 100 rem produces a temporary decrease in the number of white blood cells. If the

TABLE 16.6 Average Annual Radiation Received by a Person in the United States

Source	Dose (mrem)
Natural	
The ground	20
Air, water, food	30
Cosmic rays	40
Wood, concrete, brick	50
Medical	
Chest X-ray	20
Dental X-ray	20
Mammogram	40
Hip X-ray	60
Lumbar spine X-ray	70
Upper gastrointestinal tract X-ray	200
Other	
Television	20
Air travel	10
Radon	200[a]

[a]Varies widely.

TABLE 16.7 Lethal Doses of Radiation for Some Life-Forms

Life-Form	LD$_{50}$ (rem)
Insect	100 000
Bacterium	50 000
Rat	800
Human	500
Dog	300

exposure to radiation is greater than 100 rem, a person may suffer the symptoms of radiation sickness: nausea, vomiting, fatigue, and a reduction in white-cell count. A whole-body dosage greater than 300 rem can decrease the white-cell count to zero. The victim suffers diarrhea, hair loss, and infection. Exposure to radiation of about 500 rem is expected to cause death in 50% of the people receiving that dose. This amount of radiation to the whole body is called the *lethal dose for one half the population*, or the LD$_{50}$. The LD$_{50}$ varies for different life-forms, as Table 16.7 shows. Radiation dosages of 600 rem would be fatal to all humans within a few weeks.

QUESTIONS AND PROBLEMS

Radiation Measurement

16.23 a. How does a Geiger counter detect radiation?
   b. What SI unit and what common unit describe the activity of a radioactive sample?
   c. What SI unit and what common unit describe the radiation dose absorbed by tissue?
   d. What is meant by the term kilogray?

16.24 a. What is background radiation?
   b. What SI unit and what common unit describe the biological effect of radiation?
   c. What is meant by the term mrem?
   d. Why is a factor used to determine the equivalent dose?

16.25 The recommended dosage of iodine-131 is 4.20 μCi/kg of body mass. How many microcuries of iodine-131 are needed for a 70.0-kg person with hyperthyroidism?

16.26 a. The dosage of technetium-99m for a lung scan is 20. μCi/kg of body mass. How many millicuries of technetium-99m should be given to a 50.0-kg person (1 mCi = 1000 μCi)?
   b. A person receives 50 mrad of gamma radiation. What is that amount in grays? What would be the equivalent dose in mrem?
   c. Suppose a person absorbed 50 mrad of alpha radiation. What would be the equivalent dose in mrem? How does it compare with the mrem in part **b**?

16.27 Why would an airline pilot be exposed to more background radiation than the person who works at the ticket counter?

16.28 In radiation therapy, a person receives high doses of radiation. What symptoms of radiation sickness may occur?

16.4 Half-Life of a Radioisotope

LEARNING GOAL

Given the half-life of a radioisotope, calculate the amount of radioisotope remaining after one or more half-lives.

SELF STUDY ACTIVITY

Nuclear Chemistry

The **half-life** of a radioisotope is the amount of time it takes for one half of the atoms in a sample to decay. For example, I-131 has a half-life of 8 d. As I-131 decays, it produces the nonradioactive isotope Xe-131 and a beta particle:

$$^{131}_{53}\text{I} \longrightarrow {}^{131}_{54}\text{Xe} + {}^{0}_{-1}e$$

Suppose we have a sample that initially contains 20. mg of I-131. In 8 days, one half (10. mg) of all the I-131 nuclei will decay, which leaves 10. mg of I-131. After 16 days (two half-lives), 5.0 mg of the remaining I-131 decays, which leaves 5.0 mg of I-131. After 24 days, (three half-lives), 2.5 mg of the remaining I-131 decays, which leaves 2.5 mg of I-131.

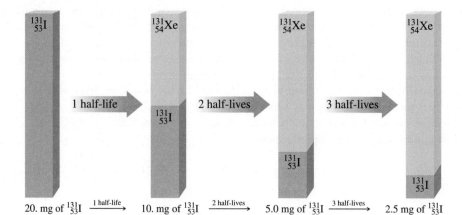

A **decay curve** is a diagram of the decay of a radioactive isotope. Figure 16.4 shows such a curve for the $^{131}_{53}$I we have discussed.

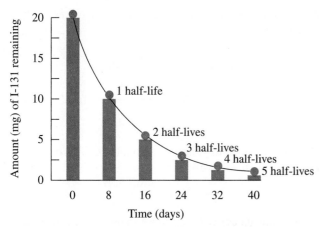

FIGURE 16.4 The decay curve for iodine-131 shows that one-half of the radioactive sample decays and one-half remains radioactive after each half-life of 8 d.

Q How many milligrams of the 20.-mg sample remain radioactive after 2 half-lives?

CONCEPT CHECK 16.5

▪ Half-Lives

Iridium-192, used to treat breast cancer, has a half-life of 74 d. What is the activity of the Ir-192 after 74 d if the activity of the initial sample of Ir-192 is 8×10^4 Bq?

ANSWER

In 74 d, which is one half-life of iridium-192, one-half of all of the iridium-192 atoms will decay. Thus, after 74 days, the activity is half of the initial activity, which is 4×10^4 Bq.

SAMPLE PROBLEM 16.7

▪ Using Half-Lives of a Radioisotope

Phosphorus-32, a radioisotope used in the treatment of leukemia, has a half-life of 14.3 d. If a treatment dose contains 8.0 mCi of phosphorus-32, what is the activity, in mCi, after 42.9 days?

SOLUTION

STEP 1 State the given and needed amounts of radioisotope.
    Given   8.0 mCi of P-32; 42.9 d; 14.3 d/half-life
    Need   mCi of P-32 remaining

STEP 2 Write a plan to calculate amount of active radioisotope.

  number of days   Half-life   number of half-lives

  mCi of P-32   Number of half-lives   mCi of P-32 remaining

STEP 3 Write the half-life equality and conversion factors.

$$1 \text{ half-life} = 14.3 \text{ d}$$

$$\frac{14.3 \text{ d}}{1 \text{ half-life}} \quad \text{and} \quad \frac{1 \text{ half-life}}{14.3 \text{ d}}$$

STEP 4 Set up problem to calculate amount of active radioisotope. We can do this problem with two calculations. First we determine the number of half-lives in the amount of time that has elapsed.

$$\text{number of half-lives} = 42.9 \cancel{\text{ d}} \times \frac{1 \text{ half-life}}{14.3 \cancel{\text{ d}}} = 3.00 \text{ half-lives}$$

TUTORIAL
Radioactive Half-Lives

Guide to Using Half-Lives

STEP 1
State the given and needed amounts of radioisotope.

STEP 2
Write a plan to calculate amount of active radioisotope.

STEP 3
Write the half-life equality and conversion factors.

STEP 4
Set up problem to calculate amount of active radioisotope.

Now we can determine how much of the sample decays in 3.00 half-lives and how many grams of the phosphorus remain.

$$8.0 \text{ mCi of P-32} \xrightarrow{\text{1 half-life}} 4.0 \text{ mCi of P-32} \xrightarrow{\text{2 half-lives}}$$

$$2.0 \text{ mCi of P-32} \xrightarrow{\text{3 half-lives}} 1.0 \text{ mCi of P-32}$$

STUDY CHECK

Iron-59 has a half-life of 44 d. If a nuclear laboratory receives a sample of 32 μg of iron-59, how many micrograms of iron-59 remain after 176 d?

Naturally occurring isotopes of the elements are typically more stable and therefore usually have long half-lives, as shown in Table 16.8. They disintegrate slowly and produce radiation over a long period of time, even hundreds of millions of years. In contrast, many of the radioisotopes used in nuclear medicine are extremely unstable and have much shorter half-lives. They disintegrate rapidly and produce almost all their radiation in a short period of time. For example, technetium-99m emits half of its radiation in the first six hours. This means that the radioisotope is essentially gone within two days. The decay products of technetium-99m are totally eliminated by the body.

TABLE 16.8 Half-Lives of Some Radioisotopes

Element	Radioisotope	Half-Life
Naturally Occurring Radioisotopes		
Carbon	$^{14}_{6}\text{C}$	5730 y
Potassium	$^{40}_{19}\text{K}$	1.3×10^9 y
Radium	$^{226}_{88}\text{Ra}$	1600 y
Uranium	$^{238}_{92}\text{U}$	4.5×10^9 y
Some Medical Radioisotopes		
Chromium	$^{51}_{24}\text{Cr}$	28 d
Iodine	$^{131}_{53}\text{I}$	8 d
Iridium	$^{192}_{77}\text{Ir}$	74 d
Iron	$^{59}_{26}\text{Fe}$	44 d
Technetium	$^{99m}_{43}\text{Tc}$	6.0 h

TUTORIAL

Radioactive Half-Lives

SAMPLE PROBLEM 16.8

■ **Dating Using Half-Lives**

In Los Angeles, the remains of ancient animals have been unearthed at the La Brea Tar Pits. Suppose a bone sample from the tar pits is subjected to the carbon-14 dating method. How long ago did the animal live if the sample shows that two half-lives have passed?

SOLUTION

We can calculate the age of the bone sample by using the half-life of carbon-14 (5730 y).

$$2 \text{ half-lives} \times \frac{5730 \text{ y}}{1 \text{ half-life}} = 11\,500 \text{ y}$$

We would estimate that the animal lived 11 500 y ago, or about 9500 B.C.E.

STUDY CHECK

Suppose that a piece of wood found in a tomb had $\frac{1}{8}$ of its original carbon-14 activity. About how many years ago was the wood part of a living tree?

CHEMISTRY AND THE ENVIRONMENT

Dating Ancient Objects

Radiological dating is a technique used by geologists, archaeologists, and historians to determine the age of ancient objects. The age of an object derived from plants or animals (such as wood, fiber, natural pigments, bone, and cotton or woolen clothing) is determined by measuring the amount of carbon-14, a naturally occurring radioactive form of carbon. In 1960, Willard Libby received the Nobel Prize for the work he did developing carbon-14 dating techniques during the 1940s. Carbon-14 is produced in the upper atmosphere by the bombardment of $^{14}_{7}N$ by high-energy neutrons from cosmic rays.

$$^{1}_{0}n \quad + \quad ^{14}_{7}N \quad \longrightarrow \quad ^{14}_{6}C \quad + \quad ^{1}_{1}H$$

Neutron from cosmic rays   Nitrogen in atmosphere   Radioactive carbon-14   Proton

The carbon-14 reacts with oxygen to form radioactive carbon dioxide, $^{14}_{6}CO_2$. Living plants continuously absorb carbon dioxide, which incorporates carbon-14 into the plant material. The uptake of carbon-14 stops when the plant dies.

As the carbon-14 undergoes beta decay, the amount of radioactive carbon-14 in the plant material steadily decreases.

$$^{14}_{6}C \longrightarrow ^{14}_{7}N + ^{0}_{-1}e$$

In a process called *carbon dating*, scientists use the half-life of carbon-14 (5730 y) to calculate the length of time since the plant died. As the plant material ages, the amount of radioactive carbon-14 that remains is less than in the original living plant. For example, a wooden beam found in an ancient Indian dwelling might have one-half of the carbon-14 found in living plants today. Because one half-life of carbon-14 is 5730 y, the dwelling was constructed about 5730 y ago. Carbon-14 dating was used to determine that the Dead Sea Scrolls are about 2000 y old.

A radiological dating method used for determining the age of much older items is based on the radioisotope uranium-238, which decays through a series of reactions to lead-206. The uranium-238 isotope has an incredibly long half-life, about 4×10^9 (4 billion) y. Measurements of the amounts of uranium-238 and lead-206 enable geologists to determine the age of rock samples. The older rocks will have a higher percentage of lead-206 because more of the uranium-238 has decayed. The age of rocks brought back from the moon by the *Apollo* missions, for example, was determined using uranium-238. They were found to be about 4×10^9 y old, approximately the same age calculated for Earth.

The age of the Dead Sea scrolls was determined using carbon-14.

QUESTIONS AND PROBLEMS

Half-Life of a Radioisotope

16.29 What is meant by the term half-life?

16.30 Why are radioisotopes with short half-lives used for diagnosis in nuclear medicine?

16.31 Technetium-99m is an ideal radioisotope for scanning organs because it has a half-life of 6.0 h and is a pure gamma emitter. Suppose that 80.0 mg were prepared in the technetium generator this morning. How many milligrams of technetium-99m would remain after the following intervals?
 a. one half-life   **b.** two half-lives
 c. 18 h   **d.** 24 h

16.32 A sample of sodium-24 with an activity of 12 mCi is used to study the rate of blood flow in the circulatory system. If sodium-24 has a half-life of 15 h, what is the activity of the sodium after 2.5 days?

16.33 Strontium-85, used for bone scans, has a half-life of 65 days. How long will it take for the radiation level of strontium-85 to drop to one-fourth of its original level?

16.34 Fluorine-18, which has a half-life of 110 min, is used in PET scans (see Section 16.5). If 100 mg of fluorine-18 is shipped at 8:00 A.M., how many milligrams of the radioisotope are still active if the sample arrives at the radiology laboratory at 1:30 P.M.?

16.5 Medical Applications Using Radioactivity

To determine the condition of an organ in the body, a radiologist may use a radioisotope that concentrates in that organ. The cells in the body do not differentiate between a nonradioactive atom and a radioactive one. However, radioactive atoms can be detected because they emit radiation. Some radioisotopes used in nuclear medicine are listed in Table 16.9.

LEARNING GOAL

Describe the use of radioisotopes in medicine.

TABLE 16.9 Medical Applications of Radioisotopes

Isotope	Half-Life	Medical Application
Au-198	2.7 d	Liver imaging; treatment of abdominal carcinoma
Ce-141	32.5 d	Gastrointestinal tract diagnosis; measuring blood flow to the heart
Cs-131	10 d	Brachytherapy
Ga-67	78 h	Abdominal imaging; tumor detection
Ga-68	68 min	Detection of pancreatic cancer
I-125	60 d	Treatment of brain and prostate cancer
I-131	8 d	Imaging of thyroid; treatment of Graves' disease, goiter, and hyperthyroidism; treatment of thyroid and prostate cancer
Ir-192	74 d	Treatment of breast and prostate cancer
P-32	14.3 d	Treatment of leukemia, excess red blood cells, pancreatic cancer
Pd-103	17 d	Brachytherapy
Sr-85	65 d	Detection of bone lesions; brain scans
Tc-99m	6 h	Imaging of skeleton and heart muscle, brain, liver, heart, lungs, bone, spleen, kidney, and thyroid; most widely used radioisotope in nuclear medicine

Scans with Radioisotopes

After a person receives a radioisotope, the radiologist determines the level and location of radioactivity emitted by the radioisotope. An apparatus called a scanner is used to produce an image of the organ. The scanner moves slowly across the body above the region where the organ containing the radioisotope is located. The gamma rays emitted from the radioisotope in the organ can be used to expose a photographic plate, producing a scan of the organ. On a scan, an area of decreased or increased radiation can indicate conditions such as a disease of the organ, a tumor, a blood clot, or edema.

A common method of determining thyroid function is the use of *radioactive iodine uptake* (RAIU). Taken orally, the radioisotope iodine-131 mixes with the iodine already present in the body. Twenty-four hours later, the amount of iodine taken up by the thyroid is determined. A detection tube held up to the area of the thyroid gland detects the radiation coming from the iodine-131 that has located there (see Figure 16.5).

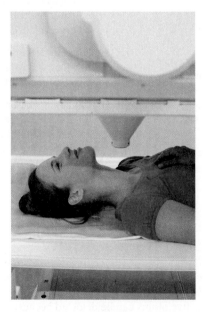

(b)

FIGURE 16.5 **(a)** A scanner is used to detect radiation from a radioisotope that has accumulated in an organ. **(b)** A scan of the thyroid shows the accumulation of radioactive iodine-131 in the thyroid.

Q What type of radiation would move through body tissues to create a scan?

(a)

A person with a hyperactive thyroid will have a higher than normal level of radioactive iodine, whereas a person with a hypoactive thyroid will record lower values. If a person has hyperthyroidism, treatment is begun to lower the activity of the thyroid. One treatment involves giving a therapeutic dosage of radioactive iodine, which has a higher radiation level than the diagnostic dose. The radioactive iodine goes to the thyroid where its radiation destroys some of the thyroid cells. The thyroid produces less thyroid hormone, bringing the hyperthyroid condition under control.

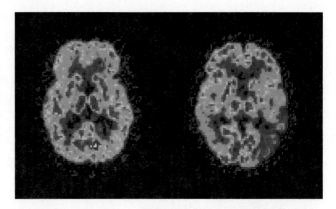

Positron Emission Tomography (PET)

Positron emitters with short half-lives such as carbon-11, oxygen-15, nitrogen-13, and fluorine-18 are used in an imaging method called *positron emission tomography* (PET). A positron-emitting isotope such as fluorine-18 combined with substances in the body such as glucose is used to study brain function, metabolism, and blood flow.

$$^{18}_{9}F \longrightarrow {}^{18}_{8}O + {}^{0}_{+1}e$$

As positrons are emitted, they combine with electrons to produce gamma rays that are detected by computerized equipment to create a three-dimensional image of the organ (see Figure 16.6).

FIGURE 16.6 These PET scans of the brain show a normal brain on the left and a brain affected by Alzheimer's disease on the right.

Q When positrons collide with electrons, what type of radiation is produced that gives an image of an organ?

SAMPLE PROBLEM $\boxed{16.9}$

■ Medical Application of Radioactivity

In the treatment of abdominal carcinoma, a person is treated with seeds of gold-198, which is a beta emitter. Write the nuclear equation for the beta decay of gold-198.

SOLUTION

We can write the incomplete nuclear equation starting with gold-198.

$$^{198}_{79}Au \longrightarrow ? + {}^{0}_{-1}e$$

In beta decay, the mass number, 198, does not change, but the atomic number of the new nucleus increases by one. The new atomic number is 80, which is mercury, Hg.

$$^{198}_{79}Au \longrightarrow {}^{198}_{80}Hg + {}^{0}_{-1}e$$

STUDY CHECK

In an experimental treatment, a person is given boron-10, which is taken up by malignant tumors. When bombarded with neutrons, boron-10 decays by emitting alpha particles that destroy the surrounding tumor cells. Write the nuclear equation for this experimental procedure.

Computed Tomography (CT)

Another imaging method used to detect changes within the body is computed tomography (CT). A computer monitors the degree of absorption of 30 000 X-ray beams directed at the brain at successive layers. The differences in absorption based upon the densities of the tissues and fluids in the brain provide a series of images of the brain. This technique is successful in the identification of brain hemorrhages, tumors, and atrophy.

Magnetic Resonance Imaging (MRI)

Magnetic resonance imaging (MRI) is a powerful imaging technique that does not involve radiation. It is the least invasive imaging method available. MRI is based on the absorption of energy when the protons in hydrogen atoms are excited by a strong magnetic field.

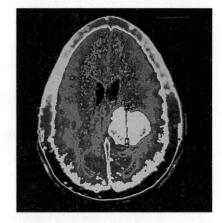

A CT scan shows a brain tumor (yellow) on the right side of the brain.

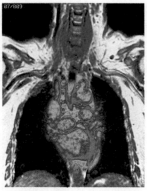

An MRI scan provides images of the heart and lungs.

Hydrogen atoms make up 63 percent of all the atoms in the body. In the hydrogen nuclei, the protons act like tiny bar magnets. With no external field, the protons have random orientations. However, when placed within a large magnet, the protons align with the magnetic field. A proton aligned with the field has a lower energy than one that is aligned against the field. As the MRI scan proceeds, radiofrequency pulses of energy are applied. When a nucleus absorbs certain energy, its proton "flips" and becomes aligned against the field. Because hydrogen atoms in the body are in different chemical environments, energies of different frequencies are absorbed. The energies absorbed are calculated and converted to color images of the body. MRI is particularly useful in obtaining images of soft tissues because these tissues contain large amounts of water.

CHEMISTRY AND HEALTH

Brachytherapy

The process called *brachytherapy*, or seed implantation, is an internal form of radiation therapy. The prefix *brachy* is from the Greek word for short distance. With internal radiation, a high dose of radiation is delivered to a cancerous area, while normal tissue sustains minimal damage. Because higher doses are used, fewer treatments of shorter duration are needed. Conventional external treatment delivers a lower dose per treatment, but requires six to eight weeks of treatments.

Permanent Brachytherapy

One of the most common forms of cancer in males is prostate cancer. In addition to surgery and chemotherapy, one treatment option is to place 40 or more titanium capsules, or "seeds", in the malignant area. Each seed, which is the size of a small grain of rice, contains radioactive iodine-125, palladium-103, or cesium-131, which decays by gamma emission. The radiation from the seeds destroys the cancer by interfering with the reproduction of cancer cells. Because the radiation targets the cancer cells, there is minimal damage to normal tissues. Ninety percent (90%) of the radioisotopes decay within a few months because they have short half-lives.

Isotope	I-125	Pd-103	Cs-131
Half-life	60 d	17 d	10 d
Time required to deliver 90% of radiation	7 months	2 months	1 month

Almost no radiation passes out of the patient's body. The amount of radiation received by a family member is no greater than that received on a long plane flight. The titanium capsules are left in the body permanently, but the products of decay are not radioactive and cause no further damage.

Temporary Brachytherapy

In another type of treatment for prostate cancer, long needles containing iridium-192 are placed in the tumor. However, the needles are removed after 5 to 10 min, depending on the activity of the iridium isotope. Compared to permanent brachytherapy, temporary brachytherapy can deliver a higher dose of radiation over a shorter time. The procedure may be repeated in a few days.

Brachytherapy is also used following breast cancer lumpectomy. An iridium-192 isotope is inserted into the catheter implanted in the space left by the removal of the tumor. The isotope is removed after 5 to 10 min, depending on the activity of the iridium source. Radiation is delivered primarily to the tissue surrounding the cavity that contained the tumor and where the cancer is most likely to reoccur. The procedure is repeated twice a day for five days to give an absorbed dose of 34 Gy (3400 rads). The catheter is removed, and no radioactive material remains in the body.

In conventional external beam therapy for breast cancer, a patient receives 2 Gy/treatment for seven weeks, which gives a total absorbed dose of about 100 Gy or 10 000 rads. The external beam therapy irradiates the entire breast including the tumor cavity.

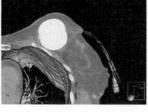

A catheter is placed temporarily in the breast for radiation from Ir-192.

QUESTIONS AND PROBLEMS

Medical Applications Using Radioactivity

16.35 Bone and bony structures contain calcium and phosphorus.
    a. Why would the radioisotopes calcium-47 and phosphorus-32 be used in the diagnosis and treatment of bone diseases?
    b. During nuclear tests, scientists were concerned that strontium-85, a radioactive product, would be harmful to the growth of bone in children. Explain.

16.36 **a.** Technetium-99m emits only gamma radiation. Why would this type of radiation be used in diagnostic imaging rather than an isotope that also emits beta or alpha radiation?
    b. A person with *polycythemia vera* (excess production of red blood cells) receives radioactive phosphorus-32. Why would this treatment reduce the production of red blood cells in the bone marrow?

16.37 In a diagnostic test for leukemia, a person receives 4.0 mL of a solution containing selenium-75. If the activity of the selenium-75 is 45 μCi/mL, what dose, in μCi, is given?

16.38 A vial contains radioactive iodine-131 with an activity of 2.0 mCi/mL. If the thyroid test requires 3.0 mCi in an "atomic cocktail," how many milliliters are used to prepare the iodine-131 solution?

16.6 Nuclear Fission and Fusion

LEARNING GOAL

Describe the processes of nuclear fission and fusion.

During the 1930s, scientists bombarding uranium-235 with neutrons discovered that the U-235 nucleus splits into two smaller nuclei and produces a great amount of energy. This was the discovery of nuclear **fission**. The energy generated by splitting the atom was called atomic energy. A typical equation for nuclear fission is

$$\, _{0}^{1}n \;+\; _{92}^{235}U \;\longrightarrow\; _{92}^{236}U \;\longrightarrow\; _{36}^{91}Kr \;+\; _{56}^{142}Ba \;+\; 3\,_{0}^{1}n \;+\; \text{energy}$$

If we could determine the masses of the initial reactants and the products with great accuracy, we would find that the total mass of the products is slightly less than the total mass of the starting materials. The missing mass has been converted into energy, consistent with the famous equation derived by Albert Einstein:

$$E = mc^2$$

where E is the energy released, m is the mass lost, and c is the speed of light, 3×10^8 m/s. Even though the mass loss is very small, when it is multiplied by the speed of light squared the result is a large value for the energy released. The fission of one gram of uranium-235 produces about as much energy as the burning of three tons of coal.

Chain Reaction

Fission begins when a neutron collides with the nucleus of a uranium atom. The resulting nucleus is unstable and splits into smaller nuclei. This fission process also releases several neutrons and large amounts of gamma radiation and energy. The neutrons emitted have high energies and bombard more uranium-235 nuclei. As fission continues, there is a rapid increase in the number of high-energy neutrons capable of splitting more uranium atoms, a process called a **chain reaction**. To sustain a nuclear chain reaction, sufficient quantities of uranium-235 must be brought together to provide a critical mass in which almost all the neutrons immediately collide with more uranium-235 nuclei. So much heat and energy build up that an atomic explosion can occur (see Figure 16.7).

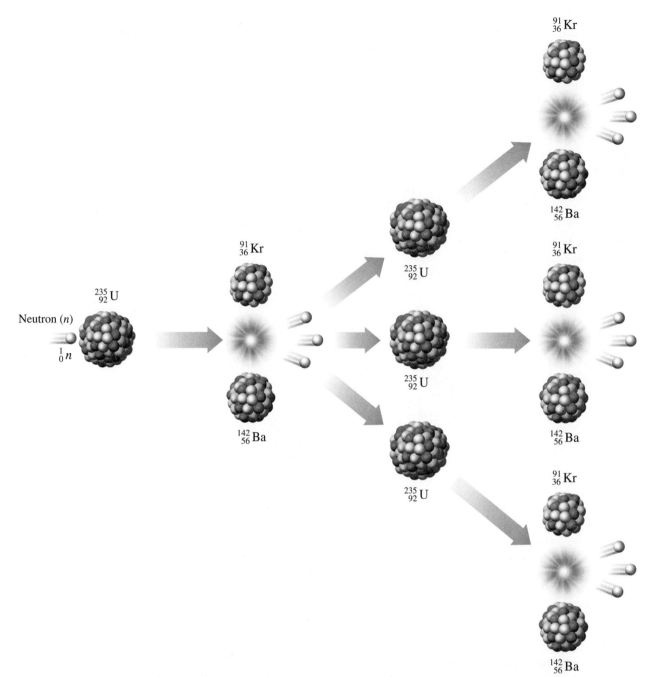

FIGURE 16.7 In a nuclear chain reaction, the fission of each uranium-235 atom produces three neutrons that cause the nuclear fission of more and more uranium-235 atoms.

Q Why is the fission of uranium-235 called a chain reaction?

Nuclear Fusion

In **fusion**, two small nuclei combine to form a larger nucleus. Mass is lost, and a tremendous amount of energy is released, even more than the energy released from nuclear fission. However, a fusion reaction requires a temperature of 100 000 000 °C to overcome the repulsion of the hydrogen nuclei and cause them to undergo fusion. Fusion reactions occur continuously in the Sun and other stars, providing us with heat and light. The huge amounts of energy produced by our Sun come from the fusion of 6×10^{11} kilograms of hydrogen every second. In a fusion reaction, isotopes of hydrogen combine to form helium and large amounts of energy.

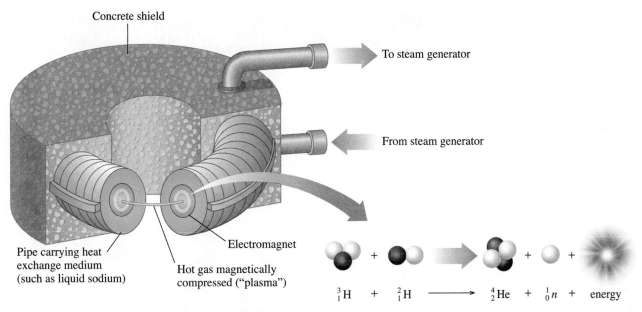

In a fusion reactor, high temperatures are needed to combine hydrogen atoms.

Scientists expect less radioactive waste from fusion reactors. However, fusion is still in the experimental stage because the extremely high temperatures needed have been difficult to reach and even more difficult to maintain. Research groups around the world are attempting to develop the technology needed to make the harnessing of the fusion reaction for energy a reality in our lifetime.

CONCEPT CHECK 16.6

■ Identifying Fission and Fusion

Classify the following as pertaining to fission, fusion, or both:

a. A large nucleus breaks apart to produce smaller nuclei.
b. Large amounts of energy are released.
c. Very high temperatures are needed for reaction.
d. $^{3}_{1}H + ^{2}_{1}H \longrightarrow ^{4}_{2}He + ^{1}_{0}n +$ energy

ANSWER
a. When a large nucleus breaks apart to produce smaller nuclei, the process is fission.
b. Large amounts of energy are generated in both the fusion and fission processes.
c. An extremely high temperature is required for fusion.
d. When small nuclei combine to release energy, the process is fusion.

QUESTIONS AND PROBLEMS

Nuclear Fission and Fusion

16.39 What is nuclear fission?

16.40 How does a chain reaction occur in nuclear fission?

16.41 Complete the following fission reaction:

$$^{1}_{0}n + ^{235}_{92}U \longrightarrow ^{131}_{50}Sn + ? + 2^{1}_{0}n + \text{energy}$$

16.42 In another fission reaction, uranium-235 bombarded with a neutron produces strontium-94, another small nucleus, and three neutrons. Write the complete equation for the fission reaction.

16.43 Indicate whether each of the following is characteristic of the fission or fusion process or both:

a. Neutrons bombard a nucleus.
b. The nuclear process occurring in the Sun.
c. A large nucleus splits into smaller nuclei.
d. Small nuclei combine to form larger nuclei.

16.44 Indicate whether each of the following is characteristic of the fission or fusion process or both:
a. Very high temperatures are required to initiate the reaction.
b. Less radioactive waste is produced.
c. Hydrogen nuclei are the reactants.
d. Large amounts of energy are released when the nuclear reaction occurs.

CHEMISTRY AND THE ENVIRONMENT

Nuclear Power Plants

In a nuclear power plant, the quantity of uranium-235 is held below a critical mass, so it cannot sustain a chain reaction. The fission reactions are slowed by placing control rods, which absorb some of the fast-moving neutrons, among the uranium samples. In this way, less fission occurs, and there is a slower, controlled production of energy. The heat from the controlled fission is used to produce steam. The steam drives a generator, which produces electricity. Approximately 10% of the electrical energy produced in the United States is generated in nuclear power plants.

Although nuclear power plants help meet some of our energy needs, there are some problems. One of the most serious problems is the production of radioactive by-products that have very long half-lives. It is essential that these waste products be stored safely for a very long time in a place where they do not contaminate the environment.

Early in 1990, the Environmental Protection Agency gave its approval for the storage of radioactive hazardous wastes in chambers 2150 ft underground. In 1998, the Waste Isolation Pilot Plant (WIPP) repository site in New Mexico was ready to receive plutonium waste from former U.S. bomb factories. Although authorities have determined the caverns are safe, some people are concerned with the safe transport of the radioactive waste by trucks on the highways.

Nuclear power plants supply about 10% of the electricity in the United States.

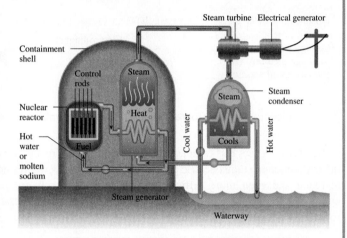

Heat from nuclear fission is used to generate electricity.

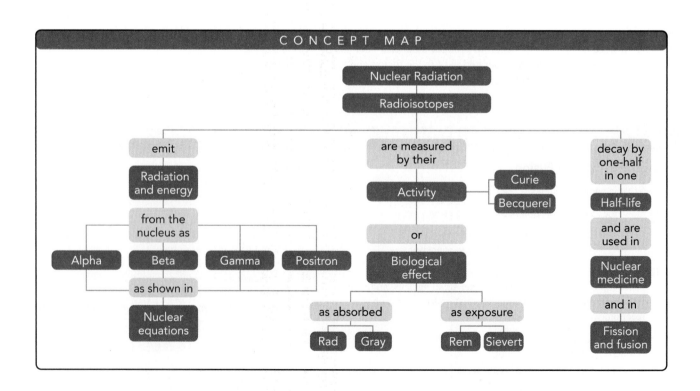

CONCEPT MAP

CHAPTER REVIEW

16.1 Natural Radioactivity

LEARNING GOAL: *Describe alpha, beta, positron, and gamma radiation.*

Radioactive isotopes have unstable nuclei that break down (decay), spontaneously emitting alpha (α), beta (β), positron (β^+), or gamma (γ) radiation. Because radiation can damage the cells in the body, proper protection must be used: shielding, limiting the time of exposure, and distance.

16.2 Nuclear Equations

LEARNING GOAL: *Write an equation showing mass numbers and atomic numbers for radioactive decay.*

A balanced equation is used to represent the changes that take place in the nuclei of the reactants and products. The new isotopes and the type of radiation emitted can be determined from the symbols that show the mass numbers and atomic numbers of the isotopes in the nuclear equation. A radioisotope is produced artificially when a nonradioactive isotope is bombarded by a small particle. Many radioactive isotopes used in nuclear medicine are produced in this way.

16.3 Radiation Measurement

LEARNING GOAL: *Describe the detection and measurement of radiation.*

In a Geiger counter, radiation produces charged particles in the gas contained in a tube, which produces an electrical current. The curie (Ci) and the becquerel (Bq) measure the activity, which is number of nuclear transformations per second. The amount of radiation absorbed by a substance is measured in rads or the gray (Gy). The rem and the sievert

(Sv) are units used to determine the biological damage from the different types of radiation.

16.4 Half-Life of a Radioisotope

LEARNING GOAL: *Given the half-life of a radioisotope, calculate the amount of radioisotope remaining after one or more half-lives.*

Every radioisotope has its own rate of emitting radiation. The time it takes for one-half of a radioactive sample to decay is called its half-life. For many medical radioisotopes, such as Tc-99m and I-131, half-lives are short. For other naturally occurring isotopes such as C-14, Ra-226, and U-238, half-lives are extremely long.

16.5 Medical Applications Using Radioactivity

LEARNING GOAL: *Describe the use of radioisotopes in medicine.*

In nuclear medicine, radioisotopes are administered that go to specific sites in the body. By detecting the radiation they emit, an evaluation can be made about the location and extent of an injury, disease, tumor, or the level of function of a particular organ. Higher levels of radiation are used to treat or destroy tumors.

16.6 Nuclear Fission and Fusion

LEARNING GOAL: *Describe the processes of nuclear fission and fusion.*

In fission, the bombardment of a large nucleus breaks it into smaller nuclei, releasing one or more types of radiation and a great amount of energy. In fusion, small nuclei combine to form larger nuclei releasing great amounts of energy.

KEY TERMS

alpha particle A nuclear particle identical to a helium (^4_2He, α) nucleus (two protons and two neutrons).

becquerel (Bq) A unit of activity of a radioactive sample equal to one disintegration per second.

beta particle A particle identical to an electron ($^{\,\,0}_{-1}e$, β) that forms in the nucleus when a neutron changes to a proton and an electron.

chain reaction A fission reaction that will continue once it has been initiated by a high-energy neutron bombarding a heavy nucleus such as uranium-235.

curie (Ci) A unit of activity of a radioactive sample equal to 3.7×10^{10} disintegrations/s.

decay curve A diagram of the decay of a radioactive element.

equivalent dose The measure of biological damage from an absorbed dose that has been adjusted for the type of radiation.

fission A process in which large nuclei are split into smaller pieces, releasing large amounts of energy.

fusion A reaction in which large amounts of energy are released when small nuclei combine to form larger nuclei.

gamma ray High-energy radiation (symbol $^0_0\gamma$) emitted to make a nucleus more stable.

gray (Gy) A unit of absorbed dose equal to 100 rad.

half-life The length of time it takes for one-half of a radioactive sample to decay.

positron A particle of radiation with no mass and a positive charge ($^{\,\,0}_{+1}e$) produced by an unstable nucleus when a proton is transformed into a neutron and a positron.

rad (radiation absorbed dose) A measure of an amount of radiation absorbed by the body.

radiation Energy or particles released by radioactive atoms.

radioactive decay The process by which an unstable nucleus breaks down with the release of high-energy radiation.

radioisotope A radioactive atom of an element.

rem (radiation equivalent in humans) A measure of the biological damage caused by the various kinds of radiation (rad $\times$ radiation biological factor).

sievert (Sv) A unit of biological damage (equivalent dose) equal to 100 rem.

UNDERSTANDING THE CONCEPTS

16.45 Consider the following nucleus of a radioactive isotope:

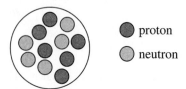

● proton
○ neutron

a. What is the nuclear symbol for this isotope?
b. Draw the resulting nucleus if this isotope emits a positron.

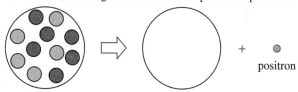

+ ● positron

16.46 Draw the nucleus that emits a beta particle to complete the following:

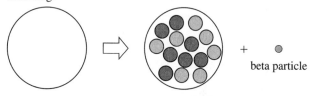

+ ● beta particle

16.47 Draw the nucleus of the atom to complete the following:

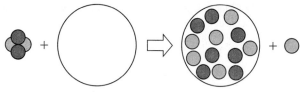

16.48 Complete the following by drawing the nucleus of the atom produced:

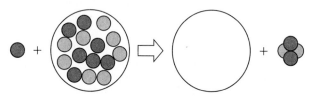

16.49 Carbon dating of small bits of charcoal used in cave paintings has determined that some of the paintings are from 10 000 to 30 000 y old. Carbon-14 has a half-life of 5730 y. In a 1 μg-sample of carbon from a live tree, the activity of carbon-14 is 6.4 μCi. If researchers determine that 1μg of charcoal from a prehistoric cave painting in France has an activity of 0.80 μCi, what is the age of the painting?

The technique of carbon dating is used to determine the age of ancient cave paintings.

16.50 Using the decay curve for I-131, determine the following:
  a. Complete the values for the mass of radioactive iodine-131 on the vertical axis.
  b. Complete the number of days on the horizontal axis.
  c. What is the half-life, in days, of iodine-131?

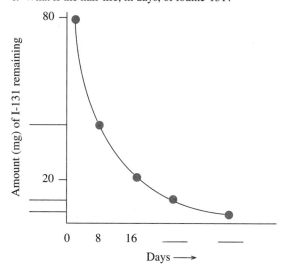

ADDITIONAL QUESTIONS AND PROBLEMS

16.51 Give the number of protons and number of neutrons in the nucleus of each the following:
  a. sodium-25   **b.** nickel-61
  c. rubidium-84   **d.** silver-110

16.52 Give the number of protons and number of neutrons in the nucleus of each of the following:
  a. boron-10   **b.** zinc-72
  c. iron-59   **d.** gold-198

16.53 Describe alpha, beta, and gamma radiation in terms of each of the following:
  a. radiation   **b.** symbols

16.54 Describe alpha, beta, and gamma radiation in terms of each of the following:
  a. depth of tissue penetration
  b. shielding needed for protection

16.55 Identify each of the following as alpha decay, beta decay, positron emission, or gamma emission:
  a. $^{27m}_{13}\text{Al} \longrightarrow ^{27}_{13}\text{Al} + ^{0}_{0}\gamma$
  b. $^{8}_{5}\text{B} \longrightarrow ^{8}_{4}\text{Be} + ^{0}_{+1}e$
  c. $^{220}_{86}\text{Rn} \longrightarrow ^{216}_{84}\text{Po} + ^{4}_{2}\text{He}$

16.56 Identify each of the following as alpha decay, beta decay, positron emission, or gamma emission:
  a. $^{127}_{55}\text{Cs} \longrightarrow ^{127}_{54}\text{Xe} + ^{0}_{+1}e$
  b. $^{90}_{38}\text{Sr} \longrightarrow ^{90}_{39}\text{Y} + ^{0}_{-1}e$
  c. $^{218}_{85}\text{At} \longrightarrow ^{214}_{83}\text{Bi} + ^{4}_{2}\text{He}$

16.57 Write a balanced nuclear equation for each of the following:
  a. Th-225 (α decay)   **b.** Bi-210 (α decay)
  c. cesium-137 (β decay) **d.** tin-126 (β decay)
  e. I-121 (β^{+}emission)

16.58 Write a balanced nuclear equation for each of the following:
a. potassium-40 (β decay)
b. sulfur-35 (β decay)
c. platinum-190 (α decay)
d. Ra-210 (α decay)
e. In-113m (γ emission)

16.59 Complete each of the following nuclear equations:
a. $^{4}_{2}\text{He} + ^{14}_{7}\text{N} \longrightarrow ? + ^{1}_{1}\text{H}$
b. $^{4}_{2}\text{He} + ^{27}_{13}\text{Al} \longrightarrow ^{30}_{14}\text{Si} + ?$
c. $^{1}_{0}n + ^{235}_{92}\text{U} \longrightarrow ^{90}_{38}\text{Sr} + 3^{1}_{0}n + ?$
d. $? \longrightarrow ^{127}_{54}\text{Xe} + ^{0}_{+1}e$

16.60 Complete each of the following nuclear equations:
a. $? + ^{59}_{27}\text{Co} \longrightarrow ^{56}_{25}\text{Mn} + ^{4}_{2}\text{He}$
b. $? \longrightarrow ^{14}_{7}\text{N} + ^{0}_{-1}e$
c. $^{0}_{-1}e + ^{76}_{36}\text{Kr} \longrightarrow ?$
d. $^{106}_{47}\text{Ag} \longrightarrow ^{106}_{46}\text{Pd} + ?$

16.61 Write the nuclear equation for each of the following:
a. When two oxygen-16 atoms collide, one of the products is an alpha particle.
b. When californium-249 is bombarded by oxygen-18, a new element, seaborgium-263, and four neutrons are produced.
c. Radon-222 emits an alpha particle, and the product emits another alpha particle. Write nuclear equations for the two reactions.
d. An atom of strontium-80 emits a positron.

16.62 Write the nuclear equation for each of the following:
a. Polonium-210 emits an alpha particle.
b. Bismuth-211 decays by emitting an alpha particle, and the product emits a beta particle. Write nuclear equations for the two reactions.
c. A radioisotope emits a positron to form titanium-48.
d. An atom of germanium-69 emits a positron.

16.63 If the amount of radioactive phosphorus-32 in a sample decreases from 1.2 mg to 0.30 mg in 28 d, what is the half-life of phosphorus-32?

16.64 If the amount of radioactive iodine-123 in a sample decreases from 0.4 mg to 0.1 mg in 26.2 h, what is the half-life of iodine-123?

16.65 Calcium-47, which decays by beta emission, has a half-life of 4.5 d.
a. Write the nuclear equation for the beta decay of calcium-47.
b. How many milligrams of an initial 16 mg of calcium-47 remain after 18 d?
c. How many days are required for 4.8 mg of calcium-47 to decay to 1.2 mg?

16.66 Cesium-137, which decays by beta emission, has a half-life of 30 y.
a. Write the nuclear equation for the beta decay of cesium-137.
b. How many milligrams of a 16-mg sample of cesium-137 remain after 90 y?
c. How many years are required for 28 mg of cesium-137 to decay to 3.5 mg?

16.67 A 120-mg sample of technetium-99m is used for a diagnostic test. If technetium-99m has a half-life of 6.0 h, how many milligrams of the technetium-99m sample remain 24 h after the test?

16.68 The half-life of oxygen-15 is 124 s. If a sample of oxygen-15 has an activity of 4000 Bq, how many minutes will elapse before it reaches an activity of 500 Bq?

16.69 What is the purpose of irradiating meats, fruits, and vegetables?

16.70 The irradiation of foods was approved in the United States in the 1980s.
a. Why have we not seen many irradiated products in our markets?
b. Would you buy foods that have been irradiated? Why or why not?

16.71 What is the difference between fission and fusion?

16.72 a. What are the products in the fission of uranium-235 that make possible a nuclear chain reaction?
b. What is the purpose of placing control rods among uranium samples in a nuclear reactor?

16.73 Where does fusion occur naturally?

16.74 Why are scientists continuing to try to build a fusion reactor even though very high temperatures needed have been difficult to reach and maintain?

CHALLENGE QUESTIONS

16.75 The half-life for the radioactive decay of calcium-47 is 4.5 d. If a sample has an activity of 1.0 μCi after 27 d, what was the initial activity of the sample?

16.76 A 16-μg sample of sodium-24 decays to 2.0 μg in 45 h. What is the half-life of sodium-24?

16.77 A nuclear technician was accidentally exposed to potassium-42 while doing brain scans for possible tumors. The error was not discovered until 36 h later when the activity of the potassium-42 sample was 2.0 μCi. If potassium-42 has a half-life of 12 h, what was the activity of the sample at the time the technician was exposed?

16.78 A wooden object from the site of an ancient temple has a carbon-14 activity of 10 counts per minute compared with a reference piece of wood cut today that has an activity of 40 counts per minute. If the half-life for carbon-14 is 5730 y, what is the age of the ancient wood object?

ANSWERS

ANSWERS TO STUDY CHECKS

16.1 Limiting the time one spends near a radioactive source and staying as far away as possible will reduce exposure to radiation.

16.2 $^{214}_{84}\text{Po} \longrightarrow \ ^{210}_{82}\text{Pb} + \ ^{4}_{2}\text{He}$

16.3 $^{131}_{53}\text{I} \longrightarrow \ ^{131}_{54}\text{Xe} + \ ^{0}_{-1}e$

16.4 $^{118}_{54}\text{Xe} \longrightarrow \ ^{118}_{53}\text{I} + \ ^{0}_{+1}e$

16.5 $^{4}_{2}\text{He} + \ ^{14}_{7}\text{N} \longrightarrow \ ^{17}_{8}\text{O} + \ ^{1}_{1}\text{H}$

16.6 5.6×10^{11} iodine-131 atoms

16.7 $2.0\ \mu\text{g}$ of iron-59

16.8 17 200 years

16.9 $^{10}_{5}\text{B} + \ ^{1}_{0}n \longrightarrow \ ^{7}_{3}\text{Li} + \ ^{4}_{2}\text{He}$

ANSWERS TO SELECTED QUESTIONS AND PROBLEMS

16.1 a. Both an alpha particle and a helium nucleus have two protons and two neutrons.

b. $\alpha, \ ^{4}_{2}\text{He}$

c. An alpha particle is emitted from an unstable nucleus during radioactive decay.

16.3 a. $^{39}_{19}\text{K}, \quad ^{40}_{19}\text{K}, \quad ^{41}_{19}\text{K}$

b. They all have 19 protons and 19 electrons, but they differ in the number of neutrons.

16.5

Medical Use	Atomic Symbol	Mass Number	Number of Protons	Number of Neutrons
Heart imaging	$^{201}_{81}\text{Tl}$	201	81	120
Radiation therapy	$^{60}_{27}\text{Co}$	60	27	33
Abdominal scan	$^{67}_{31}\text{Ga}$	67	31	36
Hyperthyroidism	$^{131}_{53}\text{I}$	131	53	78
Leukemia treatment	$^{32}_{15}\text{P}$	32	15	17

16.7 a. $\alpha, \ ^{4}_{2}\text{He}$ **b.** $n, \ ^{1}_{0}n$ **c.** $\beta, \ ^{0}_{-1}e$

d. $^{15}_{7}\text{N}$ **e.** $^{125}_{53}\text{I}$

16.9 a. β or $^{0}_{-1}e$ **b.** α or $^{4}_{2}\text{He}$ **c.** $n, \ ^{1}_{0}n$

d. $^{24}_{11}\text{Na}$ **e.** $^{14}_{6}\text{C}$

16.11 a. Because β particles are so much less massive and move faster than α particles, they can penetrate farther into body tissue.

b. Radiation produces undesirable reactions in the cells.

c. X-ray technicians leave the room to increase the distance between them and the radiation. Also, there is a wall that contains lead to shield them.

d. Wearing gloves shields the skin from α and β radiation.

16.13 a. $^{208}_{84}\text{Po} \longrightarrow \ ^{204}_{82}\text{Pb} + \ ^{4}_{2}\text{He}$

b. $^{232}_{90}\text{Th} \longrightarrow \ ^{228}_{88}\text{Ra} + \ ^{4}_{2}\text{He}$

c. $^{251}_{102}\text{No} \longrightarrow \ ^{247}_{100}\text{Fm} + \ ^{4}_{2}\text{He}$

d. $^{220}_{86}\text{Rn} \longrightarrow \ ^{216}_{84}\text{Po} + \ ^{4}_{2}\text{He}$

16.15 a. $^{25}_{11}\text{Na} \longrightarrow \ ^{25}_{12}\text{Mg} + \ ^{0}_{-1}e$

b. $^{20}_{8}\text{O} \longrightarrow \ ^{20}_{9}\text{F} + \ ^{0}_{-1}e$

c. $^{92}_{38}\text{Sr} \longrightarrow \ ^{92}_{39}\text{Y} + \ ^{0}_{-1}e$

d. $^{42}_{19}\text{K} \longrightarrow \ ^{42}_{20}\text{Ca} + \ ^{0}_{-1}e$

16.17 a. $^{26}_{14}\text{Si} \longrightarrow \ ^{26}_{13}\text{Al} + \ ^{0}_{+1}e$

b. $^{54}_{27}\text{Co} \longrightarrow \ ^{54}_{26}\text{Fe} + \ ^{0}_{+1}e$

c. $^{77}_{37}\text{Rb} \longrightarrow \ ^{77}_{36}\text{Kr} + \ ^{0}_{+1}e$

d. $^{93}_{45}\text{Rh} \longrightarrow \ ^{93}_{44}\text{Ru} + \ ^{0}_{+1}e$

16.19 a. $^{28}_{14}\text{Si}$ **b.** $^{87}_{36}\text{Kr}$ **c.** $^{0}_{-1}e$

d. $^{238}_{92}\text{U}$ **e.** $^{188}_{79}\text{Au}$

16.21 a. $^{10}_{4}\text{Be}$ **b.** $^{0}_{-1}e$

c. $^{27}_{13}\text{Al}$ **d.** $^{30}_{15}\text{P}$

16.23 a. When radiation enters the Geiger counter, it ionizes a gas in the detection tube, which produces a burst of current that is detected by the instrument.

b. becquerel (Bq), curie (Ci)

c. gray (Gy), rad

d. 1000 Gy

16.25 $294\ \mu\text{Ci}$

16.27 When pilots fly at high altitudes, there is less atmosphere to protect them from cosmic radiation.

16.29 A half-life is the time it takes for one-half of a radioactive sample to decay.

16.31 a. 40.0 mg   **b.** 20.0 mg

c. 10.0 mg   **d.** 5.00 mg

16.33 130 days

16.35 a. Since the elements Ca and P are part of the bone, the radioactive isotopes of Ca and P will become part of the bony structures of the body, where their radiation can be used to diagnose or treat bone diseases.

b. Strontium (Sr) acts much like calcium (Ca) because both are Group 2A (2) elements. The body will accumulate radioactive strontium in bones in the same way that it incorporates calcium. Radioactive strontium is harmful to children because the radiation it produces causes more damage in cells that are dividing rapidly.

16.37 $180\ \mu\text{Ci}$

16.39 Nuclear fission is the splitting of a large atom into smaller fragments with the release of large amounts of energy.

16.41 $^{103}_{42}\text{Mo}$

16.43 a. fission   **b.** fusion

c. fission   **d.** fusion

16.45 a. $^{11}_{6}\text{C}$

b.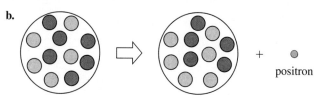

16.47

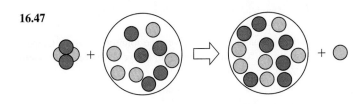

16.49 17 200 years old

16.51 a. 11 protons and 14 neutrons

b. 28 protons and 33 neutrons

c. 37 protons and 47 neutrons

d. 47 protons and 63 neutrons

16.53 a. In alpha decay, a helium nucleus is emitted from the nucleus of a radioisotope. In beta decay, a neutron in an unstable nucleus is converted to a proton and electron, which is emitted as a beta particle. In gamma emission, high-energy radiation is emitted from the nucleus of a radioisotope.

b. α, ^4_2He; β, $^{\ 0}_{-1}e$; γ, $^0_0\gamma$

16.55 a. gamma emission

b. positron emission

c. alpha decay

16.57 a. $^{225}_{90}\text{Th} \longrightarrow {}^{221}_{88}\text{Ra} + {}^4_2\text{He}$

b. $^{210}_{83}\text{Bi} \longrightarrow {}^{206}_{81}\text{Tl} + {}^4_2\text{He}$

c. $^{137}_{55}\text{Cs} \longrightarrow {}^{137}_{56}\text{Ba} + {}^{\ 0}_{-1}e$

d. $^{126}_{50}\text{Sn} \longrightarrow {}^{126}_{51}\text{Sb} + {}^{\ 0}_{-1}e$

e. $^{121}_{53}\text{I} \longrightarrow {}^{121}_{52}\text{Te} + {}^0_{+1}e$

16.59 a. $^{17}_8\text{O}$ **b.** ^1_1H

c. $^{143}_{54}\text{Xe}$ **d.** $^{127}_{55}\text{Cs}$

16.61 a. $^{16}_8\text{O} + {}^{16}_8\text{O} \longrightarrow {}^{28}_{14}\text{Si} + {}^4_2\text{He}$

b. $^{18}_8\text{O} + {}^{249}_{98}\text{Cf} \longrightarrow {}^{263}_{106}\text{Sg} + 4{}^1_0n$

c. $^{222}_{86}\text{Rn} \longrightarrow {}^{218}_{84}\text{Po} + {}^4_2\text{He}$

$^{218}_{84}\text{Po} \longrightarrow {}^{214}_{82}\text{Pb} + {}^4_2\text{He}$

d. $^{80}_{38}\text{Sr} \longrightarrow {}^{80}_{37}\text{Rb} + {}^0_{+1}e$

16.63 14 d

16.65 a. $^{47}_{20}\text{Ca} \longrightarrow {}^{47}_{21}\text{Sc} + {}^{\ 0}_{-1}e$

b. 1.0 mg of Ca-47

c. 9.0 d

16.67 7.5 mg of Tc-99m

16.69 The irradiation of meats, fruits, and vegetables kills bacteria such as *E. coli* that can cause food-borne illnesses. In addition, spoilage is deterred and shelf life is extended.

16.71 In the fission process, an atom splits into smaller nuclei. In fusion, small nuclei combine (fuse) to form a larger nucleus.

16.73 Fusion occurs naturally in the Sun and other stars.

16.75 64 μCi

16.77 16 μCi

CI.27 Consider the reaction of sodium oxalate ($Na_2C_2O_4$) and potassium permanganate ($KMnO_4$) in acidic solution. The unbalanced equation is the following:

$$MnO_4^-(aq) + C_2O_4^{2-}(aq) \longrightarrow Mn^{2+}(aq) + CO_2(g)$$

In an oxidation–reduction titration, the $KMnO_4$ in the buret reacts with $Na_2C_2O_4$.

a. What is the balanced oxidation half-reaction?

b. What is the balanced reduction half-reaction?

c. What is the balanced ionic equation for the reaction?

d. If 24.6 mL of a $KMnO_4$ solution is needed to titrate a solution containing 0.758 g of sodium oxalate ($Na_2C_2O_4$), what is the molarity of the $KMnO_4$ solution?

CI.28 A strip of magnesium metal dissolves rapidly in 6.00 mL of a 0.150 M HCl solution, producing magnesium chloride and hydrogen gas.

$$Mg(s) + HCl(aq) \longrightarrow MgCl_2(aq) + H_2(g) \text{ (unbalanced)}$$

Magnesium metal reacts vigorously with hydrochloric acid.

a. Assign oxidation numbers to all of the elements in the reactants and products.

b. What is the balanced equation for the reaction?

c. What is the oxidizing agent?

d. What is the reducing agent?

e. What is the pH of 0.150 M HCl solution?

f. How many grams of magnesium can dissolve in the HCl solution?

CI.29 A piece of magnesium with a mass of 0.121 g is added to 50.0 mL of a 1.00 M HCl solution at a temperature of 22 °C.

When the magnesium dissolves, the solution reaches a temperature of 33 °C. For the equation, see your answer to Problem CI.28 b.

a. What is the limiting reactant?

b. What volume, in mL, of hydrogen gas would be produced at 33 °C when the pressure is 750. mmHg?

c. How many joules were released by the reaction of the magnesium? Assume the density of the HCl solution is 1.00 g/mL and the specific heat of the HCl solution is the same as that of water.

d. What is the heat of reaction for magnesium in J/g? in kJ/mol?

CI.30 The iceman known as "Ötzi" was discovered in a high mountain pass on the Austrian–Italian border. Samples of his hair and bones had carbon-14 activity that was 50% of that present in new hair or bone. Carbon-14 undergoes beta decay and has a half-life of 5730 y.

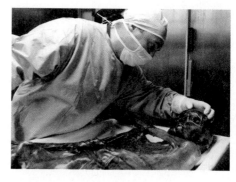

The mummified remains of "Ötzi" were discovered in 1991.

a. How long ago did "Ötzi" live?

b. Write a nuclear equation for the decay of carbon-14.

CI.31 A sample of isotopes of silicon is listed on the following table:

Isotope	% Natural Abundance	Atomic Mass	Half-Life (radioactive)	Radiation
$^{27}_{14}Si$		26.987	4.9 s	Positron
$^{28}_{14}Si$	92.230	27.977	Stable	None
$^{29}_{14}Si$	4.683	28.976	Stable	None
$^{30}_{14}Si$	3.087	29.974	Stable	None
$^{31}_{14}Si$		30.975	2.6 h	Beta

a. In the following table, indicate the number of protons, neutrons, and electrons for each isotope listed:

Isotope	Number of Protons	Number of Neutrons	Number of Electrons
$^{27}_{14}Si$			
$^{28}_{14}Si$			
$^{29}_{14}Si$			
$^{30}_{14}Si$			
$^{31}_{14}Si$			

b. What is the electron configuration and the abbreviated electron configuration of silicon?

c. Calculate the atomic mass for silicon using the isotopes that have a natural abundance.

d. Write the nuclear equations for the positron emission of Si-27 and the beta decay of Si-31.

e. Draw the electron-dot formula and predict the shape of $SiCl_4$.

f. How many hours are needed for a sample of Si-31 with an activity of 16 μCi to decay to 2.0 μCi?

CI.32 K^+ is an electrolyte required by the human body and found in many foods as well as salt substitutes. One of the isotopes of potassium is potassium-40, which has a natural abundance of 0.012% and a half-life of 1.30×10^9 y. The isotope potassium-40 decays to calcium-40 or to argon-40. A typical activity for potassium-40 is 7.0 μCi per gram.

Potassium chloride is used as a salt substitute.

a. Write a nuclear equation for each type of decay.

b. Identify the particle emitted for each type of decay.

c. How many K^+ ions are in 3.5 oz of KCl?

d. What is the activity of 25 g of KCl in becquerels?

CI.33 Uranium-238 decays in a series of nuclear changes until stable lead-206 is produced. Complete the following nuclear equations that are part of the uranium-238 decay series:

a. $^{238}_{92}U \longrightarrow {}^{234}_{90}Th + ?$

b. $^{234}_{90}Th \longrightarrow ? + {}^{0}_{-1}e$

c. $? \longrightarrow {}^{222}_{86}Rn + {}^{4}_{2}He$

CI.34 Of much concern to environmentalists is radon-222, which is a radioactive noble gas that can seep from the ground into basements of homes and buildings. Radon-222 is a product of the decay of radium-226 that occurs naturally in rocks and soil in much of the United States. Radon-222, which has a half-life of 3.8 d, decays by alpha decay. Because radon-222 is a gas, it can be inhaled and is strongly associated with lung cancer. Radon levels in a home can be measured with a home radon-detection kit. Environmental agencies have set the maximum level of radon-222 in a home at 4 picocuries per liter (pCi/L) of air.

a. Write the equation for the decay of Ra-226.

b. Write the equation for the decay of Rn-222.

c. If a room contains 24 000 atoms of radon-222, how many atoms of radon-222 remain after 15.2 d?

d. Suppose a room is 4.0 m wide, 6.0 m long, and 3.0 m high. If the radon level is the maximum allowed (4 pCi/L), how many alpha particles are emitted from Rn-222 in one day? (1 Ci = 3.7×10^{10} disintegrations per second)

A home detection kit is used to measure the level of radon-222.

■ ANSWERS

CI.27 a. $C_2O_4{}^{2-}(aq) \longrightarrow 2CO_2(g) + 2 e^-$

b. $5 e^- + MnO_4{}^-(aq) + 8H^+(aq) \longrightarrow$
$$Mn^{2+}(aq) + 4H_2O(l)$$

c. $2MnO_4{}^-(aq) + 5C_2O_4{}^{2-}(aq) + 16H^+(aq) \longrightarrow$
$$10CO_2(g) + 2Mn^{2+}(aq) + 8H_2O(l)$$

d. 0.0920 M $KMnO_4$ solution

CI.29 a. Mg is the limiting reactant.

b. 127 mL of $H_2(g)$

c. 2.30×10^3 J

d. 1.90×10^4 J/g; 462 kJ/mol

CI.31 a.

Isotope	Number of Protons	Number of Neutrons	Number of Electrons
$^{27}_{14}Si$	14	13	14
$^{28}_{14}Si$	14	14	14
$^{29}_{14}Si$	14	15	14
$^{30}_{14}Si$	14	16	14
$^{31}_{14}Si$	14	17	14

b. $1s^2 2s^2 2p^6 3s^2 3p^2$; $[Ne]3s^2 3p^2$

c. 28.09 amu

d. $^{27}_{14}Si \longrightarrow {}^{27}_{13}Al + {}^{0}_{+1}e$
$^{31}_{14}Si \longrightarrow {}^{31}_{15}P + {}^{0}_{-1}e$

e.

$$:\ddot{C}l—\underset{\underset{:\ddot{C}l:}{|}}{\overset{\overset{:\ddot{C}l:}{|}}{Si}}—\ddot{C}l: \qquad \text{Tetrahedral}$$

f. 7.8 h

CI.33 a. $^{238}_{92}U \longrightarrow {}^{234}_{90}Th + {}^{4}_{2}He$

b. $^{234}_{90}Th \longrightarrow {}^{234}_{91}Pa + {}^{0}_{-1}e$

c. $^{226}_{88}Ra \longrightarrow {}^{222}_{86}Rn + {}^{4}_{2}He$

CREDITS

Digital Vision, Pearson Education. p. 511 NASA. p. 512 Pearson Education/Pearson Science. p. 514 Corbis. p. 515 Shutterstock. p. 517 *left:* Dorling Kindersley Media Library. p. 517 *right:* Andrew Lambert Photography/Photo Researchers. p. 520 iStockphoto.

Chapter 16

p. 524 Pearson Education/Pearson Science. p. 527 iStockphoto. p. 528 Health Protection Agency. p. 532 Family Safety Products, Inc. p. 535 Australian Nuclear Science and Technology Organisation. p. 537 *top:* AP Photo/James MacPherson. p. 537 *middle:* Don Farrall/Getty Images. p. 537 *bottom:* Stanford Dosimetry, LLC. p. 538 Pearson Education/Pearson Science. p. 543 CORBIS. p. 544 *left:* Phanie/Photo Researchers. p. 544 *right:* Pasieka/Photo Researchers. p. 545 *top:* Getty Images. p. 545 *bottom:* Getty Images. p. 546 James Cavallini/Custom Medical Stock Photo/Newscom. p. 546 *bottom, left:* Cytyc Hologic Corporation. p. 546 *bottom, right:* Cytyc Hologic Corporation. p. 550 iStockphoto. p. 552 Pearson.

Combining Ideas from Chapters 15 and 16

p. 556 *top:* Andrew Lambert Photography/ SPL/Photo Researchers. p. 556 *bottom:* Richard Megna/Fundamental Photographs. p. 556 *middle:* Augustin Ochsenreite/AP.

p. 557 *left:* AlsoSalt. p. 557 *right:* Family Safety Products, Inc.

Chapter 17

p. 558 Pearson Education. p. 559 Pearson Education/Pearson Science. p. 565 *top:* iStockphoto. p. 565 *bottom:* Mark Dahners/AP. p. 569 *top:* Alastair Shay/Papilio/CORBIS. p. 569 *middle, left:* Pearson Education. p. 569 *middle, middle:* Pearson Education. p. 569 *middle, right:* Pearson Education. p. 569 *bottom, left:* Pearson Education. p. 569 *bottom, middle:* Pearson Education. p. 569 *bottom, right:* Pearson Education. p. 571 *top:* Pearson Education. p. 571 iStockphoto. p. 575 Medical-on-Line/Alamy. p. 578 Pearson Education. p. 582 *bottom, left:* Pearson Education. p. 582 *bottom, right:* Pearson Education. p. 586 Pearson Education. p. 587 *top:* Shutterstock. p. 587 *bottom:* Pearson Education. p. 592 *top:* Carlos Alvarez/ iStockphoto. p. 592 *bottom:* Michael S. Yamashita/CORBIS. p. 598 *top:* DuPont & Company. p. 598 *middle:* Getty Images. p. 598 *bottom:* Pearson Education. p. 598 *right:* Frank Greenaway-Dorling Kindersley Media Library. p. 600 Pearson Education.

Chapter 18

p. 603 Pearson Education. p. 604 Pearson Education. p. 606 Pearson Education.

p. 609 Pearson Education. p. 610 Pearson Education. p. 612 Pearson Education. p. 613 *left:* Shutterstock. p. 613 *right:* David Toase/Getty Images. p. 616 Pearson Education. p. 617 Pearson Education. p. 618 John Pitcher/iStockphoto. p. 619 Pearson Education. p. 620 Pearson Education. p. 622 L.G. Patterson/AP. p. 623 *top:* National Heart, Lung, and Blood Institute. p. 623 *bottom:* National Heart, Lung, and Blood Institute. p. 624 Pearson Education. p. 630 Pearson Education. p. 638 M. Freeman/Getty Images. p. 648 *middle, right:* Shutterstock. p. 648 *bottom, left:* Pearson Education. p. 648 *top, left:* Pearson Education. p. 648 *bottom, right:* Pearson Education. p. 649 *top, left:* iStockphoto. p. 649 *bottom, left:* iStockphoto. p. 649 *middle, left:* Pearson Education. p. 649 *top, right:* Pearson Education. p. 649 *bottom, right:* Pearson Education. p. 655 iStockphoto.

Combining Ideas from Chapters 17 and 18

p. 655 *top, left:* Arthur Glauberman/Photo Researchers, Inc. p. 655 *top, right:* David Lees/Getty Images. p. 655 *bottom, left:* Charles D. Winters/Photo Researchers. p. 655 *middle, right:* Pearson Education. p. 655 *bottom, right:* Photolibrary. p. 656 *bottom:* Pearson Education.

GLOSSARY / INDEX

METRIC AND SI UNITS AND SOME USEFUL CONVERSION FACTORS

Length SI unit meter (m)	Volume SI unit cubic meter (m³)	Mass SI unit kilogram (kg)
1 meter (m) = 100 centimeters (cm) 1 meter (m) = 1000 millimeters (mm) 1 cm = 10 mm 1 kilometer (km) = 0.6214 mile (mi) 1 inch (in.) = 2.54 cm (exact)	1 liter (L) = 1000 milliliters (mL) 1 mL = 1 cm³ 1 L = 1.057 quart (qt) 1 qt = 946.3 mL	1 kilogram (kg) = 1000 grams (g) 1 g = 1000 milligrams (mg) 1 kg = 2.205 lb 1 lb = 453.6 g 1 mol = 6.022 × 10²³ particles **Water** density = 1.00 g/mL (at 4 °C)

Temperature SI unit kelvin (K)	Pressure SI unit pascal (Pa)	Energy SI unit joule (J)
$°F = 1.8(°C) + 32$ $°C = \dfrac{(°F - 32)}{1.8}$ $K = °C + 273$	1 atm = 760 mmHg 1 atm = 101.325 kPa 1 atm = 760 torr 1 mol of gas (STP) = 22.4 L R = 0.0821 L·atm/mol·K R = 62.4 L·mmHg/mol·K	1 calorie (cal) = 4.184 J 1 kcal = 1000 cal **Water** Heat of fusion = 334 J/g; 80. cal/g Heat of vaporization = 2260 J/g; 540 cal/g Specific heat (*SH*) = 4.184 J/g °C

PREFIXES FOR METRIC (SI) UNITS

Prefix	Symbol	Power of Ten
Values greater than 1		
peta	P	10^{15}
tera	T	10^{12}
giga	G	10^{9}
mega	M	10^{6}
kilo	k	10^{3}
Values less than 1		
deci	d	10^{-1}
centi	c	10^{-2}
milli	m	10^{-3}
micro	μ	10^{-6}
nano	n	10^{-9}
pico	p	10^{-12}
femto	f	10^{-15}

FORMULAS AND MOLAR MASSES OF SOME TYPICAL COMPOUNDS

Name	Formula	Molar Mass (g/mol)	Name	Formula	Molar Mass (g/mol)
Ammonia	NH_3	17.03	Hydrogen chloride	HCl	36.46
Ammonium chloride	NH_4Cl	53.49	Iron(III) oxide	Fe_2O_3	159.70
Ammonium sulfate	$(NH_4)_2SO_4$	132.15	Magnesium oxide	MgO	40.31
Bromine	Br_2	159.80	Methane	CH_4	16.04
Butane	C_4H_{10}	58.12	Nitrogen	N_2	28.02
Calcium carbonate	$CaCO_3$	100.09	Oxygen	O_2	32.00
Calcium chloride	$CaCl_2$	110.98	Potassium carbonate	K_2CO_3	138.21
Calcium hydroxide	$Ca(OH)_2$	74.10	Potassium nitrate	KNO_3	101.11
Calcium oxide	CaO	56.08	Propane	C_3H_8	44.09
Carbon dioxide	CO_2	44.01	Sodium chloride	$NaCl$	58.44
Chlorine	Cl_2	70.90	Sodium hydroxide	$NaOH$	40.00
Copper(II) sulfide	CuS	95.62	Sulfur trioxide	SO_3	80.07
Hydrogen	H_2	2.016	Water	H_2O	18.02